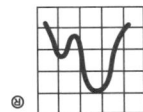

Name

Organization

Address

City, State, Zip Code

 **StatSoft**®

STATISTICS: Methods and Applications CD Offer
2300 East 14th Street
Tulsa, OK 74104

CD Order Form

Order Your CD of this Book Today!

Return this card by mail. Only original forms can be accepted.

Free CD Version!

Pay only shipping and handling.

 **StatSoft**®

CUSTOMER INFORMATION

Name (please print)

Organization

Address

City, State, Zip Code

Country

Phone*

( )　　　　　　Ext.

Fax*

( )

E-Mail Address *(We respect your privacy. This information will not be shared.)**

Where did you purchase the book?

- ☐ Amazon.com
- ☐ StatSoft, USA
- ☐ StatSoft Overseas
- ☐ Other _____

Where did you learn about the book?

☐ StatSoft.com ☐ Another user ☐ Other _____

* StatSoft will NEVER share this information with another organization

SHIPPING PAYMENT INFO

There is no charge for your CD version of the book. Pay only \$5 for domestic (US) shipping and handling.*

Payment Method:

- ☐ Check   Check # _____
- ☐ Money Order
- ☐ Credit Card   ☐ Visa   ☐ M/C   ☐ Amex   ☐ Diner's Club

Card Number

Expiration date _____ / _____

Name on Card (please sign)   By signing this, I authorize StatSoft, Inc. to charge my credit card the shipping fee.

*\$5 Fee applies only to US customers. Call StatSoft at (918) 749-1119 for international rates.

COMMENTS ON THE BOOK:

We appreciate your comments and suggestions.

I authorize StatSoft to quote me: ☐ YES (please initial _____)   ☐ NO

STATISTICS
Methods and Applications
A Comprehensive Reference for Science, Industry, and Data Mining

Thomas Hill
The University of Tulsa

Pawel Lewicki
The University of Tulsa

 **StatSoft**®

STATISTICS: METHODS AND APPLICATIONS
A COMPREHENSIVE REFERENCE FOR SCIENCE, INDUSTRY, AND DATA MINING

TABLE OF CONTENTS

STATISTICAL METHODS
(in alphabetical order)

STATISTICS: METHODS AND APPLICATIONS
A COMPREHENSIVE REFERENCE FOR
SCIENCE, INDUSTRY, AND DATA MINING

DETAILED TABLE OF CONTENTS

STATISTICAL METHODS
(in alphabetical order)

PREFACE

The content of this book was not created as a result of one organized effort to write a comprehensive manual on "how to use statistics." In fact, the idea of publishing this book came only after most of its parts had already been completed and widely distributed (in electronic format) and after we received numerous comments from readers and requests that all this material be available in the form of one "compendium of knowledge about statistics," written from the perspective of real-life users of the respective methods.

This book is a compilation of materials written, and rewritten, over the past 15 years to help "consumers" of statistics make the most of the wisdom and power that this discipline can offer to real-life data analysts. Over years of teaching graduate and undergraduate courses on quantitative methods, and consulting (both in academia and the industry), we have collected extensive notes about the most important and useful knowledge that should be presented when teaching about specific analytic methods and approaches, what common errors and misconceptions are made when trying to apply the related statistical concepts to real-life analytic problems, and what the best methods are to avoid these errors. Contacts with our colleagues in academia, consultants, and analysts at the Research and Development Department at StatSoft offices both in the U.S. and in Europe, Asia, Africa, and Australia have provided invaluable material and real-life examples that enhanced our own experience presented in this book.

Many chapters of the book benefited from specific comments by several members of the Research and Development Department at StatSoft, who are experts in the specific areas. This includes James Fultz, Ph.D.; Elizabeth Paszkiewicz, Ph.D.; Cazhaow Qazaz, Ph.D. (who wrote the chapters on *Machine Learning* and *Neural Networks*); Mark Styczen, Ph.D.; Santokh Singh, Ph.D.; and Tae-Sung Shin, Ph.D.; as well as Andrew Hunter, Ph.D. (now at the Southampton University, UK) and James Steiger, Ph.D. (now at Vanderbilt University, who wrote the chapter on *Power Analysis*). Parts of this book benefited also from comments by some of our former students: Doug Merill, Ph.D. (now at Google, Inc.); Hunter Hoffman, Ph.D. (now at the University of Washington); Rainer Neubauer, Ph.D. (now at PriceWaterhouseCoopers); Michael Lank, Ph.D. (now at Ford Motor Company); Maria Czyzewska, Ph.D. (now at University of Texas, San Marcos); Sharon Stanners, Ph.D. (now at IBM); and David Vequist, Ph.D. (now at University of Incarnate). We are also grateful to Susan Banks for her meticulous help with copyediting the manuscript and Kelly Ridgway for her thorough review and insightful comments.

Also, we would like to express our gratitude for the important contributions made by many readers of earlier versions of these chapters, who generously provided us with their feedback and requests for additional, particular topics relevant to their respective applications of statistics. These comments and requests were especially valuable because they helped us shape the coverage of specific areas in a way that is relevant for users of the respective methods. We hope that these comments will continue to be offered to us by our readers, and we promise to be responsive to them so that future editions will reflect the changing state of the discipline and the ever-changing needs of those who use statistics to convert data into knowledge.

ELEMENTARY CONCEPTS IN STATISTICS

In this chapter, we will briefly discuss those elementary statistical concepts that provide the necessary foundations for more specialized expertise in any area of statistical data analysis. The selected topics illustrate the basic assumptions of most statistical methods and/or have been demonstrated in research to be necessary components of one's general understanding of the "quantitative nature" of reality (Nisbett, et al., 1987). Because of space limitations, we will focus mostly on the functional aspects of the concepts discussed and the presentation will be very short. Further information on each of these concepts can be found in statistical textbooks. Recommended introductory textbooks are Kachigan (1986) and Runyon and Haber (1976). For a more advanced discussion of elementary theory and assumptions of statistics, see the classic books by Hays (1988) and Kendall and Stuart (1979).

What are Variables?

Variables are things that we measure, control, or manipulate in research. They differ in many respects, most notably in the role they are given in our research and in the type of measures that can be applied to them.

Correlational vs. Experimental Research

Most empirical research belongs clearly to one of these two general categories. In correlational research, we do not (or at least try not to) influence any variables but only measure them and look for relations (correlations) between some set of variables, such as blood pressure and cholesterol level. In experimental research, we manipulate some variables and then measure the effects of this manipulation on other variables. For example, a researcher might artificially increase blood pressure and then record cholesterol level. Data analysis in experimental research also comes down to calculating correlations between variables, specifically, those manipulated and those affected by the manipulation. However, experimental data may potentially provide qualitatively better information. Only experimental data can conclusively demonstrate causal relations between variables. For example, if we found that whenever we change variable A, variable B also changes, we can conclude that "A influences B." Data from correlational research can only be interpreted in causal terms based on some theories that we have, but correlational data cannot conclusively prove causality.

Dependent vs. Independent Variables

Independent variables are those that are manipulated, whereas dependent variables are only measured or registered. This distinction appears terminologically confusing to many because, as some students say, "all variables depend on something." However, once you get used to this distinction, it becomes indispensable. The terms dependent and independent variable apply mostly to experimental research where some variables are manipulated, and in this sense they are independent from the initial reaction patterns, features, intentions, etc. of the subjects. Some other variables are expected to be dependent on the manipulation or experimental conditions. That is to say, they depend on "what the subject will do" in response. Somewhat contrary to the nature of this distinction, these terms are also used in studies where we do not literally manipulate independent variables, but only assign subjects to experimental groups based on some preexisting properties of the subjects. For example, if in an experiment males are compared with females regarding their white cell count (*WCC*), *Gender* could be called the independent variable and *WCC* the dependent variable.

Measurement Scales

Variables differ in how well they can be measured, i.e., in how much measurable information their measurement scale can provide. There is obviously some measurement error involved in every measurement, which determines the amount of information that we can obtain. Another factor that determines the amount of information that can be provided by a variable is its type of measurement scale. Specifically, variables are classified as a) nominal, b) ordinal, c) interval, or d) ratio.

a. *Nominal* variables allow for only qualitative classification. That is, they can be measured only in terms of whether the individual items belong to some distinctively different categories, but we cannot quantify or even rank order those categories. For example, all we can say is that 2 individuals are different in terms of variable *A* (e.g., they are of different race), but we cannot say which one has more of the quality represented by the variable. Typical examples of nominal variables are *gender*, *race*, *color*, and *city*.

b. With *ordinal* variables, you can rank order the items you measure in terms of which has less and which has more of the quality represented by the variable, but still cannot measure how much more. A typical example of an ordinal variable is the socioeconomic status of families. For example, we know that upper-middle is higher than middle, but we cannot say that it is, for example, 18% higher. Also this very distinction between nominal, ordinal, and interval scales itself represents a good example of an ordinal variable. For example, we can say that nominal measurement provides less information than ordinal measurement, but we cannot say how much less or how this difference compares to the difference between ordinal and interval scales.

c. With *interval* variables, not only can you rank order the items that are measured, you also can quantify and compare the sizes of differences between them. For example, temperature, as measured in degrees Fahrenheit or Celsius, constitutes an interval scale. We can say that a temperature of 40 degrees is higher than a temperature of 30 degrees, and that an increase from 20 to 40 degrees is twice as much as an increase from 30 to 40 degrees.

d. *Ratio* variables are very similar to interval variables. In addition to all the properties of interval variables, they feature an identifiable absolute zero point; thus, they allow for statements such as *x* is two times more than *y*. Typical examples of ratio scales are measures of time or space. For example, as the Kelvin temperature scale is a ratio scale, not only can we say that a temperature of 200 degrees is higher than one of 100 degrees, we can correctly state that it is twice as high. Interval scales do not have the ratio property. Most statistical data analysis procedures do not distinguish between the interval and ratio properties of the measurement scales.

Relations between Variables

Regardless of their type, two or more variables are related if in a sample of observations, the values of those variables are distributed in a consistent manner. In other words, variables are related if their values systematically correspond to each other for these observations. For example, *Gender* and *WCC* would be considered to be related if most *males* had high *WCC* and most *females* low *WCC*, or vice versa. *Height* is related to *Weight* because typically tall individuals are heavier than short ones. *IQ* is related to the *Number of Errors* in a test, if people with higher IQs make fewer errors.

Why Relations between Variables are Important

Generally speaking, the ultimate goal of every research or scientific analysis is finding relations between variables. The philosophy of science teaches us that there is no other way of representing meaning except in terms of relations between some quantities or qualities; either way involves relations between variables. Thus, the advancement of science must always involve finding new relations between variables. Correlational research involves measuring such relations in the most straightforward manner. However, experimental research is not any different in this respect. For example, the above mentioned experiment comparing *WCC* in *males* and *females* can be described as looking for a correlation between two variables: *Gender* and *WCC*. Statistics does nothing else but help us evaluate relations between variables. Actually, all of the hundreds of procedures that are described in this manual can be interpreted in terms of evaluating various kinds of inter-variable relations.

Two Basic Features of Every Relation between Variables

The two most elementary formal properties of every relation between variables are the relation's a) magnitude (or strength) and b) its reliability (or truthfulness).

a. Magnitude (or strength). The magnitude is much easier to understand and measure than reliability. For example, if every *male* in our sample was found to have a higher *WCC* than any *female* in the sample, we could say that the magnitude of the relation between the two variables (*Gender* and *WCC*) is very high in our sample. In other words, we could predict one based on the other (at least among the members of our sample).

b. Reliability (or truthfulness). The reliability of a relation is a much less intuitive concept, but still extremely important. It pertains to the representativeness of the result found in our specific sample for the entire population. In other words, it says how probable it is that a similar relation would be found if the experiment was replicated with other samples drawn from the same population. Remember

that we are almost never ultimately interested only in what is going on in our sample; we are interested in the sample only to the extent it can provide information about the population. If our study meets some specific criteria (to be mentioned later), the reliability of a relation between variables observed in our sample can be quantitatively estimated and represented using a standard measure (technically called *p*-value or statistical significance level; see the next paragraph).

What is Statistical Significance (*p*-Value)?

The statistical significance of a result is the probability that the observed relationship (e.g., between variables) or a difference (e.g., between means) in a sample occurred by pure chance, and that in the population from which the sample was drawn, no such relationship or differences exist. Using less technical terms, you could say that the statistical significance of a result tells us something about the degree to which the result is true (in the sense of being representative of the population). More technically, the value of the *p*-value represents a decreasing index of the reliability of a result (see Brownlee, 1960). The higher the *p*-value, the less we can believe that the observed relation between variables in the sample is a reliable indicator of the relation between the respective variables in the population. Specifically, the *p*-value represents the probability of error that is involved in accepting our observed result as valid, that is, as representative of the population. For example, a *p*-value of .05 (i.e., 1/20) indicates that there is a 5% probability that the relation between the variables found in our sample is a "fluke." In other words, assuming that in the population

there was no relation between those variables whatsoever, and we were repeating experiments such as ours one after another, we could expect that, approximately, in every 20 replications of the experiment there would be one in which the relation between the variables in question would be equal or stronger than in ours. (Note that this is not the same as saying that, given that there *is* a relationship between the variables, we can expect to replicate the results 5% of the time or 95% of the time; when there is a relationship between the variables in the population, the probability of replicating the study and finding that relationship is related to the statistical power of the design. See also, Chapter 31 – *Power Analysis*). In many areas of research, the *p*-value of .05 is customarily treated as a borderline acceptable error level.

How to Determine that a Result is Really Significant

There is no way to avoid arbitrariness in the final decision as to what level of significance will be treated as really significant. That is, the selection of some level of significance, up to which the results will be rejected as invalid, is arbitrary. In practice, the final decision usually depends on whether the outcome was predicted *a priori* or only found post hoc in the course of many analyses and comparisons performed on the data set, on the total amount of consistent supportive evidence in the entire data set, and on traditions existing in the particular area of research. Typically, in many sciences, results that yield $p \leq .05$ are considered borderline statistically significant, but remember that this level of significance still involves a pretty high probability of error (5%). Results that are significant at the $p \leq .01$ level are commonly considered statistically significant, and $p \leq .005$ or $p \leq .001$ levels are often called highly

significant. But remember that those classifications represent nothing else but arbitrary conventions that are only informally based on general research experience.

Statistical Significance and the Number of Analyses Performed

Needless to say, the more analyses you perform on a data set, the more results will meet "by chance" the conventional significance level. For example, if you calculate correlations between 10 variables (i.e., 45 different correlation coefficients), you should expect to find by chance that about two (i.e., one in every 20) correlation coefficients are significant at the $p \leq .05$ level, even if the values of the variables were totally random and those variables do not correlate in the population. Some statistical methods that involve many comparisons, and thus a good chance for such errors, include some correction or adjustment for the total number of comparisons. However, many statistical methods (especially simple exploratory data analyses) do not offer any straightforward remedies to this problem. Therefore, it is up to the researcher to carefully evaluate the reliability of unexpected findings. Many examples in this manual offer specific advice on how to do this; relevant information can also be found in most research methods textbooks.

Magnitude vs. Reliability of a Relation between Variables

We said before that magnitude and reliability are two different features of relationships between variables. However, they are not totally independent. In general, in a sample of a particular size, the larger the magnitude of the relation between variables, the more reliable the relation (see the next paragraph).

Why Stronger Relations between Variables are More Significant

Assuming that there is no relation between the respective variables in the population, the most likely outcome would be also finding no relation between those variables in the research sample. Thus, the stronger the relation found in the sample, the less likely it is that there is no corresponding relation in the population. As you see, the magnitude and significance of a relation appear to be closely related, and we could calculate the significance from the magnitude and vice-versa; however, this is true only if the sample size is kept constant, because the relation of a given strength could be either highly significant or not significant at all, depending on the sample size (see the next paragraph).

Why Significance of a Relation between Variables Depends on the Size of the Sample

If there are very few observations, there are also few possible combinations of the values of the variables, and thus the probability of obtaining by chance a combination of those values indicative of a strong relation is relatively high. Consider the following illustration. If we are interested in two variables (*Gender: male/female* and *WCC: high/low*) and there are only four subjects in our sample (two *males* and two *females*), the probability that we will find, purely by chance, a 100% relation between the two variables can be as high as one-eighth. Specifically, there is a one-in-eight chance that both *males* will have a high *WCC* and both *females* a low *WCC*, or vice versa. Now consider the probability of obtaining such a perfect match by chance if our sample consisted of 100 subjects; the probability of obtaining such an outcome by chance would be practically zero. Let's look at a more general

example. Imagine a theoretical population in which the average value of *WCC* in *males* and *females* is exactly the same. Needless to say, if we start replicating a simple experiment by drawing pairs of samples (of *males* and *females*) of a particular size from this population and calculating the difference between the average *WCC* in each pair of samples, most of the experiments will yield results close to 0. However, from time to time, a pair of samples will be drawn where the difference between *males* and *females* will be quite different from 0. How often will it happen? The smaller the sample size in each experiment, the more likely it is that we will obtain such erroneous results, which in this case would be results indicative of the existence of a relation between *Gender* and *WCC* obtained from a population in which such a relation does not exist.

Example: Baby Boys to Baby Girls Ratio

Consider this example from research on statistical reasoning (Nisbett, et al., 1987). There are two hospitals: in the first one, 120 babies are born every day, in the other, only 12. On average, the ratio of baby boys to baby girls born every day in each hospital is 50/50. However, one day, in one of those hospitals twice as many baby girls were born as baby boys. In which hospital was it more likely to happen? The answer is obvious for a statistician, but as research shows, not so obvious for a layperson. It is much more likely to happen in the small hospital. The reason for this is that technically speaking, the probability of a random deviation of a particular size (from the population mean), decreases with the increase in the sample size.

Why Small Relations Can Be Proven Significant Only in Large Samples

The examples in the previous paragraphs indicate that if a relationship between variables in question is objectively (i.e., in the population) small, there is no way to identify such a relation in a study unless the research sample is correspondingly large. Even if our sample is in fact perfectly representative, the effect will not be statistically significant if the sample is small. Analogously, if a relation in question is objectively very large (i.e., in the population), it can be found to be highly significant even in a study based on a very small sample. Consider the following example. If a coin is slightly asymmetrical, and when tossed is somewhat more likely to produce heads than tails (e.g., 60% vs. 40%), 10 tosses would not be sufficient to convince anyone that the coin is asymmetrical, even if the outcome obtained (six heads and four tails) was perfectly representative of the bias of the coin. However, is it so that 10 tosses is not enough to prove anything? No, if the effect in question were large enough, 10 tosses could be quite enough. For instance, imagine now that the coin is so asymmetrical that no matter how you toss it, the outcome will be heads. If you toss such a coin 10 times and each toss produces heads, most people would consider it sufficient evidence that something is wrong with the coin. In other words, it would be considered convincing evidence that in the theoretical population of an infinite number of tosses of this coin there would be more heads than tails. Thus, if a relation is large, it can be found to be significant even in a small sample.

Can "No Relation" Be a Significant Result?

The smaller the relation between variables, the larger the sample size that is necessary to prove it significant. For example, imagine how many tosses would be necessary to prove that a coin is asymmetrical if its bias were only .000001%! Thus, the necessary minimum sample size increases as the magnitude of the effect to be demonstrated decreases. When the magnitude of the effect approaches 0, the necessary sample size to conclusively prove it approaches infinity. That is to say, if there is almost no relation between two variables, the sample size must be almost equal to the population size, which is assumed to be infinitely large. Statistical significance represents the probability that a similar outcome would be obtained if we tested the entire population. Thus, everything that would be found after testing the entire population would be, by definition, significant at the highest possible level, and this also includes all no-relation results.

How to Measure the Magnitude (Strength) of Relations between Variables

There are very many measures of the magnitude of relationships between variables which have been developed by statisticians; the choice of a specific measure in given circumstances depends on the number of variables involved, measurement scales used, nature of the relations, etc. Almost all of them, however, follow one general principle: they attempt to somehow evaluate the observed relation by comparing it to the maximum imaginable relation between those specific variables. Technically speaking, a common way to perform such evaluations is to look at how differentiated the values of the variables are,

and then calculate what part of this overall available differentiation is accounted for by instances when that differentiation is common in the two (or more) variables in question. Speaking less technically, we compare what is common in those variables to what potentially could have been common if the variables were perfectly related. Let's consider a simple example. Let's say that in our sample, the average index of *WCC* is *100* in *males* and *102* in *females*. Thus, we could say that on average, the deviation of each individual score from the grand mean (*101*) contains a component due to the gender of the subject; the size of this component is *1*. That value, in a sense, represents some measure of relation between *Gender* and *WCC*. However, this value is a very poor measure, because it does not tell us how relatively large this component is, given the overall differentiation of *WCC* scores. Consider two extreme possibilities:

a. If all *WCC* scores of *males* were equal to exactly *100*, and those of *females* equal to *102*, all deviations from the grand mean in our sample would be entirely accounted for by gender. We would say that in our sample, *Gender* is perfectly correlated with *WCC*, that is, 100% of the observed differences between subjects regarding their *WCC* is accounted for by their gender.

b. If *WCC* scores were in the range of *0-1000*, the same difference (of *2*) between the average *WCC* of *males* and *females* found in the study would account for such a small part of the overall differentiation of scores that most likely it would be considered negligible. For example, one more subject taken into account could change, or even reverse the direction of the difference. Therefore, every good measure of relations between variables must take into account the overall differentiation of individual scores in the sample and evaluate the relation in terms of (relatively) how much of this differentiation is accounted for by the relation in question.

Common "General Format" of Most Statistical Tests

Because the ultimate goal of most statistical tests is to evaluate relations between variables, most statistical tests follow the general format that was explained in the previous paragraph. Technically speaking, they represent a ratio of some measure of the differentiation common in the variables in question to the overall differentiation of those variables. For example, they represent a ratio of the part of the overall differentiation of the *WCC* scores that can be accounted for by gender to the overall differentiation of the *WCC* scores. This ratio is usually called a ratio of explained variation to total variation. In statistics, the term explained variation does not necessarily imply that we conceptually understand it. It is used only to denote the common variation in the variables in question, that is, the part of variation in one variable that is explained by the specific values of the other variable, and vice versa.

How the Level of Statistical Significance is Calculated

Let's assume that we have already calculated a measure of a relation between two variables (as explained above). The next question is, "how significant is this relation?" For example, is 40% of the explained variance between the two variables enough to consider the relation significant? The answer is, "it depends." Specifically, the significance depends mostly on the sample size. As explained before, in very

large samples, even very small relations between variables will be significant, whereas in very small samples even very large relations cannot be considered reliable (significant). Thus, in order to determine the level of statistical significance, we need a function that represents the relationship between magnitude and significance of relations between two variables, depending on the sample size. The function we need would tell us exactly how likely it is to obtain a relation of a given magnitude (or larger) from a sample of a given size, assuming that there is no such relation between those variables in the population. In other words, that function would give us the significance (*p*) level, and it would tell us the probability of error involved in rejecting the idea that the relation in question does not exist in the population. This alternative hypothesis (that there is no relation in the population) is usually called the null hypothesis. It would be ideal if the probability function was linear, and for example, only had different slopes for different sample sizes. Unfortunately, the function is more complex, and is not always exactly the same; however, in most cases we know its shape and can use it to determine the significance levels for our findings in samples of a particular size. Most of those functions are related to a general type of function called normal.

Why the Normal Distribution is Important

The normal distribution is important because in most cases, it well approximates the function that was introduced in the previous paragraph (for a detailed example, see *Are All Test Statistics Normally Distributed?*, page 11). The distribution of many test statistics is normal or

follows some form that can be derived from the normal distribution. In this sense, philosophically speaking, the normal distribution represents one of the empirically verified elementary truths about the general nature of reality, and its status can be compared to the one of fundamental laws of natural sciences. The exact shape of the normal distribution (the characteristic bell curve) is defined by a function that has only two parameters: mean and standard deviation.

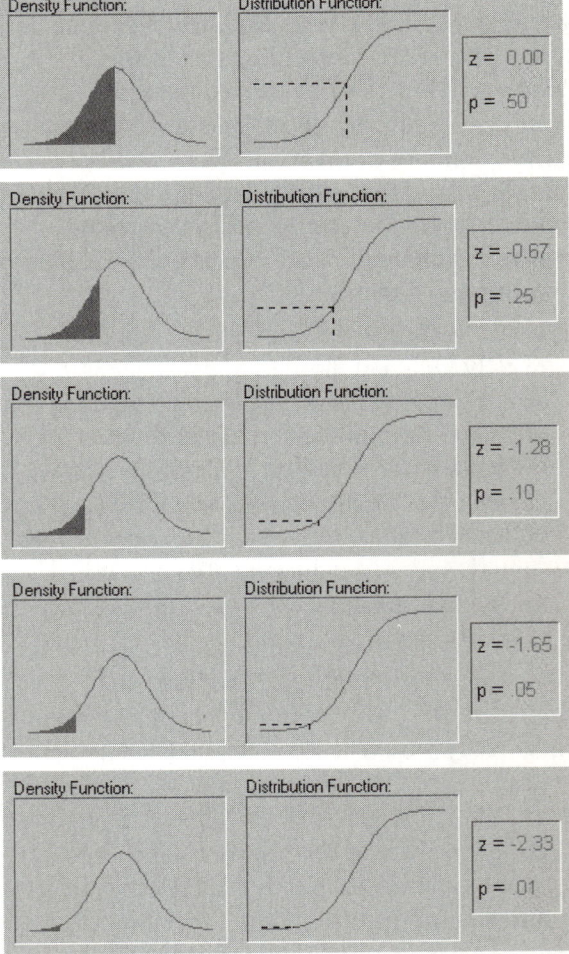

A characteristic property of the normal distribution is that 68% of all of its observations fall within a range of ±1 standard deviation from the mean, and a range of ±2 standard deviations includes 95% of the scores. In other words, in a normal distribution, observations that have a standardized value of less than -2 or more than +2 have a relative frequency of 5% or less. (Standardized value means that a value is expressed in terms of its difference from the mean, divided by the standard deviation.)

Illustration of How the Normal Distribution is Used in Statistical Reasoning (Induction)

Recall the example discussed previously, where pairs of samples of *males* and *females* were drawn from a population in which the average value of *WCC* in *males* and *females* was exactly the same. Although the most likely outcome of such experiments (one pair of samples per experiment) was that the difference between the average *WCC* in *males* and *females* in each pair is close to zero, from time to time, a pair of samples will be drawn where the difference between *males* and *females* is quite different from zero. How often does it happen? If the sample size is large enough, the results of such replications are normally distributed (this important principle is explained and illustrated in the next paragraph), and thus knowing the shape of the normal curve, we can precisely calculate the probability of obtaining by-chance outcomes representing various levels of deviation from the hypothetical population mean of zero. If such a calculated probability is so low that it meets the previously accepted criterion of statistical significance, we have only one choice: conclude that our result gives a better approximation of what is going on in the population than the null hypothesis (remember

that the null hypothesis was considered only for technical reasons as a benchmark against which our empirical result was evaluated). Note that this entire reasoning is based on the assumption that the shape of the distribution of those replications (technically, the sampling distribution) is normal.

Are All Test Statistics Normally Distributed?

Not all, but most test statistics are based either on the normal distribution directly or on distributions that are related to, and can be derived from normal, such as *t*, *F*, or *chi*-square. Typically, those tests require that the variables analyzed are themselves normally distributed in the population, that is, they meet the so-called normality assumption. Many observed variables actually are normally distributed, which is another reason why the normal distribution represents a general feature of empirical reality. The problem may occur when you try to use a normal distribution-based test to analyze data from variables that are themselves not normally distributed (see *Normality Tests* in the *Statistical Glossary*). In such cases, we have two general choices. First, we can use some alternative nonparametric test (or so-called distribution-free test; see Chapter 29 – *Nonparametric Statistics*); but this is often inconvenient because such tests are typically less powerful and less flexible in terms of types of conclusions that they can provide. Alternatively, in many cases we can still use the normal distribution-based test if we make sure that the size of our samples is large enough.

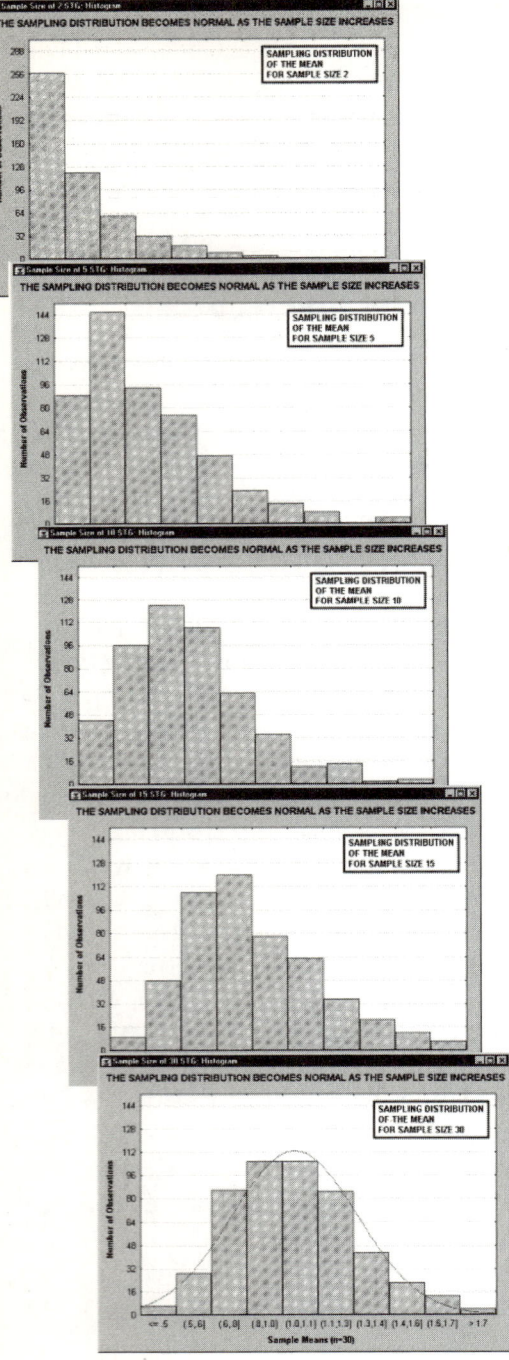

The latter option is based on an extremely important principle that is largely responsible for the popularity of tests that are based on the normal function. Namely, as the sample size increases, the shape of the sampling distribution (i.e., distribution of a statistic from the sample; this term was first used by Fisher, 1928a) approaches normal shape, even if the distribution of the variable in question is not normal. This principle is displayed in the previous illustrations, which show a series of sampling distributions (created with gradually increasing sample sizes of: 2, 5, 10, 15, and 30) using a variable that is clearly non-normal in the population, that is, the distribution of its values is clearly skewed.

However, as the sample size (of samples used to create the sampling distribution of the mean) increases, the shape of the sampling distribution becomes normal. Note that for n=30, the shape of that distribution is almost perfectly normal (see the close match of the fit).

This principle is called the central limit theorem (this term was first used by Pólya, 1920; German, "Zentraler Grenzwertsatz").

How Do We Know the Consequences of Violating the Normality Assumption?

Although many of the statements made in the preceding paragraphs can be proven mathematically, some of them do not have theoretical proofs and can be demonstrated only empirically, via so-called Monte-Carlo experiments. In these experiments, a computer following predesigned specifications generates large numbers of samples and the results from such samples are analyzed using a variety of tests. This way we can empirically evaluate the type and magnitude of errors or biases to which we are exposed when certain theoretical assumptions of the tests we are using are not met by our data. Specifically, Monte-Carlo studies were used extensively with normal distribution-based tests to determine how sensitive they are to violations of the assumption of normal distribution of the analyzed variables in the population. The general conclusion from these studies is that the consequences of such violations are less severe than previously thought. Although these conclusions should not entirely discourage anyone from being concerned about the normality assumption, they have increased the overall popularity of the distribution-dependent statistical tests in all areas of research.

BASIC STATISTICS AND TABLES

The statistics discussed in this chapter are conventionally called basic statistics and are often discussed as a group because they are usually used together in the initial, exploratory phase of data analysis. In fact, they include tests that serve different purposes. Basic statistical exploratory methods include such techniques as examining distributions of variables (e.g., to identify highly skewed or non-normal, such as bi-modal patterns), reviewing large correlation matrices for coefficients that meet certain thresholds, or examining multi-way frequency tables. Recommended introductory statistics textbooks are Kachigan (1986, and Runyon and Haber (1976); for a more advanced discussion of elementary theory of basic statistics, see the classic books by Hays (1988) and Kendall and Stuart (1979).

Descriptive Statistics

"True" Mean and Confidence Interval

Probably the most often used descriptive statistic is the mean. The mean is a particularly informative measure of the central tendency of the variable if it is reported along with its confidence intervals. As mentioned previously, usually we are interested in statistics (such as the mean) from our sample only to the extent to which they can infer information about the population. The confidence intervals for the mean give us a range of values around the mean where we expect the "true" (population) mean is located (with a given level of certainty). For example, if the mean in your sample is 23, and the lower and upper limits of the $p=.05$ confidence interval are 19 and 27 respectively, you can conclude that there is a 95% probability that the population mean is greater than 19 and lower than 27. If you set the p-level to a smaller value, the interval would become wider thereby increasing the certainty of the estimate, and vice versa; as we all know from a weather forecast, the more vague the prediction (i.e., wider the confidence interval), the more likely it will materialize. Note that the width of the confidence interval depends on the sample size and on the variation of data values. The larger the sample size, the more reliable its mean. The larger the variation, the less reliable the mean. The calculation of confidence intervals is based on the assumption that the variable is normally distributed in the population. The estimate may not be valid if this assumption is not met, unless the sample size is large, say $n=100$ or more.

Distribution Shape, Normality

An important aspect of the description of a variable is the shape of its distribution, which tells you the frequency of values from different ranges of the variable. Typically, a researcher is interested in how well the distribution can be approximated by the normal distribution (see the following illustrations for an example of this distribution).

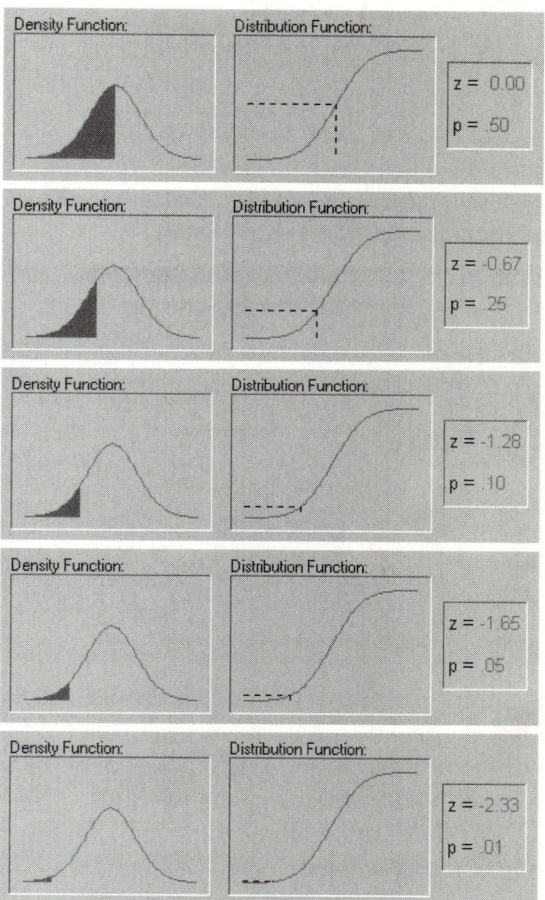

Simple descriptive statistics can provide information relevant to this issue. For example, if the skewness (which measures the deviation of the distribution from symmetry) is clearly different from 0, that distribution is asymmetrical, while normal distributions are perfectly symmetrical. If the kurtosis (which measures peakedness of the distribution) is clearly different from 0, the distribution is either flatter or more peaked than normal; the kurtosis of the normal distribution is 0.

More precise information can be obtained by performing one of the tests of normality to determine the probability that the sample came from a normally distributed population of observations (e.g., the so-called Kolmogorov-Smirnov test, or the Shapiro-Wilks' W test) . However, none of these tests can entirely substitute for the visual examination of the data using a histogram (i.e., a graph that shows the frequency distribution of a variable).

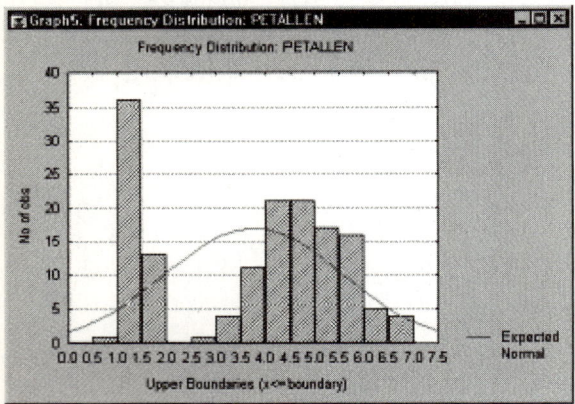

The graph enables you to evaluate the normality of the empirical distribution because it also shows the normal curve superimposed over the histogram. You can also examine various aspects of the distribution qualitatively. For example, the distribution could be bimodal (have 2 peaks). This might suggest that the sample is not homogeneous but possibly its elements came from two different populations, each more or less normally distributed. In such cases, in order to understand the nature of the variable in question, you should look for a way to quantitatively identify the two sub-samples.

Correlations

Purpose (What is Correlation?)

Correlation is a measure of the relation between two or more variables. The measurement scales used should be at least interval scales, but other correlation coefficients are available to handle other types of data. Correlation coefficients can range from -1.00 to +1.00. The value of -1.00 represents a perfect negative correlation while a value of +1.00 represents a perfect positive correlation. A value of 0.00 represents a lack of correlation.

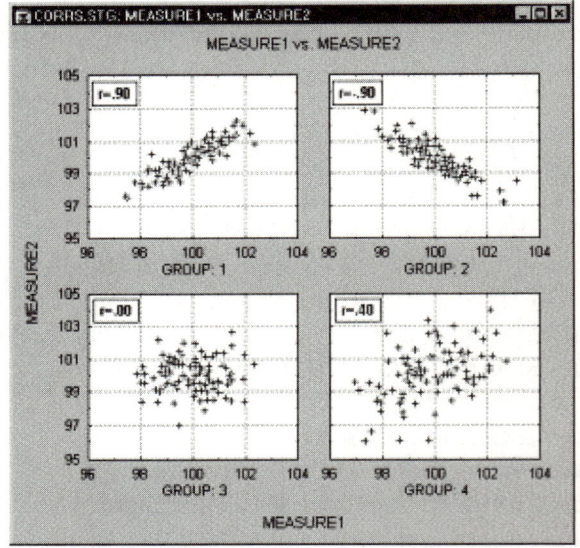

The most widely used type of correlation coefficient is Pearson r, also called linear or product-moment correlation.

Simple Linear Correlation (Pearson r)

Pearson correlation (hereafter called correlation), assumes that the two variables are measured on at least interval scales, and it determines the extent to which values of the two variables are proportional to each other. The

value of correlation (i.e., correlation coefficient) does not depend on the specific measurement units used; for example, the correlation between height and weight will be identical regardless of whether inches and pounds, or centimeters and kilograms are used as measurement units. Proportional means linearly related; that is, the correlation is high if it can be summarized by a straight line (sloped upward or downward).

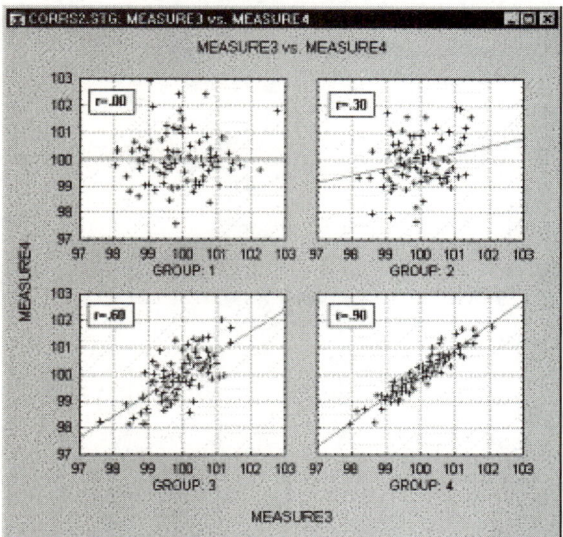

This line is called the regression line or least squares line because it is determined such that the sum of the squared distances of all the data points from the line is the lowest possible. Note that the concept of squared distances will have important functional consequences on how the value of the correlation coefficient reacts to various specific arrangements of data (as we will later see).

How to Interpret the Values of Correlations

As mentioned before, the correlation coefficient (r) represents the linear relationship between two variables. If the correlation coefficient is

squared, the resulting value (r^2, the coefficient of determination) will represent the proportion of common variation in the two variables (i.e., the strength or magnitude of the relationship). In order to evaluate the correlation between variables, it is important to know this magnitude or strength as well as the significance of the correlation.

Significance of Correlations

The significance level calculated for each correlation is a primary source of information about the reliability of the correlation. As explained previously, the significance of a correlation coefficient of a particular magnitude will change depending on the size of the sample from which it was computed. The test of significance is based on the assumption that the distribution of the residual values (i.e., the deviations from the regression line) for the dependent variable y follows the normal distribution, and that the variability of the residual values is the same for all values of the independent variable x. However, Monte Carlo studies suggest that meeting those assumptions closely is not absolutely crucial if your sample size is not very small and when the departure from normality is not very large. It is impossible to formulate precise recommendations based on those Monte Carlo results, but many researchers follow a rule of thumb that if your sample size is 50 or more then serious biases are unlikely, and if your sample size is more than 100, you should not be concerned at all with the normality assumptions. There are, however, much more common and serious threats to the validity of information that a correlation coefficient can provide; they are briefly discussed in the following paragraphs.

Outliers

Outliers are atypical (by definition), infrequent observations. Because of the way in which the regression line is determined (especially the fact that it is based on minimizing not the sum of simple distances but the sum of squares of distances of data points from the line), outliers have a profound influence on the slope of the regression line. A single outlier is capable of considerably changing the slope of the regression line and, consequently, the value of the correlation, as shown in the following illustration. Just one outlier can be entirely responsible for a high value of the correlation that otherwise (without the outlier) would be close to zero. Needless to say, never base important conclusions on the value of the correlation coefficient alone; examining the scatterplot also is recommended.

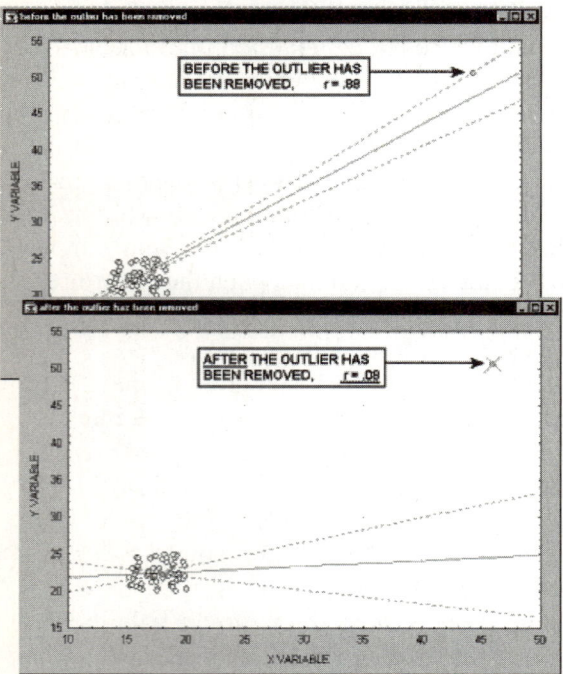

If the sample size is relatively small, including or excluding specific data points that are not

clearly outliers (as the one shown in the previous illustration) may have a profound influence on the regression line.

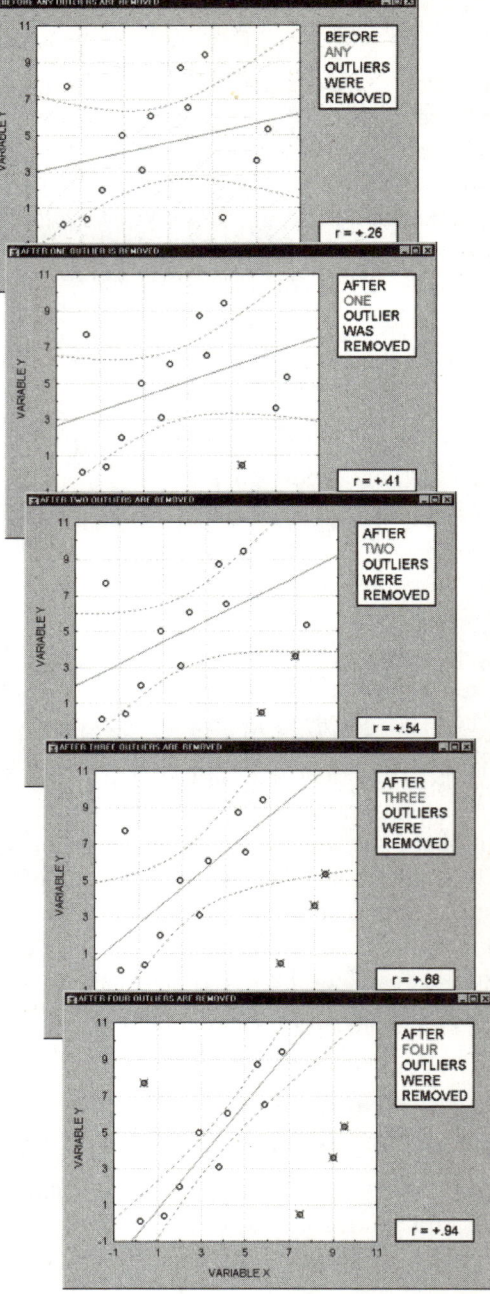

StatSoft®
Copyright © StatSoft, 2006

This is shown in the previous illustrations where we call the points being excluded outliers; you may argue, however, that they are not outliers, but rather extreme values.

Typically, we believe that outliers represent a random error that we want to be able to control. Unfortunately, there is no widely accepted method to remove outliers automatically (however, see the next paragraph), thus, what we are left with is to identify any outliers by examining a scatterplot of each important correlation. Needless to say, outliers may not only artificially increase the value of a correlation coefficient, they may also decrease the value of a legitimate correlation. See also, *Ellipse* in the *Statistical Glossary*.

Quantitative Approach to Outliers

Some researchers use quantitative methods to exclude outliers. For example, they exclude observations that are outside the range of ±2 standard deviations (or even ±1.5 sd's) around the group or design cell mean. In some areas of research, such cleaning of the data is absolutely necessary. For example, in cognitive psychology research on reaction times, even if almost all scores in an experiment are in the range of 300-700 milliseconds, just a few "distracted reactions" of 10-15 seconds will completely change the overall picture. Unfortunately, defining an outlier is subjective (as it should be), and the decisions concerning how to identify them must be made on an individual basis (taking into account specific experimental paradigms and/or accepted practice and general research experience in the respective area). It should also be noted that in some rare cases, the relative frequency of outliers across a number of groups or cells of a design can be subjected to analysis and provide interpretable results. For

example, outliers could be indicative of the occurrence of a phenomenon that is qualitatively different than the typical pattern observed or expected in the sample, thus the relative frequency of outliers could provide evidence of a relative frequency of departure from the process or phenomenon that is typical for the majority of cases in a group.

Correlations in Non-Homogeneous Groups

A lack of homogeneity in the sample from which a correlation was calculated can be another factor that biases the value of the correlation. Imagine a case where a correlation coefficient is calculated from data points that came from two different experimental groups, but this fact is ignored when the correlation is calculated. Let's assume that the experimental manipulation in one of the groups increased the values of both correlated variables and thus the data from each group form a distinctive "cloud" in the scatterplot (as shown in the next illustration).

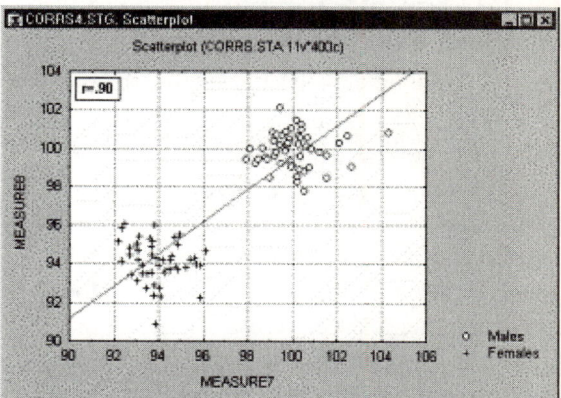

In such cases, a high correlation may result that is entirely due to the arrangement of the two groups, but which does not represent the true relation between the two variables, which may practically be equal to 0 (as could be seen if we

looked at each group separately, see the following graph).

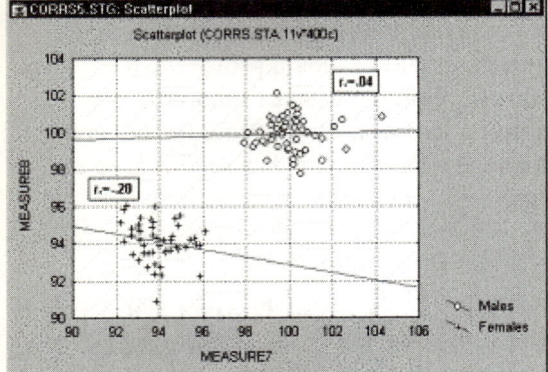

If you suspect the influence of such a phenomenon on your correlations and know how to identify such subsets of data, try to run the correlations separately in each subset of observations. If you do not know how to identify the hypothetical subsets, try to examine the data with some exploratory multivariate techniques (e.g., cluster analysis).

Nonlinear Relations between Variables

Another potential source of problems with the linear (Pearson r) correlation is the shape of the relation. As mentioned previously, Pearson r measures a relation between two variables only to the extent to which it is linear; deviations from linearity will increase the total sum of squared distances from the regression line even if they represent a true and very close relationship between two variables. The possibility of such nonlinear relationships is another reason why examining scatterplots is a necessary step in evaluating every correlation. For example, the following graph demonstrates an extremely strong correlation between the two variables that is not well described by the linear function.

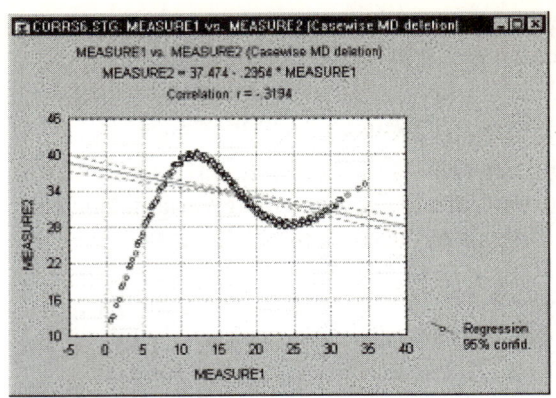

Measuring Nonlinear Relations

What do you do if a correlation is strong but clearly nonlinear (as concluded from examining scatterplots)? Unfortunately, there is no simple answer to this question, because there is no easy-to-use equivalent of Pearson r that is capable of handling nonlinear relations. If the curve is monotonous (continuously decreasing or increasing) you could try to transform one or both of the variables to remove the curvilinearity and then recalculate the correlation. For example, a typical transformation used in such cases is the logarithmic function, which will squeeze together the values at one end of the range. Another option available, if the relation is monotonous, is to try a nonparametric correlation (e.g., Spearman R, see Chapter 29 – *Nonparametric Statistics*), which is sensitive only to the ordinal arrangement of values, thus, by definition, it ignores monotonous curvilinearity. However, nonparametric correlations are generally less sensitive and sometimes this method will not produce any gains. Unfortunately, the two most precise methods are not easy to use and require a good deal of experimentation with the data. Therefore you could try to identify the specific function that best describes the curve. After a function has

been found, you can test its goodness-of-fit to your data.

Alternatively, you could experiment with dividing one of the variables into a number of segments (e.g., 4 or 5) of an equal width, treat this new variable as a grouping variable and run an analysis of variance on the data.

Exploratory Examination of Correlation Matrices

A common first step of many data analyses that involve more than a very few variables is to run a correlation matrix of all variables and then examine it for expected (and unexpected) significant relations. When this is done, you need to be aware of the general nature of statistical significance; specifically, if you run many tests (in this case, many correlations), significant results will be found surprisingly often due to pure chance. For example, by definition, a coefficient significant at the .05 level will occur by chance once in every 20 coefficients. There is no automatic way to identify the true correlations. Thus, you should treat all results that were not predicted or planned with particular caution and look for their consistency with other results; ultimately, though, the most conclusive (although costly) control for such a randomness factor is to replicate the study. This issue is general and it pertains to all analyses that involve multiple comparisons and statistical significance. This problem is also briefly discussed in *Breakdowns*: *Descriptive Statistics by Groups* (page 28) and *Post-hoc Comparisons of Means* (page 29).

Casewise vs. Pairwise Deletion of Missing Data

The default way of deleting missing data while calculating a correlation matrix is to exclude all cases that have missing data in at least one of the selected variables; that is, by casewise deletion of missing data. Only this way will you get a true correlation matrix, where all correlations are obtained from the same set of observations. However, if missing data are randomly distributed across cases, you could easily end up with no valid cases in the data set, because each of them will have at least one missing data in some variable. The most common solution used in such instances is to use so-called pairwise deletion of missing data in correlation matrices, where a correlation between each pair of variables is calculated from all cases that have valid data on those two variables. In many instances there is nothing wrong with that method, especially when the total percentage of missing data is low, say 10%, and they are relatively randomly distributed between cases and variables. However, it may sometimes lead to serious problems.

For example, a systematic bias may result from a hidden systematic distribution of missing data, causing different correlation coefficients in the same correlation matrix to be based on different subsets of subjects. In addition to the possibly biased conclusions that you could derive from such pairwise calculated correlation matrices, real problems may occur when you subject such matrices to another analysis (e.g., multiple regression, factor analysis, or cluster analysis) that expects a true correlation matrix, with a certain level of consistency and transitivity between different coefficients. Thus, if you are using the pairwise method of deleting the missing data, be sure to examine the distribution of missing data across the cells of the matrix for possible systematic patterns.

How to Identify Biases Caused by the Bias Due to Pairwise Deletion of Missing Data

If the pairwise deletion of missing data does not introduce any systematic bias to the correlation matrix, all those pairwise descriptive statistics for one variable should be very similar. However, if they differ, there are good reasons to suspect a bias. For example, if the mean (or standard deviation) of the values of variable A that were taken into account in calculating its correlation with variable B is much lower than the mean (or standard deviation) of those values of variable A that were used in calculating its correlation with variable C, we would have good reason to suspect that those two correlations (A-B and A-C) are based on different subsets of data, and thus, that there is a bias in the correlation matrix caused by a non-random distribution of missing data.

Pairwise Deletion of Missing Data vs. Mean Substitution

Another common method to avoid losing data due to casewise deletion is the so-called mean substitution of missing data (replacing all missing data in a variable by the mean of that variable). Mean substitution offers some advantages and some disadvantages as compared to pairwise deletion. Its main advantage is that it produces internally consistent sets of results (true correlation matrices). The main disadvantages are that mean substitution artificially decreases the variation of scores, and this decrease in individual variables is proportional to the number of missing data (i.e., the more missing data, the more perfectly average scores will be artificially added to the data set). Because it substitutes missing data with artificially created

average data points, mean substitution may considerably change the values of correlations.

Spurious Correlations

Although you cannot prove causal relations based on correlation coefficients (see Chapter 1 – *Elementary Concepts in Statistics*), you can still identify so-called spurious correlations; that is, correlations that are due mostly to the influences of other variables. For example, there is a correlation between the total amount of losses in a fire and the number of firemen that were putting out the fire; however, what this correlation does not indicate is that if you call fewer firemen then you would lower the losses. There is a third variable (the initial size of the fire) that influences both the amount of losses and the number of firemen. If you control for this variable (e.g., consider only fires of a fixed size), the correlation will either disappear or perhaps even change its sign. The main problem with spurious correlations is that we typically do not know what the hidden agent is. However, in cases when we know where to look, we can use partial correlations that control for (partial out) the influence of specified variables.

Are Correlation Coefficients Additive?

No, they are not. For example, an average of correlation coefficients in a number of samples does not represent an average correlation in all those samples. Because the value of the correlation coefficient is not a linear function of the magnitude of the relation between the variables, correlation coefficients cannot simply be averaged. In cases when you need to average correlations, they first have to be converted into additive measures. For example, before averaging, you can square them to obtain coefficients of determination, which are additive (as explained before in this section), or convert them into so-called Fisher z values, which are also additive.

How to Determine Whether Two Correlation Coefficients are Significant

A test is available that will evaluate the significance of differences between two correlation coefficients in two samples. The outcome of this test depends not only on the size of the raw difference between the two coefficients but also on the size of the samples and on the size of the coefficients themselves. Consistent with the previously discussed principle, the larger the sample size, the smaller the effect that can be proven significant in that sample. In general, due to the fact that the reliability of the correlation coefficient increases with its absolute value, relatively small differences between large correlation coefficients can be significant. For example, a difference of .10 between two correlations may not be significant if the two coefficients are .15 and .25, although in the same sample, the same difference of .10 can be highly significant if the two coefficients are .80 and .90.

t-Test for Independent Samples

Purpose, Assumptions

The t-test is the most commonly used method to evaluate the differences in means between two groups. For example, the t-test can be used to test for a difference in test scores between a group of patients who were given a drug and a control group who received a placebo. Theoretically, the t-test can be used even if the sample sizes are very small (e.g., as small as 10; some researchers claim that even smaller n's are possible), as long as the variables are normally distributed within

each group and the variation of scores in the two groups is not reliably different. The normality assumption can be evaluated by looking at the distribution of the data (via histograms) or by performing a normality test. The equality of variances assumption can be verified with the F test, or you can use the more robust Levene's test. If these conditions are not met, you can evaluate the differences in means between two groups using one of the nonparametric alternatives to the t-test (see Chapter 29 – *Nonparametric Statistics*).

The p-level reported with a t-test represents the probability of error involved in accepting our research hypothesis about the existence of a difference. Technically speaking, this is the probability of error associated with rejecting the hypothesis of no difference between the two categories of observations (corresponding to the groups) in the population when, in fact, the hypothesis is true. Some researchers suggest that if the difference is in the predicted direction, you can consider only one half (one tail) of the probability distribution and thus divide the standard p-level reported with a t-test (a two-tailed probability) by two. Others, however, suggest that you should always report the standard, two-tailed t-test probability. See also, *Student's t Distribution* in the *Statistical Glossary*.

Arrangement of Data

In order to perform the t-test for independent samples, one independent (grouping) variable (e.g., *Gender*: *male/female*) and at least one dependent variable (e.g., a *test score*) are required.

StatSoft®
Copyright © StatSoft, 2006

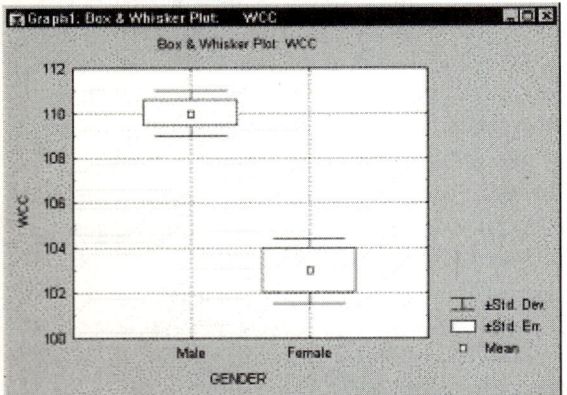

The means of the dependent variable will be compared between selected groups based on the specified values (e.g., *male* and *female*) of the independent variable. The data set shown above can be analyzed with a *t*-test comparing the average *WCC* score in *males* and *females*.

t-Test Graphs

In the *t*-test analysis, comparisons of means and measures of variation in the two groups can be visualized in box and whisker plots (for an example, see the next illustration).

These graphs help you to quickly evaluate and intuitively visualize the strength of the relation between the grouping and the dependent variable.

More Complex Group Comparisons

It often happens in research practice that you need to compare more than two groups (e.g., *drug 1*, *drug 2*, and *placebo*), or compare groups created by more than one independent variable while controlling for the separate

influence of each of them (e.g., *Gender*, *type of Drug*, and *size of Dose*). In these cases, you need to analyze the data using analysis of variance, which can be considered to be a generalization of the *t*-test. In fact, for two group comparisons, ANOVA will give results identical to a *t*-test ($t^2 [df] = F[1,df]$). However, when the design is more complex, ANOVA offers numerous advantages that *t*-tests cannot provide (even if you run a series of *t*-tests comparing various cells of the design).

t-Test for Dependent Samples

Within-Group Variation

As explained in Chapter 1 – *Elementary Concepts in Statistics*, the size of a relation between two variables, such as the one measured by a difference in means between two groups, depends to a large extent on the differentiation of values within the group. Depending on how differentiated the values are in each group, a given raw difference in-group means indicates either a stronger or weaker relationship between the independent (grouping) and dependent variable. For example, if the mean *WCC* (*White Cell Count*) was *102* in *males* and *104* in *females*, this difference of "only" *2* points would be extremely important if all values for *males* fell within a range of *101* to *103*, and all scores for *females* fell within a range of *103* to *105*; for example, we would be able to predict *WCC* pretty well based on gender. However, if the same difference of *2* were obtained from very differentiated scores (e.g., if their range was *0-200*), we would consider the difference entirely negligible. That is to say, reduction of the within-group variation increases the sensitivity of our test.

Purpose

The *t*-test for dependent samples helps us to take advantage of one specific type of design in which an important source of within-group variation (or so-called, error) can be easily identified and excluded from the analysis. Specifically, if two groups of observations (that are to be compared) are based on the same sample of subjects who were tested twice (e.g., before and after a treatment), a considerable part of the within-group variation in both groups of scores can be attributed to the initial individual differences between subjects. Note that, in a sense, this fact is not much different than in cases when the two groups are entirely independent (see *t-Test for Independent and Dependent Samples* in the *Statistical Glossary*), where individual differences also contribute to the error variance; but in the case of independent samples, we cannot do anything about it because we cannot identify (or subtract) the variation due to individual differences in subjects. However, if the same sample was tested twice, we can easily identify (or subtract) this variation. Specifically, instead of treating each group separately, and analyzing raw scores, we can look only at the differences between the two measures (e.g., pre-test and post test) in each subject. By subtracting the first score from the second for each subject and then analyzing only those pure (paired) differences, we will exclude the entire part of the variation in our data set that results from unequal base levels of individual subjects. This is precisely what is being done in the *t*-test for dependent samples, and, as compared to the *t*-test for independent samples, it always produces better results (i.e., it is always more sensitive).

Assumptions

The theoretical assumptions of the *t*-test for independent samples also apply to the dependent samples test; that is, the paired differences should be normally distributed. If these assumptions are clearly not met, one of the nonparametric alternative tests should be used. See also, *Student's t Distribution* in the *Statistical Glossary*.

Arrangement of Data

Technically, we can apply the *t*-test for dependent samples to any two variables in our data set. However, applying this test will make very little sense if the values of the two variables in the data set are not logically and methodologically comparable. For example, if you compare the average *WCC* in a sample of patients before and after a treatment, but using a different counting method or different units in the second measurement, a highly significant *t*-test value could be obtained due to an artifact; that is, to the change of units of measurement. Following is an example of a data set that can be analyzed using the *t*-test for dependent samples.

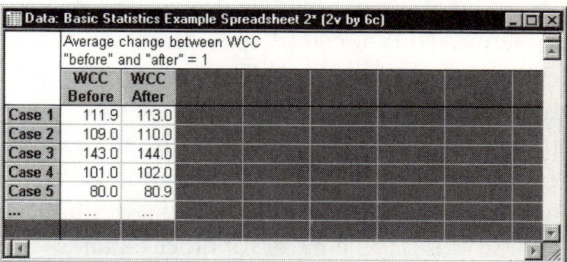

The average difference between the two conditions is relatively small (*d=1*) as compared to the differentiation (range) of the raw scores (from 80 to 143, in the first sample). However, the *t*-test for dependent samples analysis is performed only on the paired differences, ignoring the raw scores and their potential

differentiation. Thus, the size of this particular difference of *1* will be compared not to the differentiation of raw scores but to the differentiation of the individual difference scores, which is relatively small: *0.2* (from *0.9* to *1.1*). Compared to that variability, the difference of *1* is extremely large and can yield a highly significant *t* value.

Matrices of *t*-Tests

t-tests for dependent samples can be calculated for long lists of variables and reviewed in the form of matrices produced with casewise or pairwise deletion of missing data, much as the correlation matrices. Thus, the precautions discussed in the context of correlations also apply to *t*-test matrices; see the issue of artifacts caused by the pairwise deletion of missing data and the issue of randomly significant test values (see page 23).

More Complex Group Comparisons

If there are more than two correlated samples (e.g., *before treatment*, *after treatment 1*, and *after treatment 2*), analysis of variance with repeated measures should be used. The repeated measures ANOVA can be considered a generalization of the t-test for dependent samples and it offers various features that increase the overall sensitivity of the analysis. For example, it can simultaneously control not only for the base level of the dependent variable, but it can control for other factors and/or include in the design more than one interrelated dependent variable (MANOVA; for additional details refer to Chapter 3 – *ANOVA/MANOVA*).

Breakdown: Descriptive Statistics by Groups

Purpose

The breakdowns analysis calculates descriptive statistics and correlations for dependent variables in each of a number of groups defined by one or more grouping (independent) variables.

Arrangement of Data

In the following example data set (spreadsheet), the dependent variable *WCC* (*White Cell Count*) can be broken down by two independent variables: *Gender* (values: *males* and *females*), and *Height* (values: *tall* and *short*).

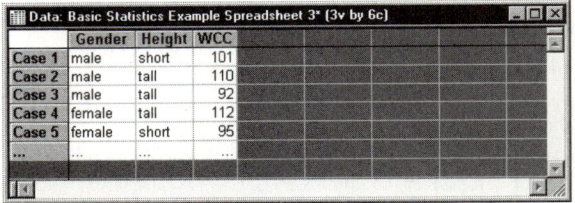

The resulting breakdowns might look as follows (we are assuming that *Gender* was specified as the first independent variable, and *Height* as the second).

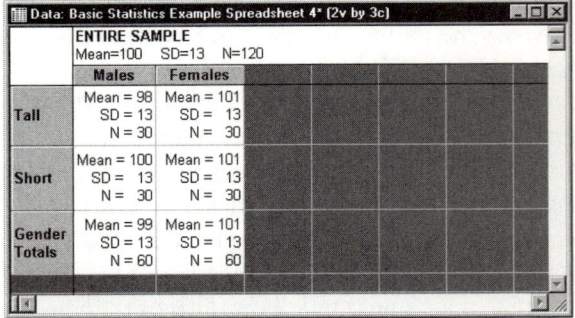

The composition of the intermediate level cells of the breakdown tree depends on the order in which independent variables are arranged. For example, in the above example, you see the means for *all males* and *all females* but you do not see the means for *all tall subjects* and *all short subjects*, which would have been produced had you specified independent variable *Height* as the first grouping variable rather than the second.

Statistical Tests in Breakdowns

Breakdowns are typically used as an exploratory data analysis technique; the typical question that this technique can help answer is very simple: Are the groups created by the independent variables different regarding the dependent variable? If you are interested in differences concerning the means, the appropriate test is the breakdowns one-way ANOVA (*F* test). If you are interested in variation differences, you should test for homogeneity of variances.

Other Related Data Analysis Techniques

Although for exploratory data analysis, breakdowns can use more than one independent variable, the statistical procedures in breakdowns assume the existence of a single grouping factor (even if, in fact, the breakdown results from a combination of a number of grouping variables). Thus, those statistics do not reveal or even take into account any possible interactions between grouping variables in the design. For example, there could be differences between the influence of one independent variable on the dependent variable at different levels of another independent variable (e.g., tall people could have lower WCC than short ones, but only if they are males; see the tree data above). You can explore such effects by examining breakdowns visually using different orders of independent variables, but the magnitude or significance of such effects cannot be estimated by the breakdown statistics.

Post-hoc Comparisons of Means

Usually, after obtaining a statistically significant *F* test from the ANOVA, you want to know which of the means contributed to the effect (i.e., which groups are particularly different from each other). You could of course perform a series of simple *t*-tests to compare all possible pairs of means. However, such a procedure would capitalize on chance. This means that the reported probability levels would actually overestimate the statistical significance of mean differences. Without going into too much detail, suppose you take 20 samples of 10 random numbers each, and compute 20 means. Then, take the group (sample) with the highest mean and compare it with that of the lowest mean. The *t*-test for independent samples will test whether or not those two means are significantly different from each other, provided they were the only two samples taken. Post-hoc comparison techniques, though, specifically take into account the fact that more than two samples were taken.

Breakdowns vs. Discriminant Function Analysis

Breakdowns can be considered a first step toward another type of analysis that explores differences between groups: discriminant function analysis. Similar to breakdowns, discriminant function analysis explores the differences between groups created by values (group codes) of an independent (grouping) variable. However, unlike breakdowns, discriminant function analysis simultaneously analyzes more than one dependent variable and it identifies patterns of values of those dependent variables. Technically, it determines a linear combination of the dependent variables that best predicts the group membership. For example, discriminant function analysis can be

used to analyze differences between three groups of persons who have chosen different professions (e.g., lawyers, physicians, and engineers) in terms of various aspects of their scholastic performance in high school. You could claim that such an analysis could explain the choice of a profession in terms of specific talents shown in high school; thus discriminant function analysis can be considered to be an exploratory extension of simple breakdowns.

Breakdowns vs. Frequency Tables

Another related type of analysis that cannot be directly performed with breakdowns is comparisons of frequencies of cases (n's) between groups. Specifically, often the n's in individual cells are not equal because the assignment of subjects to those groups typically results not from an experimenter's manipulation, but from subjects' pre-existing dispositions. If, in spite of the random selection of the entire sample, the n's are unequal, it may suggest that the independent variables are related. For example, crosstabulating levels of independent variables *Age* and *Education* most likely would not create groups of equal n, because education is distributed differently in different age groups. If you are interested in such comparisons, you can explore specific frequencies in the breakdowns tables, trying different orders of independent variables. However, in order to subject such differences to statistical tests, you should use crosstabulations and frequency tables, log-linear analysis, or correspondence analysis (for more advanced analyses on multi-way frequency tables).

Graphical Breakdowns

Graphs can often identify effects (both expected and unexpected) in the data more quickly and sometimes better than any other data analysis

method. With categorized graphs, you can plot the means, distributions, correlations, etc. across the groups of a given table (e.g., categorized histograms, categorized probability plots, categorized box and whisker plots). The graph in the next illustration shows a categorized histogram, which you can use to quickly evaluate and visualize the shape of the data for each group (group1-female, group2-female, etc.).

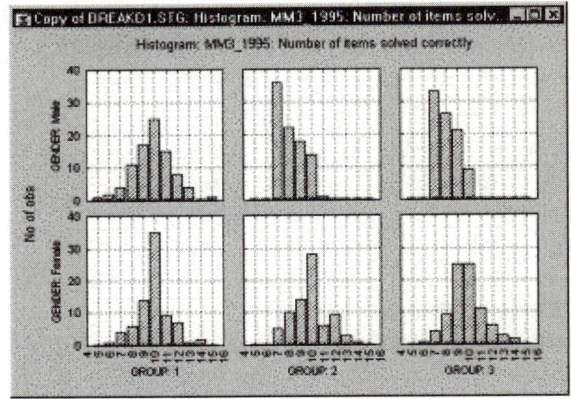

A categorized scatterplot shows the differences between patterns of correlations between dependent variables across the groups.

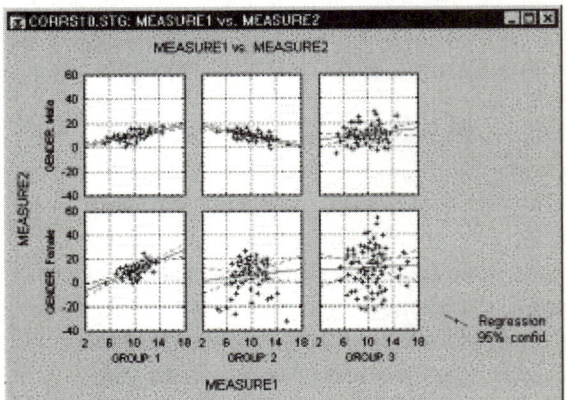

Additionally, if the software has a brushing facility that supports animated brushing, you can select (i.e., highlight) in a matrix scatterplot all data points that belong to a certain category in order to examine how those specific observations contribute to relations between other variables in the same data set.

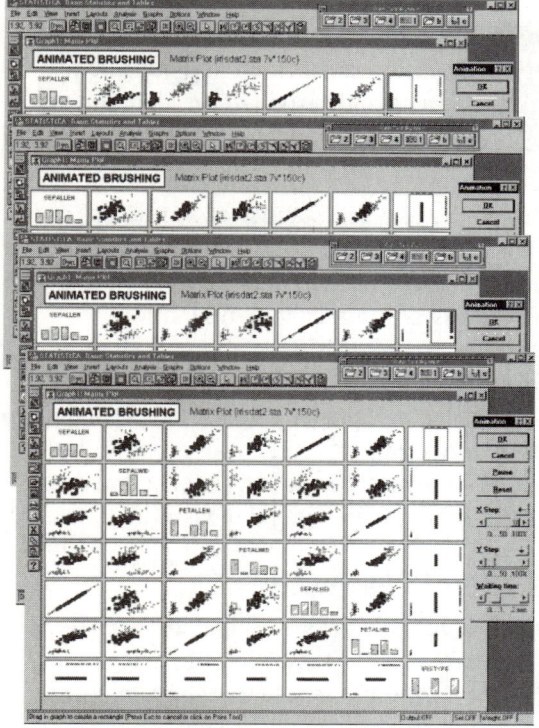

Frequency Tables

Purpose

Frequency or one-way tables represent the simplest method for analyzing categorical (nominal) data. They are often used as one of the exploratory procedures to review how different categories of values are distributed in the sample. For example, in a survey of spectator interest in different sports, we could summarize the respondents' interest in watching football in a frequency table as follows:

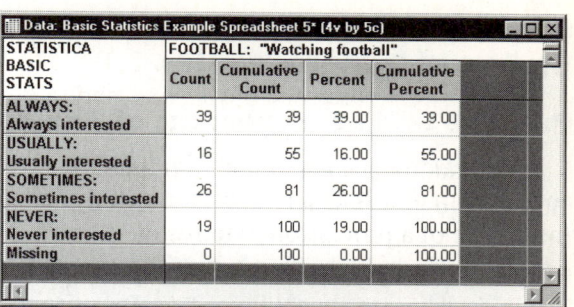

STATISTICA BASIC STATS	FOOTBALL: "Watching football"			
	Count	Cumulative Count	Percent	Cumulative Percent
ALWAYS: Always interested	39	39	39.00	39.00
USUALLY: Usually interested	16	55	16.00	55.00
SOMETIMES: Sometimes interested	26	81	26.00	81.00
NEVER: Never interested	19	100	19.00	100.00
Missing	0	100	0.00	100.00

This table shows the number, proportion, and cumulative proportion of respondents who characterized their interest in watching football as 1) *Always interested*, 2) *Usually interested*, 3) *Sometimes interested*, or 4) *Never interested*.

Applications

In practically every research project, a first look at the data usually includes frequency tables. For example, in survey research, frequency tables can show the number of males and females who participated in the survey, the number of respondents from specific ethnic and racial backgrounds, etc. Responses on some labeled attitude measurement scales (e.g., interest in watching football) can also be nicely summarized via the frequency table. In medical research, you can tabulate the number of patients displaying specific symptoms; in industrial research you can tabulate the frequency of different causes leading to catastrophic failure of products during stress tests (e.g., which parts are actually responsible for the malfunction of television sets under extreme temperatures). Customarily, if a data set includes any categorical data, one of the first steps in the data analysis is to compute a frequency table for those categorical variables.

Crosstabulation and Stub-and-Banner Tables

Purpose and Arrangement of Table

Crosstabulation is a combination of two (or more) frequency tables arranged so that each cell in the resulting table represents a unique combination of specific values of crosstabulated variables. Thus, crosstabulation enables us to examine frequencies of observations that belong to specific categories on more than one variable. By examining these frequencies, we can identify relations between crosstabulated variables. Only categorical (nominal) variables or variables with a relatively small number of different meaningful values should be crosstabulated. Note that in the cases where we do want to include a continuous variable in a crosstabulation (e.g., income), we can first *recode* it into a particular number of distinct ranges (e.g., low, medium, and high).

2x2 Table

The simplest form of crosstabulation is the 2x2 table where two variables are crossed, and each variable has only two distinct values. For example, suppose we conduct a simple study in which males and females are asked to choose one of two different brands of soda (brand *A* and brand *B*); the data file can be arranged as the following:

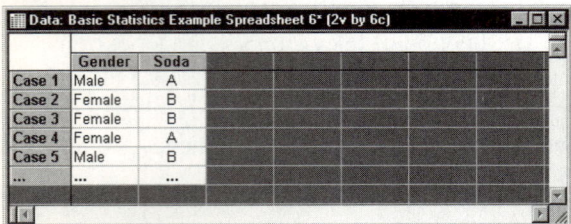

The following illustration shows how the resulting crosstabulation could appear:

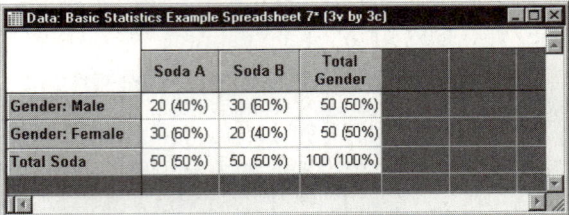

Each cell represents a unique combination of values of the two crosstabulated variables (row variable *Gender* and column variable *Soda*), and the numbers in each cell tell us how many observations fall into each combination of values. In general, this table shows us that more *females* than *males* chose the soda pop brand *A*, and that more *males* than *females* chose soda *B*. Thus, *Gender* and preference for a particular brand of soda may be related (later we will see how this relationship can be measured).

Marginal Frequencies

The values in the margins of the table are simply one-way (frequency) tables for all values in the table. They are important in that they help us to evaluate the arrangement of frequencies in individual columns or rows. For example, the frequencies of 40% and 60% of *males* and *females* (respectively) who chose soda *A* (see the first column of the previous table), would not indicate any relationship between *Gender* and *Soda* if the marginal frequencies for *Gender* were also 40% and 60%; in that case they would simply reflect the different proportions of males and females in the study. Thus, the differences between the distributions of frequencies in individual rows (or columns) and in the respective margins informs us of the relationship between the crosstabulated variables.

Column, Row, and Total Percentages

The example in the previous paragraph demonstrates that in order to evaluate relationships between crosstabulated variables, we need to compare the proportions of marginal and individual column or row frequencies. Such comparisons are easiest to perform when the frequencies are presented as percentages.

Graphical Representations of Crosstabulations

For analytic purposes, the individual rows or columns of a table can be represented as column graphs. However, often it is useful to visualize the entire table in a single graph. A two-way table can be visualized in a 3-dimensional histogram; alternatively, a categorized histogram can be produced, where one variable is represented by individual histograms that are drawn at each level (category) of the other variable in the crosstabulation. The advantage of the 3D histogram is that it produces an integrated picture of the entire table; the advantage of the categorized graph is that it enables us to precisely evaluate specific frequencies in each cell of the table.

Stub-and-Banner Tables

Stub-and-banner tables, or banners for short, are a way to display several two-way tables in a compressed form. This type of table is most easily explained with an example. Let's return to the survey of sports spectators example. (In order to simplify matters, only the response categories *Always* and *Usually* were tabulated in the following table.)

Data: Basic Statistics Example Spreadsheet 8 (3v by 6c)

Basic Statistics ---------- Factor	Stub-and-Banner Table - Row Percents		
	Football Always	Football Usually	Row Total
Baseball: Always	92.31	7.69	66.67
Baseball: Usually	61.54	38.46	33.33
BASEBALL: TOTAL	82.05	17.95	100.00
Tennis: Always	87.50	12.50	66.67
Tennis: Usually	87.50	12.50	33.33
TENNIS: TOTAL	87.50	12.50	100.00
Boxing: Always	77.78	22.22	52.94
Boxing: Usually	100.00	0.00	47.06
BOXING: TOTAL	88.24	11.76	100.00

Interpreting the Banner Table

In the previous illustration, we see the two-way tables of expressed interest in *Football* by expressed interest in *Baseball*, *Tennis*, and *Boxing*. The table entries represent percentages of rows, so that the percentages across columns will add up to 100 percent. For example, the number in the upper-left corner of the spreadsheet shows that *92.31* percent of all respondents who said they are always interested in watching football also said that they were always interested in watching baseball. Further down we can see that the percent of those always interested in watching football that were also always interested in watching tennis was *87.50* percent; for boxing this number is *77.78* percent. The percentages in the last column (*Row Total*) are always relative to the total number of cases.

Multi-Way Tables with Control Variables

When only two variables are crosstabulated, we call the resulting table a two-way table. However, the general idea of crosstabulating values of variables can be generalized to more than just two variables. For example, to return to the soda example presented earlier, a third variable could be added to the data set. This variable might contain information about the state in which the study was conducted (either *Nebraska* or *New York*).

	GENDER	SODA	STATE
case 1	Male	A	Nebraska
case 2	Female	B	New York
case 3	Female	B	Nebraska
case 4	Female	A	Nebraska
case 5	Male	B	New York
...	...	...	...

Data: Basic Statistics Example Spreadsheet 9* (3v by 6c)

The crosstabulation of these variables would result in a three-way table:

Data: Basic Statistics Example Spreadsheet 10

Gender	State	Soda A	Soda B	Row Totals
Male	Nebraska	5	45	50
Male	New York	20	30	50
Total		25	75	100
Female	Nebraska	45	5	50
Female	New York	30	20	50
Total		75	25	100
Column Total		100	100	200

Theoretically, an unlimited number of variables can be crosstabulated in a single multi-way table. However, research practice shows that it is usually difficult to examine and understand tables that involve more than four variables.

It is recommended to analyze relationships between the factors in such tables using modeling techniques such as log-linear analysis or correspondence analysis.

Graphical Representations of Multi-Way Tables

There is a variety of ways in which multi-way tables can be represented graphically. For example, you can produce double categorized histograms, 3D histograms,

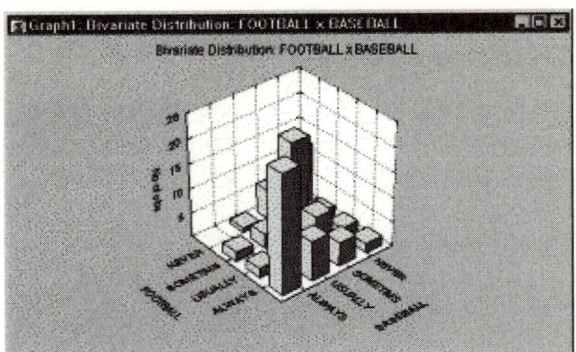

Graph1: Bivariate Distribution: FOOTBALL x BASEBALL

or line-plots that will summarize the frequencies for up to three factors in a single graph.

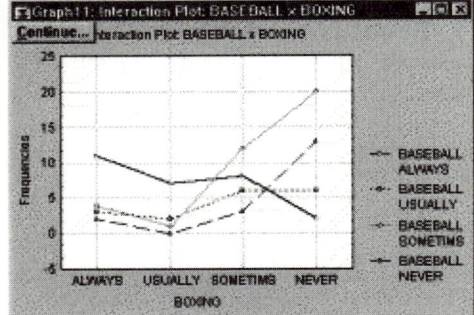

Graph11: Interaction Plot: BASEBALL x BOXING

Batches (cascades) of graphs can be used to summarize higher-way tables (as shown in the next illustration).

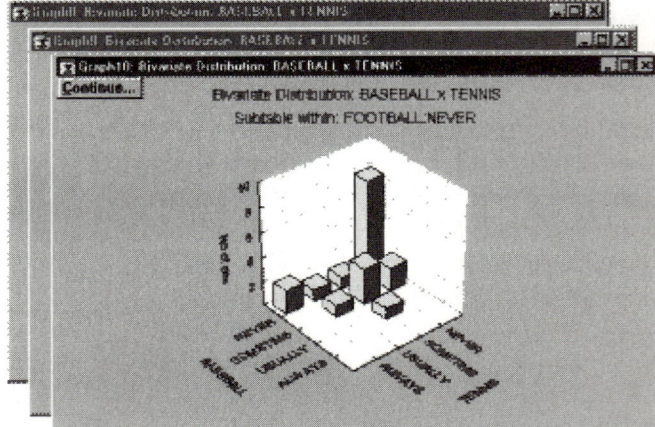

Graph10: Bivariate Distribution: BASEBALL x TENNIS

Statistics in Crosstabulation Tables

Crosstabulations generally enable us to identify relationships between the crosstabulated variables. The following table illustrates an example of a very strong relationship between two variables: variable *Age* (*Adult* vs. *Child*) and variable *Cookie* preference (*A* vs. *B*).

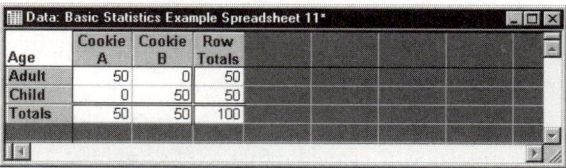

All adults chose *Cookie A*, while all children chose *Cookie B*. In this case there is little doubt about the reliability of the finding, because it is hardly conceivable that you would obtain such a pattern of frequencies by chance alone; that is, without the existence of a true difference between the cookie preferences of adults and children. However, in real-life, relations between variables are typically much weaker, and thus the question arises as to how to measure those relationships, and how to evaluate their reliability (statistical significance). The following review includes the most common measures of relationships between two categorical variables; that is, measures for two-way tables. The techniques used to analyze simultaneous relations between more than two variables in higher order crosstabulations are discussed in Chapter 11 – *Correspondence Analysis* and Chapter 22 – *Log-Linear Analysis of Frequency Tables*.

Pearson chi-square. The Pearson *chi*-square is the most common test for significance of the relationship between categorical variables. This measure is based on the fact that we can compute the expected frequencies in a two-way table (i.e., frequencies that we would expect if

there was no relationship between the variables). For example, suppose we ask 20 males and 20 females to choose between two brands of soda pop (brands *A* and *B*). If there is no relationship between preference and gender, we would expect about an equal number of choices of brand *A* and brand *B* for each sex. The *chi*-square test becomes increasingly significant as the numbers deviate further from this expected pattern; that is, the more this pattern of choices for males and females differs.

The value of the *chi*-square and its significance level depends on the overall number of observations and the number of cells in the table. Consistent with the principles discussed in Chapter 1 – *Elementary Concepts in Statistics*, relatively small deviations of the relative frequencies across cells from the expected pattern will prove significant if the number of observations is large.

The only assumption underlying the use of the *chi*-square (other than random selection of the sample) is that the expected frequencies are not very small. The reason for this is that, actually, the *chi*-square inherently tests the underlying probabilities in each cell; and when the expected cell frequencies fall, for example, below 5, those probabilities cannot be estimated with sufficient precision. For further discussion of this issue refer to Everitt (1977), Hays (1988), or Kendall and Stuart (1979).

Maximum-likelihood *chi*-square. The maximum-likelihood *chi*-square tests the same hypothesis as the Pearson chi-square statistic; however, its computation is based on Maximum-Likelihood (M-L) theory. In practice, the M-L chi-square is usually very close in magnitude to the Pearson chi-square statistic. For more details about this statistic refer to Bishop, Fienberg, and Holland (1975), or Fienberg, S. E. (1977); Chapter 22 – *Log-Linear Analysis of Frequency Tables* discusses this statistic in greater detail.

Yates correction. The approximation of the *chi*-square statistic in small 2x2 tables can be improved by reducing the absolute value of differences between expected and observed frequencies by 0.5 before squaring (Yates' correction). This correction, which makes the estimation more conservative, is usually applied when the table contains only small observed frequencies, so that some expected frequencies become less than 10 (for further discussion of this correction, see Conover, 1974; Everitt, 1977; Hays, 1988; Kendall & Stuart, 1979; and Mantel, 1974).

Fisher exact test. This test is only available for 2x2 tables; it is based on the following rationale: Given the marginal frequencies in the table, and assuming that in the population the two factors in the table are not related, how likely is it to obtain cell frequencies as uneven or worse than the ones that were observed? For small *n*, this probability can be computed exactly by counting all possible tables that can be constructed based on the marginal frequencies. Thus, the Fisher exact test computes the exact probability under the null hypothesis of obtaining the current distribution of frequencies across cells, or one that is more uneven.

McNemar *chi*-square. This test is applicable in situations where the frequencies in the 2x2 table represent dependent samples. For example, in a before-after design study, we may count the number of students who fail a test of minimal math skills at the beginning of the semester and at the end of the semester. Two *chi*-square values are reported: A/D and B/C. The *chi*-square A/D tests the hypothesis that the frequencies in cells *A* and *D* (upper-left and lower-right) are identical. The *chi*-square B/C tests the hypothesis that the frequencies in cells *B* and *C* (upper-right and lower-left) are identical.

Coefficient phi. The *phi*-square is a measure of correlation between two categorical variables in a 2x2 table. Its value can range from *0* (no relation between factors; *chi*-square=0.0) to *1* (perfect relation between the two factors in the table). For more details concerning this statistic, see Castellan and Siegel (1988, p. 232).

Tetrachoric correlation. This statistic is also only computed for (applicable to) 2x2 tables. If the 2x2 table can be thought of as the result of two continuous variables that were (artificially) forced into two categories each, the tetrachoric correlation coefficient will estimate the correlation between the two.

Coefficient of contingency. The coefficient of contingency is a *chi*-square based measure of the relation between two categorical variables (proposed by Pearson, the originator of the *chi*-square test). Its advantage over the ordinary *chi*-square is that it is more easily interpreted, since its range is always limited to *0* through *1* (where 0 means complete independence). The disadvantage of this statistic is that its specific upper limit is limited by the size of the table; *C* can reach the limit of *1* only if the number of categories is unlimited (see Siegel, 1956, p. 201).

Interpretation of contingency measures. An important disadvantage of measures of contingency (reviewed above) is that they do not lend themselves to clear interpretations in terms of probability or proportion of variance, as is the case, for example, of the Pearson *r* (see Chapter 29 – *Nonparametric Statistics*). There is no commonly accepted measure of relation between categories that has such a clear interpretation.

Statistics based on ranks. In many cases the categories used in the crosstabulation contain meaningful rank-ordering information; that is, they measure some characteristic on an ordinal scale (see Chapter 1 – *Elementary Concepts in Statistics*). Suppose we ask a sample of respondents to indicate their interest in watching different sports on a 4-point scale with the explicit labels 1) *always*, 2) *usually*, 3) *sometimes*, and 4) *never interested*. Obviously, we can assume that the response *sometimes interested* is indicative of less interest than *always interested*. Thus, we could rank the respondents with regard to their expressed interest in, for example, watching football. When categorical variables can be interpreted in this manner, there are several additional indices that can be computed to express the relationship between variables.

Spearman R. Spearman *R* can be thought of as the regular Pearson product-moment correlation coefficient (Pearson *r*); that is, in terms of the proportion of variability accounted for, except that Spearman *R* is computed from ranks. As mentioned above, Spearman *R* assumes that the variables under consideration were measured on at least an ordinal (rank order) scale; that is, the individual observations (cases) can be ranked into two ordered series. Detailed discussions of the Spearman *R* statistic, its power and efficiency can be found in Gibbons (1985), Hays (1981), McNemar (1969), Siegel (1956), Siegel and Castellan (1988), Kendall (1948), Olds (1949), or Hotelling and Pabst (1936).

Kendall tau. Kendall *tau* is equivalent to the Spearman *R* statistic with regard to the underlying assumptions. It is also comparable in terms of its statistical power. However, Spearman *R* and Kendall *tau* are usually not identical in magnitude because their underlying logic, as well as their computational formulas

are very different. Siegel and Castellan (1988) express the relationship of the two measures in terms of the inequality:

-1 < = 3 * Kendall tau – 2 * Spearman R < = 1

More importantly, Kendall *tau* and Spearman *R* imply different interpretations: While Spearman *R* can be thought of as the regular Pearson product-moment correlation coefficient as computed from ranks, Kendall *tau* rather represents a probability. Specifically, it is the difference between the probability that the observed data are in the same order for the two variables versus the probability that the observed data are in different orders for the two variables. Kendall (1948, 1975), Everitt (1977), and Siegel and Castellan (1988) discuss Kendall *tau* in greater detail. Two different variants of *tau* are computed, usually called tau_b and tau_c. These measures differ only with regard as to how tied ranks are handled. In most cases these values will be fairly similar, and when discrepancies occur, it is probably always safest to interpret the lowest value.

Sommer's d: d(X|Y), d(Y|X). Sommer's *d* is an asymmetric measure of association related to t_b (see Siegel & Castellan, 1988, p. 303-310).

Gamma. The *gamma* statistic is preferable to Spearman *R* or Kendall *tau* when the data contain many tied observations. In terms of the underlying assumptions, *gamma* is equivalent to Spearman *R* or Kendall *tau*; in terms of its interpretation and computation, it is more similar to Kendall *tau* than Spearman *R*. In short, *gamma* is also a probability; specifically, it is computed as the difference between the probability that the rank ordering of the two variables agree minus the probability that they disagree, divided by 1 minus the probability of ties. Thus, *gamma* is basically equivalent to Kendall *tau*, except that ties are explicitly taken into account. Detailed discussions of the *gamma* statistic can be found in Goodman and Kruskal (1954, 1959, 1963, 1972), Siegel (1956), and Siegel and Castellan (1988).

Uncertainty coefficients. These are indices of stochastic dependence; the concept of stochastic dependence is derived from the information theory approach to the analysis of frequency tables and you should refer to the appropriate references (see Kullback, 1959; Ku & Kullback, 1968; Ku, Varner, & Kullback, 1971; see also, Bishop, Fienberg, & Holland, 1975, p. 344-348). $S(Y,X)$ refers to symmetrical dependence, $S(X|Y)$ and $S(Y|X)$ refer to asymmetrical dependence.

Multiple Responses/Dichotomies

Multiple response variables or multiple dichotomies often arise when summarizing survey data. The nature of such variables or factors in a table is best illustrated with examples.

Multiple response variables. As part of a larger market survey, suppose you asked a sample of consumers to name their three favorite soft drinks. The specific item on the questionnaire may look as the following:

| What are your three favorite soft drinks? |
| 1:_____ 2:_____ 3:_____ |

Thus, the questionnaires returned to you will contain somewhere between 0 and 3 answers to this item. Also, a wide variety of soft drinks will most likely be named. Your goal is to summarize the responses to this item; that is, to produce a table that summarizes the percent of respondents who mentioned a respective soft drink.

The next question is how to enter the responses into a data file. Suppose 50 different soft drinks were mentioned among all of the questionnaires. You could of course set up 50 variables – one for

each soft drink – and then enter a *1* for the respective respondent and variable (soft drink) if he or she mentioned the respective soft drink (and a *0* if not); for example:

Data: Basic Statistics Example Spreadsheet 12* (4v by 4c)			
	Coke	Pepsi	Sprite	...
Case 1	0	1	0	
Case 2	1	1	0	
Case 3	0	0	1	
...	...	...	...	

This method of coding the responses would be very tedious and wasteful. Note that each respondent can only give a maximum of three responses; yet we use 50 variables to code those responses. (However, if we are only interested in these three soft drinks, this method of coding just those three variables would be satisfactory; to tabulate soft drink preferences, we could then treat the three variables as a multiple dichotomy.)

Coding multiple response variables. Alternatively, we could set up three variables, and a coding scheme for the 50 soft drinks. Then we could enter the respective codes (or alpha labels) into the three variables, in the same way that respondents wrote them down in the questionnaire.

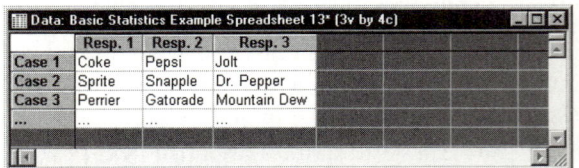

Data: Basic Statistics Example Spreadsheet 13* (3v by 4c)			
	Resp. 1	Resp. 2	Resp. 3	
Case 1	Coke	Pepsi	Jolt	
Case 2	Sprite	Snapple	Dr. Pepper	
Case 3	Perrier	Gatorade	Mountain Dew	
...	...	...	...	

To produce a table of the number of respondents by soft drink we would now treat *Resp.1* to *Resp3* as a multiple response variable. That table could look as the following:

Data: Basic Statistics Example Spreadsheet 14* (3v by 6c)			
N = 500 Category	Count	Percent of Responses	Percent of Cases
COKE: Coca Cola	44	5.23	8.80
PEPSI: Pepsi Cola	43	5.11	8.60
MOUNTAIN: Mountain Dew	81	9.62	16.20
PEPPER: Doctor Pepper	74	8.79	14.80
... : ...	...		
	842	100.00	168.40

Interpreting the multiple response frequency table. The total number of respondents was $n = 500$. Note that the counts in the first column of the table do not add up to 500, but rather to 842. That is the total number of responses; since each respondent could make up to 3 responses (write down three names of soft drinks), the total number of responses is naturally greater than the number of respondents. For example, referring back to the sample listing of the data file shown above, the first case (*Coke, Pepsi, Jolt*) contributes three times to the frequency table, once to the category *Coke*, once to the category *Pepsi*, and once to the category *Jolt*. The second and third columns in the table above report the percentages relative to the number of responses (second column) as well as respondents (third column). Thus, the entry 8.80 in the first row and last column in the table above means that 8.8% of all respondents mentioned *Coke* either as their first, second, or third soft drink preference.

Multiple dichotomies. Suppose in the previous example we were only interested in *Coke, Pepsi*, and *Sprite*. As pointed out earlier, one way to code the data in that case would be as follows:

Data: Basic Statistics Example Spreadsheet 15* (4v by 4c)			
	Coke	Pepsi	Sprite	...
Case 1		1		
Case 2	1	1		
Case 3			1	
...	...	...	...	

In other words, one variable was created for each soft drink, and a value of *1* was entered

into the respective variable whenever the respective respondent mentioned the respective drink. Note that each variable represents a dichotomy; that is, only *1*s and *not 1*s are allowed (we could have entered *1*s and *0*s, but to save typing we can also simply leave the *0*s blank or missing). When tabulating these variables, we would want to obtain a summary table very similar to the one shown earlier for multiple response variables; that is, we would want to compute the number and percent of respondents (and responses) for each soft drink. In a sense, we compact the three variables *Coke, Pepsi*, and *Sprite* into a single variable (*Soft Drink*) consisting of multiple dichotomies.

Crosstabulation of multiple responses/ dichotomies. All of these types of variables can then be used in crosstabulation tables. For example, we could crosstabulate a multiple dichotomy for *Soft Drink* (coded as described in the previous paragraph) with a multiple response variable *Favorite Fast Foods* (with many categories such as *Hamburgers, Pizza*, etc.), by the simple categorical variable *Gender*. As in the frequency table, the percentages and marginal totals in that table can be computed from the total number of respondents as well as the total number of responses. For example, consider the following hypothetical respondent:

Gender	Coke	Pepsi	Sprite	Food1	Food2
Female	1	1		Fish	Pizza

This female respondent mentioned *Coke* and *Pepsi* as her favorite drinks, and *Fish* and *Pizza* as her favorite fast foods. In the complete crosstabulation table she will be counted in the following cells of the table:

Data: Basic Statistics Example Spreadsheet 17* (7v by 6c)					
Gender	Drink	Food: Hamburger	Food: Fish	Food: Pizza ···	Total No. of Resp.
Female	Coke		X	X	2
	Pepsi		X	X	2
	Sprite				
Male	Coke				
	Pepsi				
	Sprite				

This female respondent will contribute to (i.e., be counted in) the crosstabulation table a total of 4 times. In addition, she will be counted twice in the *Female-Coke* marginal frequency column if that column is requested to represent the total number of responses; if the marginal totals are computed as the total number of respondents, this respondent will only be counted once.

Paired crosstabulation of multiple response variables. A unique option for tabulating multiple response variables is to treat the variables in two or more multiple response variables as matched pairs. Again, this method is best illustrated with a simple example. Suppose we conduct a survey of past and present home ownership. We ask the respondents to describe their last three (including the present) homes that they purchased. Naturally, for some respondents the present home is the first and only home; others have owned more than one home in the past. For each home we asked our respondents to write down the number of rooms in the respective house, and the number of occupants. Here is how the data for one respondent (say case number *112*) can be entered into a data file:

Data: Basic Statistics Example Spreadsheet 18* (9v by 1c)								
Case No.	Rooms	1	2	3	No. Occ.	1	2	3
112		3	3	4			2	3	5

This respondent owned three homes; the first had *3* rooms, the second also had 3 rooms, and the third had *4* rooms. The family apparently

also grew; there were 2 occupants in the first home, 3 in the second, and 5 in the third.

Now suppose we want to crosstabulate the number of rooms by the number of occupants for all respondents. One way to do so is to prepare three different two-way tables; one for each home. We can also treat the two factors in this study (*Number of Rooms, Number of Occupants*) as multiple response variables. However, it would obviously not make any sense to count the example respondent *112* shown above in cell *3 Rooms – 5 Occupants* of the crosstabulation table (which we would, if we simply treated the two factors as ordinary multiple response variables). In other words, we want to ignore the combination of occupants in the third home with the number of rooms in the first home. Rather, we would want to count these variables in pairs; we would want to consider the number of rooms in the first home together with the number of occupants in the first home, the number of rooms in the second home with the number of occupants in the second home, and so on. This is exactly what will be accomplished if we asked for a paired crosstabulation of these multiple response variables.

A final comment. When preparing complex crosstabulation tables with multiple responses/ dichotomies, it is sometimes difficult (in our experience) to keep track of exactly how the cases in the file are counted. The best way to verify that you understand the way in which the respective tables are constructed is to crosstabulate some simple example data, and then to trace how each case is counted. See the example in *Crosstabulation and Stub and Banner Tables* (page 32), which employs this method to illustrate how data are counted for tables involving multiple response variables and multiple dichotomies.

ANOVA/MANOVA

This chapter includes a general introduction to ANOVA and a discussion of the general topics in the analysis of variance techniques, including repeated measures designs, ANCOVA, MANOVA, unbalanced and incomplete designs, contrast effects, post-hoc comparisons, and assumptions. For related topics, see also, Chapter 39 – *Variance Components and Mixed Model ANOVA/ANCOVA* (topics related to estimation of variance components in mixed model designs), Chapter 15 – *Experimental Design/DOE* (topics related to specialized applications of ANOVA in industrial settings), and *Gage Repeatability and Reproducibility*, in Chapter 32 – *Process Analysis* (topics related to specialized designs for evaluating the reliability and precision of measurement systems). See also, Chapter 18 – *General Linear Models* and Chapter 19 – *General Regression Models*; to analyze nonlinear models, see Chapter 21 – *Generalized Linear/Nonlinear Models*.

Overview

The purpose of analysis of variance. In
general, the purpose of analysis of variance
(ANOVA) is to test for significant differences
between means. Chapter 1 – *Elementary
Concepts in Statistics* provides a brief
introduction to the basics of statistical
significance testing. If we are only comparing
two means, ANOVA will give the same results
as the *t*-test for independent samples (if we are
comparing two different groups of cases or
observations) or the *t*-test for dependent
samples (if we are comparing two variables in
one set of cases or observations). If you are not
familiar with these tests, see *t-Test for
Independent Samples*, page 25, and *t-Test for
Dependent Samples*, page 26.

Why the name analysis of variance? It may
seem odd to you that a procedure that compares
means is called analysis of variance. However,
this name is derived from the fact that in order
to test for statistical significance between
means, we are actually comparing (i.e.,
analyzing) variances.

Partitioning of Sums of Squares

At the heart of ANOVA is the fact that
variances can be divided up, that is, partitioned.
Remember that the variance is computed as the
sum of squared deviations from the overall
mean, divided by *n-1* (sample size minus one).
Thus, given a certain *n*, the variance is a
function of the sums of (deviation) squares, or
SS for short. Partitioning of variance works as
follows. Consider the following data set:

Data: ANOVA Example Spreadsheet 1 (3v by 8c)		
	Group 1	Group 2
Observation 1	2	6
Observation 2	3	7
Observation 3	1	5
Mean	2	6
Sums of Squares (SS)	2	2
Overall Mean		4
Total Sums of Squares		28

The means for the two groups are quite different
(*2* and *6*, respectively). The sums of squares
within each group are equal to *2*. Adding them
together, we get *4*. If we now repeat these
computations ignoring group membership, that is,
if we compute the total SS based on the overall
mean, we get the number *28*. In other words,
computing the variance (sums of squares) based
on the within-group variability yields a much
smaller estimate of variance than computing it
based on the total variability (the overall mean).
The reason for this in the above example is of
course that there is a large difference between
means, and it is this difference that accounts for
the difference in the SS. In fact, if we were to
perform an ANOVA on the above data, we would
get the following result:

Data: ANOVA Example Spreadsheet 2* (5v by 2c)					
	Main Effect				
	SS	df	MS	F	p
Effect	24	1	24	24	0.008
Error	4	4	1		

As you can see, in the above table, the total *SS
(28)* was partitioned into the *SS* due to within-
group variability (*2+2=4*) and variability due to
differences between means (*28-(2+2)=24*).

SS error and SS effect. The within-group
variability (SS) is usually referred to as error
variance. This term denotes the fact that we
cannot readily explain or account for it in the
current design. However, the SS effect we can
explain. Namely, it is due to the differences in
means between the groups. Put another way,
group membership explains this variability

because we know that it is due to the differences in means.

Significance testing. The basic idea of statistical significance testing is discussed in Chapter 1 – *Elementary Concepts in Statistics*. That chapter also explains why very many statistical tests represent ratios of explained to unexplained variability. ANOVA is a good example of this. Here, we base this test on a comparison of the variance due to the between-groups variability (called mean square effect, or MS_{effect}) with the within-group variability (called mean square error, or MS_{error}; this term was first used by Edgeworth, 1885). Under the null hypothesis (that there are no mean differences between groups in the population), we would still expect some minor random fluctuation in the means for the two groups when taking small samples (as in our example). Therefore, under the null hypothesis, the variance estimated based on within-group variability should be about the same as the variance due to between-groups variability. We can compare those two estimates of variance via the F test (see also, *F Distribution*, page 170), which tests whether the ratio of the two variance estimates is significantly greater than 1. In our example above, that test is highly significant, and we would in fact conclude that the means for the two groups are significantly different from each other.

Summary of the basic logic of ANOVA. To summarize the discussion up to this point, the purpose of analysis of variance is to test differences in means (for groups or variables) for statistical significance. This is accomplished by analyzing the variance, that is, by partitioning the total variance into the component that is due to true random error (i.e., within-group SS) and the components that are due to differences between means. These latter variance components are then tested for statistical significance, and, if significant, we reject the null hypothesis of no differences between means, and accept the alternative hypothesis that the means (in the population) are different from each other.

Dependent and independent variables. The variables that are measured (e.g., a test score) are called dependent variables. The variables that are manipulated or controlled (e.g., a teaching method or some other criterion used to divide observations into groups that are compared) are called factors or independent variables. For more information on this important distinction, refer to Chapter 1, *Elementary Concepts in Statistics*.

Multi-Factor ANOVA

In the simple example above, it may have occurred to you that we could have simply computed a t test for independent samples to arrive at the same conclusion. And, indeed, we would get the identical result if we were to compare the two groups using this test. However, ANOVA is a much more flexible and powerful technique that can be applied to much more complex research issues.

Multiple factors. The world is complex and multivariate in nature, and instances when a single variable completely explains a phenomenon are rare. For example, when trying to explore how to grow a bigger tomato, we would need to consider factors that have to do with the plants' genetic makeup, soil conditions, lighting, temperature, etc. Thus, in a typical experiment, many factors are taken into account. One important reason for using ANOVA methods rather than multiple two-group studies analyzed via t tests is that the former method is more efficient, and with fewer observations we can gain more information. Let's expand on this statement.

Controlling for factors. Suppose that in the above two-group example we introduce another grouping factor, for example, *Gender*. Imagine that in each group we have 3 males and 3 females. We could summarize this design in a 2x2 table:

Data: ANOVA Example Spreadsheet 3* (2v by 8c)		
2 by 2 Table	Experimental Group 1	Experimental Group 2
Males	2	6
	3	7
	1	5
Mean	2	6
Females	4	8
	5	9
	3	7
Mean	4	8

Before performing any computations, it appears that we can partition the total variance into at least 3 sources: 1) error (within-group) variability, 2) variability due to experimental group membership, and 3) variability due to gender. (Note that there is an additional source – interaction – that we will discuss shortly.) What would have happened had we not included *Gender* as a factor in the study but rather computed a simple *t* test? If you compute the SS ignoring the *Gender* factor (use the within-group means ignoring or collapsing across *gender*; the result is SS=10+10=20), you will see that the resulting within-group SS is larger than it is when we include gender (use the within-group, within-gender means to compute those SS; they will be equal to 2 in each group, thus the combined SS-within is equal to 2+2+2+2=8). This difference is due to the fact that the means for *males* are systematically lower than those for *females*, and this difference in means adds variability if we ignore this factor. Controlling for error variance increases the sensitivity (power) of a test. This

example demonstrates another principal of ANOVA that makes it preferable over simple two-group *t* test studies: In ANOVA we can test each factor while controlling for all others; this is actually the reason why ANOVA is more statistically powerful (i.e., we need fewer observations to find a significant effect) than the simple *t* test.

Interaction Effects

There is another advantage of ANOVA over simple *t*-tests. ANOVA enables us to detect interaction effects between variables and, therefore, to test more complex hypotheses about reality. Let's consider another example to illustrate this point. (The term interaction was first used by Fisher, 1926.)

Main effects, two-way interaction. Imagine that we have a sample of highly achievement-oriented students and another of achievement avoiders. We now create two random halves in each sample, and give one half of each sample a challenging test, the other an easy test. We measure how hard the students work on the test. The means of this (fictitious) study are as follows:

Data: ANOVA Example Spreadsheet 4* (2v by 2c)		
	Achievement-oriented	Achievement-avoiders
Challenging Test	10	5
Easy Test	5	10

How can we summarize these results? Is it appropriate to conclude that 1) challenging tests make students work harder, and 2) achievement-oriented students work harder than achievement-avoiders? None of these statements captures the essence of this clearly systematic pattern of means. The appropriate way to summarize the result would be to say that challenging tests make only achievement-oriented students work harder, while easy tests make only achievement-avoiders work harder.

In other words, the type of achievement orientation and test difficulty interact in their effect on effort; specifically, this is an example of a two-way interaction between achievement orientation and test difficulty. Note that statements 1 and 2 above describe so-called main effects.

Higher order interactions. While the previous two-way interaction can be put into words relatively easily, higher order interactions are increasingly difficult to verbalize. Imagine that we had included factor *Gender* in the achievement study above, and we had obtained the following pattern of means:

Data: ANOVA Example Spreadsheet 5* (3v by 6c)		
Females	**Achievement-oriented**	**Achievement-avoiders**
Challenging Test	10	5
Easy Test	5	10
Males	**Achievement-oriented**	**Achievement-avoiders**
Challenging Test	1	6
Easy Test	6	1

How could we now summarize the results of our study? Graphs of means for all effects greatly facilitate the interpretation of complex effects. The pattern shown in the previous table (and in the graph below) represents a three-way cross-over interaction between factors.

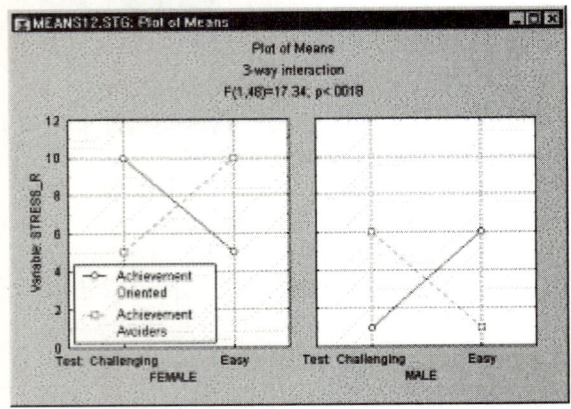

Thus, we can summarize this pattern by saying that for *females* there is a two-way interaction between achievement-orientation type and test difficulty: Achievement-oriented *females* work harder on challenging tests than on easy tests, achievement-avoiding *females* work harder on easy tests than on difficult tests. For *males*, this interaction is reversed. As you can see, the description of the interaction has become much more involved.

A general way to express interactions. A general way to express all interactions is to say that an effect is modified (qualified) by another effect. Let's try this with the two-way interaction above. The main effect for test difficulty is modified by achievement orientation. For the three-way interaction in the previous paragraph, we can summarize that the two-way interaction between test difficulty and achievement orientation is modified (qualified) by *Gender*. If we have a four-way interaction, we can say that the three-way interaction is modified by the fourth variable, that is, that there are different types of interactions in the different levels of the fourth variable. As it turns out, in many areas of research five- or higher-way interactions are not that uncommon.

Complex Designs

Following is a review of the basic building blocks of complex designs.

Between-Groups and Repeated-Measures

When we want to compare two groups, we would use the *t*-test for independent samples; when we want to compare two variables given the same subjects (observations), we would use the *t*-test for dependent samples. This distinction – dependent and independent samples – is important for ANOVA as well. Basically, if we have repeated measurements of the same variable (under different conditions or at different points in time) on the same subjects, the factor is a repeated measures factor (also called a within-subjects factor, because to estimate its significance we compute the within-subjects SS). If we compare different groups of subjects (e.g., *males* and *females*; *three strains of bacteria*, etc.), we refer to the factor as a between-groups factor. The computations of significance tests are different for these different types of factors; however, the logic of computations and interpretations is the same.

Between-within designs. In many instances, experiments call for the inclusion of between-groups and repeated measures factors. For example, we can measure math skills in male and female students (*Gender*, a between-groups factor) at the beginning and the end of the semester. The two measurements on each student would constitute a within-subjects (repeated measures) factor. The interpretation of main effects and interactions is not affected by whether a factor is between-groups or repeated measures, and both factors may obviously interact with each other (e.g., females improve over the semester while males deteriorate).

Incomplete (Nested) Designs

There are instances where we may decide to ignore interaction effects. This happens when 1) we know that in the population the interaction effect is negligible, or 2) when a complete factorial design (this term was first introduced by Fisher, 1935a) cannot be used for economic reasons. Imagine a study where we want to evaluate the effect of four fuel additives on gas mileage. For our test, our company has provided us with four cars and four drivers. A complete factorial experiment, that is, one in

which each combination of driver, additive, and car appears at least once, would require 4 x 4 x 4 = 64 individual test conditions (groups). However, we may not have the resources (time) to run all of these conditions; moreover, it seems unlikely that the type of driver would interact with the fuel additive to an extent that would be of practical relevance. Given these considerations, you could actually run a so-called Latin square design and get away with only 16 individual groups (the four additives are denoted by letters A, B, C, and D):

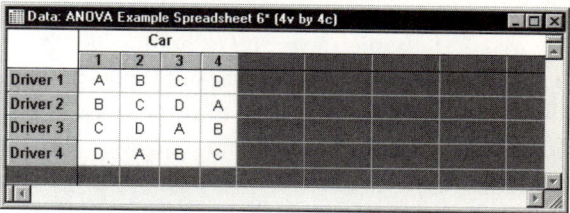

Latin square designs (this term was first used by Euler, 1782) are described in most textbooks on experimental methods (e.g., Hays, 1988; Lindman, 1974; Milliken & Johnson, 1984; Winer, 1962), and we do not want to discuss here the details of how they are constructed. Suffice it to say that this design is incomplete insofar as not all combinations of factor levels occur in the design. For example, Driver 1 will only drive Car 1 with additive A, while Driver 3 will drive that car with additive C. In a sense, the levels of the additives factor (A, B, C, and D) are placed into the cells of the *Car* by *Driver* matrix as "eggs into a nest." This mnemonic device is sometimes useful for remembering the nature of nested designs.

Note that there are several other statistical procedures that can be used to analyze these types of designs; see *Methods for Analysis of Variance*, page 54, for details. In particular the methods discussed in Chapter 39 – *Variance Components and Mixed Model ANOVA/*

ANCOVA are very efficient for analyzing designs with unbalanced nesting (when the nested factors have different numbers of levels within the levels of the factors in which they are nested), very large nested designs (e.g., with more than 200 levels overall), or hierarchically nested designs (with or without random factors).

Analysis of Covariance (ANCOVA)

General idea. The *Overview*, page 43, discussed briefly the idea of controlling for factors and how the inclusion of additional factors can reduce the error SS and increase the statistical power (sensitivity) of our design. This idea can be extended to continuous variables, and when such continuous variables are included as factors in the design they are called covariates.

Fixed Covariates

Suppose that we want to compare the math skills of students who were randomly assigned to one of two alternative textbooks. Imagine that we also have data about the general intelligence (IQ) for each student in the study. We would suspect that general intelligence is related to math skills, and we can use this information to make our test more sensitive. Specifically, imagine that in each one of the two groups we can compute the correlation coefficient (see Chapter 2 – *Basic Statistics and Tables*) between IQ and math skills. Remember that once we have computed the correlation coefficient we can estimate the amount of variance in math skills that is accounted for by IQ, and the amount of (residual) variance that we cannot explain with IQ (refer also to Chapter 1 – *Elementary Concepts in Statistics* and Chapter 2 – *Basic Statistics and Tables*). We

can use this residual variance in the ANOVA as an estimate of the true error SS after controlling for IQ. If the correlation between IQ and math skills is substantial, a large reduction in the error SS can be achieved.

Effect of a covariate on the *F*-test. In the *F*-test (see also, *F Distribution*, page 170), to evaluate the statistical significance of between-groups differences, we compute the ratio of the between-groups variance (MS_{effect}) over the error variance (MS_{error}). If MS_{error} becomes smaller, due to the explanatory power of IQ, the overall *F* value will become larger.

Multiple covariates. The logic described above for the case of a single covariate (IQ) can easily be extended to the case of multiple covariates. For example, in addition to IQ, we might include measures of motivation, spatial reasoning, etc., and instead of a simple correlation, compute the multiple correlation coefficient (see Chapter 26 – *Multiple Linear Regression*).

When the *F* value gets smaller. In some studies with covariates it happens that the *F* value actually becomes smaller (less significant) after including covariates in the design. This is usually an indication that the covariates are not only correlated with the dependent variable (e.g., math skills), but also with the between-groups factors (e.g., the two different textbooks). For example, imagine that we measured IQ at the end of the semester, after the students in the different experimental groups had used the respective textbook for almost one year. It is possible that, even though students were initially randomly assigned to one of the two textbooks, the different books were so different that both math skills and IQ improved differentially in the two groups. In that case, the covariate will not only partition variance away from the error variance, but also from the

variance due to the between-groups factor. Put another way, after controlling for the differences in IQ that were produced by the two textbooks, the math skills are not that different. Put in yet a third way, by eliminating the effects of IQ, we have inadvertently eliminated the true effect of the textbooks on students' math skills.

Adjusted means. When the latter case happens, that is, when the covariate is affected by the between-groups factor, it is appropriate to compute so-called adjusted means. These are the means that you would get after removing all differences that can be accounted for by the covariate.

Interactions between covariates and factors. Just as we can test for interactions between factors, we can also test for the interactions between covariates and between-groups factors. Specifically, imagine that one of the textbooks is particularly suited for intelligent students, while the other actually bores those students but challenges the less intelligent ones. As a result, we may find a positive correlation in the first group (the more intelligent, the better the performance), but a zero or slightly negative correlation in the second group (the more intelligent the student, the less likely he or she is to acquire math skills from the particular textbook). In some older statistics textbooks this condition is discussed as a case where the assumptions for analysis of covariance are violated (see *Assumptions and Effects of Violating Assumptions*, page 51). However, because ANOVA/MANOVA uses a very general approach to analysis of covariance, you can specifically estimate the statistical significance of interactions between factors and covariates.

Changing Covariates

While fixed covariates are commonly discussed in textbooks on ANOVA, changing covariates are discussed less frequently. In general, when we have repeated measures, we are interested in testing the differences in repeated measurements on the same subjects. Thus, we are actually interested in evaluating the significance of changes. If we have a covariate that is also measured at each point when the dependent variable is measured, we can compute the correlation between the changes in the covariate and the changes in the dependent variable. For example, we could study math anxiety and math skills at the beginning and at the end of the semester. It would be interesting to see whether any changes in math anxiety over the semester correlate with changes in math skills.

Multivariate Designs: MANOVA/MANCOVA

Between-Groups Designs

All examples discussed so far have involved only one dependent variable. Even though the computations become increasingly complex, the logic and nature of the computations do not change when there is more than one dependent variable at a time. For example, we may conduct a study where we try two different textbooks, and we are interested in the students' improvements in math and physics. In that case, we have two dependent variables, and our hypothesis is that both together are affected by the difference in textbooks. We could now perform a multivariate analysis of variance (MANOVA) to test this hypothesis. Instead of a univariate *F* value, we would obtain a multivariate *F* value (Wilks' *lambda*) based on a comparison of the error variance/covariance matrix and the effect variance/covariance matrix. The covariance here is included because the two measures are probably correlated and we must take this correlation into account when performing the significance test. Obviously, if we were to take the same measure twice, we would really not learn anything new. If we take a correlated measure, we gain some new information, but the new variable will also contain redundant information that is expressed in the covariance between the variables.

Interpreting results. If the overall multivariate test is significant, we conclude that the respective effect (e.g., textbook) is significant. However, our next question would of course be whether only math skills improved, only physics skills improved, or both. In fact, after obtaining a significant multivariate test for a particular main effect or interaction, customarily you would examine the univariate *F*-tests (see also, *F Distribution*, page 170) for each variable to interpret the respective effect. In other words, you would identify the specific dependent variables that contributed to the significant overall effect.

Repeated Measures Designs

If we were to measure math and physics skills at the beginning of the semester and the end of the semester, we would have a multivariate repeated measure. Again, the logic of significance testing in such designs is simply an extension of the univariate case. Note that MANOVA methods are also commonly used to test the significance of univariate repeated measures factors with more than two levels; this application will be discussed later in this section.

Sum Scores vs. MANOVA

Even experienced users of ANOVA and MANOVA techniques are often puzzled by the differences in results that sometimes occur when performing a MANOVA on, for example, three variables as compared to a univariate

ANOVA on the sum of the three variables. The logic underlying the summing of variables is that each variable contains some true value of the variable in question, as well as some random measurement error. Therefore, by summing up variables, the measurement error will sum to approximately 0 across all measurements, and the sum score will become more and more reliable (increasingly equal to the sum of true scores). In fact, under these circumstances, ANOVA on sums is appropriate and represents a very sensitive (powerful) method. However, if the dependent variable is truly multi-dimensional in nature, summing is inappropriate. For example, suppose that the dependent measure consists of four indicators of success in society, and each indicator represents a completely independent way in which a person could make it in life (e.g., successful professional, successful entrepreneur, successful homemaker, etc.). Now, summing up the scores on those variables would be like adding apples to oranges, and the resulting sum score will not be a reliable indicator of a single underlying dimension. Thus, you should treat such data as multivariate indicators of success in a MANOVA.

Contrast Analysis and Post-hoc Tests

Why Compare Individual Sets of Means?

Usually, experimental hypotheses are stated in terms that are more specific than simply main effects or interactions. We may have the specific hypothesis that a particular textbook will improve math skills in males, but not in females, while another book would be about equally effective for both genders, but less effective overall for males. Now generally, we are predicting an interaction here: the effectiveness of the book is modified (qualified) by the student's gender. However, we have a particular prediction concerning the nature of the interaction: we expect a significant difference between genders for one book, but not the other. This type of specific prediction is usually tested via contrast analysis.

Contrast Analysis

Briefly, contrast analysis enables us to test the statistical significance of predicted specific differences in particular parts of our complex design. It is a major and indispensable component of the analysis of every complex ANOVA design.

Post-hoc Comparisons

Sometimes we find effects in our experiment that were not expected. Even though in most cases a creative experimenter will be able to explain almost any pattern of means, it would not be appropriate to analyze and evaluate that pattern as if you had predicted it all along. The problem here is one of capitalizing on chance when performing multiple tests post hoc, that is, without *a priori* hypotheses. To illustrate this point, let's consider the following experiment. Imagine we were to write down a number between 1 and 10 on 100 pieces of paper. We then put all of those pieces into a hat and draw 20 samples (of pieces of paper) of 5 observations each, and compute the means (from the numbers written on the pieces of paper) for each group. How likely do you think it is that we will find two sample means that are significantly different from each other? It is very likely! Selecting the extreme means obtained from 20 samples is very different from taking only 2 samples from the hat in the first place, which is what the test via the contrast

analysis implies. Without going into further detail, there are several so-called post hoc tests that are explicitly based on the first scenario (taking the extremes from 20 samples), that is, they are based on the assumption that we have chosen for our comparison the most extreme (different) means out of *k* total means in the design. Those tests apply corrections that are designed to offset the advantage of post-hoc selection of the most extreme comparisons.

Assumptions and Effects of Violating Assumptions

Deviation from Normal Distribution

Assumptions. It is assumed that the dependent variable is measured on at least an interval scale level (see Chapter 1 – *Elementary Concepts in Statistics*). Moreover, the dependent variable should be normally distributed within groups.

Effects of violations. Overall, the *F* test is remarkably robust to deviations from normality (see Lindman, 1974, for a summary). If the kurtosis (see Chapter 2 – *Basic Statistics and Tables*) is greater than 0, the *F* tends to be too small and we cannot reject the null hypothesis even though it is incorrect. The opposite is the case when the kurtosis is less than 0. The skewness of the distribution usually does not have a sizable effect on the *F* statistic. If the *n* per cell is fairly large, deviations from normality do not matter much at all because of the central limit theorem, according to which the sampling distribution of the mean approximates the normal distribution, regardless of the distribution of the variable in the population. A detailed discussion of the robustness of the *F* statistic can be found in Box and Anderson (1955), or Lindman (1974).

Homogeneity of Variances

Assumptions. It is assumed that the variances in the different groups of the design are identical; this assumption is called the homogeneity of variances assumption. Remember that at the beginning of this section we computed the error variance (SS error) by adding up the sums of squares within each group. If the variances in the two groups are different from each other, adding the two together is not appropriate, and will not yield an estimate of the common within-group variance (since no common variance exists).

Effects of violations. Lindman (1974, p. 33) shows that the F statistic is quite robust against violations of this assumption (heterogeneity of variances; see also, Box, 1954a, 1954b; Hsu, 1938).

Special case: correlated means and variances. One instance when the F statistic is very misleading is when the means are correlated with variances across cells of the design. A scatterplot of variances or standard deviations against the means will detect such correlations. The reason why this is a dangerous violation is the following: Imagine that you have 8 cells in the design, 7 with about equal means but one with a much higher mean. The F statistic may suggest to you a statistically significant effect. However, suppose that there also is a much larger variance in the cell with the highest mean, that is, the means and the variances are correlated across cells (the higher the mean the larger the variance). In that case, the high mean in the one cell is actually quite unreliable, as is indicated by the large variance. However, because the overall F statistic is based on a pooled within-cell variance estimate, the high mean is identified as significantly different from the others, when in fact it is not

at all significantly different if you based the test on the within-cell variance in that cell alone.

This pattern – a high mean and a large variance in one cell – frequently occurs when there are outliers present in the data. One or two extreme cases in a cell with only 10 cases can greatly bias the mean, and will dramatically increase the variance.

Homogeneity of Variances and Covariances

Assumptions. In multivariate designs, with multiple dependent measures, the homogeneity of variances assumption described earlier also applies. However, since there are multiple dependent variables, it is also required that their intercorrelations (covariances) are homogeneous across the cells of the design. There are various specific tests of this assumption.

Effects of violations. The multivariate equivalent of the F-test is Wilks' *lambda*. Not much is known about the robustness of Wilks' *lambda* to violations of this assumption. However, because the interpretation of MANOVA results usually rests on the interpretation of significant univariate effects (after the overall test is significant), the above discussion concerning univariate ANOVA basically applies, and important significant univariate effects should be carefully scrutinized.

Special case: ANCOVA. A special, serious violation of the homogeneity of variances/ covariances assumption may occur when covariates are involved in the design. Specifically, if the correlations of the covariates with the dependent measure(s) are very different in different cells of the design, gross misinterpretations of results may occur.

Remember that in ANCOVA, we in essence perform a regression analysis within each cell to partition out the variance component due to the covariates. The homogeneity of variances/covariances assumption implies that we perform this regression analysis subject to the constraint that all regression equations (slopes) across the cells of the design are the same. If this is not the case, serious biases may occur. There are specific tests of this assumption, and it is advisable to look at those tests to ensure that the regression equations in different cells are approximately the same.

Sphericity and Compound Symmetry

Reasons for using the multivariate approach to repeated measures ANOVA.

In repeated measures ANOVA containing repeated measures factors with more than two levels, additional special assumptions enter the picture: The compound symmetry assumption and the assumption of sphericity. Because these assumptions rarely hold, the MANOVA approach to repeated measures ANOVA has gained popularity in recent years (both tests are automatically computed in ANOVA/MANOVA). The compound symmetry assumption requires that the variances (pooled within-group) and covariances (across subjects) of the different repeated measures are homogeneous (identical). This is a sufficient condition for the univariate F test for repeated measures to be valid (i.e., for the reported F values to actually follow the F distribution). However, it is not a necessary condition. The sphericity assumption is a necessary and sufficient condition for the F-test to be valid; it states that the within-subject model consists of independent (orthogonal) components. The nature of these assumptions, and the effects of

violations are usually not well-described in ANOVA textbooks; in the following paragraphs we will try to clarify this matter and explain what it means when the results of the univariate approach differ from the multivariate approach to repeated measures ANOVA.

The necessity of independent hypotheses.

One general way of looking at ANOVA is to consider it a model fitting procedure. In a sense, we bring to our data a set of *a priori* hypotheses; we then partition the variance (test main effects, interactions) to test those hypotheses. Computationally, this approach translates into generating a set of contrasts (comparisons between means in the design) that specify the main effect and interaction hypotheses. However, if these contrasts are not independent of each other, the partitioning of variances runs afoul. For example, if two contrasts A and B are identical to each other and we partition out their components from the total variance, we take the same thing out twice. Intuitively, specifying the two (not independent) hypotheses "the mean in Cell 1 is higher than the mean in Cell 2" and "the mean in Cell 1 is higher than the mean in Cell 2" is silly and simply makes no sense. Thus, hypotheses must be independent of each other, or orthogonal (the term orthogonality was first used by Yates, 1933).

Independent hypotheses in repeated measures.

The general algorithm implemented will attempt to generate, for each effect, a set of independent (orthogonal) contrasts. In repeated measures ANOVA, these contrasts specify a set of hypotheses about differences between the levels of the repeated measures factor. However, if these differences are correlated across subjects, the resulting contrasts are no longer independent. For example, in a study where we measured learning at three times during the experimental session, it may happen that the changes from time 1 to time 2 are negatively correlated with the changes from time 2 to time 3: subjects who learn most of the material between time 1 and time 2 improve less from time 2 to time 3. In fact, in most instances where a repeated measures ANOVA is used, you would probably suspect that the changes across levels are correlated across subjects. However, when this happens, the compound symmetry and sphericity assumptions have been violated, and independent contrasts cannot be computed.

Effects of violations and remedies. When the compound symmetry or sphericity assumptions have been violated, the univariate ANOVA table will give erroneous results. Before multivariate procedures were well understood, various approximations were introduced to compensate for the violations (e.g., Greenhouse & Geisser, 1959; Huynh & Feldt, 1970), and these techniques are still widely used.

MANOVA approach to repeated measures. To summarize, the problem of compound symmetry and sphericity pertains to the fact that multiple contrasts involved in testing repeated measures effects (with more than two levels) are not independent of each other. However, they do not need to be independent of each other if we use multivariate criteria to simultaneously test the statistical significance of the two or more repeated measures contrasts. This insight is the reason why MANOVA methods are increasingly applied to test the significance of univariate repeated measures factors with more than two levels. We wholeheartedly endorse this approach because it simply bypasses the assumption of compound symmetry and sphericity altogether.

Cases when the MANOVA approach cannot be used. There are instances (designs) when the MANOVA approach cannot be applied; specifically, when there are few subjects in the design and many levels on the repeated measures factor, there may not be enough degrees of freedom to perform the multivariate analysis. For example, if we have 12 subjects and $p = 4$ repeated measures factors, each at $k = 3$ levels, the four-way interaction would "consume" $(k-1)^p = 2^4 = 16$ degrees of freedom. However, we have only 12 subjects, so in this instance the multivariate test cannot be performed.

Differences in univariate and multivariate results. Anyone whose research involves extensive repeated measures designs has seen cases when the univariate approach to repeated measures ANOVA gives clearly different results from the multivariate approach. To repeat the point, this means that the differences between the levels of the respective repeated measures factors are in some way correlated across subjects. Sometimes, this insight by itself is of considerable interest.

Methods for Analysis of Variance

Several chapters in this textbook discuss methods for performing analysis of variance. Although many of the available statistics overlap in the different chapters, each is best suited for particular applications.

ANOVA/MANOVA. Chapter 3 – *ANOVA/ MANOVA*, includes discussions of full factorial designs, repeated measures designs, mutivariate design (MANOVA), designs with balanced nesting (designs can be unbalanced, i.e., have unequal n), for evaluating planned and post-hoc comparisons, etc.

General linear models. Chapter 18, an extremely comprehensive chapter, discusses a complete implementation of the general linear model, and describes the *sigma*-restricted as well as the overparameterized approach. The chapter includes information on incomplete designs, complex analysis of covariance designs, nested designs (balanced or unbalanced), mixed model ANOVA designs (with random effects), and huge balanced ANOVA designs (efficiently). It also contains descriptions of six types of Sums of Squares.

General regression models. Chapter 19 discusses the between subject designs and multivariate designs that are appropriate for stepwise regression as well as discussing how to perform stepwise and best-subset model building (for continuous as well as categorical predictors).

Variance components and mixed model ANOVA/ANCOVA. Chapter 39 includes discussions of experiments with random effects (mixed model ANOVA), estimating variance components for random effects, or large main effect designs (e.g., with factors with more than 100 levels) with or without random effects, or large designs with many factors, when you do not need to estimate all interactions.

Experimental design (DOE). Chapter 15 includes discussions of standard experimental designs for industrial/manufacturing applications, including $2^{(k-p)}$ and $3^{(k-p)}$ designs, central composite and non-factorial designs, designs for mixtures, D and A optimal designs, and designs for arbitrarily constrained experimental regions.

Repeatability and reproducibility analysis. This section, in Chapter 32 – *Process Analysis*, includes a discussion of specialized designs for evaluating the reliability and precision of measurement systems; these designs usually include two or three random factors, and specialized statistics can be computed for evaluating the quality of a measurement system (typically in industrial/manufacturing applications).

Breakdown tables. This section, on page 28 of Chapter 2 – *Basic Statistics and Tables*, includes discussions of experiments with only one factor (and many levels), or with multiple factors, when a complete ANOVA table is not required.

ASSOCIATION RULES

The goal of the techniques described in this chapter is to detect relationships or associations between specific values of categorical variables in large data sets. This is a common task in many data mining projects and in the subcategory text mining. These powerful exploratory techniques have a wide range of applications in many areas of business practice and also research – from the analysis of consumer preferences or human resource management, to the history of language. These techniques enable analysts and researchers to uncover hidden patterns in large data sets, such as "customers who order product *A*, often also order product *B* or *C*" or "employees who said positive things about initiative *X*, also frequently complain about issue *Y* but are happy with issue *Z*." The implementation of the so-called *a priori* algorithm (see Agrawal and Swami, 1993; Agrawal and Srikant, 1994; Han and Lakshmanan, 2001; see also, Witten and Frank, 2000) enables you to process huge data sets for such associations rapidly, based on predefined threshold values for detection.

Overview

How association rules work. The usefulness of this technique to address unique data mining problems is best illustrated in a simple example. Suppose you are collecting data at the checkout cash registers at a large bookstore. Each customer transaction is logged in a database and consists of the titles of the books purchased by the customer, and perhaps additional magazine titles and other gift items that were purchased. Hence, each record in the database will represent one customer, and may consist of a single book purchased by that customer, or it may consist of many (perhaps hundreds of) different items that were purchased, arranged in an arbitrary order depending on the order in which the different items came down the conveyor belt at the cash register.

The purpose of the analysis is to find associations between the items that were purchased, i.e., to derive association rules that identify the items and co-occurrences of different items that appear with the greatest frequencies. For example, you want to learn which books are likely to be purchased by a customer who you know already purchased (or is about to purchase) a particular book. This type of information could then quickly be used to suggest to the customer those additional titles.

You may already be familiar with the results of these types of analyses if you are a customer of various on-line (Web-based) retail businesses. Many times when making a purchase on-line, the vendor will suggest similar items (to the ones purchased by you) at the time of check-out, based on rules such as "customers who buy book title *A* are also likely to purchase book title *B*."

Unique data analysis requirements. Crosstabulation tables, and in particular multiple response tables, can be used to analyze

data of this kind. However, in cases where the number of different items (categories) in the data is very large (and not known ahead of time), and the factorial degree of important association rules is not known ahead of time, these tabulation facilities may be too cumbersome to use, or simply not applicable.

Consider once more the simple bookstore example. First, the number of book titles is practically unlimited. In other words, if we were to create a table where each book title represents one dimension, and the purchase of a book (yes/no) was the class or category for each dimension, the complete crosstabulation table would be huge and sparse (consisting mostly of empty cells). Alternatively, we could construct all possible two-way tables from all items available in the store; this would enable us to detect two-way associations (association rules) between items. However, the number of tables that would have to be constructed would again be huge, most of the two-way tables would be sparse, and worse, if there were any three-way association rules hiding in the data, we would miss them completely.

The *a priori* algorithm implemented in association rules not only automatically detects the relationships (cross-tabulation tables) that are important (i.e., cross-tabulation tables that are not sparse, not containing mostly zeros), but also determines the factorial degree of the tables that contain the important association rules.

To summarize, with association rules, you can find rules of the kind *If X then (likely) Y* where *X* and *Y* can be single values, items, words, etc., or conjunctions of values, items, words, etc. (e.g., *if (Car=Porsche and Gender=Male and Age<20) then (Risk=High and Insurance=High)*). Association rules can be used to analyze simple categorical variables, dichotomous variables,

and/or multiple response variables. The algorithm will determine association rules without requiring you to specify the number of distinct categories present in the data, or any prior knowledge regarding the maximum factorial degree or complexity of the important associations. In a sense, the algorithm will construct cross-tabulation tables without the need to specify the number of dimensions for the tables or the number of categories for each dimension. Hence, this technique is particularly well suited for data and text mining of huge databases.

Computational Procedures and Terminology

Categorical or class variables. Categorical variables are single variables that contain codes or text values to denote distinct classes; for example, a variable *Gender* would have the categories *male* and *female*.

Multiple-response variables. Multiple-response variables usually consist of multiple variables (i.e., a list of variables) that can contain, for each observation, codes or text values describing a single dimension or transaction. A good example of a multiple response variable would be if a vendor recorded the purchases made by a customer in a single record, where each record could contain one or more items purchased, in arbitrary order. This is a typical format in which customer transaction data would be kept.

Multiple dichotomies. In this data format, each variable would represent one item or category, and the dichotomous data in each variable would indicate whether or not the respective item or category applies to the respective case. For example, suppose a vendor created a data spreadsheet where each column represented one of the products available for purchase. Each transaction (row of the data spreadsheet) would record whether or not the respective customer did or did not purchase that product, i.e., whether or not the respective transaction involved each item.

Association rules: If body, then head. The *a priori* algorithm attempts to derive from the data association rules of the form: *If "Body" then "Head,"* where *Body* and *Head* stand for simple codes or text values (items), or the conjunction of codes and text values (items; e.g., *if (Car=Porsche and Age<20) then (Risk=High and Insurance=High)*; here the logical conjunction before the then would be the *Body*, and the logical conjunction following the then would be the *Head* of the association rule).

Initial pass through the data: The support value. First, an association rules program will scan all variables to determine the unique codes or text values (items) found in the variables selected for the analysis. In this initial pass, the relative frequencies with which the individual codes or text values occur in each transaction will also be computed. The probability that a transaction contains a particular code or text value is called support; the support value is also computed in consecutive passes through the data as the joint probability (relative frequency of co-occurrence) of pairs, triplets, etc., of codes or text values (items), i.e., separately for the *Body* and *Head* of each association rule.

Second pass through the data: The confidence value; correlation value. After the initial pass through the data, all items with a support value greater than some predefined minimum support value will be remembered for subsequent passes through the data. Specifically, the conditional probabilities will be computed for all pairs of codes or text values that have support values greater than the minimum support value. This conditional probability – that an observation (transaction) that contains a code or text value X also contains a code or text value Y – is called the confidence value. In general (in later passes through the data) the confidence value denotes the conditional probability of the *Head* of the association rule, given the *Body* of the association rule.

In addition, the support value will be computed for each pair of codes or text values and a correlation value based on the support values. The correlation value for a pair of codes or text values $\{X, Y\}$ is computed as the support value for that pair, divided by the square root of the product of the support values for X and Y. After the second pass through the data those pairs of codes or text values that 1) have a confidence value that is greater than some user-defined minimum confidence value, 2) have a support value that is greater than some user-defined minimum support value, and 3) have a correlation value that is greater than some minimum correlation value will be retained.

Subsequent passes through the data: Maximum item size in body, head. The data in subsequent steps, the data will be further scanned computing support, confidence, and correlation values for pairs of codes or text values (associations between single codes or text values), triplets of codes or text values, and so on. To reiterate, in general, at each association rules will be derived of the general form if "*Body*" then "*Head*," where *Body* and *Head* stand for simple codes or text values (items), or the conjunction of codes and text values (items).

Unless the process stops because no further associations can be found that satisfy the minimum support, confidence, and correlation conditions, the process could continue to build very complex association rules (e.g., *if X1 and*

X2 .. and X20 then Y1 and Y2 ... and Y20). To avoid excessive complexity, additionally, you can specify the maximum number of codes or text values (items) in the *Body* and *Head* of the association rules; this value is referred to as the maximum item set size in the *Body* and *Head* of an association rule.

Tabular Representation of Associations

Association rules are generated of the general form *if Body then Head*, where *Body* and *Head* stand for single codes or text values (items) or conjunctions of codes or text values (items; e.g., *if (Car=Porsche and Age<20) then (Risk=High and Insurance= High)*. The major statistics computed for the association rules are support (relative frequency of the *Body* or *Head* of the rule), confidence (conditional probability of the *Head* given the *Body* of the rule), and correlation (support for *Body* and *Head*, divided by the square root of the product of the support for the *Body* and the support for the *Head*). These statistics can be summarized in a spreadsheet, as shown in the next illustration.

Data: Summary of association rules (Scene 1)*
Summary of association rules (Scene 1.sta)
Min. support = 5.0%, Min. confidence = 5.0%, Min. correlation = 5.0%
Max. size of body = 10, Max. size of head = 10

	Body	==>	Head	Support(%)	Confidence(%)	Correlation(%)
154	and, that	==>	like	6.94444	83.3333	91.28709
126	like	==>	and, that	6.94444	100.0000	91.28709
163	and, PAROLLES	==>	will	5.55556	80.0000	73.02967
148	will	==>	and, PAROLLES	5.55556	66.6667	73.02967
155	and, you	==>	your	5.55556	80.0000	67.61234
122	your	==>	and, virginity	5.55556	57.1429	67.61234
164	and, virginity	==>	your	5.55556	80.0000	67.61234
121	your	==>	and, you	5.55556	57.1429	67.61234
73	that	==>	like	6.94444	41.6667	64.54972
75	that	==>	and, like	6.94444	41.6667	64.54972
161	and, like	==>	that	6.94444	100.0000	64.54972

This results spreadsheet shows an example of how association rules can be applied to text mining tasks. This analysis was performed on the paragraphs (dialog spoken by the characters in the play) in the first scene of Shakespeare's *All's Well That Ends Well*, after removing a few very frequent words such as "is," "of," etc. The values for support, confidence, and correlation are expressed in percent.

Graphical Representation of Associations

As a result of applying association rules data mining techniques to large data sets rules of the form *if "Body" then "Head"* will be derived, where *Body* and *Head* stand for simple codes or text values (items), or the conjunction of codes and text values (items; e.g., *if (Car=Porsche and Age<20) then (Risk=High and Insurance= High))*. These rules can be reviewed in textual format or tables, or in graphical format (see the next illustration).

Association rules networks, 2D. For example, consider the data that describe a (fictitious) survey of 100 patrons of sports bars and their preferences for watching various sports on television. This is an example of simple categorical variables, where each variable represents one sport. The respondents indicate how frequently they watch each sport on television. The association rules derived from these data could be summarized as follows:

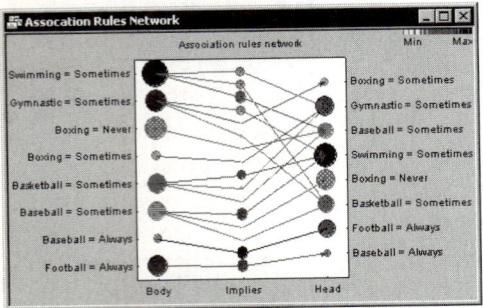

In this graph, the support values for the *Body* and *Head* portions of each association rule are indicated by the sizes and colors of each. The thickness of each line indicates the confidence value

(conditional probability of *Head* given *Body*) for the respective association rule; the sizes and colors of the circles in the center, above the *Implies* label, indicate the joint support (for the co-occurrences) of the respective *Body* and *Head* components of the respective association rules. Hence, in this graphical summary, the strongest support value was found for *Swimming=Sometimes*, which was associated with *Gymnastic=Sometimes*, *Baseball=Sometimes*, and *Basketball= Sometimes*.

Unlike simple frequency and crosstabulation tables, the absolute frequencies with which individual codes or text values (items) occur in the data are often not reflected in the association rules; instead, only those codes or text values (items) are retained that show sufficient values for support, confidence, and correlation, i.e., that co-occur with other codes or text values (items) with sufficient relative (co-) frequency.

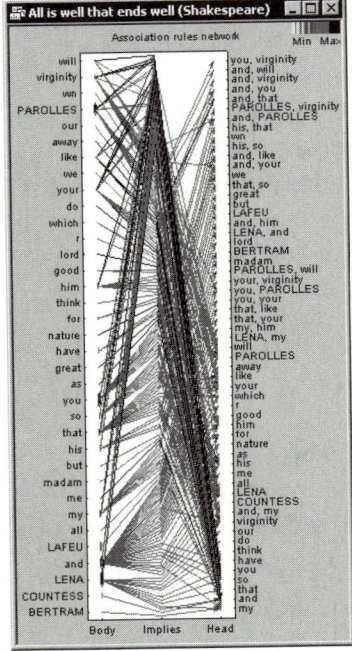

The results that can be summarized in 2D association rules networks can be relatively simple, or complex as illustrated in the network shown in the previous illustration. This is an example of how association rules can be applied to text mining tasks. This analysis was performed on the paragraphs (dialog spoken by the characters in the play) in the first scene of Shakespeare's *All's Well That Ends Well*, after removing a few very frequent words such as "is," "of," etc. Of course, the specific words and phrases removed during the data preparation phase of text (or data) mining projects will depend on the purpose of the research.

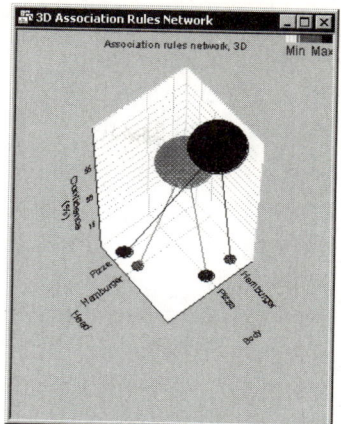

Association rules networks, 3D. Association rules can be graphically summarized in 2D association networks, as well as 3D association networks. Shown in the next illustration are some (very clear) results from an analysis. Respondents in a survey were asked to list their (up to) 3 favorite fast foods. The association rules derived from those data are summarized in a 3D association network display.

As in the 2D association network, the support values for the *Body* and *Head* portions of each association rule are indicated by the sizes and colors of each circle in the 2D. The thickness of each line indicates the confidence value (joint probability) for the respective association rule; the sizes and colors of the floating circles plotted against the (vertical) z-axis indicate the joint support (for the co-occurrences) of the respective *Body* and *Head* components of the association rules. The plot position of each circle along the vertical z-axis indicates the respective confidence value. Hence, this particular graphical summary clearly shows two simple rules: Respondents who name *Pizza* as a preferred fast food also mention *Hamburger*, and vice versa.

Interpreting and Comparing Results

When comparing the results of applying association rules to those from simple frequency or cross-tabulation tables, you may notice that in some cases very high-frequency codes or text values (items) are not part of any association rule. This can sometimes be perplexing.

To illustrate how this pattern of findings can occur, consider this example: Suppose you analyzed data from a survey of insurance rates for different makes of automobiles in America. Simple tabulation would very likely show that many people drive automobiles manufactured by Ford, GM, and Chrysler; however, none of these makes can be associated with particular patterns in insurance rates, i.e., none of these brands can be involved in high-confidence, high-correlation association rules linking them to particular categories of insurance rates. However, when applying association rules methods, automobile makes that occur in the sample with relatively low frequency (e.g., Porsche) may be found to be associated with high insurance rates (allowing you to infer, for example, a rule that *if Car=Porsche then*

Insurance=High). If you only reviewed a
simple cross-tabulation table (make of car by
insurance rate) this high-confidence association
rule may well have gone unnoticed.

BOOSTING TREES

Boosting Trees for Regression and Classification (Stochastic Gradient Boosting). The general computational approach of stochastic gradient boosting is also known by the names TreeNet (TM Salford Systems, Inc.) and MART (TM Jerill, Inc.). Over the past few years, this technique has emerged as one of the most powerful methods for predictive data mining. Some implementations of these powerful algorithms enable them to be used for regression as well as classification problems, with continuous and/or categorical predictors. Detailed technical descriptions of these methods can be found in Friedman (1999a, b) as well as Hastie, Tibshirani, & Friedman (2001).

Gradient Boosting Trees

The algorithm for boosting trees evolved from the application of boosting methods to regression trees. The general idea is to compute a sequence of (very) simple trees, where each successive tree is built for the prediction residuals of the preceding tree. As described in the *Classification and Regression Trees Overview*, page 83, this method will build binary trees, i.e., partition the data into two samples at each split node. Now suppose that you were to limit the complexities of the trees to 3 nodes only: a root node and two child nodes, i.e., a single split. Thus, at each step of the boosting (boosting trees algorithm), a simple (best) partitioning of the data is determined, and the deviations of the observed values from the respective means (residuals for each partition) are computed. The next 3-node tree will then be fitted to those residuals, to find another partition that will further reduce the residual (error) variance for the data, given the preceding sequence of trees.

It can be shown that such "additive weighted expansions" of trees can eventually produce an excellent fit of the predicted values to the observed values, even if the specific nature of the relationships between the predictor variables and the dependent variable of interest is very complex (nonlinear in nature). Hence, the method of gradient boosting – fitting a weighted additive expansion of simple trees – represents a very general and powerful machine learning algorithm.

The Problem of Overfitting; Stochastic Gradient Boosting

One of the major problems of all machine learning algorithms is to "know when to stop," i.e., how to prevent the learning algorithm to fit esoteric aspects of the training data that are not likely to improve the predictive validity of the respective model. This issue is also known as the problem of overfitting. To reiterate, this is a general problem applicable to most machine learning algorithms used in predictive data mining. A general solution to this problem is to evaluate the quality of the fitted model by predicting observations in a test-sample of data that have not been used before to estimate the respective model(s). In this manner, one hopes to gage the predictive accuracy of the solution, and to detect when overfitting has occurred (or is starting to occur).

A similar approach is for each consecutive simple tree to be built for only a randomly selected subsample of the full data set. In other words, each consecutive tree is built for the prediction residuals (from all preceding trees) of an independently drawn random sample. The introduction of a certain degree of randomness into the analysis in this manner can serve as a powerful safeguard against overfitting (since each consecutive tree is built for a different sample of observations), and yield models (additive weighted expansions of simple trees) that generalize well to new observations, i.e., exhibit good predictive validity. This technique, i.e., performing consecutive boosting computations on independently drawn samples of observations, is knows as stochastic gradient boosting.

The following illustration is a plot of the prediction error function for the training data over successive trees and also an independently sampled testing data set at each stage.

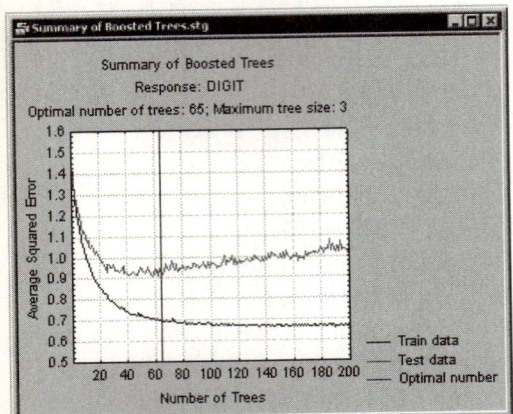

With this graph, you can identify very quickly the point where the model (consisting of a certain number of successive trees) begins to overfit the data. Notice how the prediction error for the training data steadily decreases as more and more additive terms (trees) are added to the model. However, somewhere past 35 trees, the performance for independently sampled testing data actually begins to deteriorate, clearly indicating the point where the model begins to overfit the data.

Stochastic Gradient Boosting Trees and Classification

So far, the discussion of boosting trees has exclusively focused on regression problems, i.e., on the prediction of a continuous dependent variable. The technique can easily be expanded to handle classification problems as well (this is described in detail in Friedman, 1999a, section 4.6; in particular, see Algorithm 6).

First, different boosting trees are built for (fitted to) each category or class of the categorical dependent variable, after creating a coded variable (vector) of values for each class with the values 1 or 0 to indicate whether or not an observation does or does not belong to the

respective class. In successive boosting steps, the algorithm will apply the logistic transformation (see also Chapter 28 – *Nonlinear Estimation*) to compute the residuals for subsequent boosting steps. To compute the final classification probabilities, the logistic transformation is again applied to the predictions for each 0/1 coded vector (class). This algorithm is described in detail in Friedman (1999a; see also Hastie, Tibshirani, and Freedman, 2001, for a description of this general procedure).

Large numbers of categories. Note that the procedure for applying this method to classification problems requires that separate sequences of (boosted) trees be built for each category or class. Hence, the computational effort generally becomes larger by a multiple of what it takes to solve a simple regression prediction problem (for a single continuous dependent variable). Therefore, it is not prudent to analyze categorical dependent variables (class variables) with more than, approximately, 100 or so classes; past that point, the computations performed may require an unreasonable amount of effort and time. (For example, a problem with 200 boosting steps and 100 categories or classes for the dependent variable would yield 200 * 100 = 20,000 individual trees!)

CANONICAL ANALYSIS

There are several measures of correlation to express the relationship between two or more variables. For example, the standard Pearson product moment correlation coefficient (r) measures the extent to which two variables are related. There are various nonparametric measures of relationships that are based on the similarity of ranks in two variables; with multiple regression, you can assess the relationship between a dependent variable and a set of independent variables; multiple correspondence analysis is useful for exploring the relationships between a set of categorical variables.

Canonical correlation is an additional procedure for assessing the relationship between variables. Specifically, with this analysis, you can investigate the relationship between two sets of variables. For example, an educational researcher may want to compute the (simultaneous) relationship among three measures of scholastic ability with five measures of success in school. A sociologist may want to investigate the relationship between two predictors of social mobility based on interviews, with actual subsequent social mobility as measured by four different indicators. A medical researcher may want to study the relationship of various risk factors to the development of a group of symptoms. In all of these cases, the researcher is interested in the relationship between two sets of variables, and canonical correlation is the appropriate method of analysis.

This chapter, briefly introduces the major concepts and statistics in canonical correlation analysis. You should be familiar with the correlation coefficient described in Chapter 2 – *Basic Statistics and Tables* and the basics of multiple regression described in Chapter 26 – *Multiple Linear Regression*.

General Ideas

Suppose you conduct a study in which you measure satisfaction at work with three questionnaire items, and satisfaction in various other domains with an additional seven items. The general question that you may want to answer is how satisfaction at work relates to the satisfaction in those other domains.

Sum Scores

A first approach that you might take is simply to add up the responses to the work satisfaction items, and to correlate that sum with the responses to all other satisfaction items. If the correlation between the two sums is statistically significant, we could conclude that work satisfaction is related to satisfaction in other domains.

In a way, this is a rather crude conclusion. We still know nothing about the particular domains of satisfaction that are related to work satisfaction. In fact, we could potentially have lost important information by simply adding up items. For example, suppose there were two items, one measuring satisfaction with one's relationship with the spouse, the other measuring satisfaction with one's financial situation. Adding the two together is, obviously, like adding apples to oranges. Doing so implies that a person who is dissatisfied with her finances but happy with her spouse is comparable overall to a person who is satisfied financially but not happy in the relationship with her spouse. Most likely, people's psychological makeup is not that simple.

The problem then with simply correlating two sums is that you might lose important information in the process, and, in the worst case, actually destroy important relationships between variables by adding apples to oranges.

Using a weighted sum. It seems reasonable to correlate some kind of a weighted sum instead so that the structure of the variables in the two sets is reflected in the weights. For example, if satisfaction with one's spouse is only marginally related to work satisfaction, but financial satisfaction is strongly related to work satisfaction, we could assign a smaller weight to the first item and a greater weight to the second item. We can express this general idea in the following equation:

$$a_1{}^*y_1 + a_2{}^*y_2 + \ldots + a_p{}^*y_p = b_1{}^*x_1 + b_2{}^*x_2 + \ldots + b_q{}^*x_q$$

If we have two sets of variables, the first one containing p variables and the second one containing q variables, we would want to correlate the weighted sums on each side of the equation with each other.

Determining the weights. We have now formulated the general model equation for canonical correlation. The only problem that remains is how to determine the weights for the two sets of variables. It seems to make little sense to assign weights so that the two weighted sums do not correlate with each other. A reasonable approach to take is to impose the condition that the two weighted sums shall correlate maximally with each other.

Canonical Roots/Variates

In the terminology of canonical correlation analysis, the weighted sums define a canonical root or variate. You can think of those canonical variates (weighted sums) as describing some underlying latent variables. For example, if for a set of diverse satisfaction items we were to obtain a weighted sum marked by large weights for all items having to do with work, we could conclude that the respective canonical variate measures satisfaction with work.

Number of Roots

So far we have pretended as if there is only one set of weights (weighted sum) that can be extracted from the two sets of variables. However, suppose that we had among our work satisfaction items particular questions regarding satisfaction with pay, and questions pertaining to satisfaction with one's social relationships with other employees. It is possible that the pay satisfaction items correlate with satisfaction with one's finances, and that the social relationship satisfaction items correlate with the reported satisfaction with one's spouse. If so, we should really derive two weighted sums to reflect this complexity in the structure of satisfaction.

In fact, the computations involved in canonical correlation analysis will lead to more than one set of weighted sums. To be precise, the number of roots extracted will be equal to the minimum number of variables in either set. For example, if we have three work satisfaction items and seven general satisfaction items, three canonical roots will be extracted.

Extraction of Roots

As mentioned before, you can extract roots so that the resulting correlation between the canonical variates is maximal. When extracting more than one root, each successive root will explain a unique additional proportion of variability in the two sets of variables. Therefore, successively extracted canonical roots will be uncorrelated with each other, and account for less and less variability.

Computational Methods and Results

Some of the computational issues involved in canonical correlation and the major results that are commonly reported will now be reviewed.

Eigenvalues. When extracting the canonical roots, you will compute the eigenvalues. These can be interpreted as the proportion of variance accounted for by the correlation between the respective canonical variates. Note that the proportion here is computed relative to the variance of the canonical variates, that is, of the weighted sum scores of the two sets of variables; the eigenvalues do not tell how much variability is explained in either set of variables. You will compute as many eigenvalues as there are canonical roots, that is, as many as the minimum number of variables in either of the two sets.

Successive eigenvalues will be of smaller and smaller size. First, compute the weights that maximize the correlation of the two sum scores. After this first root has been extracted, you will find the weights that produce the second largest correlation between sum scores, subject to the constraint that the next set of sum scores does not correlate with the previous one, and so on.

Canonical correlations. If the square root of the eigenvalues is taken, the resulting numbers can be interpreted as correlation coefficients. Because the correlations pertain to the canonical variates, they are called canonical correlations. As eigenvalues, the correlations between successively extracted canonical variates are smaller and smaller. Therefore, as an overall index of the canonical correlation between two sets of variables, it is customary to report the largest correlation, that is, the one for the first root. However, the other canonical variates can also be correlated in a meaningful and interpretable manner.

Significance of roots. The significance test of the canonical correlations is straightforward in principle. Simply stated, the different canonical correlations are tested, one by one, beginning with the largest one. Only those roots that are statistically significant are then retained for subsequent interpretation. Actually, the nature of the significance test is somewhat different. First, evaluate the significance of all roots combined, then of the roots remaining after removing the first root, the second root, etc.

Some authors have criticized this sequential testing procedure for the significance of canonical roots (e.g., Harris, 1976). However, this procedure was rehabilitated in a subsequent Monte Carlo study by Mendoza, Markos, and Gonter (1978).

In short, the results of that study showed that this testing procedure would detect strong canonical correlations most of the time, even with samples of relatively small size (e.g., $n = 50$). Weaker canonical correlations (e.g., $R = .3$) require larger sample sizes ($n > 200$) to be detected at least 50% of the time. Note that canonical correlations of small magnitude are often of little practical value, as they account for very little actual variability in the data. This issue, as well as the sample size issue, will be discussed shortly.

Canonical weights. After determining the number of significant canonical roots, the question arises as to how to interpret each (significant) root. Remember that each root actually represents two weighted sums, one for each set of variables. One way to interpret the meaning of each canonical root would be to look at the weights for each set. These weights are called the canonical weights.

In general, the larger the weight (i.e., the absolute value of the weight), the greater is the respective variable's unique positive or negative contribution to the sum. To facilitate comparisons between weights, the canonical weights are usually reported for the standardized variables, that is, for the z transformed variables with a mean of 0 and a standard deviation of 1.

If you are familiar with multiple regression, you can interpret the canonical weights in the same manner as you would interpret the beta weights in a multiple regression equation. In a sense, they represent the partial correlations of the variables with the respective canonical root. If you are familiar with factor analysis, you can interpret the canonical weights in the same manner as you would interpret the factor score coefficients. To summarize, the canonical weights enable you to understand the makeup of each canonical root, that is, it lets you see how each variable in each set uniquely contributes to the respective weighted sum (canonical variate).

Canonical scores. Canonical weights can also be used to compute actual values of the canonical variates; that is, you can simply use the weights to compute the respective sums. Again, remember that the canonical weights are customarily reported for the standardized (z transformed) variables.

Factor structure. Another way of interpreting the canonical roots is to look at the simple correlations between the canonical variates (or factors) and the variables in each set. These correlations are also called canonical factor loadings. The logic here is that variables that are highly correlated with a canonical variate have more in common with it. Therefore, you should weigh them more heavily when deriving a meaningful interpretation of the respective canonical variate. This method of interpreting canonical variates is identical to the manner in which factors are interpreted in factor analysis.

Factor structure vs. canonical weights. Sometimes the canonical weights for a variable

are nearly zero, but the respective loading for the variable is very high. The opposite pattern of results may also occur. At first, such a finding may seem contradictory; however, remember that the canonical weights pertain to the unique contribution of each variable, while the canonical factor loadings represent simple overall correlations.

For example, suppose you included in your satisfaction survey two items that measure basically the same thing, namely: 1) Are you satisfied with your supervisors? and 2) Are you satisfied with your bosses? Obviously, these items are redundant. When the weights for the weighted sums (canonical variates) in each set is computed so that they correlate maximally, it only needs to include one of the items to capture the essence of what they measure. Once a large weight is assigned to the first item, the contribution of the second item is redundant; consequently, it will receive a zero or negligibly small canonical weight. Nevertheless, if you then look at the simple correlations between the respective sum score with the two items (i.e., the factor loadings), those may be substantial for both. To reiterate, the canonical weights pertain to the unique contributions of the respective variables with a particular weighted sum or canonical variate; the canonical factor loadings pertain to the overall correlation of the respective variables with the canonical variate.

Variance extracted. As discussed earlier, the canonical correlation coefficient refers to the correlation between the weighted sums of the two sets of variables. It tells nothing about how much variability (variance) each canonical root explains in the variables. However, you can infer the proportion of variance extracted from each set of variables by a particular root by looking at the canonical factor loadings. Remember that those loadings represent correlations between the canonical variates and the variables in the respective set. If you square those correlations, the resulting numbers reflect the proportion of variance accounted for in each variable. For each root, you can take the average of those proportions across variables to get an indication of how much variability is explained, on the average, by the respective canonical variate in that set of variables. Put another way, you can compute in this manner the average proportion of variance extracted by each root.

Redundancy. The canonical correlations can be squared to compute the proportion of variance shared by the sum scores (canonical variates) in each set. If you multiply this proportion by the proportion of variance extracted, you arrive at a measure of redundancy, that is, of how redundant one set of variables is, given the other set of variables. In equation form, you can express the redundancy as:

$$\text{Redundancy}_{left} = [\Sigma(\text{loadings}_{left}^2)/p]*R_c^2$$

$$\text{Redundancy}_{right} = [\Sigma(\text{loadings}_{right}^2)/q]*R_c^2$$

In these equations, p denotes the number of variables in the first set of variables, and q denotes the number of variables in the second set of variables; R_c^2 is the respective squared canonical correlation.

Note that you can compute the redundancy of the first set of variables given the second set, and the redundancy of the second set of variables, given the first set. Because successively extracted canonical roots are uncorrelated, you could sum up the redundancies across all (or only the first significant) roots to arrive at a single index of redundancy (as proposed by Stewart and Love, 1968).

Practical significance. The measure of redundancy is also useful for assessing the practical significance of canonical roots. With large sample sizes, canonical correlations of magnitude $R = .30$ may become statistically significant (see above). If you square this coefficient (R-square $= .09$) and use it in the redundancy formula shown above, it becomes clear that such canonical roots account for only very little variability in the variables. Of course, the final assessment of what does and does not constitute a finding of practical significance is subjective by nature. However, to maintain a realistic appraisal of how much actual variance (in the variables) is accounted for by a canonical root, it is important to always keep in mind the redundancy measure, that is, how much of the actual variability in one set of variables is explained by the other.

Assumptions

The following discussion provides only a list of the most important assumptions of canonical correlation analysis, and the major threats to the reliability and validity of results.

Distributions. The test of significance of the canonical correlations is based on the assumption that the distributions of the variables in the population (from which the sample was drawn) are multivariate normal. Little is known about the effects of violations of the multivariate normality assumption. However, with a sufficiently large sample size the results from canonical correlation analysis are usually quite robust.

Sample sizes. Stevens (1986) provides a very thorough discussion of the sample sizes that should be used in order to obtain reliable results. As mentioned earlier, if there are strong canonical correlations in the data (e.g., $R > .7$), even relatively small samples (e.g., $n = 50$) will detect them most of the time. However, in order to arrive at reliable estimates of the canonical

factor loadings (for interpretation), Stevens recommends that there should be at least 20 times as many cases as variables in the analysis if you want to interpret the most significant canonical root only. To arrive at reliable estimates for two canonical roots, Barcikowski and Stevens (1975) recommend, based on a Monte Carlo study, to include 40 to 60 times as many cases as variables.

Outliers. Outliers can greatly affect the magnitudes of correlation coefficients. Since canonical correlation analysis is based on (computed from) correlation coefficients, they can also seriously affect the canonical correlations. Of course, the larger the sample size, the smaller is the impact of one or two outliers. However, it is a good idea to examine various scatterplots to detect possible outliers (as shown in the next illustrations).

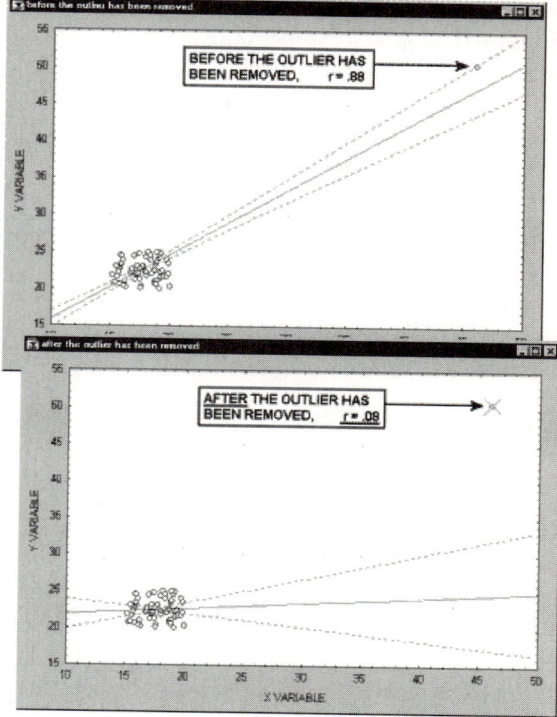

See also, *Ellipse* in the *Statistical Glossary*.

Matrix ill-conditioning. One assumption is that the variables in the two sets should not be completely redundant. For example, if you included the same variable twice in one of the sets, it is not clear how to assign different weights to each of them. Computationally, such complete redundancies will upset the canonical correlation analysis. When there are perfect correlations in the correlation matrix, or if any of the multiple correlations between one variable and the others is perfect ($R = 1.0$), the correlation matrix cannot be inverted, and the computations for the canonical analysis cannot be performed. Such correlation matrices are said to be ill-conditioned.

Once again, this assumption appears trivial on the surface; however, it often is almost violated when the analysis includes very many highly redundant measures, as is often the case when analyzing questionnaire responses.

StatSoft®
Copyright © StatSoft, 2006

CHAID ANALYSIS

The acronym CHAID stands for *chi*-squared Automatic Interaction Detector. It is one of the oldest tree classification methods originally proposed by Kass (1980; according to Ripley, 1996, the CHAID algorithm is a descendent of THAID developed by Morgan and Messenger, 1973). CHAID will build non-binary trees (i.e., trees where more than two branches can attach to a single root or node), based on a relatively simple algorithm that is particularly well suited for the analysis of larger data sets. Also, because the CHAID algorithm will often effectively yield many multi-way frequency tables (e.g., when classifying a categorical response variable with many categories, based on categorical predictors with many classes), it has been particularly popular in marketing research, in the context of market segmentation studies.

Both CHAID and CART techniques will construct trees where each (non-terminal) node identifies a split condition to yield optimum prediction (of continuous dependent or response variables) or classification (for categorical dependent or response variables). Hence, both types of algorithms can be applied to analyze regression-type or classification-type problems.

Basic Tree-Building Algorithm: CHAID and Exhaustive CHAID

The acronym CHAID stands for Chi-squared Automatic Interaction Detector. This name derives from the basic algorithm that is used to construct (non-binary) trees, which for classification problems (when the dependent variable is categorical in nature) relies on the *chi*-square test to determine the best next split at each step; for regression-type problems (continuous dependent variable), *F*-tests are computed. Specifically, the algorithm proceeds as follows:

Preparing predictors. The first step is to create categorical predictors out of any continuous predictors by dividing the respective continuous distributions into a number of categories with an approximately equal number of observations. For categorical predictors, the categories (classes) are naturally defined.

Merging categories. The next step is to cycle through the predictors to determine for each predictor the pair of (predictor) categories that is least significantly different with respect to the dependent variable; for classification problems (where the dependent variable is categorical as well), it will compute a *chi*-square test (Pearson *chi*-square); for regression problems (where the dependent variable is continuous), it will compute F tests. If the respective test for a given pair of predictor categories is not statistically significant as defined by an *alpha*-to-merge value, it will merge the respective predictor categories and repeat this step (i.e., find the next pair of categories, which now may include previously merged categories). If the statistical significance for the respective pair of predictor categories is significant (less than the respective alpha-to-merge value), then (optionally) it will compute a Bonferroni adjusted *p*-value for the set of categories for the respective predictor.

Selecting the split variable. The next step is to choose the variable to split. The algorithm selects the predictor variable with the smallest adjusted *p*-value, i.e., the predictor variable that will yield the most significant split. If the smallest (Bonferroni) adjusted *p*-value for any predictor is greater than some *alpha*-to-split value, no further splits will be performed, and the respective node is a terminal node.

Continue this process until no further splits can be performed (given the alpha-to-merge and alpha-to-split values).

CHAID and exhaustive CHAID algorithms. A modification to the basic CHAID algorithm, exhaustive CHAID, performs a more thorough merging and testing of predictor variables, and hence requires more computing time. Specifically, the merging of categories continues (without reference to any alpha-to-merge value) until only two categories remain for each predictor. The algorithm then proceeds as described in the previous step – *Selecting the split variable* – and selects among the predictors the one that yields the most significant split. For large data sets, and with many continuous predictor variables, this modification of the simpler CHAID algorithm may require significant computing time.

General Computation Issues of CHAID

Reviewing large trees: Unique analysis management tools. A general issue that arises when applying tree classification or regression methods is that the final trees can become very large. In practice, when the input data are complex and, for example, contain many different categories for classification problems and many possible predictors for performing the classification, the resulting trees can become very large. This is not so much a computational problem as it is a problem of presenting the trees in a manner that is easily accessible to the data analyst, or for presentation to the consumers of the research.

Analyzing ANCOVA-like designs. The classic CHAID algorithms can accommodate both continuous and categorical predictors. However, in practice, it is not uncommon to combine such variables into analysis of variance/covariance (ANCOVA)-like predictor designs with main effects or interaction effects for categorical and continuous predictors. This method of analyzing coded ANCOVA-like designs is relatively new. However, it is easy to see how the use of coded predictor designs expands these powerful classification and regression techniques to the analysis of data from experimental designs.

CHAID, CART, and QUEST

For classification-type problems (categorical dependent variable), all three algorithms can be used to build a tree for prediction. QUEST is generally faster than the other two algorithms, however, for very large data sets, the memory requirements are usually larger, so using the QUEST algorithms for classification with very large input data sets may be impractical.

For regression-type problems (continuous dependent variable), the QUEST algorithm is not applicable, so only CHAID and CART can be used. CHAID will build non-binary trees that tend to be wider. This has made the CHAID method particularly popular in market research applications: CHAID often yields many terminal nodes connected to a single branch, which can be conveniently summarized in a simple two-way table with multiple categories for each variable or dimension of the table. This type of display matches well the requirements for research on market segmentation, for example, it may yield a split on a variable *Income*, dividing that variable into 4 categories and groups of individuals belonging to those categories that are different with respect to some important consumer-behavior related variable (e.g., types of cars most likely to be purchased). CART will always yield binary trees, which sometimes can not be summarized as efficiently for interpretation and/or presentation.

As far as predictive accuracy is concerned, it is difficult to derive general recommendations, and this issue is still the subject of active research. As a practical matter, it is best to apply different algorithms, perhaps compare them with user-defined interactively derived trees, and decide on the most reasonable and best performing model based on the prediction errors. For a discussion of various schemes for combining predictions from different models, see, for example, Witten and Frank, 2000.

CLASSIFICATION AND REGRESSION TREES (CART)

CART builds classification and regression trees for predicting continuous dependent variables (regression) and categorical predictor variables (classification). The classic CART algorithm was popularized by Breiman et al. (Breiman, Friedman, Olshen, & Stone, 1984; see also, Ripley, 1996). A general introduction to tree-classifiers, specifically to the QUEST (Quick, Unbiased, Efficient Statistical Trees) algorithm, is also presented in Chapter 9 – *Classification Trees*, and much of the following discussion presents the same information in only a slightly different context. Another similar type of tree building algorithm is CHAID (Chi-square Automatic Interaction Detector; see Kass, 1980).

Overview

Classification and Regression Problems

There are numerous algorithms for predicting continuous variables or categorical variables from a set of continuous predictors and/or categorical factor effects. For example, in GLM (general linear models) and GRM (general regression models), you can specify a linear combination (design) of continuous predictors and categorical factor effects (e.g., with two-way and three-way interaction effects) to predict a continuous dependent variable. In GDA (general discriminant function analysis), you can specify such designs for predicting categorical variables, i.e., to solve classification problems.

Regression-type problems. Regression-type problems are generally those for which you attempt to predict the values of a continuous variable from one or more continuous and/or categorical predictor variables. For example, you may want to predict the selling prices of single family homes (a continuous dependent variable) from various other continuous predictors (e.g., square footage) as well as categorical predictors (e.g., style of home, such as ranch, two-story, etc.; zip code or telephone area code where the property is located, etc.; note that this latter variable would be categorical in nature, even though it would contain numeric values or codes).

If you use simple multiple regression or some general linear model to predict the selling prices of single family homes, you would determine a linear equation for these variables that could be used to compute predicted selling prices. There are many different analytic procedures for fitting linear models (e.g., GLM, GRM, regression), various types of nonlinear models (e.g., generalized linear/nonlinear models, generalized

additive models, etc.), or completely custom-defined nonlinear models (nonlinear estimation), where you can type in an arbitrary equation containing parameters to be estimated. CHAID also analyzes regression-type problems, and produces results that are similar (in nature) to those computed by CART. Note that various neural network architectures are also applicable to solve regression-type problems.

Classification-type problems. Classification-type problems are generally those for which you attempt to predict values of a categorical dependent variable (class, group membership, etc.) from one or more continuous and/or categorical predictor variables. For example, you may want to predict who will graduate from college, or who will renew a subscription. These are examples of simple binary classification problems, where the categorical dependent variable can only assume two distinct and mutually exclusive values.

In other cases, you may want to predict which one of multiple different alternative consumer products (e.g., makes of cars) a person decides to purchase, or what type of failure occurs with different types of engines. In these cases, there are multiple categories or classes for the categorical dependent variable. There are a number of methods for analyzing classification-type problems and to compute predicted classifications, either from simple continuous predictors (e.g., binomial or multinomial logit regression in GLZ), from categorical predictors (e.g., log-linear analysis of multi-way frequency tables), or both (e.g., via ANCOVA-like designs in GLZ or GDA). CHAID also analyzes classification-type problems, and produces results that are similar (in nature) to those computed by CART. Note that various neural network architectures are also applicable to solve classification-type problems.

Classification and Regression Trees (CART)

In general terms, the purpose of the analyses via tree-building algorithms is to determine a set of *if-then* logical (split) conditions that permit accurate prediction or classification of cases.

Classification Trees

For example, consider the widely referenced Iris data classification problem introduced by Fisher 1936; see also, Chapter 13 – *Discriminant Function Analysis* and Chapter 17 – *General Discriminant Analysis (GDA)*. The data file *Irisdat* reports the lengths and widths of sepals and petals of three types of irises (Setosa, Versicol, and Virginic). The purpose of the analysis is to learn how to discriminate between the three types of flowers based on the four measures of width and length of petals and sepals. Discriminant function analysis will estimate several linear combinations of predictor variables for computing classification scores (or probabilities) that enable you to determine the predicted classification for each observation. A classification tree will determine a set of logical if-then conditions (instead of linear equations) for predicting or classifying cases instead.

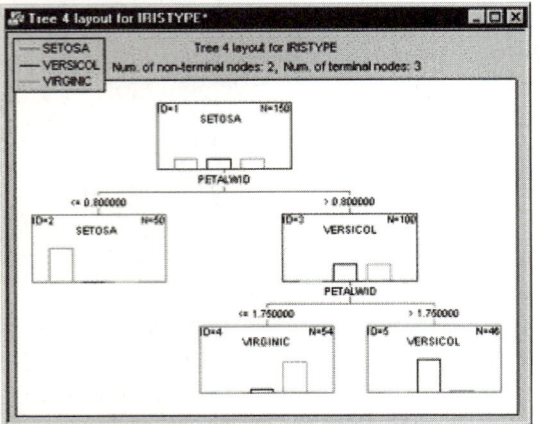

The interpretation of this tree is straightforward: If the petal width is less than or equal to 0.8, the flower would be classified as Setosa; if the petal width is greater than 0.8 and less than or equal to 1.75, the flower would be classified as Virginic; others belong to class Versicol.

Regression Trees

The general approach to derive predictions from a few simple if-then conditions can be applied to regression problems as well. This example is based on the data file *Poverty*, which contains 1960 and 1970 Census figures for a random selection of 30 counties. The research question (for that example) was to determine the correlates of poverty, that is, the variables that best predict the percent of families below the poverty line in a county. A re-analysis of those data using the regression tree analysis (and v-fold cross-validation), yields the following results:

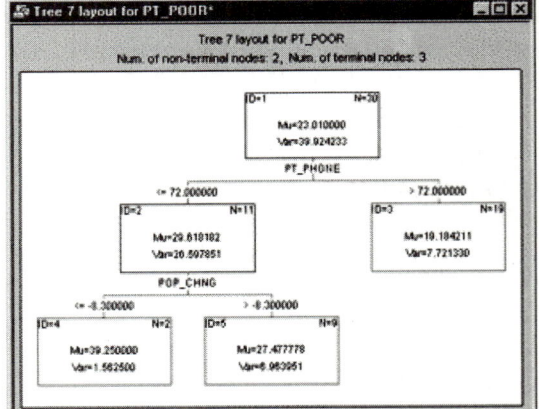

Again, the interpretation of these results is straightforward. Counties where the percentage of households with a phone is greater than 72% generally have a lower poverty rate. The greatest poverty rate is evident in those counties that show less than (or equal to) 72% of households with a phone, and where the population change (from the 1960 census to the

170 census) is less than -8.3 (minus 8.3). These results are easily presented and intuitively clear as well. There are some affluent counties (where most households have a telephone), and those generally have little poverty. Then there are counties that are generally less affluent, and among those the ones that shrunk most showed the greatest poverty rate.

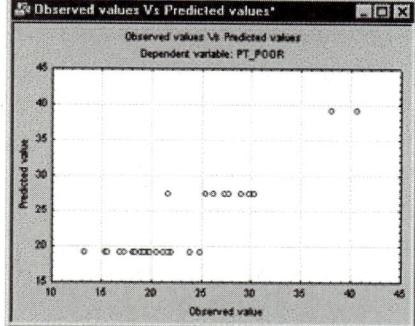

A quick review of the scatterplot of observed vs. predicted values shows how the discrimination between the latter two groups is particularly well explained by the tree model.

Advantages of Classification and Regression Trees (CART) Methods

As mentioned previously, there are a large number of methods from which an analyst can choose when analyzing classification or regression problems. Tree classification techniques, when they work and produce accurate predictions or predicted classifications based on few logical *if-then* conditions, have a number of advantages over many of those alternative techniques.

Simplicity of results. In most cases, the interpretation of results summarized in a tree is very simple.

This simplicity is useful not only for purposes of rapid classification of new observations (it is much easier to evaluate just one or two logical conditions, than to compute classification scores for each possible group, or predicted values, based on all predictors and using possibly some complex nonlinear model equations), but can also often yield a much simpler model for explaining why observations are classified or predicted in a particular manner (e.g., when analyzing business problems, it is much easier to present a few simple if-then statements to management, than some elaborate equations).

Tree methods are nonparametric and nonlinear. The final results of using tree methods for classification or regression can be summarized in a series of (usually few) logical *if-then* conditions (tree nodes). Therefore, there is no implicit assumption that the underlying relationships between the predictor variables and the dependent variable are linear, follow some specific nonlinear link function [e.g., see Chapter 21 – *Generalized Linear/Nonlinear Models (GLZ)*], or that they are even monotonic in nature. For example, some continuous outcome variable of interest could be positively related to a variable Income if the income is less than some certain amount, but negatively related if it is more than that amount (i.e., the tree could reveal multiple splits based on the same variable Income, revealing such a non-monotonic relationship between the variables). Thus, tree methods are particularly well suited for data mining tasks, where there is often little *a priori* knowledge nor any coherent set of theories or predictions regarding which variables are related and how. In those types of data analyses, tree methods can often reveal simple relationships between just a few variables that could have easily gone unnoticed using other analytic techniques.

General Computation Issues and Unique Solutions of CART

The computational details involved in determining the best split conditions to construct a simple yet useful and informative tree are quite complex. Refer to Breiman et al. (1984) for a discussion of their CART® to learn more about the general theory of and specific computational solutions for constructing classification and regression trees. An excellent general discussion of tree classification and regression methods, and comparisons with other approaches to pattern recognition and neural networks, is provided in Ripley (1996).

Avoiding Over-Fitting: Pruning, Cross-Validation, and V-Fold Cross-Validation

A major issue that arises when applying regression or classification trees to real data with much random error noise concerns the decision when to stop splitting. For example, if you have a data set with 10 cases, and perform 9 splits (determined 9 *if-then* conditions), you could perfectly predict every single case. In general, if you only split a sufficient number of times, eventually you will be able to predict (reproduce would be the more appropriate term here) your original data (from which you determined the splits). Of course, it is far from clear whether such complex results (with many splits) will replicate in a sample of new observations; most likely they will not.

This general issue is also discussed in the literature on tree classification and regression methods, as well as neural networks, under the topic of "overlearning" or "overfitting." If not stopped, the tree algorithm will ultimately extract all information from the data, including information that is not and cannot be predicted in the population with the current set of predictors, i.e., random or noise variation. The general approach to addressing this issue is first to stop generating new split nodes when subsequent splits only result in very little overall improvement of the prediction. For example, if you can predict 90% of all cases correctly from 10 splits, and 90.1% of all cases from 11 splits, it obviously makes little sense to add that 11th split to the tree. There are many such criteria for automatically stopping the splitting (tree-building) process.

Once the tree-building algorithm has stopped, it is always useful to further evaluate the quality of the prediction of the current tree in samples of observations that did not participate in the original computations. These methods are used to prune back the tree, i.e., to eventually (and ideally) select a simpler tree than the one obtained when the tree-building algorithm stopped, but one that is equally as accurate for predicting or classifying new observations.

Cross-validation. One approach is to apply the tree computed from one set of observations (learning sample) to another completely independent set of observations (testing sample). If most or all of the splits determined by the analysis of the learning sample are essentially based on random noise, the prediction for the testing sample will be very poor. Hence, you can infer that the selected tree is not very good (useful), and not of the right size.

V-fold cross-validation. Continuing further along this line of reasoning (described in the context of cross-validation above), why not repeat the analysis many times over with different randomly drawn samples from the data, for every tree size starting at the root of the tree, and applying it to the prediction of observations from randomly selected testing samples. Then use (interpret, or accept as your final result) the tree that shows the best average accuracy for cross-validated predicted classifications or predicted values. In most cases, this tree will not be the one with the most terminal nodes, i.e., the most complex tree. This method for pruning a tree, and for selecting a smaller tree from a sequence of trees, can be very powerful, and is particularly useful for smaller data sets. It is an essential step for generating useful (for prediction) tree models, and because it can be computationally difficult to do, this method is often not found in tree classification or regression software.

Reviewing Large Trees: Unique Analysis Management Tools

Another general issue that arises when applying tree classification or regression methods is that the final trees can become very large. In practice, when the input data are complex and, for example, contain many different categories for classification problems and many possible predictors for performing the classification, the resulting trees can become very large. This is not so much a computational problem as it is a problem of presenting the trees in a manner that is easily accessible to the data analyst, or for presentation to the consumers of the research.

Analyzing ANCOVA-Like Designs

The classic (Breiman et. al., 1984) classification and regression trees algorithms can accommodate both continuous and categorical predictors. However, in practice, it is not uncommon to combine such variables into analysis of variance/covariance (ANCOVA)-like predictor designs with main effects or interaction effects for categorical and continuous predictors. This method of analyzing coded ANCOVA-like designs is relatively new

and. However, it is easy to see how the use of coded predictor designs expands these powerful classification and regression techniques to the analysis of data from experimental designs [e.g., see the detailed discussion of experimental design methods for quality improvement in Chapter 15 – *Experimental Design (DOE)*].

Computational Details

The process of computing classification and regression trees can be characterized as involving four basic steps:

- Specifying the criteria for predictive accuracy
- Selecting splits
- Determining when to stop splitting
- Selecting the right-sized tree.

These steps are very similar to those discussed in Chapter 9 – *Classification Trees* (see also, Breiman et al., 1984, for more details). See also, *Computational Formulas*, page 92.

Specifying the Criteria for Predictive Accuracy

The classification and regression trees (CART) algorithms are generally aimed at achieving the best possible predictive accuracy. Operationally, the most accurate prediction is defined as the prediction with the minimum costs. The notion of costs was developed as a way to generalize, to a broader range of prediction situations, the idea that the best prediction has the lowest misclassification rate. In most applications, the cost is measured in terms of proportion of misclassified cases, or variance. In this context, it follows, therefore, that a prediction would be considered best if it has the lowest misclassification rate or the smallest variance. The need for minimizing costs, rather than just

the proportion of misclassified cases, arises when some predictions that fail are more catastrophic than others, or when some predictions that fail occur more frequently than others.

Priors. In the case of a categorical response (classification problem), minimizing costs amounts to minimizing the proportion of misclassified cases when priors are taken to be proportional to the class sizes and when misclassification costs are taken to be equal for every class.

The *a priori* probabilities used in minimizing costs can greatly affect the classification of cases or objects. Therefore, care has to be taken while using the priors. If differential base rates are not of interest for the study, or if you know that there are about an equal number of cases in each class, you would use equal priors. If the differential base rates are reflected in the class sizes (as they would be, if the sample is a probability sample), you would use priors estimated by the class proportions of the sample. Finally, if you have specific knowledge about the base rates (for example, based on previous research), you would specify priors in accordance with that knowledge The general point is that the relative size of the priors assigned to each class can be used to adjust the importance of misclassifications for each class. However, no priors are required when you are building a regression tree.

Misclassification costs. Sometimes more accurate classification of the response is desired for some classes than others for reasons not related to the relative class sizes. If the criterion for predictive accuracy is misclassification costs, minimizing costs would amount to minimizing the proportion of misclassified cases when priors are considered proportional to the class sizes and misclassification costs are taken to be equal for every class.

Case weights. Case weights are treated strictly as case multipliers. For example, the misclassification rates from an analysis of an aggregated data set using case weights will be identical to the misclassification rates from the same analysis where the cases are replicated the specified number of times in the data file.

However, note that the use of case weights for aggregated data sets in classification problems is related to the issue of minimizing costs. Interestingly, as an alternative to using case weights for aggregated data sets, you could specify appropriate priors and/or misclassification costs and produce the same results while avoiding the additional processing required to analyze multiple cases with the same values for all variables. Suppose that in an aggregated data set with two classes having an equal number of cases, there are case weights of 2 for all cases in the first class, and case weights of 3 for all cases in the second class. If you specified priors of .4 and .6, respectively, specified equal misclassification costs, and analyzed the data without case weights, you will get the same misclassification rates as you would get if you specified priors estimated by the class sizes, specified equal misclassification costs, and analyzed the aggregated data set using the case weights. You would also get the same misclassification rates if you specified priors to be equal, specified the costs of misclassifying class 1 cases as class 2 cases to be 2/3 of the costs of misclassifying class 2 cases as class 1 cases, and analyzed the data without case weights.

Selecting Splits

The second basic step in classification and regression trees is to select the splits on the predictor variables that are used to predict membership in classes of the categorical dependent variables, or to predict values of the continuous dependent (response) variable. In general terms, the algorithm will find at each node the split that will generate the greatest improvement in predictive accuracy. This is usually measured with some type of node impurity measure, which provides an indication of the relative homogeneity (the inverse of impurity) of cases in the terminal nodes. If all cases in each terminal node show identical values, node impurity is minimal, homogeneity is maximal, and prediction is perfect (at least for the cases used in the computations; predictive validity for new cases is of course a different matter).

For classification problems using CART, you have the choice of several impurity measures: The Gini index, *chi*-square, or G-square. The Gini index of node impurity is the measure most commonly chosen for classification-type problems. As an impurity measure, it reaches a value of zero when only one class is present at a node. With priors estimated from class sizes and equal misclassification costs, the Gini measure is computed as the sum of products of all pairs of class proportions for classes present at the node; it reaches its maximum value when class sizes at the node are equal; the Gini index is equal to zero if all cases in a node belong to the same class. The *chi*-square measure is similar to the standard *chi*-square value computed for the expected and observed classifications (with priors adjusted for misclassification cost), and the G-square measure is similar to the maximum-likelihood *chi*-square (as, for example, computed in log-linear analyses). For

regression-type problems, a least-squares deviation criterion (similar to what is computed in least squares regression) is automatically used. *Computational Formulas*, page 92, provides further computational details.

Determining When to Stop Splitting

As discussed in *Basic Ideas* at the beginning of this chapter, in principal, splitting could continue until all cases are perfectly classified or predicted. However, this wouldn't make much sense since you would likely end up with a tree structure that is as complex and tedious as the original data file (with many nodes possibly containing single observations), and that would most likely not be very useful or accurate for predicting new observations. What is required is some reasonable stopping rule. In CART, two options are available that can be used to keep a check on the splitting process; namely minimum *n* and fraction of objects.

Minimum n. One way to control splitting is to allow splitting to continue until all terminal nodes are pure or contain no more than a specified minimum number of cases or objects. This enables you to specify the desired minimum number of cases as a check on the splitting process.

Fraction of objects. Another way to control splitting is to allow splitting to continue until all terminal nodes are pure or contain no more cases than a specified minimum fraction of the sizes of one or more classes (in the case of classification problems, or all cases in regression problems). This is used in conjunction with FACT-style direct stopping (see below). The desired minimum fraction can be specified. For classification problems, if the priors used in the analysis are equal and class sizes are equal as well, splitting will stop when all terminal nodes containing more than one class have no more cases than the specified fraction of the class sizes for one or more classes. Alternatively, if the priors used in the analysis are not equal, splitting will stop when all terminal nodes containing more than one class have no more cases than the specified fraction for one or more classes. See Loh and Vanichestakul, 1988 for details.

Pruning and Selecting the "Right-Sized" Tree

The size of a tree in the classification and regression trees analysis is an important issue, since an unreasonably big tree can only make the interpretation of results more difficult. Some generalizations can be offered about what constitutes the "right-sized" tree. It should be sufficiently complex to account for the known facts, but at the same time it should be as simple as possible. It should exploit information that increases predictive accuracy and ignore information that does not. It should, if possible, lead to greater understanding of the phenomena it describes. With CART, you can use either, or both, of two different strategies for selecting the right-sized tree from among all the possible trees. One strategy is to grow the tree to just the right size, where the right size is determined by the user, based on the knowledge from previous research, diagnostic information from previous analyses, or even intuition. The other strategy is to use a set of well-documented, structured procedures developed by Breiman et al. (1984) for selecting the right-sized tree. These procedures are not foolproof, as Breiman et al. (1984) readily acknowledge, but at least they take subjective judgment out of the process of selecting the right-sized tree.

FACT-style direct stopping. We will begin by describing the first strategy, in which the user

specifies the size to grow the tree. In FACT-style direct stopping, *Fraction of objects* (see above) is used to grow the tree to the desired size. Then cross-validation of the selected tree is performed by Test sample, V-fold, or Minimal cost-complexity.

Test sample cross-validation. The first, and most preferred type of cross-validation is the test sample cross-validation. In this type of cross-validation, the tree is computed from the learning sample, and its predictive accuracy is tested by applying it to predict the class membership in the test sample. If the costs for the test sample exceed the costs for the learning sample, this is an indication of poor cross-validation. In that case, a different sized tree might cross-validate better. The test and learning samples can be formed by collecting two independent data sets, or if a large learning sample is available, by reserving a randomly selected proportion of the cases, say a third or a half, for use as the test sample.

In CART, test sample cross-validation is performed by specifying a sample identifier variable which contains codes for identifying the sample (learning or test) to which each case or object belongs.

V-fold cross-validation. The second type of cross-validation available in CART is v-fold cross-validation. This type of cross-validation is useful when no test sample is available and the learning sample is too small to have the test sample taken from it. The user-specified v value for v-fold cross-validation (its default value is 3) determines the number of random subsamples, as equal in size as possible, that are formed from the learning sample. A tree of the specified size is computed v times, each time leaving out one of the subsamples from the computations, and using that subsample as a test

sample for cross-validation, so that each subsample is used (v – 1) times in the learning sample and just once as the test sample. The CV costs (cross-validation cost) computed for each of the v test samples are then averaged to give the v-fold estimate of the CV costs.

Minimal cost-complexity cross-validation pruning. In CART, two ways to prune are: minimal cost-complexity cross-validation pruning or minimal deviance-complexity cross-validation pruning. The only difference in the two options is the measure of prediction error that is used. Prune on misclassification error uses the cost that equals the misclassification rate when priors are estimated and misclassification costs are equal, while Prune on deviance uses a measure, based on maximum-likelihood principles, called the deviance (see Ripley, 1996). For details about the algorithms used in CART to implement Minimal cost-complexity cross-validation pruning, see the *Overview*, page 95, and *Computational Methods*, page 105, of Chapter 9 – *Classification Trees*.

The sequence of trees obtained by this algorithm has a number of interesting properties. They are nested, because the successively pruned trees contain all the nodes of the next smaller tree in the sequence. Initially, many nodes are often pruned going from one tree to the next smaller tree in the sequence, but fewer nodes tend to be pruned as the root node is approached. The sequence of largest trees is also optimally pruned, because for every size of tree in the sequence, there is no other tree of the same size with lower costs. Proofs and/or explanations of these properties can be found in Breiman et al. (1984).

Tree selection after pruning. The pruning, as discussed above, often results in a sequence of optimally pruned trees. So the next task is to use

an appropriate criterion to select the right-sized tree from this set of optimal trees. A natural criterion would be the CV costs (cross-validation costs). While there is nothing wrong with choosing the tree with the minimum CV costs as the right-sized tree, oftentimes there will be several trees with CV costs close to the minimum. Following Breiman et al. (1984) you could use the automatic tree selection procedure and choose as the right-sized tree the smallest-sized (least complex) tree whose CV costs do not differ appreciably from the minimum CV costs. In particular, they proposed a "1 SE rule" for making this selection, i.e., choose as the right-sized tree the smallest-sized tree whose CV costs do not exceed the minimum CV costs plus 1 times the standard error of the CV costs for the minimum CV costs tree. In CART, a multiple other than the one (the default) can also be specified for the SE rule. Thus, specifying a value of 0.0 would result in the minimal CV cost tree being selected as the right-sized tree. Values greater than 1.0 could lead to trees much smaller than the minimal CV cost tree being selected as the right-sized tree. One distinct advantage of the automatic tree selection procedure is that it helps to avoid overfitting and underfitting of the data.

As can be seen, minimal cost-complexity cross-validation pruning and subsequent right-sized tree selection is a truly automatic process. The algorithms make all the decisions leading to the selection of the right-sized tree, except for, perhaps, specification of a value for the SE rule. V-fold cross-validation enables you to evaluate how well each tree performs when repeatedly cross-validated in different samples randomly drawn from the data.

Computational Formulas

In classification and regression trees, estimates of accuracy are computed by different formulas for categorical and continuous dependent variables (classification and regression-type problems). For classification-type problems (categorical dependent variable) accuracy is measured in terms of the true classification rate of the classifier, while in the case of regression (continuous dependent variable) accuracy is measured in terms of mean squared error of the predictor.

In addition to measuring accuracy, the following measures of node impurity are used for classification problems: The Gini measure, generalized *chi*-square measure, and generalized G-square measure. The *chi*-square measure is similar to the standard *chi*-square value computed for the expected and observed classifications (with priors adjusted for misclassification cost), and the G-square measure is similar to the maximum-likelihood *chi*-square (as for example computed in log-linear analysis). The Gini measure is the one most often used for measuring purity in the context of classification problems, and it is described below.

For continuous dependent variables (regression-type problems), the least squared deviation (LSD) measure of impurity is automatically applied.

Estimation of Accuracy in Classification

In classification problems (categorical dependent variable), three estimates of the accuracy are used: resubstitution estimate, test sample estimate, and v-fold cross-validation. These estimates are defined here.

Resubstitution estimate. Resubstitution estimate is the proportion of cases that are misclassified by the classifier constructed from the entire sample. This estimate is computed in the following manner:

$$R(d) = \frac{1}{N} \sum_{n=1}^{N} X(d(x_n) \neq j_n)$$

where X is the indicator function;

$X = 1$ if the statement $X(d(x_n) \neq j_n)$ is true

$X = 0$ if the statement $X(d(x_n) \neq j_n)$ is false

and $d(x)$ is the classifier.

The resubstitution estimate is computed using the same data as used in constructing the classifier d.

Test sample estimate. The total number of cases is divided into two subsamples Z_1 and Z_2. The test sample estimate is the proportion of cases in the subsample Z_2 that are misclassified by the classifier constructed from the subsample Z_1. This estimate is computed in the following way.

Let the learning sample Z of size N be partitioned into subsamples Z_1 and Z_2 of sizes N and N_2, respectively.

$$R^{ts}(d) = \frac{1}{N_2} \sum_{(x_n, j_n) \in Z_2} X(d(x_n) \neq j_n)$$

where Z_2 is the sub sample that is not used for constructing the classifier.

V-fold cross-validation. The total number of cases are divided into v subsamples Z_1, Z_2, ..., Z_v of almost equal sizes. V-fold cross validation estimate is the proportion of cases in the subsample Z that are misclassified by the classifier constructed from the subsample $Z - Z_v$. This estimate is computed in the following way.

Let the learning sample Z of size N be partitioned into v sub samples $Z_1, Z_2, ..., Z_v$ of almost sizes $N_1, N_2, ..., N_v$, respectively.

$$R^{ts}(d^{(v)}) = \frac{1}{N_v} \sum_{(x_n, j_n) \in Z_v} X(d^{(v)}(x_n) \neq j_n)$$

where $d^{(v)}(x)$ is computed from the sub sample $Z - Z_v$.

Estimation of Accuracy in Regression

In the regression problem (continuous dependent variable) three estimates of the accuracy are used: resubstitution estimate, test sample estimate, and v-fold cross-validation. These estimates are defined here.

Resubstitution estimate. The resubstitution estimate is the estimate of the expected squared error using the predictor of the continuous dependent variable. This estimate is computed in the following way:

$$R(d) = \frac{1}{N} \sum_{i=1}^{N} (y_i - d(x_i))^2$$

where the learning sample Z consists of (x_i, y_i), $i = 1, 2, ..., N$. The resubstitution estimate is computed using the same data as used in constructing the predictor d.

Test sample estimate. The total number of cases is divided into two subsamples Z_1 and Z_2. The test sample estimate of the mean squared error is computed in the following way:

Let the learning sample Z of size N be partitioned into subsamples Z_1 and Z_2 of sizes N and N_2, respectively.

$$R^{ts}(d) = \frac{1}{N_2} \sum_{(x_i, y_i) \in Z_2}^{N} (y_i - d(x_i))^2$$

where Z_2 is the subsample that is not used for constructing the predictor.

v-fold cross-validation. The total number of cases are divided into v sub samples $Z_1, Z_2, ..., Z_v$ of almost equal sizes. The subsample $Z - Z_v$ is used to construct the predictor d. Then, v-fold cross validation estimate is computed from the subsample Z_v in the following way:

Let the learning sample Z of size N be partitioned into v sub samples $Z_1, Z_2, ..., Z_v$ of almost sizes $N_1, N_2, ..., N_v$, respectively.

$$R^{CV}(d) = \frac{1}{N_v} \sum_{v} \sum_{(x_n, y_n) \in Z_v} (y_i - d^{(v)}(x_n))^2$$

where $d^{(v)}(x)$ is computed from the sub sample $Z - Z_v$.

Estimation of Node Impurity: Gini Measure

The Gini measure is the measure of impurity of a node and is commonly used when the dependent variable is a categorical variable, defined as:

$$g(t) = \sum_{j \neq i} p(j/t) p(i/t)$$

if costs of misclassification are not specified,

$$= \sum_{j \neq i} C(i/j) p(j/t) \, p(i/t)$$

if costs of misclassification are specified,

where the sum extends over all k categories. $p(j/t)$ is the probability of category j at the node t and $C(i/j)$ is the probability of misclassifying a category j case as category i.

Estimation of Node Impurity: Least-Squared Deviation

Least-squared deviation (LSD) is used as the measure of impurity of a node when the response variable is continuous, and is computed as:

$$R(t) = \frac{1}{N_w(t)} \sum_{i \in t} w_i \, f_i (y_i - \bar{y}(t))^2$$

where $N_w(t)$ is the weighted number of cases in node t, w_i is the value of the weighting variable for case i, f_i is the value of the frequency variable, y_i is the value of the response variable, and $y(t)$ is the weighted mean for node t.

CLASSIFICATION TREES

Classification trees are used to predict membership of cases or objects in the classes of a categorical dependent variable from their measurements on one or more predictor variables. Classification tree analysis is one of the main techniques used in data mining.

The goal of classification trees is to predict or explain responses on a categorical dependent variable, and as such, the available techniques have much in common with the techniques used in the more traditional methods of discriminant analysis, cluster analysis, nonparametric statistics, and nonlinear estimation. The flexibility of classification trees make them a very attractive analysis option, but this is not to say that their use is recommended to the exclusion of more traditional methods. Indeed, when the typically more stringent theoretical and distributional assumptions of more traditional methods are met, the traditional methods may be preferable. But as an exploratory technique, or as a technique of last resort when traditional methods fail, classification trees are, in the opinion of many researchers, unsurpassed.

Overview

What are classification trees? Imagine that you want to devise a system for sorting a collection of coins into different classes (perhaps pennies, nickels, dimes, quarters). Suppose that there is a measurement by which the coins differ, say width, that can be used to devise a hierarchical system for sorting coins. You might roll the coins on edge down a narrow track in which a slot the width of a dime is cut. If the coin falls through the slot, it is classified as a dime; otherwise, it continues down the track to where a slot the width of a penny is cut. If the coin falls through the slot, it is classified as a penny; otherwise, it continues down the track to where a slot the width of a nickel is cut, and so on. You have just constructed a classification tree. The decision process used by your classification tree provides an efficient method for sorting a pile of coins, and more generally, can be applied to a wide variety of classification problems.

The study and use of classification trees are not widespread in the fields of probability and statistical pattern recognition (Ripley, 1996), but classification trees are widely used in applied fields as diverse as medicine (diagnosis), computer science (data structures), botany (classification), and psychology (decision theory). Classification trees readily lend themselves to being displayed graphically, helping to make them easier to interpret than they would be if only a strict numerical interpretation were possible.

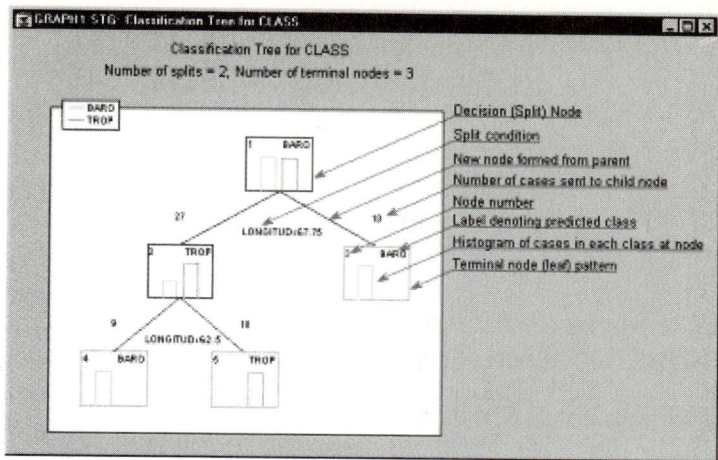

Classification trees can be and sometimes are quite complex. However, graphical procedures can be developed to help simplify interpretation even for complex trees. If your interest is mainly in the conditions that produce a particular class of response, perhaps a high response, a 3D contour plot can be produced to identify which terminal node of the classification tree classifies most of the cases with high responses.

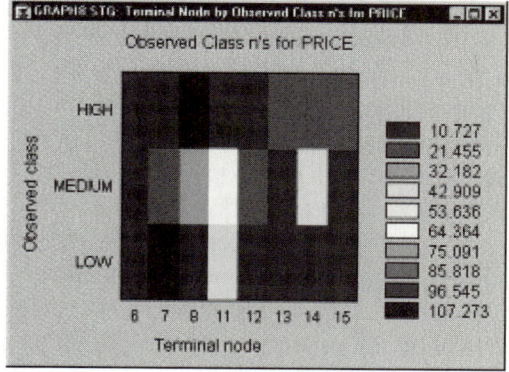

In the example illustrated by this 3D contour plot, you could follow the branches leading to terminal node 8 to obtain an understanding of the conditions leading to high responses.

Amenability to graphical display and ease of interpretation are perhaps partly responsible for the popularity of classification trees in applied fields, but two features that characterize classification trees more generally are their hierarchical nature and their flexibility.

For information on techniques and issues in computing classification trees, see *Computational Methods*, page 105. See also, Chapter 12 – *Data Mining Techniques*.

Characteristics of Classification Trees

Hierarchical Nature

Breiman et al. (1984) give a number of examples of the use of classification trees. As one example, when heart attack patients are admitted to a hospital, dozens of tests are often performed to obtain physiological measures such as heart rate, blood pressure, and so on. A wide variety of other information is also obtained, such as the patient's age and medical history. Patients subsequently can be tracked to see if they survive the heart attack, say, at least 30 days. It would be useful in developing treatments for heart attack patients, and in advancing medical theory on heart failure, if measurements taken soon after hospital admission could be used to identify high-risk patients (those who are not likely to survive at least 30 days). One classification tree that Breiman et al. (1984) developed to address this problem was a simple, three question decision tree. Verbally, the binary classification tree can be described by the statement, "If the patient's minimum systolic blood pressure over the initial 24 hour period is greater than 91, then if the patient's age is over 62.5 years, then if the patient displays sinus tachycardia, then and only then the patient is predicted not to survive for at least 30 days." It is easy to conjure up the image of a decision tree from such a statement. A

hierarchy of questions is asked and the final decision that is made depends on the answers to all the previous questions. Similarly, the relationship of a leaf to the tree on which it grows can be described by the hierarchy of splits of branches (starting from the trunk) leading to the last branch from which the leaf hangs. The hierarchical nature of classification trees is one of their most basic features (but the analogy with trees in nature should not be taken too far; most decision trees are drawn downward on paper, so the more exact analogy in nature would be a decision root system leading to the root tips, hardly a poetic image).

The hierarchical nature of classification trees is illustrated by a comparison to the decision-making procedure employed in discriminant analysis. A traditional linear discriminant analysis of the heart attack data would produce a set of coefficients defining the single linear combination of blood pressure, patient age, and sinus tachycardia measurements that best differentiates low-risk from high-risk patients. A score for each patient on the linear discriminant function would be computed as a composite of each patient's measurements on the three predictor variables, weighted by the respective discriminant function coefficients. The predicted classification of each patient as a low-risk or a high-risk patient would be made by simultaneously considering the patient's scores on the three predictor variables. That is, suppose P (minimum systolic blood pressure over the 24 hour period), A (age in years), and T (presence of sinus tachycardia: $0 =$ not present; $1 =$ present) are the predictor variables, $p, a,$ and $t,$ are the corresponding linear discriminant function coefficients, and c is the "cut point" on the discriminant function for separating the two classes of heart attack patients. The decision equation for each patient would be of the form, "if $pP + aA + tT - c$ *is less than or equal to zero, the patient is low risk, else the patient is in high risk."*

In comparison, the decision tree developed by Breiman et al. (1984) would have the following hierarchical form, where *p, a,* and *t* would be -91, -62.5, and 0, respectively, *"If p + P is less than or equal to zero, the patient is low risk, else if a + A is less than or equal to zero, the patient is low risk, else if t + T is less than or equal to zero, the patient is low risk, else the patient is high risk."* Superficially, the discriminant analysis and classification tree decision processes might appear similar, because both involve coefficients and decision equations. But the difference of the simultaneous decisions of discriminant analysis from the hierarchical decisions of classification trees cannot be emphasized enough.

The distinction between the two approaches can perhaps be made most clear by considering how each analysis would be performed in regression. Because risk in the example of Breiman et al. (1984) is a dichotomous dependent variable, the discriminant analysis predictions could be reproduced by a simultaneous multiple regression of risk on the three predictor variables for all patients. The classification tree predictions could only be reproduced by three separate simple regression analyses, where risk is first regressed on *P* for all patients, then risk is regressed on *A* for patients not classified as low risk in the first regression, and finally, risk is regressed on *T* for patients not classified as low risk in the second regression. This clearly illustrates the simultaneous nature of discriminant analysis decisions as compared to the recursive, hierarchical nature of classification trees decisions, a characteristic of classification trees that has far-reaching implications.

Flexibility

Another distinctive characteristic of classification trees is their flexibility. The ability of classification trees to examine the effects of the predictor variables one at a time, rather than just all at once, has already been described, but there are a number of other ways in which classification trees are more flexible than traditional analyses. The ability of classification trees to perform univariate splits, examining the effects of predictors one at a time, has implications for the variety of types of predictors that can be analyzed. In the Breiman et al. (1984) heart attack example, blood pressure and age were continuous predictors, but presence of sinus tachycardia was a categorical (two-level) predictor. Even if sinus tachycardia was measured as a three-level categorical predictor (perhaps coded as 0 = not present; 1 = present; 3 = unknown or unsure), without any underlying continuous dimension represented by the values assigned to its levels, univariate splits on the predictor variables could still be easily performed. Additional decisions would be added to the decision tree to exploit any additional information on risk provided by the additional category. To summarize, classification trees can be computed for categorical predictors, continuous predictors, or any mix of the two types of predictors when univariate splits are used.

Traditional linear discriminant analysis requires that the predictor variables be measured on at least an interval scale. For classification trees based on univariate splits for ordinal scale predictor variables, it is interesting that any monotonic transformation of the predictor variables (i.e., any transformation that preserves the order of values on the variable) will produce splits yielding the same predicted classes for the cases or objects (if the CART-style univariate split selection method is used, see Breimen et al., 1984). Therefore, classification trees based on univariate splits can be computed without concern for whether a unit change on a continuous predictor represents a unit change on the dimension underlying the values on the predictor variable; it need only be assumed that predictors are measured on at least an ordinal scale. In short, assumptions regarding the level of measurement of predictor variables are less stringent.

Classification trees are not limited to univariate splits on the predictor variables. When continuous predictors are indeed measured on at least an interval scale, linear combination splits, similar to the splits for linear discriminant analysis, can be computed for classification trees. However, the linear combination splits computed for classification trees do differ in important ways from the linear combination splits computed for discriminant analysis. In linear discriminant analysis the number of linear discriminant functions that can be extracted is the lesser of the number of predictor variables or the number of classes on the dependent variable minus one. The recursive approach implemented for classification trees module does not face this limitation. For example, dozens of recursive, linear combination splits potentially could be performed when there are dozens of predictor variables but only two classes on the dependent variable. This compares with the single linear combination split that could be performed using traditional, non-recursive linear discriminant analysis, which could leave a substantial amount of the information in the predictor variables unused.

Now consider the situation in which there are many categories but few predictors. Suppose you were trying to sort coins into classes (perhaps pennies, nickels, dimes, and quarters) based only on thickness and diameter measurements. Using traditional linear discriminant analysis, at most two linear discriminant functions could be extracted, and the coins could be successfully sorted only if there were no more than two dimensions represented by linear combinations of thickness and diameter on which the coins differ. Again, the approach implemented for classification trees does not face a limitation on the number of linear combination splits that can be formed.

The approach implemented for classification trees for linear combination splits can also be used as the analysis method for constructing classification trees using univariate splits. Actually, a univariate split is just a special case of a linear combination split. Imagine a linear combination split in which the coefficients for creating the weighted composite were zero for all predictor variables except one. Since scores on the weighted composite would depend only on the scores on the one predictor variable with the nonzero coefficient, the resulting split would be a univariate split.

The approach implemented for classification trees for the discriminant-based univariate split selection method for categorical and ordered predictors and for the discriminant-based linear combination split selection method for ordered predictors is an adaptation of the algorithms used in QUEST (Quick, Unbiased, Efficient Statistical Trees). QUEST is a classification tree program developed by Loh and Shih (1997) that employs a modification of recursive quadratic discriminant analysis and includes a number of innovative features for improving the reliability and efficiency of the classification trees that it computes.

The algorithms used in QUEST are fairly technical, but classification trees analysis also offers a split selection method option based on a conceptually simpler approach. The CART-style univariate split selection method is an adaptation of the algorithms used in CART, as described by Breiman et al. (1984). CART (Classification and Regression Trees) is a classification tree program that uses an exhaustive grid search of all possible univariate splits to find the splits for a classification tree.

The QUEST and CART analysis options compliment each other nicely. CART searches can be lengthy when there are a large number of predictor variables with many levels, and they are biased toward choosing predictor variables with more levels for splits, but because they employ an exhaustive search, they are guaranteed to find the splits producing the best classification (in the learning sample, but not necessarily in cross-validation samples).

QUEST is fast and unbiased. The speed advantage of QUEST over CART is particularly dramatic when the predictor variables have dozens of levels (Loh & Shih, 1997, report an analysis completed by QUEST in 1 CPU second that took CART 30.5 CPU hours to complete). QUEST's lack of bias in variable selection for splits is also a distinct advantage when some predictor variables have few levels and other predictor variables have many levels (predictors with many levels are more likely to produce fluke theories that fit the data well but have low predictive accuracy; see Doyle, 1973, and Quinlan & Cameron-Jones, 1995). Finally, QUEST does not sacrifice predictive accuracy for speed (Lim, Loh, & Shih, 1997). Together, the QUEST and CART options enable you to fully exploit the flexibility of classification trees.

Power and Pitfalls

The advantages of classification trees over traditional methods such as linear discriminant analysis, at least in some applications, can be illustrated using a simple, fictitious data set. To keep the presentation even-handed, other situations in which linear discriminant analysis would outperform classification trees are illustrated using a second data set.

Suppose you have records of the *Longitude* and *Latitude* coordinates at which 37 storms reached hurricane strength for two classifications of hurricanes – *Baro* hurricanes and *Trop* hurricanes. The fictitious data (partially shown in the next illustration) were presented for illustrative purposes by Elsner, Lehmiller, and Kimberlain (1996), who investigated the differences between baroclinic and tropical North Atlantic hurricanes.

A linear discriminant analysis of hurricane *Class* (*Baro* or *Trop*) using *Longitude* and *Latitude* as predictors correctly classifies only 20 of the 37 hurricanes (54%). A classification tree for *Class* using the CART-style exhaustive search for univariate splits option correctly classifies all 37 hurricanes. The tree graph for the classification tree is shown in the next illustration.

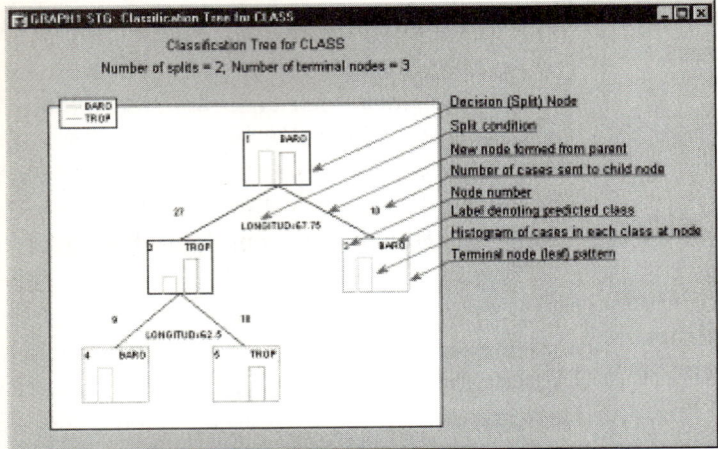

The headings of the graph give the summary information that the classification tree has 2 splits and 3 terminal nodes. Terminal nodes, or terminal leaves as they are sometimes called, are points on the tree beyond which no further decisions are made. In the graph itself, terminal nodes are outlined with dotted red lines, while the remaining decision nodes or split nodes are outlined with solid black lines. The tree starts with the top decision node, sometimes called the root node. In the graph it is labeled as node 1 in the upper-left corner. Initially, all 37 hurricanes are assigned to the root node and tentatively classified as *Baro* hurricanes, as indicated by the *Baro* label in the upper-right corner of the root node. *Baro* is chosen as the initial classification because there are slightly more *Baro* than *Trop* hurricanes, as indicated by the histogram plotted within the root node. The legend identifying which bars in the node histograms correspond to *Baro* and *Trop* hurricanes is located in the upper-left corner of the graph.

The root node is split, forming two new nodes. The text below the root node describes the split. It indicates that hurricanes with *Longitude* coordinate values of less than or equal to 67.75 are sent to node number 2 and tentatively

classified as *Trop* hurricanes, and that hurricanes with *Longitude* coordinate values of greater than 67.75 are assigned to node number 3 and classified as *Baro* hurricanes. The values of 27 and 10 printed above nodes 2 and 3, respectively, indicate the number of cases sent to each of these two child nodes from their parent, the root node. Similarly, node 2 is subsequently split. The split is such that the 9 hurricanes with *Longitude* coordinate values of less than or equal to 62.5 are sent to node number 4 and classified as *Baro* hurricanes, and the remaining 18 hurricanes with *Longitude* coordinate values of greater than 62.5 are sent to node number 5 and classified as *Trop* hurricanes.

The tree graph presents all this information in a simple, straightforward way, and probably enables you to digest the information in much less time than it takes to read the two preceding paragraphs. Getting to the bottom line, the histograms plotted within the tree's terminal nodes show that the classification tree classifies the hurricanes perfectly. Each of the terminal nodes is pure, containing no misclassified hurricanes. All the information in the tree graph is also available in the tree structure spreadsheet shown in the next illustration.

Data: Classification Trees Example Spreadsheet 1

Tree Structure (Barotrop.sta)
Child nodes, observed class n's,
predicted class, and split condition for each node

Node	Left branch	Right branch	n in cls BARO	n in cls TROP	Predict. class	Split constant	Split variable
1	2	3	19	18	BARO	-67.7500	LONGITUD
2	4	5	9	18	TROP	-62.5000	LONGITUD
3			10	0	BARO		
4			9	0	BARO		
5			0	18	TROP		

Note that in the spreadsheet, nodes 3 through 5 are identified as terminal nodes because no split

is performed at those nodes. Also note the signs of the split constants displayed in the spreadsheet, for example, *-67.75* for the split at node 1. In the tree graph, the split condition at node 1 is described as *LONGITUD 67.75* rather than as (the equivalent) *-67.75 + LONGITUD 0.* This is done simply to save space on the graph.

When univariate splits are performed, the predictor variables can be ranked on a 0 - 100 scale in terms of their potential importance in accounting for responses on the dependent variable. In this example, *Longitude* is clearly very important and *Latitude* is relatively unimportant.

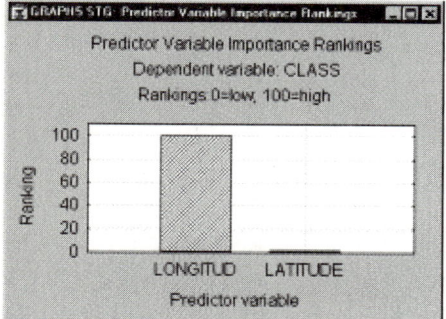

A classification tree *Class* using the discriminant-based univariate split selection method option produces similar results.

Node	Left branch	Right branch	n in cls BARO	n in cls TROP	Predict. class	Split constant	Split variable
1	2	3	19	18	BARO	-63.4716	LONGITUD
2			9	1	BARO		
3	4	5	10	17	TROP	-67.7516	LONGITUD
4			0	17	TROP		
5			10	0	BARO		

Tree Structure (Barotrop.sta)
Child nodes, observed class n's, predicted class, and split condition for each node

The tree structure spreadsheet for this analysis shows that the splits of *-63.4716* and *-67.7516* are quite similar to the splits found using the CART-style exhaustive search for univariate

splits option, although 1 *Trop* hurricane in terminal node 2 is misclassified as *Baro*.

A categorized scatterplot for *Longitude* and *Latitude* clearly shows why linear discriminant analysis fails so miserably at predicting *Class,* and why the classification tree succeeds so well.

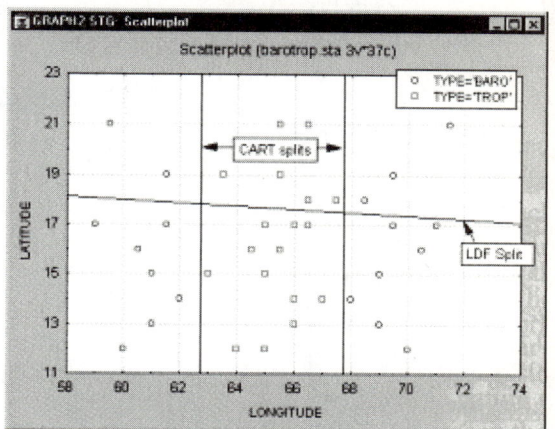

The plot clearly shows that there is no strong linear relationship of longitude or latitude coordinates with *Class,* or of any possible linear combination of longitude and latitude with *Class.* *Class* is not functionally related to longitude or latitude, at least in the linear sense. The LDF (Linear Discriminant Function) Split shown on the graph is almost a "shot in the dark" at trying to separate predicted *Trop* hurricanes (above the split line) from predicted *Baro* hurricanes (below the split line). The CART univariate splits, because they are not restricted to a single linear combination of longitude and latitude scores, find the cut points on the *Longitude* dimension that allow the best possible (in this case, perfect) classification of hurricane *Class.*

Now we can examine a situation illustrating the pitfalls of classification tree. Suppose that the hurricane data shown in the next illustration were available.

	1	2	3
	LONGITUD	LATITUDE	CLASS
1	59.00	17.00	BARO
2	59.50	21.00	BARO
3	60.00	12.00	TROP
4	60.50	16.00	BARO
5	61.00	13.00	TROP
6	61.00	15.00	TROP
7	61.50	17.00	BARO
8	61.50	19.00	BARO
9	62.00	14.00	TROP
10	63.00	15.00	TROP
11	63.50	19.00	BARO
12	64.00	12.00	TROP
13	64.50	16.00	TROP
14	65.00	12.00	TROP
15	65.00	15.00	TROP
16	65.00	17.00	BARO
17	65.50	16.00	TROP
18	65.50	19.00	BARO
19	65.50	21.00	BARO
20	66.00	13.00	TROP
21	66.00	14.00	TROP
22	66.00	17.00	BARO
23	66.50	17.00	BARO
24	66.50	18.00	BARO
25	66.50	21.00	BARO
26	67.00	14.00	TROP
27	67.50	18.00	BARO
28	68.00	14.00	TROP
29	68.50	18.00	BARO
30	69.00	13.00	TROP
31	69.00	15.00	TROP
32	69.50	17.00	TROP
33	69.50	19.00	BARO
34	70.00	12.00	TROP
35	70.50	16.00	TROP
36	71.00	17.00	TROP
37	71.50	21.00	BARO

Data: Barotro2 (3v by 37c)
Example data for Classification Tr

that hurricanes in the western Atlantic at low latitudes are likely to be *Trop* hurricanes, and that hurricanes further east in the Atlantic at higher latitudes are likely to be *Baro* hurricanes.

The tree graph for the classification tree analysis using the CART-style exhaustive search for univariate splits option is shown in the next illustration.

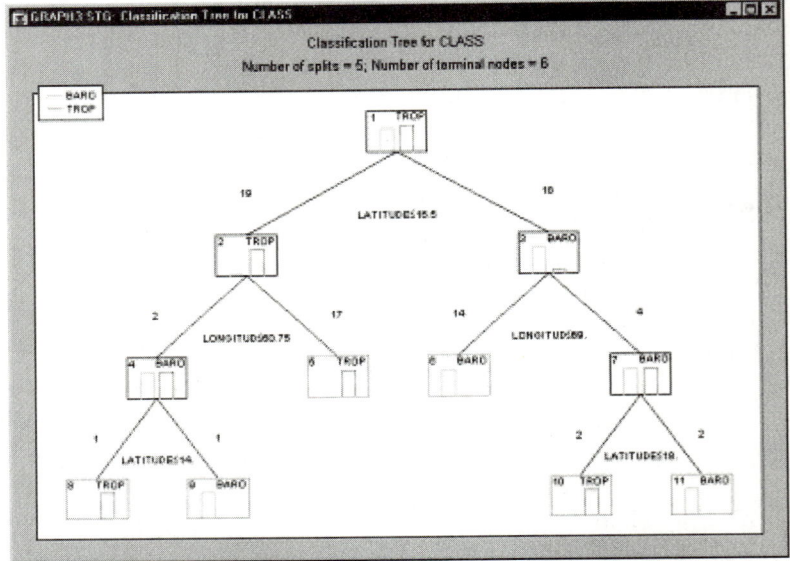

A linear discriminant analysis of hurricane *Class* (*Baro* or *Trop*) using *Longitude* and *Latitude* as predictors correctly classifies all 37 of the hurricanes. A classification tree analysis for *Class* using the CART-style exhaustive search for univariate splits option also correctly classifies all 37 hurricanes, but the tree requires 5 splits producing 6 terminal nodes. Which results are easier to interpret? In the linear discriminant analysis, the raw canonical discriminant function coefficients for *Longitude* and *Latitude* on the (single) discriminant function are *.122073* and *-.633124*, respectively, and hurricanes with higher longitude and lower latitude coordinates are classified as *Trop*. The interpretation would be

You could methodically describe the splits in this classification tree, exactly as was done in the previous example, but because there are so many splits, the interpretation would necessarily be more complex than the simple interpretation provided by the single discriminant function from the linear discrimination analysis.

However, recall that in describing the flexibility of classification trees, it was noted that an option exists for discriminant-based linear combination splits for ordered predictors using algorithms from QUEST. The tree graph for the classification tree analysis using linear combination splits is shown in the next illustration.

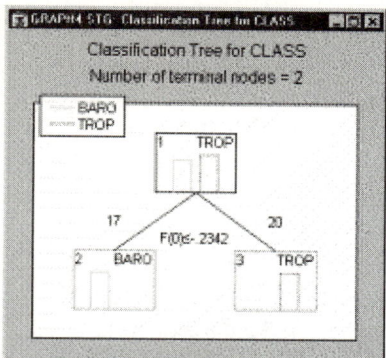

Note that in this tree, just one split yields perfect prediction. Each of the terminal nodes is "pure," containing no misclassified hurricanes. The linear combination split used to split the root node into its left child node and right child node is summarized by the description "*F(0) -.2342.*" This indicates that if a hurricane has a score of less than or equal to -.*2342* on the split function – abbreviated as *F(0)* – then it is sent to the left child node and classified as *Baro*, otherwise it is sent to the right child node and classified as *Trop*. The split function coefficients (.*011741* for *Longitude* and -.*060896* for *Latitude*) have the same signs and are similar in their relative magnitude to the corresponding linear discriminant function coefficients from the linear discriminant analysis, so the two analyses are functionally identical, at least in terms of their predictions of hurricane *Class*.

The moral of this story of the power and pitfalls of classification trees is that classification trees are only as good as the choice of analysis options used to produce them. For finding models that predict well, there is no substitute for a thorough understanding of the nature of the relationships between the predictor and dependent variables.

We have seen that classification trees analysis can be characterized as a hierarchical, highly flexible set of techniques for predicting membership of cases or objects in the classes of a categorical dependent variable from their measurements on one or more predictor variables. With this groundwork behind us, we now are ready to look at the methods for computing classification trees in greater detail.

For information on the basic purpose of classification trees, see the *Overview* (page 97). See also, Chapter 12 – *Data Mining Techniques*.

Computational Methods

The process of computing classification trees can be characterized as involving four basic steps:

- Specifying the criteria for predictive accuracy,
- Selecting splits,
- Determining when to stop splitting, and
- Choosing the right-sized tree.

Specifying the Criteria for Predictive Accuracy

The goal of classification tree analysis, simply stated, is to obtain the most accurate prediction possible. Unfortunately, an operational definition of accurate prediction is hard to come by. To solve the problem of defining predictive accuracy, the problem is "stood on its head," and the most accurate prediction is operationally defined as the prediction with the minimum costs. The term *costs* need not seem mystifying. In many typical applications, costs simply correspond to the proportion of misclassified cases. The notion of costs was developed as a way to generalize, to a broader range of prediction situations, the idea that the best prediction has the lowest misclassification rate.

The need for minimizing costs, rather than just the proportion of misclassified cases, arises when some predictions that fail are more catastrophic than others, or when some predictions that fail occur more frequently than others. The costs to a gambler of losing a single bet (or prediction) on which the gambler's whole fortune is at stake are greater than the costs of losing many bets (or predictions) on which a tiny part of the gambler's fortune is at stake. Conversely, the costs of losing many small bets can be larger than the costs of losing just a few bigger bets. You should spend proportionately more effort in minimizing losses on bets where losing (making errors in prediction) costs you more.

Priors. Minimizing costs, however, does correspond to minimizing the proportion of misclassified cases when priors are taken to be proportional to the class sizes and when misclassification costs are taken to be equal for every class. We will address priors first. Priors, or, *a priori* probabilities, specify how likely it is, without using any prior knowledge of the values for the predictor variables in the model, that a case or object will fall into one of the classes. For example, in an educational study of high school drop-outs, it may happen that, overall, there are fewer drop-outs than students who stay in school (i.e., there are different base rates); thus, the *a priori* probability that a student drops out is lower than that a student remains in school.

The *a priori* probabilities used in minimizing costs can greatly affect the classification of cases or objects. If differential base rates are not of interest for the study, or if you know that there are about an equal number of cases in each class, you would use equal priors. If the differential base rates are reflected in the class sizes (as they would be, if the sample is a probability sample), you would use priors estimated by the class proportions of the sample. Finally, if you have specific knowledge about the base rates (for example, based on previous research), you would specify priors in

accordance with that knowledge. For example, *a priori* probabilities for carriers of a recessive gene could be specified as twice as high as for individuals who display a disorder caused by the recessive gene. The general point is that the relative size of the priors assigned to each class can be used to adjust the importance of misclassifications for each class. Minimizing costs corresponds to minimizing the overall proportion of misclassified cases when priors are taken to be proportional to the class sizes (and misclassification costs are taken to be equal for every class), because prediction should be better in larger classes to produce an overall lower misclassification rate.

Misclassification costs. Sometimes more accurate classification is desired for some classes than others for reasons unrelated to relative class sizes. Regardless of their relative frequency, carriers of a disease who are contagious to others might need to be more accurately predicted than carriers of the disease who are not contagious to others. If you assume that little is lost in avoiding a non-contagious person but much is lost in not avoiding a contagious person, higher misclassification costs could be specified for misclassifying a contagious carrier as non-contagious than for misclassifying a non-contagious person as contagious. But to reiterate, minimizing costs corresponds to minimizing the proportion of misclassified cases when priors are taken to be proportional to the class sizes and when misclassification costs are taken to be equal for every class.

Case weights. A little less conceptually, the use of case weights on a weighting variable as case multipliers for aggregated data sets is also related to the issue of minimizing costs. Interestingly, as an alternative to using case weights for aggregated data sets, you could specify appropriate priors and/or misclassification costs and produce the same results while avoiding the additional processing required to analyze multiple cases with the same values for all variables. Suppose that in an aggregated data set with two classes having an equal number of cases, there are case weights of 2 for all the cases in the first class, and case weights of 3 for all the cases in the second class. If you specify priors of .4 and .6, respectively, specify equal misclassification costs, and analyze the data without case weights, you will get the same misclassification rates as you would get if you specify priors estimated by the class sizes, specify equal misclassification costs, and analyze the aggregated data set using the case weights. You would also get the same misclassification rates if you specify priors to be equal, specify the costs of misclassifying class 1 cases as class 2 cases to be 2/3 of the costs of misclassifying class 2 cases as class 1 cases, and analyze the data without case weights.

The relationships between priors, misclassification costs, and case weights become quite complex in all but the simplest situations (for discussions, see Breiman et al, 1984; Ripley, 1996). In analyses where minimizing costs corresponds to minimizing the misclassification rate, however, these issues need not cause any concern. Priors, misclassification costs, and case weights are brought up here, however, to illustrate the wide variety of prediction situations that can be handled using the concept of minimizing costs, as compared to the rather limited (but probably typical) prediction situations that can be handled using the narrower (but simpler) idea of minimizing misclassification rates. Furthermore, minimizing costs is an underlying goal of classification tree analysis, and is explicitly addressed in the fourth and final basic step in classification tree analysis, where in trying to select the right-sized tree, you choose the tree with

the minimum estimated costs. Depending on the type of prediction problem you are trying to solve, understanding the idea of reduction of estimated costs may be important for understanding the results of the analysis.

Selecting Splits

The second basic step in classification tree analysis is to select the splits on the predictor variables that are used to predict membership in the classes of the dependent variables for the cases or objects in the analysis. Not surprisingly, given the hierarchical nature of classification trees, these splits are selected one at a time, starting with the split at the root node, and continuing with splits of resulting child nodes until splitting stops, and the child nodes that have not been split become terminal nodes. Three split selection methods are discussed here.

Discriminant-based univariate splits. The first step in split selection when the discriminant-based univariate splits option is chosen is to determine the best terminal node to split in the current tree, and which predictor variable to use to perform the split. For each terminal node, *p*-levels are computed for tests of the significance of the relationship of class membership with the levels of each predictor variable. For categorical predictors, the *p*-levels are computed for *chi*-square tests of independence of the classes and the levels of the categorical predictor that are present at the node. For ordered predictors, the *p*-levels are computed for ANOVAs of the relationship of the classes to the values of the ordered predictor that are present at the node. If the smallest computed *p*-level is smaller than the default Bonferoni-adjusted *p*-level for multiple comparisons of .05 (a different threshold value can be used), the predictor variable producing that smallest *p*-level is chosen to split the corresponding node. If no *p*-level smaller than the threshold *p*-level is found, *p*-levels are computed for statistical tests that are robust to distributional violations, such as Levene's *F*. Details concerning node and predictor variable selection when no *p*-level is smaller than the specified threshold are described in Loh and Shih (1997).

The next step is to determine the split. For ordered predictors, the 2-means clustering algorithm of Hartigan and Wong (1979; see also, Chapter 10 – *Cluster Analysis*) is applied to create two super classes for the node. The two roots are found for a quadratic equation describing the difference in the means of the super classes on the ordered predictor, and the values for a split corresponding to each root are computed. The split closest to a super class mean is selected. For categorical predictors, dummy-coded variables representing the levels of the categorical predictor are constructed, and then singular value decomposition methods are applied to transform the dummy-coded variables into a set of non-redundant ordered predictors. The procedures for ordered predictors are then applied and the obtained split is mapped back onto the original levels of the categorical variable and represented as a contrast between two sets of levels of the categorical variable. Again, further details about these procedures are described in Loh and Shih (1997). Although complicated, these procedures reduce a bias in split selection that occurs when using the CART-style exhaustive search method for selecting splits. This is the bias toward selecting variables with more levels for splits, a bias which can skew the interpretation of the relative importance of the predictors in explaining responses on the dependent variable (Breiman et. al., 1984).

Discriminant-based linear combination splits. The second split selection method is the discriminant-based linear combination split option for ordered predictor variables (however,

the predictors are assumed to be measured on at least interval scales). Surprisingly, this method works by treating the continuous predictors from which linear combinations are formed in a manner that is similar to the way categorical predictors are treated in the previous method. Singular value decomposition methods are used to transform the continuous predictors into a new set of non-redundant predictors. The procedures for creating super classes and finding the split closest to a super class mean are then applied, and the results are mapped back onto the original continuous predictors and represented as a univariate split on a linear combination of predictor variables.

CART-style exhaustive search for univariate splits. The third split-selection method is the CART-style exhaustive search for univariate splits method for categorical or ordered predictor variables. With this method, all possible splits for each predictor variable at each node are examined to find the split producing the largest improvement in goodness of fit (or equivalently, the largest reduction in lack of fit). What determines the domain of possible splits at a node? For categorical predictor variables with *k* levels present at a node, there are $2^{(k-1)} - 1$ possible contrasts between two sets of levels of the predictor. For ordered predictors with *k* distinct levels present at a node, there are *k* -1 midpoints between distinct levels. Thus it can be seen that the number of possible splits that must be examined can become very large when there are large numbers of predictors with many levels that must be examined at many nodes.

How is improvement in goodness of fit determined? Three choices of goodness of fit measures are discussed here. The Gini measure of node impurity is a measure which reaches a value of zero when only one class is present at a node (with priors estimated from class sizes and equal misclassification costs, the Gini measure is computed as the sum of products of all pairs of class proportions for classes present at the node; it reaches its maximum value when class sizes at the node are equal). The Gini measure was the measure of goodness of fit preferred by the developers of CART (Breiman et. al., 1984). The two other indices are the *chi*-square measure, which is similar to Bartlett's *chi*-square (Bartlett, 1948), and the G-square measure, which is similar to the maximum-likelihood *chi*-square used in structural equation modeling. The CART-style exhaustive search for univariate splits method works by searching for the split that maximizes the reduction in the value of the selected goodness of fit measure. When the fit is perfect, classification is perfect.

Determining When to Stop Splitting

The third step in classification tree analysis is to determine when to stop splitting. One characteristic of classification trees is that if no limit is placed on the number of splits that are performed, eventually pure classification will be achieved, with each terminal node containing only one class of cases or objects. However, pure classification is usually unrealistic. Even a simple classification tree such as a coin sorter can produce impure classifications for coins whose sizes are distorted or if wear changes the lengths of the slots cut in the track. This potentially could be remedied by further sorting of the coins that fall into each slot, but to be practical, at some point the sorting would have to stop and you would have to accept that the coins have been reasonably well sorted.

Likewise, if the observed classifications on the dependent variable or the levels on the predicted variable in a classification tree analysis are measured with error or contain noise, it is unrealistic to continue to sort until every terminal node is pure. Two options for controlling when splitting stops will be discussed here. These two options are linked to the choice of the stopping rule specified for the analysis.

Minimum n. One option for controlling when splitting stops is to allow splitting to continue until all terminal nodes are pure or contain no more than a specified minimum number of cases or objects. The desired minimum number of cases can be specified as the Minimum n, and splitting will stop when all terminal nodes containing more than one class have no more than the specified number of cases or objects.

Fraction of objects. Another option for controlling when splitting stops is to allow splitting to continue until all terminal nodes are pure or contain no more cases than a specified minimum fraction of the sizes of one or more classes. The desired minimum fraction can be specified as the fraction of objects and, if the priors used in the analysis are equal and class sizes are equal, splitting will stop when all terminal nodes containing more than one class have no more cases than the specified fraction of the class sizes for one or more classes. If the priors used in the analysis are not equal, splitting will stop when all terminal nodes containing more than one class have no more cases than the specified fraction for one or more classes.

Selecting the Right-Sized Tree

After a night at the horse track, a studious gambler computes a huge classification tree with numerous splits that perfectly account for the win, place, show, and no show results for every horse in every race. Expecting to become rich, the gambler takes a copy of the tree graph to the races the next night, sorts the horses racing that night using the classification tree, makes his or her predictions and places his or her bets, and leaves the race track later much less rich than had been expected. The poor gambler has foolishly assumed that a classification tree computed from a learning sample in which the outcomes are already known will perform equally well in predicting outcomes in a second, independent test sample. The gambler's classification tree performed poorly during cross-validation. The gambler's payoff might have been larger using a smaller classification tree that did not classify perfectly in the learning sample, but which was expected to predict equally well in the test sample.

Some generalizations can be offered about what constitutes the right-sized classification tree. It should be sufficiently complex to account for the known facts, but at the same time it should be as simple as possible. It should exploit information that increases predictive accuracy and ignore information that does not. It should, if possible, lead to greater understanding of the phenomena that it describes. Of course, these same characteristics apply to any scientific theory, so we must try to be more specific about what constitutes the right-sized classification tree. One strategy is to grow the tree to just the right size, where the right size is determined by the user from knowledge from previous research, diagnostic information from previous analyses, or even intuition. The other strategy is to use a set of well-documented, structured procedures developed by Breiman et al. (1984) for selecting the right-sized tree. These procedures are not foolproof, as Breiman et al. (1984) readily acknowledge, but at least they take subjective judgment out of the process of selecting the right-sized tree.

FACT-style direct stopping. We'll begin by describing the first strategy, in which the researcher specifies the size to grow the classification tree. This strategy is followed by using FACT-style direct stopping as the stopping rule for the analysis, and by specifying the fraction of objects that allows the tree to grow to the desired size. There are several options for obtaining diagnostic information to determine the reasonableness of the choice of size for the tree. Three options for performing cross-validation of the selected classification tree are discussed below.

Test sample cross-validation. The first, and most preferred type of cross-validation is test sample cross-validation. In this type of cross-validation, the classification tree is computed from the learning sample, and its predictive accuracy is tested by applying it to predict class membership in the test sample. If the costs for the test sample exceed the costs for the learning sample (remember, costs equal the proportion of misclassified cases when priors are estimated and misclassification costs are equal), this indicates poor cross-validation and that a different sized tree might cross-validate better. The test and learning samples can be formed by collecting two independent data sets, or if a large learning sample is available, by reserving a randomly selected proportion of the cases, say a third or a half, for use as the test sample.

V-fold cross-validation. This type of cross-validation is useful when no test sample is available and the learning sample is too small to have the test sample taken from it. A specified V value for V-fold cross-validation determines the number of random subsamples, as equal in size as possible, that are formed from the learning sample. The classification tree of the specified size is computed V times, each time leaving out one of the subsamples from the computations, and using that subsample as a test sample for cross-validation, so that each subsample is used V − 1 times in the learning sample and just once as the test sample. The CV costs computed for each of the V test samples are then averaged to give the V-fold estimate of the CV costs.

Global cross-validation. In global cross-validation, the entire analysis is replicated a specified number of times holding out a fraction of the learning sample equal to 1 over the specified number of times, and using each hold-out sample in turn as a test sample to cross-validate the selected classification tree. This type of cross-validation is probably no more useful than V-fold cross-validation when FACT-style direct stopping is used, but can be quite useful as a method validation procedure when automatic tree selection techniques are used (for discussion, see Breiman et. al., 1984). This brings us to the second of the two strategies that can used to select the right-sized tree, an automatic tree selection method based on a technique developed by Breiman et al. (1984) called minimal cost-complexity cross-validation pruning.

Minimal cost-complexity cross-validation pruning. Two methods of pruning can be used depending on the stopping rule you choose to use. Minimal cost-complexity cross-validation pruning is performed when you decide to prune on misclassification error (as a stopping rule), and minimal deviance-complexity cross-validation pruning is performed when you choose to prune on deviance (as a stopping rule). The only difference in the two options is the measure of prediction error that is used. Prune on misclassification error uses the costs that we have discussed repeatedly (which equal the misclassification rate when priors are estimated and misclassification costs are equal). Prune on deviance uses a measure, based on maximum-likelihood principles, called the deviance (see Ripley, 1996). We will focus on cost-complexity cross-validation pruning (as originated by Breiman et. al., 1984), since deviance-

complexity pruning merely involves a different measure of prediction error.

The costs needed to perform cost-complexity pruning are computed as the tree is being grown, starting with the split at the root node up to its maximum size, as determined by the specified Minimum n. The learning sample costs are computed as each split is added to the tree, so that a sequence of generally decreasing costs (reflecting better classification) are obtained corresponding to the number of splits in the tree. The learning sample costs are called resubstitution costs to distinguish them from CV costs, because V-fold cross-validation is also performed as each split is added to the tree. Use the estimated CV costs from V-fold cross-validation as the costs for the root node. Note that tree size can be taken to be the number of terminal nodes, because for binary trees the tree size starts at one (the root node) and increases by one with each added split. Now, define a parameter called the complexity parameter whose initial value is zero, and for every tree (including the first, containing only the root node), compute the value for a function defined as the costs for the tree plus the complexity parameter times the tree size. Increase the complexity parameter continuously until the value of the function for the largest tree exceeds the value of the function for a smaller-sized tree. Take the smaller-sized tree to be the new largest tree, continue increasing the complexity parameter continuously until the value of the function for the largest tree exceeds the value of the function for a smaller-sized tree, and continue the process until the root node is the largest tree. (Those who are familiar with numerical analysis will recognize the use of a penalty function in this algorithm. The function is a linear combination of costs, which generally decrease with tree size, and tree size, which increases linearly. As the complexity parameter is increased, larger trees are penalized for their complexity more and more, until a discrete threshold is reached at which a smaller-sized tree's higher costs are outweighed by the largest tree's higher complexity).

The sequence of largest trees obtained by this algorithm has a number of interesting properties. They are nested, because successively pruned trees contain all the nodes of the next smaller tree in the sequence. Initially, many nodes are often pruned going from one tree to the next smaller tree in the sequence, but fewer nodes tend to be pruned as the root node is approached. The sequence of largest trees is also optimally pruned, because for every size of tree in the sequence, there is no other tree of the same size with lower costs. Proofs and/or explanations of these properties can be found in Breiman et al. (1984).

Tree selection after pruning. We now select the right-sized tree from the sequence of optimally pruned trees. A natural criterion is the CV costs. While there is nothing wrong with choosing the tree with the minimum CV costs as the right-sized tree, oftentimes there will be several trees with CV costs close to the minimum. Breiman et al. (1984) make the reasonable suggestion to choose as the right-sized tree the smallest-sized (least complex) tree whose CV costs do not differ appreciably from the minimum CV costs. They proposed a "1 SE rule" for making this selection, i.e., choose as the right-sized tree the smallest-sized tree whose CV costs do not exceed the minimum CV costs plus 1 times the standard error of the CV costs for the minimum CV costs tree.

One distinct advantage of the automatic tree selection procedure is that it helps to avoid overfitting and underfitting of the data. The graph in the next illustration shows a typical plot of the Resubstitution costs and CV costs for the sequence of successively pruned trees.

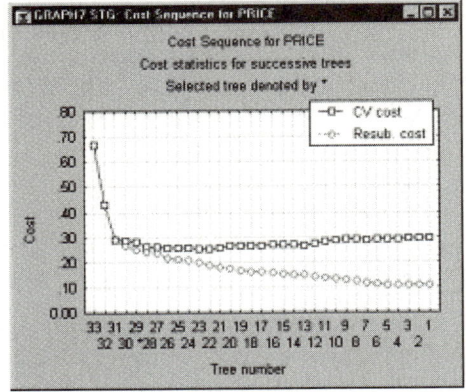

As shown in this graph, the Resubstitution costs (e.g., the misclassification rate in the learning sample) rather consistently decrease as tree size increases. The CV costs, on the other hand, approach the minimum quickly as tree size initially increases, but actually start to rise as tree size becomes very large. Note that the selected right-sized tree is close to the inflection point in the curve, that is, close to the point where the initial sharp drop in CV costs with increased tree size starts to level out. The automatic tree selection procedure is designed to select the simplest (smallest) tree with close to minimum CV costs, and thereby avoid the loss in predictive accuracy produced by underfitting or overfitting the data (note the similarity to the logic underlying the use of a scree plot to determine the number of factors to retain in factor analysis; see also, *Reviewing the results of a principal components analysis*, page 234).

As has been seen, minimal cost-complexity cross-validation pruning and subsequent right-sized tree selection is a truly automatic process. The algorithms make all the decisions leading to selection of the right-sized tree, except for, perhaps, specification of a value for the SE rule. One issue that arises with the use of such automatic procedures is how well the results replicate, where replication might involve the selection of trees of quite different sizes across replications, given the automatic selection process that is used. This is where global cross-validation can be very useful. As explained previously, in global cross-validation, the entire analysis is replicated a specified number of times holding out a fraction of the cases to use as a test sample to cross-validate the selected classification tree. If the average of the costs for the test samples, called the global CV costs, exceeds the CV costs for the selected tree, or if the standard error of the global CV costs exceeds the standard error of the CV costs for the selected tree, this indicates that the automatic tree selection procedure is allowing too much variability in tree selection rather than consistently selecting a tree with minimum estimated costs.

Classification trees and traditional methods.

As can be seen in the methods used in computing classification trees, in a number of respects classification trees are decidedly different from traditional statistical methods for predicting class membership on a categorical dependent variable. They employ a hierarchy of predictions, with many predictions sometimes being applied to particular cases, to sort the cases into predicted classes. Traditional methods use simultaneous techniques to make one and only one class membership prediction for each and every case. In other respects, such as having as its goal accurate prediction, classification tree analysis is indistinguishable from traditional methods. Time will tell if classification tree analysis has enough to commend itself to become as accepted as the traditional methods.

For information on the basic purpose of classification trees, see *Basic Ideas* (page 98). For information on the hierarchical nature and flexibility of classification trees, see

Characteristics of Classification Trees (page 98). See also, *Data Mining Techniques*, page 141.

A Brief Comparison of Classification Tree Programs

A variety of classification tree programs have been developed to predict membership of cases or objects in the classes of a categorical dependent variable from their measurements on one or more predictor variables. In the previous section, *Computational Methods*, we discussed the QUEST (Loh & Shih, 1997) and CART (Breiman et. al., 1984) programs for computing binary classification trees based on univariate splits for categorical predictor variables, ordered predictor variables (measured on at least an ordinal scale), or a mix of both types of predictors. We have also discussed computing classification trees based on linear combination splits for interval scale predictor variables.

Some classification trees programs, such as FACT (Loh & Vanichestakul, 1988) and THAID (Morgan & Messenger, 1973, as well as the related programs AID, for Automatic Interaction Detection, Morgan & Sonquist, 1963, and CHAID, for Chi-Square Automatic Interaction Detection, Kass, 1980) perform multi-level splits rather than binary splits when computing classification trees. A multi-level split performs $k - 1$ splits (where k is the number of levels of the splitting variable), as compared to a binary split, which performs one split (regardless of the number of levels of the splitting variable). However, it should be noted that there is no inherent advantage of multi-level splits, because any multi-level split can be represented as a series of binary splits, and there may be disadvantages of using multi-level splits. With multi-level splits, predictor variables can be used for splitting only once, so

the resulting classification trees may be unrealistically short and uninteresting (Loh & Shih, 1997). A more serious problem is bias in variable selection for splits. This bias is possible in any program such as THAID (Morgan & Sonquist, 1963) that employs an exhaustive search for finding splits (for a discussion, see Loh & Shih, 1997). Bias in variable selection is the bias toward selecting variables with more levels for splits, a bias that can skew the interpretation of the relative importance of the predictors in explaining responses on the dependent variable (Breiman et. al., 1984).

Bias in variable selection can be avoided by using the Discriminant-based (univariate or linear combination) split options. These options make use of the algorithms in QUEST (Loh & Shih, 1997) to prevent bias in variable selection. The CART-style exhaustive search for univariate splits option is useful if your goal is to find splits producing the best possible classification in the learning sample (but not necessarily in independent cross-validation samples). For reliable splits, as well as computational speed, the Discriminant-based split options are recommended. For information on techniques and issues in computing classification trees, see *Computational Methods*, page 105.

Building trees interactively. In contrast, another method for building trees that has proven popular in applied research and data exploration is based on experts' knowledge about the domain or area under investigation, and relies on interactive choices (for how to grow the tree) by such experts to arrive at "good" (valid) models for prediction or predictive classification. In other words, instead of building trees automatically, using sophisticated algorithms for choosing good predictors and splits (for growing the branches of the tree), a user may want to determine manually which variables to include in the tree, and how to split those variables to create the branches of the tree. This enables the user to experiment with different variables and scenarios, and ideally to derive a better understanding of the phenomenon under investigation by combining her or his expertise with the analytic capabilities and options for building the tree. In practice, it may often be most useful to combine the automatic methods for building trees with "educated guesses" and domain-specific expertise. You may want to grow some portions of the tree using automatic methods and refine and modify the tree based on your expertise. Another common situation where this type of combined automatic and interactive tree building is called for is when some variables that are chosen automatically for some splits are not easily observable because they cannot be measured reliably or economically (i.e., obtaining such measurements would be too expensive). For example, suppose the automatic analysis at some point selects a variable *Income* as a good predictor for the next split; however, you may not be able to obtain reliable data on income from the new sample to which you want to apply the results of the current analysis (e.g., for predicting some behavior of interest, such as whether or not the person will purchase something from your catalog). In this case, you may want to select a "surrogate" variable, i.e., a variable that you can observe easily and that is likely related or similar to variable Income (with respect to its predictive power; for example, a variable *Number of years of education* may be related to *Income* and have similar predictive power; while most people are reluctant to reveal their level of income, they are more likely to report their level of education, and hence, this latter variable is more easily measured).

CLUSTER ANALYSIS

The term cluster analysis (first used by Tryon, 1939) encompasses a number of different algorithms and methods for grouping objects of similar kind into respective categories. A general question facing researchers in many areas of inquiry is how to organize observed data into meaningful structures, that is, to develop taxonomies. In other words, cluster analysis is an exploratory data analysis tool that aims to sort different objects into groups in a way that the degree of association between two objects is maximal if they belong to the same group and minimal otherwise. Given the above, cluster analysis can be used to discover structures in data without providing an explanation/interpretation. In other words, cluster analysis simply discovers structures in data without explaining why they exist.

We deal with clustering in almost every aspect of daily life. For example, a group of diners sharing the same table in a restaurant may be regarded as a cluster of people. In grocery stores items of similar nature, such as different types of meat or vegetables, are displayed in the same or nearby locations. There is a countless number of examples in which clustering plays an important role. For instance, biologists have to organize the different species of animals before a meaningful description of the differences between animals is possible. According to the modern system employed in biology, man belongs to the primates, the mammals, the amniotes, the vertebrates, and the animals. Note how in this classification, the higher the level of aggregation the less similar are the members in the respective class. Man has more in common with all other primates (e.g., apes) than it does with the more distant members of the mammals (e.g., dogs), etc. For a review of the general categories of cluster analysis methods, see *Joining (Tree Clustering)*, *Two-Way Joining (Block Clustering)*, and *k-Means Clustering*. In short, whatever the nature of your business is, sooner or later you will run into a clustering problem of one form or another.

122 EM (Expectation Maximization) Clustering

123 Finding the Right Number of Clusters in *k*-Means and EM Clustering: V-Fold Cross-Validation

Statistical Significance Testing

Note that the chapter introduction (page 115) refers to clustering algorithms and does not mention anything about statistical significance testing. In fact, cluster analysis is not as much a typical statistical test as it is a collection of different algorithms that "put objects into clusters according to well defined similarity rules." The point here is that, unlike many other statistical procedures, cluster analysis methods are mostly used when we do not have any *a priori* hypotheses, but are still in the exploratory phase of our research. In a sense, cluster analysis finds the most significant solution possible. Therefore, statistical significance testing is really not appropriate here, even in cases when *p*-levels are reported (as in *k*-means clustering).

Area of Application

Clustering techniques have been applied to a wide variety of research problems. Hartigan (1975) provides an excellent summary of the many published studies reporting the results of cluster analyses. For example, in the field of medicine, clustering diseases, cures for diseases, or symptoms of diseases can lead to very useful taxonomies. In the field of psychiatry, the correct diagnosis of clusters of symptoms such as paranoia, schizophrenia, etc. is essential for successful therapy. In archeology, researchers have attempted to establish taxonomies of stone tools, funeral objects, etc., by applying cluster analytic techniques. In general, whenever you need to classify a mountain of information into manageable meaningful piles, cluster analysis is of great utility.

Joining (Tree Clustering)

General Logic

The example in the chapter introduction (page 115) illustrates the goal of the joining or tree clustering algorithm. The purpose of this algorithm is to join together objects (e.g., animals) into successively larger clusters, using some measure of similarity or distance. A typical result of this type of clustering is the hierarchical tree.

Hierarchical Tree

Consider a horizontal hierarchical tree plot (see the next illustration), on the left of the plot, we begin with each object in a class by itself. Now imagine that, in very small steps, we relax our criterion as to what is and is not unique. Put another way, we lower our threshold regarding the decision when to declare two or more objects to be members of the same cluster.

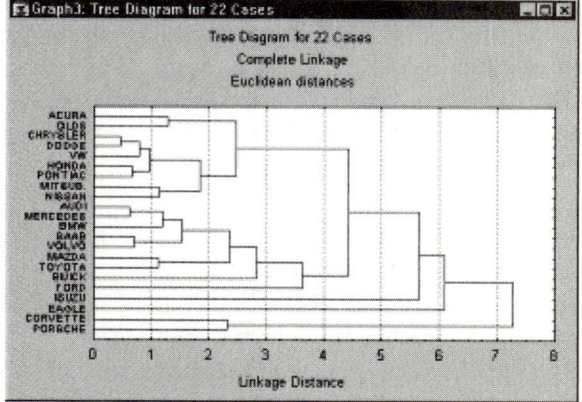

As a result we link more and more objects together and aggregate (amalgamate) larger and larger clusters of increasingly dissimilar elements. Finally, in the last step, all objects are joined together. In these plots, the horizontal axis denotes the linkage distance (in vertical icicle plots, the vertical axis denotes the linkage distance). Thus, for each node in the graph (where a new cluster is formed) we can read off the criterion distance at which the respective elements were linked together into a new single cluster. When the data contain a clear structure in terms of clusters of objects that are similar to each other, this structure will often be reflected in the hierarchical tree as distinct branches. As the result of a successful analysis with the joining method, you are able to detect clusters (branches) and interpret those branches.

Distance Measures

The joining or tree clustering method uses the dissimilarities (similarities) or distances between objects when forming the clusters. Similarities are a set of rules that serve as criteria for grouping or separating items. In the previous example the rule for grouping a number of diners was whether they shared the same table or not. These distances (similarities) can be based on a single dimension or multiple dimensions, with each dimension representing a rule or condition for grouping objects. For example, if we were to cluster fast foods, we could take into account the number of calories they contain, their price, subjective ratings of taste, etc. The most straightforward way of computing distances between objects in a multi-dimensional space is to compute Euclidean distances. If we had a two- or three-dimensional space this measure is the actual geometric distance between objects in the space (i.e., as if measured with a ruler). However, the joining algorithm does not care whether the distances that are fed to it are actual real distances, or some other derived measure of distance that is more meaningful to the researcher; and it is up to the researcher to select the right method for his/her specific application.

Euclidean distance. This is probably the most commonly chosen type of distance. It simply is the geometric distance in the multidimensional space. It is computed as:

$$\text{distance}(x,y) = \{\textstyle\sum_i (x_i - y_i)^2\}^{\frac{1}{2}}$$

Note that Euclidean (and squared Euclidean) distances are usually computed from raw data, and not from standardized data. This method has certain advantages (e.g., the distance between any two objects is not affected by the addition of new objects to the analysis, which may be outliers). However, the distances can be greatly affected by differences in scale among the dimensions from which the distances are computed. For example, if one of the dimensions denotes a measured length in centimeters, and you then convert it to millimeters (by multiplying the values by 10), the resulting Euclidean or squared Euclidean distances (computed from multiple dimensions) can be greatly affected (i.e. biased by those dimensions which have a larger scale), and consequently, the results of cluster analyses may be very different. Generally, it is good practice to transform the dimensions so they have similar scales.

Squared Euclidean distance. You may want to square the standard Euclidean distance in order to place progressively greater weight on objects that are further apart. This distance is computed as (see also, the note in the previous paragraph):

$$\text{distance}(x,y) = \textstyle\sum_i (x_i - y_i)^2$$

City-block (Manhattan) distance. This distance is simply the average difference across dimensions. In most cases, this distance measure yields results similar to the simple Euclidean distance. However, note that in this measure, the effect of single large differences (outliers) is dampened (since they are not squared). The city-block distance is computed as:

$$\text{distance}(x,y) = \sum_i |x_i - y_i|$$

Chebychev distance. This distance measure may be appropriate in cases when you want to define two objects as different if they are different on any one of the dimensions. The Chebychev distance is computed as:

$$\text{distance}(x,y) = \text{Maximum}|x_i - y_i|$$

Power distance. Sometimes you may want to increase or decrease the progressive weight that is placed on dimensions on which the respective objects are very different. This can be accomplished via the power distance. The power distance is computed as:

$$\text{distance}(x,y) = \left(\sum_i |x_i - y_i|^p\right)^{1/r}$$

where r and p are user-defined parameters. A few example calculations may demonstrate how this measure behaves. Parameter p controls the progressive weight that is placed on differences on individual dimensions, parameter r controls the progressive weight that is placed on larger differences between objects. If r and p are equal to 2, this distance is equal to the Euclidean distance.

Percent disagreement. This measure is particularly useful if the data for the dimensions included in the analysis are categorical in nature. This distance is computed as:

$$\text{distance}(x,y) = (\text{Number of } x_i \neq y_i)/ i$$

Amalgamation or Linkage Rules

At the first step, when each object represents its own cluster, the distances between those objects are defined by the chosen distance measure. However, once several objects have been linked together, how do we determine the distances between those new clusters? In other words, we need a linkage or amalgamation rule to determine when two clusters are sufficiently similar to be linked together. There are various possibilities: for example, we could link two clusters together when *any* two objects in the two clusters are closer together than the respective linkage distance. Put another way, we use the nearest neighbors across clusters to determine the distances between clusters; this method is called single linkage. This rule produces "stringy" types of clusters, that is, clusters "chained together" by only single objects that happen to be close together. Alternatively, we can use the neighbors across clusters that are furthest away from each other; this method is called complete linkage. There are numerous other linkage rules such as these that have been proposed.

Single linkage (nearest neighbor). As described above, in this method the distance between two clusters is determined by the distance of the two closest objects (nearest neighbors) in the different clusters. This rule will, in a sense, string objects together to form clusters, and the resulting clusters tend to represent long chains.

Complete linkage (furthest neighbor). In this method, the distances between clusters are determined by the greatest distance between any two objects in the different clusters (i.e., by the furthest neighbors). This method usually performs quite well in cases when the objects actually form naturally distinct clumps. If the clusters tend to be elongated or of a "chain" type nature, this method is inappropriate.

Unweighted pair-group average. In this method, the distance between two clusters is calculated as the average distance between all pairs of objects in the two different clusters. This method is also very efficient when the objects form natural distinct clumps, however, it performs equally well with elongated, chain type clusters. Note that in their book, Sneath and Sokal (1973) introduced the abbreviation UPGMA to refer to this method as unweighted pair-group method using arithmetic averages.

Weighted pair-group average. This method is identical to the unweighted pair-group average method, except that in the computations, the size of the respective clusters (i.e., the number of objects contained in them) is used as a weight. Thus, this method (rather than the previous method) should be used when the cluster sizes are suspected to be greatly uneven. Note that in their book, Sneath and Sokal (1973) introduced the abbreviation WPGMA to refer to this method as weighted pair-group method using arithmetic averages.

Unweighted pair-group centroid. The centroid of a cluster is the average point in the multidimensional space defined by the dimensions. In a sense, it is the center of gravity for the respective cluster. In this method, the distance between two clusters is determined as the difference between centroids. Sneath and Sokal (1973) use the abbreviation UPGMC to refer to this method as unweighted pair-group method using the centroid average.

Weighted pair-group centroid (median). This method is identical to the previous one, except that weighting is introduced into the computations to take into consideration differences in cluster sizes (i.e., the number of objects contained in them). Thus, when there are (or you suspect there to be) considerable differences in cluster sizes, this method is preferable to the previous one. Sneath and Sokal (1973) use the abbreviation WPGMC to refer to this method as weighted pair-group method using the centroid average.

Ward's method. This method is distinct from all other methods because it uses an analysis of variance approach to evaluate the distances between clusters. In short, this method attempts to minimize the Sum of Squares (SS) of any two (hypothetical) clusters that can be formed at each step. Refer to Ward (1963) for details concerning this method. In general, this method is regarded as very efficient, however, it tends to create clusters of small size.

For an overview of the other two methods of clustering, see *Two-Way Joining*, page 120, and *k-Means Clustering*, page 121.

Two-Way Joining

Previously, we have discussed this method in terms of objects that are to be clustered [see *Joining (Tree Clustering)*, page 117]. In all other types of analyses the research question of interest is usually expressed in terms of cases (observations) or variables. It turns out that the clustering of both may yield useful results. For example, imagine a study where a medical researcher has gathered data on different measures of physical fitness (variables) for a sample of heart patients (cases). The researcher may want to cluster cases (patients) to detect clusters of patients with similar syndromes. At the same time, the researcher may want to cluster variables (fitness measures) to detect clusters of measures that appear to tap similar physical abilities.

Given the discussion in the paragraph above concerning whether to cluster cases or variables, you may wonder why not cluster both simultaneously? Two-way joining is useful in (the relatively rare) circumstances when you expect that both cases and variables will simultaneously contribute to the uncovering of meaningful patterns of clusters.

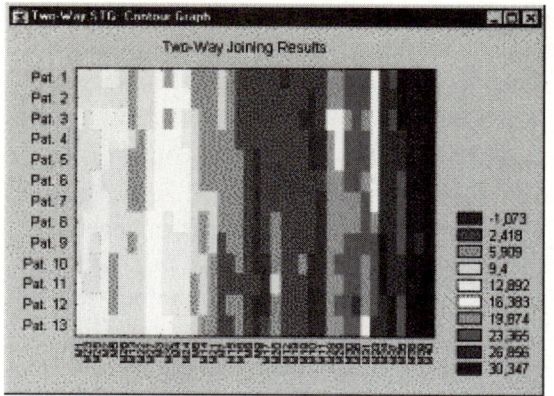

For example, returning to the previous example, the medical researcher may want to identify clusters of patients that are similar with regard to particular clusters of similar measures of physical fitness. The difficulty with interpreting these results may arise from the fact that the similarities between different clusters may pertain to (or be caused by) somewhat different subsets of variables. Thus, the resulting structure (clusters) is by nature not homogeneous. This may seem a bit confusing at first, and, indeed, compared to the other clustering methods described [see *Joining (Tree Clustering)* and *k-Means Clustering*], two-way joining is probably the one least commonly used. However, some researchers believe that this method offers a powerful exploratory data analysis tool (for more information, refer to the detailed description of this method in Hartigan, 1975).

k-Means Clustering

General Logic

This method of clustering is very different from the joining (tree clustering) and two-way joining. Suppose that you already have hypotheses concerning the number of clusters in your cases or variables. You may want to specify that the computer forms exactly 3 clusters that are to be as distinct as possible. This is the type of research question that can be addressed by the *k*-means clustering algorithm. In general, the *k*-means method will produce exactly *k* different clusters of greatest possible distinction. It should be mentioned that the best number of clusters *k* leading to the greatest separation (distance) is not known as *a priori* and must be computed from the data (see *Finding the Right Number of Clusters*, page 123).

Example

In the physical fitness example (see *Two-Way Joining*), the medical researcher may have a hunch from clinical experience that her heart patients fall basically into three different categories with regard to physical fitness. She might wonder whether this intuition can be quantified, that is, whether a k-means cluster analysis of the physical fitness measures would indeed produce the three clusters of patients as expected. If so, the means on the different measures of physical fitness for each cluster would represent a quantitative way of expressing the researcher's hypothesis or intuition (i.e., patients in cluster 1 are high on measure 1, low on measure 2, etc.).

Computations

Computationally, you can think of this method as analysis of variance (ANOVA) in reverse. Starting with *k* random clusters, move objects between those clusters with the goal to 1) minimize variability within clusters and 2) maximize variability between clusters. In other words, the similarity rules will apply maximally to the members of one cluster and minimally to members belonging to the rest of the clusters. This is analogous to "ANOVA in reverse" in the sense that the significance test in ANOVA evaluates the between group variability against the within-group variability when computing the significance test for the hypothesis that the means in the groups are different from each other. In *k*-means clustering, objects (e.g., cases) are moved in and out of groups (clusters) to get the most significant ANOVA results.

Interpretation of Results

Usually, as the result of a *k*-means clustering analysis, we would examine the means for each cluster on each dimension to assess how distinct our *k* clusters are. Ideally, we would obtain very different means for most, if not all dimensions, used in the analysis. The magnitude of the *F* values from the analysis of variance performed on each dimension is another indication of how well the respective dimension discriminates between clusters.

EM (Expectation Maximization) Clustering

The methods described here are similar to the *k*-Means algorithm described above, and you may want to review that section for a general overview of these techniques and their applications. The general purpose of these techniques is to detect clusters in observations (or variables) and to assign those observations to the clusters. A typical example application for this type of analysis is a marketing research study in which a number of consumer behavior related variables are measured for a large sample of respondents. The purpose of the study is to detect "market segments," i.e., groups of respondents that are somehow more similar to each other (to all other members of the same cluster) when compared to respondents that "belong to" other clusters. In addition to identifying such clusters, it is usually equally of

interest to determine how the clusters are different, i.e., determine the specific variables or dimensions that vary and how they vary in regard to members in different clusters.

***k*-means clustering.** To reiterate, the classic *k*-means algorithm was popularized and refined by Hartigan (1975; see also Hartigan and Wong, 1978). The basic operation of that algorithm is relatively simple: Given a fixed number of (desired or hypothesized) *k* clusters, assign observations to those clusters so that the means across clusters (for all variables) are as different from each other as possible.

Extensions and generalizations. The EM (expectation maximization) algorithm extends this basic approach to clustering in two important ways:

1. Instead of assigning cases or observations to clusters to maximize the differences in means for continuous variables, the *EM* clustering algorithm computes probabilities of cluster memberships based on one or more probability distributions. The goal of the clustering algorithm then is to maximize the overall probability or likelihood of the data, given the (final) clusters.

2. Unlike the classic implementation of *k*-means clustering, the general EM algorithm can be applied to both continuous and categorical variables (note that the classic *k*-means algorithm can also be modified to accommodate categorical variables).

EM Algorithm

The EM algorithm for clustering is described in detail in Witten and Frank (2001). The basic approach and logic of this clustering method is as follows. Suppose you measure a single continuous variable in a large sample of observations. Further, suppose that the sample consists of two clusters of observations with different means (and perhaps different standard deviations); within each sample, the distribution of values for the continuous variable follows the normal distribution. The resulting distribution of values (in the population) may look like this:

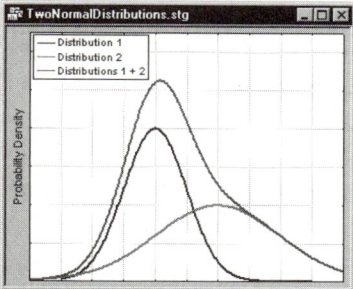

Mixtures of distributions. The previous illustration shows two normal distributions with different means and different standard deviations, and the sum of the two distributions. Only the mixture (sum) of the two normal distributions (with different means and standard deviations) would be observed. The goal of EM clustering is to estimate the means and standard deviations for each cluster so as to maximize the likelihood of the observed data (distribution). Put another way, the EM algorithm attempts to approximate the observed distributions of values based on mixtures of different distributions in different clusters.

With the implementation of the EM algorithm in some computer programs, you may be able to select (for continuous variables) different distributions such as the *Normal* (see page 173), *Lognormal* (see page 172), and *Poisson* (see page 174) distributions. You can select different distributions for different variables and, thus, derive clusters for mixtures of different types of distributions.

Categorical variables. The EM algorithm can also accommodate categorical variables. The

method will at first randomly assign different probabilities (weights, to be precise) to each class or category, for each cluster. In successive iterations, these probabilities are refined (adjusted) to maximize the likelihood of the data given the specified number of clusters.

Classification probabilities instead of classifications. The results of EM clustering are different from those computed by *k*-means clustering. The latter will assign observations to clusters to maximize the distances between clusters. The EM algorithm does not compute actual assignments of observations to clusters, but classification probabilities. In other words, each observation belongs to each cluster with a certain probability. Of course, as a final result you can usually review an actual assignment of observations to clusters, based on the (largest) classification probability.

Finding the Right Number of Clusters in *k*-Means and EM Clustering: V-Fold Cross-Validation

An important question that needs to be answered before applying the *k*-means or EM clustering algorithms is how many clusters there are in the data. This is not known *a priori* and, in fact, there might be no definite or unique answer as to what value *k* should take. In other words, *k* is a nuisance parameter of the clustering model. Luckily, an estimate of *k* can be obtained from the data using the method of cross-validation. Remember that the *k*-means and EM methods will determine cluster solutions for a particular user-defined number of clusters. The *k*-means and EM clustering techniques (described above) can be optimized and enhanced for typical applications in data mining (see Chapter 12 – *Data Mining*

Techniques). The general metaphor of data mining implies the situation in which an analyst searches for useful structures and "nuggets" in the data, usually without any strong *a priori* expectations of what the analyst might find (in contrast to the hypothesis-testing approach of scientific research). In practice, the analyst usually does not know ahead of time how many clusters there might be in the sample. For that reason, some programs include an implementation of a v-fold cross-validation algorithm (see the *Statistical Glossary*) for automatically determining the number of clusters in the data.

This unique algorithm is immensely useful in all general "pattern-recognition" tasks to determine the number of market segments in a marketing research study, the number of distinct spending patterns in studies of consumer behavior, the number of clusters of different medical symptoms, the number of different types (clusters) of documents in text mining, the number of weather patterns in meteorological research, the number of defect patterns on silicon wafers, and so on.

The v-fold cross-validation algorithm applied to clustering. The v-fold cross-validation algorithm is described in some detail in Chapter 9 – *Classification Trees* and Chapter 8 – *Classification and Regression Trees (CART)*. The general idea of this method is to divide the overall sample into a number of *v* folds. The same type of analysis is then successively applied to the observations belonging to the *v-1* folds (training sample), and the results of the analyses are applied to sample *v* (the sample or fold that was not used to estimate the parameters, build the tree, determine the clusters, etc.; this is the testing sample) to compute some index of predictive validity. The results for the *v* replications are aggregated (averaged) to yield a single measure of the stability of the respective model, i.e., the validity of the model for predicting new observations.

Cluster analysis is an unsupervised learning technique (see the *Statistical Glossary*), and we cannot observe the (real) number of clusters in the data. However, it is reasonable to replace the usual notion (applicable to supervised learning, see the *Statistical Glossary*) of "accuracy" with that of "distance." In general, we can apply the v-fold cross-validation method to a range of numbers of clusters in *k*-means or EM clustering, and observe the resulting average distance of the observations (in the cross-validation or testing samples) from their cluster centers (for *k*-means clustering); for EM clustering, an appropriate equivalent measure would be the average negative (log-) likelihood computed for the observations in the testing samples.

Reviewing the results of v-fold cross-validation. The results of v-fold cross-validation are best reviewed in a simple line graph.

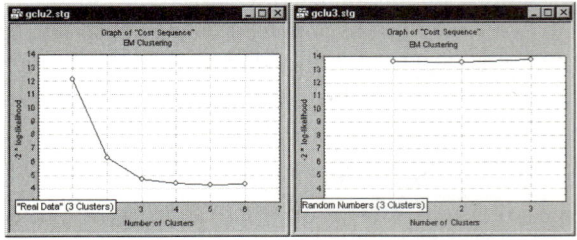

The previous illustration shows the result of analyzing a data set widely known to contain three clusters of observations (specifically, the well-known *Iris* data file reported by Fisher, 1936, and widely referenced in the literature on discriminant function analysis). Also shown (in the graph to the right) are the results for analyzing simple normal random numbers. The real data (shown to the left) exhibit the characteristic scree-plot (see the *Statistical*

Glossary) pattern (see also, Chapter 16 – *Factor Analysis*), where the cost function (in this case, 2 times the log-likelihood of the cross-validation data, given the estimated parameters) quickly decreases as the number of clusters increases, but then levels off (past 3 clusters), and even increases as the data are overfitted. Alternatively, the random numbers show no such pattern, in fact, there is basically no decrease in the cost function at all, and it quickly begins to increase as the number of clusters increases and overfitting occurs.

It is easy to see from this simple illustration how useful the v-fold cross-validation technique, applied to *k*-means and EM clustering can be for determining the right number of clusters in the data.

CORRESPONDENCE ANALYSIS

Correspondence analysis is a descriptive/exploratory technique designed to analyze simple two-way and multi-way tables containing some measure of correspondence between the rows and columns. The results provide information that is similar in nature to those produced by factor analysis techniques, and they enable you to explore the structure of categorical variables included in the table. The most common table of this type is the two-way frequency crosstabulation table (see, for example, Chapter 2 – *Basic Statistics* or Chapter 22 – *Log-Linear*).

In a typical correspondence analysis, a crosstabulation table of frequencies is first standardized so that the relative frequencies across all cells sum to 1.0. One way to state the goal of a typical analysis is to represent the entries in the table of relative frequencies in terms of the distances between individual rows and/or columns in a low-dimensional space. This is best illustrated by a simple example, which is described in this chapter. There are several parallels in interpretation between correspondence analysis and factor analysis, and some similar concepts will be pointed out in this chapter.

For a comprehensive description of this method, computational details, and its applications (in the English language), refer to the classic text by Greenacre (1984). These methods were originally developed primarily in France by Jean-Paul Benzérci in the early 1960s and 1970s (e.g., see Benzérci, 1973; see also, Lebart, Morineau, and Tabard, 1977), but have only more recently gained increasing popularity in English-speaking countries (see, for example, Carrol, Green, and Schaffer, 1986; Hoffman and Franke, 1986). (Note that similar techniques were developed independently in several countries, where they were known as optimal scaling, reciprocal averaging, optimal scoring, quantification method, or homogeneity analysis). In the following paragraphs, a general introduction to correspondence analysis will be presented.

Overview

Suppose you collected data on the smoking habits of different employees in a company. The following data set is presented in Greenacre (1984, p. 55).

STAFF GROUP	SMOKING CATEGORY				
	None (1)	Light (2)	Medium (3)	Heavy (4)	Row Totals
Senior Managers	4	2	3	2	11
Junior Managers	4	3	7	4	18
Senior Employees	25	10	12	4	51
Junior Employees	18	24	33	13	88
Secretaries	10	6	7	2	25
Column Totals	61	45	62	25	193

Data: Correspondance Analysis Example Spreadsheet 1* (5v by 6c)

You can think of the 4 column values in each row of the table as coordinates in a 4-dimensional space, and you could compute the (Euclidean) distances between the 5 row points in the 4-dimensional space. The distances between the points in the 4-dimensional space summarize all information about the similarities between the rows in the table above. Now suppose you could find a lower-dimensional space, in which to position the row points in a manner that retains all, or almost all, of the information about the differences between the rows. You could then present all information about the similarities between the rows (types of employees in this case) in a simple 1, 2, or 3-dimensional graph. While this may not appear to be particularly useful for small tables such as the one shown above, you can easily imagine how the presentation and interpretation of very large tables (e.g., differential preference for 10 consumer items among 100 groups of respondents in a consumer survey) could greatly benefit from the simplification that can be achieved via correspondence analysis (e.g., represent the 10 consumer items in a two-dimensional space).

Mass. To continue with the simpler example of the two-way table presented above, computationally, the relative frequencies for the frequency table are computed so that the sum of all table entries is equal to 1.0 (each element will be divided by the total, i.e., *193*). You could say that this table now shows how one unit of mass is distributed across the cells. In the terminology of correspondence analysis, the row and column totals of the matrix of relative frequencies are called the row mass and column mass, respectively.

Inertia. The term inertia in correspondence analysis is used by analogy with the definition in applied mathematics of "moment of inertia," which stands for the integral of mass times the squared distance to the centroid (e.g., Greenacre, 1984, p. 35). Inertia is defined as the total Pearson *chi*-square for the two-way divided by the total sum (*193* in the present example).

Inertia and row and column profiles. If the rows and columns in a table are completely independent of each other, the entries in the table (distribution of mass) can be reproduced from the row and column totals alone, or row and column profiles in the terminology of correspondence analysis. According to the well-known formula for computing the *chi*-square statistic for two-way tables, the expected frequencies in a table, where the column and rows are independent of each other, are equal to the respective column total times the row total, divided by the grand total. Any deviations from the expected values (expected under the hypothesis of complete independence of the row and column variables) will contribute to the overall *chi*-square. Thus, another way of looking at correspondence analysis is to consider it a method for decomposing the overall *chi*-square statistic (or *Inertia=Chi-square/Total N*) by identifying a small number of dimensions in which the deviations from the expected values can be represented. This is similar to the goal of factor analysis, where the total variance is decomposed to arrive at a lower-dimensional representation of the variables that enable you to reconstruct most of the variance/covariance matrix of variables.

Analyzing rows and columns. This simple example began with a discussion of the row-points in the table shown above. However, you may rather be interested in the column totals, in which case you could plot the column points in a small-dimensional space, which satisfactorily reproduces the similarity (and distances) between the relative frequencies for the columns, across the rows, in the table shown above. In fact it is customary to simultaneously plot the column points and the row points in a single graph, to summarize the information contained in a two-way table.

Reviewing results. Let's now look at some of the results for the table shown in the previous illustration. First, shown in the next illustration, are the so-called singular values, eigenvalues, percentages of inertia explained, cumulative percentages, and the contribution to the overall *chi*-square.

Data: Correspondance Analysis Example Spreadsheet 2* (6v by 3c)

Eigenvalues and Inertia for all Dimensions
Input Table (Rows x Columns): 5 x 4
Total Inertia = .08519 Chi² = 16.442

	No. of Dims	Singular Values	Eigen-Values	Perc. of Inertia	Cumulatv Percent	Chi Squares
1	1	273421	74759	87.75587	87.7559	14.42851
2	2	0.100086	0.01	11.75865	99.5145	1.93332
3	3	0.020337	0.0004	0.48547	100	0.07982

Note that the dimensions are "extracted" so as to maximize the distances between the row or column points, and successive dimensions (which are independent of or orthogonal to each other) will explain less and less of the overall *chi*-square value (and, thus, inertia). Thus, the extraction of

the dimensions is similar to the extraction of principal components in factor analysis.

First, it appears that, with a single dimension, 87.76% of the inertia can be explained, that is, the relative frequency values that can be reconstructed from a single dimension can reproduce 87.76% of the total *chi*-square value (and, thus, of the inertia) for this two-way table; two dimensions enable you to explain 99.51%.

Maximum number of dimensions. Since the sums of the frequencies across the columns must be equal to the row totals, and the sums across the rows equal to the column totals, there are in a sense only (*no. of columns-1*) independent entries in each row, and (*no. of rows-1*) independent entries in each column of the table (once you know what these entries are, you can fill in the rest based on your knowledge of the column and row marginal totals). Thus, the maximum number of eigenvalues that can be extracted from a two-way table is equal to the minimum of the number of columns minus 1, and the number of rows minus 1. If you choose to extract (i.e., interpret) the maximum number of dimensions that can be extracted, you can reproduce exactly all information contained in the table.

Row and column coordinates. Next, look at the coordinates for the two-dimensional solution.

Data: Correspondance Analysis Example Spreadsheet 3* (2v by 5c)

Row Name	Dim 1	Dim 2
1 Senior Managers	-0.065768	193737
2 Junior Managers	0.258958	0.243305
3 Senior Employees	-0.380595	0.01066
4 Junior Employees	0.232952	-0.057744
5 Secretaries	-0.201089	-0.078911

Of course, you can plot these coordinates in a two-dimensional scatterplot. Remember that the purpose of correspondence analysis is to

reproduce the distances between the row and/or column points in a two-way table in a lower-dimensional display; note that, as in factor analysis, the actual rotational orientation of the axes is arbitrarily chosen so that successive dimensions explain less and less of the overall *chi*-square value (or inertia). You could, for example, reverse the signs in each column in the table shown above, thereby effectively rotating the respective axis in the plot by 180 degrees.

What is important are the distances of the points in the two-dimensional display, which are informative in that row points that are close to each other are similar with regard to the pattern of relative frequencies across the columns. If you have produced this plot you will see that, along the most important first axis in the plot, the *Senior employees* and *Secretaries* are relatively close together on the left side of the origin (scale position 0). If you looked at the table of relative row frequencies (i.e., frequencies standardized, so that their sum in each row is equal to 100%), you will see that these two groups of employees indeed show very similar patterns of relative frequencies across the categories of smoking intensity.

	Percentages of Row Total				
	Smoking Category None (1)	Smoking Category Light (2)	Smoking Category Medium (3)	Smoking Category Heavy (4)	Row Totals
1 Senior Managers	36.36	18.18	27.27	18.18	100
2 Junior Managers	22.22	16.67	38.89	22.22	100
3 Senior Employees	49.02	19.61	23.53	7.84	100
4 Junior Employees	20.45	27.27	37.50	14.77	100
5 Secretaries	40.00	24.00	28.00	8.00	100

Data: Correspondance Analysis Example Spreadsheet 4" (5v by 5c)

Obviously the final goal of correspondence analysis is to find theoretical interpretations (i.e., meaning) for the extracted dimensions. One method that may aid in interpreting extracted dimensions is to plot the column points. Shown in the next illustration are the column coordinates for the first and second dimension.

Smoking Category	Dim. 1	Dim. 2
None	-0.393308	30492
Light	0.099456	-0.141064
Medium	0.196321	-0.007359
Heavy	0.293776	0.197766

Data: Correspondance Analysis Example Spreadsheet 5" (2v by 4c)

It appears that the first dimension distinguishes mostly between the different degrees of smoking, and in particular between category None and the others. Thus, you can interpret the greater similarity of *Senior Managers* with *Secretaries*, with regard to their position on the first axis, as mostly deriving from the relatively large numbers of *None* smokers in these two groups of employees.

Compatibility of row and column coordinates. It is customary to summarize the row and column coordinates in a single plot. However, it is important to remember that in such plots, you can only interpret the distances between row points, and the distances between column points, but not the distances between row points and column points.

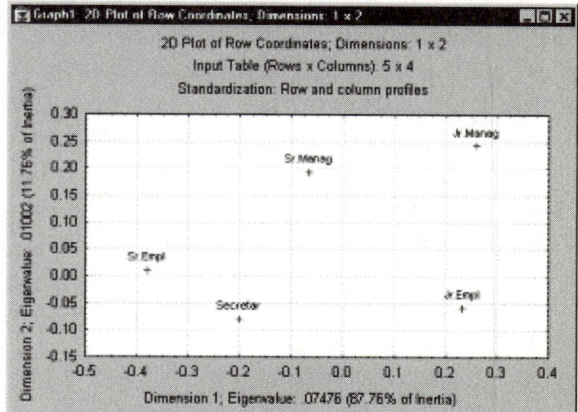

To continue with this example, it would not be appropriate to say that the category *None* is similar to *Senior Employees* (the two points are very close in the simultaneous plot of row and column coordinates). However, as was indicated earlier, it is appropriate to make general statements about the nature of the dimensions, based on which side of the origin particular points fall. For example, because category *None* is the only column point on the left side of the origin for the first axis, and since employee group *Senior Employees* also falls onto that side of the first axis, you can conclude that the first axis separates *None* smokers from the other categories of smokers, and that *Senior Employees* are different from, for example, *Junior Employees*, in that there are relatively more non-smoking *Senior Employees*.

Scaling of the coordinates (standardization options). Another important decision that the analyst must make concerns the scaling of the coordinates. The nature of the choice pertains to whether or not you want to analyze the relative row percentages, column percentages, or both. In the context of the example described above, the row percentages were shown to illustrate how the patterns of those percentages across the columns are similar for points that appear more closely together in the graphical display of the row coordinates. Put another way, the coordinates are based on the analysis of the row profile matrix, where the sum of the table entries in a row, across all columns, is equal to 1.0 (each entry r_{ij} in the row profile matrix can be interpreted as the conditional probability that a case belongs to column *j*, given its membership in row i). Thus, the coordinates are computed so as to maximize the differences between the points with respect to the row profiles (row percentages). The row coordinates are computed from the row profile matrix, the column coordinates are computed from the column profile matrix.

A fourth option, canonical standardization (see Gifi, 1981), is also provided, and it amounts to a standardization of the columns and rows of the matrix of relative frequencies. This standardization amounts to a rescaling of the coordinates based on the row profile standardization and the column profile standardization, and this type of standardization is not widely used. Note also that a variety of other custom standardizations can be easily performed if you have the raw eigenvalues and eigenvector matrices.

Metric of coordinate system. In several places in this introduction, the term distance is used (loosely) to refer to the differences between the pattern of relative frequencies for the rows across the columns and columns across the rows that are to be reproduced in a lower-dimensional solution as a result of the correspondence analysis. Actually, these distances represented by the coordinates in the respective space are not simple Euclidean distances computed from the relative row or column frequencies but, rather, are weighted distances. Specifically, the weighting that is applied is such that the metric in the lower-dimensional space is a *chi*-square metric, provided that 1) you are comparing row points and chose either row-profile standardization or both row- and column-profile standardization, or 2) you are comparing column points and chose either column-profile standardization or both row- and column-profile standardization.

In that case (but *not* if you chose the canonical standardization), the squared Euclidean distance between, for example, two row points *i* and *i'* in the respective coordinate system of a given number of dimensions actually approximates a weighted (i.e., *chi*-square) distance between the relative frequencies (see Hoffman and Franke, 1986, formula 21):

$$d_{ii\cdot}^{2} = \sum_j (1/c_j \, (p_{ij}/r_i - p^2_{i\cdot j}/r_{i\cdot}))$$

In this formula, $d_{ii\cdot}$ stands for the squared distance between the two points, c_j stands for the column total for the j'th column of the standardized frequency table (where the sum of all entries or mass is equal to *1.0*), p_{ij} stands for the individual cell entries in the standardized frequency table (row i, column j), r_i stands for the row total for the i'th column of the relative frequency table, and the summation $\sum$ is over the columns of the table. To reiterate, only the distances between row points, and correspondingly, between column points are interpretable in this manner; the distances between row points and column points cannot be interpreted.

Judging the quality of a solution. A number of auxiliary statistics are reported to aid in the evaluation of the quality of the respective chosen numbers of dimensions. The general concern here is that all (or at least most) points are properly represented by the respective solution, that is, that their distances to other points can be approximated to a satisfactory degree. Shown in the next spreadsheet are all statistics reported for the row coordinates for the example table discussed so far, based on a one-dimensional solution only (i.e., only one dimension is used to reconstruct the patterns of relative frequencies across the columns).

Data: Correspondance Analysis Example Spreadsheet 6* (6v by 5c)						
ROW COORDINATES AND CONTRIBUTIONS TO INERTIA						
	Dim.1	Mass	Quality	Relative Inertia	Inertia Dim.1	Cosine² Dim.1
1 Sr. Managers	-0.065768	0.056995	92232	31376	0.003298	0.092232
2 Jr. Managers	0.258958	0.093264	0.5264	0.139467	0.083659	0.5264
3 Sr. Employees	-0.380595	0.264249	0.999033	0.44975	0.512006	0.999033
4 Jr. Employees	0.232952	0.455959	0.941934	0.308354	0.330974	0.941934
5 Secretaries	-0.201089	0.129534	0.865346	0.071053	0.070064	0.865346

Coordinates. The first numeric column shown in the table above contains the coordinates, as discussed in the previous paragraphs. To reiterate, the specific interpretation of these coordinates depends on the standardization chosen for the solution (see above). The number of dimensions is chosen by the user (in this case we chose only one dimension), and coordinate values will be shown for each dimension (i.e., there will be one column with coordinate values for each dimension).

Mass. The *Mass* column contains the row totals (since these are the row coordinates) for the table of relative frequencies (i.e., for the table where each entry is the respective mass, as discussed earlier in this section). Remember that the coordinates are computed based on the matrix of conditional probabilities shown in the *Mass* column.

Quality. The *Quality* column contains information concerning the quality of representation of the respective row point in the coordinate system defined by the respective numbers of dimensions, as chosen by the user. In the table shown above, only one dimension was chosen, and the numbers in the *Quality* column pertain to the quality of representation in the one-dimensional space. To reiterate, computationally, the goal of the correspondence analysis is to reproduce the distances between points in a low-dimensional space. If you extracted (i.e., interpreted) the maximum number of dimensions (which is equal to the minimum of the number of rows and the number of columns, minus 1), you could reconstruct all distances exactly. The quality of a point is defined as the ratio of the squared distance of the point from the origin in the chosen number of dimensions, over the squared distance from the origin in the space defined by the maximum number of dimensions (remember that the metric here is *chi*-square, as described earlier). By analogy to factor analysis, the quality of a point is similar in its interpretation to the communality for a variable in factor analysis.

Note that the quality measure reported is independent of the chosen method of standardization, and always pertains to the default standardization (i.e., the distance metric is *chi*-square, and the quality measure can be interpreted as the "proportion of *chi*-square accounted for" for the respective row, given the respective number of dimensions). A low quality means that the current number of dimensions does not well represent the respective row (or column). In the table shown above, the quality for the first row (*Senior Managers*) is less than *.1*, indicating that this row point is not well represented by the one-dimensional representation of the points.

Relative inertia. The quality of a point (see above) represents the proportion of the contribution of that point to the overall inertia (*chi*-square) that can be accounted for by the chosen number of dimensions. However, it does not indicate whether or not, and to what extent, the respective point does in fact contribute to the overall inertia (*chi*-square value). The relative inertia represents the proportion of the total inertia accounted for by the respective point, and it is independent of the number of dimensions chosen by the user. Note that a particular solution may represent a point very well (high quality), but the same point may not contribute much to the overall inertia (e.g., a row point with a pattern of relative frequencies across the columns that is similar to the average pattern across all rows).

Relative inertia for each dimension. This column contains the relative contribution of the respective (row) point to the inertia accounted for by the respective dimension. Thus, this

value will be reported for each (row or column) point, for each dimension.

Cosine² (quality or squared correlations with each dimension). This column contains the quality for each point, by dimension. The sum of the values in these columns across the dimensions is equal to the total *Quality* value discussed above (since in the example table above, only one dimension was chosen, the values in this column are identical to the values in the overall *Quality* column). This value can also be interpreted as the correlation of the respective point with the respective dimension. The term *Cosine*² refers to the fact that this value is also the squared cosine value of the angle the point makes with the respective dimension (refer to Greenacre, 1984, for details concerning the geometric aspects of correspondence analysis).

A note about statistical significance. It should be noted at this point that correspondence analysis is an exploratory technique. Actually, the method was developed based on a philosophical orientation that emphasizes the development of models that fit the data, rather than the rejection of hypotheses based on the lack of fit (Benzecri's "second principle" states that "The model must fit the data, not vice versa;" see Greenacre, 1984, p. 10). Therefore, there are no statistical significance tests that are customarily applied to the results of a correspondence analysis; the primary purpose of the technique is to produce a simplified (low-dimensional) representation of the information in a large frequency table (or tables with similar measures of correspondence).

Supplementary Points

The *Overview* at the beginning of this chapter (page 129) explains how to interpret the

coordinates and related statistics computed in a correspondence analysis. An important aid in the interpretation of the results from a correspondence analysis is to include supplementary row or column points, that were not used to perform the original analyses. For example, consider the following results, which are based on the example given in the introductory (based on Greenacre, 1984).

Row Name	Dim 1	Dim 2
1 Senior Managers	-0.065768	193737
2 Junior Managers	0.258958	0.243305
3 Senior Employees	-0.380595	0.01066
4 Junior Employees	0.232952	-0.057744
5 Secretaries	-0.201089	-0.078911
6 National Average	-0.258368	-0.117648

Data: Correspondance Analysis Example Spreadsheet 7* (2v by 6c)

The table in the previous illustration shows the coordinate values (for two dimensions) computed for a frequency table of different types of employees by type of smoking habit. The row labeled *National Average* contains the coordinate values for the supplementary point, which is the national average (percentages) for the different smoking categories (which make up the columns of the table; those fictitious percentages reported in Greenacre (1984) are: Nonsmokers: 42%, light smokers: 29%, medium smokers, 20%; heavy smokers: 9%). If you plotted these coordinates in a two-dimensional scatterplot, along with the column coordinates, it would be apparent that the *National Average* supplementary row point is plotted close to the point representing the Secretaries group, and on the same side of the horizontal axis (first dimension) as the *Nonsmokers* column point. If you refer back to the original two-way table shown in the overview, this finding is consistent with the entries in the table of row frequencies, that is, there are relatively more nonsmokers among the *Secretaries*, and in the *National Average*. Put

another way, the sample represented in the original frequency table contains more smokers than the national average.

While this type of information could have been easily gleaned from the original frequency table (that was used as the input to the analysis), in the case of very large tables, such conclusions may not be as obvious.

Quality of representation of supplementary points. Another interesting result for supplementary points concerns the quality of their representation in the chosen number of dimensions (see page 133 for a more detailed discussion of the concept of quality of representation). To reiterate, the goal of the correspondence analysis is to reproduce the distances between the row or column coordinates (patterns of relative frequencies across the columns or rows, respectively) in a low-dimensional solution. Given such a solution, you may ask whether particular supplementary points of interest can be represented equally well in the final space, that is, whether or not their distances from the other points in the table can also be represented in the chosen numbers of dimensions. Shown in the next illustration are the summary statistics for the original points, and the supplementary row point *National Average*, for the two-dimensional solution.

STAFF GROUP	Quality	Cosine^2 Dim. 1	Cosine^2 Dim. 2
1 Senior Managers	0.892568	0.092232	0.800336
2 Junior Managers	0.991082	0.5264	0.464682
3 Senior Employees	0.999817	0.999033	0.000784
4 Junior Employees	0.99981	0.941934	0.057876
5 Secretaries	0.998603	0.865346	0.133257
National Average	0.761324	0.630578	0.130746

Data: Correspondence Analysis Example Spreadsheet 8 (3v by 6c)

The statistics reported in the table above are discussed in the *Overview* of this chapter. In short, the *Quality* of a row or column point is

defined as the ratio of the squared distance of the point from the origin in the chosen number of dimensions, over the squared distance from the origin in the space defined by the maximum number of dimensions (remember that the metric here is *chi*-square, as described in the *Overview* of this chapter). In a sense, the overall quality is the "proportion of squared distance-from-the-overall-centroid accounted for." The supplementary row point *National Average* has a quality of .76, indicating that it is reasonably well represented in the two-dimensional solution. The *Cosine*² statistic is the quality accounted for by the respective row point, by the respective dimension (the sum of the *Cosine*² values over the respective number of dimensions is equal to the total *Quality*, see the *Overview* of this chapter).

Multiple Correspondence Analysis (MCA)

Multiple correspondence analysis (MCA) can be considered to be an extension of simple correspondence analysis to more than two variables. For an introductory overview of simple correspondence analysis, refer to the *Overview* at the beginning of this chapter. Multiple correspondence analysis is a simple correspondence analysis carried out on an indicator (or design) matrix with cases as rows and categories of variables as columns. Actually, you usually analyze the inner product of such a matrix, called the Burt table in an MCA; this will be discussed later. However, to clarify the interpretation of the results from a multiple correspondence analysis, it is easier to discuss the simple correspondence analysis of an indicator or design matrix.

Indicator or design matrix. Consider again the simple two-way table presented in the *Overview* at the beginning of this chapter:

STAFF GROUP	SMOKING CATEGORY				
	None (1)	Light (2)	Medium (3)	Heavy (4)	Row Totals
Senior Managers	4	2	3	2	11
Junior Managers	4	3	7	4	18
Senior Employees	25	10	12	4	51
Junior Employees	18	24	33	13	88
Secretaries	10	6	7	2	25
Column Totals	61	45	62	25	193

Suppose you had entered the data for this table in the following manner, as an indicator or design matrix:

CASE NO.	Sr. Man- ager	Jr. Man- ager	Sr. Em- ployee	Jr. Em- ployee	Sec- retary	Smoke None	Smoke Light	Smoke Medium	Smoke Heavy
1	1	0	0	0	0	1	0	0	0
2	1	0	0	0	0	1	0	0	0
3	1	0	0	0	0	1	0	0	0
4	1	0	0	0	0	1	0	0	0
5	1	0	0	0	0	0	1	0	0
...	0	0	0	0	0	0	0	0	0
...	0	0	0	0	0	0	0	0	0
...	0	0	0	0	0	0	0	0	0
191	0	0	0	0	1	0	0	1	0
192	0	0	0	0	1	0	0	0	1
193	0	0	0	0	1	0	0	0	1

Each one of the 193 total cases in the table is represented by one case in this data file. For each case a *1* is entered into the category where the respective case belongs, and a *0* otherwise. For example, case *1* represents a *Senior Manager* who is a *None* smoker. As can be seen in the table above, there are a total of 4 such cases in the two-way table and, thus, there will be four cases like this in the indicator matrix. In all, there will be *193* cases in the indicator or design matrix.

Analyzing the design matrix. If you now analyze this data file (design or indicator matrix) shown above as if it were a two-way frequency table, the results of the correspondence analysis would provide column

coordinates that would allow you to relate the different categories to each other, based on the distances between the row points, i.e., between the individual cases. In fact, the two-dimensional display you would obtain for the column coordinates would look very similar to the combined display for row and column coordinates, if you had performed the simple correspondence analysis on the two-way frequency table (note that the metric will be different, but the relative positions of the points will be very similar).

More than two variables. The approach to analyzing categorical data outlined above can easily be extended to more than two categorical variables. For example, the indicator or design matrix could contain two additional variables – *Male* and *Female*, again coded *0* and *1* – to indicate the subjects' gender; and three variables could be added to indicate to which one of three age groups a case belongs. Thus, in the final display, you could represent the relationships (similarities) between *Gender, Age, Smoking habits*, and *Occupation (Staff Groups)*.

Fuzzy coding. It is not necessary that each case is assigned exclusively to only one category of each categorical variable. Rather than the *0-or-1* coding scheme, you could enter probabilities for membership in a category, or some other measure that represents a fuzzy rule for group membership. Greenacre (1984) discusses different types of coding schemes of this kind. For example, suppose in the example design matrix shown earlier, you had missing data for a few cases regarding their smoking habits. Instead of discarding those cases entirely from the analysis (or creating a new category *Missing data*), you could assign to the different smoking categories proportions (which should add to 1.0) to represent the probabilities that the respective case belongs to the respective category (e.g., you

could enter proportions based on your knowledge about estimates for the national averages for the different categories).

Interpretation of coordinates and other results. To reiterate, the results of a multiple correspondence analysis are identical to the results you would obtain for the column coordinates from a simple correspondence analysis of the design or indicator matrix. Therefore, the interpretation of coordinate values, quality values, cosine²s, and other statistics reported as the results from a multiple correspondence analysis can be interpreted in the same manner as described in the context of the simple correspondence analysis (see the *Overview* of this chapter), however, these statistics pertain to the total inertia associated with the entire design matrix.

Supplementary column points and multiple regression for categorical variables. Another application of the analysis of design matrices via correspondence analysis techniques is that it enables you to perform the equivalent of a multiple regression for categorical variables, by adding supplementary columns to the design matrix. For example, suppose you added to the design matrix shown earlier two columns to indicate whether or not the respective subject had or had not been ill over the past year (i.e., you could add one column *Ill* and another column *Not ill*, and again enter *0*s and *1*s to indicate each subject's health status). If, in a simple correspondence analysis of the design matrix, you added those columns as supplementary columns to the analysis, then 1) the summary statistics for the quality of representation (see the *Overview* of this chapter) for those columns would give you an indication of how well you can explain illness as a function of the other variables in the design matrix, and 2) the display of the column points in the final coordinate system would provide an indication of the nature (e.g., direction) of the relationships between the columns in the design matrix and the column points indicating illness; this technique (adding supplementary points to an MCA analysis) is also sometimes called predictive mapping.

Burt table. The actual computations in multiple correspondence analysis are not performed on a design or indicator matrix (which, potentially, may be very large if there are many cases), but on the inner product of this matrix; this matrix is also called the Burt matrix. With frequency tables, this amounts to tabulating the stacked categories against each other; for example the Burt table for the two-way frequency table presented earlier would look as follows:

Data: Correspondance Analysis Example Spreadsheet 10* (9v by 9c)								
	EMPLOYEE					**SMOKING**		
	1	**2**	**3**	**4**	**5**	**1**	**2**	**3**	**4**
1 Sr. Managers	11	0	0	0	0	4	2	3	2
2 Jr. Managers	0	18	0	0	0	4	3	7	4
3 Sr. Employees	0	0	51	0	0	25	10	12	4
4 Jr. Employees	0	0	0	88	0	18	24	33	13
5 Secretaries	0	0	0	0	25	10	6	7	2
1 Smoking: None	4	4	25	18	10	61	0	0	0
2 Smoking: Light	2	3	10	24	6	0	45	0	0
3 Smoking: Medium	3	7	12	33	7	0	0	62	0
4 Smoking: Heavy	2	4	4	13	2	0	0	0	25

The Burt table has a clearly defined structure. In the case of two categorical variables (shown above), it consists of 4 partitions: 1) the crosstabulation of variable *Employee* against itself, 2) the crosstabulation of variable *Employee* against variable *Smoking*, 3), the crosstabulation of variable *Smoking* against variable *Employee*, and 4) the crosstabulation of variable *Smoking* against itself. Note that the matrix is symmetrical, and that the sum of the diagonal elements in each partition representing the crosstabulation of a variable against itself must be the same (e.g., there were a total of 193 observations in the present example, and hence, the diagonal elements in the crosstabulation

tables of variable *Employee* against itself, and *Smoking* against itself must also be equal to 193).

Note that the off-diagonal elements in the partitions representing the crosstabulations of a variable against itself are equal to *0* in the table shown above. However, this is not necessarily always the case, for example, when the Burt table was derived from a design or indicator matrix that included fuzzy coding of category membership (see above).

Burt Tables

The Burt table is the result of the inner product of a design or indicator matrix, and the multiple correspondence analysis results are identical to the results you would obtain for the column points from a simple correspondence analysis of the indicator or design matrix (see also, *Multiple Correspondence Analysis*, page 136).

For example, suppose you had entered data concerning the *Survival* for different *Age* groups in different *Locations* such as the following:

Data: Correspondance Analysis Example 11 (8v by 10c)								
CASE NO.	**SURVIVAL**		**AGE**			**LOCATION**		
	No	Yes	< 50	50 - 69	> 69	Tokyo	Boston	Glamorgn	
1	0	1	0	1	0	0	0	1	
2	1	0	1	0	0	1	0	0	
3	0	1	0	1	0	0	1	0	
4	0	1	0	0	1	0	0	1	
...	...	...	...	...	...	...	...	...	
...	...	...	...	...	...	...	...	...	
...	...	...	...	...	...	...	...	...	
762	1	0	0	1	0	1	0	0	
763	0	1	1	0	0	0	1	0	
764	0	1	0	1	0	0	0	1	

In this data arrangement, for each case a *1* was entered to indicate to which category, of a particular set of categories, a case belongs (e.g., *Survival*, with the categories *No* and *Yes*). For example, case *1* survived (a *0* was entered for variable *No*, and a 1 was entered for variable *Yes*), case *1* is between age 50 and 69 (a *1* was entered for variable *A50to69*), and was observed in *Glamorgn*). Overall there are *764* observations in the data set.

If you denote the data (design or indicator matrix) shown above as matrix **X**, then matrix product **X'X** is a Burt table; shown in the next illustration is an example of a Burt table that you might obtain in this manner.

Data: Correspondance Analysis Example Spreadsheet 12* (8v by 10c)		SURVIVAL		AGE			LOCATION		
		No	Yes	< 50	50 - 69	> 69	Tokyo	Boston	Glamorgn
Survival: No		210	0	68	93	49	60	82	68
Survival: Yes		0	554	212	258	84	230	171	153
Age: < 50		68	212	280	0	0	151	58	71
Age: 50 - 69		93	258	0	351	0	120	122	109
Age: < 69		49	84	0	0	133	19	73	41
Location: Tokyo		60	230	151	120	19	290	0	0
Location: Boston		82	171	58	122	73	0	253	0
Location: Glamorgn		68	153	71	109	41	0	0	221

The Burt table has a clearly defined structure. Overall, the data matrix is symmetrical. In the case of 3 categorical variables (as shown above), the data matrix consists 3 x 3 = 9 partitions, created by each variable being tabulated against itself, and against the categories of all other variables. Note that the sum of the diagonal elements in each diagonal partition (i.e., where the respective variables are tabulated against themselves) is constant (equal to 764 in this case).

The off-diagonal elements in each diagonal partition in this example are all *0*. If the cases in the design or indicator matrix are assigned to categories via fuzzy coding (i.e., if probabilities are used to indicate likelihood of membership in a category, rather than 0/1 coding to indicate actual membership), then the off-diagonal elements of the diagonal partitions are not necessarily equal to 0.

DATA MINING TECHNIQUES

Data mining is an analytic process designed to explore data (usually large amounts of data – typically business or market related) in search of consistent patterns and/or systematic relationships between variables, and then to validate the findings by applying the detected patterns to new data. The ultimate goal of data mining is prediction, and predictive data mining is the most common type of data mining and one that has the most direct business applications. Data mining has emerged as a result of the fusion of three types of technologies, namely 1) computing power, 2) statistical learning algorithms and tools and finally 3) advances in data gathering and management.

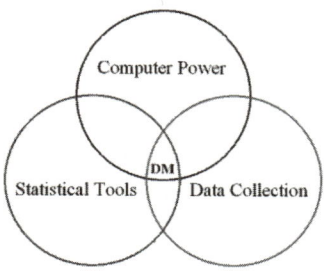

The increase in computing power has made it possible to process and model large amounts of datasets within an affordable amount of time. This significant advantage was brought about by the quick pace of advancements in computer technology and affordability of computer products, such as powerful servers and workstations. Improvements in access to data sources has also gone hand in hand with the emergence of a variety of state of the art statistical and machine learning technologies and algorithms, capable of analyzing and extracting hidden patterns from large datasets in an efficient and accurate manner.

The techniques used by the data mining can be divided into a number of categories depending on the nature of the problem (and indeed the data set) in hand. These include: regression, classification, time series, association and sequence analysis, and clustering. For details of these statistical tasks, see these chapters of this book.

Overview

The process of data mining consists of three stages: 1) the initial exploration of data, 2) model building or pattern identification with validation/verification, and 3) deployment (i.e., the application of the model to new data in order to generate predictions).

Stage 1: Exploration

This stage usually starts with data preparation, which may involve cleaning data, data transformations, selecting subsets of records and, for data sets with large numbers of variables (fields), performing preliminary feature selection operations to bring the number of variables to a manageable range (depending on the statistical methods being considered).

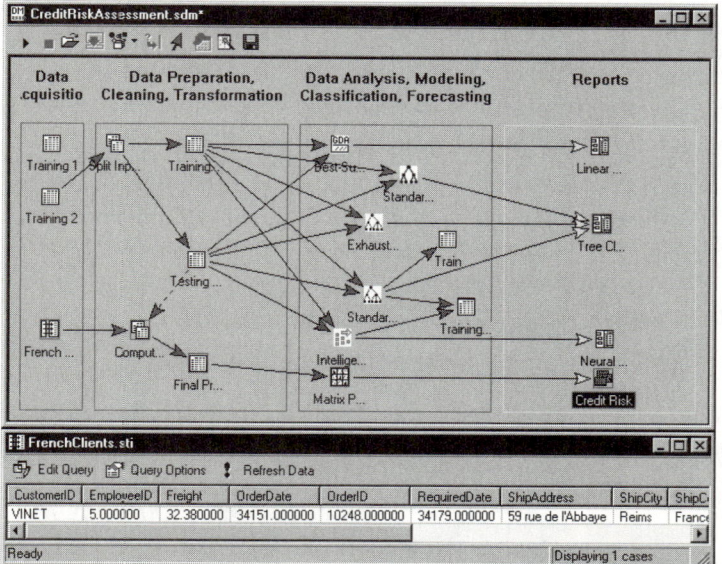

Then, depending on the nature of the analytic problem, this first stage of the data mining process may involve anywhere between a simple choice of straightforward predictors for a regression model, to elaborate exploratory analyses using a wide variety of graphical and statistical methods [see *Exploratory Data Analysis (EDA)*, page 150] in order to identify the most relevant variables and determine the complexity and/or the general nature of models that can be taken into account in the next stage.

Stage 2: Model Building and Validation

This stage involves considering various models and choosing the best one based on their predictive performance (i.e., explaining the variability in question and producing stable results across samples). This may sound like a simple operation, but in fact, it sometimes involves a very elaborate process. There are a variety of techniques developed to achieve that goal, many of which are based on so-called "competitive evaluation of models," that is, applying different models to the same data set and then comparing their performance to choose the best. These techniques – which are often considered the core of predictive data mining – include: bagging (voting, averaging), boosting, stacking (stacked generalizations), and meta-learning.

Stage 3: Deployment

This final stage involves using the model selected as best in the previous stage and applying it to new data in order to generate predictions or estimates of the expected outcome.

The concept of data mining is becoming increasingly popular as a business information management tool where it is expected to reveal knowledge structures that can guide decisions in conditions of limited certainty. Recently, there has been increased interest in developing new analytic techniques specifically designed to address the issues relevant to business data mining (e.g., classification trees), but data mining is still based on the conceptual principles of statistics including the traditional exploratory data analysis and modeling and it shares with them both some components of its general approaches and specific techniques.

However, an important general difference in the focus and purpose between data mining and the traditional exploratory data analysis is that data mining is more oriented toward applications than the basic nature of the underlying phenomena. In other words, data mining is relatively less concerned with identifying the specific relations between the involved variables. For example, uncovering the nature of the underlying functions or the specific types of interactive, multivariate dependencies between variables are not the main goal of data mining. Instead, the focus is on producing a solution that can generate useful predictions. Therefore, data mining accepts, among others, a "black box" approach to data exploration or knowledge discovery and uses not only the traditional exploratory data analysis techniques, but also techniques such as neural networks, tree methods, and support vector machines which can generate valid predictions but are not capable of identifying the specific nature of the interrelations between the variables on which the predictions are based.

Data mining is often considered to be "*a blend of statistics, AI [artificial intelligence], and data base research*" (Pregibon, 1997, p. 8), which until very recently was not commonly recognized as a field of interest for statisticians, and was even considered by some "*a dirty word in Statistics*" (Pregibon, 1997, p. 8). Due to its applied importance, however, the field has emerged as a rapidly growing and major area (also in statistics) where important theoretical advances are being made (see, for example, the recent annual *International Conferences on Knowledge Discovery and Data Mining*, cohosted by the American Statistical Association).

There are numerous books that review the theory and practice of data mining; the following list of books offers a representative sample of influential general books on data mining, representing a variety of approaches and perspectives:

Berry, M., J., A., & Linoff, G., S., (2000). *Mastering data mining*. New York: Wiley.

Edelstein, H., A. (1999). *Introduction to data mining and knowledge discovery (3rd ed)*. Potomac, MD: Two Crows Corp.

Fayyad, U. M., Piatetsky-Shapiro, G., Smyth, P., & Uthurusamy, R. (1996). *Advances in knowledge discovery & data mining*. Cambridge, MA: MIT Press.

Han, J., Kamber, M. (2000). *Data mining: Concepts and Techniques*. New York: Morgan-Kaufman.

Hastie, T., Tibshirani, R., & Friedman, J. H. (2001). *The elements of statistical learning : Data mining, inference, and prediction*. New York: Springer.

Pregibon, D. (1997). *Data Mining*. Statistical Computing and Graphics, 7, 8.

Weiss, S. M., & Indurkhya, N. (1997). *Predictive data mining: A practical guide*. New York: Morgan-Kaufman.

Westphal, C., Blaxton, T. (1998). *Data mining solutions*. New York: Wiley.

Witten, I. H., & Frank, E. (2000). *Data mining*. New York: Morgan-Kaufmann.

Crucial Concepts in Data Mining

Bagging (voting, averaging). The concept of bagging (voting for classification, averaging for regression-type problems with continuous dependent variables of interest) applies to the area of predictive data mining, to combine the predicted classifications (prediction) from multiple models, or from the same type of model for different learning data. It is also used to address the inherent instability of results when applying complex models to relatively small data sets.

Suppose your data mining task is to build a model for predictive classification, and the data set from which to train the model (learning data set, which contains observed classifications) is relatively small. You could repeatedly sub-sample (with replacement) from the data set, and apply, for example, a tree classifier (e.g., CART and CHAID) to the successive samples. In practice, very different trees will often be grown for the different samples, illustrating the instability of models often evident with small data sets. One method of deriving a single prediction (for new observations) is to use all trees found in the different samples, and to apply some simple voting. The final classification is the one most often predicted by the different trees.

Note that some weighted combination of predictions (weighted vote, weighted average) is also possible, and commonly used. A sophisticated (machine learning) algorithm for generating weights for weighted prediction or voting is the boosting procedure.

Boosting. The concept of boosting applies to the area of predictive data mining, to generate multiple models or classifiers (for prediction or classification), and to derive weights to combine the predictions from those models into a single prediction or predicted classification.

A simple algorithm for boosting works like this: Start by applying some method (e.g., a tree classifier such as CART or CHAID) to the learning data, where each observation is assigned an equal weight. Compute the predicted classifications, and apply weights to the observations in the learning sample that are inversely proportional to the accuracy of the classification. In other words, assign greater weight to those observations that were difficult to classify (where the misclassification rate was high), and lower weights to those that were easy to classify (where the misclassification rate was low). In the context of CART for example, different misclassification costs (for the different classes) can be applied, inversely proportional to the accuracy of prediction in each class. Then apply the classifier again to the weighted data (or with different misclassification costs), and continue with the next iteration (application of the analysis method for classification to the re-weighted data).

Boosting will generate a sequence of classifiers, where each consecutive classifier in the sequence is an expert in classifying observations that were not well classified by those preceding it. During deployment (for prediction or classification of new cases), the predictions from the different classifiers can then be combined (e.g., via voting, or some weighted voting procedure) to derive a single best prediction or classification.

Note that boosting can also be applied to learning methods that do not explicitly support weights or misclassification costs. In that case, random sub-sampling can be applied to the learning data in the successive steps of the iterative boosting procedure, where the probability for selection of an observation into the subsample is inversely proportional to the accuracy of the prediction for that observation in the previous iteration (in the sequence of iterations of the boosting procedure).

CRISP. See *Models for data mining*, page 147.

Data preparation (in data mining). Data preparation and cleaning is an often neglected but extremely important step in the data mining process. The old saying "garbage-in-garbage-out" is particularly applicable to the typical data mining projects where large data sets collected via some automatic methods (e.g., via the Web) serve as the input into the analyses. Often, the method by which the data were gathered was not tightly controlled, and so the data may contain out-of-range values (e.g., *Income: -100*), impossible data combinations (e.g., *Gender: Male, Pregnant: Yes*), etc. Analyzing data that has not been carefully screened for such problems can produce highly misleading results, in particular in predictive data mining.

Data reduction (for data mining). The term data reduction in the context of data mining is usually applied to projects where the goal is to aggregate or amalgamate the information contained in large data sets into manageable (smaller) information nuggets. Data reduction methods can include simple tabulation, aggregation (computing descriptive statistics) or more sophisticated techniques such as clustering, principal components analysis, etc. See also, *Predictive data mining*, page 148, and *Drill-down analysis*, page 146.

Deployment. The concept of deployment in predictive data mining refers to the application of a model for prediction or classification to new data. After a satisfactory model or set of models has been identified (trained) for a particular application, you usually want to deploy those models so that predictions or predicted classifications can quickly be obtained for new data. For example, a credit card company may want to deploy a trained model or set of models (e.g., neural networks, meta-learner) to quickly identify transactions that have a high probability of being fraudulent.

Drill-down analysis. The concept of drill-down analysis applies to the area of data mining to denote the interactive exploration of data of large databases in particular. The process of drill-down analyses begins by considering some simple breakdowns of the data by a few variables of interest (e.g., gender, geographic region, etc.). Various statistics, tables, histograms, and other graphical summaries can be computed for each group. Next you may want to drill-down to expose and further analyze the data underneath one of the categorizations, for example, you might want to further review the data for males from the mid-west. Again, various statistical and graphical summaries can be computed for those cases only, which might suggest further breakdowns by other variables (e.g., income, age, etc.). At the lowest (bottom) level are the raw data. For example, you may want to review the addresses of male customers from one region, for a certain income group, etc., and to offer to those customers some particular services of particular utility to that group.

Feature selection. One of the preliminary stages in predictive data mining, when the data set includes more variables than could be included (or would be efficient to include) in the actual model building phase (or even in initial exploratory operations), is to select predictors from a large list of candidates. For example, when data are collected via automated (computerized) methods, it is not uncommon that measurements are recorded for thousands or hundreds of thousands (or more) of predictors. The standard analytic methods for predictive data mining, such as neural networks, generalized linear models, or general linear models become impractical when the number of predictors is large.

Feature selection selects a subset of predictors from a large list of candidate predictors without assuming that the relationships between the predictors and the dependent or outcome variables of interest are linear, or even monotone. Therefore, this is used as a pre-processor for predictive data mining, to select manageable sets of predictors that are likely related to the dependent (outcome) variables of interest, for further analyses with any of the other methods for regression and classification.

Machine learning. Machine learning, computational learning theory, and similar terms are often used in the context of data mining, to denote the application of generic model-fitting or classification algorithms for predictive data mining. Unlike traditional statistical data analysis, which is usually concerned with the estimation of population parameters by statistical inference, the emphasis in data mining (and machine learning) is usually on the accuracy of prediction (predicted classification), regardless of whether or not the models or techniques that are used to generate the prediction are interpretable or open to simple explanation.

Good examples of this type of technique often applied to predictive data mining are neural networks or meta-learning techniques such as boosting, etc. These methods usually involve the fitting of very complex generic models, that are not related to any reasoning or theoretical understanding of underlying causal processes; instead, these techniques can be shown to generate accurate predictions or classification in cross-validation samples.

Meta-learning. The concept of meta-learning applies to the area of predictive data mining, to combine the predictions from multiple models.

It is particularly useful when the types of models included in the project are very different. In this context, this procedure is also referred to as stacking (stacked generalization).

Suppose your data mining project includes tree classifiers, such as CART and CHAID, linear discriminant analysis [e.g., see Chapter 17 – *General Discriminant Analysis (GDA)*], and neural networks. Each computes predicted classifications for a cross-validation sample, from which overall goodness-of-fit statistics (e.g., misclassification rates) can be computed. Experience has shown that combining the predictions from multiple methods often yields more accurate predictions than can be derived from any one method (e.g., see Witten and Frank, 2000). The predictions from different classifiers can be used as input into a meta-learner, which will attempt to combine the predictions to create a final best-predicted classification. For example, the predicted classifications from a tree classifier, a linear model, and a neural network can be used as input variables into a neural network meta-classifier, which will attempt to "learn" from the data how to combine the predictions from the different models to yield maximum classification accuracy.

You can apply meta-learners to the results from different meta-learners to create "meta-meta"-learners, and so on; however, in practice such exponential increase in the amount of data processing, in order to derive an accurate prediction, will yield less and less marginal utility.

Models for data mining. In the business environment, complex data mining projects may require the coordinate efforts of various experts, stakeholders, or departments throughout an entire organization. In the data mining literature, various general frameworks have been proposed to serve as blueprints for how to organize the process of gathering data, analyzing data, disseminating results, implementing results, and monitoring improvements.

One such model, CRISP (Cross-Industry Standard Process for data mining) was proposed in the mid-1990s by a European consortium of companies to serve as a non-proprietary standard process model for data mining. This general approach postulates the following (perhaps not particularly controversial) general sequence of steps for data mining projects:

Business Understanding   ↔   Data Understanding
↓
Data Preparation   ↔   Modeling
↓
Evaluation
↓
Deployment

Another approach – the Six Sigma methodology – is a well-structured, data-driven methodology for eliminating defects, waste, or quality control problems of all kinds in manufacturing, service delivery, management, and other business activities. This model has recently become very popular (due to its successful implementations) in various American industries, and it appears to gain favor worldwide. It postulated a sequence of, so-called, DMAIC steps –

Define → Measure → Analyze → Improve → Control

– that grew up from the manufacturing, quality improvement, and process control traditions and is particularly well suited to production environments (including production of services, i.e., service industries).

Another framework of this kind (actually somewhat similar to Six Sigma) is the approach proposed by SAS Institute called SEMMA –

Sample → Explore → Modify → Model → Assess

– which is focused more on the technical activities typically involved in a data mining project.

All of these models are concerned with the process of how to integrate data mining methodology into an organization, how to convert data into information, how to involve important stake-holders, and how to disseminate the information in a form that can easily be converted by stake-holders into resources for strategic decision making.

Some software tools for data mining are specifically designed and documented to fit into one of these specific frameworks. Some data mining software has the general underlying philosophy to provide a flexible data mining workbench that can be integrated into any organization, industry, or organizational culture, regardless of the general data mining process-model that the organization chooses to adopt.

Predictive data mining. The term predictive data mining is usually applied to identify data mining projects with the goal to identify a statistical model or set of models that can be used to predict some response of interest. For example, a credit card company may want to engage in predictive data mining to derive a (trained) model or set of models (e.g., neural networks, meta-learner) that can quickly identify transactions that have a high probability of being fraudulent. Other types of data mining projects may be more exploratory in nature (e.g., to identify cluster or segments of customers), in which case drill-down descriptive and exploratory methods would be applied. Data reduction is another possible objective for data mining (e.g., to aggregate or amalgamate information contained in large data sets into useful and manageable chunks).

SEMMA. See *Models for data mining*, page 147.

Stacked generalization. See *Stacking* below.

Stacking (stacked generalization). The concept of stacking (short for stacked generalization) applies to the area of predictive data mining, to combine the predictions from multiple models. It is particularly useful when the types of models included in the project are very different.

Suppose a data mining project includes tree classifiers, such as CART or CHAID, linear discriminant analysis (e.g., see GDA), and neural networks. Each computes predicted classifications for a cross-validation sample, from which overall goodness-of-fit statistics (e.g., misclassification rates) can be computed. Experience has shown that combining the predictions from multiple methods often yields more accurate predictions than can be derived from any one method (e.g., see Witten and Frank, 2000). In stacking, the predictions from different classifiers are used as input into a meta-learner, which attempts to combine the predictions to create a final best-predicted classification. So, for example, the predicted classifications from the tree classifiers, linear model, and the neural network classifier(s) can be used as input variables into a neural network meta-classifier, which will attempt to learn from the data how to combine the predictions from the different models to yield maximum classification accuracy.

Other methods for combining the prediction from multiple models or methods (e.g., from multiple data sets used for learning) are boosting and bagging (voting).

Text mining. While data mining is typically concerned with the detection of patterns in numeric data, important (e.g., critical to business) information very often is stored in the form of text. Unlike numeric data, text is often amorphous, and difficult to deal with. Text mining generally consists of the analysis of (multiple) text documents by extracting key phrases, concepts, etc., and the preparation of the text processed in that manner for further analyses with numeric data mining techniques (e.g., to determine co-occurrences of concepts, key phrases, names, addresses, product names, etc.).

Voting. See *Bagging*, page 144.

Data Warehousing

Data warehousing is the process of organizing the storage of large, multivariate data sets in a way that facilitates the retrieval of information for analytic purposes.

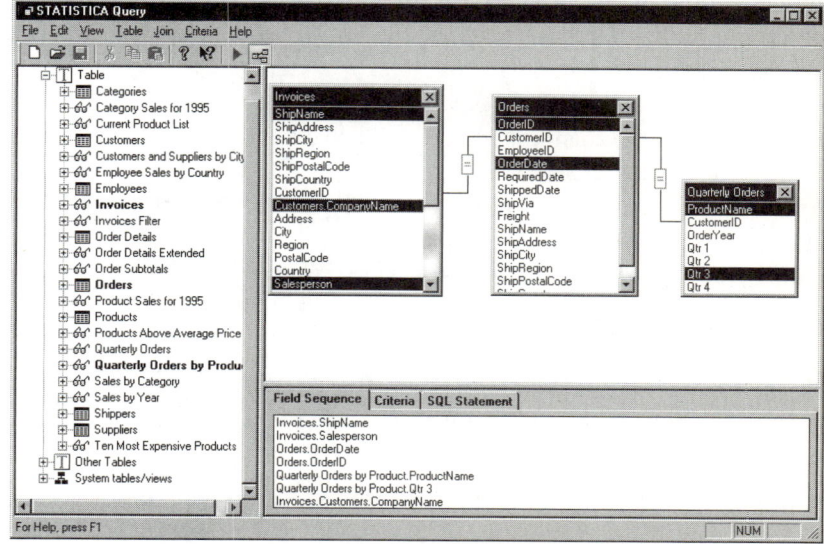

The most efficient data warehousing architecture will be capable of incorporating or at least referencing all data available in the relevant enterprise-wide information management systems, using designated technology suitable for corporate data base management (e.g., Oracle, Sybase, or MS SQL Server).

On-Line Analytic Processing (OLAP)

The term on-line analytic processing – OLAP (or Fast Analysis of Shared Multidimensional Information – FASMI) refers to technology that enables users of multidimensional databases to generate on-line descriptive or comparative summaries (views) of data and other analytic queries. Note that despite its name, analyses referred to as OLAP do not need to be performed truly on-line (or in real-time); the term applies to analyses of multidimensional databases (that may, obviously, contain dynamically updated information) through efficient multidimensional queries that reference various types of data. OLAP facilities can be integrated into corporate (enterprise-wide) database systems and they enable analysts and managers to monitor the performance of the business (e.g., such as various aspects of the manufacturing process or numbers and types of completed transactions at different locations) or the market.

The final result of OLAP techniques can be very simple (e.g., frequency tables, descriptive statistics, simple cross-tabulations) or more complex (e.g., they may involve seasonal adjustments, removal of outliers,

and other forms of cleaning the data). Although data mining techniques can operate on any kind of unprocessed or even unstructured information, they can also be applied to the data views and summaries generated by OLAP to provide more in-depth and often more multidimensional knowledge. In this sense, data mining techniques could be considered to represent either a different analytic approach (serving different purposes than OLAP) or as an analytic extension of OLAP.

Exploratory Data Analysis (EDA) and Data Mining Techniques

EDA vs. Hypothesis Testing

As opposed to traditional hypothesis testing designed to verify *a priori* hypotheses about relations between variables (e.g., "There is a positive correlation between the age of a person and his/her risk-taking disposition"), exploratory data analysis is used to identify systematic relations between variables when there are no (or not complete) *a priori* expectations as to the nature of those relations.

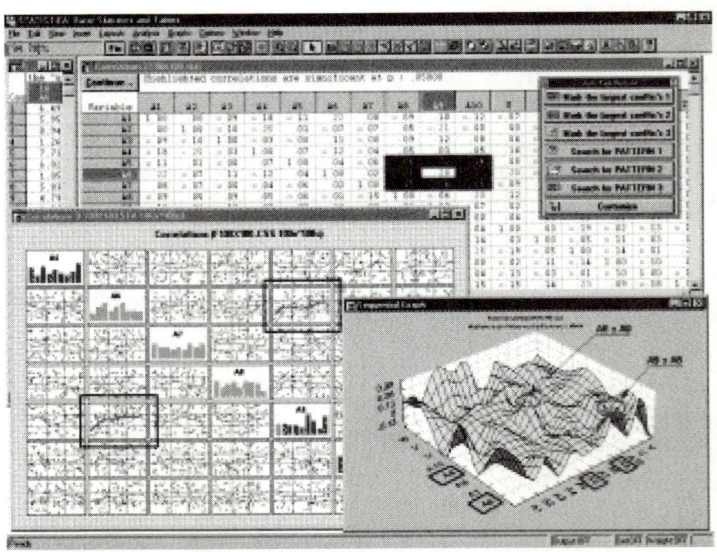

In a typical exploratory data analysis process, many variables are taken into account and compared, using a variety of techniques in the search for systematic patterns.

Computational EDA Techniques

Computational exploratory data analysis methods include both simple basic statistics and more advanced, designated multivariate exploratory techniques designed to identify patterns in multivariate data sets.

Basic statistical exploratory methods. The basic statistical exploratory methods include such techniques as examining distributions of variables (e.g., to identify highly skewed or non-normal, such as bi-modal patterns), reviewing large correlation matrices for coefficients that meet certain thresholds (see example above), or examining multi-way frequency tables (e.g., slice by slice systematically reviewing combinations of levels of control variables).

Multivariate exploratory techniques. Multivariate exploratory techniques designed specifically to identify patterns in multivariate (or univariate, such as sequences of measurements) data sets include: cluster analysis, factor analysis, discriminant function analysis, multidimensional scaling, log-linear analysis, canonical correlation, stepwise linear and nonlinear (e.g., logit) regression, correspondence analysis, time series analysis, and classification trees.

Graphical (Data Visualization) EDA Techniques

A large selection of powerful exploratory data analytic techniques is also offered by graphical data visualization methods that can identify relations, trends, and biases hidden in unstructured data sets.

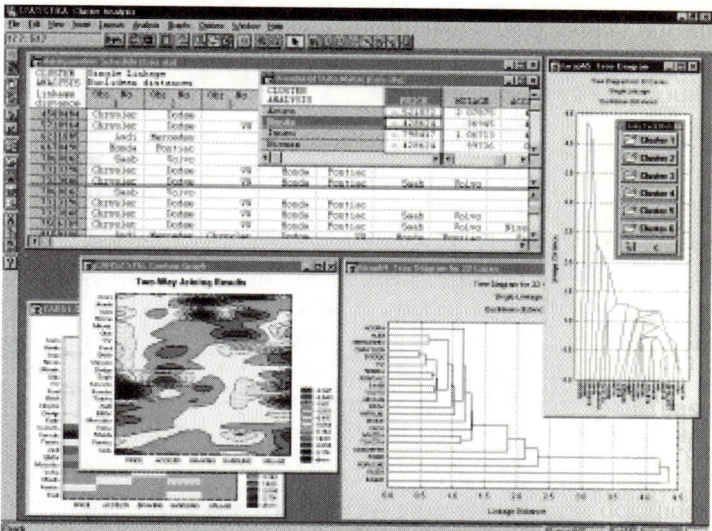

Brushing. Perhaps the most common and historically first widely used technique explicitly identified as graphical exploratory data analysis is brushing, an interactive method that enables you to select on-screen specific data points or subsets of data and identify their (e.g., common) characteristics, or to examine their effects on relations between relevant variables.

Those relations between variables can be visualized by fitted functions (e.g., 2D lines or 3D surfaces) and their confidence intervals, thus, for example, you can examine changes in those functions by interactively (temporarily) removing or adding specific subsets of data.

For example, one of many applications of the brushing technique is to select (i.e., highlight) in a matrix scatterplot all data points that belong to a certain category (e.g., a medium income level, see the highlighted subset in the fourth component graph of the first row in the next illustration) in order to examine how those specific observations contribute to relations between other variables in the same data set (e.g., the correlation between the *debt* and *assets* in the current example).

If the brushing facility supports features such as animated brushing or automatic function re-fitting, you can define a dynamic brush that would move over the consecutive ranges of a criterion variable [e.g., *income* measured on a continuous scale or a discrete (3-level) scale as in the following illustration] and examine the dynamics of the contribution of the criterion variable to the relations between other relevant variables in the same
data set.

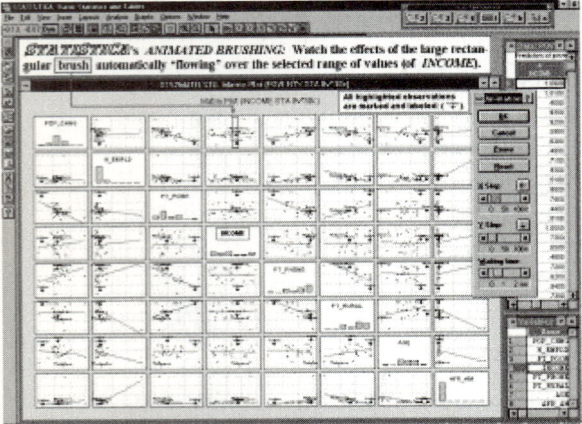

Other graphical EDA techniques. Other graphical exploratory analytic techniques include function fitting and plotting, data smoothing, overlaying and merging of multiple displays, categorizing data, splitting/merging subsets of

data in graphs, aggregating data in graphs, identifying and marking subsets of data that meet specific conditions, icon plots,

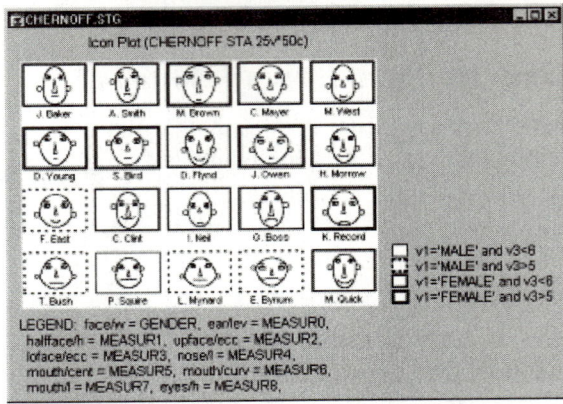

shading, plotting confidence intervals and confidence areas (e.g., ellipses),

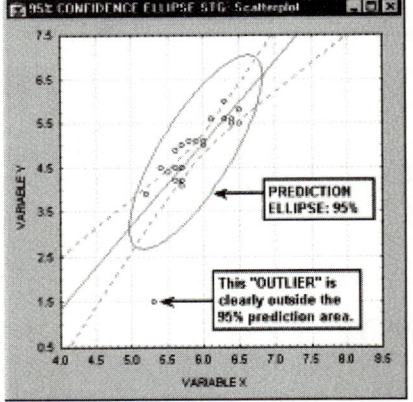

generating tessellations, spectral planes,

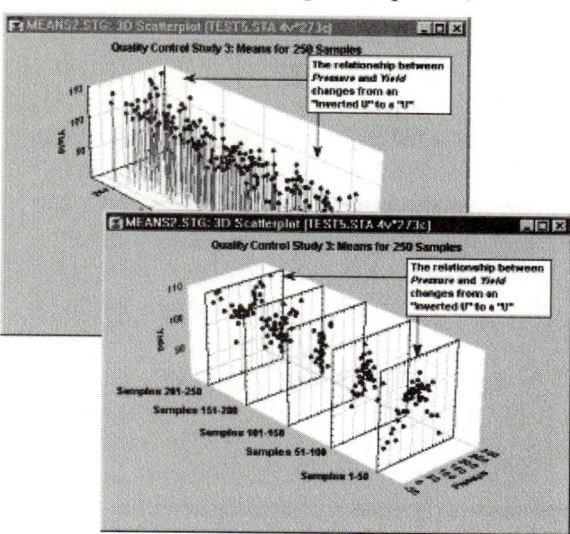

integrated layered compressions,

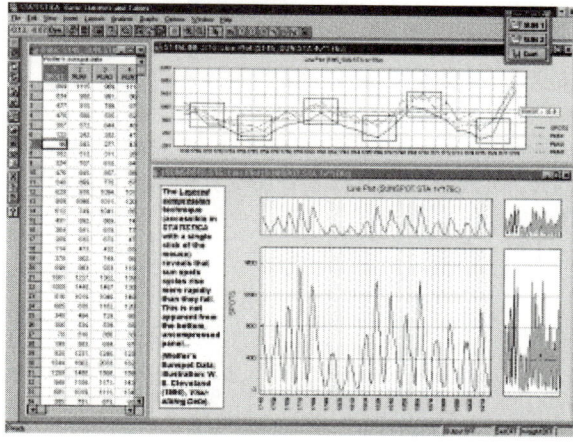

and projected contours, data image reduction techniques, interactive (and continuous) rotation,

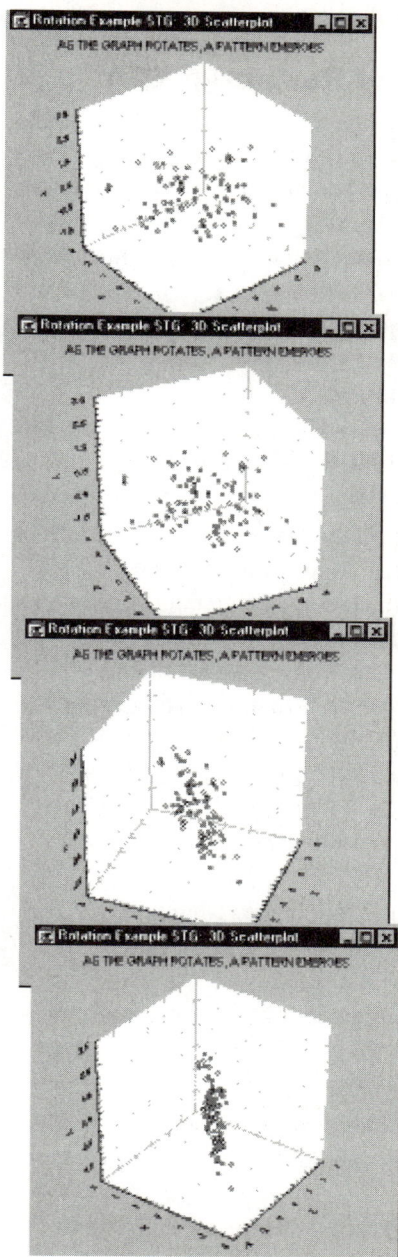

with animated stratification (cross-sections) of 3D displays, and selective highlighting of specific series and blocks of data.

Verification of Results of EDA

The exploration of data can only serve as the first stage of data analysis and its results can be treated as tentative at best as long as they are not confirmed, e.g., cross validated, using a different data set (or and independent subset). If the result of the exploratory stage suggests a particular model, its validity can be verified by applying it to a new data set and testing its fit (e.g., testing its predictive validity).

Popular Techniques Used in Data Mining

Data mining takes advantage of a wide variety of statistical tools and algorithms for detecting and extracting patterns hidden in the data. These tools include a broad range of methods and techniques with capabilities ranging from performing regression tasks to classification, time series, clustering, and rule extraction. The most commonly used of these statistical tools include:

- Neural Networks (see Chapter 27 – *Neural Networks*)
- Decision Trees Models [see Chapter 7 – *CHAID* and Chapter 8 – *Classification and Regression Trees (CART)*]
- Nearest Neighbor Methods (see Chapter 10 – *Cluster Analysis* and Chapter 23 – *Machine Learning*)
- Association and Sequence Models
- *k*-Means Clustering (see Chapter 10 – *Cluster Analysis*)
- Multivariate Adaptive Splines [see Chapter 24 – *Multivariate Adaptive Regression Splines (MARSplines)*]

- Support Vector Machines (see Chapter 23 – *Machine Learning*)
- Naïve Bayes (see Chapter 23 – *Machine Learning*)

Although any of these data mining tools can be applied to a wide variety of analysis, research and problem solving, it should be noted that the effectiveness and suitability of each model depends very much on the nature of the problem (data set) at hand. While some models might be particularly suited for analyzing certain types of data sets, others may not perform so well on the same problem. For example, neural networks might be a suitable choice when the relationship between the variables of the data set is highly nonlinear. Alternately, for localized type of inference and handling data sets with exceptionally large number of variables, decision trees and multivariate adaptive splines may be better choices. It is the task of data mining not only to perform explanatory data analysis but also to select and find model(s) that explain and predict the data best. For further details of the above statistical models, refer to the appropriate chapters of this book.

DISCRIMINANT FUNCTION ANALYSIS

Discriminant function analysis is used to determine which variables discriminate between two or more naturally occurring groups. For example, an educational researcher may want to investigate which variables discriminate between high school graduates who decide 1) to go to college, 2) to attend a trade or professional school, or 3) to seek no further training or education. For that purpose the researcher could collect data on numerous variables prior to students' graduation. After graduation, most students will naturally fall into one of the three categories. Discriminant analysis could then be used to determine which variable(s) are the best predictors of students' subsequent educational choice.

A medical researcher may record different variables relating to patients' backgrounds in order to learn which variables best predict whether a patient is likely to recover completely (group 1), partially (group 2), or not at all (group 3). A biologist could record different characteristics of similar types (groups) of flowers, and then perform a discriminant function analysis to determine the set of characteristics that allows for the best discrimination between the types.

Computational Approach

Computationally, discriminant function analysis is very similar to analysis of variance (*ANOVA*). Let's consider a simple example. Suppose we measure height in a random sample of 50 males and 50 females. Females are, on the average, not as tall as males, and this difference will be reflected in the difference in means (for the variable *Height*). Therefore, variable height allows us to discriminate between males and females with a better than chance probability: if a person is tall, then he is likely to be a male, if a person is short, then she is likely to be a female.

We can generalize this reasoning to groups and variables that are less trivial. For example, suppose we have two groups of high school graduates: Those who choose to attend college after graduation and those who do not. We could have measured students' stated intention to continue on to college one year prior to graduation. If the means for the two groups (those who actually went to college and those who did not) are different, we can say that intention to attend college as stated one year prior to graduation allows us to discriminate between those who are and are not college bound (and this information can be used by career counselors to provide the appropriate guidance to the respective students).

To summarize the discussion so far, the basic idea underlying discriminant function analysis is to determine whether groups differ with regard to the mean of a variable, and then to use that variable to predict group membership (e.g., of new cases).

Analysis of variance. Stated in this manner, the discriminant function problem can be rephrased as a one-way analysis of variance (ANOVA) problem. Specifically, you can ask whether two or more groups are significantly different from

each other with respect to the mean of a particular variable. To learn more about how you can test for the statistical significance of differences between means in different groups, you may want to read Chapter 3 – *ANOVA/ MANOVA*, page 41. However, it should be clear that, if the means for a variable are significantly different in different groups, then we can say that this variable discriminates between the groups.

In the case of a single variable, the final significance test of whether or not a variable discriminates between groups is the *F* test. As described in Chapter 1 – *Elementary Concepts in Statistics* and Chapter 3 – *ANOVA /MANOVA*, *F* is essentially computed as the ratio of the between-groups variance in the data over the pooled (average) within-group variance. If the between-group variance is significantly larger then there must be significant differences between means.

Multiple variables. Usually, you include several variables in a study in order to see which one(s) contribute to the discrimination between groups. In that case, we have a matrix of total variances and covariances; likewise, we have a matrix of pooled within-group variances and covariances. We can compare those two matrices via multivariate *F* tests in order to determined whether or not there are any significant differences (with regard to all variables) between groups. This procedure is identical to multivariate analysis of variance, or MANOVA. As in MANOVA, you could first perform the multivariate test and, if statistically significant, proceed to see which of the variables have significantly different means across the groups. Thus, even though the computations with multiple variables are more complex, the principal reasoning still applies, namely, that we are looking for variables that discriminate between groups, as evident in observed mean differences.

Stepwise Discriminant Analysis

Probably the most common application of discriminant function analysis is to include many measures in the study, in order to determine the ones that discriminate between groups. For example, an educational researcher interested in predicting high school graduates' choices for further education would probably include as many measures of personality, achievement motivation, academic performance, etc. as possible in order to learn which one(s) offer the best prediction.

Model. Put another way, we want to build a model of how we can best predict to which group a case belongs. In the following discussion we will use the term "in the model" in order to refer to variables that are included in the prediction of group membership, and we will refer to variables as being "not in the model" if they are not included.

Forward stepwise analysis. In stepwise discriminant function analysis, a model of discrimination is built step-by-step. Specifically, at each step all variables are reviewed and evaluated to determine which one will contribute most to the discrimination between groups. That variable will then be included in the model, and the process starts again.

Backward stepwise analysis. You can also step backwards; in that case all variables are included in the model and then, at each step, the variable that contributes least to the prediction of group membership is eliminated. Thus, as the result of a successful discriminant function analysis, you would only keep the important variables in the model, that is, those variables

that contribute the most to the discrimination between groups.

F to enter, F to remove. The stepwise procedure is guided by the respective F to enter and F to remove values. The F value for a variable indicates its statistical significance in the discrimination between groups, that is, it is a measure of the extent to which a variable makes a unique contribution to the prediction of group membership. If you are familiar with stepwise multiple regression procedures, you can interpret the F to enter/remove values in the same way as in stepwise regression.

Capitalizing on chance. A common misinterpretation of the results of stepwise discriminant analysis is to take statistical significance levels at face value. By nature, the stepwise procedures will capitalize on chance because they pick and choose the variables to be included in the model so as to yield maximum discrimination. Thus, when using the stepwise approach the researcher should be aware that the significance levels do not reflect the true *alpha* error rate, that is, the probability of erroneously rejecting H_0 (the null hypothesis that there is no discrimination between groups).

Interpreting a Two-Group Discriminant Function

In the two-group case, discriminant function analysis can also be thought of as (and is analogous to) multiple regression (see Chapter 26 – *Multiple Linear Regression*); the two-group discriminant analysis is also called Fisher linear discriminant analysis after Fisher, 1936; computationally all of these approaches are analogous). If we code the two groups in the analysis as *1* and *2*, and use that variable as the dependent variable in a multiple regression analysis, we would get results that are

analogous to those we would obtain via discriminant analysis. In general, in the two-group case we fit a linear equation of the type:

Group = a + b$_1$*x$_1$ + b$_2$*x$_2$ + ... + b$_m$*a$_m$

where *a* is a constant and *b$_1$* through *b$_m$* are regression coefficients. The interpretation of the results of a two-group problem is straightforward and closely follows the logic of multiple regression: Those variables with the largest (standardized) regression coefficients are the ones that contribute most to the prediction of group membership.

Discriminant Functions for Multiple Groups

When there are more than two groups, we can estimate more than one discriminant function such as the one presented above. For example, when there are three groups, we could estimate 1) a function for discriminating between group 1 and groups 2 and 3 combined, and 2) another function for discriminating between group 2 and group 3. For example, we could have one function that discriminates between those high school graduates that go to college and those who do not (but rather get a job or go to a professional or trade school), and a second function to discriminate between those graduates that go to a professional or trade school versus those who get a job. The *b* coefficients in those discriminant functions could then be interpreted as before.

Canonical analysis. When actually performing a multiple group discriminant analysis, we do not have to specify how to combine groups so as to form different discriminant functions. Rather, you can automatically determine some optimal combination of variables so that the first function provides the most overall discrimination between groups, the second provides second most, and so on. Moreover, the functions will be independent or orthogonal, that is, their contributions to the discrimination between groups will not overlap. Computationally, you will perform a canonical correlation analysis (see Chapter 6 – *Canonical Analysis*) that will determine the successive functions and canonical roots (the term root refers to the eigenvalues that are associated with the respective canonical function). The maximum number of functions will be equal to the number of groups minus one, or the number of variables in the analysis, whichever is smaller.

Interpreting the discriminant functions. As before, we will get *b* (and standardized *beta*) coefficients for each variable in each discriminant (now also called canonical) function, and they can be interpreted as usual: the larger the standardized coefficient, the greater is the contribution of the respective variable to the discrimination between groups. (Note that we could also interpret the structure coefficients; see below.) However, these coefficients do not tell us between which of the groups the respective functions discriminate. We can identify the nature of the discrimination for each discriminant (canonical) function by looking at the means for the functions across groups. We can also visualize how the two functions discriminate between groups by plotting the individual scores for the two discriminant functions (see the example graph in the next illustration).

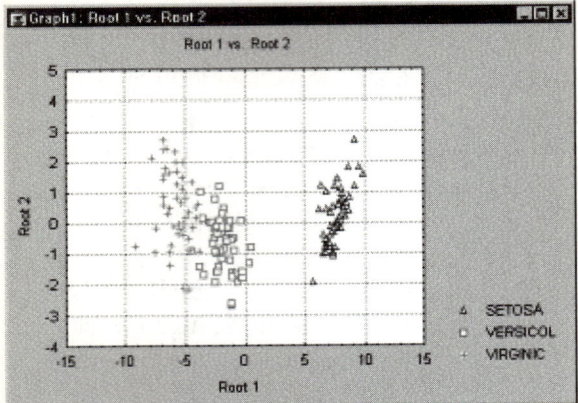

In this example, *Root* (function) *1* seems to discriminate mostly between groups *Setosa*, and *Virginic* and *Versicol* combined. In the vertical direction (*Root 2*), a slight trend of *Versicol* points to fall below the center line (*0*) is apparent.

Factor structure matrix. Another way to determine which variables mark or define a particular discriminant function is to look at the factor structure. The factor structure coefficients are the correlations between the variables in the model and the discriminant functions; if you are familiar with factor analysis (see Chapter 16 – *Factor Analysis*), you can think of these correlations as factor loadings of the variables on each discriminant function.

Some authors have argued that these structure coefficients should be used when interpreting the substantive meaning of discriminant functions. The reasons given by those authors are that 1) supposedly the structure coefficients are more stable, and 2) they allow for the interpretation of factors (discriminant functions) in the manner that is analogous to factor analysis. However, subsequent Monte Carlo research (Barcikowski & Stevens, 1975; Hubert, 1975) has shown that the discriminant function coefficients and the structure coefficients are about equally unstable, unless

the n is fairly large (e.g., if there are 20 times more cases than there are variables). The most important thing to remember is that the discriminant function coefficients denote the unique (partial) contribution of each variable to the discriminant function(s), while the structure coefficients denote the simple correlations between the variables and the function(s). If you want to assign substantive meaningful labels to the discriminant functions (akin to the interpretation of factors in factor analysis), the structure coefficients should be used (interpreted); if you want to learn what each variable's unique contribution to the discriminant function is, use the discriminant function coefficients (weights).

Significance of discriminant functions. You can test the number of roots that add significantly to the discrimination between group. Only those found to be statistically significant should be used for interpretation; non-significant functions (roots) should be ignored.

Summary. To summarize, when interpreting multiple discriminant functions, which arise from analyses with more than two groups and more than one variable, you would first test the different functions for statistical significance, and only consider the significant functions for further examination. Next, we would look at the standardized *b* coefficients for each variable for each significant function. The larger the standardized *b* coefficient, the larger is the respective variable's unique contribution to the discrimination specified by the respective discriminant function. In order to derive substantive meaningful labels for the discriminant functions, you can also examine the factor structure matrix with the correlations between the variables and the discriminant functions. Finally, we would look at the means for the significant discriminant functions in

order to determine between which groups the respective functions seem to discriminate.

Assumptions

As mentioned earlier, discriminant function analysis is computationally very similar to MANOVA, and all assumptions for MANOVA mentioned in ANOVA/MANOVA apply. In fact, you can use the wide range of diagnostics and statistical tests of assumption that are available to examine your data for the discriminant analysis.

Normal distribution. It is assumed that the data (for the variables) represent a sample from a multivariate normal distribution. You can examine whether or not variables are normally distributed with histograms of frequency distributions. However, note that violations of the normality assumption are usually not fatal, meaning, that the resultant significance tests are still trustworthy. You can use specific tests for normality in addition to graphs.

Homogeneity of variances/covariances. It is assumed that the variance/covariance matrices of variables are homogeneous across groups. Again, minor deviations are not that important; however, before accepting final conclusions for an important study it is probably a good idea to review the within-groups variances and correlation matrices. In particular a scatterplot matrix can be produced and can be very useful for this purpose. When in doubt, try re-running the analyses excluding one or two groups that are of less interest. If the overall results (interpretations) hold up, you probably do not have a problem. You can also use the numerous tests available to examine whether or not this assumption is violated in your data. However, as mentioned in Chapter 3 – *ANOVA/MANOVA*, the multivariate box *M* test for homogeneity of

variances/covariances is particularly sensitive to deviations from multivariate normality, and should not be taken too seriously.

Correlations between means and variances. The real threat to the validity of significance tests occurs when the means for variables across groups are correlated with the variances (or standard deviations). Intuitively, if there is large variability in a group with particularly high means on some variables, those high means are not reliable. However, the overall significance tests are based on pooled variances, that is, the average variance across all groups. Thus, the significance tests of the relatively larger means (with the large variances) would be based on the relatively smaller pooled variances, resulting erroneously in statistical significance. In practice, this pattern may occur if one group in the study contains a few extreme outliers, who have a large impact on the means, and also increase the variability. To guard against this problem, inspect the descriptive statistics, that is, the means and standard deviations or variances for such a correlation.

The matrix ill-conditioning problem. Another assumption of discriminant function analysis is that the variables that are used to discriminate between groups are not completely redundant. As part of the computations involved in discriminant analysis, you will invert the variance/covariance matrix of the variables in the model. If any one of the variables is completely redundant with the other variables then the matrix is said to be ill-conditioned, and it cannot be inverted. For example, if a variable is the sum of three other variables that are also in the model, the matrix is ill-conditioned.

Tolerance values. In order to guard against matrix ill-conditioning, constantly check the so-called tolerance value for each variable. This tolerance value is computed as *1 minus R-square* of the respective variable with all other variables included in the current model. Thus, it is the proportion of variance that is unique to the respective variable. You can also refer to Chapter 26 – *Multiple Linear Regression*, to learn more about multiple regression and the interpretation of the tolerance value. In general, when a variable is almost completely redundant (and, therefore, the matrix ill-conditioning problem is likely to occur), the tolerance value for that variable will approach 0.

Classification

Another major purpose to which discriminant analysis is applied is the issue of predictive classification of cases. Once a model has been finalized and the discriminant functions have been derived, how well can we predict to which group a particular case belongs?

A priori and post hoc predictions. Before going into the details of different estimation procedures, we want to make sure that this difference is clear. Obviously, if we estimate, based on some data set, the discriminant functions that best discriminate between groups, and then use the same data to evaluate how accurate our prediction is, we are very much capitalizing on chance. In general, you will always get a worse classification when predicting cases that were not used for the estimation of the discriminant function. Put another way, post hoc predictions are always better than *a priori* predictions. (The trouble with predicting the future *a priori* is that you do not know what will happen; it is much easier to find ways to predict what we already know has happened.) Therefore, you should never base your confidence regarding the correct classification of future observations on the same data set from which the discriminant functions were derived; rather, if you want to classify cases productively, it is necessary to collect new data to try out (cross-validate) the utility of the discriminant functions.

Classification functions. These are not to be confused with the discriminant functions. The classification functions can be used to determine to which group each case most likely belongs. There are as many classification functions as there are groups. Each function enables us to compute classification scores for each case for each group, by applying the formula:

$$S_i = c_i + w_{i1}{}^*x_1 + w_{i2}{}^*x_2 + \ldots + w_{im}{}^*x_m$$

In this formula, the subscript i denotes the respective group; the subscripts *1, 2, ..., m* denote the m variables; c_i is a constant for the i'th group, w_{ij} is the weight for the j'th variable in the computation of the classification score for the i'th group; x_j is the observed value for the respective case for the j'th variable. S_i is the resultant classification score.

We can use the classification functions to directly compute classification scores for some new observations.

Classification of cases. Once we have computed the classification scores for a case, it is easy to decide how to classify the case. In general, we classify the case as belonging to the group for which it has the highest classification score (unless the *a priori* classification probabilities are widely disparate). Thus, if we were to study high school students' post-graduation career/educational choices (e.g., attending college, attending a professional or trade school, or getting a job) based on several variables assessed one year prior to graduation, we could use the classification functions to predict what each student is most likely to do

after graduation. However, we would also want to know the probability that the student will make the predicted choice. Those probabilities are called posterior probabilities, and can also be computed. However, to understand how those probabilities are derived, let's first consider the so-called Mahalanobis distances.

Mahalanobis distances. In general, the Mahalanobis distance is a measure of distance between two points in the space defined by two or more correlated variables. For example, if there are two variables that are uncorrelated, we could plot points (cases) in a standard two-dimensional scatterplot; the Mahalanobis distances between the points would then be identical to the Euclidean distance; that is, the distance as, for example, measured by a ruler. If there are three uncorrelated variables, we could also simply use a ruler (in a 3D plot) to determine the distances between points. If there are more than 3 variables, we cannot represent the distances in a plot any more. Also, when the variables are correlated, the axes in the plots can be thought of as being non-orthogonal; that is, they would not be positioned in right angles to each other. In those cases, the simple Euclidean distance is not an appropriate measure, while the Mahalanobis distance will adequately account for the correlations.

Mahalanobis distances and classification. For each group in our sample, we can determine the location of the point that represents the means for all variables in the multivariate space defined by the variables in the model. These points are called group centroids. For each case we can then compute the Mahalanobis distances (of the respective case) from each of the group centroids. Again, we would classify the case as belonging to the group to which it is closest, that is, where the Mahalanobis distance is smallest.

Posterior classification probabilities. Using the Mahalanobis distances to do the classification, we can now derive probabilities. The probability that a case belongs to a particular group is basically proportional to the Mahalanobis distance from that group centroid (it is not exactly proportional because we assume a multivariate normal distribution around each centroid). Because we compute the location of each case from our prior knowledge of the values for that case on the variables in the model, these probabilities are called posterior probabilities. In summary, the posterior probability is the probability, based on our knowledge of the values of other variables, that the respective case belongs to a particular group. Some software packages will automatically compute those probabilities for all cases (or for selected cases only for cross-validation studies).

A priori classification probabilities. There is one additional factor that needs to be considered when classifying cases. Sometimes, we know ahead of time that there are more observations in one group than in any other; thus, the *a priori* probability that a case belongs to that group is higher. For example, if we know ahead of time that 60% of the graduates from our high school usually go to college (20% go to a professional school, and another 20% get a job), we should adjust our prediction accordingly: *a priori*, and all other things being equal, it is more likely that a student will attend college than choose either of the other two options. You can specify different *a priori* probabilities, which will then be used to adjust the classification of cases (and the computation of posterior probabilities) accordingly.

In practice, the researcher needs to ask him or herself whether the unequal number of cases in different groups in the sample is a reflection of the true distribution in the population, or whether it is only the (random) result of the sampling procedure. In the former case, we would set the *a priori* probabilities to be proportional to the sizes of the groups in our sample, in the latter case we would specify the *a priori* probabilities as being equal in each group. The specification of different *a priori* probabilities can greatly affect the accuracy of the prediction.

Summary of the prediction. A common result that you look at in order to determine how well the current classification functions predict group membership of cases is the classification matrix. The classification matrix shows the number of cases that were correctly classified (on the diagonal of the matrix) and those that were misclassified.

Another word of caution. To reiterate, post-hoc predicting of what has happened in the past is not that difficult. It is not uncommon to obtain very good classification if you use the same cases from which the classification functions were computed. In order to get an idea of how well the current classification functions perform, you must classify (*a priori*) different cases, that is, cases that were not used to estimate the classification functions. You can include or exclude cases from the computations; thus, the classification matrix can be computed for old cases as well as new cases. Only the classification of new cases allows us to assess the predictive validity of the classification functions (see also, cross-validation); the classification of old cases only provides a useful diagnostic tool to identify outliers or areas where the classification function seems to be less adequate.

Summary. In general, discriminant analysis is a very useful tool 1) for detecting the variables that enable the researcher to discriminate between different (naturally occurring) groups, and 2) for classifying cases into different groups with a better than chance accuracy.

DISTRIBUTION FITTING

In some research applications, you can formulate hypotheses about the specific distribution of the variable of interest. For example, variables whose values are determined by an infinite number of independent random events will be distributed following the normal distribution. You can think of a person's height as being the result of very many independent factors such as numerous specific genetic predispositions, early childhood diseases, nutrition, etc. As a result, height tends to be normally distributed in the U.S. population. Alternatively, if the values of a variable are the result of very rare events, the variable will be distributed according to the Poisson distribution (sometimes called the distribution of rare events). For example, industrial accidents can be thought of as the result of the intersection of a series of unfortunate (and unlikely) events, and their frequency tends to be distributed according to the Poisson distribution.

Another common application where distribution fitting procedures are useful is when you want to verify the assumption of normality before using a parametric test (see *General Purpose* in Chapter 29 – *Nonparametric Statistics*). For example, you may want to use the Kolmogorov-Smirnov Test or the Shapiro-Wilks' W Test (see the *Statistical Glossary*) to test for normality.

Fit of the Observed Distribution

For predictive purposes, it is often desirable to understand the shape of the underlying distribution of the population. To determine the underlying distribution, it is common to fit the observed distribution to a theoretical distribution by comparing the frequencies observed in the data to the expected frequencies of the theoretical distribution (i.e., a *chi*-square goodness of fit test). In addition to this type of test, some software packages also enable you to compute maximum likelihood tests and method of matching moments tests (see *Fitting Distributions by Moments* in Chapter 32 – *Process Analysis*).

Which distribution to use. Certain types of variables follow specific distributions. Variables whose values are determined by an infinite number of independent random events will be distributed following the normal distribution, whereas variables whose values are the result of an extremely rare event would follow the Poisson distribution. The major distributions that have been proposed for modeling survival or failure times are the exponential (and linear exponential) distribution, the Weibull distribution of extreme events, and the Gompertz distribution. The next section, *Types of Distributions*, contains a number of distributions with brief examples of what type of data would most commonly follow a specific distribution as well as the probability density function (pdf) for each distribution.

Types of Distributions

Bernoulli

This distribution best describes all situations where a trial is made resulting in either success or failure, such as when tossing a coin, or when modeling the success or failure of a surgical procedure. The Bernoulli distribution is defined as:

$$f(x) = p^x * (1-p)^{1-x} \quad \text{for } x \in \{0,1\}$$

where p is the probability that a particular event (e.g., success) will occur.

Beta

The *beta* distribution arises from a transformation of the F distribution and is typically used to model the distribution of order statistics. Because the beta distribution is bounded on both sides, it is often used for representing processes with natural lower and upper limits [refer to Hahn and Shapiro (1967)]. The *beta* distribution is defined as:

$$f(x) = \Gamma(\nu+\omega)/[\Gamma(\nu)\Gamma(\omega)] * x^{\nu-1}*(1-x)^{\omega-1}$$

$$\text{for } 0 < x < 1, \nu > 0, \omega > 0$$

where Γ is the *gamma* function and ν and ω are the shape parameters (Shape1 and Shape2).

The following illustrations show the *beta* distribution as the two shape parameters change.

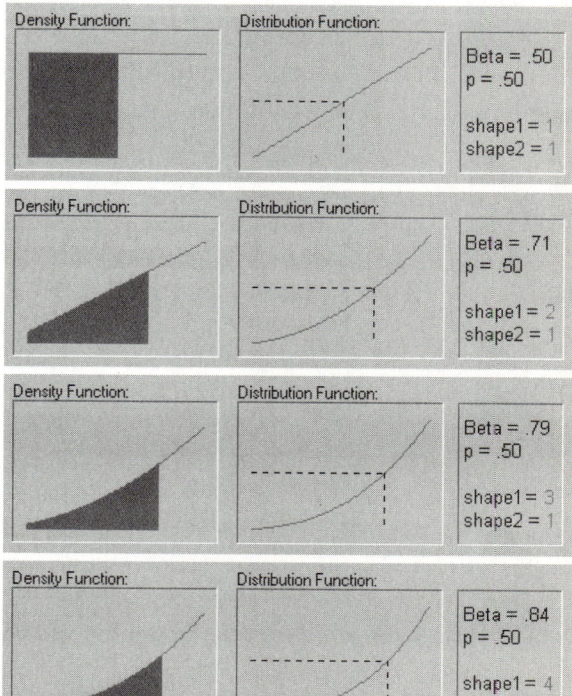

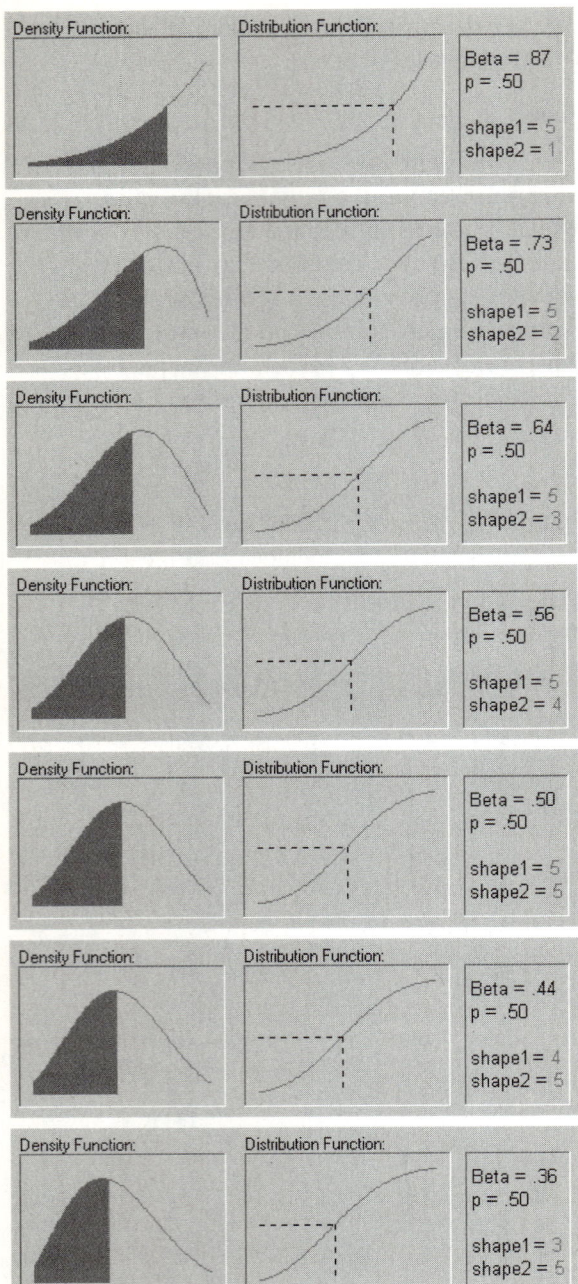

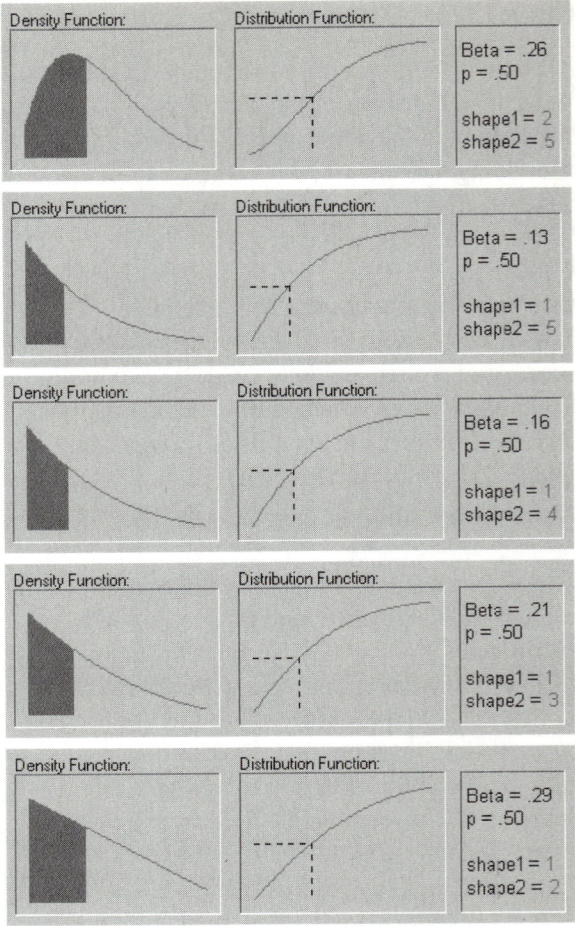

Binomial

The binomial distribution is useful for describing distributions of binomial events, such as the number of males and females in a random sample of companies, or the number of defective components in samples of 20 units taken from a production process. The binomial distribution is defined as:

$$f(x) = [n!/(x!*(n-x)!)]*p^x * q^{n-x}$$

for $x = 0,1,2,...,n$

where p is the probability that the respective event will occur, q is equal to 1-p, and n is the maximum number of independent trials.

Cauchy

The Cauchy distribution is interesting for theoretical reasons. Although its mean can be taken as zero, since it is symmetrical about zero, the expectation, variance, higher moments, and moment generating function do not exist. The Cauchy distribution is defined as:

$$f(x) = 1/(\theta*\pi*\{1+[(x-\eta)/\theta]^2\}) \quad \text{for } 0 < \theta$$

where η is the location parameter (median), θ is the scale parameter, and π is the constant *pi*.

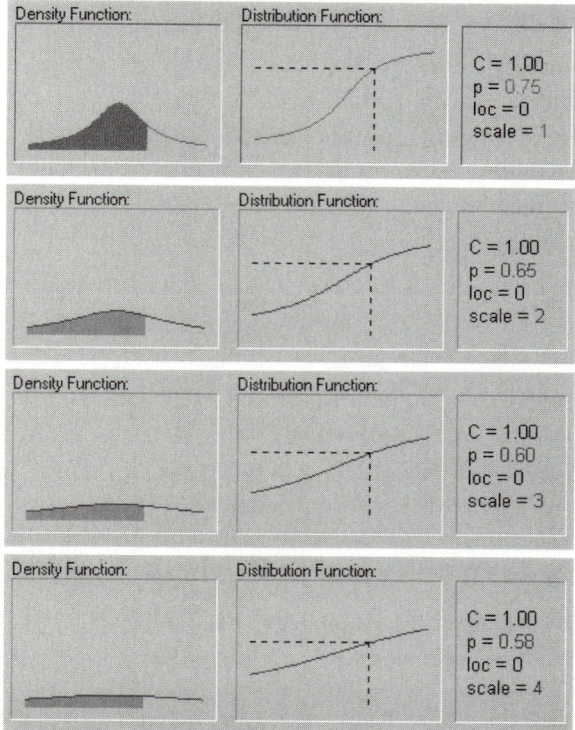

The previous illustrations show the changing shape of the Cauchy distribution when the location parameter equals 0 and the scale parameter equals 1, 2, 3, and 4.

Chi-Square

The sum of ν independent squared random variables, each distributed following the standard normal distribution, is distributed as *chi*-square

with ν degrees of freedom. The following illustrations show the shape of the *chi*-square distribution as the degrees of freedom increase (1, 2, 5, 10, 25 and 50).

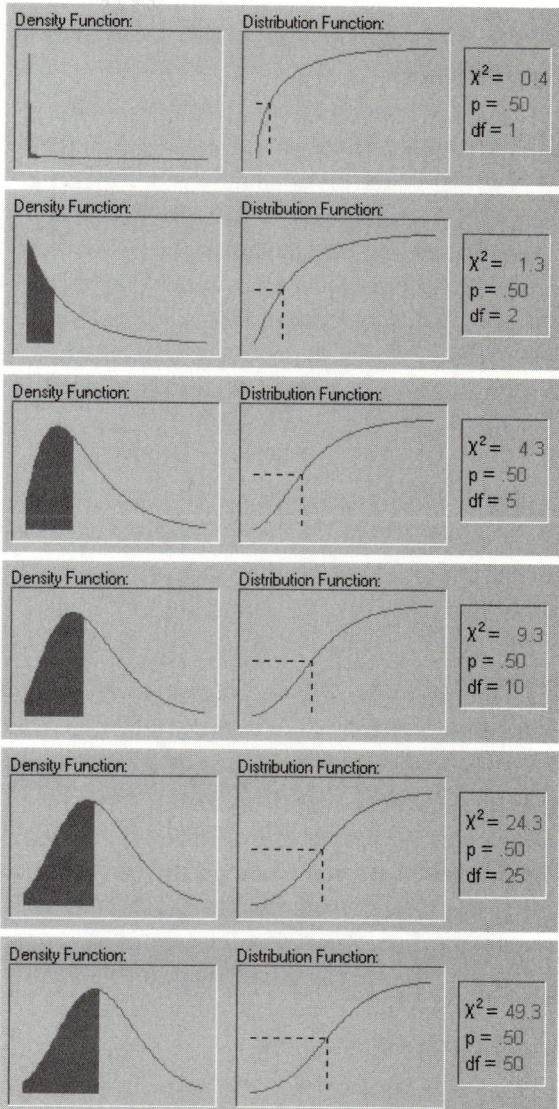

This distribution is most frequently used in the modeling of random variables (e.g., representing frequencies) in statistical applications, and is defined by:

$$f(x) = \{1/[2^{v/2} * \Gamma(v/2)]\} * [x^{(v/2)-1} * e^{-x/2}]$$
for $v = 1, 2, ..., 0 < x$

where v is the degrees of freedom, e is the base of the natural logarithm, sometimes called Euler's e (2.71...), and Γ (*gamma*) is the *gamma* function.

Exponential

If T is the time between occurrences of rare events that happen on the average with a rate l per unit of time, T is distributed exponentially with parameter λ (*lambda*). Thus, the exponential distribution is frequently used to model the time interval between successive random events. Examples of variables distributed in this manner would be the gap length between cars crossing an intersection, lifetimes of electronic devices, or arrivals of customers at the checkout counter in a grocery store. The exponential distribution function is defined as:

$$f(x) = \lambda \, e^{-\lambda x}, \text{ for } 0 \le x < \infty, \lambda > 0$$

where λ is the parameter of the exponential distribution function (an alternative parameterization is scale parameter b=1/λ) and e is the base of the natural logarithm, sometimes called Euler's e (2.71...).

Extreme Value

The extreme value distribution is often used to model extreme events, such as the size of floods, gust velocities encountered by airplanes, maxima of stock marked indices over a given year, etc.; it is also often used in reliability testing, e.g., to represent the distribution of failure times for electric circuits (see Hahn and Shapiro, 1967). The extreme value (Type I) distribution has the probability density function:

$$f(x) = 1/b * e^{-(x-a)/b} * e^{-e^{-(x-a)/b}}$$

for $-\infty < x < \infty, b > 0$

where a is the location parameter, b is the scale parameter, and e is the base of the natural logarithm, sometimes called Euler's e (2.71...).

F

Snedecor's F distribution is most commonly used in tests of variance (e.g., ANOVA). The ratio of two *chi*-squares divided by their respective degrees of freedom is said to follow an F distribution. The F distribution (for x > 0) has the probability density function (for $v = 1, 2, ...; \omega = 1, 2, ...$):

$$f(x) = \{\Gamma[(v+\omega)/2]\}/[\,\Gamma(v/2) * \Gamma(\omega/2)] * (v/\omega)^{(v/2)} * x^{[(v/2)-1]} * \{1+[(v/\omega)*x]\}^{[-(v+\omega)/2]}$$

for $0 \le x < \infty \; v=1,2,..., \; \omega=1,2,...$

where v and ω are the shape parameters, degrees of freedom and Γ is the *Gamma* function.

The following illustrations show various tail areas (*p*-values) for an F distribution with both degrees of freedom equal to 10.

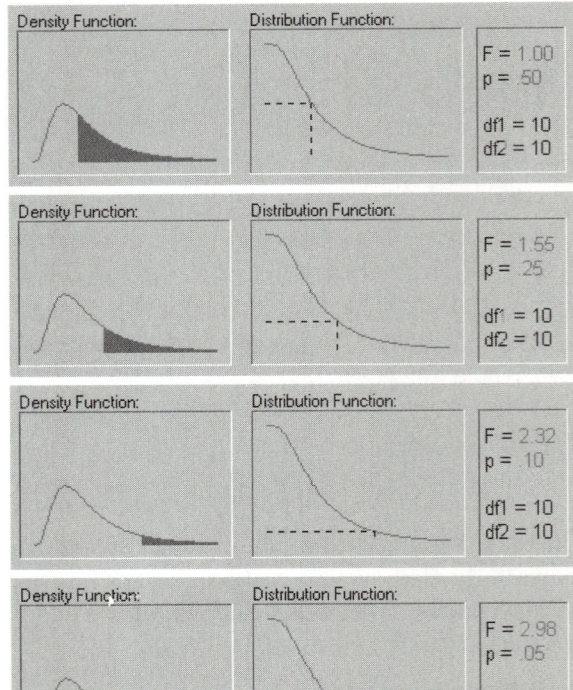

StatSoft
Copyright © StatSoft, 2006

Gamma

The *gamma* distribution is defined as:

$$f(x) = \{1/[b\Gamma(c)]\}*[x/b]^{c-1}*e^{-x/b}, \text{ for } 0 \leq x, c > 0$$

where Γ is the *gamma* function, c is the shape parameter, b is the scale parameter, and e is the base of the natural logarithm, sometimes called Euler's e (2.71...).

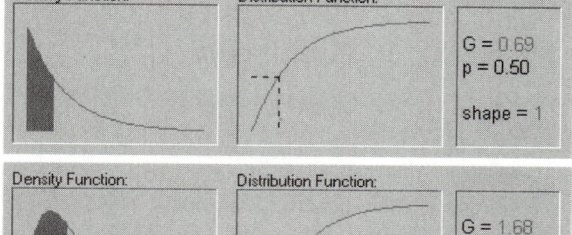

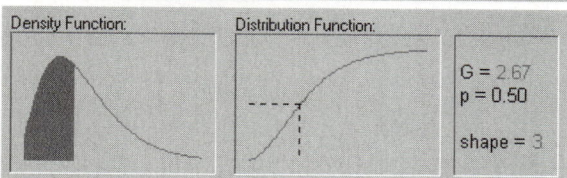

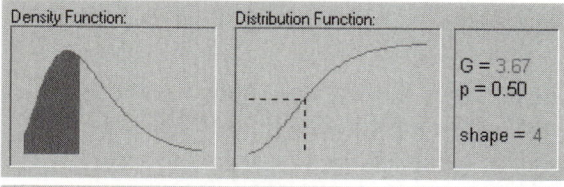

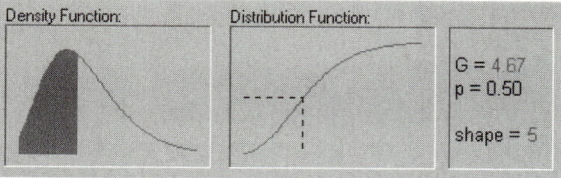

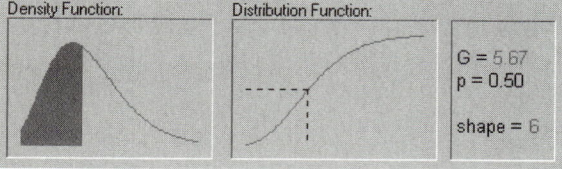

The probability density function of the exponential distribution has a mode of zero. In many instances, it is known *a priori* that the mode of the distribution of a particular random variable of interest is not equal to zero (e.g., when modeling the distribution of the lifetimes of a product such as an electric light bulb, or the serving time taken at a ticket booth at a baseball game). In these cases, the *gamma* distribution is more appropriate for describing the underlying distribution.

The previous illustrations show the *gamma* distribution as the shape parameter changes from 1 to 6.

Geometric

If independent Bernoulli trials are made until a success occurs, the total number of trials required is a geometric random variable. The geometric distribution is defined as:

$$f(x) = p*(1-p)^x, \text{ for } x = 0,1,2,...$$

where p is the probability that a particular event (e.g., success) will occur.

Gompertz

The Gompertz distribution is a theoretical distribution of survival times. Gompertz (1825) proposed a probability model for human mortality, based on the assumption that the "average exhaustion of a man's power to avoid death to be such that at the end of equal infinitely small intervals of time he lost equal portions of his remaining power to oppose destruction which he had at the commencement of these intervals" (Johnson, Kotz, Blakrishnan, 1995, p. 25). The resultant hazard function:

$$r(x)=Bc^x, \text{ for } x \leq 0, B > 0, c \leq 1$$

is often used in survival analysis. See Johnson, Kotz, Blakrishnan (1995) for additional details.

Laplace

For interesting mathematical applications of the Laplace distribution see Johnson and Kotz (1995). The Laplace (or Double Exponential) distribution is defined as:

$$f(x) = 1/(2b) * e^{[-(|x-a|/b)]}, \text{ for } -\infty < x < \infty$$

where a is the location parameter (mean), b is the scale parameter, and e is the base of the natural logarithm, sometimes called Euler's e (2.71...).

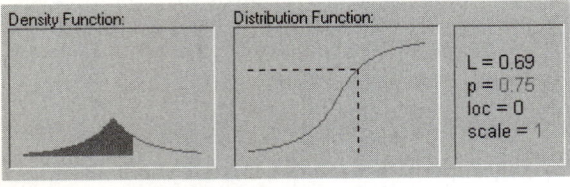

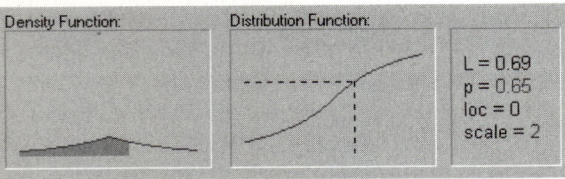

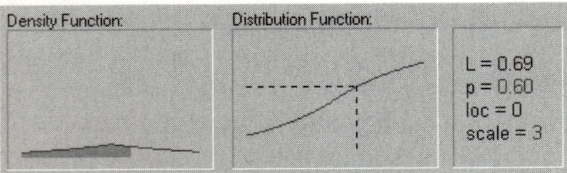

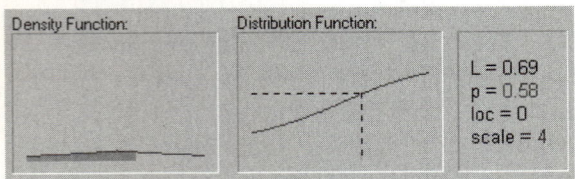

The previous illustrations show the changing shape of the Laplace distribution when the location parameter equals 0 and the scale parameter equals 1, 2, 3, and 4.

Logistic

The logistic distribution is used to model binary responses (e.g., Gender) and is commonly used in logistic regression. The logistic distribution is defined as:

$$f(x) = (1/b) * e^{[-(x-a)/b]} * [1 + e^{[-(x-a)/b]-2}]$$

for $-\infty < x < \infty$, $0 < b$

where a is the location parameter (mean), b is the scale parameter, and e is the base of the natural logarithm, sometimes called Euler's e (2.71...).

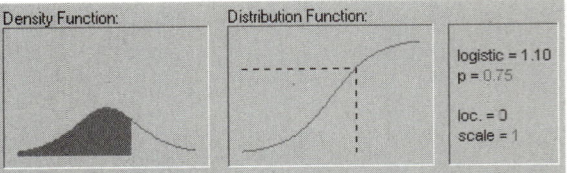

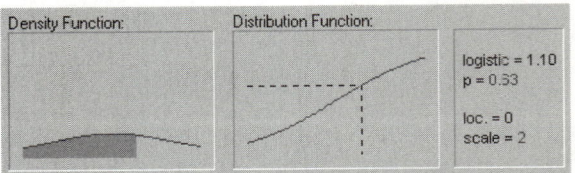

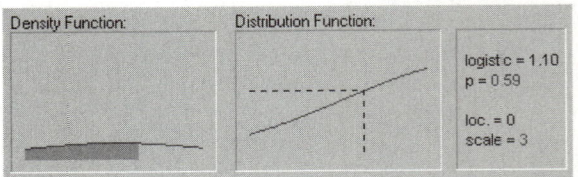

The previous illustrations show the changing shape of the logistic distribution when the location parameter equals 0 and the scale parameter equals 1, 2, and 3.

Lognormal

The lognormal distribution is often used in simulations of variables such as personal incomes, age at first marriage, or tolerance to poison in animals. In general, if x is a sample from a normal distribution, then $y = e^x$ is a sample from a lognormal distribution. Thus, the lognormal distribution is defined as:

$$f(x) = 1/[x\sigma(2\pi)^{1/2}] * \exp(-[\log(x)-\mu]^2/2\sigma^2)$$

for $0 < x < \infty$, $\mu > 0$, $\sigma > 0$

where μ is the scale parameter, σ is the shape parameter, and e is the base of the natural logarithm, sometimes called Euler's e (2.71...).

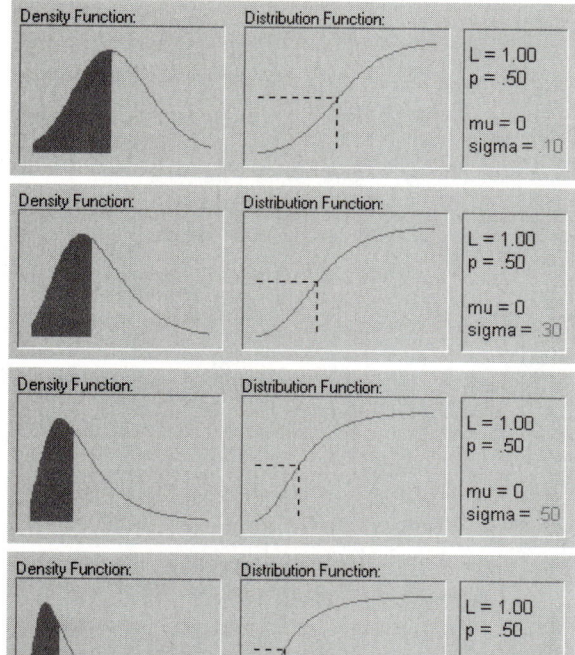

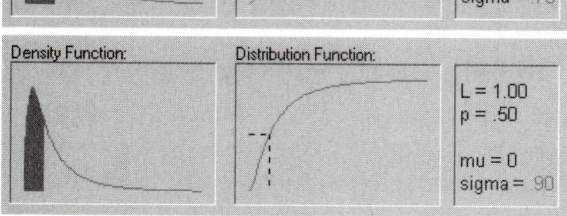

The previous illustrations show the lognormal distribution with *mu* equal to 0 and *sigma* equals .10, .30, .50, .70, and .90.

Normal

The normal distribution (the bell-shaped curve, which is symmetrical about the mean) is a theoretical function commonly used in inferential statistics as an approximation to

sampling distributions (see also, Chapter 1 – *Elementary Concepts in Statistics*). In general, the normal distribution provides a good model for a random variable, when:

1. There is a strong tendency for the variable to take a central value;

2. Positive and negative deviations from this central value are equally likely;

3. The frequency of deviations falls off rapidly as the deviations become larger.

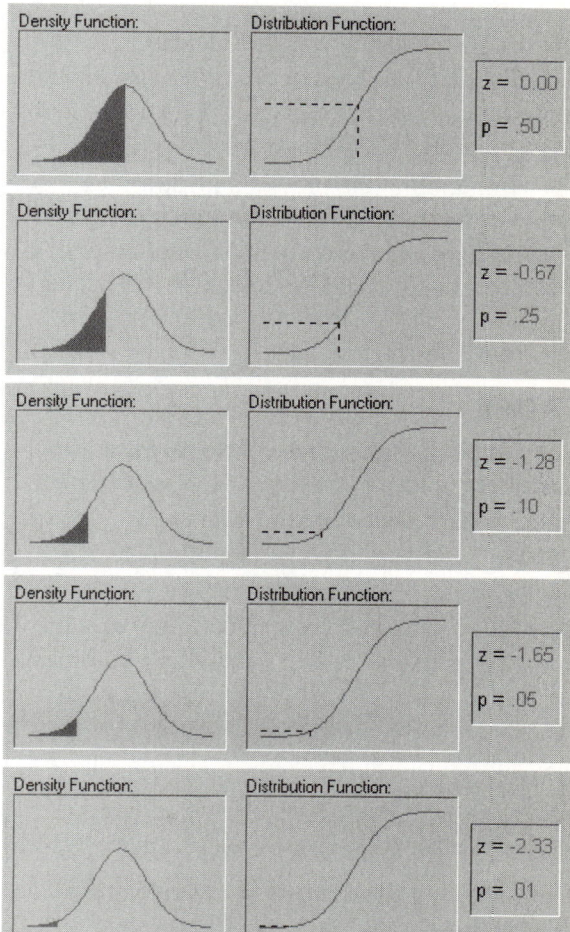

As an underlying mechanism that produces the normal distribution, you can think of an infinite number of independent random (binomial) events that bring about the values of a particular variable. For example, there are probably a nearly infinite number of factors that determine a person's height (thousands of genes, nutrition, diseases, etc.). Thus, height can be expected to be normally distributed in the population. The normal distribution function is determined by the following formula:

$$f(x) = \frac{1}{\sigma\sqrt{2\pi}} e^{\frac{-(x-\mu)^2}{2\sigma^2}} \quad for -\infty < x < \infty$$

where μ is the mean, σ is the standard deviation, e is the base of the natural logarithm, sometimes called Euler's e (2.71...), and π is the constant *pi* (3.14...).

The animation in the previous illustrations shows several tail areas of the standard normal distribution (i.e., the normal distribution with a mean of 0 and a standard deviation of 1). The standard normal distribution is often used in hypothesis testing.

Pareto

The Pareto distribution is commonly used in monitoring production processes (see Chapter 32 – *Process Analysis* and Chapter 33 – *Quality Control Charts*).

For example, a machine that produces copper wire will occasionally generate a flaw at some point along the wire. The Pareto distribution can be used to model the length of wire between successive flaws. The standard Pareto distribution is defined as:

$f(x) = c/x^{c+1}$, for $1 \leq x$, $c < 0$

where c is the shape parameter.

The following illustrations show the Pareto distribution for the shape parameter equal to 1, 2, 3, 4, and 5.

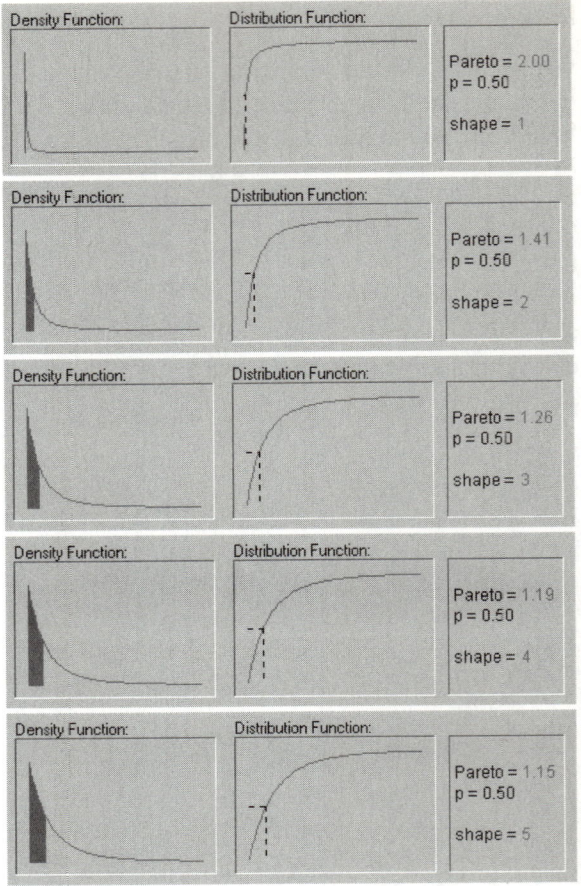

Poisson

The Poisson distribution is sometimes referred to as the distribution of rare events; e.g., Poisson distributed variables could be number of sweepstakes won per person or the number of catastrophic defects found in a production process. It is defined as:

$f(x) = (\lambda^x * e^{-\lambda})/x!$, for $x = 0,1,2,..., 0 < \lambda$

where λ (*lambda*) is the expected value of x (the mean) and e is the base of the natural logarithm, sometimes called Euler's e (2.71...).

Rayleigh

If two independent variables y_1 and y_2 are independent from each other and normally distributed with equal variance, the variable $x = \sqrt{(y_1^2 + y_2^2)}$ will follow the Rayleigh distribution. An example for such a variable is the distance of darts from the target in a dart-throwing game, where the errors in the two dimensions of the target plane are independent and normally distributed. The Rayleigh distribution is defined as:

$$f(x) = x/b^2 * e^{\wedge}[-(x^2/2b^2)], \text{ for } 0 \le x < \infty, b > 0$$

where b is the scale parameter and e is the base of the natural logarithm, sometimes called Euler's e (2.71...).

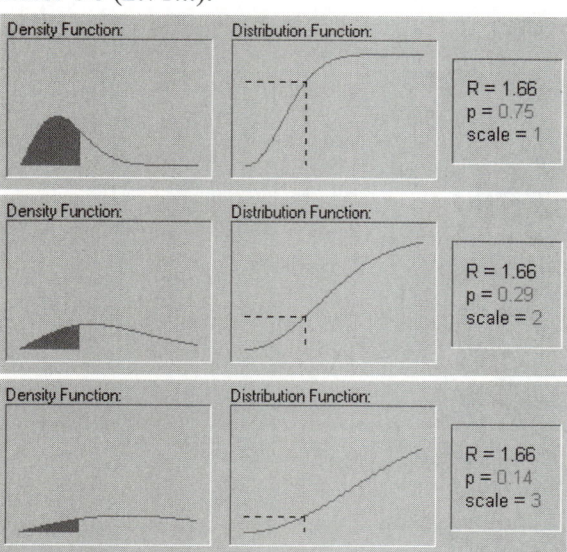

The previous illustrations show the changing shape of the Rayleigh distribution when the scale parameter equals 1, 2, and 3.

Rectangular

The rectangular distribution is useful for describing random variables with a constant probability density over the defined range a<b.

$$f(x) = 1/(b-a), \text{ for } a<x<b$$
$$= 0, \text{ elsewhere}$$

where $a<b$ are constants.

Student's t

The student's t distribution is symmetric about zero, and its general shape is similar to that of the standard normal distribution. It is most commonly used in testing hypothesis about the mean of a particular population.

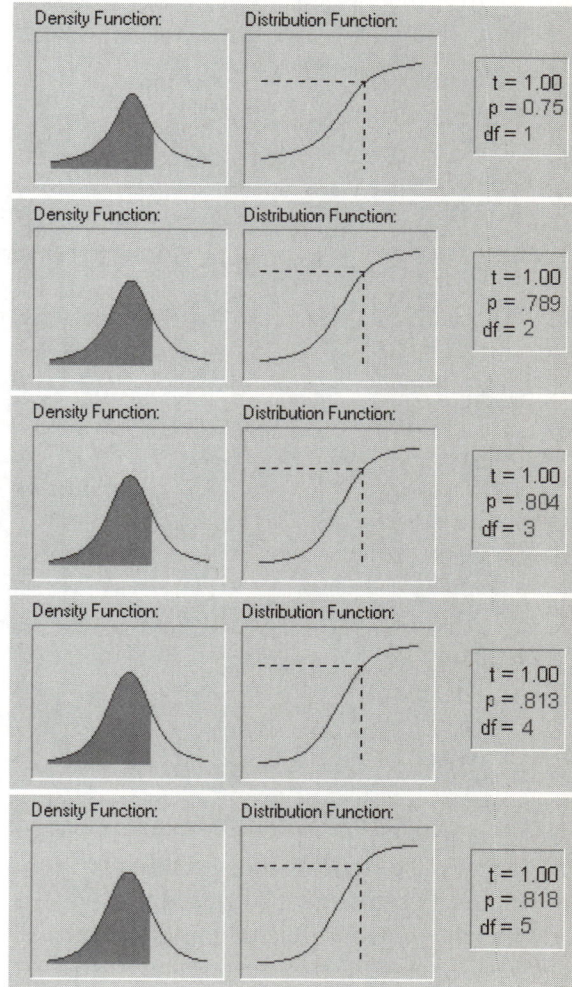

The student's t distribution is defined as (for n = 1, 2, ...):

$f(x) = \{ \Gamma[(\nu+1)/2] / \Gamma(\nu/2)\} * (\nu*\pi)^{-1/2} * [1 + (x^2/\nu)]^{-(\nu+1)/2}$

where ν is the shape parameter, degrees of freedom, Γ is the Gamma function, and π is the constant Pi (3.14 . . .).

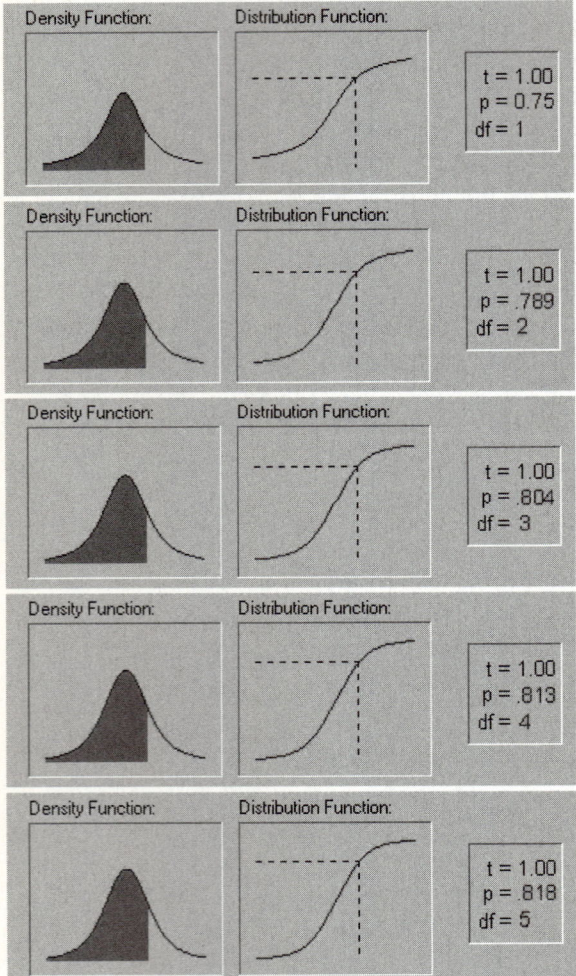

The shape of the student's t distribution is determined by the degrees of freedom. As shown in the previous illustrations, its shape changes as the degrees of freedom increase.

Weibull

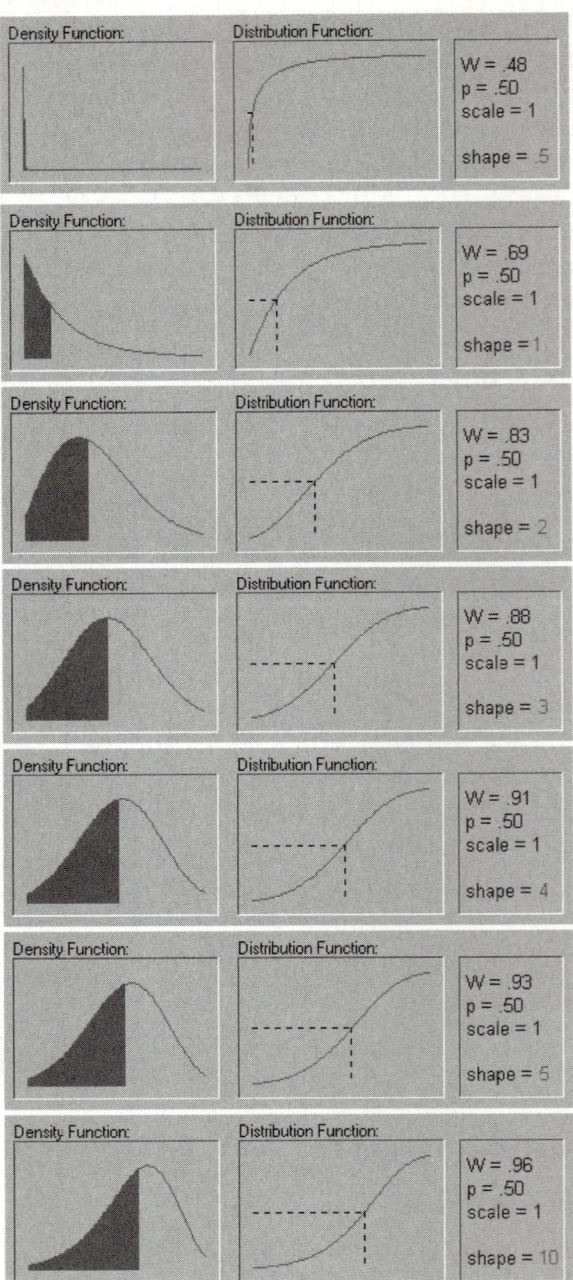

The exponential distribution is often used as a model of time-to-failure measurements, when the failure (hazard) rate is constant over time. When the failure probability varies over time, the Weibull distribution is appropriate. Thus, the Weibull distribution is often used in reliability testing (e.g., of electronic relays, ball bearings, etc.; see Hahn and Shapiro, 1967).

The Weibull distribution is defined as:

$$f(x) = c/b * (x/b)^{(c-1)} * e^{[-(x/b)^{\wedge}c]}$$

for $0 \leq x < \infty$, $b > 0$, $c > 0$

where b is the scale parameter, c is the shape parameter, and e is the base of the natural logarithm, sometimes called Euler's e (2.71...).

The previous illustrations show the Weibull distribution as the shape parameter increases (.5, 1, 2, 3, 4, 5, and 10).

EXPERIMENTAL DESIGN (INDUSTRIAL DOE)

The general ideas and principles on which experimentation in industry is based and the types of designs used are discussed in this chapter. The chapter is meant to be introductory in nature. However, it is assumed that you are familiar with the basic ideas of analysis of variance and the interpretation of main effects and interactions in ANOVA. Otherwise, it is recommended that you read the *Basic Ideas* section of *ANOVA/MANOVA* and Chapter 18 – *General Linear Models*.

Overview

Experiments in Science and Industry

Experimental methods are widely used in research as well as in industrial settings, however, sometimes for very different purposes. The primary goal in scientific research is usually to show the statistical significance of an effect that a particular factor exerts on the dependent variable of interest (for more information on statistical significance, see Chapter 1 – *Elementary Concepts in Statistics*).

In industrial settings, the primary goal is usually to extract the maximum amount of unbiased information regarding the factors affecting a production process from as few (costly) observations as possible. While in the former application (in science) analysis of variance (*ANOVA*) techniques are used to uncover the interactive nature of reality, as manifested in higher-order interactions of factors, in industrial settings interaction effects are often regarded as a nuisance (they are often of no interest; they only complicate the process of identifying important factors).

Differences in Techniques

These differences in purpose have a profound effect on the techniques that are used in the two settings. If you review a standard ANOVA text for the sciences, for example the classic texts by Winer (1962) or Keppel (1982), you will find that they will primarily discuss designs with up to, perhaps, five factors (designs with more than six factors are usually impractical; see *ANOVA/MANOVA*, page 41). The focus of these discussions is how to derive valid and robust statistical significance tests. However, if you review standard texts on experimentation in industry (Box, Hunter, and Hunter, 1978; Box

and Draper, 1987; Mason, Gunst, and Hess, 1989; Taguchi, 1987) you will find that they will primarily discuss designs with many factors (e.g., 16 or 32) in which interaction effects cannot be evaluated, and the primary focus of the discussion is how to derive unbiased main effect (and, perhaps, two-way interaction) estimates with a minimum number of observations.

This comparison can be expanded further, however, a more detailed description of experimental design in industry will now be discussed and other differences will become clear. Note that Chapter 18 – *General Linear Models* and Chapter 3 – *ANOVA/MANOVA* contain detailed discussions of typical design issues in scientific research; the general linear model procedure is a very comprehensive implementation of the general linear model approach to ANOVA/MANOVA (univariate and multivariate ANOVA). There are of course applications in industry where general ANOVA designs, as used in scientific research, can be immensely useful. You may want to read the *General Linear Models* and *ANOVA/MANOVA* chapters to gain a more general appreciation of the range of methods encompassed by the term experimental design.

General Ideas

In general, every machine used in a production process allows its operators to adjust various settings, affecting the resultant quality of the product manufactured by the machine. Experimentation allows the production engineer to adjust the settings of the machine in a systematic manner and to learn which factors have the greatest impact on the resultant quality. Using this information, the settings can be constantly improved until optimum quality is obtained. To illustrate this reasoning, here are a few examples:

Example 1: Dyestuff manufacture. Box and Draper (1987, page 115) report an experiment concerned with the manufacture of certain dyestuff. Quality in this context can be described in terms of a desired (specified) hue and brightness and maximum fabric strength. Moreover, it is important to know what to change in order to produce a different hue and brightness should the consumers' taste change. Put another way, the experimenter would want to identify the factors that affect the brightness, hue, and strength of the final product. In the example described by Box and Draper, there are 6 different factors that are evaluated in a $2^{(6-0)}$ design (the $2^{(k-p)}$ notation is explained below). The results of the experiment show that the three most important factors determining fabric strength are the *Polysulfide index, Time*, and *Temperature* (see Box and Draper, 1987, page 116). You can summarize the expected effect (predicted means) for the variable of interest (i.e., fabric strength in this case) in a so-called cube-plot. This plot shows the expected (predicted) mean fabric strength for the respective low and high settings for each of the three variables (factors).

simultaneously evaluated. It is not uncommon, that there are very many (e.g., 100) different factors that may potentially be important. Special designs (e.g., Plackett-Burman designs, see Plackett and Burman, 1946) have been developed to screen such large numbers of factors in an efficient manner, that is, with the least number of observations necessary. For example, you can design and analyze an experiment with 127 factors and only 128 runs (observations); still, you will be able to estimate the main effects for each factor, and thus, you can quickly identify which ones are important and most likely to yield improvements in the process under study.

Example 2: 3^3 design. Montgomery (1976, page 204) describes an experiment conducted in order to identify the factors that contribute to the loss of soft drink syrup due to frothing during the filling of five-gallon metal containers. Three factors where considered: a) the nozzle configuration, b) the operator of the machine, and c) the operating pressure. Each factor was set at three different levels, resulting in a complete $3^{(3-0)}$ experimental design (the $3^{(k-p)}$ notation is explained below).

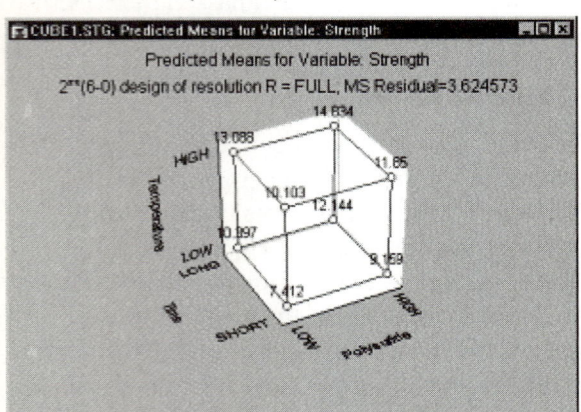

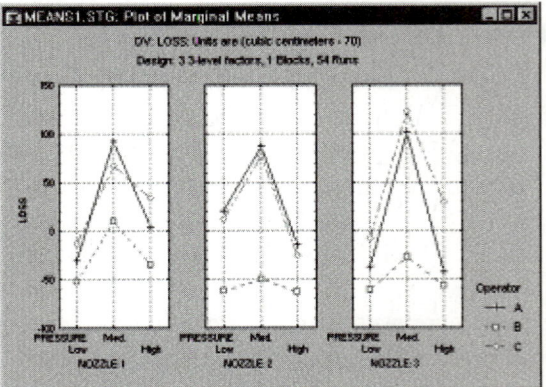

Example 1.1: Screening designs. In the previous example, 6 different factors were

Moreover, two measurements were taken for each combination of factor settings, that is, the $3^{(3-0)}$ design was completely replicated once.

Example 3: Maximizing yield of a chemical reaction. The yield of many chemical reactions is a function of time and temperature. Unfortunately, these two variables often do not affect the resultant yield in a linear fashion. In other words, it is not so that "the longer the time, the greater the yield" and "the higher the temperature, the greater the yield." Rather, both of these variables are usually related in a curvilinear fashion to the resultant yield.

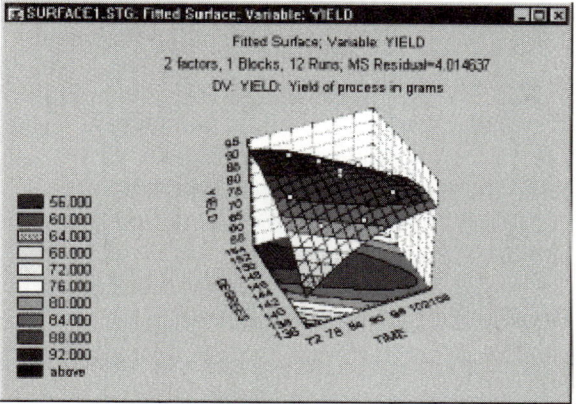

Thus, in this example your goal as experimenter is to optimize the yield surface that is created by the two variables: *time* and *temperature*.

Example 4: Testing the effectiveness of four fuel additives. Latin square designs are useful when the factors of interest are measured at more than two levels, and the nature of the problem suggests some blocking. For example, imagine a study of 4 fuel additives on the reduction in oxides of nitrogen (see Box, Hunter, and Hunter, 1978, page 263). You may have 4 drivers and 4 cars at your disposal. You are not particularly interested in any effects of particular cars or drivers on the resultant oxide reduction; however, you do not want the results for the fuel additives to be biased by the particular driver or car. With Latin square designs, you can estimate the main effects of all factors in the design in an unbiased manner. With regard to the example, the arrangement of treatment levels in a Latin square design assures that the variability among drivers or cars does not affect the estimation of the effect due to different fuel additives.

Example 5: Improving surface uniformity in the manufacture of polysilicon wafers. The manufacture of reliable microprocessors requires very high consistency in the manufacturing process. Note that in this instance, it is equally, if not more important to control the variability of certain product characteristics than it is to control the average for a characteristic. For example, with regard to the average surface thickness of the polysilicon layer, the manufacturing process may be perfectly under control; yet, if the variability of the surface thickness on a wafer fluctuates widely, the resultant microchips will not be reliable. Phadke (1989) describes how different characteristics of the manufacturing process (such as deposition temperature, deposition pressure, nitrogen flow, etc.) affect the variability of the polysilicon surface thickness on wafers. However, no theoretical model exists that would enable the engineer to predict how these factors affect the uniformness of wafers. Therefore, systematic experimentation with the factors is required to optimize the process. This is a typical example where Taguchi robust design methods would be applied.

Example 6: Mixture designs. Cornell (1990, page 9) reports an example of a typical (simple) mixture problem. Specifically, a study was conducted to determine the optimum texture of fish patties as a result of the relative proportions of different types of fish (Mullet, Sheepshead, and Croaker) that made up the patties. Unlike in non-mixture experiments, the total sum of the proportions must be equal to a constant, for example, to 100%. The results of such experiments are usually graphically represented in so-called triangular (or ternary) graphs.

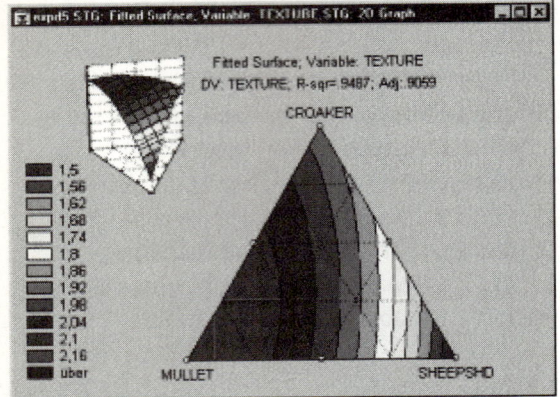

In general, the overall constraint – that the three components must sum to a constant – is reflected in the triangular shape of the graph (see the previous illustration).

Example 6.1: Constrained mixture designs.

It is particularly common in mixture designs that the relative amounts of components are further constrained (in addition to the constraint that they must sum to, for example, 100%). For example, suppose we wanted to design the best-tasting fruit punch consisting of a mixture of juices from five fruits. Since the resulting mixture is supposed to be a fruit punch, pure blends consisting of the pure juice of only one fruit are necessarily excluded. Additional constraints can be placed on the "universe" of mixtures due to cost constraints or other considerations, so that one particular fruit cannot, for example, account for more than 30% of the mixtures (otherwise the fruit punch would be too expensive, the shelf-life would be compromised, the punch could not be produced in large enough quantities, etc.). Such so-called constrained experimental regions present numerous problems, which, however, can be addressed.

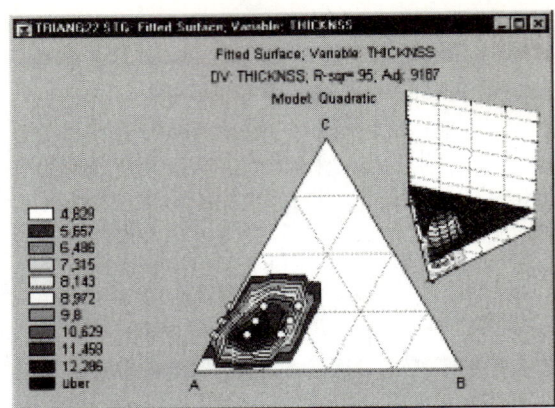

In general, under those conditions, you seek to design an experiment that can potentially extract the maximum amount of information about the respective response function (e.g., taste of the fruit punch) in the experimental region of interest.

Computational Problems

There are basically two general issues to which experimental design is addressed:

1. How to design an optimal experiment, and

2. How to analyze the results of an experiment.

With regard to the first issue, there are different considerations that enter into the different types of designs, and they will be discussed shortly. In the most general terms, the goal is always to allow the experimenter to evaluate in an unbiased (or least biased) way, the consequences of changing the settings of a particular factor, that is, regardless of how other factors were set. In more technical terms, you attempt to generate designs where main effects are unconfounded among themselves, and in some cases, even unconfounded with the interaction of factors.

Components of Variance, Denominator Synthesis

There are several statistical methods for analyzing designs with random effects (see *Methods for Analysis of Variance*, page 54). Chapter 39 – *Variance Components and Mixed Model ANOVA/ANCOVA* discusses numerous options for estimating variance components for random effects, and for performing approximate F tests based on synthesized error terms.

Summary

Experimental methods are finding increasing use in manufacturing to optimize the production process. Specifically, the goal of these methods is to identify the optimum settings for the different factors that affect the production process. In the discussion so far, the major classes of designs that are typically used in industrial experimentation have been introduced: $2^{(k-p)}$ (two-level, multi-factor) designs, screening designs for large numbers of factors, $3^{(k-p)}$ (3-level, multi-factor) designs (mixed designs with 2 and 3-level factors are also supported), central composite (or response surface) designs, Latin square designs, Taguchi robust design analysis, mixture designs, and special procedures for constructing experiments in constrained experimental regions. Interestingly, many of these experimental techniques have made their way from the production plant into management, and successful implementations have been reported in profit planning in business, cash-flow optimization in banking, etc. (e.g., see Yokyama and Taguchi, 1975). These techniques will be described in greater detail in the remainder of the chapter.

$2^{(k-p)}$ Fractional Factorial Designs at 2 Levels

Overview

In many cases, it is sufficient to consider the factors affecting the production process at two levels. For example, the temperature for a chemical process can either be set a little higher or a little lower, the amount of solvent in a dyestuff manufacturing process can either be slightly increased or decreased, etc. The experimenter would want to determine whether any of these changes affect the results of the production process. The most intuitive approach to study those factors would be to vary the factors of interest in a full factorial design, that is, to try all possible combinations of settings. This would work fine, except that the number of necessary runs in the experiment (observations) will increase geometrically. For example, if you want to study 7 factors, the necessary number of runs in the experiment would be $2^7 = 128$. To study 10 factors you would need $2^{10} = 1,024$ runs in the experiment. Because each run may require time-consuming and costly setting and resetting of machinery, it is often not feasible to require that many different production runs for the experiment. In these conditions, fractional factorials are used that sacrifice interaction effects so that main effects can still be computed correctly.

Generating the Design

A technical description of how fractional factorial designs are constructed is beyond the scope of this introduction. Detailed accounts of how to design $2^{(k-p)}$ experiments can be found, for example, in Bayne and Rubin (1986), Box and Draper (1987), Box, Hunter, and Hunter (1978), Montgomery (1991), Daniel (1976), Deming and Morgan (1993), Mason, Gunst, and Hess (1989), or Ryan (1989), to name only a few of the many text books on this subject. In general, it will successively use the highest-order interactions to generate new factors. For example, consider the following design that includes 11 factors but requires only 16 runs (observations).

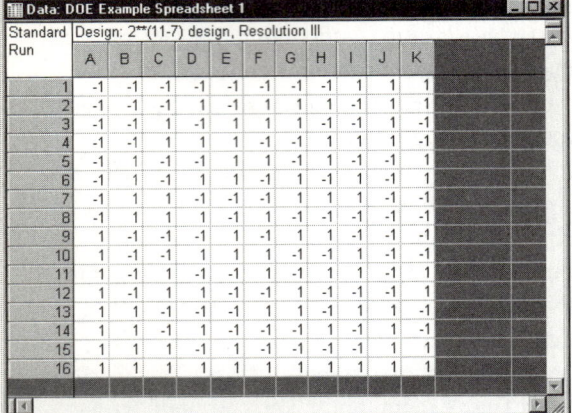

Reading the design. The design displayed above should be interpreted as follows. Each column contains +1s or -1s to indicate the setting of the respective factor (high or low, respectively). So, for example, in the first run of the experiment, set all factors *A* through *K* to the plus setting (e.g., a little higher than before); in the second run, set factors *A, B*, and *C* to the positive setting, factor *D* to the negative setting, and so on. Note that there are numerous options provided to display (and save) the design using notation other than ±1 to denote factor settings. For example, you can use actual values of factors (e.g., *90* degrees Celsius and *100* degrees Celsius) or text labels (*Low* temperature, *High* temperature).

Randomizing the runs. Because many other things may change from production run to production run, it is always a good practice to randomize the order in which the systematic runs of the designs are performed.

The Concept of Design Resolution

The design above is described as a $2^{(11-7)}$ design of resolution III (three). This means that you study overall $k = 11$ factors (the first number in parentheses); however, $p = 7$ of those factors (the second number in parentheses) were generated from the interactions of a full $2^{[(11-7)=4]}$ factorial design. As a result, the design does not give full resolution; that is, there are certain interaction effects that are confounded with (identical to) other effects. In general, a design of resolution R is one where no l-way interactions are confounded with any other interaction of order less than R-l. In the current example, R is equal to 3. Here, no l = 1 level interactions (i.e., main effects) are confounded with any other interaction of order less than R-l = 3-1 = 2. Thus, main effects in this design are confounded with 2-way interactions; and consequently, all higher-order interactions are equally confounded. If you had included 64 runs, and generated a $2^{(11-5)}$ design, the resultant resolution would have been R = IV (4). You would have concluded that no l = 1-way interaction (main effect) is confounded with any other interaction of order less than R-l = 4-1 = 3. In this design then, main effects are not confounded with 2-way interactions, but only with 3-way interactions. What about the 2-way interactions? No l = 2-way interaction is confounded with any other interaction of order less than R-l = 4-2 = 2. Thus, the 2-way interactions in that design are confounded with each other.

Plackett-Burman (Hadamard Matrix) Designs for Screening

When you need to screen a large number of factors to identify those that may be important (i.e., those that are related to the dependent variable of interest), you want to employ a design that enables you to test the largest number of factor main effects with the least number of observations, that is to construct a resolution III design with as few runs as possible. One way to design such experiments is to confound all interactions with new main effects. Such designs are also sometimes called saturated designs, because all information in those designs is used to estimate the parameters, leaving no degrees of freedom to estimate the error term for the ANOVA. Because the added factors are created by equating (aliasing, see below) the new factors with the interactions of a full factorial design, these designs always will have 2^k runs (e.g., 4, 8, 16, 32, and so on). Plackett and Burman (1946) showed how full factorial design can be fractionalized in a different manner to yield saturated designs where the number of runs is a multiple of 4 rather than a power of 2. These designs are also sometimes called Hadamard matrix designs. Of course, you do not have to use all available factors in those designs, and, in fact, sometimes you want to generate a saturated design for one more factor than you are expecting to test. This will enable you to estimate the random error variability, and test for the statistical significance of the parameter estimates.

Enhancing Design Resolution via Foldover

One way in which a resolution III design can be enhanced and turned into a resolution IV design is via foldover (e.g., see Box and Draper, 1987, Deming and Morgan, 1993).

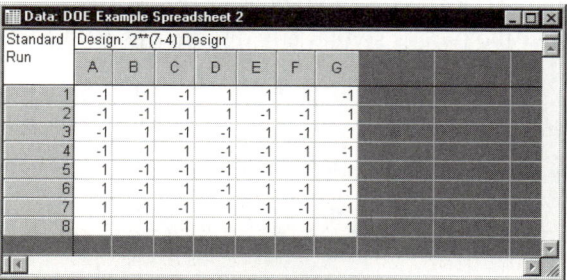

Suppose you have a 7-factor design in 8 runs (see previous illustration). This is a resolution III design, that is, the two-way interactions will be confounded with the main effects. You can turn this design into a resolution IV design via the Foldover (enhance resolution) option. The foldover method copies the entire design and appends it to the end, reversing all signs:

Standard Run	Design: 2**(7-4) design (+Foldover)							New: H
	A	B	C	D	E	F	G	
1	-1	-1	-1	1	1	1	-1	1
2	-1	-1	1	1	-1	-1	1	1
3	-1	1	-1	-1	1	-1	1	1
4	-1	1	1	-1	-1	1	-1	1
5	1	-1	-1	-1	-1	1	1	1
6	1	-1	1	-1	1	-1	-1	1
7	1	1	-1	1	-1	-1	-1	1
8	1	1	1	1	1	1	1	1
9	1	1	1	-1	-1	-1	1	-1
10	1	1	-1	-1	1	1	-1	-1
11	1	-1	1	1	-1	1	-1	-1
12	1	-1	-1	1	1	-1	1	-1
13	-1	1	1	1	1	-1	-1	-1
14	-1	1	-1	1	-1	1	1	-1
15	-1	-1	1	-1	1	1	1	-1
16	-1	-1	-1	-1	-1	-1	-1	-1

Thus, the standard run number 1 was -1, -1, -1, 1, 1, 1, -1; the new run number 9 (the first run of the folded-over portion) has all signs reversed: 1, 1, 1, -1, -1, -1, 1. In addition to enhancing the resolution of the design, we also have gained an 8th factor (factor H), which contains all +1s for the first 8 runs, and -1s for the folded-over portion of the new design. Note that the resultant design is actually a $2^{(8-4)}$ design of resolution IV (see also, Box and Draper, 1987, page 160).

Aliases of Interactions: Design Generators

To return to the example of the resolution R = III design, now that you know that main effects are confounded with 2-way interactions, you may ask the question, "Which interaction is confounded with which main effect?"

Factor	Fractional Design Generators 2**(11-7) design (Factors are denoted by numbers)
	Alias
5	123
6	234
7	134
8	124
9	1234
10	12
11	13

Data: DOE Example Spreadsheet 4

Design generators. The design generators shown above are the key to how factors *5* through *11* were generated by assigning them to particular interactions of the first 4 factors of the full factorial 2^4 design. Specifically, factor *5* is identical to the *123* (factor 1 by factor 2 by factor 3) interaction. Factor *6* is identical to the *234* interaction, and so on. Remember that the design is of resolution III (three), and you expect some main effects to be confounded with some two-way interactions; indeed, factor *10* (ten) is identical to the *12* (factor 1 by factor 2) interaction, and factor *11* (eleven) is identical to the *13* (factor 1 by factor 3) interaction. Another way in which these equivalencies are often expressed is by saying that the main effect for factor 10 is an alias for the interaction of 1 by 2. (The term alias was first used by Finney, 1945).

To summarize, whenever you want to include fewer observations (runs) in your experiment than would be required by the full factorial 2^k design, you sacrifice interaction effects and assign them to the levels of factors. The resulting design is no longer a full factorial but a fractional factorial.

The fundamental identity. Another way to summarize the design generators is in a simple equation. Namely, if, for example, factor *5* in a fractional factorial design is identical to the *123* (factor *1* by factor *2* by factor *3*) interaction, then it follows that multiplying the coded values for the *123* interaction by the coded values for

factor *5* will always result in *+1* (if all factor levels are coded ±*1*); or:

I = 1235

where *I* stands for *+1* (using the standard notation as, for example, found in Box and Draper, 1987). Thus, we also know that factor *1* is confounded with the *235* interaction, factor *2* with the *135*, interaction, and factor *3* with the *125* interaction, because, in each instance their product must be equal to *1*. The confounding of two-way interactions is also defined by this equation, because the *12* interaction multiplied by the *35* interaction must yield *1* and, hence, they are identical or confounded. Therefore, you can summarize all confounding in a design with such a fundamental identity equation.

Blocking

In some production processes, units are produced in natural "chunks" or blocks. You want to make sure that these blocks do not bias your estimates of main effects. For example, you may have a kiln to produce special ceramics, but the size of the kiln is limited so that you cannot produce all runs of your experiment at once. In that case you need to break up the experiment into blocks. However, you do not want to run positive settings of all factors in one block, and all negative settings in the other. Otherwise, any incidental differences between blocks would systematically affect all estimates of the main effects of the factors of interest. Rather, you want to distribute the runs over the blocks so that any differences between blocks (i.e., the blocking factor) do not bias your results for the factor effects of interest. This is accomplished by treating the blocking factor as another factor in the design. Consequently, you lose another interaction effect to the blocking factor, and the resultant design will be of lower resolution. However, these designs often have the advantage of being statistically more powerful because they enable you to estimate and control the variability in the production process that is due to differences between blocks.

Replicating the Design

It is sometimes desirable to replicate the design, i.e., to run each combination of factor levels in the design more than once. This will enable you to later estimate the so-called pure error in the experiment. The analysis of experiments is further discussed below; however, it should be clear that, when replicating the design, you can compute the variability of measurements within each unique combination of factor levels. This variability will give an indication of the random error in the measurements (e.g., due to uncontrolled factors, unreliability of the measurement instrument, etc.), because the replicated observations are taken under identical conditions (settings of factor levels). Such an estimate of the pure error can be used to evaluate the size and statistical significance of the variability that can be attributed to the manipulated factors.

Partial replications. When it is not possible or feasible to replicate each unique combination of factor levels (i.e., the full design), you can still gain an estimate of pure error by replicating only some of the runs in the experiment. However, you must consider the possible bias that may be introduced by selectively replicating only some runs. If you only replicate those runs that are most easily repeated (e.g., gather information at the points where it is "cheapest"), you may inadvertently only choose those combinations of factor levels that happen to produce very little (or very much) random variability, causing you to underestimate (or overestimate) the true amount

of pure error. Thus, you should carefully consider, typically based on your knowledge about the process that is being studied, which runs should be replicated, that is, which runs will yield a good (unbiased) estimate of pure error.

Adding Center Points

Designs with factors that are set at two levels implicitly assume that the effect of the factors on the dependent variable of interest (e.g., fabric *Strength*) is linear. It is impossible to test whether or not there is a nonlinear (e.g., quadratic) component in the relationship between a factor *A* and a dependent variable if *A* is only evaluated at two points (.i.e., at the *low* and *high* settings). If you suspect that the relationship between the factors in the design and the dependent variable is curve-linear instead, you should include one or more runs where all (continuous) factors are set at their midpoint. Such runs are called center-point runs (or center points), since they are, in a sense, in the center of the design (see the following graph).

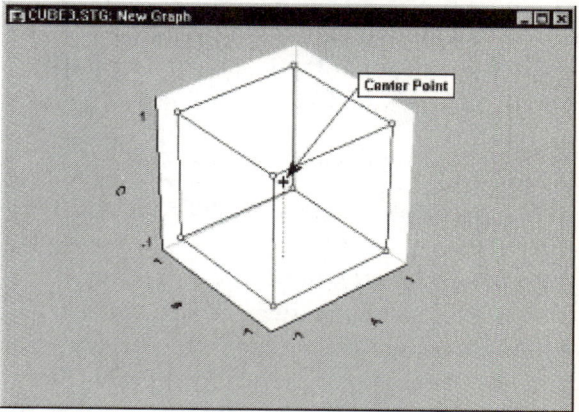

Later in the analysis, you can compare the measurements for the dependent variable at the center point with the average for the rest of the design. This provides a check for curvature (see Box and Draper, 1987). If the mean for the dependent variable at the center of the design is significantly different from the overall mean at all other points of the design, you have good reason to believe that the simple assumption that the factors are linearly related to the dependent variable, does not hold.

Analyzing the Results of a $2^{(k-p)}$ Experiment

Analysis of variance. Next, you need to determine exactly which of the factors significantly affected the dependent variable of interest. For example, in the study reported by Box and Draper (1987, page 115), it is desired to learn which of the factors involved in the manufacture of dyestuffs affected the strength of the fabric. In this example, factors 1 (*Polysulfide*), 4 (*Time*), and 6 (*Temperature*) significantly affected the strength of the fabric. Note that to simplify matters, only main effects are shown in the illustration.

Factor	ANOVA; Var.:STRENGTH; R-sqr=.60614; Adj:.56469 (Fabric.sta) 2**(6-0) design; MS Residual=3.62509 DV: STRENGTH				
Data: DOE Example Spreadsheet 5	SS	df	MS	F	p
(1) Polysulfide	48.8252	1	48.8252	13.46867	0.000536
(2) Reflux	7.9102	1	7.9102	2.18206	0.145132
(3) Moles	0.1702	1	0.1702	0.04694	0.829252
(4) Time	142.5039	1	142.5039	39.31044	0.000000
(5) Solvent	2.7639	1	2.7639	0.76244	0.386230
(6) Temperature	115.8314	1	115.8314	31.95269	0.000001
Error	206.6302	57	3.6251		
Total SS	524.6348	63			

Pure error and lack of fit. If the experimental design is at least partially replicated, you can estimate the error variability for the experiment from the variability of the replicated runs. Since those measurements were taken under identical conditions, that is, at identical settings of the factor levels, the estimate of the error variability from those runs is independent of whether or not the true model is linear or nonlinear in nature, or includes higher-order interactions.

The error variability so estimated represents pure error, that is, it is entirely due to unreliabilities in the measurement of the dependent variable. If available, you can use the estimate of pure error to test the significance of the residual variance, that is, all remaining variability that cannot be accounted for by the factors and their interactions that are currently in the model. If, in fact, the residual variability is significantly larger than the pure error variability, you can conclude that there is still some statistically significant variability left that is attributable to differences between the groups, and hence, that there is an overall lack of fit of the current model.

Factor	ANOVA; Var.:STRENGTH; R-sqr=.58547; Adj:.56475 (Fabric.sta) 2**(3-0) design; MS Pure Error=3.594844 DV: STRENGTH				
Data: DOE Example Spreadsheet 6	SS	df	MS	F	p
(1) Polysulfide	48.8252	1	48.8252	13.58200	0.000517
(2) Time	142.5039	1	142.5039	39.64120	0.000000
(3) Temperature	115.8314	1	115.8314	32.22154	0.000001
Lack of Fit	16.1631	4	4.0408	1.12405	0.354464
Pure Error	201.3113	56	3.5948		
Total SS	524.6348	63			

For example, the table above shows the results for the three factors that were previously identified as most important in their effect on fabric strength; all other factors where ignored in the analysis. As you can see in the row with the label *Lack of Fit*, when the residual variability for this model (i.e., after removing the three main effects) is compared against the pure error estimated from the within-group variability, the resulting *F* test is not statistically significant. Therefore, this result additionally supports the conclusion that, indeed, factors *Polysulfide, Time*, and *Temperature* significantly affected resultant fabric strength in an additive manner (i.e., there are no interactions). Or, put another way, all differences between the means obtained in the different experimental conditions can be sufficiently explained by the simple additive model for those three variables.

Parameter or effect estimates. Now, look at how these factors affected the strength of the fabrics.

Data: DOE Example Spreadsheet 7				
Factor	Effect Estimates; Var.:STRENGTH; R-sqr=.60614; Adj:.56469 2**(6-0) design; MS Residual=3.62509 DV: STRENGTH			
	Effect	Std.Err.	t(57)	p
Mean/Interc.	11.12344	0.237996	46.73794	0.000000
(1) Polysulfide	1.74687	0.475992	3.66997	0.000536
(2) Reflux	0.70312	0.475992	1.47718	0.145132
(3) Moles	0.10313	0.475992	0.21665	0.829252
(4) Time	2.98438	0.475992	6.26980	0.000000
(5) Solvent	-0.41562	0.475992	-0.87318	0.386230
(6) Temperature	2.69062	0.475992	5.65267	0.000001

The numbers in the previous spreadsheet are the effect or parameter estimates. With the exception of the overall *Mean/Intercept*, these estimates are the deviations of the mean of the negative settings from the mean of the positive settings for the respective factor. For example, if you change the setting of factor *Time* from *low* to *high*, you can expect an improvement in *Strength* by *2.98*; if you set the value for factor *Polysulfide* to its high setting, you can expect a further improvement by *1.75*, and so on.

As you can see, the same three factors that were statistically significant show the largest parameter estimates; thus the settings of these three factors were most important for the resultant strength of the fabric.

For analyses including interactions, the interpretation of the effect parameters is a bit more complicated. Specifically, the two-way interaction parameters are defined as half the difference between the main effects of one factor at the two levels of a second factor (see Mason, Gunst, and Hess, 1989, page 127); likewise, the three-way interaction parameters are defined as half the difference between the

two-factor interaction effects at the two levels of a third factor, and so on.

Regression coefficients. You can also look at the parameters in the multiple regression model (see Chapter 26 – *Multiple Linear Regression*). To continue this example, consider the following prediction equation:

Strength = const + b_1 *x_1 +... + b_6 *x_6

Here, x_1 through x_6 stand for the 6 factors in the analysis. The Effect Estimates shown earlier also contains these parameter estimates:

Data: DOE Example Spreadsheet 8				
Factor	Effect Estimates; Var.:STRENGTH; R-sqr=.60614; Adj:.56469 2**(6-0) design; MS Residual=3.62509 DV: STRENGTH			
	Coeff.	Std.Err. Coeff.	-95.% Cnf.Limt	+95.% Cnf.Limt
Mean/Interc.	11.12344	0.237996	10.64686	11.60002
(1) Polysulfide	0.87344	0.237996	0.39686	1.35002
(2) Reflux	0.35156	0.237996	-0.12502	0.82814
(3) Moles	0.05156	0.237996	-0.42502	0.52814
(4) Time	1.49219	0.237996	1.01561	1.96877
(5) Solvent	-0.20781	0.237996	-0.68439	0.26877
(6) Temperature	1.34531	0.237996	0.86873	1.82189

Actually, these parameters contain little new information, as they simply are one-half of the parameter values (except for the *Mean/Intercept*) shown earlier. This makes sense since now; the coefficient can be interpreted as the deviation of the high setting for the respective factors from the center. However, note that this is only the case if the factor values (i.e., their levels) are coded as -1 and +1, respectively. Otherwise, the scaling of the factor values will affect the magnitude of the parameter estimates. In the example data reported by Box and Draper (1987, page 115), the settings or values for the different factors were recorded on very different scales:

Data: DOE Example Spreadsheet 9* (9v by 64c)

2**(6-0) Design (Box & Draper, 1987; page 114-115)

	Poly-sulfide	Reflux	Moles	Time	Solvent	Temper-ature	Strength	Hue	Bright-ness
1	6	150	1.8	24	30	120	3.4	15.0	36.0
2	7	150	1.8	24	30	120	9.7	5.0	35.0
3	6	170	1.8	24	30	120	7.4	23.0	37.0
4	7	170	1.8	24	30	120	10.6	8.0	34.0
5	6	150	2.4	24	30	120	6.5	20.0	30.0
6	7	150	2.4	24	30	120	7.9	9.0	32.0
7	6	170	2.4	24	30	120	10.3	13.0	28.0
8	7	170	2.4	24	30	120	9.5	5.0	38.0
9	6	150	1.8	36	30	120	14.3	23.0	40.0
10	7	150	1.8	36	30	120	10.5	1.0	32.0
11	6	170	1.8	36	30	120	7.8	11.0	32.0
12	7	170	1.8	36	30	120	17.2	5.0	28.0
13	6	150	2.4	36	30	120	9.4	15.0	34.0
14	7	150	2.4	36	30	120	12.1	8.0	26.0
15	6	170	2.4	36	30	120	9.5	15.0	30.0
16	7	170	2.4	36	30	120	15.8	1.0	28.0

Shown in the next illustration are the regression coefficient estimates based on the uncoded original factor values:

Data: DOE Example Spreadsheet 10

Regr. Coefficients; Var.:STRENGTH; R-sqr=.60614; Adj:.5646
2**(6-0) design; MS Residual=3.62509
DV: STRENGTH

Factor	Regressn Coeff.	Std.Err.	t(57)	p
Mean/Interc.	-46.0641	8.109341	-5.68037	0.000000
(1) Polysulfide	1.7469	0.475992	3.66997	0.000536
(2) Reflux	0.0352	0.023800	1.47718	0.145132
(3) Moles	0.1719	0.793320	0.21665	0.829252
(4) Time	0.2487	0.039666	6.26980	0.000000
(5) Solvent	-0.0346	0.039666	-0.87318	0.386230
(6) Temperature	0.2691	0.047599	5.65267	0.000001

Because the metric for the different factors is no longer compatible, the magnitudes of the regression coefficients are not compatible either. This is why it is usually more informative to look at the ANOVA parameter estimates (for the coded values of the factor levels), as shown before. However, the regression coefficients can be useful when you want to make predictions for the dependent variable, based on the original metric of the factors.

Graph Options

Diagnostic plots of residuals. To start with, before accepting a particular model that includes a particular number of effects (e.g., main effects for *Polysulfide, Time,* and *Temperature* in the current example), you should always examine the

distribution of the residual values. These are computed as the difference between the predicted values (as predicted by the current model) and the observed values. You can compute the histogram for these residual values, as well as probability plots (as shown in the next illustration).

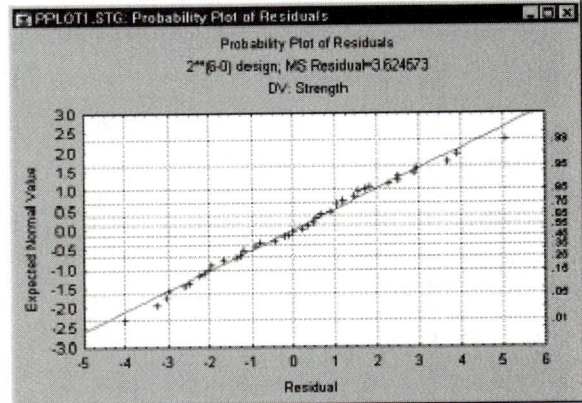

The parameter estimates and ANOVA table are based on the assumption that the residuals are normally distributed (see also, Chapter 1 – *Elementary Concepts in Statistics*). The histogram provides one way to check (visually) whether this assumption holds. The so-called normal probability plot is another common tool to assess how closely a set of observed values (residuals in this case) follows a theoretical distribution. In this plot the actual residual values are plotted along the horizontal x-axis; the vertical y-axis shows the expected normal values for the respective values, after they were rank-ordered. If all values fall onto a straight line, you can be satisfied that the residuals follow the normal distribution.

Pareto chart of effects. The Pareto chart of effects is often an effective tool for communicating the results of an experiment, in particular to laymen.

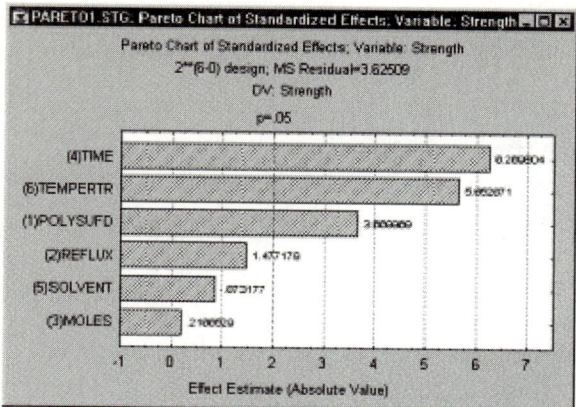

In this graph, the ANOVA effect estimates are sorted from the largest absolute value to the smallest absolute value. The magnitude of each effect is represented by a column, and often, a line going across the columns indicates how large an effect has to be (i.e., how long a column must be) to be statistically significant.

Normal probability plot of effects. Another useful, albeit more technical summary graph, is the normal probability plot of the estimates. As in the normal probability plot of the residuals, first the effect estimates are rank ordered, and then a normal z score is computed based on the assumption that the estimates are normally distributed. This z score is plotted on the Y-axis; the observed estimates are plotted on the X-axis.

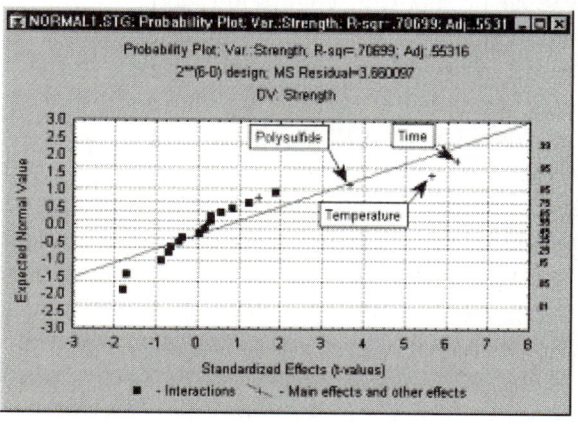

Square and cube plots. These plots are often used to summarize predicted values for the dependent variable, given the respective high and low setting of the factors. The square plot (see the next illustration) will show the predicted values (and, optionally, their confidence intervals) for two factors at a time. The cube plot will show the predicted values (and, optionally, confidence intervals) for three factors at a time.

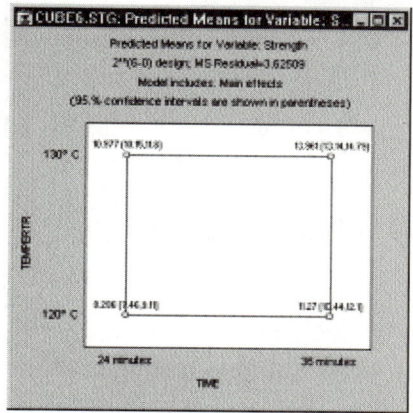

Interaction plots. A general graph for showing the means is the standard interaction plot, where the means are indicated by points connected by lines. This plot (see the next illustration) is particularly useful when there are significant interaction effects in the model.

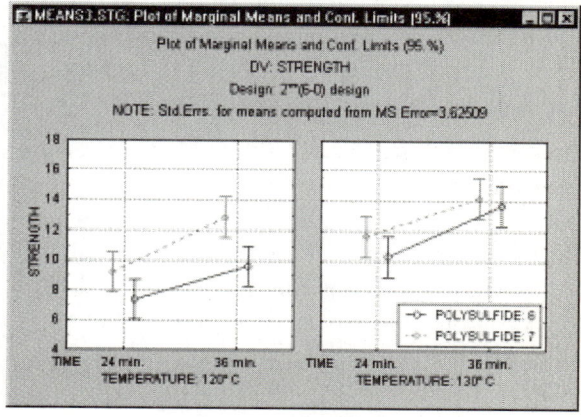

Surface and contour plots. When the factors in the design are continuous in nature, it is often also useful to look at surface and contour plots of the dependent variable as a function of the factors.

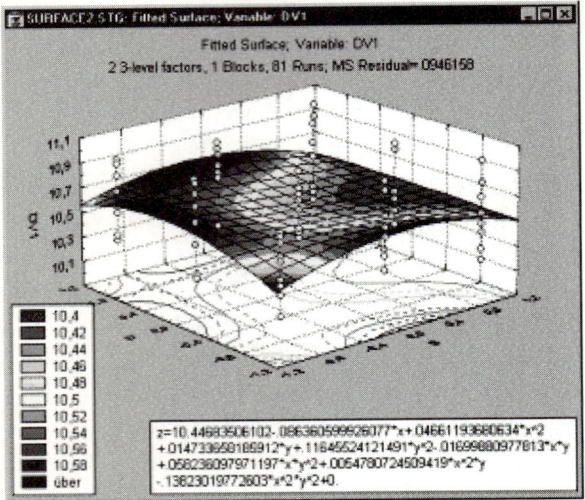

These types of plots will be discussed further later in this section, in the context of $3^{(k-p)}$ and central composite and response surface designs.

Summary

$2^{(k-p)}$ designs are the "workhorse" of industrial experiments. The impact of a large number of factors on the production process can simultaneously be assessed with relative efficiency (i.e., with few experimental runs). The logic of these types of experiments is straightforward (each factor has only two settings).

Disadvantages. The simplicity of these designs is also their major flaw. As mentioned previously, underlying the use of two-level factors is the belief that the resultant changes in the dependent variable (e.g., fabric strength) are basically linear in nature. This is often not the case, and many variables are related to quality characteristics in a nonlinear fashion. In the example above, if you were to continuously

increase the temperature factor (which was significantly related to fabric strength), you would of course eventually hit a peak, and from there on the fabric strength would decrease as the temperature increases. While this type of curvature in the relationship between the factors in the design and the dependent variable can be detected if the design included center point runs, you cannot fit explicit nonlinear (e.g., quadratic) models with $2^{(k-p)}$ designs (however, central composite designs will do exactly that).

Another problem of fractional designs is the implicit assumption that higher-order interactions do not matter; but sometimes they do, for example, when some other factors are set to a particular level, temperature may be negatively related to fabric strength. Again, in fractional factorial designs, higher-order interactions (greater than two-way) particularly will escape detection.

$2^{(k-p)}$ Maximally Unconfounded and Minimum Aberration Designs

Overview

$2^{(k-p)}$ fractional factorial designs are often used in industrial experimentation because of the economy of data collection that they provide. For example, suppose an engineer needed to investigate the effects of varying 11 factors, each with 2 levels, on a manufacturing process. Let's call the number of factors k, which would be 11 for this example. An experiment using a full factorial design, where the effects of every combination of levels of each factor are studied, would require $2^{(k)}$ experimental runs, or 2,048 runs for this example. To minimize the data collection effort, the engineer might decide to forego investigation of higher-order interaction effects of the 11 factors, and focus instead on identifying the main effects of the 11 factors and any low-order interaction effects that could be estimated from an experiment using a smaller, more reasonable number of experimental runs. There is another, more theoretical reason for not conducting huge, full factorial 2-level experiments. In general, it is not logical to be concerned with identifying higher-order interaction effects of the experimental factors, while ignoring lower-order nonlinear effects, such as quadratic or cubic effects, which cannot be estimated if only 2 levels of each factor are employed. So although practical considerations often lead to the need to design experiments with a reasonably small number of experimental runs, there is a logical justification for such experiments.

The alternative to the $2^{(k)}$ full factorial design is the $2^{(k-p)}$ fractional factorial design, which requires only a fraction of the data collection effort required for full factorial designs. For our example with $k=11$ factors, if only 64 experimental runs can be conducted, a $2^{(11-5)}$ fractional factorial experiment would be designed with $2^6 = 64$ experimental runs. In essence, a k-p = 6 way full factorial experiment is designed, with the levels of the p factors being generated by the levels of selected higher order interactions of the other 6 factors. Fractional factorials sacrifice higher order interaction effects so that lower order effects can still be computed correctly. However, different criteria can be used in choosing the higher order interactions to be used as generators, with different criteria sometimes leading to different "best" designs.

$2^{(k-p)}$ fractional factorial designs can also include blocking factors. In some production processes, units are produced in natural chunks or blocks. To make sure that these blocks do not bias your estimates of the effects for the k factors, blocking factors can be added as additional factors in the design. Consequently, you may sacrifice additional interaction effects to generate the blocking factors, but these designs often have the advantage of being statistically more powerful because they enable you to estimate and control the variability in the production process that is due to differences between blocks.

Design Criteria

Many of the concepts discussed in this section are also addressed in $2^{(k-p)}$ *Fractional Factorial Designs* (page 185). However, a technical description of how fractional factorial designs are constructed is beyond the scope of either section. Detailed accounts of how to design $2^{(k-p)}$ experiments can be found, for example, in Bayne and Rubin (1986), Box and Draper (1987), Box, Hunter, and Hunter (1978), Montgomery (1991), Daniel (1976), Deming and Morgan (1993), Mason, Gunst, and Hess (1989), or Ryan (1989), to name only a few of the many textbooks on this subject.

In general, the $2^{(k-p)}$ maximally unconfounded and minimum aberration designs techniques will successively select which higher-order interactions to use as generators for the p factors. For example, consider the following design that includes 11 factors but requires only 16 runs (observations).

Data: DOE Example Spreadsheet 11

Design: 2**(11-7), Resolution III

Standard Run	A	B	C	D	E	F	G	H	I	J	K
1	-1	-1	-1	-1	-1	-1	-1	-1	1	1	1
2	-1	-1	-1	1	-1	1	1	1	-1	1	1
3	-1	-1	1	-1	1	1	1	-1	-1	1	-1
4	-1	-1	1	1	1	-1	-1	1	1	1	-1
5	-1	1	-1	-1	1	1	-1	1	-1	-1	1
6	-1	1	-1	1	1	-1	1	-1	1	-1	1
7	-1	1	1	-1	-1	-1	1	1	1	-1	-1
8	-1	1	1	1	-1	1	-1	-1	-1	-1	-1
9	1	-1	-1	-1	1	-1	1	-1	1	-1	-1
10	1	-1	-1	1	1	1	-1	1	-1	-1	-1
11	1	-1	1	-1	-1	1	-1	-1	1	-1	1
12	1	-1	1	1	-1	-1	1	1	-1	-1	1
13	1	1	-1	-1	-1	1	1	-1	-1	1	-1
14	1	1	-1	1	-1	-1	-1	1	1	1	-1
15	1	1	1	-1	1	-1	-1	-1	-1	1	1
16	1	1	1	1	1	1	1	1	1	1	1

Interpreting the design. The design displayed in the previous spreadsheet should be interpreted as follows. Each column contains +1s or -1s to indicate the setting of the respective factor (high or low, respectively). So for example, in the first run of the experiment, all factors A through K are set to the higher level, and in the second run, factors A, B, and C are set to the higher level, but factor D is set to the lower level, and so on. Notice that the settings for each experimental run for factor E can be produced by multiplying the respective settings for factors A, B, and C. The $A \times B \times C$ interaction effect therefore cannot be estimated independently of the factor E effect in this design because these two effects are confounded. Likewise, the settings for factor F can be produced by multiplying the respective settings for factors B, C, and D. We say that ABC and BCD are the generators for factors E and F, respectively.

The maximum resolution design criterion. In the spreadsheet shown above, the design is described as a $2^{(11-7)}$ design of resolution III (three). This means that you study overall $k = 11$ factors, but $p = 7$ of those factors were generated from the interactions of a full $2^{(11-7)=4}$ factorial design. As a result, the design does not give full resolution; that is, there are certain interaction effects that are confounded with (identical to) other effects. In general, a design of resolution R is one where no l-way interactions are confounded with any other interaction of order less than R – 1. In the current example, R is equal to 3. Here, no l = 1-way interactions (i.e., main effects) are confounded with any other interaction of order less than R – 1 = 3 -1 = 2. Thus, main effects in this design are unconfounded with each other, but are confounded with two-factor interactions; and consequently, with other higher-order

interactions. One obvious, but nevertheless very important overall design criterion is that the higher-order interactions to be used as generators should be chosen such that the resolution of the design is as high as possible.

The maximum unconfounding design criterion. Maximizing the resolution of a design, however, does not by itself ensure that the selected generators produce the best design. Consider, for example, two different resolution IV designs. In both designs, main effects would be unconfounded with each other and 2-factor interactions would be unconfounded with main effects, i.e, no l = 2-way interactions are confounded with any other interaction of order less than R – 1 = 4 – 2 = 2. The two designs might be different, however, with regard to the degree of confounding for the 2-factor interactions. For resolution IV designs, the crucial order, in which confounding of effects first appears, is for 2-factor interactions. In one design, none of the crucial order, 2-factor interactions might be unconfounded with all other 2-factor interactions, while in the other design, virtually all of the 2-factor interactions might be unconfounded with all of the other 2-factor interactions. The second "almost resolution V" design would be preferable to the first "just barely resolution IV" design. This suggests that even though the maximum resolution design criterion should be the primary criterion, a subsidiary criterion might be that generators should be chosen such that the maximum number of interactions of less than or equal to the crucial order, given the resolution, are unconfounded with all other interactions of the crucial order. This is called the maximum unconfounding design criterion, and is one of the optional, subsidiary design criterion to use in a search for a $2^{(k-p)}$ design.

The minimum aberration design criterion.
The minimum aberration design criterion is another optional, subsidiary criterion to use in a search for a $2^{(k-p)}$ design. In some respects, this criterion is similar to the maximum unconfounding design criterion. Technically, the minimum aberration design is defined as the design of maximum resolution "which minimizes the number of words in the defining relation that are of minimum length" (Fries & Hunter, 1980). Less technically, the criterion apparently operates by choosing generators that produce the smallest number of pairs of confounded interactions of the crucial order. For example, the minimum aberration resolution IV design would have the minimum number of pairs of confounded 2-factor interactions.

To illustrate the difference between the maximum unconfounding and minimum aberration criteria, consider the maximally unconfounded $2^{(9-4)}$ design and the minimum aberration $2^{(9-4)}$ design, as for example, listed in Box, Hunter, and Hunter (1978). If you compare these two designs, you will find that in the maximally unconfounded design, 15 of the 36 2-factor interactions are unconfounded with any other 2-factor interactions, while in the minimum aberration design, only 8 of the 36 2-factor interactions are unconfounded with any other 2-factor interactions. The minimum aberration design, however, produces 18 pairs of confounded interactions, while the maximally unconfounded design produces 21 pairs of confounded interactions. So, the two criteria lead to the selection of generators producing different best designs.

Fortunately, the choice of whether to use the maximum unconfounding criterion or the minimum aberration criterion makes no difference in the design that is selected (except for, perhaps, relabeling of the factors) when

there are 11 or fewer factors, with the single exception of the $2^{(9-4)}$ design described above (see Chen, Sun, & Wu, 1993). For designs with more than 11 factors, the two criteria can lead to the selection of very different designs, and for lack of better advice, we suggest using both criteria, comparing the designs that are produced, and choosing the design that best suits your needs. We will add, editorially, that maximizing the number of totally unconfounded effects often makes more sense than minimizing the number of pairs of confounded effects.

Summary

$2^{(k-p)}$ fractional factorial designs are probably the most frequently used type of design in industrial experimentation. Things to consider in designing any $2^{(k-p)}$ fractional factorial experiment include the number of factors to be investigated, the number of experimental runs, and whether there will be blocks of experimental runs. Beyond these basic considerations, you should also take into account whether the number of runs will allow a design of the required resolution and degree of confounding for the crucial order of interactions, given the resolution.

$3^{(k-p)}$, Box-Behnken, and Mixed 2- and 3-Level Factorial Designs

Overview

In some cases, factors that have more than 2 levels have to be examined. For example, if you suspect that the effect of the factors on the dependent variable of interest is not simply linear, then, as discussed earlier (see $2^{(k-p)}$ designs), you need at least 3 levels in order to test for the linear and quadratic effects (and interactions) for those factors. Also, sometimes some factors may be categorical in nature, with more than 2 categories. For example, you may have three different machines that produce a particular part.

Designing $3^{(k-p)}$ Experiments

The general mechanism of generating fractional factorial designs at 3 levels ($3^{(k-p)}$ designs) is very similar to that described in the context of $2^{(k-p)}$ designs. Specifically, you start with a full factorial design, and then uses the interactions of the full design to construct new factors (or blocks) by making their factor levels identical to those for the respective interaction terms (i.e., by making the new factors aliases of the respective interactions).

For example, consider the following simple $3^{(3-1)}$ factorial design:

Data: DOE Example Spreadsheet 12

Standard Run	3**(3-1) Fractional Factorial Design, 1 Block , 9 Runs		
	A	B	C
1	0	0	0
2	0	1	2
3	0	2	1
4	1	0	2
5	1	1	1
6	1	2	0
7	2	0	1
8	2	1	0
9	2	2	2

As in the case of $2^{(k-p)}$ designs, the design is constructed by starting with the full *3-1=2* factorial design; those factors are listed in the first two columns (factors *A* and *B*). Factor *C* is constructed from the interaction *AB* of the first two factors. Specifically, the values for factor *C* are computed as

$$C = 3 - mod_3 (A+B)$$

Here, $mod_3(x)$ stands for the so-called *modulo-3* operator, which will first find a number *y* that is less than or equal to *x*, and that is evenly divisible by 3, and then compute the difference (remainder) between number *y* and *x*. For example, $mod_3(0)$ is equal to *0*, $mod_3(1)$ is equal to *1*, $mod_3(3)$ is equal to *0*, $mod_3(5)$ is equal to *2* (*3* is the largest number that is less than or equal to *5*, and that is evenly divisible by *3*; finally, *5-3=2*), and so on.

Fundamental identity. If you apply this function to the sum of columns *A* and *B* shown above, you will obtain the third column C. Similar to the case of $2^{(k-p)}$ designs (see $2^{(k-p)}$ designs for a discussion of the fundamental identity in the context of $2^{(k-p)}$ designs), this confounding of interactions with "new" main effects can be summarized in an expression:

$$0 = mod_3 (A+B+C)$$

If you look back at the $3^{(3-1)}$ design shown earlier, you will see that, indeed, if you add the numbers in the three columns they will all sum to either *0*, *3*, or *6*, that is, values that are evenly divisible by 3 (and hence: $mod_3(A+B+C)=0$). Thus, you could write as a shortcut notation ABC=0, in order to summarize the confounding of factors in the fractional $3^{(k-p)}$ design.

Some of the designs will have fundamental identities that contain the number 2 as a multiplier; e.g.,

$$0 = mod_3 (B+C*2+D+E*2+F)$$

This notation can be interpreted exactly as before, that is, the $modulo_3$ of the sum $B+2*C+D+2*E+F$ must be equal to 0. The next example shows such an identity.

An Example $3^{(4-1)}$ Design in 9 Blocks

Here is the summary for a 4-factor, 3-level fractional factorial design in 9 blocks that requires only 27 runs.

SUMMARY: $3^{(4-1)}$ fractional factorial
Design generators: ABCD
Block generators: AB,AC2
Number of factors (independent variables): 4
Number of runs (cases, experiments): 27
Number of blocks: 9

With this design, you can test for linear and quadratic main effects for 4 factors in 27 observations, which can be gathered in 9 blocks of 3 observations each. The fundamental identity or design generator for the design is ABCD, thus the modulo$_3$ of the sum of the factor levels across the four factors is equal to 0. The fundamental identity also enables you to determine the confounding of factors and interactions in the design (see McLean and Anderson, 1984, for details).

Data: DOE Table 13

Unconfounded Effects (Spreadsheet8)
List of uncorrelated factors and interactions
3**(4-1) fractional factorial design, 9 blocks, 27 runs

	Unconf. Effects (excl. blocks)	Unconfounded if blocks included?			
1	(1) A (L)	Yes			
2	A (Q)	Yes			
3	(2) B (L)	Yes			
4	B (Q)	Yes			
5	(3) C (L)	Yes			
6	C (Q)	Yes			
7	(4) D (L)	Yes			
8	D (Q)	Yes			

As you can see, in this $3^{(4-1)}$ design the main effects are not confounded with each other, even when the experiment is run in 9 blocks.

Box-Behnken Designs

In the case of $2^{(k-p)}$ designs, Plackett and Burman (1946) developed highly fractionalized designs to screen the maximum number of (main) effects in the least number of experimental runs. The equivalent in the case of $3^{(k-p)}$ designs are the so-called Box-Behnken designs (Box and Behnken, 1960; see also, Box

and Draper, 1984). These designs do not have simple design generators (they are constructed by combining two-level factorial designs with incomplete block designs), and have complex confounding of interaction. However, the designs are economical and therefore particularly useful when it is expensive to perform the necessary experimental runs.

Analyzing the $3^{(k-p)}$ Design

The analysis of these types of designs proceeds basically in the same way as was described in the context of $2^{(k-p)}$ designs. However, for each effect, you can now test for the linear effect and the quadratic (nonlinear effect). For example, when studying the yield of a chemical process, the temperature may be related in a nonlinear fashion, that is, the maximum yield may be attained when the temperature is set at the medium level. Thus, nonlinearity often occurs when a process performs near its optimum.

ANOVA Parameter Estimates

To estimate the ANOVA parameters, the factors levels for the factors in the analysis are internally recoded so that you can test the linear and quadratic components in the relationship between the factors and the dependent variable. Thus, regardless of the original metric of factor settings (e.g., *100 degrees C, 110 degrees C, 120 degrees C*), you can always recode those values to *-1, 0,* and *+1* to perform the computations. The resultant ANOVA parameter estimates can be interpreted analogously to the parameter estimates for $2^{(k-p)}$ designs.

For example, consider the following ANOVA results:

Data: DOE Example Spreadsheet 14 (4v by 13c)

Factor	Effect	Std. Err.	t(69)	p
Mean/Interc.	103.6942	0.390591	265.4805	0
Blocks (1)	0.8028	1.360542	0.5901	0.557055
Blocks (2)	-1.2307	1.291511	-0.9529	0.343952
(1) Temperature (L)	-0.3245	0.977778	-0.3319	0.740991
Temperature (Q)	-0.5111	0.809946	-0.6311	0.530091
(2) Time (L)	0.0017	0.977778	0.0018	0.998589
Time (Q)	0.0045	0.809946	0.0056	0.995541
(3) Speel (L)	-10.3073	0.977778	-10.5415	0
Speed (Q)	-3.7915	0.809946	-4.6812	0.000014
1 L by 2 L	3.9256	1.540235	2.5487	0.013041
1 L by 2 Q	0.4384	1.371941	0.3195	0.750297
1 Q by 2 L	0.4747	1.371941	0.346	0.730403
1 Q by 2 Q	-2.7499	0.995575	-2.7621	0.007353

Main-effect estimates. By default, the effect estimate for the linear effects (marked by an *L* next to the factor name) can be interpreted as the difference between the average response at the low and high settings for the respective factors. The estimate for the quadratic (nonlinear) effect (marked by a *Q* next to the factor name) can be interpreted as the difference between the average response at the center (medium) settings and the combined high and low settings for the respective factors.

Interaction effect estimates. As in the case of $2^{(k-p)}$ designs, the linear-by-linear interaction effect can be interpreted as half the difference between the linear main effect of one factor at the high and low settings of another. Analogously, the interactions by the quadratic components can be interpreted as half the difference between the quadratic main effect of one factor at the respective settings of another; that is, either the high or low setting (quadratic by linear interaction), or the medium or high and low settings combined (quadratic by quadratic interaction).

In practice, and from the standpoint of interpretability of results, you would usually try to avoid quadratic interactions. For example, a quadratic-by-quadratic *A*-by-*B* interaction indicates that the nonlinear effect of factor *A* is modified in a nonlinear fashion by the setting of *B*. This means

that there is a fairly complex interaction between factors present in the data that will make it difficult to understand and optimize the process. Sometimes, performing nonlinear transformations (e.g., a *log* transformation) of the dependent variable values can remedy the problem.

Centered and non-centered polynomials. As mentioned above, the interpretation of the effect estimates applies only when you use the default parameterization of the model. In that case, you would code the quadratic factor interactions so that they become maximally "untangled" from the linear main effects.

Graphical Presentation of Results

The same diagnostic plots (e.g., of residuals) are available for $3^{(k-p)}$ designs as were described in the context of $2^{(k-p)}$ designs. Thus, before interpreting the final results, you should always first look at the distribution of the residuals for the final fitted model. The ANOVA assumes that the residuals (errors) are normally distributed.

Plot of means. When an interaction involves categorical factors (e.g., type of machine, specific operator of machine, and distinct setting of the machine), the best way to understand interactions is to look at the interaction plot of means.

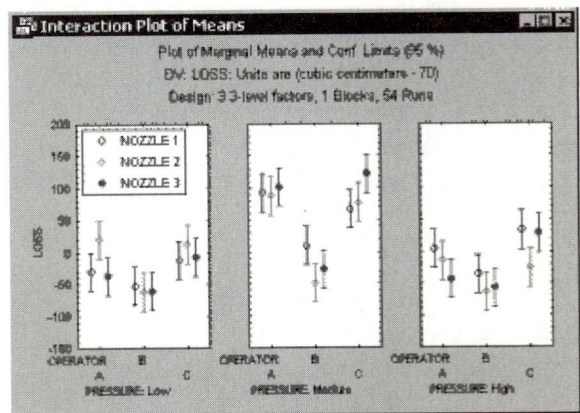

Surface plot. When the factors in an interaction are continuous in nature, you may want to look at the surface plot that shows the response surface applied by the fitted model. Note that this graph also contains the prediction equation (in terms of the original metric of factors) that produces the respective response surface.

Designs for Factors at 2 and 3 Levels

You can also generate standard designs with 2 and 3 level factors. Specifically, you can generate the standard designs as enumerated by Connor and Young for the US National Bureau of Standards (see McLean and Anderson, 1984). The technical details of the method used to generate these designs are beyond the scope of this introduction. However, in general the technique is, in a sense, a combination of the procedures described in the context of $2^{(k-p)}$ and $3^{(k-p)}$ designs. It should be noted however, that, while all of these designs are very efficient, they are not necessarily orthogonal with respect to all main effects. This is, however, not a problem, if you use a general algorithm for estimating the ANOVA parameters and sums of squares that does not require orthogonality of the design.

The design and analysis of these experiments proceeds along the same lines as discussed in the context of $2^{(k-p)}$ and $3^{(k-p)}$ experiments.

Central Composite and Non-Factorial Response Surface Designs

Overview

The $2^{(k-p)}$ and $3^{(k-p)}$ designs all require that the levels of the factors are set at, for example, 2 or 3 levels. In many instances, such designs are not feasible, because, for example, some factor

combinations are constrained in some way (e.g., factors A and B cannot be set at their high levels simultaneously). Also, for reasons related to efficiency, which will be discussed shortly, it is often desirable to explore the experimental region of interest at particular points that cannot be represented by a factorial design.

The designs (and how to analyze them) discussed in this section all pertain to the estimation (fitting) of response surfaces, following the general model equation:

$$y = b_0 + b_1 {}^*x_1 + ... + b_k {}^*x_k + b_{12} {}^*x_1 {}^*x_2 + b_{13} {}^*x_1 {}^*x_3 + ... + b_{k-1,k} {}^*x_{k-1} {}^*x_k + b_{11} {}^*x_1{}^2 + ... + b_{kk} {}^*x_k{}^2$$

Put into words, you are fitting a model to the observed values of the dependent variable y, that include 1) main effects for factors x_1, ..., x_k, 2) their interactions ($x_1{}^*x_2$, $x_1{}^*x_3$, ... ,$x_{k-1}{}^*x_k$), and 3) their quadratic components ($x_1{}^{**}2$, ..., $x_k{}^{**}2$). No assumptions are made concerning the "levels" of the factors, and you can analyze any set of continuous values for the factors.

There are some considerations concerning design efficiency and biases, which have led to standard designs that are ordinarily used when attempting to fit these response surfaces, and those standard designs will be discussed shortly (e.g., see Box, Hunter, and Hunter, 1978; Box and Draper, 1987; Khuri and Cornell, 1987; Mason, Gunst, and Hess, 1989; Montgomery, 1991). But, as will be discussed later, in the context of constrained surface designs and D- and A-optimal designs, these standard designs can sometimes not be used for practical reasons. However, the central composite design analysis options do not make any assumptions about the structure of your data file, that is, the number of distinct factor values, or their combinations across the runs of the experiment, and, hence, these options can be used to analyze any type of design, to fit to the data the general model described above.

Design Considerations

Orthogonal designs. One desirable characteristic of any design is that the main effect and interaction estimates of interest are independent of each other. For example, suppose you have a two-factor experiment, with both factors at two levels. Your design consists of four runs:

	A	B
1	1	1
2	1	1
3	-1	-1
4	-1	-1

Data: DOE Example Spreadsheet 15

For the first two runs, both factors A and B are set at their high levels ($+1$). In the last two runs, both are set at their low levels (-1). Suppose you want to estimate the independent contributions of factors A and B to the prediction of the dependent variable of interest. Clearly this is a silly design, because there is no way to estimate the A main effect and the B main effect. You can only estimate one effect – the difference between *Runs 1+2* vs. *Runs 3+4* – which represents the combined effect of A and B.

The point here is that, in order to assess the independent contributions of the two factors, the factor levels in the four runs must be set so that the columns in the design (under A and B in the previous illustration) are independent of each other. Another way to express this requirement is to say that the columns of the design matrix (with as many columns as there are main effect and interaction parameters that you want to estimate) should be orthogonal (this term was first used by Yates, 1933). For example, if the four runs in the design are arranged as follows:

	A	B
1	1	1
2	1	-1
3	-1	1
4	-1	-1

Data: DOE Example Spreadsheet 16

the A and B columns are orthogonal. Now you can estimate the A main effect by comparing the high level for A within each level of B, with the low level for A within each level of B; the B main effect can be estimated in the same way.

Technically, two columns in a design matrix are orthogonal if the sum of the products of their elements within each row is equal to zero. In practice, you often encounter situations, for example due to loss of some data in some runs or other constraints, where the columns of the design matrix are not completely orthogonal. In general, the rule here is that the more orthogonal the columns are, the better the design, that is, the more independent information can be extracted from the design regarding the respective effects of interest. Therefore, one consideration for choosing standard central composite designs is to find designs that are orthogonal or near orthogonal.

Rotatable designs. The second consideration is related to the first requirement, in that it also has to do with how best to extract the maximum amount of (unbiased) information from the design, or specifically, from the experimental region of interest. Without going into details (see Box, Hunter, and Hunter, 1978; Box and Draper, 1987, Chapter 14; see also, Deming and Morgan, 1993, Chapter 13), it can be shown that the standard error for the prediction of dependent variable values is proportional to:

$$(1 + f(x)' * (X'X)^{-1} * f(x))^{**\frac{1}{2}}$$

where $\mathbf{f(x)}$ stands for the (coded) factor effects for the respective model [$\mathbf{f(x)}$ is a vector, $\mathbf{f(x)'}$ is the transpose of that vector], and $\mathbf{X}$ is the design matrix for the experiment, that is, the matrix of coded factor effects for all runs; $\mathbf{X'X^{-1}}$ is the inverse of the crossproduct matrix. Deming and Morgan (1993) refer to this expression as the normalized uncertainty; this function is also related to the variance function as defined by Box and Draper (1987). The amount of uncertainty in the prediction of dependent variable values depends on the variability of the design points, and their covariance over the runs. (Note that it is inversely proportional to the determinant of $\mathbf{X'X}$; this issue is further discussed in *Constructing D- and A-Optimal Designs*, page 221).

The point here is that, again, you want to choose a design that extracts the most information regarding the dependent variable and leaves the least amount of uncertainty for the prediction of future values. It follows that the amount of information (or normalized information according to Deming and Morgan, 1993) is the inverse of the normalized uncertainty.

For the simple 4-run orthogonal experiment shown earlier, the information function is equal to:

$$I_x = 4/(1 + x_1^2 + x_2^2)$$

where x_1 and x_2 stand for the factor settings for factors A and B, respectively (see Box and Draper, 1987).

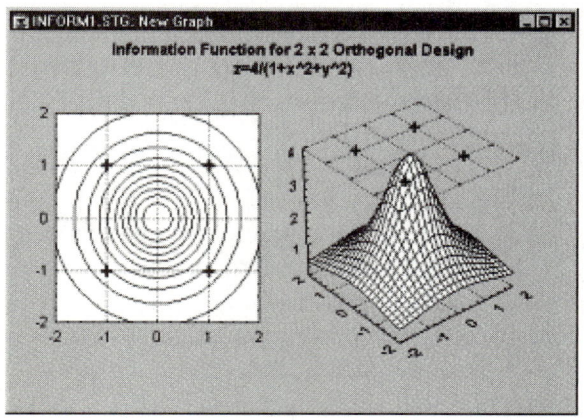

Information Function for 2 x 2 Orthogonal Design
z=4/(1+x^2+y^2)

Inspection of this function in a plot (see the previous illustration) shows that it is constant on circles centered at the origin. Thus any kind of rotation of the original design points will generate the same amount of information, that is, generate the same information function. Therefore, the 2-by-2 orthogonal design in 4 runs shown earlier is said to be rotatable.

As pointed out before, in order to estimate the second order, quadratic, or nonlinear component of the relationship between a factor and the dependent variable, you need at least 3 levels for the respective factors. What does the information function look like for a simple 3-by-3 factorial design, for the second-order quadratic model as shown at the beginning of this section?

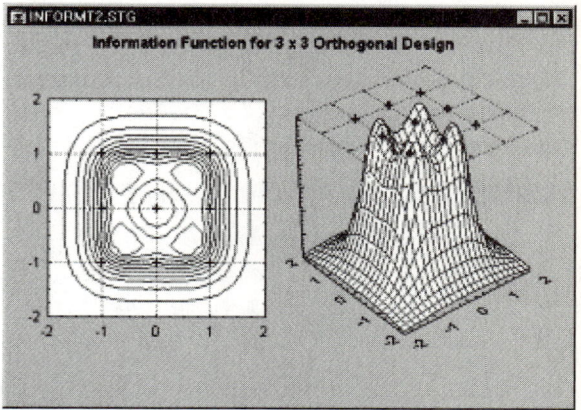

Information Function for 3 x 3 Orthogonal Design

As it turns out (see Box and Draper, 1987 and Montgomery, 1991; refer also to the manual), this function looks more complex, contains "pockets" of high-density information at the edges (which are probably of little particular interest to the experimenter), and clearly it is not constant on circles around the origin. Therefore, it is not rotatable, meaning different rotations of the design points will extract different amounts of information from the experimental region.

Star-points and rotatable second-order designs.

It can be shown that by adding so-called star-points to the simple (square or cube) 2-level factorial design points, you can achieve rotatable and often orthogonal or nearly orthogonal designs. For example, adding to the simple 2-by-2 orthogonal design shown earlier the following points, will produce a rotatable design.

	A	B
Run 1	1	1
Run2	1	-1
Run 3	-1	1
Run 4	-1	-1
Run 5	-1.414	0
Run 6	1.414	0
Run 7	0	-1.414
Run 8	0	1.414
Run 9	0	0
Run 10	0	0

Data: DOE Example Spreadsheet 17 (2v by 10c)

The first four runs in this design are the previous 2-by-2 factorial design points (or square points or cube points); runs 5 through 8 are the so-called star points or axial points, and runs 9 and 10 are center points.

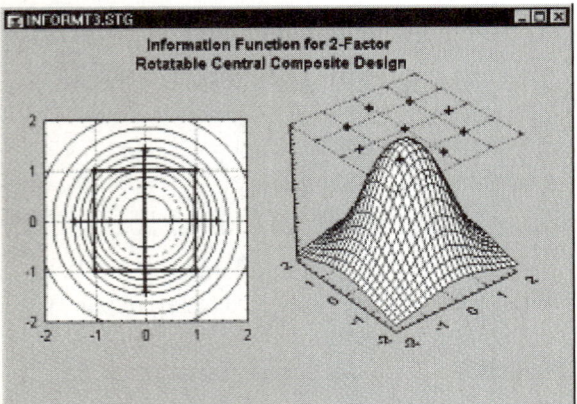

Information Function for 2-Factor Rotatable Central Composite Design

INFORMT3.STG

The information function for this design for the second-order (quadratic) model is rotatable, that is, it is constant on the circles around the origin.

Alpha for Rotatability and Orthogonality

The two design characteristics orthogonality and rotatability depend on the number of center points in the design and on the so-called axial distance α (*alpha*), which is the distance of the star points from the center of the design (i.e., *1.414* in the design shown above). It can be shown (e.g., see Box, Hunter, and Hunter, 1978; Box and Draper, 1987, Khuri and Cornell, 1987; Montgomery, 1991) that a design is rotatable if:

$$\alpha = (n_c)^{\frac{1}{4}}$$

where n_c stands for the number of cube points in the design (i.e., points in the factorial portion of the design).

A central composite design is orthogonal if you choose the axial distance so that:

$$\alpha = \{[(n_c + n_s + n_0)^{\frac{1}{2}} - n_c^{\frac{1}{2}}]^2 * n_c/4\}^{\frac{1}{4}}$$

where n_c is the number of cube points in the design, n_s is the number of star points in the design, and n_0 is the number of center points in the design.

To make a design both (approximately) orthogonal and rotatable, you would first choose the axial distance for rotatability, and then add center points (see Kkuri and Cornell, 1987), so that:

$$n_0 \approx 4*n_c^{\frac{1}{2}} + 4 - 2k$$

where k stands for the number of factors in the design.

Finally, if blocking is involved, Box and Draper (1987) give the following formula for computing the axial distance to achieve orthogonal blocking, and in most cases also reasonable information function contours, that is, contours that are close to spherical:

$$\alpha = [k*(l+n_{s0}/n_s)/(1+n_{c0}/n_c)]^{\frac{1}{2}}$$

where n_{s0} is the number of center points in the star portion of the design, n_s is the number of non-center star points in the design, n_{c0} is the number of center points in the cube portion of the design, and n_c is the number of non-center cube points in the design

Available Standard Designs

The standard central composite designs are usually constructed from a $2^{(k-p)}$ design for the cube portion of the design, which is augmented with center points and star points. Box and Draper (1987) list a number of such designs.

Small composite designs. In the standard designs, the cube portion of the design is typically of resolution V (or higher). This is, however, not necessary, and in cases when the experimental runs are expensive, or when it is not necessary to perform a statistically powerful test of model adequacy, you could choose for the cube portion designs of resolution III. For example, it could be constructed from highly fractionalized Plackett-Burman designs. Hartley (1959) described such designs.

Analyzing Central Composite Designs

The analysis of central composite designs proceeds in much the same way as for the analysis of $3^{(k-p)}$ designs. You fit to the data the general model described above; for example, for two variables you would fit the model:

$$y = b_0 + b_1*x_1 + b_2*x_2 + b_{12}*x_1*x_2 + b_{11}*x_1{}^2 + b_{22}*x_2{}^2$$

Fitted Response Surface

The shape of the fitted overall response can best be summarized in graphs and you can generate both contour plots and response surface plots (see the next illustration) for the fitted model.

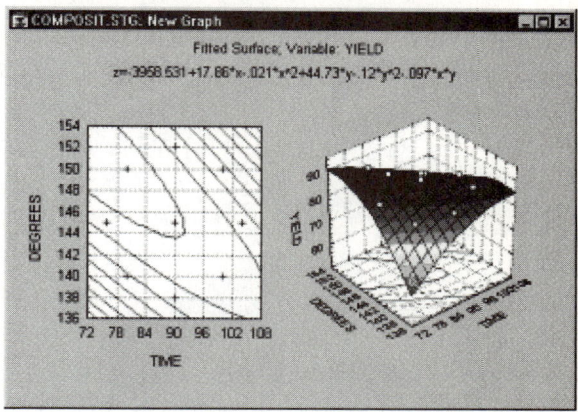

Categorized Response Surface

You can fit 3D surfaces to your data, categorized by some other variable. For example, if you replicated a standard central composite design 4 times, it may be very informative to see how similar the surfaces are when fitted to each replication.

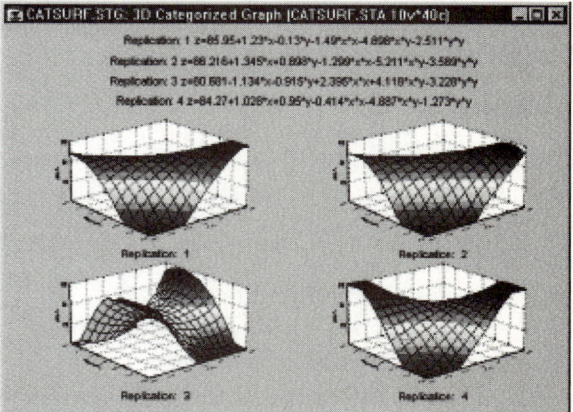

This would give you a graphical indication of the reliability of the results and where (e.g., in which region of the surface) deviations occur.

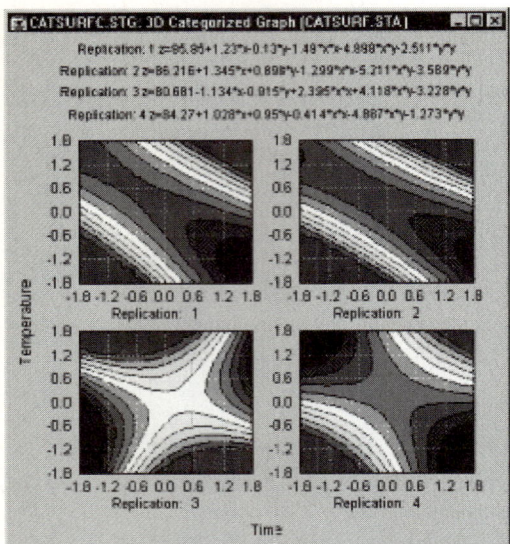

Clearly, the third replication produced a different surface. In replications *1*, *2*, and *4*, the fitted surfaces are very similar to each other. Thus, you should investigate what could have caused this noticeable difference in the third replication of the design.

Latin Square Designs

Overview

Latin square designs (the term *Latin square* was first used by Euler, 1782) are used when the factors of interest have more than two levels and you know ahead of time that there are no (or only negligible) interactions between factors. For example, if you wanted to examine the effect of 4 fuel additives on reduction in oxides of nitrogen and had 4 cars and 4 drivers at your disposal, you could of course run a full 4 x 4 x 4 factorial design, resulting in 64 experimental runs. However, you are not really interested in any (minor) interactions between the fuel additives and drivers, fuel additives and cars, or cars and drivers. You are mostly

interested in estimating main effects, in particular the one for the fuel additives factor. At the same time, you want to make sure that the main effects for drivers and cars do not affect (bias) your estimate of the main effect for the fuel additive.

If you labeled the additives with the letters A, B, C, and D, the Latin square design that would enable you to derive unconfounded main effects estimates could be summarized as follows (see also, Box, Hunter, and Hunter, 1978, page 263):

DRIVER	CAR			
	1	2	3	4
1	A	B	D	C
2	D	C	A	B
3	B	D	C	A
4	C	A	B	D

The example shown above is actually only one of the three possible arrangements in effect estimates. These arrangements are also called Latin square. The example above constitutes a 4 x 4 Latin square; and rather than requiring the 64 runs of the complete factorial, you can complete the study in only 16 runs.

Greco-Latin square. A nice feature of Latin squares is that they can be superimposed to form what are called Greco-Latin squares (this term was first used by Fisher and Yates, 1934). For example, the following two 3 x 3 Latin squares can be superimposed to form a Greco-Latin square:

a b c   α β τ   aα bβ cτ
b c a and τ α β results in bτ cα aβ
c a b   β τ α   cβ aτ bα

In the resultant Greco-Latin square design, you can evaluate the main effects of four 3-level factors (row factor, column factor, Roman letters, Greek letters) in only 9 runs.

Hyper-Greco Latin square. For some numbers of levels, there are more than two possible Latin

square arrangements. For example, there are three possible arrangements for 4-level Latin squares. If all three of them are superimposed, you get a Hyper-Greco Latin square design. In that design you can estimate the main effects of all five 4-level factors with only 16 runs in the experiment.

Analyzing the Design

Analyzing Latin square designs is straightforward. Also, plots of means can be produced to aid in the interpretation of results.

Very Large Designs, Random Effects, Unbalanced Nesting

Note that there are several other statistical methods that can also analyze these types of designs; see *Methods for Analysis of Variance*, page 54, for details. In particular, Chapter 39 – *Variance Components and Mixed Model ANOVA/ANCOVA* discusses very efficient methods for analyzing designs with unbalanced nesting (when the nested factors have different numbers of levels within the levels of the factors in which they are nested), very large nested designs (e.g., with more than 200 levels overall), or hierarchically nested designs (with or without random factors).

Taguchi Methods: Robust Design Experiments

Overview

Applications. Taguchi methods have become increasingly popular in recent years. The documented examples of sizable quality improvements that resulted from implementations of these methods (see, for example, Phadke, 1989; Noori, 1989) have added to the curiosity among American manufacturers. In fact, some of the leading manufacturers in this country have begun to use these methods with usually great success. For example, AT&T is using these methods in the manufacture of very large-scale integrated (VLSI) circuits; also, Ford Motor Company has gained significant quality improvements due to these methods (American Supplier Institute, 1984 to 1988). However, as the details of these methods are becoming more widely known, critical appraisals are also beginning to appear (for example, Bhote, 1988; Tribus and Szonyi, 1989).

Overview. Taguchi robust design methods are set apart from traditional quality control procedures (see Chapter 32 – *Process Analysis* and Chapter 33 – *Quality Control*) and industrial experimentation in various respects. Of particular importance are:

1. The concept of quality loss functions,

2. The use of signal-to-noise (S/N) ratios, and

3. The use of orthogonal arrays.

These basic aspects of robust design methods will be discussed in the following sections. Several books have recently been published on these methods, for example, Peace (1993), Phadke (1989), Ross (1988), and Roy (1990), to name a few, and it is recommended that you refer to those books for further specialized discussions. Introductory overviews of Taguchi's ideas about quality and quality improvement can also be found in Barker (1986), Garvin (1987), Kackar (1986), and Noori (1989).

Quality and Loss Functions

What is quality? Taguchi's analysis begins with the question of how to define quality. It is not easy to formulate a simple definition of what constitutes quality; however, when your new car stalls in the middle of a busy intersection – putting yourself and other motorists at risk – you know that your car is not of high quality. Put another way, the definition of the inverse of quality is rather straightforward: it is the total loss to you and society due to functional variations and harmful side effects associated with the respective product. Thus, as an operational definition, you can measure quality in terms of this loss, and the greater the quality loss the lower the quality.

Discontinuous (step-shaped) loss function.

You can formulate hypotheses about the general nature and shape of the loss function. Assume a specific ideal point of highest quality; for example, a perfect car with no quality problems. It is customary in statistical process control (SPC; see also, Chapter 32 – *Process Analysis*) to define tolerances around the nominal ideal point of the production process. According to the traditional view implied by common SPC methods, as long as you are within the manufacturing tolerances you do not have a problem. Put another way, within the tolerance limits the quality loss is zero; once you move outside the tolerances, the quality loss is declared to be unacceptable. Thus, according to traditional views, the quality loss function is a discontinuous step function: as long as you are within the tolerance limits, quality loss is negligible; when you step outside those tolerances, quality loss becomes unacceptable.

Quadratic loss function.

Is the step function implied by common SPC methods a good model of quality loss? Return to the "perfect automobile" example. Is there a difference between a car that, within one year after purchase, has nothing wrong with it, and a car where minor rattles develop, a few fixtures fall off, and the clock in the dashboard breaks (all in-warranty repairs)? If you ever bought a new car of the latter kind, you know very well how annoying those admittedly minor quality

problems can be. The point here is that it is not realistic to assume that, as you move away from the nominal specification in your production process, the quality loss is zero as long as you stay within the set tolerance limits. Rather, if you are not exactly on target, loss will result, for example, in terms of customer satisfaction. Moreover, this loss is probably not a linear function of the deviation from nominal specifications, but rather a quadratic function (inverted U). A rattle in one place in your new car is annoying, but you would probably not get too upset about it; add two more rattles, and you might declare the car junk. Gradual deviations from the nominal specifications do not produce proportional increments in loss, but rather squared increments.

Conclusion: Controlling variability.

If, in fact, quality loss is a quadratic function of the deviation from a nominal value, the goal of your quality improvement efforts should be to minimize the squared deviations or variance of the product around nominal (ideal) specifications, rather than the number of units within specification limits (as is done in traditional SPC procedures).

Signal-to-Noise (S/N) Ratios

Measuring quality loss.

Even though you have concluded that the quality loss function is probably quadratic in nature, you still do not know precisely how to measure quality loss. However, you know that whatever measure you decide upon should reflect the quadratic nature of the function.

Signal, noise, and control factors.

The product of ideal quality should always respond in exactly the same manner to the signals provided by the user. When you turn the key in the ignition of your car you expect that the

starter motor turns and the engine starts. In the ideal quality car, the starting process would always proceed in exactly the same manner – for example, after three turns of the starter motor the engine comes to life. If, in response to the same signal (turning the ignition key) there is random variability in this process, you have less than ideal quality. For example, due to such uncontrollable factors as extreme cold, humidity, engine wear, etc., sometimes the engine may start only after turning over 20 times and finally not start at all. This example illustrates the key principle in measuring quality according to Taguchi: You want to minimize the variability in the product's performance in response to noise factors while maximizing the variability in response to signal factors.

Noise factors are those that are not under the control of the operator of a product. In the car example, those factors include temperature changes, different qualities of gasoline, engine wear, etc. Signal factors are those factors that are set or controlled by the operator of the product to make use of its intended functions (turning the ignition key to start the car).

Finally, the goal of your quality improvement effort is to find the best settings of factors under your control that are involved in the production process, in order to maximize the S/N ratio; thus, the factors in the experiment represent control factors.

S/N ratios. The conclusion of the previous paragraph is that quality can be quantified in terms of the respective product's response to noise factors and signal factors. The ideal product will only respond to the operator's signals and will be unaffected by random noise factors (weather, temperature, humidity, etc.). Therefore, the goal of your quality improvement effort can be stated as attempting to maximize the signal-to-noise (S/N) ratio for the respective product. The S/N ratios described in the following paragraphs were proposed by Taguchi (1987).

Smaller-the-better. In cases where you want to minimize the occurrences of some undesirable product characteristics, you would compute the following S/N ratio:

$$\text{Eta} = -10 * \log_{10} [(1/n) * \Sigma (y_i^2)]$$

for i = 1 to no. vars

See *Outer Arrays* in the *Statistical Glossary*.

Here, *eta* is the resultant S/N ratio; *n* is the number of observations on the particular product, and *y* is the respective characteristic. For example, the number of flaws in the paint on an automobile could be measured as the *y* variable and analyzed via this S/N ratio. The effect of the signal factors is zero, since zero flaws is the only intended or desired state of the paint on the car. Note how this S/N ratio is an expression of the assumed quadratic nature of the loss function. The factor *10* ensures that this ratio measures the inverse of "bad quality;" the more flaws in the paint, the greater is the sum of the squared number of flaws, and the smaller (i.e., more negative) the S/N ratio. Thus, maximizing this ratio will increase quality.

Nominal-the-best. Here, you have a fixed signal value (nominal value), and the variance around this value can be considered the result of noise factors:

$$\text{Eta} = 10 * \log_{10} (\text{Mean}^2/\text{Variance})$$

This signal-to-noise ratio could be used whenever ideal quality is equated with a particular nominal value. For example, the size of piston rings for an automobile engine must be as close to specification as possible to ensure high quality.

Larger-the-better. Examples of this type of engineering problem are fuel economy (miles per gallon) of an automobile, strength of concrete, resistance of shielding materials, etc. The following S/N ratio should be used:

$$Eta = -10 * \log_{10} [(1/n) * \Sigma (1/y_i^2)]$$

for i = 1 to no. vars

See *Outer Arrays* in the *Statistical Glossary*.

Signed target. This type of S/N ratio is appropriate when the quality characteristic of interest has an ideal value of 0 (zero), and both positive and negative values of the quality characteristic may occur. For example, the dc offset voltage of a differential operational amplifier can be positive or negative (see Phadke, 1989). The following S/N ratio should be used for these types of problems:

$$Eta = -10 * \log_{10}(s^2)$$

for i = 1 to no. vars

where s^2 stands for the variance of the quality characteristic across the measurements (variables). See *Outer Arrays* in the *Statistical Glossary*.

Fraction defective. This S/N ratio is useful for minimizing scrap, minimizing the percent of patients who develop side effects to a drug, etc. Taguchi also refers to the resultant *eta* values as *omegas*; note that this S/N ratio is identical to the familiar logit transformation (see also, Chapter 28 – *Nonlinear Estimation*):

$$Eta = -10 * \log_{10}[p/(1-p)]$$

where p is the proportion defective

Ordered categories (the accumulation analysis). In some cases, measurements on a quality characteristic can only be obtained in terms of categorical judgments. For example, consumers may rate a product as excellent, good, average, or below average. In that case, you would attempt to maximize the number of excellent or good ratings. Typically, the results of an accumulation analysis are summarized graphically in a stacked bar plot.

Orthogonal Arrays

The third aspect of Taguchi robust design methods is the one most similar to traditional techniques. Taguchi has developed a system of tabulated designs (arrays) that allow for the maximum number of main effects to be estimated in an unbiased (orthogonal) manner, with a minimum number of runs in the experiment. Latin square designs, $2^{(k-p)}$ designs (Plackett-Burman designs, in particular), and Box-Behnken designs main are also aimed at accomplishing this goal. In fact, many of the standard orthogonal arrays tabulated by Taguchi are identical to fractional two-level factorials, Plackett-Burman designs, Box-Behnken designs, Latin square, Greco-Latin squares, etc.

Analyzing Designs

Most analyses of robust design experiments amount to a standard ANOVA of the respective S/N ratios, ignoring two-way or higher-order interactions. However, when estimating error variances, you customarily pool together main effects of negligible size.

Analyzing S/N ratios in standard designs. It should be noted at this point that, of course, all of the designs discussed up to this point (e.g., $2^{(k-p)}$, $3^{(k-p)}$, mixed 2 and 3 level factorials, Latin squares, central composite designs) can be used to analyze S/N ratios that you computed. In fact, the many additional diagnostic plots and other options available for those designs (e.g., estimation of quadratic components, etc.) may prove very useful when analyzing the variability (S/N ratios) in the production process.

Plot of means. A visual summary of the experiment is the plot of the average *eta* (S/N ratio) by factor levels. In this plot, the optimum setting (i.e., largest S/N ratio) for each factor can easily be identified.

Verification experiments. For prediction purposes, you can compute the expected S/N ratio given a user-defined combination of settings of factors (ignoring factors that were pooled into the error term). These predicted S/N ratios can then be used in a verification experiment, where the engineer actually sets the machine accordingly and compares the resultant observed S/N ratio with the predicted S/N ratio from the experiment. If major deviations occur, you must conclude that the simple main effect model is not appropriate.

In these cases, Taguchi (1987) recommends transforming the dependent variable to accomplish additivity of factors, that is, to make the main effects model fit. Phadke (1989, Chapter 6) also discusses in detail methods for achieving additivity of factors.

Accumulation Analysis

When analyzing ordered categorical data, ANOVA is not appropriate. Rather, you produce a cumulative plot of the number of observations in a particular category. For each level of each factor, you plot the cumulative proportion of the number of defectives. Thus, this graph provides valuable information concerning the distribution of the categorical counts across the different factor settings.

Summary

To briefly summarize, when using Taguchi methods you first need to determine the design or control factors that can be set by the designer or engineer. Those are the factors in the experiment for which you will try different levels. Next, you decide to select an appropriate orthogonal array for the experiment. Next, you need to decide on how to measure the quality characteristic of interest. Remember that most S/N ratios require that multiple measurements are taken in each run of the experiment; for example, the variability around the nominal value cannot otherwise be assessed. Finally, you conduct the experiment and identify the factors that most strongly affect the chosen S/N ratio, and you reset your machine or production process accordingly.

Mixture Designs and Triangular Surfaces

Overview

Special issues arise when analyzing mixtures of components that must sum to a constant. For example, if you wanted to optimize the taste of a fruit-punch consisting of the juices of 5 fruits, the sum of the proportions of all juices in each mixture must be 100%. Thus, the task of optimizing mixtures commonly occurs in food processing, refining, or the manufacturing of chemicals. A number of designs have been developed to address specifically the analysis and modeling of mixtures (see, for example, Cornell, 1990a, 1990b; Cornell and Khuri, 1987; Deming and Morgan, 1993; Montgomery, 1991).

Triangular Coordinates

The common manner in which mixture proportions can be summarized is via triangular (ternary) graphs. For example, suppose you have a mixture that consists of 3 components *A, B*, and *C*. Any mixture of the three components can be summarized by a point in the triangular coordinate system defined by the three variables.

For example, take the following 6 different mixtures of the 3 components.

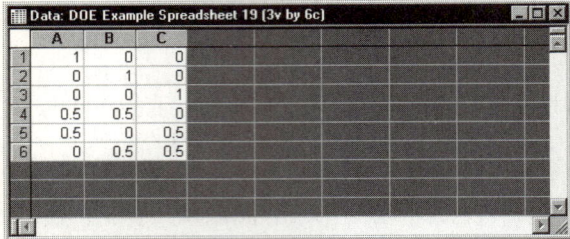

The sum for each mixture is 1.0, so the values for the components in each mixture can be interpreted as proportions. If you graph these data in a regular 3D scatterplot, it becomes apparent that the points form a triangle in the 3D space. Only the points inside the triangle where the sum of the component values is equal to 1 are valid mixtures. Therefore, you can simply plot only the triangle to summarize the component values (proportions) for each mixture.

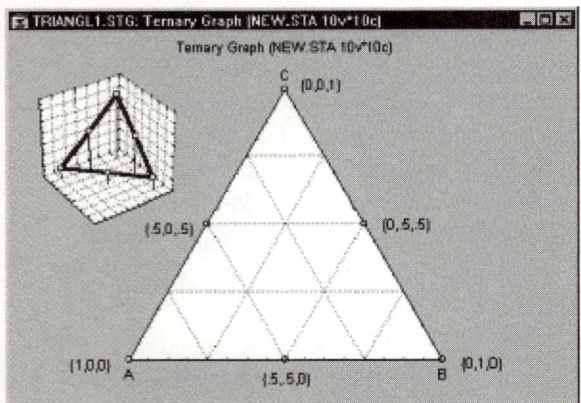

To read off the coordinates of a point in the triangular graph, you would simply drop a line from each respective vertex to the side of the triangle.

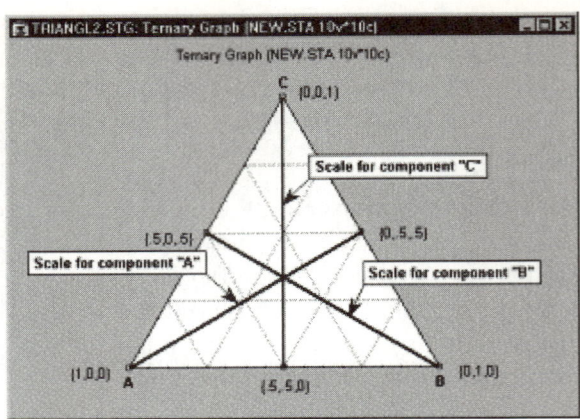

At the vertex for the particular factor, there is a pure blend, that is, one that only contains the respective component. Thus, the coordinates for the vertex point is *1* (or *100%*, or however else the mixtures are scaled) for the respective component, and *0* (zero) for all other components. At the side opposite to the respective vertex, the value for the respective component is *0* (zero), and *.5* (or *50%*, etc.) for the other components.

Triangular Surfaces and Contours

You can now add to the triangle a fourth dimension that is perpendicular to the first three. Using that dimension, you could plot the values for a dependent variable, or function (surface) that was fit to the dependent variable. Note that the response surface can be shown in 3D, where the predicted response (*Taste* rating) is indicated by the distance of the surface from the triangular plane, or it can be indicated in a contour plot where the contours of constant height are plotted on the 2D triangle.

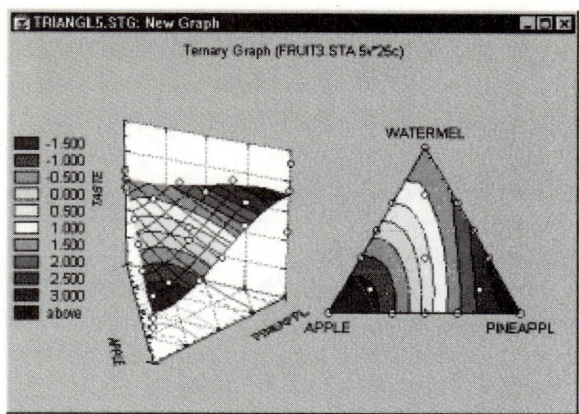

It should be mentioned at this point that you can produce categorized ternary graphs. These are very useful, because they enable you to fit to a dependent variable (e.g., Taste) a response surface, for different levels of a fourth component.

Canonical Form of Mixture Polynomials

Fitting a response surface to mixture data is, in principle, done in the same manner as fitting surfaces to, for example, data from central composite designs. However, there is the issue that mixture data are constrained, that is, the sum of all component values must be constant.

Consider the simple case of two factors A and B. You may want to fit the simple linear model:

$$y = b_0 + b_A{}^*x_A + b_B{}^*x_B$$

Here y stands for the dependent variable values, b_A and b_B stand for the regression coefficients, x_A and x_B stand for the values of the factors. Suppose that x_A and x_B must sum to 1; you can multiple b_0 by $1=(x_A + x_B)$:

$$y = (b_0{}^*x_A + b_0{}^*x_B) + b_A{}^*x_A + b_B{}^*x_B$$

or:

$$y = b'_A{}^*x_A + b'_B{}^*x_B$$

where $b'_A = b_0 + b_A$ and $b'_B = b_0 + b_B$. Thus, the estimation of this model comes down to fitting a no-intercept multiple regression model. (See also, Chapter 26 – *Multiple Linear Regression* for details concerning multiple regression.)

Common Models for Mixture Data

The quadratic and cubic model can be similarly simplified (as illustrated for the simple linear model above), yielding four standard models that are customarily fit to the mixture data. Here are the formulas for the 3-variable case for those models (see Cornell, 1990, for additional details).

Linear model:

$$y = b_1{}^*x_1 + b_2{}^*x_2 + b_3{}^*x_3$$

Quadratic model:

$$y = b_1{}^*x_1 + b_2{}^*x_2 + b_3{}^*x_3 + b_{12}{}^*x_1{}^*x_2 + b_{13}{}^*x_1{}^*x_3 + b_{23}{}^*x_2{}^*x_3$$

Special cubic model:

$$y = b_1{}^*x_1 + b_2{}^*x_2 + b_3{}^*x_3 + b_{12}{}^*x_1{}^*x_2 + b_{13}{}^*x_1{}^*x_3 + b_{23}{}^*x_2{}^*x_3 + b_{123}{}^*x_1{}^*x_2{}^*x_3$$

Full cubic model:

$$y = b_1{}^*x_1 + b_2{}^*x_2 + b_3{}^*x_3 + b_{12}{}^*x_1{}^*x_2 + b_{13}{}^*x_1{}^*x_3 + b_{23}{}^*x_2{}^*x_3 + d_{12}{}^*x_1{}^*x_2{}^*(x_1 - x_2) + d_{13}{}^*x_1{}^*x_3{}^*(x_1 - x_3) + d_{23}{}^*x_2{}^*x_3{}^*(x_2 - x_3) + b_{123}{}^*x_1{}^*x_2{}^*x_3$$

(Note that the d_{ij}'s are also parameters of the model.)

Standard Designs for Mixture Experiments

Two different types of standard designs are commonly used for experiments with mixtures. Both of them will evaluate the triangular response surface at the vertices (i.e., the corners of the triangle) and the centroids (sides of the

triangle). Sometimes, those designs are enhanced with additional interior points.

Simplex-lattice designs. In this arrangement of design points, *m+1* equally spaced proportions are tested for each factor or component in the model:

$$x_i = 0, 1/m, 2/m, ..., 1 \quad i = 1,2,...,q$$

and all combinations of factor levels are tested. The resulting design is called a *{q,m} simplex lattice* design. For example, a *{q=3, m=2}* simplex lattice design will include the following mixtures:

Data: DOE Example Spreadsheet 20 (3v by 6c)

	A	B	C
1	1	0	0
2	0	1	0
3	0	0	1
4	0.5	0.5	0
5	0.5	0	0.5
6	0	0.5	0.5

A *{q=3, m=3}* simplex lattice design will include the points:

Data: DOE Example Spreadsheet 21 (3v by 10c)

	A	B	C
1	1.0000	0.0000	0.0000
2	0.0000	1.0000	0.0000
3	0.0000	0.0000	1.0000
4	0.3333	0.6667	0.0000
5	0.3333	0.0000	0.6667
6	0.0000	0.3333	0.6667
7	0.6667	0.3333	0.0000
8	0.6667	0.0000	0.3333
9	0.0000	0.6667	0.3333
10	0.3333	0.3333	0.3333

Simplex-centroid designs. An alternative arrangement of settings introduced by Scheffé (1963) is the so-called simplex-centroid design. Here the design points correspond to all permutations of the pure blends (e.g., *1 0 0; 0 1 0; 0 0 1*), the permutations of the binary blends (*½ ½ 0; ½ 0 ½; 0 ½ ½*), the permutations of the blends involving three components, and so on. For example, for 3 factors the simplex centroid design consists of the points:

Data: DOE Example Spreadsheet 22 (3v by 7c)

	A	B	C
1	1.0000	0.0000	0.0000
2	0.0000	1.0000	0.0000
3	0.0000	0.0000	1.0000
4	0.5000	0.5000	0.0000
5	0.5000	0.0000	0.5000
6	0.0000	0.5000	0.5000
7	0.3333	0.3333	0.3333

Adding interior points. These designs are sometimes augmented with interior points (see Khuri and Cornell, 1987, page 343; Mason, Gunst, Hess; 1989; page 230). For example, for 3 factors one could add the interior points:

Data: DOE Example Spreadsheet 23 (3v by 3c)

	A	B	C
1	0.6667	0.1667	0.1667
2	0.1667	0.6667	0.1667
3	0.1667	0.1667	0.6667

If you plot these points in a scatterplot with triangular coordinates; you can see how these designs evenly cover the experimental region defined by the triangle.

Lower Constraints

The designs described above all require vertex points, that is, pure blends consisting of only one ingredient. In practice, those points may often not be valid, that is, pure blends cannot be produced because of cost or other constraints. For example, suppose you want to study the effect of a food additive on the taste of the fruit punch. The additional ingredient can be varied only within small limits; for example, it cannot exceed a certain percentage of the total. Clearly, a fruit punch that is a pure blend, consisting only of the additive, would not be a fruit punch at all, or worse, may be toxic. These types of constraints are very common in many applications of mixture experiments.

Let's consider a 3-component example, where component *A* is constrained so that $x_A \geq .3$. The total of the 3-component mixture must be equal

to 1. This constraint can be visualized in a triangular graph by a line at the triangular coordinate for $x_A = .3$, that is, a line that is parallel to the triangle's edge opposite to the A vertex point.

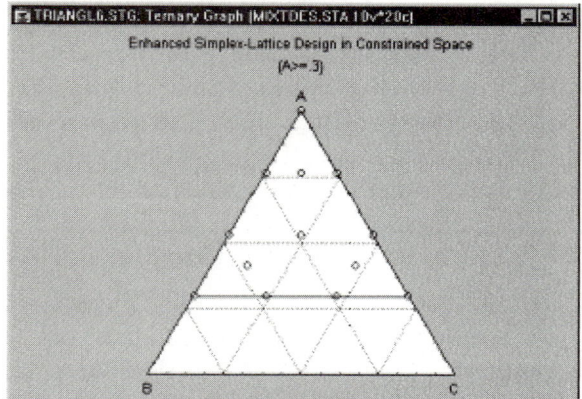

You can now construct the design as before, except that one side of the triangle is defined by the constraint. Later, in the analysis, you can review the parameter estimates for the so-called pseudo-components, treating the constrained triangle as if it were a full triangle.

Multiple constraints. Multiple lower constraints can be treated analogously, that is, you can construct the sub-triangle within the full triangle, and then place the design points in that sub-triangle according to the chosen design.

Upper and Lower Constraints

When there are both upper and lower constraints (as is often the case in experiments involving mixtures), the standard simplex-lattice and simplex-centroid designs can no longer be constructed, because the subregion defined by the constraints is no longer a triangle. There is a general algorithm for finding the vertex and centroid points for such constrained designs.

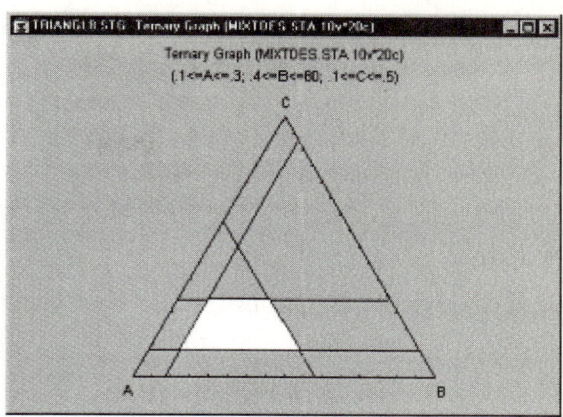

Note that you can still analyze such designs by fitting the standard models to the data.

Analyzing Mixture Experiments

The analysis of mixture experiments amounts to a multiple regression with the intercept set to zero. As explained previously, the mixture constraint – that the sum of all components must be constant – can be accommodated by fitting multiple regression models that do not include an intercept term. If you are not familiar with multiple regression, you may want to review Chapter 26 – *Multiple Linear Regression* at this point.

The specific models that are usually considered were described earlier. To summarize, you fit to the dependent variable response surfaces of increasing complexity, that is, starting with the linear model, then the quadratic model, special cubic model, and full cubic model. Shown in the next illustration is a table with the number of terms or parameters in each model, for a selected number of components (see also, Table 4, Cornell, 1990):

Data: DOE Example Spreadsheet 24 (4v by 7c)

Number of Components	Model (Degree of Polynomial)			
	Linear	Quadratic	Special Cubic	Full Cubic
2	2	3		
3	3	6	7	10
4	4	10	14	20
5	5	15	25	35
6	6	21	41	56
7	7	28	63	84
8	8	36	92	120

Analysis of Variance

To decide which of the models of increasing complexity provides a sufficiently good fit to the observed data, you usually compare the models in a hierarchical, stepwise fashion. For example, consider a 3-component mixture to which the full cubic model was fitted.

Data: DOE Example Spreadsheet 25 (10v by 5c)

Model	ANOVA Var.: DV (mixt4.sta) 3-Factor mixture design; Mixture total = 1.0, 14 Runs Sequential fit of models of increasing complexity									
	SS Effect	df Effect	MS Effect	SS Error	df Error	MS Error	F	p	R-square	R-square (adjusted)
Linear	44.76	2	22.38	46.87	11	4.26	5.25	0.03	0.49	0.40
Quadratic	30.56	3	10.19	16.31	8	2.04	4.99	0.03	0.82	0.71
Special Cubic	0.72	1	0.72	15.60	7	2.23	0.32	0.59	0.83	0.68
Cubic	8.23	3	2.74	7.37	4	1.84	1.49	0.35	0.92	0.74
Total Adjusted	91.63	13	7.05							

First, the linear model was fit to the data. Even though this model has 3 parameters, one for each component, this model has only 2 degrees of freedom. This is because of the overall mixture constraint, that the sum of all component values is constant. The simultaneous test for all parameters of this model is statistically significant $(F(2,11)=5.25; p<.05)$. The addition of the 3 quadratic model parameters $(b_{12}*x_1*x_2, b_{13}*x_1*x_3, b_{23}*x_2*x_3)$ further significantly improves the fit of the model $(F(3,8)=4.99; p<.05)$. However, adding the parameters for the special cubic and cubic models does not significantly improve the fit of the surface. Thus, you could conclude that the quadratic model provides an adequate fit to the data (of course, pending further examination of the residuals for outliers, etc.).

R-square. The R-square value can be interpreted as the proportion of variability around the mean for the dependent variable that can be accounted for by the respective model. (Note that for non-intercept models, some multiple regression programs will only compute the R-square value pertaining to the proportion of variance around 0 (zero) accounted for by the independent variables; for more information, see Kvalseth, 1985; Okunade, Chang, and Evans, 1993.

Pure error and lack of fit. The usefulness of the estimate of pure error for assessing the overall lack of fit was discussed in the context of central composite designs. If some runs in the design were replicated, you can compute an estimate of error variability based only on the variability between replicated runs. This variability provides a good indication of the unreliability in the measurements, independent of the model that was fit to the data, since it is based on identical factor settings (or blends in this case). You can test the residual variability after fitting the current model against this estimate of pure error. If this test is statistically significant, that is, if the residual variability is significantly larger than the pure error variability, you can conclude that, most likely, there are additional significant differences between blends that cannot be accounted for by the current model. Thus, there may be an overall lack of fit of the current model. In that case, try a more complex model, perhaps by only adding individual terms of the next higher-order model (e.g., only the $b_{13}*x_1*x_3$ to the linear model).

Parameter Estimates

Usually, after fitting a particular model, you would next review the parameter estimates. Remember that the linear terms in mixture models are constrained, that is, the sum of the components must be constant. Hence, independent statistical significance tests for the linear components cannot be performed.

Pseudo-Components

To allow for scale-independent comparisons of the parameter estimates, during the analysis, the component settings are customarily recoded to so-called pseudo-components so that (see also, Cornell, 1993, Chapter 3):

$$x'_i = (x_i - L_i)/(Total - L)$$

Here, x'_i stands for the i'th pseudo-component, x_i stands for the original component value, L_i stands for the lower constraint (limit) for the i'th component, L stands for the sum of all lower constraints (limits) for all components in the design, and $Total$ is the mixture total.

The issue of lower constraints was also discussed earlier in this section. If the design is a standard simplex-lattice or simplex-centroid design, this transformation amounts to a rescaling of factors so as to form a sub-triangle (sub-simplex) as defined by the lower constraints. However, you can compute the parameter estimates based on the original (untransformed) metric of the components in the experiment. If you want to use the fitted parameter values for prediction purposes (i.e., to predict dependent variable values), the parameters for the untransformed components are often more convenient to use. Note that the results dialog for mixture experiments contains options to make predictions for the dependent variable for user-defined values of the components, in their original metric.

Graph Options

Surface and contour plots. The respective fitted model can be visualized in triangular surface plots or contour plots, which, optionally, can also include the respective fitted function.

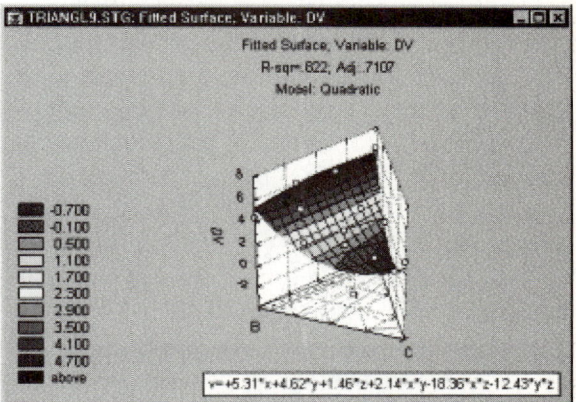

Note that the fitted function displayed in the surface and contour plots always pertains to the parameter estimates for the pseudo-components.

Categorized surface plots. If your design involves replications (and the replications are coded in your data file), you can use 3D ternary plots to look at the respective fit, replication by replication.

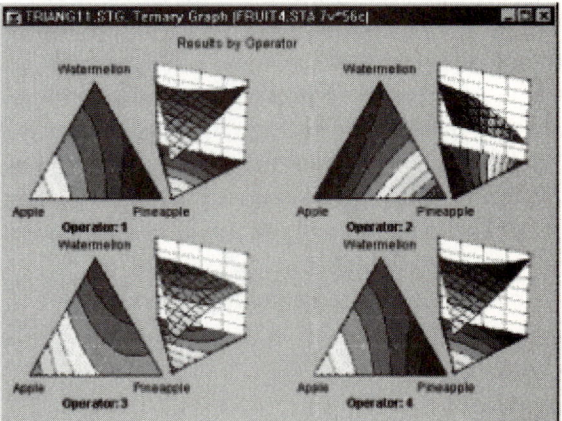

Of course, if you have other categorical variables in your study (e.g., operator or experimenter; machine, etc.) you can also categorize the 3D surface plot by those variables.

Trace plots. One aid for interpreting the triangular response surface is the so-called trace plot. Suppose you looked at the contour plot of the response surface for three components. Then, determine a reference blend for two of the components, for example, hold the values for *A* and *B* at *1/3* each. Keeping the relative proportions of *A* and *B* constant (i.e., equal proportions in this case), you can then plot the estimated response (values for the dependent variable) for different values of *C*.

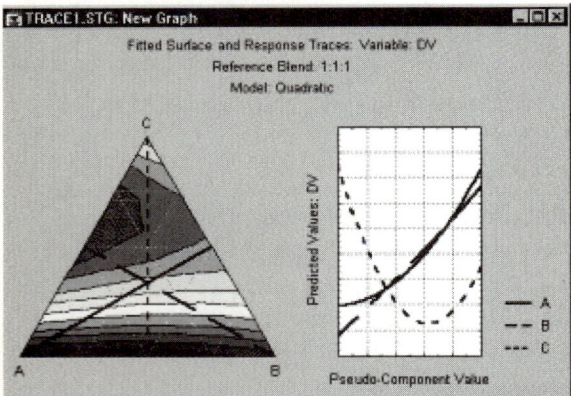

If the reference blend for *A* and *B* is *1:1*, the resulting line or response trace is the axis for factor *C*; that is, the line from the C vertex point connecting with the opposite side of the triangle at a right angle. However, trace plots for other reference blends can also be produced. Typically, the trace plot contains the traces for all components, given the current reference blend.

Residual plots. Finally, it is important, after deciding on a model, to review the prediction residuals, in order to identify outliers or regions of misfit-fit. In addition, you should review the standard normal probability plot of residuals and

the scatterplot of observed versus predicted values. Remember that the multiple regression analysis (i.e., the process of fitting the surface) assumes that the residuals are normally distributed, and you should carefully review the residuals for any apparent outliers.

Designs for Constrained Surfaces and Mixtures

Overview

As mentioned in the context of mixture designs, it often happens in real-world studies that the experimental region of interest is constrained, that is, that not all factors settings can be combined with all settings for the other factors in the study. There is an algorithm suggested by Piepel (1988) and Snee (1985) for finding the vertices and centroids for such constrained regions.

Designs for Constrained Experimental Regions

When in an experiment with many factors, there are constraints concerning the possible values of those factors and their combinations, it is not clear how to proceed. A reasonable approach is to include in the experiments runs at the extreme vertex points and centroid points of the constrained region, which should usually provide good coverage of the constrained experimental region (e.g., see Piepel, 1988; Snee, 1975). In fact, the mixture designs reviewed in the previous section provide examples for such designs, since they are typically constructed to include the vertex and centroid points of the constrained region that consists of a triangle (simplex).

Linear Constraints

One general way in which you can summarize most constraints that occur in real world

experimentation is in terms of a linear equation (see Piepel, 1988):

$$A_1x_1 + A_2x_2 + ... + A_qx_q + A_0 \geq 0$$

Here, $A_0, .., A_q$ are the parameters for the linear constraint on the q factors, and $x_1,.., x_q$ stands for the factor values (levels) for the q factors. This general formula can accommodate even very complex constraints. For example, suppose that in a two-factor experiment the first factor must always be set at least twice as high as the second, that is, $x_1 \geq 2*x_2$. This simple constraint can be rewritten as $x_1-2*x_2 \geq 0$. The ratio constraint $2*x_1/x_2 \geq 1$ can be rewritten as $2*x_1 - x_2 \geq 0$, and so on.

The problem of multiple upper and lower constraints on the component values in mixtures was discussed earlier, in the context of mixture experiments. For example, suppose in a three-component mixture of fruit juices, the upper and lower constraints on the components are (see example 3.2, in Cornell 1993):

$40\% \leq$ Watermelon $(x_1) \leq 80\%$
$10\% \leq$ Pineapple $(x_2) \leq 50\%$
$10\% \leq$ Orange $(x_3) \leq 30\%$

These constraints can be rewritten as linear constraints into the form:

Watermelon: $x_1-40 \geq 0$
                $-x_1+80 \geq 0$

Pineapple: $x_2-10 \geq 0$
               $-x_2+50 \geq 0$

Orange: $x_3-10 \geq 0$
            $-x_3+30 \geq 0$

Thus, the problem of finding design points for mixture experiments with components with multiple upper and lower constraints is only a special case of general linear constraints.

Piepel & Snee Algorithm

For the special case of constrained mixtures, algorithms such as the XVERT algorithm (see, for example, Cornell, 1990) are often used to find the vertex and centroid points of the constrained region (inside the triangle of three components, tetrahedron of four components, etc.). The general algorithm proposed by Piepel (1988) and Snee (1979) for finding vertices and centroids can be applied to mixtures as well as non-mixtures. The general approach of this algorithm is described in detail by Snee (1979).

Specifically, it will consider each constraint one by one, written as a linear equation as described above. Each constraint represents a line (or plane) through the experimental region. For each successive constraint, you will evaluate whether or not the current (new) constraint crosses into the current valid region of the design. If so, new vertices will be computed which define the new valid experimental region, updated for the most recent constraint. It will then check whether or not any of the previously processed constraints have become redundant, that is, define lines or planes in the experimental region that are now entirely outside the valid region. After all constraints have been processed, it will then compute the centroids for the sides of the constrained region (of the order specified by the user). For the two-dimensional (two-factor) case, you can easily recreate this process by simply drawing lines through the experimental region, one for each constraint; what is left is the valid experimental region.

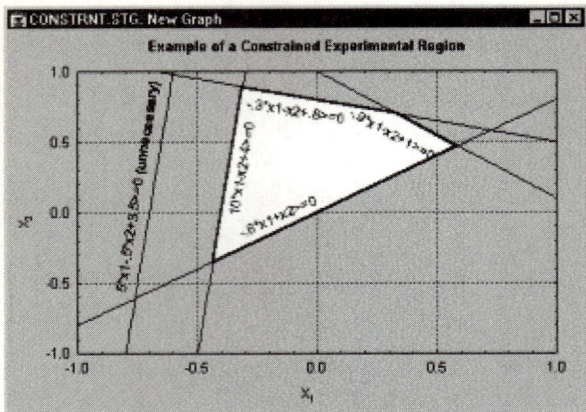

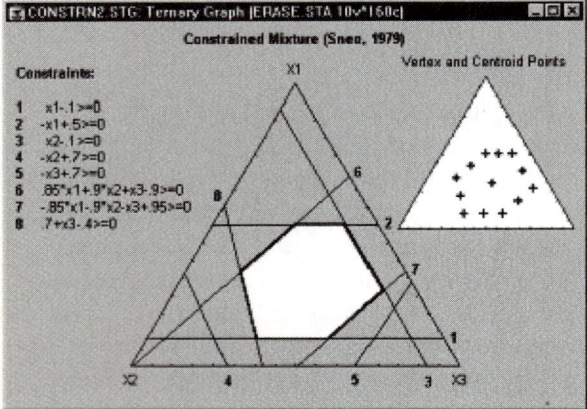

For more information, see Piepel (1988) or Snee (1979).

Choosing Points for the Experiment

Once the vertices and centroids have been computed, you may face the problem of having to select a subset of points for the experiment. If each experimental run is costly, it may not be feasible to simply run all vertex and centroid points. In particular, when there are many factors and constraints, the number of centroids can quickly get very large.

If you are screening a large number of factors, and are not interested in nonlinear effects,

choosing the vertex points only will usually yield good coverage of the experimental region. To increase statistical power (to increase the degrees of freedom for the ANOVA error term), you may also want to include a few runs with the factors set at the overall centroid of the constrained region.

If you are considering a number of different models that you might fit once the data have been collected, you may want to use the D- and A-optimal design options. Those options will help you select the design points that will extract the maximum amount of information from the constrained experimental region, given your models.

Analyzing Designs for Constrained Surfaces and Mixtures

As mentioned in the section on central composite designs and mixture designs, once the constrained design points have been chosen for the final experiment, and the data for the dependent variables of interest have been collected, the analysis of these designs can proceed in the standard manner.

For example, Cornell (1990, page 68) describes an experiment of three plasticizers, and their effect on resultant vinyl thickness (for automobile seat covers). The constraints for the three plasticizers components x_1, x_2, and x_3 are:

$.409 \leq x_1 \leq .849$
$.000 \leq x_2 \leq .252$
$.151 \leq x_3 \leq .274$

(Note that these values are already rescaled, so that the total for each mixture must be equal to 1.) The vertex and centroid points generated are:

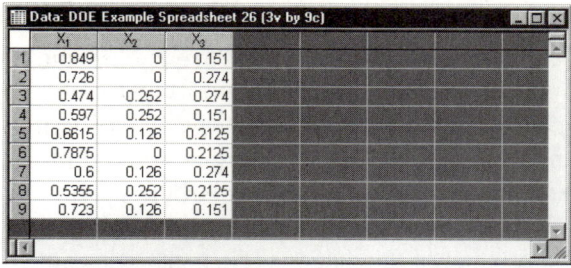

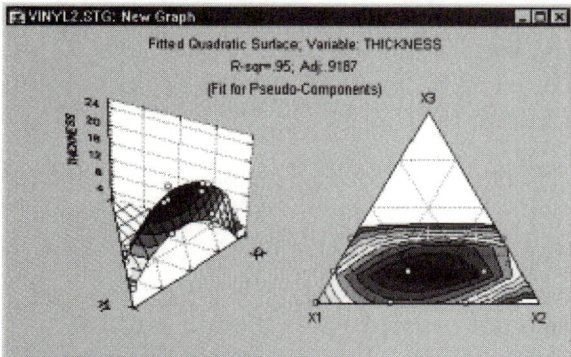

Constructing D- and A-Optimal Designs for Surfaces and Mixtures

Overview

In the sections on standard factorial designs (see $2^{(k-p)}$ *Fractional Factorial Designs*, page 185, and $3^{(k-p)}$, *Box Behnken, and Mixed 2 and 3 Level Factorial Designs*, page 197) and *Central Composite Designs*, the property of orthogonality of factor effects was discussed. In short, when the factor level settings for two factors in an experiment are uncorrelated, that is, when they are varied independently of each other, they are said to be orthogonal to each other. (If you are familiar with matrix and vector algebra, two column vectors X_1 and X_2 in the design matrix are orthogonal if $X_1'^*X_2 = 0$). Intuitively, it should be clear that you can extract the maximum amount of information regarding a dependent variable from the experimental region (the region defined by the settings of the factor levels), if all factor effects are orthogonal to each other. Conversely, suppose you run a four-run experiment for two factors as follows:

	X_1	X_2
Run 1	1	1
Run 2	1	1
Run 3	-1	-1
Run 4	-1	-1

Now the columns of factor settings for X_1 and X_2 are identical to each other (their correlation is *1*), and there is no way in the results to distinguish between the main effect for X_1 and X_2.

The D- and A-optimal design procedures provide various options to select from a list of valid (candidate) points (i.e., combinations of factor settings) those points that will extract the maximum amount of information from the experimental region, given the respective model that you expect to fit to the data. You need to supply the list of candidate points, for example the vertex and centroid points, and specify the type of model you expect to fit to the data and the number of runs for the experiment. A design will be constructed with the desired number of cases that will provide as much orthogonality between the columns of the design matrix as possible.

The reasoning behind D- and A-optimality is discussed, for example, in Box and Draper (1987, Chapter 14). The different algorithms used for searching for optimal designs are described in Dykstra (1971), Galil and Kiefer (1980), and Mitchell (1974a, 1974b). A detailed comparison study of the different algorithms is discussed in Cook and Nachtsheim (1980).

Basic Ideas

A technical discussion of the reasoning (and limitations) of D- and A-optimal designs is

beyond the scope of this introduction. However, the general ideas are fairly straightforward. Consider again the simple two-factor experiment in four runs.

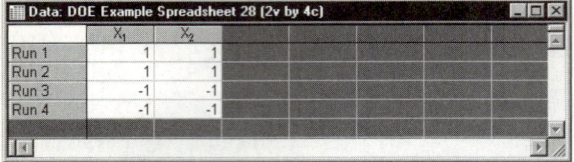

As mentioned previously, this design, of course, does not enable you to test, independently, the statistical significance of the two variables' contribution to the prediction of the dependent variable. If you computed the correlation matrix for the two variables, they would correlate at *1*:

	X_1	X_2
X_1	1	1
X_2	1	1

Data: DOE Example Spreadsheet 29 (2v by 2c)

Normally, you would run this experiment so that the two factors are varied independently of each other:

	X_1	X_2
Run 1	1	1
Run 2	1	-1
Run 3	-1	1
Run 4	-1	-1

Data: DOE Example Spreadsheet 30 (2v by 4c)

Now the two variables are uncorrelated, that is, the correlation matrix for the two factors is:

	X_1	X_2
X_1	1	0
X_2	0	1

Data: DOE Example Spreadsheet 31 (2v by 2c)

Another term that is customarily used in this context is that the two factors are orthogonal. Technically, if the sum of the products of the elements of two columns (vectors) in the design (design matrix) is equal to 0 (zero), the two columns are orthogonal.

The determinant of the design matrix. The determinant D of a square matrix (such as the 2-by-2 correlation matrices shown above) is a specific numerical value that reflects the amount of independence or redundancy between the columns and rows of the matrix. For the 2-by-2 case, it is simply computed as the product of the diagonal elements minus the off-diagonal elements of the matrix (for larger matrices the computations are more complex). For example, for the two matrices shown above, the determinant D is:

$$D_1 = \begin{vmatrix} 1.0 & 1.0 \\ 1.0 & 1.0 \end{vmatrix} = 1*1 - 1*1 = 0$$

$$D_2 = \begin{vmatrix} 1.0 & 0.0 \\ 0.0 & 1.0 \end{vmatrix} = 1*1 - 0*0 = 1$$

Thus, the determinant for the first matrix computed from completely redundant factor settings is equal to *0*. The determinant for the second matrix, when the factors are orthogonal, is equal to *1*.

D-optimal designs. This basic relationship extends to larger design matrices, that is, the more redundant the vectors (columns) of the design matrix, the closer to 0 (zero) is the determinant of the correlation matrix for those vectors; the more independent the columns, the larger is the determinant of that matrix. Thus, finding a design matrix that maximizes the determinant D of this matrix means finding a design where the factor effects are maximally independent of each other. This criterion for selecting a design is called the D-optimality criterion.

Matrix notation. Actually, the computations are commonly not performed on the correlation matrix of vectors, but on the simple cross-product matrix. In matrix notation, if the design matrix is denoted by **X**, then the quantity of interest here is the determinant of $\mathbf{X'X}$ (*X*-transposed times *X*). Thus, the search for *D-optimal* designs aims to maximize $|X'X|$, where the vertical lines ($|..|$) indicate the determinant.

A-optimal designs. Looking back at the computations for the determinant, another way to look at the issue of independence is to maximize the diagonal elements of the $X'X$ matrix, while minimizing the off-diagonal elements. The so-called *trace criterion* or *A-optimality* criterion expresses this idea. Technically, the *A*-criterion is defined as:

$$A = \text{trace}(X'X)^{-1}$$

where *trace* stands for the sum of the diagonal elements [of the $\mathbf{(X'X)^{-1}}$ matrix].

The information function. It should be mentioned at this point that *D*-optimal designs minimize the expected prediction error for the dependent variable, that is, those designs will maximize the precision of prediction, and thus the information (which is defined as the inverse of the error) that is extracted from the experimental region of interest.

Measuring Design Efficiency

A number of standard measures have been proposed to summarize the efficiency of a design.

D-efficiency. This measure is related to the *D*-optimality criterion:

$$D\text{-efficiency} = 100 * (|X'X|^{1/p}/N)$$

Here, *p* is the number of factor effects in the design (columns in *X*), and *N* is the number of requested runs. This measure can be interpreted as the relative number of runs (in percent) that would be required by an orthogonal design to achieve the same value of the determinant $|X'X|$. However, remember that an orthogonal

design may not be possible in many cases, that is, it is only a theoretical "yard-stick." Therefore, you should use this measure rather as a relative indicator of efficiency, to compare other designs of the same size, and constructed from the same design points candidate list. Also note that this measure is only meaningful (and will only be reported) if you chose to recode the factor settings in the design (i.e., the factor settings for the design points in the candidate list), so that they have a minimum of *-1* and a maximum of *+1*.

A-efficiency. This measure is related to the A-optimality criterion:

$$A\text{-efficiency} = 100 * p/\text{trace}(N*(X'X)^{-1})$$

Here, *p* stands for the number of factor effects in the design, *N* is the number of requested runs, and *trace* stands for the sum of the diagonal elements [of $(N*(X'X)^{-1})$]. This measure can be interpreted as the relative number of runs (in percent) that would be required by an orthogonal design to achieve the same value of the trace of $(X'X)^{-1}$. However, again you should use this measure as a relative indicator of efficiency, to compare other designs of the same size and constructed from the same design points candidate list; also this measure is only meaningful if you chose to recode the factor settings in the design to the *-1* to *+1* range.

G-efficiency. This measure is computed as:

$$G\text{-efficiency} = 100 * \text{square root}(p/N)/\sigma_M$$

Again, *p* stands for the number of factor effects in the design and *N* is the number of requested runs; σ_M (*sigma$_M$*) stands for the maximum standard error for prediction across the list of candidate points. This measure is related to the so-called G-optimality criterion; G-optimal designs are defined as those that will minimize the maximum value of the standard error of the predicted response.

Constructing Optimal Designs

The optimal design facilities will search for optimal designs, given a list of candidate points. Put another way, given a list of points that specifies which regions of the design are valid or feasible, and given a user-specified number of runs for the final experiment, it will select points to optimize the respective criterion. This searching for the best design is not an exact method, but rather an algorithmic procedure that employs certain search strategies to find the best design (according to the respective optimality criterion).

The search procedures or algorithms that have been proposed are described below (for a review and detailed comparison, see Cook and Nachtsheim, 1980). They are reviewed here in the order of speed, that is, the Sequential or Dykstra method is the fastest method, but often most likely to fail, that is, to yield a design that is not optimal (e.g., only locally optimal; this issue will be discussed shortly).

Sequential or Dykstra method. This algorithm is due to Dykstra (1971). Starting with an empty design, it will search through the candidate list of points, and choose in each step the one that maximizes the chosen criterion. There are no iterations involved, they will simply pick the requested number of points sequentially. Thus, this method is the fastest of the ones discussed. Also, by default, this method is used to construct the initial designs for the remaining methods.

Simple exchange (Wynn-Mitchell) method. This algorithm is usually attributed to Mitchell and Miller (1970) and Wynn (1972). The method starts with an initial design of the requested size (by default constructed via the sequential search

algorithm described above). In each iteration, one point (run) in the design will be dropped from the design and another added from the list of candidate points. The choice of points to be dropped or added is sequential, that is, at each step the point that contributes least with respect to the chosen optimality criterion (D or A) is dropped from the design; then the algorithm chooses a point from the candidate list so as to optimize the respective criterion. The algorithm stops when no further improvement is achieved with additional exchanges.

DETMAX algorithm (exchange with excursions). This algorithm, due to Mitchell (1974b), is probably the best known and most widely used optimal design search algorithm. Like the simple exchange method, first an initial design is constructed (by default, via the sequential search algorithm described above). The search begins with a simple exchange as described above. However, if the respective criterion (D or A) does not improve, the algorithm will undertake excursions. Specifically, the algorithm will add or subtract more than one point at a time, so that, during the search, the number of points in the design may vary between $N_D + N_{excursion}$ and $N_D - N_{excursion}$, where N_D is the requested design size, and $N_{excursion}$ refers to the maximum allowable excursion, as specified by the user. The iterations will stop when the chosen criterion (D or A) no longer improves within the maximum excursion.

Modified Fedorov (simultaneous switching). This algorithm represents a modification (Cook and Nachtsheim, 1980) of the basic Fedorov algorithm described below. It also begins with an initial design of the requested size (by default constructed via the sequential search algorithm). In each iteration, the algorithm will exchange each point in the design with one chosen from the candidate list, so as to optimize the design according to the chosen criterion (*D* or *A*). Unlike the simple exchange algorithm described above, the exchange is not sequential, but simultaneous. Thus, in each iteration each point in the design is compared with each point in the candidate list, and the exchange is made for the pair that optimizes the design. The algorithm terminates when there are no further improvements in the respective optimality criterion.

Fedorov (simultaneous switching). This is the original simultaneous switching method proposed by Fedorov (see Cook and Nachtsheim, 1980). The difference between this procedure and the one described above (modified Fedorov) is that in each iteration only a single exchange is performed, that is, in each iteration all possible pairs of points in the design and those in the candidate list are evaluated. The algorithm will then exchange the pair that optimizes the design (with regard to the chosen criterion). Thus, it is easy to see that this algorithm potentially can be somewhat slow, since in each iteration $N_D * N_C$ comparisons are performed, in order to exchange a single point.

General Recommendations

If you think about the basic strategies represented by the different algorithms described above, it should be clear that there are usually no exact solutions to the optimal design problem. Specifically, the determinant of the **X'X** matrix (and trace of its inverse) are complex functions of the list of candidate points. In particular, there are usually several local minima with regard to the chosen optimality criterion; for example, at any point during the search, a design may appear optimal unless you simultaneously discard half of the points in the design and choose certain other

points from the candidate list; but, if you only exchange individual points or only a few points (via DETMAX), no improvement occurs.

Therefore, it is important to try a number of different initial designs and algorithms. If, after repeating the optimization several times with random starts, the same or very similar final optimal design results, you can be reasonably sure that you are not caught in a local minimum or maximum.

Also, the methods described above vary greatly with regard to their ability to get trapped in local minima or maxima. As a general rule, the slower the algorithm (i.e., the further down on the list of algorithms described above), the more likely is the algorithm to yield a truly optimal design. However, note that the modified Fedorov algorithm will practically perform just as well as the unmodified algorithm (see Cook and Nachtsheim, 1980); therefore, if time is not a consideration, we recommend the modified Fedorov algorithm as the best method to use.

D-optimality and A-optimality. For computational reasons (see Galil and Kiefer, 1980), updating the trace of a matrix (for the *A*-optimality criterion) is much slower than updating the determinant (for *D*-optimality). Thus, when you choose the *A*-optimality criterion, the computations may require significantly more time as compared to the *D*-optimality criterion. Since in practice, there are many other factors that will affect the quality of an experiment (e.g., the measurement reliability for the dependent variable), we generally recommend that you use the *D* optimality criterion. However, in difficult design situations, for example, when there appear to be many local maxima for the *D* criterion, and repeated trials yield very different results, you may want to run several optimization trials using the *A* criterion to learn more about the different types of designs that are possible.

Avoiding Matrix Singularity

It may happen during the search process that it cannot compute the inverse of the **X'X** matrix (for *A*-optimality), or that the determinant of the matrix becomes almost 0 (zero). At that point, the search can usually not continue. To avoid this situation, perform the optimization based on an augmented **X'X** matrix:

$$X'X_{augmented} = X'X + \alpha \, {}^*(X_0'X_0/N_0)$$

where X_0 stands for the design matrix constructed from the list of all N_0 candidate points, and α (*alpha*) is a user-defined small constant. Thus, you can turn off this feature by setting α to 0 (zero).

"Repairing" Designs

The optimal design features can be used to "repair" designs. For example, suppose you ran an orthogonal design, but some data were lost (e.g., due to equipment malfunction), and now some effects of interest can no longer be estimated. You could of course make up the lost runs, but suppose you do not have the resources to redo them all. In that case, you can set up the list of candidate points from among all valid points for the experimental region, add to that list all the points that you have already run, and instruct it to always force those points into the final design (and never to drop them out; you can mark points in the candidate list for such forced inclusion). It will then only consider excluding those points from the design that you did not actually run. In this manner you can, for example, find the best single run to add to an existing experiment that would optimize the respective criterion.

Constrained Experimental Regions and Optimal Design

A typical application of the optimal design features is to situations when the experimental region of interest is constrained. As described earlier in this section, there are facilities for finding vertex and centroid points for linearly constrained regions and mixtures. Those points can then be submitted as the candidate list for constructing an optimal design of a particular size for a particular model. Thus, these two facilities combined provide a very powerful tool to cope with the difficult design situation when the design region of interest is subject to complex constraints, and you want to fit particular models with the least number of runs.

Special Topics

The following topics introduce several analysis techniques. Response/desirability profiling, conducting residual analyses, and performing Box-Cox transformations of the dependent variable are described.

See also, Chapter 3 – *ANOVA/MANOVA, Methods for Analysis of Variance*, page 54, and Chapter 39 – *Variance Components and Mixed Model ANOVA/ANCOVA*.

Profiling Predicted Responses and Response Desirability

Basic Idea. A typical problem in product development is to find a set of conditions, or levels of the input variables, that produces the most desirable product in terms of its characteristics, or responses on the output variables. The procedures used to solve this problem generally involve two steps:
1) predicting responses on the dependent, or *Y* variables, by fitting the observed responses using an equation based on the levels of the

independent, or X variables, and 2) finding the levels of the X variables which simultaneously produce the most desirable predicted responses on the Y variables. Derringer and Suich (1980) give, as an example of these procedures, the problem of finding the most desirable tire tread compound. There are a number of Y variables, such as PICO Abrasion Index, 200 percent modulus, elongation at break, and hardness. The characteristics of the product in terms of the response variables depend on the ingredients, the X variables, such as hydrated silica level, silane coupling agent level, and sulfur. The problem is to select the levels for the X's that will maximize the desirability of the responses on the Y's. The solution must take into account the fact that the levels for the X's that maximize one response may not maximize a different response.

When analyzing $2^{(k-p)}$ (two-level factorial) designs, 2-level screening designs, $2^{(k-p)}$ maximally unconfounded and minimum aberration designs, $3^{(k-p)}$ and Box Behnken designs, mixed 2 and 3 level designs, central composite designs, and mixture designs, response/desirability profiling enables you to inspect the response surface produced by fitting the observed responses using an equation based on levels of the independent variables.

Prediction profiles. When you analyze the results of any of the designs listed above, a separate prediction equation for each dependent variable (containing different coefficients but the same terms) is fitted to the observed responses on the respective dependent variable. Once these equations are constructed, predicted values for the dependent variables can be computed at any combination of levels of the predictor variables. A prediction profile for a dependent variable consists of a series of graphs, one for each independent variable, of the predicted values for the dependent variable at different levels of one independent variable, holding the levels of the other independent variables constant at specified values, called current values. If appropriate current values for the independent variables have been selected, inspecting the prediction profile can show which levels of the predictor variables produce the most desirable predicted response on the dependent variable.

You might be interested in inspecting the predicted values for the dependent variables only at the actual levels at which the independent variables were set during the experiment. Alternatively, you also might be interested in inspecting the predicted values for the dependent variables at levels other than the actual levels of the independent variables used during the experiment, to see if there might be intermediate levels of the independent variables that could produce even more desirable responses. Also, returning to the Derringer and Suich (1980) example, for some response variables, the most desirable values may not necessarily be the most extreme values; for example, the most desirable value of elongation may fall within a narrow range of the possible values.

Response desirability. Different dependent variables might have different kinds of relationships between scores on the variable and the desirability of the scores. Less filling beer may be more desirable, but better tasting beer can also be more desirable – lower "fillingness" scores and higher "taste" scores are both more desirable. The relationship between predicted responses on a dependent variable and the desirability of responses is called the desirability function. Derringer and Suich (1980) developed a procedure for specifying the relationship between predicted responses on a dependent variable and the desirability of the responses, a procedure that provides for up to three inflection points in the function. Returning to the tire tread compound example described above, their procedure involved transforming scores on each of the four tire tread compound outcome variables into desirability scores that could range from 0.0 for undesirable to 1.0 for very desirable. For example, their desirability function for hardness of the tire tread compound was defined by assigning a desirability value of 0.0 to hardness scores below 60 or above 75, a desirability value of 1.0 to mid-point hardness scores of 67.5, a desirability value that increased linearly from 0.0 up to 1.0 for hardness scores between 60 and 67.5 and a desirability value that decreased linearly from 1.0 down to 0.0 for hardness scores between 67.5 and 75.0. More generally, they suggested that procedures for defining desirability functions should accommodate curvature in the falloff of desirability between inflection points in the functions.

After transforming the predicted values of the dependent variables at different combinations of levels of the predictor variables into individual desirability scores, the overall desirability of the outcomes at different combinations of levels of the predictor variables can be computed. Derringer and Suich (1980) suggested that overall desirability be computed as the geometric mean of the individual desirabilities (which makes intuitive sense, because if the individual desirability of any outcome is 0.0, or unacceptable, the overall desirability will be 0.0, or unacceptable, no matter how desirable the other individual outcomes are – the geometric mean takes the product of all of the values, and raises the product to the power of the reciprocal of the number of values). Derringer and Suich's procedure provides a straightforward way for transforming predicted

values for multiple dependent variables into a single overall desirability score. The problem of simultaneously optimization of several response variables then comes down to selecting the levels of the predictor variables that maximize the overall desirability of the responses on the dependent variables.

Summary. When you are developing a product whose characteristics are known to depend on the ingredients of which it is constituted, producing the best product possible requires determining the effects of the ingredients on each characteristic of the product, and then finding the balance of ingredients that optimizes the overall desirability of the product. In data analytic terms, the procedure that is followed to maximize product desirability is to 1) find adequate models (i.e., prediction equations) to predict characteristics of the product as a function of the levels of the independent variables, and 2) determine the optimum levels of the independent variables for overall product quality. These two steps, if followed faithfully, will likely lead to greater success in product improvement than the fabled, but statistically dubious technique of hoping for accidental breakthroughs and discoveries that radically improve product quality.

Residuals Analysis

Basic idea. Extended residuals analysis is a collection of methods for inspecting different residual and predicted values, and thus to examine the adequacy of the prediction model, the need for transformations of the variables in the model, and the existence of outliers in the data.

Residuals are the deviations of the observed values on the dependent variable from the predicted values, given the current model. The ANOVA models used in analyzing responses on the dependent variable make certain assumptions about the distributions of residual (but not predicted) values on the dependent variable. These assumptions can be summarized by saying that the ANOVA model assumes normality, linearity, homogeneity of variances and covariances, and independence of residuals. All of these properties of the residuals for a dependent variable can be inspected using residuals analysis.

Box-Cox Transformations of Dependent Variables

Basic Idea. It is assumed in analysis of variance that the variances in the different groups (experimental conditions) are homogeneous, and that they are uncorrelated with the means. If the distribution of values within each experimental condition is skewed, and the means are correlated with the standard deviations, you can often apply an appropriate power transformation to the dependent variable to stabilize the variances, and to reduce or eliminate the correlation between the means and standard deviations. The Box-Cox transformation is useful for selecting an appropriate (power) transformation of the dependent variable. Selecting the Box-Cox transformation usually provides a plot of the residual sum of squares, given the model, as a function of the value of *lambda*, where *lambda* is used to define a transformation of the dependent variable:

$y' = (y^{**}(lambda) - 1) / (g^{**}(lambda-1) * lambda)$ if lambda $\geq$ 0

$y' = g *$ natural log(y) if lambda = 0

in which g is the geometric mean of the dependent variable and all values of the dependent variable are non-negative. The value of *lambda* for which the residual sum of squares is a minimum is the maximum likelihood estimate for this parameter. It produces the variance stabilizing transformation of the dependent variable that reduces or eliminates the correlation between the group means and standard deviations.

In practice, it is not important that you use the exact estimated value of *lambda* for transforming the dependent variable. Rather, as a rule of thumb, you should consider the following transformations:

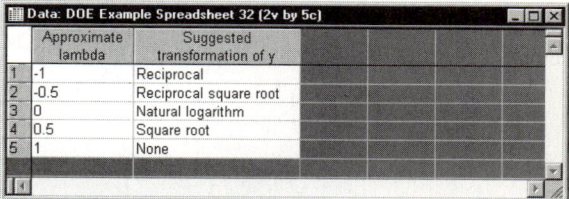

	Approximate lambda	Suggested transformation of y			
1	-1	Reciprocal			
2	-0.5	Reciprocal square root			
3	0	Natural logarithm			
4	0.5	Square root			
5	1	None			

For additional information regarding this family of transformations, see Box and Cox (1964), Box and Draper (1987), and Maddala (1977).

FACTOR ANALYSIS AND PRINCIPAL COMPONENTS

The main applications of factor analytic techniques are 1) to reduce the number of variables and 2) to detect structure in the relationships between variables, that is to classify variables. Therefore, factor analysis is applied as a data reduction or structure detection method (the term factor analysis was first introduced by Thurstone, 1931). This chapter describes the principles of factor analysis and how it can be applied toward these two purposes. We will assume that you are familiar with the basic logic of statistical reasoning as described in Chapter 1 – *Elementary Concepts in Statistics*. Moreover, we will also assume that you are familiar with the concepts of variance and correlation; if not, we advise that you read Chapter 2 – *Basic Statistics and Tables*.

There are many excellent books on factor analysis. For example, a hands-on, how-to approach can be found in Stevens (1986); more detailed technical descriptions are provided in Cooley and Lohnes (1971); Harman (1976); Kim and Mueller, (1978a, 1978b); Lawley and Maxwell (1971); Lindeman, Merenda, and Gold (1980); Morrison (1967); or Mulaik (1972). The interpretation of secondary factors in hierarchical factor analysis, as an alternative to traditional oblique rotational strategies, is explained in detail by Wherry (1984).

Confirmatory factor analysis. With structural equation modeling, you can test specific hypotheses about the factor structure for a set of variables in one or several samples (i.e., you can compare factor structures across samples). For more information, refer to Chapter 35 – *Structural Equation Modeling*.

Correspondence analysis. Correspondence analysis is a descriptive/exploratory technique designed to analyze two-way and multi-way tables containing some measure of correspondence between the rows and columns. The results provide information that is similar in nature to those produced by factor analysis techniques, and they enable you to explore the structure of categorical variables included in the table. For more information regarding these methods, refer to Chapter 11 – *Correspondence Analysis*.

239 Other Issues and Statistics

Factor Analysis as a Data Reduction Method

Suppose we conduct a (rather silly) study in which we measure 100 people's height in inches and centimeters. Thus, we would have two variables that measure height. If in future studies, we want to research, for example, the effect of different nutritional food supplements on height, would we continue to use both measures? Probably not; height is one characteristic of a person, regardless of how it is measured.

Let's now extrapolate from this silly study to something that you might actually do as a researcher. Suppose we want to measure people's satisfaction with their lives. We design a satisfaction questionnaire with various items; among other things we ask our subjects how satisfied they are with their hobbies (item 1) and how intensely they are pursuing a hobby (item 2). Most likely, the responses to the two items are highly correlated with each other. (If you are not familiar with the correlation coefficient, see *Correlations* in Chapter 2 – *Basic Statistics and Tables*.) Given a high correlation between the two items, we can conclude that they are quite redundant.

Combining two variables into a single factor. You can summarize the correlation between two variables in a scatterplot. A regression line can then be fitted that represents the best summary of the linear relationship between the variables. If we could define a variable that would approximate the regression line in such a plot, that variable would capture most of the "essence" of the two items. Subjects' single scores on that new factor, represented by the regression line, could then be used in future data analyses to represent that essence of the two items. In a sense we have reduced the two

variables to one factor. Note that the new factor is actually a linear combination of the two variables.

Principal components analysis. The example described above, combining two correlated variables into one factor, illustrates the basic idea of factor analysis, or of principal components analysis to be precise (we will return to this later). If we extend the two-variable example to multiple variables, the computations become more involved but the basic principle of expressing two or more variables by a single factor remains the same.

Extracting principal components. We do not want to go into the details about the computational aspects of principal components analysis here, which can be found elsewhere (references were provided at the beginning of this section). However, basically, the extraction of principal components amounts to a variance maximizing (varimax) rotation of the original variable space. For example, in a scatterplot we can think of the regression line as the original X axis, rotated so that it approximates the regression line. This type of rotation is called variance maximizing because the criterion for (goal of) the rotation is to maximize the variance (variability) of the new variable (factor), while minimizing the variance around the new variable (see *Rotational Strategies*, page 237).

Generalizing to the case of multiple variables. When there are more than two variables, we can think of them as defining a "space," just as two variables defined a plane. Thus, when we have three variables, we could plot a three-dimensional scatterplot, and, again we could fit a plane through the data.

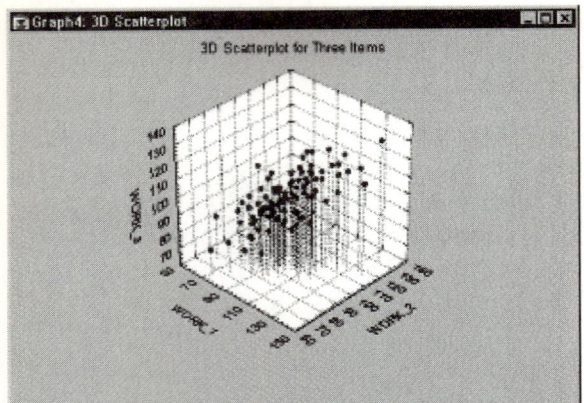

With more than three variables, it becomes impossible to illustrate the points in a scatterplot; however, the logic of rotating the axes so as to maximize the variance of the new factor remains the same.

Multiple orthogonal factors. After we have found the line on which the variance is maximal, there remains some variability around this line. In principal components analysis, after the first factor has been extracted, that is, after the first line has been drawn through the data, we continue and define another line that maximizes the remaining variability, and so on. In this manner, consecutive factors are extracted. Because each consecutive factor is defined to maximize the variability that is not captured by the preceding factor, consecutive factors are independent of each other. Put another way, consecutive factors are uncorrelated or orthogonal to each other.

How many factors to extract? Remember that, so far, we are considering principal components analysis as a data reduction method, that is, as a method for reducing the number of variables. The question then is, how many factors do we want to extract. Note that as we extract consecutive factors, they account for less and less variability. The decision of when to stop

extracting factors basically depends on when there is only very little "random" variability left. The nature of this decision is arbitrary; however, various guidelines have been developed, and they are reviewed in *Eigenvalues and the Number-of-Factors Problem*, page 235.

Reviewing the results of a principal components analysis. Without further ado, let's now look at some of the standard results from a principal components analysis. To reiterate, we are extracting factors that account for less and less variance. To simplify matters, you usually start with the correlation matrix, where the variances of all variables are equal to 1.0. Therefore, the total variance in that matrix is equal to the number of variables. For example, if we have 10 variables each with a variance of 1 then the total variability that can potentially be extracted is equal to 10 times 1. Suppose that in the satisfaction study introduced earlier we included 10 items to measure different aspects of satisfaction at home and at work. The variance accounted for by successive factors would be summarized as follows:

Value	Eigenvalue	% Total variance	Cumulative Eigenvalue	Cumulative %
	Eigenvalues (Factor.sta) Extraction: Principal components			
1	6.118369	61.18369	6.11837	61.1837
2	1.800682	18.00682	7.91905	79.1905
3	0.472888	4.72888	8.39194	83.9194
4	0.407996	4.07996	8.79993	87.9993
5	0.317222	3.17222	9.11716	91.1716
6	0.293300	2.93300	9.41046	94.1046
7	0.195808	1.95808	9.60626	96.0626
8	0.170431	1.70431	9.77670	97.7670
9	0.137970	1.37970	9.91467	99.1467
10	0.085334	0.85334	10.00000	100.0000

Eigenvalues. In the first column (*Eigenvalue*) above, we find the variance on the new factors that were successively extracted. In the second column, these values are expressed as a percent of the total variance (in this example, 10). As we can see, factor 1 accounts for 61 percent of the variance, factor 2 for 18 percent, and so on. As expected, the sum of the eigenvalues is equal to the number of variables. The third column contains the cumulative variance extracted. The variances extracted by the factors are called the eigenvalues. This name derives from the computational issues involved.

Eigenvalues and the number-of-factors problem. Now that we have a measure of how much variance each successive factor extracts, we can return to the question of how many factors to retain. As mentioned earlier, by its nature this is an arbitrary decision. However, there are some guidelines that are commonly used, and that, in practice, seem to yield the best results.

The Kaiser criterion. First, we can retain only factors with eigenvalues greater than 1. In essence, this is like saying that, unless a factor extracts at least as much as the equivalent of one original variable, we drop it. This criterion was proposed by Kaiser (1960), and is probably the one most widely used. In our example above, using this criterion, we would retain 2 factors (principal components).

The scree test. A graphical method is the scree test first proposed by Cattell (1966). We can plot the eigenvalues shown above in a simple line plot.

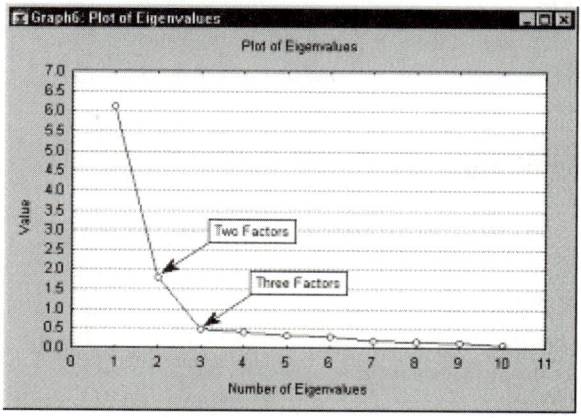

Cattell suggests finding the place where the smooth decrease of eigenvalues appears to level off to the right of the plot. To the right of this point, presumably, you find only factorial scree – scree is the geological term referring to the debris that collects on the lower part of a rocky slope. According to this criterion, we would probably retain 2 or 3 factors in our example.

Which criterion to use. Both criteria have been studied in detail (Browne, 1968; Cattell & Jaspers, 1967; Hakstian, Rogers, & Cattell, 1982; Linn, 1968; Tucker, Koopman & Linn, 1969). Theoretically, you can evaluate those criteria by generating random data based on a particular number of factors. You can then see whether the number of factors is accurately detected by those criteria. Using this general technique, the first method (Kaiser criterion) sometimes retains too many factors, while the second technique (scree test) sometimes retains too few; however, both do quite well under normal conditions, that is, when there are relatively few factors and many cases. In practice, an additional important aspect is the extent to which a solution is interpretable. Therefore, you usually examine several solutions with more or fewer factors, and chooses the one that makes the best sense. We'll discuss this issue in the section on factor rotations, page 236.

Principal factors analysis. Before we continue to examine the different aspects of the typical output from a principal components analysis, let's now introduce principal factors analysis. Let's return to our satisfaction questionnaire example to conceive of another "mental model" for factor analysis. We can think of subjects' responses as being dependent on two components. First, there are some underlying common factors, such as the satisfaction-with-hobbies factor we looked at before. Each item measures some part of this common aspect of satisfaction. Second, each item also captures a unique aspect of satisfaction that is not addressed by any other item.

Communalities. If this model is correct, we should not expect that the factors will extract all variance from our items; rather, only that proportion that is due to the common factors and shared by several items. In the language of factor analysis, the proportion of variance of a particular item that is due to common factors (shared with other items) is called communality. Therefore, an additional task facing us when applying this model is to estimate the communalities for each variable, that is, the proportion of variance that each item has in common with other items. The proportion of variance that is unique to each item is then the respective item's total variance minus the communality. A common starting point is to use the squared multiple correlation of an item with all other items as an estimate of the communality (refer to Chapter 26 – *Multiple Linear Regression* for details about multiple regression). Some authors have suggested various iterative post-solution improvements to the initial multiple regression communality estimate; for example, the so-called MINRES method (minimum residual factor method; Harman & Jones, 1966) will try various modifications to the factor loadings with the goal to minimize the residual (unexplained) sums of squares.

Principal factors vs. principal components. The defining characteristic then that distinguishes between the two factor analytic models is that in principal components analysis we assume that all variability in an item should be used in the analysis, while in principal factors analysis we only use the variability in an item that it has in common with the other items. A detailed discussion of the pros and cons of

each approach is beyond the scope of this introduction (refer to the general references provided in the *Overview* of this chapter, page 231). In most cases, these two methods usually yield very similar results. However, principal components analysis is often preferred as a method for data reduction, while principal factors analysis is often preferred when the goal of the analysis is to detect structure (see the next section, *Factor Analysis as a Classification Method*).

Factor Analysis as a Classification Method

Let's now return to the interpretation of the standard results from a factor analysis. We will henceforth use the term factor analysis generically to encompass both principal components and principal factors analysis. Let's assume that we are at the point in our analysis where we basically know how many factors to extract. We may now want to know the meaning of the factors, that is, whether and how we can interpret them in a meaningful manner. To illustrate how this can be accomplished, let's work backwards, that is, begin with a meaningful structure and then see how it is reflected in the results of a factor analysis. Let's return to our satisfaction example; shown in the next illustration is the correlation matrix for items pertaining to satisfaction at work and items pertaining to satisfaction at home.

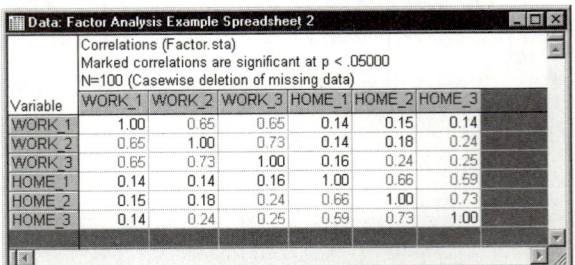

Variable	WORK_1	WORK_2	WORK_3	HOME_1	HOME_2	HOME_3
WORK_1	1.00	0.65	0.65	0.14	0.15	0.14
WORK_2	0.65	1.00	0.73	0.14	0.18	0.24
WORK_3	0.65	0.73	1.00	0.16	0.24	0.25
HOME_1	0.14	0.14	0.16	1.00	0.66	0.59
HOME_2	0.15	0.18	0.24	0.66	1.00	0.73
HOME_3	0.14	0.24	0.25	0.59	0.73	1.00

Data: Factor Analysis Example Spreadsheet 2
Correlations (Factor.sta)
Marked correlations are significant at p < .05000
N=100 (Casewise deletion of missing data)

The work satisfaction items are highly correlated amongst themselves, and the home satisfaction items are highly intercorrelated amongst themselves. The correlations across these two types of items (work satisfaction items with home satisfaction items) is comparatively small. It thus seems that there are two relatively independent factors reflected in the correlation matrix, one related to satisfaction at work, the other related to satisfaction at home.

Factor loadings. Let's now perform a principal components analysis and look at the two-factor solution. Specifically, let's look at the correlations between the variables and the two factors (or new variables), as they are extracted by default; these correlations are also called factor loadings.

Data: Factor Analysis Example Spreadsheet 3		
Factor Loadings (Unrotated) (Factor.sta) Extraction: Principal components (Marked loadings are > .700000)		
Variable	Factor 1	Factor 2
WORK_1	-0.654384	-0.564143
WORK_2	-0.715256	-0.541444
WORK_3	-0.741688	-0.508212
HOME_1	-0.634120	0.563123
HOME_2	-0.706267	0.572658
HOME_3	-0.707446	0.525602
Expl.Var	2.891313	1.791000
Prp.Totl	0.481885	0.298500

Apparently, the first factor is generally more highly correlated with the variables than the second factor. This is to be expected because, as previously described, these factors are extracted successively and will account for less and less variance overall.

Rotating the factor structure. We could plot the factor loadings shown above in a scatterplot. In that plot, each variable is represented as a point. In this plot we could rotate the axes in any direction without changing the *relative* locations of the points to each other; however, the actual coordinates of the points, that is, the

factor loadings would of course change. In this example, if you produce the plot it will be evident that if we were to rotate the axes by about 45 degrees we might attain a clear pattern of loadings identifying the work satisfaction items and the home satisfaction items.

Rotational strategies. There are various rotational strategies that have been proposed. The goal of all of these strategies is to obtain a clear pattern of loadings, that is, factors that are somehow clearly marked by high loadings for some variables and low loadings for others. This general pattern is also sometimes referred to as simple structure (a more formalized definition can be found in most standard textbooks). Typical rotational strategies are varimax, quartimax, and equamax.

We have described the idea of the varimax rotation before (see *Extracting Principal Components*, page 233), and it can be applied to this problem as well. As before, we want to find a rotation that maximizes the variance on the new axes; put another way, we want to obtain a pattern of loadings on each factor that is as diverse as possible, lending itself to easier interpretation. Shown in the next illustration is the table of rotated factor loadings.

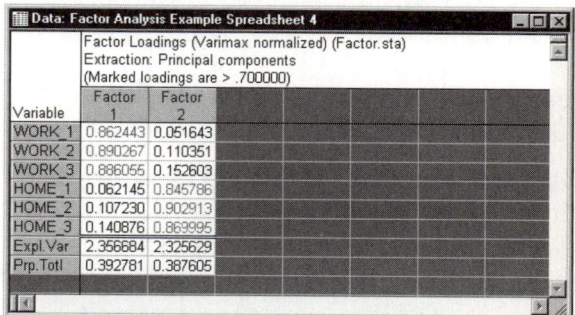

Data: Factor Analysis Example Spreadsheet 4		
Factor Loadings (Varimax normalized) (Factor.sta) Extraction: Principal components (Marked loadings are > .700000)		
Variable	Factor 1	Factor 2
WORK_1	0.862443	0.051643
WORK_2	0.890267	0.110351
WORK_3	0.886055	0.152603
HOME_1	0.062145	0.845786
HOME_2	0.107230	0.902913
HOME_3	0.140876	0.869995
Expl.Var	2.356684	2.325629
Prp.Totl	0.392781	0.387605

Interpreting the factor structure. Now the pattern is much clearer. As expected, the first factor is marked by high loadings on the work satisfaction items, the second factor is marked by high loadings on the home satisfaction items. We would thus conclude that satisfaction, as measured by our questionnaire, is composed of those two aspects; hence we have arrived at a classification of the variables.

Consider another example, this time with four additional Hobby/Misc variables added to our earlier example.

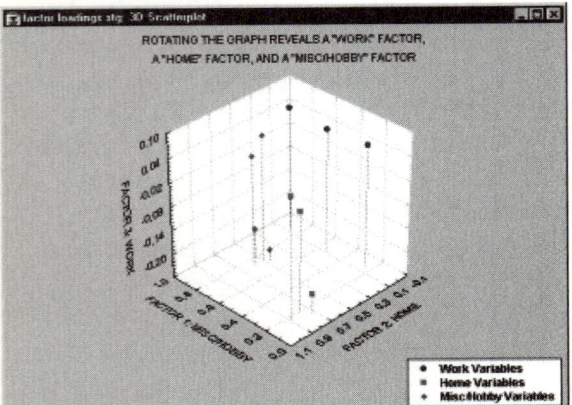

In the plot of factor loadings in the previous illustration, 10 variables were reduced to three specific factors, a work factor, a home factor and a hobby/misc. factor. Note that factor loadings for each factor are spread out over the values of the other two factors but are high for its own values. For example, the factor loadings for the hobby/misc variables have both high and low "work" and "home" values, but all four of these variables have high factor loadings on the "hobby/misc" factor.

Oblique factors. Some authors (e.g., Catell & Khanna; Harman, 1976; Jennrich & Sampson, 1966; Clarkson & Jennrich, 1988) have discussed in some detail the concept of oblique (non-orthogonal) factors, in order to achieve more interpretable simple structure. Specifically, computational strategies have been developed to rotate factors so as to best represent clusters of variables, without the constraint of orthogonality

of factors. However, the oblique factors produced by such rotations are often not easily interpreted. To return to the example, suppose we had included in the satisfaction questionnaire four items that measured other miscellaneous types of satisfaction. Let's assume that people's responses to those items were affected about equally by their satisfaction at home (*Factor 1*) and at work (*Factor 2*). An oblique rotation will likely produce two correlated factors with less-than-obvious meaning, that is, with many cross-loadings.

Hierarchical factor analysis. Instead of computing loadings for often difficult to interpret oblique factors, you can use a strategy first proposed by Thompson (1951) and Schmid and Leiman (1957), which has been elaborated and popularized in the detailed discussions by Wherry (1959, 1975, 1984). In this strategy, you first identify clusters of items and rotate axes through those clusters; next the correlations between those (oblique) factors is computed, and that correlation matrix of oblique factors is further factor-analyzed to yield a set of orthogonal factors that divide the variability in the items into that due to shared or common variance (secondary factors), and unique variance due to the clusters of similar variables (items) in the analysis (primary factors). To return to the example above, such a hierarchical analysis might yield the following factor loadings:

Data: Factor Analysis Example Spreadsheet 5					
Secondary & Primary (Unique) Factor Loadings (Factor.sta) Marked loadings are > .700000					
Factor	Second. 1	Primary 1	Primary 2		
WORK_1	0.483178	0.649499	-0.187074		
WORK_2	0.570953	0.687056	-0.140627		
WORK_3	0.565624	0.656790	-0.115461		
HOBBY_1	0.776013	0.439010	0.303672		
HOBBY_2	0.714183	0.455157	0.228351		
HOME_1	0.535812	-0.117278	0.630076		
HOME_2	0.615403	-0.079910	0.668880		
HOME_3	0.586405	-0.065512	0.626730		
MISCEL_1	0.780488	0.466823	0.280141		
MISCEL_2	0.734854	0.464779	0.238512		

Careful examination of these loadings would lead to the following conclusions:

1. There is a general (secondary) satisfaction factor that likely affects all types of satisfaction measured by the 10 items;

2. There appear to be two primary unique areas of satisfaction that can best be described as satisfaction with work and satisfaction with home life.

Wherry (1984) discusses in great detail examples of such hierarchical analyses, and how meaningful and interpretable secondary factors can be derived.

Confirmatory factor analysis. Over the past 15 years, so-called confirmatory methods have become increasingly popular (e.g., see Jöreskog and Sörbom, 1979). In general, you can specify *a priori*, a pattern of factor loadings for a particular number of orthogonal or oblique factors, and then test whether the observed correlation matrix can be reproduced given these specifications. Confirmatory factor analyses can be performed via structural equation modeling.

Other Issues and Statistics

Factor scores. We can estimate the actual values of individual cases (observations) for the factors. These factor scores are particularly useful when you want to perform further analyses involving the factors that you have identified in the factor analysis.

Reproduced and residual correlations. An additional check for the appropriateness of the respective number of factors that were extracted is to compute the correlation matrix that would result if those were indeed the only factors. That matrix is called the reproduced correlation

matrix. To see how this matrix deviates from the observed correlation matrix, you can compute the difference between the two; that matrix is called the matrix of residual correlations. The residual matrix may point to misfits, that is, to particular correlation coefficients that cannot be reproduced appropriately by the current number of factors.

Matrix ill-conditioning. If there are variables that are 100% redundant in the correlation matrix, the inverse of the matrix cannot be computed. For example, if a variable is the sum of two other variables selected for the analysis, the correlation matrix of those variables cannot be inverted, and the factor analysis can basically not be performed. In practice this happens when you are attempting to factor analyze a set of highly intercorrelated variables, as it, for example, sometimes occurs in correlational research with questionnaires. Then you can artificially lower all correlations in the correlation matrix by adding a small constant to the diagonal of the matrix, and then restandardizing it. This procedure will usually yield a matrix that now can be inverted and thus factor-analyzed; moreover, the factor patterns should not be affected by this procedure. However, note that the resulting estimates are not exact.

Data: Factor Analysis Example Spreadsheet 6

Secondary & Primary (Unique) Factor Loadings (Factor.sta)
Marked loadings are > .700000

Factor	Second. 1	Primary 1	Primary 2
WORK_1	0.483178	0.649499	-0.187074
WORK_2	0.570953	0.687056	-0.140627
WORK_3	0.565624	0.656790	-0.115461
HOBBY_1	0.776013	0.439010	0.303672
HOBBY_2	0.714183	0.455157	0.228351
HOME_1	0.535812	-0.117278	0.630076
HOME_2	0.615403	-0.079910	0.668880
HOME_3	0.586405	-0.065512	0.626730
MISCEL_1	0.780488	0.466823	0.280141
MISCEL_2	0.734854	0.464779	0.238512

Careful examination of these loadings would lead to the following conclusions:

1. There is a general (secondary) satisfaction factor that likely affects all types of satisfaction measured by the 10 items;

2. There appear to be two primary unique areas of satisfaction that can best be described as satisfaction with work and satisfaction with home life.

Wherry (1984) discusses in great detail, examples of such hierarchical analyses and how meaningful and interpretable secondary factors can be derived.

GENERAL DISCRIMINANT ANALYSIS (GDA)

General discriminant analysis (GDA) is called a "general" discriminant analysis because it applies the methods of the general linear model [see also, Chapter 18 – *General Linear Models (GLM)*] to the discriminant function analysis problem. A general overview of discriminant function analysis, and the traditional methods for fitting linear models with categorical dependent variables and continuous predictors is provided in Chapter 13 – *Discriminant Function Analysis*. In GDA, the discriminant function analysis problem is "recast" as a general multivariate linear model, where the dependent variables of interest are (dummy-) coded vectors that reflect the group membership of each case. The remainder of the analysis is then performed as described in general regression models (GRM) with a few additional features, which are noted in this chapter.

Specifying Models for Predictor Variables and Predictor Effects

One advantage of applying the general linear model to the discriminant analysis problem is that you can specify complex models for the set of predictor variables. For example, you can specify for a set of continuous predictor variables, a polynomial regression model, response surface model, factorial regression, or mixture surface regression (without an intercept). Thus, you could analyze a constrained mixture experiment (where the predictor variable values must sum to a constant), where the dependent variable of interest is categorical in nature. In fact, GDA does not impose any particular restrictions on the type of predictor variable (categorical or continuous) that can be used, or the models that can be specified. However, when using categorical predictor variables, caution should be used (see *A note of caution for models with categorical predictors, and other advanced techniques* below).

Stepwise and Best-Subset Analyses

In addition to the traditional stepwise analyses for single continuous predictors provided in discriminant analysis, general discriminant analysis makes available the options for stepwise and best-subset analyses provided in general regression models (GRM*)*. Specifically, you can request stepwise and best-subset selection of predictors or sets of predictors (in multiple-degree of freedom effects, involving categorical predictors), based on the F-to-enter and p-to-enter statistics (associated with the multivariate Wilks' *lambda* test statistic). In addition, when a cross-validation sample is specified, best-subset selection can also be based on the misclassification rates for the cross-validation sample; in other words, after estimating the discriminant functions for a given set of predictors, the misclassification rates for the cross-validation sample are computed, and the model (subset of predictors) that yields the lowest misclassification rate for the cross-validation sample is chosen. This is a powerful technique for choosing models that may yield good predictive validity, while avoiding overfitting of the data (see also, Chapter 27 – *Neural Networks*).

Desirability Profiling of Posterior Classification Probabilities

Another unique option of general discriminant analysis is the inclusion of Response/desirability profiler options. These options are described in some detail in Chapter 15 – *Experimental Design (DOE)*. In short, the predicted response values for each dependent variable are computed, and those values can be combined into a single desirability score. A graphical summary can then be produced to show the behavior of the predicted responses and the desirability score over the ranges of values for the predictor variables. In GDA, you can profile both simple predicted values (such as in general regression models) for the coded dependent variables (i.e., dummy-coded categories of the categorical dependent variable), and you can also profile posterior prediction probabilities. This unique latter option enables you to evaluate how different values for the predictor variables affect the predicted classification of cases, and is particularly useful when interpreting the results for complex models that involve categorical and continuous predictors and their interactions.

A Note of Caution for Models with Categorical Predictors, and Other Advanced Techniques

General discriminant analysis provides functionality that makes this technique a general tool for classification and data mining. However, most, if not all, textbook treatments of discriminant function analysis are limited to simple and stepwise analyses with single degree of freedom continuous predictors. No experience (in the literature) exists regarding issues of robustness and effectiveness of these techniques, when they are generalized in the manner provided in this very powerful analysis. The use of best-subset methods, in particular when used in conjunction with categorical predictors or when using the misclassification rates in a cross-validation sample for choosing the best subset of predictors, should be considered a heuristic search method, rather than a statistical analysis technique.

The Use of Categorical Predictor Variables

The use of categorical predictor variables or effects in a discriminant function analysis model may be (statistically) questionable. For example, you can use GDA to analyze a 2x2 frequency table, by specifying one variable in the 2x2 table as the dependent variable, and the other as the predictor. Clearly, the (ab)use of GDA in this manner would be silly (although, interestingly, in most cases you will get results that are generally compatible with those you would get by computing a simple *chi*-square test for the 2x2 table). Alternatively, if you only consider the parameter estimates computed

by GDA as the least squares solution to a set of linear (prediction) equations, the use of categorical predictors in GDA is fully justified; moreover, it is not uncommon in applied research to be confronted with a mixture of continuous and categorical predictors (e.g., income or age which are continuous, along with occupational status, which is categorical) for predicting a categorical dependent variable. In those cases, it can be very instructive to consider specific models involving the categorical predictors, and possibly interactions between categorical and continuous predictors for classifying observations. However, to reiterate, the use of categorical predictor variables in discriminant function analysis is not widely documented, and you should proceed cautiously before accepting the results of statistical significance tests, and before drawing final conclusions from your analyses. Also remember that there are alternative methods available to perform similar analyses, namely, the multinomial logit models available in generalized linear/nonlinear models (GLZ), and the methods for analyzing multi-way frequency tables in log-linear.

GENERAL LINEAR MODELS (GLM)

This chapter describes the use of the general linear model in a wide variety of statistical analyses. If you are unfamiliar with the basic methods of ANOVA and regression in linear models, it may be useful to first review the information on these topics in Chapter 1 – *Elementary Concepts in Statistics*. A detailed discussion of univariate and multivariate ANOVA techniques can also be found in Chapter 3 – *ANOVA/MANOVA*

Overview: The General Linear Model

The following topics summarize the historical, mathematical, and computational foundations for the general linear model. For a basic introduction to ANOVA (MANOVA, ANCOVA) techniques, refer to Chapter 3 – *ANOVA/MANOVA*; for an introduction to multiple regression, see Chapter 26 – *Multiple Linear Regression*; for an introduction to the design and analysis of experiments in applied (industrial) settings, see Chapter 15 – *Experimental Design (DOE)*.

Historical Background

The roots of the general linear model surely go back to the origins of mathematical thought, but it is the emergence of the theory of algebraic invariants in the 1800's that made the general linear model, as we know it today, possible. The theory of algebraic invariants developed from the groundbreaking work of 19th century mathematicians such as Gauss, Boole, Cayley, and Sylvester. The theory seeks to identify those quantities in systems of equations that remain unchanged under linear transformations of the variables in the system. Stated more imaginatively (but in a way in which the originators of the theory would not consider an overstatement), the theory of algebraic invariants searches for the eternal and unchanging amongst the chaos of the transitory and the illusory. That is no small goal for any theory, mathematical or otherwise.

The wonder of it all is the theory of algebraic invariants was successful far beyond the hopes of its originators. Eigenvalues, eigenvectors, determinants, matrix decomposition methods; all derive from the theory of algebraic invariants. The contributions of the theory of algebraic invariants to the development of statistical theory and methods are numerous, but a simple example familiar to even the most casual student of statistics is illustrative. The correlation between two variables is unchanged by linear transformations of either or both variables. We probably take this property of correlation coefficients for granted, but what would data analysis be like if we did not have statistics that are invariant to the scaling of the variables involved? Some thought on this question should convince you that without the theory of algebraic invariants, the development of useful statistical techniques would be nigh impossible.

The development of the linear regression model in the late 19th century, and the development of correlational methods shortly thereafter, are clearly direct outgrowths of the theory of algebraic invariants. Regression and correlational methods, in turn, serve as the basis for the general linear model. Indeed, the general linear model can be seen as an extension of linear multiple regression for a single dependent variable. Understanding the multiple regression model is fundamental to understanding the general linear model, so we will look at the purpose of multiple regression, the computational algorithms used to solve regression problems, and how the regression model is extended in the case of the general linear model. A basic introduction to multiple regression methods and the analytic problems to which they are applied is provided in Chapter 26 – *Multiple Linear Regression*.

The Purpose of Multiple Regression

The general linear model can be seen as an extension of linear multiple regression for a single dependent variable, and understanding the multiple regression model is fundamental to understanding the general linear model. The general purpose of multiple regression (the term was first used by Pearson, 1908) is to quantify the relationship between several independent or predictor variables and a dependent or criterion variable. For example, a real estate agent might record for each listing the size of the house (in square feet), the number of bedrooms, the average income in the respective neighborhood according to census data, and a subjective rating of appeal of the house. Once this information has been compiled for various houses it would be interesting to see whether and how these measures relate to the price for which a house is sold. For example, you might learn that the number of bedrooms is a better predictor of the price for which a house sells in a particular neighborhood than how pretty the house is (subjective rating). You can also detect outliers; for example, houses that should really sell for more, given their location and characteristics.

Personnel professionals customarily use multiple regression procedures to determine equitable compensation. You can determine a number of factors or dimensions such as amount of responsibility (*Resp*) or number of people to supervise (*No_Super*) that you believe to contribute to the value of a job. The personnel analyst then usually conducts a salary survey among comparable companies in the market, recording the salaries and respective characteristics (i.e., values on dimensions) for different positions. This information can be used in a multiple regression analysis to build a regression equation of the form:

Salary = .5*Resp + .8*No_Super

Once this so-called regression equation has been determined, the analyst can now easily construct a graph of the expected (predicted) salaries and the actual salaries of job incumbents in his or her company. Thus, the analyst is able to determine which position is underpaid (below the regression line) or overpaid (above the regression line), or paid equitably.

In the social and natural sciences, multiple regression procedures are very widely used in research. In general, multiple regression allows the researcher to ask (and hopefully answer) the general question "what is the best predictor of ...". For example, educational researchers might want to learn what the best predictors of success in high school are. Psychologists may want to determine which personality variable best predicts social adjustment. Sociologists may want to find out which of the multiple social indicators best predict whether or not a new immigrant group will adapt and be absorbed into society.

Computations for Solving the Multiple Regression Equation

A one-dimensional surface in a two-dimensional or two-variable space is a line defined by the equation $Y = b_0 + b_1 X$. According to this equation, the Y variable can be expressed in terms of or as a function of a constant (b_0) and a slope (b_1) times the X variable. The constant is also referred to as the intercept, and the slope as the regression coefficient. For example, *GPA* can best be predicted as *1+.02*IQ*. Thus, knowing that a student has an *IQ* of 130 would lead us to predict that her *GPA* would be 3.6 (since, 1+.02*130=3.6). In the multiple regression case, when there are multiple predictor variables, the regression surface usually cannot be visualized in a two dimensional space, but the computations are a straightforward extension of the computations in the single predictor case. For example, if in addition to *IQ* we had additional predictors of achievement (e.g., *Motivation, Self-*

discipline) we could construct a linear equation containing all those variables. In general then, multiple regression procedures will estimate a linear equation of the form:

$$Y = b_0 + b_1X_1 + b_2X_2 + ... + b_kX_k$$

where k is the number of predictors. Note that in this equation, the regression coefficients (or b_1 ... b_k coefficients) represent the independent contributions of each in dependent variable to the prediction of the dependent variable. Another way to express this fact is to say that, for example, variable X_1 is correlated with the Y variable, after controlling for all other independent variables. This type of correlation is also referred to as a partial correlation (this term was first used by Yule, 1907). Perhaps the following example will clarify this issue. You would probably find a significant negative correlation between hair length and height in the population (i.e., short people have longer hair). At first this may seem odd; however, if we were to add the variable Gender into the multiple regression equation, this correlation would probably disappear. This is because women, on the average, have longer hair than men; they also are shorter on the average than men. Thus, after we remove this gender difference by entering Gender into the equation, the relationship between hair length and height disappears because hair length does *not* make any unique contribution to the prediction of height, above and beyond what it shares in the prediction with variable *Gender*. Put another way, after controlling for the variable *Gender*, the partial correlation between hair length and height is zero.

The regression surface (a line in simple regression, a plane or higher-dimensional surface in multiple regression) expresses the best prediction of the dependent variable (*Y*),

given the independent variables (*X*'s). However, nature is rarely (if ever) perfectly predictable, and usually there is substantial variation of the observed points from the fitted regression surface. The deviation of a particular point from the nearest corresponding point on the predicted regression surface (its predicted value) is called the residual value. Since the goal of linear regression procedures is to fit a surface, which is a linear function of the X variables, as closely as possible to the observed Y variable, the residual values for the observed points can be used to devise a criterion for the best fit. Specifically, in regression problems the surface is computed for which the sum of the squared deviations of the observed points from that surface are minimized. Thus, this general procedure is sometimes also referred to as least squares estimation. (See also, the description of weighted least squares estimation.)

The actual computations involved in solving regression problems can be expressed compactly and conveniently using matrix notation. Suppose that there are n observed values of Y and n associated observed values for each of k different X variables. Then Y_i, X_{ik}, and e_i can represent the ith observation of the Y variable, the ith observation of each of the X variables, and the ith unknown residual value, respectively. Collecting these terms into matrices, we have:

$$Y = \begin{bmatrix} Y_1 \\ \vdots \\ \vdots \\ \vdots \\ Y_n \end{bmatrix}, \quad X = \begin{bmatrix} 1 & X_{11} & \cdots & \cdots & X_{1k} \\ \vdots & \vdots & \vdots & \vdots & \vdots \\ \vdots & \vdots & \vdots & \vdots & \vdots \\ \vdots & \vdots & \vdots & \vdots & \vdots \\ 1 & X_{n1} & \cdots & \cdots & X_{nk} \end{bmatrix}, \quad e = \begin{bmatrix} e_1 \\ \vdots \\ \vdots \\ e_n \end{bmatrix}$$

The multiple regression model in matrix notation then can be expressed as:

$$Y = Xb + e$$

where b is a column vector of 1 (for the intercept) + k unknown regression coefficients. Recall that the goal of multiple regression is to minimize the sum of the squared residuals. Regression coefficients that satisfy this criterion are found by solving the set of normal equations

X'Xb = X'Y

When the X variables are linearly independent (i.e., they are nonredundant, yielding an $X'X$ matrix which is of full rank) there is a unique solution to the normal equations. Premultiplying both sides of the matrix formula for the normal equations by the inverse of $X'X$ gives

$(X'X)^{-1}X'Xb = (X'X)^{-1}X'Y$

or

b = $(X'X)^{-1}X'Y$

This last result is very satisfying in view of its simplicity and its generality. With regard to its simplicity, it expresses the solution for the regression equation in terms just 2 matrices (**X** and **Y**) and 3 basic matrix operations, 1) matrix transposition, which involves interchanging the elements in the rows and columns of a matrix, 2) matrix multiplication, which involves finding the sum of the products of the elements for each row and column combination of two conformable (i.e., multipliable) matrices, and 3) matrix inversion, which involves finding the matrix equivalent of a numeric reciprocal, that is, the matrix that satisfies

$A^{-1}AA=A$

for a matrix **A**.

It took literally centuries for the ablest mathematicians and statisticians to find a satisfactory method for solving the linear least square regression problem. But their efforts have paid off, for it is hard to imagine a simpler solution.

With regard to the generality of the multiple regression model, its only notable limitations are that 1) it can be used to analyze only a single dependent variable, 2) it cannot provide a solution for the regression coefficients when the X variables are not linearly independent and the inverse of X'X therefore does not exist. These restrictions, however, can be overcome, and in doing so the multiple regression model is transformed into the general linear model.

Extension of Multiple Regression to the General Linear Model

One way in which the general linear model differs from the multiple regression model is in terms of the number of dependent variables that can be analyzed. The **Y** vector of n observations of a single Y variable can be replaced by a **Y** matrix of n observations of m different Y variables. Similarly, the **b** vector of regression coefficients for a single Y variable can be replaced by a **b** matrix of regression coefficients, with one vector of b coefficients for each of the m dependent variables. These substitutions yield what is sometimes called the multivariate regression model, but it should be emphasized that the matrix formulations of the multiple and multivariate regression models are identical, except for the number of columns in the Y and b matrices. The method for solving for the b coefficients is also identical, that is, m different sets of regression coefficients are separately found for the m different dependent variables in the multivariate regression model.

The general linear model goes a step beyond the multivariate regression model by allowing for linear transformations or linear combinations of multiple dependent variables. This extension gives the general linear model important

advantages over the multiple and the so-called multivariate regression models, both of which are inherently univariate (single dependent variable) methods. One advantage is that multivariate tests of significance can be employed when responses on multiple dependent variables are correlated. Separate univariate tests of significance for correlated dependent variables are not independent and may not be appropriate. Multivariate tests of significance of independent linear combinations of multiple dependent variables also can give insight into which dimensions of the response variables are, and are not, related to the predictor variables. Another advantage is the ability to analyze effects of repeated measure factors. Repeated measure designs, or within-subject designs, have traditionally been analyzed using ANOVA techniques. Linear combinations of responses reflecting a repeated measure effect (for example, the difference of responses on a measure under differing conditions) can be constructed and tested for significance using either the univariate or multivariate approach to analyzing repeated measures in the general linear model.

A second important way in which the general linear model differs from the multiple regression model is in its ability to provide a solution for the normal equations when the X variables are not linearly independent and the inverse of $X'X$ does not exist. Redundancy of the X variables may be incidental (e.g., two predictor variables might happen to be perfectly correlated in a small data set), accidental (e.g., two copies of the same variable might unintentionally be used in an analysis) or designed (e.g., indicator variables with exactly opposite values might be used in the analysis, as when both *Male* and *Female* predictor variables are used in representing *Gender*). Finding the

regular inverse of a non-full-rank matrix is reminiscent of the problem of finding the reciprocal of 0 in ordinary arithmetic. No such inverse or reciprocal exists because division by 0 is not permitted. This problem is solved in the general linear model by using a generalized inverse of the $X'X$ matrix in solving the normal equations. A generalized inverse is any matrix that satisfies

$$AA^-A = A$$

for a matrix A.

A generalized inverse is unique and is the same as the regular inverse only if the matrix A is full rank. A generalized inverse for a non-full-rank matrix can be computed by the simple expedient of zeroing the elements in redundant rows and columns of the matrix. Suppose that an $X'X$ matrix with r non-redundant columns is partitioned as:

$$X'X = \begin{bmatrix} A_{11} & A_{12} \\ A_{21} & A_{22} \end{bmatrix}$$

where A_{11} is an r by r matrix of rank r. Then the regular inverse of A_{11} exists and a generalized inverse of $X'X$ is

$$(X'X)^- = \begin{bmatrix} A_{11}^{-1} & 0_{12} \\ 0_{21} & 0_{22} \end{bmatrix}$$

where each 0 (null) matrix is a matrix of 0s (zeroes) and has the same dimensions as the corresponding A matrix.

In practice, however, a particular generalized inverse of $X'X$ for finding a solution to the normal equations is usually computed using the sweep operator (Dempster, 1960). This generalized inverse, called a g2 inverse, has two important properties. One is that zeroing of the elements in redundant rows is unnecessary. Another is that partitioning or reordering of the columns of $\mathbf{X'X}$ is unnecessary, so that the matrix can be inverted "in place."

There are infinitely many generalized inverses of a non-full-rank $\mathbf{X'X}$ matrix, and thus, infinitely many solutions to the normal equations. This can make it difficult to understand the nature of the relationships of the predictor variables to responses on the dependent variables, because the regression coefficients can change depending on the particular generalized inverse chosen for solving the normal equations. It is not cause for dismay, however, because of the invariance properties of many results obtained using the general linear model.

A simple example may be useful for illustrating one of the most important invariance properties of the use of generalized inverses in the general linear model. If both *Male* and *Female* predictor variables with exactly opposite values are used in an analysis to represent *Gender*, it is essentially arbitrary as to which predictor variable is considered to be redundant (e.g., *Male* can be considered to be redundant with *Female*, or vice versa). No matter which predictor variable is considered to be redundant, no matter which corresponding generalized inverse is used in solving the normal equations, and no matter which resulting regression equation is used for computing predicted values on the dependent variables, the predicted values and the corresponding residuals for males and females will be unchanged. In using the general linear model, you must keep in mind that finding a particular arbitrary solution to the normal equations is primarily a means to the end of accounting for responses on the dependent variables, and not necessarily an end in itself.

Sigma-Restricted vs. Overparameterized Model

Unlike the multiple regression model, which is usually applied to cases where the X variables are continuous, the general linear model is frequently applied to analyze any ANOVA or MANOVA design with categorical predictor variables and any ANCOVA or MANCOVA design with both categorical and continuous predictor variables, as well as any multiple or multivariate regression design with continuous predictor variables. To illustrate, *Gender* is clearly a nominal level variable (anyone who attempts to rank order the sexes on any dimension does so at his or her own peril in today's world). There are two basic methods by which *Gender* can be coded into one or more (non-offensive) predictor variables, and analyzed using the general linear model.

Sigma-restricted model (coding of categorical predictors). Using the first method, males and females can be assigned any two arbitrary, but distinct values on a single predictor variable. The values on the resulting predictor variable will represent a quantitative contrast between males and females. Typically, the values corresponding to group membership are chosen not arbitrarily but rather to facilitate interpretation of the regression coefficient associated with the predictor variable. In one widely used strategy, cases in the two groups are assigned values of 1 and -1 on the predictor variable, so that if the regression coefficient for the variable is positive, the group coded as 1 on the predictor variable will have a higher predicted value (i.e., a higher group mean) on the dependent variable, and if the regression coefficient is negative, the group coded as -1 on the predictor variable will have a higher predicted value on the dependent variable. An additional advantage is that since each group is coded with a value one unit from zero, this helps in interpreting the magnitude of differences in predicted values between groups, because regression coefficients reflect the units of change in the dependent variable for each unit change in the predictor variable. This coding strategy is aptly called the sigma-restricted parameterization, because the values used to represent group membership (1 and -1) sum to zero.

Note that the sigma-restricted parameterization of categorical predictor variables usually leads to $X'X$ matrices that do not require a generalized inverse for solving the normal equations. Potentially redundant information, such as the characteristics of maleness and femaleness, is literally reduced to full rank by creating quantitative contrast variables representing differences in characteristics.

Overparameterized model (coding of categorical predictors). The second basic method for recoding categorical predictors is the indicator variable approach. In this method a separate predictor variable is coded for each group identified by a categorical predictor variable. To illustrate, females might be assigned a value of 1 and males a value of 0 on a first predictor variable identifying membership in the female *Gender* group, and males would then be assigned a value of 1 and females a value of 0 on a second predictor variable identifying membership in the male *Gender* group. Note that this method of recoding categorical predictor variables will almost always lead to $X'X$ matrices with redundant columns, and thus require a generalized inverse for solving the normal equations. As such, this method is often called the overparameterized model for representing categorical predictor variables, because it results in more columns in the $X'X$ than are

necessary for determining the relationships of categorical predictor variables to responses on the dependent variables.

True to its description as general, the general linear model can be used to perform analyses with categorical predictor variables that are coded using either of the two basic methods that have been described.

Summary of Computations

To conclude this discussion of the ways in which the general linear model extends and generalizes regression methods, the general linear model can be expressed as

YM = Xb + e

Here Y, X, b, and e are as described for the multivariate regression model and **M** is an m x s matrix of coefficients defining s linear transformation of the dependent variables. The normal equations are

X'Xb = X'YM

and a solution for the normal equations is given by

b = (X'X)⁻X'YM

Here the inverse of **X'X** is a generalized inverse if **X'X** contains redundant columns.

Add a provision for analyzing linear combinations of multiple dependent variables, add a method for dealing with redundant predictor variables and recoded categorical predictor variables, and the major limitations of multiple regression are overcome by the general linear model.

Types of Analyses

A wide variety of types of designs can be analyzed using the general linear model. In fact, the flexibility of the general linear model enables it to handle so many different types of designs that it is difficult to develop simple typologies of the ways in which these designs might differ. Some general ways in which designs might differ can be suggested, but keep in mind that any particular design can be a "hybrid" in the sense that it could have combinations of features of a number of different types of designs.

In the following discussion, references will be made to the design matrix **X**, as well as sigma-restricted and overparameterized model coding. For an explanation of this terminology, refer to *Basic Ideas: The General Linear Model*, page 247, or, for a brief summary, to *Summary of Computations*, page 253.

A basic discussion to univariate and multivariate ANOVA techniques can also be found in Chapter 3 – *ANOVA/MANOVA*; a discussion of multiple regression methods is also provided in Chapter 26 – *Multiple Linear Regression*.

Between-Subject Designs

Overview. The levels or values of the predictor variables in an analysis describe the differences between the *n* subjects or the *n* valid cases that are analyzed. Thus, when we speak of the between subject design (or simply the between design) for an analysis, we are referring to the nature, number, and arrangement of the predictor variables.

Concerning the nature or type of predictor variables, between designs which contain only categorical predictor variables can be called ANOVA (analysis of variance) designs, between designs which contain only continuous predictor variables can be called regression designs, and between designs which contain

both categorical and continuous predictor variables can be called ANCOVA (analysis of covariance) designs. Further, continuous predictors are always considered to have fixed values, but the levels of categorical predictors can be considered to be fixed or to vary randomly. Designs that contain random categorical factors are called mixed-model designs (see Chapter 39 – *Variance Components and Mixed Model ANOVA/ANCOVA*).

Between designs can involve only a single predictor variable and therefore be described as simple (e.g., simple regression) or can employ numerous predictor variables (e.g., multiple regression).

Concerning the arrangement of predictor variables, some between designs employ only "main effect" or first-order terms for predictors, that is, the values for different predictor variables are independent and raised only to the first power. Other between designs may employ higher-order terms for predictors by raising the values for the original predictor variables to a power greater than 1 (e.g., in polynomial regression designs), or by forming products of different predictor variables (i.e., interaction terms). A common arrangement for ANOVA designs is the full-factorial design, in which every combination of levels for each of the categorical predictor variables is represented in the design. Designs with some but not all combinations of levels for each of the categorical predictor variables are aptly called fractional factorial designs. Designs with a hierarchy of combinations of levels for the different categorical predictor variables are called nested designs.

These basic distinctions about the nature, number, and arrangement of predictor variables can be used in describing a variety of different types of between designs. Some of the more common between designs can now be described.

One-way ANOVA. A design with a single categorical predictor variable is called a one-way ANOVA design. For example, a study of 4 different fertilizers used on different individual plants could be analyzed via one-way ANOVA, with four levels for the factor *Fertilizer*.

In general, consider a single categorical predictor variable A with 1 case in each of its 3 categories. Using the sigma-restricted coding of A into 2 quantitative contrast variables, the matrix $\mathbf{X}$ defining the between design is

$$\mathbf{X} = \begin{array}{c} \\ A_1 \\ A_2 \\ A_3 \end{array} \begin{array}{ccc} X_0 & X_1 & X_2 \\ \left[\begin{array}{ccc} 1 & 1 & 0 \\ 1 & 0 & 1 \\ 1 & -1 & -1 \end{array}\right] \end{array}$$

That is, cases in groups A_1, A_2, and A_3 are all assigned values of 1 on X_0 (the intercept), the case in group A_1 is assigned a value of 1 on X_1 and a value 0 on X_2, the case in group A_2 is assigned a value of 0 on X_1 and a value 1 on X_2, and the case in group A_3 is assigned a value of -1 on X_1 and a value -1 on X_2. Of course, any additional cases in any of the 3 groups would be coded similarly. If there were 1 case in group A_1, 2 cases in group A_2, and 1 case in group A_3, the $\mathbf{X}$ matrix would be:

$$\mathbf{X} = \begin{array}{c} \\ A_{11} \\ A_{12} \\ A_{22} \\ A_{13} \end{array} \begin{array}{ccc} X_0 & X_1 & X_2 \\ \left[\begin{array}{ccc} 1 & 1 & 0 \\ 1 & 0 & 1 \\ 1 & 0 & 1 \\ 1 & -1 & -1 \end{array}\right] \end{array}$$

where the first subscript for A gives the replicate number for the cases in each group. For brevity, replicates usually are not shown when describing ANOVA design matrices.

Note that in one-way designs with an equal number of cases in each group, sigma-restricted coding yields $X_1 \ldots X_k$ variables all of which have means of 0.

Using the overparameterized model to represent A, the $\mathbf{X}$ matrix defining the between design is simply:

$$\mathbf{X} = \begin{array}{c} \\ A_1 \\ A_2 \\ A_3 \end{array} \begin{array}{cccc} X_0 & X_1 & X_2 & X_3 \\ \left[\begin{array}{cccc} 1 & 1 & 0 & 0 \\ 1 & 0 & 1 & 0 \\ 1 & 0 & 0 & 1 \end{array}\right] \end{array}$$

These simple examples show that the $\mathbf{X}$ matrix actually serves two purposes. It specifies 1) the coding for the levels of the original predictor variables on the X variables used in the analysis as well as 2) the nature, number, and arrangement of the X variables, that is, the between design.

Main Effect ANOVA. Main effect ANOVA designs contain separate one-way ANOVA designs for 2 or more categorical predictors. A good example of main effect ANOVA would be the typical analysis performed on screening designs as described in Chapter 15 – *Experimental Design (DOE)*.

Consider 2 categorical predictor variables A and B each with 2 categories. Using the sigma-restricted coding, the $\mathbf{X}$ matrix defining the between design is:

$$\mathbf{X} = \begin{array}{c} \\ A_1B_1 \\ A_1B_2 \\ A_2B_1 \\ A_2B_2 \end{array} \begin{array}{ccc} X_0 & X_1 & X_2 \\ \left[\begin{array}{ccc} 1 & 1 & 1 \\ 1 & 1 & -1 \\ 1 & -1 & 1 \\ 1 & -1 & -1 \end{array}\right] \end{array}$$

Note that if there are equal numbers of cases in each group, the sum of the cross-products of values for the X_1 and X_2 columns is 0, for example, with 1 case in each group (1*1)+(1*-1)+(-1*1)+(-1*-1)=0. Using the overparameterized model, the matrix $\mathbf{X}$ defining the between design is:

$$\mathbf{X} = \begin{array}{c} \\ A_1B_1 \\ A_1B_2 \\ A_2B_1 \\ A_2B_2 \end{array} \begin{array}{ccccc} X_0 & X_1 & X_2 & X_3 & X_4 \\ \left[\begin{array}{ccccc} 1 & 1 & 0 & 1 & 0 \\ 1 & 1 & 0 & 0 & 1 \\ 1 & 0 & 1 & 1 & 0 \\ 1 & 0 & 1 & 0 & 1 \end{array}\right] \end{array}$$

Comparing the two types of coding, it can be seen that the overparameterized coding takes almost twice as many values as the sigma-restricted coding to convey the same information.

Factorial ANOVA. Factorial ANOVA designs contain X variables representing combinations of the levels of 2 or more categorical predictors (e.g., a study of boys and girls in four age groups, resulting in a *2 (Gender) x 4 (Age Group)* design). In particular, full-factorial designs represent all possible combinations of the levels of the categorical predictors. A full-factorial design with 2 categorical predictor variables A and B each with 2 levels each would be called a 2x2 full-factorial design. Using the sigma-restricted coding, the **X** matrix for this design would be:

$$\mathbf{X} = \begin{array}{c} \\ A_1B_1 \\ A_1B_2 \\ A_2B_1 \\ A_2B_2 \end{array} \begin{array}{cccc} X_0 & X_1 & X_2 & X_3 \\ \left[\begin{array}{cccc} 1 & 1 & 1 & 1 \\ 1 & 1 & -1 & -1 \\ 1 & -1 & 1 & -1 \\ 1 & -1 & -1 & 1 \end{array}\right] \end{array}$$

Several features of this **X** matrix deserve comment. Note that the X_1 and X_2 columns represent main effect contrasts for one variable, (i.e., A and B, respectively) collapsing across the levels of the other variable. The X_3 column instead represents a contrast between different combinations of the levels of A and B. Note also that the values for X_3 are products of the corresponding values for X_1 and X_2. Product variables such as X_3 represent the multiplicative or interaction effects of their factors, so X_3 would be said to represent the 2-way interaction of A and B. The relationship of such product variables

to the dependent variables indicate the interactive influences of the factors on responses above and beyond their independent (i.e., main effect) influences on responses. Thus, factorial designs provide more information about the relationships between categorical predictor variables and responses on the dependent variables than is provided by corresponding one-way or main effect designs.

When many factors are being investigated, however, full-factorial designs sometimes require more data than reasonably can be collected to represent all possible combinations of levels of the factors, and high-order interactions between many factors can become difficult to interpret. With many factors, a useful alternative to the full-factorial design is the fractional factorial design. As an example, consider a 2x2 x 2 fractional factorial design to degree 2 with 3 categorical predictor variables each with 2 levels. The design would include the main effects for each variable, and all 2-way interactions between the three variables, but would not include the 3-way interaction between all three variables. Using the overparameterized model, the **X** matrix for this design is:

$$\mathbf{X} = \begin{array}{c} A_1B_1C_1 \\ A_1B_1C_2 \\ A_1B_2C_1 \\ A_1B_2C_2 \\ A_2B_1C_1 \\ A_2B_1C_2 \\ A_2B_2C_1 \\ A_2B_2C_2 \end{array} \left[\begin{array}{ccccccc|cccccccccccc} 1&1&0&1&0&1&0&1&0&0&0&1&0&0&0&1&0&0&0 \\ 1&1&0&1&0&0&1&1&0&0&0&0&1&0&0&0&1&0&0 \\ 1&1&0&0&1&1&0&0&1&0&0&1&0&0&0&0&0&1&0 \\ 1&1&0&0&1&0&1&0&1&0&0&0&1&0&0&0&0&0&1 \\ 1&0&1&1&0&1&0&0&0&1&0&0&0&1&0&1&0&0&0 \\ 1&0&1&1&0&0&1&0&0&1&0&0&0&0&1&0&1&0&0 \\ 1&0&1&0&1&1&0&0&0&0&1&0&0&1&0&0&0&1&0 \\ 1&0&1&0&1&0&1&0&0&0&1&0&0&0&1&0&0&0&1 \end{array}\right]$$

(...... main effects .... 2 - way interactions ....)

The 2-way interactions are the highest degree effects included in the design. These types of designs are discussed in *2^(k-p) Fractional Factorial Designs*, page 185.

Nested ANOVA designs. Nested designs are similar to fractional factorial designs in that all possible combinations of the levels of the categorical predictor variables are not represented in the design. In nested designs, however, the omitted effects are lower-order effects. Nested effects are effects in which the nested variables never appear as main effects. Suppose that for 2 variables A and B with 3 and 2 levels, respectively, the design includes the main effect for A and the effect of B nested within the levels of A. The **X** matrix for this design using the overparameterized model is:

$$\mathbf{X} = \begin{array}{c} \\ A_1B_1 \\ A_1B_2 \\ A_2B_1 \\ A_2B_2 \\ A_3B_1 \\ A_3B_2 \end{array} \begin{array}{cccccccccc} X_0 & X_1 & X_2 & X_3 & X_4 & X_5 & X_6 & X_7 & X_8 & X_9 \\ \left[\begin{array}{cccccccccc} 1 & 1 & 0 & 0 & 1 & 0 & 0 & 0 & 0 & 0 \\ 1 & 1 & 0 & 0 & 0 & 1 & 0 & 0 & 0 & 0 \\ 1 & 0 & 1 & 0 & 0 & 0 & 1 & 0 & 0 & 0 \\ 1 & 0 & 1 & 0 & 0 & 0 & 0 & 1 & 0 & 0 \\ 1 & 0 & 0 & 1 & 0 & 0 & 0 & 0 & 1 & 0 \\ 1 & 0 & 0 & 1 & 0 & 0 & 0 & 0 & 0 & 1 \end{array}\right] \end{array}$$

Note that if the sigma-restricted coding were used, there would be only 2 columns in the **X** matrix for the B nested within A effect instead of the 6 columns in the **X** matrix for this effect when the overparameterized model coding is used (i.e., columns X_4 through X_9). The sigma-restricted coding method is overly-restrictive for nested designs, so only the overparameterized model is used to represent nested designs.

Balanced ANOVA. Most of the between designs discussed in this section can be analyzed much more efficiently, when they are balanced, i.e., when all cells in the ANOVA design have equal n, when there are no missing cells in the design, and, if nesting is present, when the nesting is balanced so that equal numbers of levels of the factors that are nested appear in the levels of the factor(s) that they are nested in. In that case, the **X'X** matrix (where **X** stands for the design matrix) is a diagonal matrix, and many of the computations necessary to compute the ANOVA results (such as matrix inversion) are greatly simplified.

Simple regression. Simple regression designs involve a single continuous predictor variable. If there were 3 cases with values on a predictor variable P of, say, 7, 4, and 9, and the design is for the first-order effect of P, the **X** matrix would be:

$$\mathbf{X} = \begin{array}{c} \\ \end{array} \begin{array}{cc} X_0 & X_1 \\ \left[\begin{array}{cc} 1 & 7 \\ 1 & 4 \\ 1 & 9 \end{array}\right] \end{array}$$

and using P for X_1 the regression equation would be:

$$Y = b_0 + b_1 P$$

If the simple regression design is for a higher-order effect of P, say the quadratic effect, the values in the X_1 column of the design matrix would be raised to the 2nd power, that is, squared:

$$\mathbf{X} = \begin{array}{c} \\ \end{array} \begin{array}{cc} X_0 & X_1 \\ \left[\begin{array}{cc} 1 & 49 \\ 1 & 16 \\ 1 & 81 \end{array}\right] \end{array}$$

and using P^2 for X_1 the regression equation would be

$$Y = b_0 + b_1 P^2$$

The sigma-restricted and overparameterized coding methods do not apply to simple regression designs and any other design containing only continuous predictors (since there are no categorical predictors to code). Regardless of which coding method is chosen, values on the continuous predictor variables are raised to the desired power and used as the values for the X variables. No recoding is performed. It is therefore sufficient, in describing regression designs, to simply describe the regression equation without explicitly describing the design matrix **X**.

Multiple regression. Multiple regression designs are to continuous predictor variables as main effect ANOVA designs are to categorical predictor variables, that is, multiple regression designs contain the separate simple regression designs for 2 or more continuous predictor variables. The regression equation for a multiple regression design for the first-order effects of 3 continuous predictor variables P, Q, and R would be:

$$Y = b_0 + b_1 P + b_2 Q + b_3 R$$

Factorial regression. Factorial regression designs are similar to factorial ANOVA designs, in which combinations of the levels of the factors are represented in the design. In factorial regression designs, however, there may be many more such possible combinations of distinct levels for the continuous predictor variables than there are cases in the data set. To simplify matters, full-factorial regression designs are defined as designs in which all possible products of the continuous predictor variables are represented in the design. For example, the full-factorial regression design for two continuous predictor variables P and Q would include the main effects (i.e., the first-order effects) of P and Q and their 2-way P by Q interaction effect, which is represented by the product of P and Q scores for each case. The regression equation would be:

$$Y = b_0 + b_1 P + b_2 Q + b_3 P*Q$$

Factorial regression designs can also be fractional, that is, higher-order effects can be omitted from the design. A fractional factorial design to degree 2 for 3 continuous predictor variables P, Q, and R would include the main effects and all 2-way interactions between the predictor variables.

$$Y = b_0 + b_1 P + b_2 Q + b_3 R + b_4 P*Q + b_5 P*R + b_6 Q*R$$

Polynomial regression. Polynomial regression designs are designs that contain main effects and higher-order effects for the continuous predictor variables but do not include interaction effects between predictor variables. For example, the polynomial regression design to degree 2 for three continuous predictor variables P, Q, and R would include the main effects (i.e., the first-order effects) of P, Q, and R and their quadratic (i.e., second-order) effects, but not the 2-way interaction effects or the P by Q by R 3-way interaction effect.

$$Y = b_0 + b_1 P + b_2 P^2 + b_3 Q + b_4 Q^2 + b_5 R + b_6 R^2$$

Polynomial regression designs do not have to contain all effects up to the same degree for every predictor variable. For example, main, quadratic, and cubic effects could be included in the design for some predictor variables, and effects up the fourth degree could be included in the design for other predictor variables.

Response surface regression. Quadratic response surface regression designs are a hybrid type of design with characteristics of both polynomial regression designs and fractional factorial regression designs. Quadratic response surface regression designs contain all the same effects of polynomial regression designs to degree 2 and additionally the 2-way interaction effects of the predictor variables. The regression equation for a quadratic response surface regression design for 3 continuous predictor variables P, Q, and R would be:

$$Y = b_0 + b_1 P + b_2 P^2 + b_3 Q + b_4 Q^2 + b_5 R + b_6 R^2 + b_7 P*Q + b_8 P*R + b_9 Q*R$$

These types of designs are commonly employed in applied research (e.g., in industrial experimentation), and a detailed discussion of these types of designs is also presented in Chapter 15 – *Experimental Design (DOE)* (see *Central Composite Designs*, page 201).

Mixture surface regression. Mixture surface regression designs are identical to factorial regression designs to degree 2 except for the omission of the intercept. Mixtures, as the name implies, add up to a constant value; the sum of the proportions of ingredients in different recipes for some material all must add up 100%. Thus, the proportion of one ingredient in a material is redundant with the remaining ingredients. Mixture surface regression designs deal with this redundancy by omitting the intercept from the design. The design matrix for a mixture surface regression design for 3 continuous predictor variables *P, Q,* and *R* would be

$$Y = b_1P + b_2Q + b_3R + b_4P*Q + b_5P*R + b_6Q*R$$

These types of designs are commonly employed in applied research (e.g., in industrial experimentation), and a detailed discussion of these types of designs is also presented in Chapter 15 – *Experimental Design (DOE)* (see *Mixture Designs and Triangular Surfaces*, page 211).

Analysis of covariance. In general, between designs that contain both categorical and continuous predictor variables can be called ANCOVA designs. Traditionally, however, ANCOVA designs have referred more specifically to designs in which the first-order effects of one or more continuous predictor variables are taken into account when assessing the effects of one or more categorical predictor variables. A basic introduction to analysis of covariance can also be found in *Analysis of Covariance (ANCOVA)*, page 48.

To illustrate, suppose a researcher wants to assess the influences of a categorical predictor variable *A* with 3 levels on some outcome, and that measurements on a continuous predictor variable *P*, known to covary with the outcome, are available. If the data for the analysis are:

$$
\begin{array}{cc}
P & \text{Group} \\
\begin{bmatrix} 7 \\ 4 \\ 9 \\ 3 \\ 6 \\ 8 \end{bmatrix} &
\begin{bmatrix} A_1 \\ A_1 \\ A_2 \\ A_2 \\ A_3 \\ A_3 \end{bmatrix}
\end{array}
$$

then the sigma-restricted **X** matrix for the design that includes the separate first-order effects of *P* and *A* would be:

$$
\mathbf{X} =
\begin{array}{cccc}
X_0 & X_1 & X_2 & X_3 \\
\begin{bmatrix} 1 & 7 & 1 & 0 \\ 1 & 4 & 1 & 0 \\ 1 & 9 & 0 & 1 \\ 1 & 3 & 0 & 1 \\ 1 & 6 & -1 & -1 \\ 1 & 8 & -1 & -1 \end{bmatrix}
\end{array}
$$

The b_2 and b_3 coefficients in the regression equation

$$Y = b_0 + b_1X_1 + b_2X_2 + b_3X_3$$

represent the influences of group membership on the *A* categorical predictor variable, controlling for the influence of scores on the *P* continuous predictor variable. Similarly, the b_1 coefficient represents the influence of scores on *P* controlling for the influences of group membership on *A*. This traditional ANCOVA analysis gives a more sensitive test of the influence of *A* to the extent that *P* reduces the prediction error, that is, the residuals for the outcome variable.

The **X** matrix for the same design using the overparameterized model would be:

$$\begin{array}{ccccc} X_0 & X_1 & X_2 & X_3 & X_4 \end{array}$$

$$\mathbf{X} = \begin{bmatrix} 1 & 7 & 1 & 0 & 0 \\ 1 & 4 & 1 & 0 & 0 \\ 1 & 9 & 0 & 1 & 0 \\ 1 & 3 & 0 & 1 & 0 \\ 1 & 6 & 0 & 0 & 1 \\ 1 & 8 & 0 & 0 & 1 \end{bmatrix}$$

The interpretation is unchanged except that the influences of group membership on the A categorical predictor variables are represented by the b_2, b_3 and b_4 coefficients in the regression equation

$$Y = b_0 + b_1X_1 + b_2X_2 + b_3X_3 + b_4X_4$$

Separate slope designs. The traditional analysis of covariance (ANCOVA) design for categorical and continuous predictor variables is inappropriate when the categorical and continuous predictors interact in influencing responses on the outcome. The appropriate design for modeling the influences of the predictors in this situation is called the separate slope design. For the same example data used to illustrate traditional ANCOVA, the overparameterized **X** matrix for the design that includes the main effect of the three-level categorical predictor A and the 2-way interaction of P by A would be:

$$\begin{array}{ccccccc} X_0 & X_1 & X_2 & X_3 & X_4 & X_5 & X_6 \end{array}$$

$$\mathbf{X} = \begin{bmatrix} 1 & 1 & 0 & 0 & 7 & 0 & 0 \\ 1 & 1 & 0 & 0 & 4 & 0 & 0 \\ 1 & 0 & 1 & 0 & 0 & 9 & 0 \\ 1 & 0 & 1 & 0 & 0 & 3 & 0 \\ 1 & 0 & 0 & 1 & 0 & 0 & 6 \\ 1 & 0 & 0 & 1 & 0 & 0 & 8 \end{bmatrix}$$

The b_4, b_5, and b_6 coefficients in the regression equation

$$Y = b_0 + b_1X_1 + b_2X_2 + b_3X_3 + b_4X_4 + b_5X_5 + b_6X_6$$

give the separate slopes for the regression of the outcome on P within each group on A, controlling for the main effect of A.

As with nested ANOVA designs, the sigma-restricted coding of effects for separate slope designs is overly restrictive, so only the overparameterized model is used to represent separate slope designs. In fact, separate slope designs are identical in form to nested ANOVA designs, since the main effects for continuous predictors are omitted in separate slope designs.

Homogeneity of slopes. The appropriate design for modeling the influences of continuous and categorical predictor variables depends on whether the continuous and categorical predictors interact in influencing the outcome. The traditional analysis of covariance (ANCOVA) design for continuous and categorical predictor variables is appropriate when the continuous and categorical predictors do not interact in influencing responses on the outcome, and the separate slope design is appropriate when the continuous and categorical predictors do interact in influencing responses. The homogeneity of slopes designs can be used to test whether the continuous and categorical predictors interact in influencing responses, and thus, whether the traditional ANCOVA design or the separate slope design is appropriate for modeling the effects of the predictors. For the same example data used to illustrate the traditional ANCOVA and separate slope designs, the overparameterized **X** matrix for the design that includes the main effect of P, the main effect of the three-level categorical predictor A, and the 2-way interaction of P by A would be:

$$X = \begin{bmatrix} X_0 & X_1 & X_2 & X_3 & X_4 & X_5 & X_6 & X_7 \\ 1 & 7 & 1 & 0 & 0 & 7 & 0 & 0 \\ 1 & 4 & 1 & 0 & 0 & 4 & 0 & 0 \\ 1 & 9 & 0 & 1 & 0 & 0 & 9 & 0 \\ 1 & 3 & 0 & 1 & 0 & 0 & 3 & 0 \\ 1 & 6 & 0 & 0 & 1 & 0 & 0 & 6 \\ 1 & 8 & 0 & 0 & 1 & 0 & 0 & 8 \end{bmatrix}$$

If the b_5, b_6, or b_7 coefficient in the regression equation

$$Y = b_0 + b_1X_1 + b_2X_2 + b_3X_3 + b_4X_4 + b_5X_5 + b_6X_6 + b_7X_7$$

is non-zero, the separate slope model should be used. If instead all 3 of these regression coefficients are zero the traditional ANCOVA design should be used.

The sigma-restricted **X** matrix for the homogeneity of slopes design would be:

$$X = \begin{bmatrix} X_0 & X_1 & X_2 & X_3 & X_4 & X_5 \\ 1 & 7 & 1 & 0 & 7 & 0 \\ 1 & 4 & 1 & 0 & 4 & 0 \\ 1 & 9 & 0 & 1 & 0 & 9 \\ 1 & 3 & 0 & 1 & 0 & 3 \\ 1 & 6 & -1 & -1 & -6 & -6 \\ 1 & 8 & -1 & -1 & -8 & -8 \end{bmatrix}$$

Using this **X** matrix, if the b_4, or b_5 coefficient in the regression equation

$$Y = b_0 + b_1X_1 + b_2X_2 + b_3X_3 + b_4X_4 + b_5X_5$$

is non-zero, the separate slope model should be used. If instead both of these regression coefficients are zero the traditional ANCOVA design should be used.

Mixed model ANOVA and ANCOVA. Designs that contain random effects for one or more categorical predictor variables are called mixed-model designs. Random effects are classification effects where the levels of the effects are assumed to be randomly selected from an infinite population of possible levels. The solution for the normal equations in mixed-

model designs is identical to the solution for fixed-effect designs (i.e., designs which do not contain Random effects. Mixed-model designs differ from fixed-effect designs only in the way in which effects are tested for significance. In fixed-effect designs, between effects are always tested using the mean squared residual as the error term. In mixed-model designs, between effects are tested using relevant error terms based on the covariation of random sources of variation in the design. Specifically, this is done using Satterthwaite's method of denominator synthesis (Satterthwaite, 1946), which finds the linear combinations of sources of random variation that serve as appropriate error terms for testing the significance of the respective effect of interest. A basic discussion of these types of designs, and methods for estimating variance components for the random effects can also be found in Chapter 39 – *Variance Components and Mixed Model ANOVA/ ANCOVA*.

Mixed-model designs, like nested designs and separate slope designs, are designs in which the sigma-restricted coding of categorical predictors is overly restrictive. Mixed-model designs require estimation of the covariation between the levels of categorical predictor variables, and the sigma-restricted coding of categorical predictors suppresses this covariation. Thus, only the overparameterized model is used to represent mixed-model designs (some programs will use the sigma-restricted approach and a so-called restricted model for random effects; however, only the overparameterized model as described in General Linear Models applies to both balanced and unbalanced designs, as well as designs with missing cells; see Searle, Casella, & McCullock, 1992, p. 127). It is important to recognize, however, that sigma-restricted coding can be used to represent any between design, with the exceptions of mixed-model, nested, and separate slope designs. Furthermore, some types of hypotheses can only be tested using the sigma-restricted coding (i.e., the effective hypothesis, Hocking, 1996), thus the greater generality of the overparameterized model for representing between designs does not justify it being used exclusively for representing categorical predictors in the general linear model.

Within-Subject (Repeated Measures) Designs

Overview. It is quite common for researchers to administer the same test to the same subjects repeatedly over a period of time or under varying circumstances. In essence, they are interested in examining differences within each subject, for example, subjects' improvement over time. Such designs are referred to as within-subject designs or repeated measures designs. A basic introduction to repeated measures designs is also provided in *Between-within designs*, page 47, in Chapter 3 – *ANOVA/MANOVA*.

For example, imagine that you want to monitor the improvement of students' algebra skills over two months of instruction. A standardized algebra test is administered after one month (level 1 of the repeated measures factor), and a comparable test is administered after two months (level 2 of the repeated measures factor). Thus, the repeated measures factor (*Time*) has 2 levels.

Now, suppose that scores for the 2 algebra tests (i.e., values on the Y_1 and Y_2 variables at *Time 1* and *Time 2*, respectively) are transformed into scores on a new composite variable (i.e., values on the T_1), using the linear transformation

$$T = YM$$

where **M** is an orthonormal contrast matrix. Specifically, if:

$$\begin{bmatrix} T_{11} \\ \vdots \\ \vdots \\ \vdots \\ T_{n1} \end{bmatrix} = \begin{bmatrix} Y_{11} & Y_{12} \\ \vdots & \vdots \\ \vdots & \vdots \\ \vdots & \vdots \\ Y_{n1} & Y_{n2} \end{bmatrix} \begin{bmatrix} \sqrt{.5} \\ -\sqrt{.5} \end{bmatrix}$$

then the difference of the mean score on T_1 from *0* indicates the improvement (or deterioration) of scores across the 2 levels of *Time*.

One-way within-subject designs. The example algebra skills study with the *Time* repeated measures factor [see also, *Within-Subject (Repeated Measures) Designs Overview* above] illustrates a one-way within-subject design. In such designs, orthonormal contrast transformations of the scores on the original dependent *Y* variables are performed via the *M* transformation (orthonormal transformations correspond to orthogonal rotations of the original variable axes). If any b_0 coefficient in the regression of a transformed *T* variable on the intercept is non-zero, this indicates a change in responses across the levels of the repeated measures factor, that is, the presence of a main effect for the repeated measure factor on responses.

What if the between design includes effects other than the intercept? If any of the b_1 through b_k coefficients in the regression of a transformed *T* variable on *X* are non-zero, this indicates a different change in responses across the levels of the repeated measures factor for different levels of the corresponding between effect, i.e., the presence of a within by between interaction effect on responses.

The same between-subject effects that can be tested in designs with no repeated-measures factors can also be tested in designs that do include repeated-measures factors. This is accomplished by creating a transformed dependent variable that is the sum of the original dependent variables divided by the square root of the number of original dependent variables. The same tests of between-subject effects that are performed in designs with no repeated-measures factors (including tests of the between intercept) are performed on this transformed dependent variable.

Multi-way within-subject designs. Suppose that in the example algebra skills study with the *Time* repeated measures factor [see *Within-Subject (Repeated Measures) Designs Overview*, page 262], students were given a number problem test and then a word problem test on each testing occasion. *Test* could then be considered as a second repeated measures factor, with scores on the number problem tests representing responses at level 1 of the *Test* repeated measure factor, and scores on the word problem tests representing responses at level 2 of the *Test* repeated measure factor. The within subject design for the study would be a 2 (*Time*) by 2 (*Test*) full-factorial design, with effects for *Time, Test,* and the *Time* by *Test* interaction.

To construct transformed dependent variables representing the effects of *Time, Test,* and the *Time* by *Test* interaction, three respective *M* transformations of the original dependent *Y* variables are performed. Assuming that the original *Y* variables are in the order *Time 1 – Test 1, Time 1 – Test 2, Time 2 – Test 1,* and *Time 2 – Test 2,* the *M* matrices for the *Time, Test,* and the *Time* by *Test* interaction would be:

$$\mathbf{M}_{Time} = \begin{bmatrix} .5 \\ .5 \\ -.5 \\ -.5 \end{bmatrix}, \quad \mathbf{M}_{Test} = \begin{bmatrix} .5 \\ -.5 \\ .5 \\ -.5 \end{bmatrix}, \quad \mathbf{M}_{Time \times Test} = \begin{bmatrix} .5 \\ -.5 \\ -.5 \\ .5 \end{bmatrix}$$

The differences of the mean scores on the transformed *T* variables from 0 are then used to interpret the corresponding within-subject

effects. If the b_0 coefficient in the regression of a transformed *T* variable on the intercept is non-zero, this indicates a change in responses across the levels of a repeated measures effect, that is, the presence of the corresponding main or interaction effect for the repeated measure factors on responses.

Interpretation of within by between interaction effects follow the same procedures as for one-way within designs, except that now within by between interactions are examined for each within effect by between effect combination.

Multivariate approach to repeated measures. When the repeated measures factor has more than 2 levels, the **M** matrix will have more than a single column. For example, for a repeated measures factor with 3 levels (e.g., *Time 1, Time 2, Time 3*), the **M** matrix will have 2 columns (e.g., the two transformations of the dependent variables could be 1) *Time 1* vs. *Time 2* and *Time 3* combined, and 2) *Time 2* vs. *Time 3*). Consequently, the nature of the design is really multivariate, that is, there are two simultaneous dependent variables, which are transformations of the original dependent variables. Therefore, when testing repeated measures effects involving more than a single degree of freedom (e.g., a repeated measures main effect with more than 2 levels), you can compute multivariate test statistics to test the respective hypotheses. This is a different (and usually the preferred) approach than the univariate method that is still widely used. For a further discussion of the multivariate approach to testing repeated measures effects and a comparison to the traditional univariate approach, see *Sphericity and Compound Symmetry*, page 53.

Doubly multivariate designs. If the product of the number of levels for each within-subject factor is equal to the number of original dependent variables, the within-subject design is called a univariate repeated measures design. The within design is univariate because there is one dependent variable representing each combination of levels of the within-subject factors. Note that this use of the term univariate design is not to be confused with the univariate and multivariate approach to the analysis of repeated measures designs, both of which can be used to analyze such univariate (single-dependent-variable-only) designs. When there are two or more dependent variables for each combination of levels of the within-subject factors, the within-subject design is called a multivariate repeated measures design, or more commonly, a doubly multivariate within-subject design. This term is used because the analysis for each dependent measure can be done via the multivariate approach; so when there is more than one dependent measure, the design can be considered doubly-multivariate.

Doubly multivariate design are analyzed using a combination of univariate repeated measures and multivariate analysis techniques. To illustrate, suppose in an algebra skills study, tests are administered three times (repeated measures factor *Time* with 3 levels). Two test scores are recorded at each level of *Time*: a *Number Problem* score and a *Word Problem* score. Thus, scores on the two types of tests could be treated as multiple measures on which improvement (or deterioration) across *Time* could be assessed. *M* transformed variables could be computed for each set of test measures, and multivariate tests of significance could be performed on the multiple transformed measures, as well as on the each individual test measure.

Multivariate Designs

Overview. When there are multiple dependent variables in a design, the design is said to be multivariate. Multivariate measures of association are by nature more complex than their univariate counterparts (such as the correlation coefficient, for example). This is because multivariate measures of association must take into account not only the relationships of the predictor variables with responses on the dependent variables, but also the relationships among the multiple dependent variables. By doing so, however, these measures of association provide information about the strength of the relationships between predictor and dependent variables independent of the dependent variable interrelationships. A basic discussion of multivariate designs is also presented in *Multivariate Designs*, page 49.

The most commonly used multivariate measures of association all can be expressed as functions of the eigenvalues of the product matrix:

$$E^{-1}H$$

where **E** is the error SSCP matrix (i.e., the matrix of sums of squares and cross-products for the dependent variables that are not accounted for by the predictors in the between design), and **H** is a hypothesis SSCP matrix (i.e., the matrix of sums of squares and cross-products for the dependent variables that are accounted for by all the predictors in the between design, or the sums of squares and cross-products for the dependent variables that are accounted for by a particular effect). If

λ_i = the ordered eigenvalues of $E^{-1}H$, if E^{-1} exists

the 4 commonly used multivariate measures of association are:

Wilks' lambda = $\Pi[1/(1+\lambda_i)]$

Pillai's trace = $\Sigma\lambda_i/(1+\lambda_i)$

Hotelling-Lawley trace = $\Sigma\lambda_i$

Roy's largest root = λ_1

These 4 measures have different upper and lower bounds, with Wilks' *lambda* perhaps being the most easily interpretable of the 4 measures. Wilks' *lambda* can range from 0 to 1, with 1 indicating no relationship of predictors to responses and 0 indicating a perfect relationship of predictors to responses. 1 – Wilks' *lambda* can be interpreted as the multivariate counterpart of a univariate R-squared, that is, it indicates the proportion of generalized variance in the dependent variables that is accounted for by the predictors.

The 4 measures of association are also used to construct multivariate tests of significance. These multivariate tests are covered in detail in a number of sources (e.g., Finn, 1974; Tatsuoka, 1971).

Estimation and Hypothesis Testing

The following sections discuss details concerning hypothesis testing, for example, how the test for the overall model fit is computed, the options for computing tests for categorical effects in unbalanced or incomplete designs, how and when custom-error terms can be chosen, and the logic of testing custom-hypotheses in factorial or regression designs.

Whole Model Tests

Partitioning sums of squares. A fundamental principle of least squares methods is that variation on a dependent variable can be partitioned, or divided into parts, according to the sources of the variation. Suppose that a

dependent variable is regressed on one or more predictor variables, and that, for convenience, the dependent variable is scaled so that its mean is 0. Then a basic least squares identity is that the total sum of squared values on the dependent variable equals the sum of squared predicted values plus the sum of squared residual values. Stated more generally,

$$\Sigma(y - \text{y-bar})^2 = \Sigma(\text{y-hat} - \text{y-bar})^2 + \Sigma(y - \text{y-hat})^2$$

where the term on the left is the total sum of squared deviations of the observed values on the dependent variable from the dependent variable mean, and the respective terms on the right are 1) the sum of squared deviations of the predicted values for the dependent variable from the dependent variable mean, and 2) the sum of the squared deviations of the observed values on the dependent variable from the predicted values, that is, the sum of the squared residuals. Stated yet another way,

Total SS = Model SS + Error SS

Note that the *Total SS* is always the same for any particular data set, but that the *Model SS* and the *Error SS* depend on the regression equation. Assuming again that the dependent variable is scaled so that its mean is 0, the Model SS and the Error SS can be computed using:

Model SS = b'X'Y

Error SS = Y'Y – b'X'Y

Testing the whole model. Given the Model SS and the Error SS, you can perform a test that all the regression coefficients for the X variables ($b1$ through bk) are zero. This test is equivalent to a comparison of the fit of the regression surface defined by the predicted values (computed from the whole model regression equation) to the fit of the regression surface defined solely by the dependent variable mean (computed from the reduced regression equation containing only the intercept). Assuming that $X'X$ is full rank, the whole model hypothesis mean square

MSH = (Model SS)/k

is an estimate of the variance of the predicted values. The error mean square

s^2 = MSE = (Error SS)/(n-k-1)

is an unbiased estimate of the residual or error variance. The test statistic is:

F = MSH/MSE

where F has $(k, n - k - 1)$ degrees of freedom.

If $X'X$ is not full rank, $r + 1$ is substituted for k, where r is the rank or the number of non-redundant columns of $X'X$.

Note that in the case of non-intercept models, some multiple regression programs will compute the full model test based on the proportion of variance around 0 (zero) accounted for by the predictors; for more information (see Kvålseth, 1985; Okunade, Chang, and Evans, 1993), while others will actually compute both values (i.e., based on the residual variance around 0, and around the respective dependent variable means).

Limitations of whole model tests. For designs such as one-way ANOVA or simple regression, the whole model test by itself may be sufficient for testing general hypotheses about whether or not the single predictor variable is related to the outcome. In more complex designs, however, hypotheses about specific X variables or subsets of X variables are usually of interest. For example, you might want to make inferences about whether a subset of regression coefficients are 0, or you might want to test whether subpopulation means corresponding to combinations of specific X

variables differ. The whole model test is usually insufficient for such purposes.

A variety of methods have been developed for testing specific hypotheses. Like whole model tests, many of these methods rely on comparisons of the fit of different models (e.g., Type I, Type II, and the effective hypothesis sums of squares). Other methods construct tests of linear combinations of regression coefficients in order to test mean differences (e.g., Type III, Type IV, and Type V sums of squares). For designs that contain only first-order effects of continuous predictor variables (i.e., multiple regression designs), many of these methods are equivalent (i.e., Type II through Type V sums of squares all test the significance of partial regression coefficients). However, there are important distinctions between the different hypothesis testing techniques for certain types of ANOVA designs (i.e., designs with unequal cell n's and/or missing cells).

All methods for testing hypotheses, however, involve the same hypothesis testing strategy employed in whole model tests, that is, the sums of squares attributable to an effect (using a given criterion) is computed, and then the mean square for the effect is tested using an appropriate error term.

Six Types of Sums of Squares

When there are categorical predictors in the model arranged in a factorial ANOVA design, you are typically interested in the main effects for and interaction effects between the categorical predictors. However, when the design is not balanced (has unequal cell n's, and consequently, the coded effects for the categorical factors are usually correlated), or when there are missing cells in a full factorial ANOVA design, there is ambiguity regarding the specific comparisons between the (population, or least-squares) cell means that constitute the main effects and interactions of interest. These issues are discussed in great detail in Milliken and Johnson (1986), and if you routinely analyze incomplete factorial designs, you should consult their discussion of various problems and approaches to solving them.

In addition to the widely used methods that are commonly labeled *Type I, II, III*, and *IV* sums of squares (see Goodnight, 1980), we also offer different methods for testing effects in incomplete designs, that are widely used in other areas (and traditions) of research.

Type V sums of squares. Specifically, we propose the term *Type V sums of squares* to denote the approach that is widely used in industrial experimentation, to analyze fractional factorial designs; these types of designs are discussed in detail in $2^{(k-p)}$ *Fractional Factorial Designs*, page 185. In effect, for those effects for which tests are performed all population marginal means (least squares means) are estimable.

Type VI sums of squares. Second, in keeping with the *Type I* labeling convention, we propose the term *Type VI sums of squares* to denote the approach that is often used in programs that only implement the *sigma*-restricted model (which is not well suited for certain types of designs; we offer a choice between the sigma-restricted and overparameterized model models). This approach is identical to what is described as the *effective hypothesis* method in Hocking (1996).

Contained effects. The following descriptions will use the term contained effect. An effect *E1* (e.g., *A * B* interaction) is contained in another effect *E2* if:

- Both effects involve the same continuous predictor variable (if included in the model; e.g., *A * B * X* would be contained in *A * C * X*, where *A*, *B*, and *C* are categorical predictors, and *X* is a continuous predictor); or

- *E2* has more categorical predictors than does *E1*, and, if *E1* includes any categorical predictors, they also appear in *E2* (e.g., *A * B* would be contained in the *A * B * C* interaction).

Type I sums of squares. Type I sums of squares involve a sequential partitioning of the whole model sums of squares. A hierarchical series of regression equations are estimated, at each step adding an additional effect into the model. In Type I sums of squares, the sums of squares for each effect are determined by subtracting the predicted sums of squares with the effect in the model from the predicted sums of squares for the preceding model not including the effect. Tests of significance for each effect are then performed on the increment in the predicted sums of squares accounted for by the effect. Type I sums of squares are therefore sometimes called sequential or hierarchical sums of squares.

Type I sums of squares are appropriate to use in balanced (equal *n*) ANOVA designs in which effects are entered into the model in their natural order (i.e., any main effects are entered before any two-way interaction effects, any two-way interaction effects are entered before any three-way interaction effects, and so on). Type I sums of squares are also useful in polynomial regression designs in which any lower-order effects are entered before any higher-order effects. A third use of Type I sums of squares is to test hypotheses for hierarchically nested designs, in which the first effect in the design is nested within the second effect, the second effect is nested within the third, and so on.

One important property of Type I sums of squares is that the sums of squares attributable to each effect add up to the whole model sums of squares. Thus, Type I sums of squares provide a complete decomposition of the predicted sums of squares for the whole model. This is not generally true for any other type of sums of squares. An important limitation of Type I sums of squares, however, is that the sums of squares attributable to a specific effect will generally depend on the order in which the effects are entered into the model. This lack of invariance to order of entry into the model limits the usefulness of Type I sums of squares for testing hypotheses for certain designs (e.g., fractional factorial designs).

Type II sums of squares. Type II sums of squares are sometimes called partially sequential sums of squares. As Type I sums of squares, Type II sums of squares for an effect controls for the influence of other effects. Which other effects to control for, however, is determined by a different criterion. In Type II sums of squares, the sums of squares for an effect is computed by controlling for the influence of all other effects of equal or lower degree. Thus, sums of squares for main effects control for all other main effects, sums of squares for two-way interactions control for all main effects and all other two-way interactions, and so on.

Unlike Type I sums of squares, Type II sums of squares are invariant to the order in which effects are entered into the model. This makes Type II sums of squares useful for testing hypotheses for multiple regression designs, for main effect ANOVA designs, for full-factorial

ANOVA designs with equal cell ns, and for hierarchically nested designs.

There is a drawback to the use of Type II sums of squares for factorial designs with unequal cell ns. In these situations, Type II sums of squares test hypotheses that are complex functions of the cell ns that ordinarily are not meaningful. Thus, a different method for testing hypotheses is usually preferred.

Type III sums of squares. Type I and Type II sums of squares usually are not appropriate for testing hypotheses for factorial ANOVA designs with unequal ns. For ANOVA designs with unequal ns, however, Type III sums of squares test the same hypothesis that would be tested if the cell ns were equal, provided that there is at least one observation in every cell. Specifically, in no-missing-cell designs, Type III sums of squares test hypotheses about differences in subpopulation (or marginal) means. When there are no missing cells in the design, these subpopulation means are least squares means, which are the best linear-unbiased estimates of the marginal means for the design (see, Milliken and Johnson, 1986).

Tests of differences in least squares means have the important property that they are invariant to the choice of the coding of effects for categorical predictor variables (e.g., the use of the sigma-restricted or overparameterized model) and to the choice of the particular g2 inverse of $X'X$ used to solve the normal equations. Thus, tests of linear combinations of least squares means in general, including Type III tests of differences in least squares means, are said to not depend on the parameterization of the design. This makes Type III sums of squares useful for testing hypotheses for any design for which Type I or Type II sums of squares are appropriate, as well as for any unbalanced ANOVA design with no missing cells.

The Type III sums of squares attributable to an effect is computed as the sums of squares for the effect controlling for any effects of equal or lower degree and orthogonal to any higher-order interaction effects (if any) that contain it. The orthogonality to higher-order containing interactions is what gives Type III sums of squares the desirable properties associated with linear combinations of least squares means in ANOVA designs with no missing cells. But for ANOVA designs with missing cells, Type III sums of squares generally do not test hypotheses about least squares means, but instead test hypotheses that are complex functions of the patterns of missing cells in higher-order containing interactions and that are ordinarily not meaningful. In this situation Type V sums of squares or tests of the effective hypothesis (Type VI sums of squares) are preferred.

Type IV sums of squares. Type IV sums of squares were designed to test balanced hypotheses for lower-order effects in ANOVA designs with missing cells. Type IV sums of squares are computed by equitably distributing cell contrast coefficients for lower-order effects across the levels of higher-order containing interactions.

Type IV sums of squares are not recommended for testing hypotheses for lower-order effects in ANOVA designs with missing cells, even though this is the purpose for which they were developed. This is because Type IV sum-of-squares are invariant to some but not all g2 inverses of $X'X$ that could be used to solve the normal equations. Specifically, Type IV sums of squares are invariant to the choice of a g2 inverse of $X'X$ given a particular ordering of the levels of the categorical predictor variables, but

are not invariant to different orderings of levels. Furthermore, as with Type III sums of squares, Type IV sums of squares test hypotheses that are complex functions of the patterns of missing cells in higher-order containing interactions and that are ordinarily not meaningful.

Statisticians who have examined the usefulness of Type IV sums of squares have concluded that Type IV sums of squares are not up to the task for which they were developed:

- Milliken & Johnson (1992, p. 204) write: "It seems likely that few, if any, of the hypotheses tested by the Type IV analysis of *[some programs]* will be of particular interest to the experimenter."

- Searle (1987, p. 463-464) writes: "In general, [Type IV] hypotheses determined in this nature are not necessarily of any interest."; and (p. 465) "This characteristic of Type IV sums of squares for rows depending on the sequence of rows establishes their non-uniqueness, and this in turn emphasizes that the hypotheses they are testing are by no means necessarily of any general interest."

- Hocking (1985, p. 152), in an otherwise comprehensive introduction to general linear models, writes: "For the missing cell problem, *[some programs]* offers a fourth analysis, Type IV, which we shall not discuss."

So, we recommend that you use the Type IV sums of squares solution with caution, and that you understand fully the nature of the (often non-unique) hypotheses that are being testing, before attempting interpretations of the results. Furthermore, in ANOVA designs with no missing cells, Type IV sums of squares are always equal to Type III sums of squares, so the use of Type IV sums of squares is either (potentially) inappropriate, or unnecessary, depending on the presence of missing cells in the design.

Type V sums of squares. Type V sums of squares were developed as an alternative to Type IV sums of squares for testing hypotheses in ANOVA designs in missing cells. Also, this approach is widely used in industrial experimentation, to analyze fractional factorial designs; these types of designs are discussed in detail in the $2^{(k-p)}$ *Fractional Factorial Designs*, page 185. In effect, for effects for which tests are performed all population marginal means (least squares means) are estimable.

Type V sums of squares involve a combination of the methods employed in computing Type I and Type III sums of squares. Specifically, whether or not an effect is eligible to be dropped from the model is determined using Type I procedures, and then hypotheses are tested for effects not dropped from the model using Type III procedures. Type V sums of squares can be illustrated by using a simple example. Suppose that the effects considered are A, B, and A by B, in that order, and that A and B are both categorical predictors with, say, 3 and 2 levels, respectively. The intercept is first entered into the model. Then A is entered into the model, and its degrees of freedom are determined (i.e., the number of non-redundant columns for A in $X'X$, given the intercept). If A's degrees of freedom are less than 2 (i.e., its number of levels minus 1), it is eligible to be dropped. Then B is entered into the model, and its degrees of freedom are determined (i.e., the number of non-redundant columns for B in $X'X$, given the intercept and A). If B's degrees of freedom are less than 1 (i.e., its number of levels minus 1), it is eligible to be dropped. Finally, A by B is entered into the model, and its degrees of freedom are determined (i.e., the number of non-redundant columns for A by B in

$X'X$, given the intercept, A, and B). If B's degrees of freedom are less than 2 (i.e., the product of the degrees of freedom for its factors if there were no missing cells), it is eligible to be dropped. Type III sums of squares are then computed for the effects that were not found to be eligible to be dropped, using the reduced model in which any eligible effects are dropped. Tests of significance, however, use the error term for the whole model prior to dropping any eligible effects.

Note that Type V sums of squares involve determining a reduced model for which all effects remaining in the model have at least as many degrees of freedom as they would have if there were no missing cells. This is equivalent to finding a subdesign with no missing cells such that the Type III sums of squares for all effects in the subdesign reflect differences in least squares means.

Appropriate caution should be exercised when using Type V sums of squares. Dropping an effect from a model is the same as assuming that the effect is unrelated to the outcome (see, e.g., Hocking, 1996). The reasonableness of the assumption does not necessarily insure its validity, so when possible the relationships of dropped effects to the outcome should be inspected. It is also important to note that Type V sums of squares are not invariant to the order in which eligibility for dropping effects from the model is evaluated. Different orders of effects could produce different reduced models.

In spite of these limitations, Type V sums of squares for the reduced model have all the same properties of Type III sums of squares for ANOVA designs with no missing cells. Even in designs with many missing cells (such as fractional factorial designs, in which many high-order interaction effects are assumed to be zero), Type V sums of squares provide tests of meaningful hypotheses, and sometimes hypotheses that cannot be tested using any other method.

Type VI (effective hypothesis) sums of squares. Type I through Type V sums of squares can all be viewed as providing tests of hypotheses that subsets of partial regression coefficients (controlling for or orthogonal to appropriate additional effects) are zero. Effective hypothesis tests (developed by Hocking, 1996) are based on the philosophy that the only unambiguous estimate of an effect is the proportion of variability on the outcome that is uniquely attributable to the effect. The overparameterized coding of effects for categorical predictor variables generally cannot be used to provide such unique estimates for lower-order effects. Effective hypothesis tests, which we propose to call Type VI sums of squares, use the sigma-restricted coding of effects for categorical predictor variables to provide unique effect estimates even for lower-order effects.

The method for computing Type VI sums of squares is straightforward. The sigma-restricted coding of effects is used, and for each effect, its Type VI sums of squares is the difference of the model sums of squares for all other effects from the whole model sums of squares. As such, the Type VI sums of squares provide an unambiguous estimate of the variability of predicted values for the outcome uniquely attributable to each effect.

In ANOVA designs with missing cells, Type VI sums of squares for effects can have fewer degrees of freedom than they would have if there were no missing cells, and for some missing cell designs, can even have zero degrees of freedom. The philosophy of Type VI

sums of squares is to test as much as possible of the original hypothesis given the observed cells. If the pattern of missing cells is such that no part of the original hypothesis can be tested, so be it. The inability to test hypotheses is simply the price you pay for having no observations at some combinations of the levels of the categorical predictor variables. The philosophy is that it is better to admit that a hypothesis cannot be tested than it is to test a distorted hypothesis that may not meaningfully reflect the original hypothesis.

Type VI sums of squares cannot generally be used to test hypotheses for nested ANOVA designs, separate slope designs, or mixed-model designs, because the sigma-restricted coding of effects for categorical predictor variables is overly restrictive in such designs. This limitation, however, does not diminish the fact that Type VI sums of squares can be computed for any other design that can be analyzed using the general linear model.

Error Terms for Tests

Lack-of-fit tests using pure error. Whole model tests and tests based on the 6 types of sums of squares use the mean square residual as the error term for tests of significance. For certain types of designs, however, the residual sum of squares can be further partitioned into meaningful parts that are relevant for testing hypotheses. One such type of design is a simple regression design in which there are subsets of cases all having the same values on the predictor variable. For example, performance on a task could be measured for subjects who work on the task under several different room temperature conditions. The test of significance for the *Temperature* effect in the linear regression of *Performance* on *Temperature* would not necessarily provide complete information on

how *Temperature* relates to *Performance*; the regression coefficient for *Temperature* only reflects its linear effect on the outcome.

One way to glean additional information from this type of design is to partition the residual sums of squares into lack-of-fit and pure error components. In the example just described, this would involve determining the difference between the sum of squares that cannot be predicted by *Temperature* levels, given the linear effect of *Temperature* (residual sums of squares) and the pure error; this difference would be the sums of squares associated with the lack-of-fit (in this example, of the linear model). The test of lack-of-fit, using the mean square pure error as the error term, would indicate whether nonlinear effects of *Temperature* are needed to adequately model *Temperature's* influence on the outcome. Further, the linear effect could be tested using the pure error term, thus providing a more sensitive test of the linear effect independent of any possible nonlinear effect.

Designs with zero degrees of freedom for error. When the model degrees of freedom equal the number of cases or subjects, the residual sums of squares will have zero degrees of freedom and preclude the use of standard hypothesis tests. This sometimes occurs for overfitted designs (designs with many predictors, or designs with categorical predictors having many levels). However, in some designed experiments, such as experiments using split-plot designs or highly fractionalized factorial designs as commonly used in industrial experimentation, it is no accident that the residual sum of squares has zero degrees of freedom. In such experiments, mean squares for certain effects are planned to be used as error terms for testing other effects, and the experiment is designed with this in

mind. It is entirely appropriate to use alternatives to the mean square residual as error terms for testing hypotheses in such designs.

Tests in mixed model designs. Designs that contain random effects for one or more categorical predictor variables are called mixed-model designs. These types of designs, and the analysis of those designs, is also described in detail in Chapter 39 – *Variance Components and Mixed Model ANOVA/ANCOVA*. Random effects are classification effects where the levels of the effects are assumed to be randomly selected from an infinite population of possible levels. The solution for the normal equations in mixed-model designs is identical to the solution for fixed-effect designs (i.e., designs which do not contain random effects). Mixed-model designs differ from fixed-effect designs only in the way in which effects are tested for significance. In fixed-effect designs, between effects are always tested using the mean square residual as the error term. In mixed-model designs, between effects are tested using relevant error terms based on the covariation of sources of variation in the design. Also, only the overparameterized model is used to code effects for categorical predictors in mixed-models, because the sigma-restricted model is overly restrictive.

The covariation of sources of variation in the design is estimated by the elements of a matrix called the expected mean squares (**EMS**) matrix. This non-square matrix contains elements for the covariation of each combination of pairs of sources of variation and for each source of variation with *Error*. Specifically, each element is the mean square for one effect (indicated by the column) that is expected to be accounted by another effect (indicated by the row), given the observed covariation in their levels. Note that expected mean squares can be computing using any type of sums of squares from Type I through

Type V. Once the **EMS** matrix is computed, it is used to the solve for the linear combinations of sources of random variation that are appropriate to use as error terms for testing the significance of the respective effects. This is done using Satterthwaite's method of denominator synthesis (Satterthwaite, 1946). Detailed discussions of methods for testing effects in mixed-models, and related methods for estimating variance components for random effects, can be found in Chapter 39 – *Variance Components and Mixed Model ANOVA/ANCOVA.*

Testing Specific Hypotheses

Whole model tests and tests based on sums of squares attributable to specific effects illustrate two general types of hypotheses that can be tested using the general linear model. Still, there may be other types of hypotheses the researcher wants to test that do not fall into either of these categories. For example, hypotheses about subsets of effects may be of interest, or hypotheses involving comparisons of specific levels of categorical predictor variables may be of interest.

Estimability of hypotheses. Before considering tests of specific hypotheses of this sort, it is important to address the issue of estimability. A test of a specific hypothesis using the general linear model must be framed in terms of the regression coefficients for the solution of the normal equations. If the **X'X** matrix is less than full rank, the regression coefficients depend on the particular g2 inverse used for solving the normal equations, and the regression coefficients will not be unique. When the regression coefficients are not unique, linear functions (*f*) of the regression coefficients having the form:

$$f = Lb$$

where **L** is a vector of coefficients, will also in general not be unique. However, *Lb* for an *L* which satisfies:

$$L = L(X'X)^{-}X'X$$

is invariant for all possible g2 inverses, and is therefore called an estimable function.

The theory of estimability of linear functions is an advanced topic in the theory of algebraic invariants (Searle, 1987, provides a comprehensive introduction), but its implications are clear enough. One instance of non-estimability of a hypothesis has been encountered in tests of the effective hypothesis that have zero degrees of freedom. On the other hand, Type III sums of squares for categorical predictor variable effects in ANOVA designs with no missing cells (and the least squares means in such designs) provide an example of estimable functions which do not depend on the model parameterization (i.e., the particular g2 inverse used to solve the normal equations). The general implication of the theory of estimability of linear functions is that hypotheses which cannot be expressed as linear combinations of the rows of *X* (i.e., the combinations of observed levels of the categorical predictor variables) are not estimable, and therefore cannot be tested. Stated another way, we simply cannot test specific hypotheses that are not represented in the data. The notion of estimability is valuable because the test for estimability makes explicit which specific hypotheses can be tested and which cannot.

Linear combinations of effects. In multiple regression designs, it is common for hypotheses of interest to involve subsets of effects. In mixture designs, for example, you might be interested in simultaneously testing whether the main effect and any of the two-way interactions involving a particular predictor variable are non-zero. It is also common in multiple regression designs for hypotheses of interest to involves comparison of slopes. For example, you might be interested in whether the regression coefficients for two predictor variables differ. In both factorial regression and factorial ANOVA designs with many factors, it is often of interest whether sets of effects, say, all three-way and higher-order interactions, are nonzero.

Tests of these types of specific hypotheses involve 1) constructing one or more Ls reflecting the hypothesis, 2) testing the estimability of the hypothesis by determining whether:

$$L = L(X'X)^{-}X'X$$

and if so, using 3):

$$(Lb)' < L(X'X)^{-}L')^{-1}(Lb)$$

to estimate the sums of squares accounted for by the hypothesis. Finally, (4) the hypothesis is tested for significance using the usual mean square residual as the error term. To illustrate this 4-step procedure, suppose that a test of the difference in the regression slopes is desired for the (intercept plus) 2 predictor variables in a first-order multiple regression design. The coefficients for L would be:

$$L = [0\ 1\ -1]$$

(note that the first coefficient 0 excludes the intercept from the comparison) for which Lb is estimable if the 2 predictor variables are not redundant with each other. The hypothesis sums of squares reflect the difference in the partial regression coefficients for the 2 predictor variables, which is tested for significance using the mean square residual as the error term.

Planned comparisons of least square means. Usually, experimental hypotheses are stated in terms that are more specific than simply main effects or interactions. We may have the specific hypothesis that a particular textbook will improve math skills in males, but not in females, while another book would be about equally effective for both genders, but less effective overall for males. Now generally, we are predicting an interaction here: the effectiveness of the book is modified (qualified) by the student's gender. However, we have a particular prediction concerning the nature of the interaction: we expect a significant difference between genders for one book, but not the other. This type of specific prediction is usually tested by testing planned comparisons of least squares means (estimates of the population marginal means), or as it is sometimes called, contrast analysis.

Briefly, with contrast analysis you can test the statistical significance of predicted specific differences in particular parts of our complex design. The 4-step procedure for testing specific hypotheses is used to specify and test specific predictions. Contrast analysis is a major and indispensable component of the analysis of many complex experimental designs.

To learn more about the logic and interpretation of contrast analysis refer to Chapter 3 – *ANOVA/MANOVA*.

Post-hoc comparisons. Sometimes we find effects in an experiment that were not expected. Even though in most cases a creative experimenter will be able to explain almost any pattern of means, it would not be appropriate to analyze and evaluate that pattern as if you had predicted it all along. The problem here is one of capitalizing on chance when performing multiple tests post-hoc, that is, without *a priori* hypotheses. To illustrate this point, let's consider the following experiment. Imagine we were to write down a number between 1 and 10 on 100 pieces of paper. We then put all of those pieces into a hat and draw 20 samples (of pieces of paper) of 5 observations each, and compute the means (from the numbers written on the pieces of paper) for each group. How likely do you think it is that we will find two sample means that are significantly different from each other? It is very likely! Selecting the extreme means obtained from 20 samples is very different from taking only 2 samples from the hat in the first place, which is what the test via the contrast analysis implies. Without going into further detail, there are several so-called post-hoc tests that are explicitly based on the first scenario (taking the extremes from 20 samples), that is, they are based on the assumption that we have chosen for our comparison the most extreme (different) means out of k total means in the design. Those tests apply corrections that are designed to offset the advantage of post-hoc selection of the most extreme comparisons. Whenever you find unexpected results in an experiment, you should use those post-hoc procedures to test their statistical significance.

Testing Hypotheses for Repeated Measures and Dependent Variables

In the discussion of different hypotheses that can be tested using the general linear model, the tests have been described as tests for the dependent variable or the outcome. This has been done solely to simplify the discussion. When there are multiple dependent variables reflecting the levels of repeated measure factors, the general linear model performs tests using orthonormalized M-transformations of the dependent variables. When there are multiple dependent variables but no repeated measure factors, the general linear model performs tests using the hypothesis sums of squares and cross-products for the multiple dependent variables, which are tested against the residual sums of squares and cross-products for the multiple dependent variables. Thus, the same hypothesis testing procedures that apply to univariate designs with a single dependent variable also apply to repeated measure and multivariate designs.

GENERAL REGRESSION MODELS (GRM)

This chapter describes the use of the general linear model for finding the best linear model from a number of possible models. If you are unfamiliar with the basic methods of ANOVA and regression in linear models, it may be useful to first review the information on these topics in Chapter 1 – *Elementary Concepts in Statistics*. A detailed discussion of univariate and multivariate ANOVA techniques can be found in Chapter 3 – *ANOVA/MANOVA*; see also, Chapter 26 – *Multiple Linear Regression*. Discussion of the ways in which the linear regression model is extended by the general linear model can be found in Chapter 18 – *General Linear Models*.

Overview: The Need for Simple Models

A good theory is the end result of a winnowing process. We start with a comprehensive model that includes all conceivable, testable influences on the phenomena under investigation. Then we test the components of the initial comprehensive model, to identify the less comprehensive submodels that adequately account for the phenomena under investigation. Finally, from these candidate submodels, we single out the simplest submodel, which by the principle of parsimony we take to be the best explanation for the phenomena under investigation.

We prefer simple models not just for philosophical but also for practical reasons. Simple models are easier to put to test again in replication and cross-validation studies. Simple models are less costly to put into practice in predicting and controlling the outcome in the future. The philosophical reasons for preferring simple models should not be downplayed, however. Simpler models are easier to understand and appreciate, and therefore have a "beauty" that their more complicated counterparts often lack.

The entire winnowing process described above is encapsulated in the model-building techniques of stepwise and best-subset regression. The use of these model-building techniques begins with the specification of the design for a comprehensive whole model. Less comprehensive submodels are then tested to determine if they adequately account for the outcome under investigation. Finally, the simplest of the adequate is adopted as the best.

Model Building in General Stepwise Regression

Unlike the multiple regression model, which is used to analyze designs with continuous predictor variables, the general linear model can be used to analyze any ANOVA design with categorical predictor variables, any ANCOVA design with both categorical and continuous predictor variables, as well as any regression design with continuous predictor variables. Effects for categorical predictor variables can be coded in the design matrix $\mathbf{X}$ using either the overparameterized model or the *sigma*-restricted model.

Only the sigma-restricted parameterization can be used for model-building. True to its description as general, the general linear model can be used to analyze designs with effects for categorical predictor variables that are coded using either parameterization method. In many uses of the general linear model, it is arbitrary whether categorical predictors are coded using the *sigma*-restricted or the overparameterized coding. When you desire to build models, however, the use of the overparameterized model is unsatisfactory; lower-order effects for categorical predictor variables are redundant with higher-order containing interactions, and therefore cannot be fairly evaluated for inclusion in the model when higher-order containing interactions are already in the model.

This problem does not occur when categorical predictors are coded using the *sigma*-restricted parameterization, so only the *sigma*-restricted parameterization is necessary in general stepwise regression.

Designs that cannot be represented using the sigma-restricted parameterization. The *sigma*-restricted parameterization can be used to represent most, but not all types of designs.

Specifically, the designs that cannot be represented using the *sigma*-restricted parameterization are designs with nested effects, such as nested ANOVA and separate slope, and random effects. Any other type of ANOVA, ANCOVA, or regression design can be represented using the *sigma*-restricted parameterization, and can therefore be analyzed with general stepwise regression.

Model building for designs with multiple dependent variables.

Stepwise and best-subset model-building techniques are well developed for regression designs with a single dependent variable (e.g., see Cooley and Lohnes, 1971; Darlington, 1990; Hocking Lindeman, Merenda, and Gold, 1980; Morrison, 1967; Neter, Wasserman, and Kutner, 1985; Pedhazur, 1973; Stevens, 1986; Younger, 1985). Using the *sigma*-restricted parameterization and general linear model methods, these model-building techniques can be readily applied to any ANOVA design with categorical predictor variables, any ANCOVA design with both categorical and continuous predictor variables, as well as any regression design with continuous predictor variables. Building models for designs with multiple dependent variables, however, involves considerations that are not typically addressed by the general linear model. Model-building techniques for designs with multiple dependent variables are available with structural equation modeling.

Types of Analyses

A wide variety of types of designs can be represented using the *sigma*-restricted coding of the design matrix **X**, and any such design can be analyzed using the general linear model. The following topics describe these different types of designs and how they differ. Some general ways in which designs might differ can be suggested, but keep in mind that any particular design can be a "hybrid" in the sense that it could have combinations of features of a number of different types of designs.

Between-Subject Designs

Overview. The levels or values of the predictor variables in an analysis describe the differences between the *n* subjects or the *n* valid cases that are analyzed. Thus, when we speak of the between subject design (or simply the between design) for an analysis, we are referring to the nature, number, and arrangement of the predictor variables.

Concerning the nature or type of predictor variables, between designs that contain only categorical predictor variables can be called ANOVA (analysis of variance) designs, between designs that contain only continuous predictor variables can be called regression designs, and between designs that contain both categorical and continuous predictor variables can be called ANCOVA (analysis of covariance) designs.

Between designs may involve only a single predictor variable and therefore be described as simple (e.g., simple regression) or may employ numerous predictor variables (e.g., multiple regression).

Concerning the arrangement of predictor variables, some between designs employ only main effect or first-order terms for predictors, that is, the values for different predictor variables are independent and raised only to the first power. Other between designs may employ higher-order terms for predictors by raising the values for the original predictor variables to a power greater than 1 (e.g., in polynomial regression designs), or by forming products of different predictor variables (i.e., interaction terms). A common arrangement for ANOVA designs is the full-factorial design, in which every combination of levels for each of the categorical predictor variables is represented in the design. Designs with some but not all combinations of levels for each of the categorical predictor variables are aptly called fractional factorial designs.

These basic distinctions about the nature, number, and arrangement of predictor variables can be used in describing a variety of different types of between designs. Some of the more common between designs can now be described.

Simple regression. Simple regression designs involve a single continuous predictor variable. If there are 3 cases with values on a predictor variable P of, say, 7, 4, and 9, and the design is for the first-order effect of P, the **X** matrix would be:

$$\mathbf{X} = \begin{matrix} X_0 & X_1 \\ \begin{bmatrix} 1 & 7 \\ 1 & 4 \\ 1 & 9 \end{bmatrix} \end{matrix}$$

and using P for X_1 the regression equation would be:

$$Y = b_0 + b_1 P$$

If the simple regression design is for a higher-order effect of P, say the quadratic effect, the values in the X_1 column of the design matrix would be raised to the 2nd power, that is, squared:

$$\mathbf{X} = \begin{matrix} X_0 & X_1 \\ \begin{bmatrix} 1 & 49 \\ 1 & 16 \\ 1 & 81 \end{bmatrix} \end{matrix}$$

and using P^2 for X_1 the regression equation is:

$$Y = b_0 + b_1 P^2$$

In regression designs, values on the continuous predictor variables are raised to the desired power and used as the values for the X variables. No recoding is performed. It is therefore sufficient, in describing regression designs, to simply describe the regression equation without explicitly describing the design matrix **X**.

Multiple regression. Multiple regression designs are to continuous predictor variables as main effect ANOVA designs are to categorical predictor variables, that is, multiple regression designs contain the separate simple regression designs for 2 or more continuous predictor variables. The regression equation for a multiple regression design for the first-order effects of 3 continuous predictor variables P, Q, and R would be:

$$Y = b_0 + b_1 P + b_2 Q + b_3 R$$

A discussion of multiple regression methods is also provided in Chapter 26 – *Multiple Linear Regression*.

Factorial regression. Factorial regression designs are similar to factorial ANOVA designs, in which combinations of the levels of the factors are represented in the design. In factorial regression designs, however, there may be many more such possible combinations of distinct levels for the continuous predictor variables than there are cases in the data set. To simplify matters, full-factorial regression designs are defined as designs in which all possible products of the continuous predictor variables are represented in the design. For example, the full-factorial regression design for two continuous predictor variables P and Q would include the main effects (i.e., the first-order effects) of P and Q and their 2-way P by Q interaction effect, which is represented by the product of P and Q

scores for each case. The regression equation would be:

$$Y = b_0 + b_1P + b_2Q + b_3P*Q$$

Factorial regression designs can also be fractional, that is, higher-order effects can be omitted from the design. A fractional factorial design to degree 2 for 3 continuous predictor variables P, Q, and R would include the main effects and all 2-way interactions between the predictor variables:

$$Y = b_0 + b_1P + b_2Q + b_3R + b_4P*Q + b_5P*R + b_6Q*R$$

Polynomial regression. Polynomial regression designs are designs that contain main effects and higher-order effects for the continuous predictor variables but do not include interaction effects between predictor variables. For example, the polynomial regression design to degree 2 for three continuous predictor variables P, Q, and R would include the main effects (i.e., the first-order effects) of P, Q, and R and their quadratic (i.e., second-order) effects, but not the 2-way interaction effects or the P by Q by R 3-way interaction effect.

$$Y = b_0 + b_1P + b_2P^2 + b_3Q + b_4Q^2 + b_5R + b_6R^2$$

Polynomial regression designs do not have to contain all effects up to the same degree for every predictor variable. For example, main, quadratic, and cubic effects could be included in the design for some predictor variables, and effects up the fourth degree could be included in the design for other predictor variables.

Response surface regression. Quadratic response surface regression designs are a hybrid type of design with characteristics of both polynomial regression designs and fractional factorial regression designs. Quadratic response surface regression designs contain all the same effects of polynomial regression designs to degree 2 and additionally the 2-way interaction effects of the predictor variables. The regression equation for a quadratic response surface regression design for 3 continuous predictor variables P, Q, and R is:

$$Y = b_0 + b_1P + b_2P^2 + b_3Q + b_4Q^2 + b_5R + b_6R^2 + b_7P*Q + b_8P*R + b_9Q*R$$

These types of designs are commonly employed in applied research (e.g., in industrial experimentation), and a detailed discussion of these types of designs is also presented in Chapter 15 – *Experimental Design* (see *Central Composite Designs*).

Mixture surface regression. Mixture surface regression designs are identical to factorial regression designs to degree 2 except for the omission of the intercept. Mixtures, as the name implies, add up to a constant value; the sum of the proportions of ingredients in different recipes for some material all must add up 100%. Thus, the proportion of one ingredient in a material is redundant with the remaining ingredients. Mixture surface regression designs deal with this redundancy by omitting the intercept from the design. The design matrix for a mixture surface regression design for 3 continuous predictor variables P, Q, and R would be:

$$Y = b_1P + b_2P^2 + b_3Q + b_4P*Q + b_5P*R + b_6Q*R$$

These types of designs are commonly employed in applied research (e.g., in industrial experimentation), and a detailed discussion of these types of designs is also presented in Chapter 15 – *Experimental Design* (see *Mixture Designs and Triangular Surfaces*).

One-way ANOVA. A design with a single categorical predictor variable is called a one-way ANOVA design. For example, a study of 4 different fertilizers used on different individual plants could be analyzed via one-way ANOVA, with four levels for the factor *Fertilizer*.

Consider a single categorical predictor variable *A* with 1 case in each of its 3 categories. Using the *sigma*-restricted coding of *A* into 2 quantitative contrast variables, the matrix **X** defining the between design is:

$$\mathbf{X} = \begin{array}{c} \\ A_1 \\ A_2 \\ A_3 \end{array} \begin{array}{ccc} X_0 & X_1 & X_2 \\ \begin{bmatrix} 1 & 1 & 0 \\ 1 & 0 & 1 \\ 1 & -1 & -1 \end{bmatrix} \end{array}$$

That is, cases in groups A_1, A_2, and A_3 are all assigned values of 1 on X_0 (the intercept), the case in group A_1 is assigned a value of 1 on X_1 and a value 0 on X_2, the case in group A_2 is assigned a value of 0 on X_1 and a value 1 on X_2, and the case in group A_3 is assigned a value of -1 on X_1 and a value -1 on X_2. Of course, any additional cases in any of the 3 groups would be coded similarly. If there were 1 case in group A_1, 2 cases in group A_2, and 1 case in group A_3, the **X** matrix would be:

$$\mathbf{X} = \begin{array}{c} \\ A_{11} \\ A_{12} \\ A_{22} \\ A_{13} \end{array} \begin{array}{ccc} X_0 & X_1 & X_2 \\ \begin{bmatrix} 1 & 1 & 0 \\ 1 & 0 & 1 \\ 1 & 0 & 1 \\ 1 & -1 & -1 \end{bmatrix} \end{array}$$

where the first subscript for *A* gives the replicate number for the cases in each group. For brevity, replicates usually are not shown when describing ANOVA design matrices.

Note that in one-way designs with an equal number of cases in each group, *sigma*-restricted coding yields $X_1 \ldots X_k$ variables all of which have means of 0.

These simple examples show that the **X** matrix actually serves two purposes. It specifies (1) the coding for the levels of the original predictor variables on the *X* variables used in the analysis as well as (2) the nature, number, and arrangement of the X variables, that is, the between design.

Main effect ANOVA. Main effect ANOVA designs contain separate one-way ANOVA designs for 2 or more categorical predictors. A good example of main effect ANOVA would be the typical analysis performed on screening designs as described in Chapter 15 – *Experimental Design (DOE)*.

Consider 2 categorical predictor variables *A* and *B* each with 2 categories. Using the *sigma*-restricted coding, the **X** matrix defining the between design is:

$$\mathbf{X} = \begin{array}{c} \\ A_1B_1 \\ A_1B_2 \\ A_2B_1 \\ A_2B_2 \end{array} \begin{array}{ccc} X_0 & X_1 & X_2 \\ \begin{bmatrix} 1 & 1 & 1 \\ 1 & 1 & -1 \\ 1 & -1 & 1 \\ 1 & -1 & -1 \end{bmatrix} \end{array}$$

Note that if there are equal numbers of cases in each group, the sum of the cross-products of values for the X_1 and X_2 columns is 0, for example, with 1 case in each group $(1*1)+(1*-1)+(-1*1)+(-1*-1)=0$.

Factorial ANOVA. Factorial ANOVA designs contain *X* variables representing combinations of the levels of 2 or more categorical predictors (e.g., a study of boys and girls in four age groups, resulting in a *2 (Gender) x 4 (Age Group)* design). In particular, full-factorial designs represent all possible combinations of the levels of the categorical predictors. A full-factorial design with 2 categorical predictor variables *A* and *B* each with 2 levels would be called a 2x2 full-factorial design. Using the *sigma*-restricted coding, the **X** matrix for this design would be:

$$
\mathbf{X} = \begin{array}{c} \\ A_1B_1 \\ A_1B_2 \\ A_2B_1 \\ A_2B_2 \end{array} \begin{array}{cccc} X_0 & X_1 & X_2 & X_3 \\ \left[\begin{array}{cccc} 1 & 1 & 1 & 1 \\ 1 & 1 & -1 & -1 \\ 1 & -1 & 1 & -1 \\ 1 & -1 & -1 & 1 \end{array}\right] \end{array}
$$

Several features of this **X** matrix deserve comment. Note that the X_1 and X_2 columns represent main effect contrasts for one variable, (i.e., *A* and *B*, respectively) collapsing across the levels of the other variable. The X_3 column instead represents a contrast between different combinations of the levels of *A* and *B*. Note also that the values for X_3 are products of the corresponding values for X_1 and X_2. Product variables such as X_3 represent the multiplicative or interaction effects of their factors, so X_3 would be said to represent the 2-way interaction of *A* and *B*. The relationship of such product variables to the dependent variables indicates the interactive influences of the factors on responses above and beyond their independent (i.e., main effect) influences on responses. Thus, factorial designs provide more information about the relationships between categorical predictor variables and responses on the dependent variables than is provided by corresponding one-way or main effect designs.

When many factors are being investigated, however, full-factorial designs sometimes require more data than reasonably can be collected to represent all possible combinations of levels of the factors, and high-order interactions between many factors can become difficult to interpret. With many factors, a useful alternative to the full-factorial design is the fractional factorial design. As an example, consider a 2x2 x 2 fractional factorial design to degree 2 with 3 categorical predictor variables each with 2 levels. The design would include the main effects for each variable, and all 2-way interactions between the three variables, but

would not include the 3-way interactions between all three variables. These types of designs are discussed in detail in $2^{(k-p)}$ *Fractional Factorial Designs* in Chapter 15 – *Experimental Design (Industrial DOE)*.

Analysis of covariance. In general, between designs that contain both categorical and continuous predictor variables can be called ANCOVA designs. Traditionally, however, ANCOVA designs have referred more specifically to designs in which the first-order effects of one or more continuous predictor variables are taken into account when assessing the effects of one or more categorical predictor variables. A basic introduction to analysis of covariance can also be found in the *Analysis of Covariance (ANCOVA)* topic in Chapter 3 – *ANOVA/MANOVA*.

To illustrate, suppose a researcher wants to assess the influences of a categorical predictor variable *A* with 3 levels on some outcome, and that measurements on a continuous predictor variable *P*, known to covary with the outcome, are available. If the data for the analysis are:

$$
\begin{array}{cc} P & Group \\ \left[\begin{array}{c} 7 \\ 4 \\ 9 \\ 3 \\ 6 \\ 8 \end{array}\right] & \left[\begin{array}{c} A_1 \\ A_1 \\ A_2 \\ A_2 \\ A_3 \\ A_3 \end{array}\right] \end{array}
$$

the *sigma*-restricted **X** matrix for the design that includes the separate first-order effects of *P* and *A* would be:

$$
\mathbf{X} = \begin{array}{cccc} X_0 & X_1 & X_2 & X_3 \\ \left[\begin{array}{cccc} 1 & 7 & 1 & 0 \\ 1 & 4 & 1 & 0 \\ 1 & 9 & 0 & 1 \\ 1 & 3 & 0 & 1 \\ 1 & 6 & -1 & -1 \\ 1 & 8 & -1 & -1 \end{array}\right] \end{array}
$$

The b_2 and b_3 coefficients in the regression equation:

$$Y = b_0 + b_1X_1 + b_2X_2 + b_3X_3$$

represent the influences of group membership on the A categorical predictor variable, controlling for the influence of scores on the P continuous predictor variable. Similarly, the b_1 coefficient represents the influence of scores on P controlling for the influences of group membership on A. This traditional ANCOVA analysis gives a more sensitive test of the influence of A to the extent that P reduces the prediction error, that is, the residuals for the outcome variable.

Homogeneity of slopes. The appropriate design for modeling the influences of continuous and categorical predictor variables depends on whether the continuous and categorical predictors interact in influencing the outcome. The traditional analysis of covariance (ANCOVA) design for continuous and categorical predictor variables is appropriate when the continuous and categorical predictors do not interact in influencing responses on the outcome. The homogeneity of slopes designs can be used to test whether the continuous and categorical predictors interact in influencing responses. For the same example data used to illustrate the traditional ANCOVA design, the *sigma*-restricted $\mathbf{X}$ matrix for the homogeneity of slopes design would be:

$$\mathbf{X} = \begin{bmatrix}
X_0 & X_1 & X_2 & X_3 & X_4 & X_5 \\
1 & 7 & 1 & 0 & 7 & 0 \\
1 & 4 & 1 & 0 & 4 & 0 \\
1 & 9 & 0 & 1 & 0 & 9 \\
1 & 3 & 0 & 1 & 0 & 3 \\
1 & 6 & -1 & -1 & -6 & -6 \\
1 & 8 & -1 & -1 & -8 & -8
\end{bmatrix}$$

Using this design matrix $\mathbf{X}$, if the b_4 and b_5 coefficients in the regression equation:

$$Y = b_0 + b_1X_1 + b_2X_2 + b_3X_3 + b_4X_4 + b_5X_5$$

are zero, the simpler traditional ANCOVA design should be used.

Multivariate Designs Overview

When there are multiple dependent variables in a design, the design is said to be multivariate. Multivariate measures of association are by nature more complex than their univariate counterparts (such as the correlation coefficient, for example). This is because multivariate measures of association must take into account not only the relationships of the predictor variables with responses on the dependent variables, but also the relationships among the multiple dependent variables. By doing so, however, these measures of association provide information about the strength of the relationships between predictor and dependent variables independent of the dependent variables interrelationships. For a basic discussion of multivariate designs, see *Multivariate Designs* in Chapter 3 – *ANOVA/MANOVA*.

The most commonly used multivariate measures of association all can be expressed as functions of the eigenvalues of the product matrix:

$$E^{-1}H$$

where $\mathbf{E}$ is the error SSCP matrix (i.e., the matrix of sums of squares and cross-products for the dependent variables that are not accounted for by the predictors in the between design), and $\mathbf{H}$ is a hypothesis SSCP matrix (i.e., the matrix of sums of squares and cross-products for the dependent variables that are accounted for by all the predictors in the between design, or the sums of squares and cross-products for the dependent variables that are accounted for by a particular effect). If

λ_i = the ordered eigenvalues of $E^{-1}H$, if E^{-1} exists the 4 commonly used multivariate measures of association are:

Wilks' *lambda* = $\Pi[1/(1+\lambda_i)]$

Pillai's trace = $\Sigma\lambda_i/(1+\lambda_i)$

Hotelling-Lawley trace = $\Sigma\lambda_i$

Roy's largest root = λ_1

These 4 measures have different upper and lower bounds, with Wilks' *lambda* perhaps being the most easily interpretable of the four measures. Wilks' *lambda* can range from 0 to 1, with 1 indicating no relationship of predictors to responses and 0 indicating a perfect relationship of predictors to responses. 1 – Wilks' *lambda* can be interpreted as the multivariate counterpart of a univariate R-squared, that is, it indicates the proportion of generalized variance in the dependent variables that is accounted for by the predictors.

The 4 measures of association are also used to construct multivariate tests of significance. These multivariate tests are covered in detail in a number of sources (e.g., Finn, 1974; Tatsuoka, 1971).

Building the Whole Model

The following sections discuss details for building and testing hypotheses about the whole model, for example, how sums of squares are partitioned and how the overall fit for the whole model is tested.

Partitioning Sums of Squares

A fundamental principle of least squares methods is that variation on a dependent variable can be partitioned, or divided into parts, according to the sources of the variation. Suppose that a dependent variable is regressed on one or more predictor variables, and that for convenience the dependent variable is scaled so that its mean is 0. Then a basic least squares identity is that the total sum of squared values on the dependent variable equals the sum of squared predicted values plus the sum of squared residual values. Stated more generally,

$$\Sigma(y - \bar{y})^2 = \Sigma(\hat{y} - \bar{y})^2 + \Sigma(y - \hat{y})^2$$

where the term on the left is the total sum of squared deviations of the observed values on the dependent variable from the dependent variable mean, and the respective terms on the right are 1) the sum of squared deviations of the predicted values for the dependent variable from the dependent variable mean and 2) the sum of the squared deviations of the observed values on the dependent variable from the predicted values, that is, the sum of the squared residuals. Stated yet another way,

Total SS = Model SS + Error SS

Note that the *Total SS* is always the same for any particular data set, but that the *Model SS* and the *Error SS* depend on the regression equation. Assuming again that the dependent variable is scaled so that its mean is 0, the *Model SS* and the *Error SS* can be computed using:

Model SS = b'X'Y

Error SS = Y'Y – b'X'Y

Testing the Whole Model

Given the Model SS and the Error SS, you can perform a test that all the regression coefficients for the X variables (b_1 through b_k, excluding the b_0 coefficient for the intercept) are zero. This test is equivalent to a comparison of the fit of the regression surface defined by the predicted values (computed from the whole model regression equation) to the fit of the regression surface defined solely by the dependent variable mean (computed from the reduced regression equation containing only the intercept). Assuming that $X'X$ is full rank, the whole model hypothesis mean square:

MSH = (Model SS)/k

where k is the number of columns of X (excluding the intercept column), is an estimate of the variance of the predicted values. The error mean square:

s^2 = MSE = (Error SS)/(n-k-1)

where n is the number of observations, is an unbiased estimate of the residual or error variance. The test statistic is:

F = MSH/MSE

where F has $(k, n - k - 1)$ degrees of freedom.

If $X'X$ is not full rank, $r + 1$ is substituted for k, where r is the rank or the number of non-redundant columns of $X'X$.

If the whole model test is not significant the analysis is complete; the whole model is concluded to fit the data no better than the reduced model using the dependent variable mean alone. It is futile to seek a submodel that adequately fits the data when the whole model is inadequate.

Note that in the case of non-intercept models, some multiple regression programs will only compute the full model test based on the proportion of variance around 0 (zero) accounted for by the predictors; for more information (see Kvålseth, 1985; Okunade, Chang, and Evans, 1993). Other programs will actually compute both values (i.e., based on the residual variance around 0, and around the respective dependent variable means.

Limitations of Whole Models

For designs such as one-way ANOVA or simple regression designs, the whole model test by itself may be sufficient for testing general hypotheses about whether or not the single predictor variable is related to the outcome. In complex designs, however, finding a statistically significant test of whole model fit is often just the first step in the analysis; you then seek to identify simpler submodels that fit the data equally well (see *Basic Ideas: The Need for Simple Models*, page 279). It is to this task, the search for submodels that fit the data well, that stepwise and best-subset regression are devoted.

Building Models via Stepwise Regression

Stepwise model-building techniques for regression designs with a single dependent variable are described in numerous sources (e.g., see Darlington, 1990; Hocking, 1966; Lindeman, Merenda, and Gold, 1980; Morrison, 1967; Neter, Wasserman, and Kutner, 1985; Pedhazur, 1973; Stevens, 1986; Younger, 1985). The basic procedures involve 1) identifying an initial model, 2) iteratively stepping, that is, repeatedly altering the model at the previous step by adding or removing a predictor variable in accordance with the stepping criteria, and 3) terminating the search when stepping is no longer possible given the stepping criteria, or when a specified maximum number of steps has been reached. The following topics provide details on the use of stepwise model-building procedures.

The initial model in stepwise regression. The initial model is designated the model at *Step 0*. The initial model always includes the regression intercept (unless a *No intercept* option has been specified). For the backward stepwise and backward removal methods, the initial model also includes all effects specified to be included in the design for the analysis. The initial model for these methods is therefore the whole model.

For the forward stepwise and forward entry methods, the initial model always includes the regression intercept (unless a *No intercept* option has been specified). The initial model can also include 1 or more effects specified to be forced into the model. If j is the number of effects specified to be forced into the model, the first j effects specified to be included in the design are entered into the model at step 0. Any such effects are not eligible to be removed from the model during subsequent steps.

Effects can also be specified to be forced into the model when the backward stepwise and backward removal methods are used. As in the forward stepwise and forward entry methods, any such effects are not eligible to be removed from the model during subsequent steps.

The forward entry method. The forward entry method is a simple model-building procedure. At each step after step 0, the entry statistic is computed for each effect eligible for entry in the model. If no effect has a value on the entry statistic that exceeds the specified critical value for model entry, stepping is terminated; otherwise the effect with the largest value on the entry statistic is entered into the model. Stepping is also terminated if the maximum number of steps is reached.

The backward removal method. The backward removal method is also a simple model-building procedure. At each step after step 0, the removal statistic is computed for each effect eligible to be removed from the model. If no effect has a value on the removal statistic that is less than the critical value for removal from the model, stepping is terminated;

otherwise the effect with the smallest value on the removal statistic is removed from the model. Stepping is also terminated if the maximum number of steps is reached.

The forward stepwise method. The forward stepwise method employs a combination of the procedures used in the forward entry and backward removal methods. At Step 1 the procedures for forward entry are performed. At any subsequent step where 2 or more effects have been selected for entry into the model, forward entry is performed if possible, and backward removal is performed if possible, until neither procedure can be performed and stepping is terminated. Stepping is also terminated if the maximum number of steps is reached.

The backward stepwise method. The backward stepwise method employs a combination of the procedures used in the forward entry and backward removal methods. At step 1 the procedures for backward removal are performed. At any subsequent step where 2 or more effects have been selected for entry into the model, forward entry is performed if possible, and backward removal is performed if possible, until neither procedure can be performed and stepping is terminated. Stepping is also terminated if the maximum number of steps is reached.

Entry and removal criteria. Either critical F values or critical p values can be specified to be used to control entry and removal of effects from the model. If p values are specified, the actual values used to control entry and removal of effects from the model are 1 minus the specified p values. The critical value for model entry must exceed the critical value for removal from the model. A maximum number of steps can also be specified. If not previously terminated, stepping stops when the specified maximum number of steps is reached.

Building Models via Best-Subset Regression

All-possible-subset regression can be used as an alternative to or in conjunction with stepwise methods for finding the best possible submodel.

Neter, Wasserman, and Kutner (1985) discuss the use of all-possible-subset regression in conjunction with stepwise regression:

> "A limitation of the stepwise regression search approach is that it presumes there is a single "best" subset of X variables and seeks to identify it. As noted earlier, there is often no unique best subset. Hence, some statisticians suggest that all possible regression models with a similar number of X variables as in the stepwise regression solution be fitted subsequently to study whether some other subsets of X variables might be better." (p. 435).

This reasoning suggests that after finding a stepwise solution, the best of all the possible subsets of the same number of effects should be examined to determine if the stepwise solution is among the best. If not, the stepwise solution is suspect.

All-possible-subset regression can also be used as an alternative to stepwise regression. Using this approach, you first decide on the range of subset sizes that could be considered to be useful. For example, you might expect that inclusion of at least 3 effects in the model is necessary to adequately account for responses, and also might expect there is no advantage to considering models with more than 6 effects. Only the "best" of all possible subsets of 3, 4, 5, and 6 effects are then considered.

Note that several different criteria can be used for ordering subsets in terms of goodness. The most often used criteria are the subset multiple R-square, adjusted R-square, and Mallow's Cp statistics. When all-possible-subset regression is used in conjunction with stepwise methods, the subset multiple R-square statistic allows direct comparisons of the best subsets identified using each approach.

The number of possible submodels increases very rapidly as the number of effects in the whole model increases, and as subset size approaches half of the number of effects in the whole model. The amount of computation required to perform all-possible-subset regression increases as the number of possible submodels increases, and holding all else constant, also increases very rapidly as the number of levels for effects involving categorical predictors increases, thus resulting in more columns in the design matrix $\mathbf{X}$. For example, all possible subsets of up to a dozen or so effects could certainly theoretically be computed for a design that includes two dozen or so effects all of which have many levels, but the computation would be very time consuming (e.g., there are about 2.7 million different ways to select 12 predictors from 24 predictors, i.e., 2.7 million models to evaluate just for subset size 12). Simpler is generally better when using all-possible-subset regression.

GENERALIZED ADDITIVE MODELS (GAM)

Generalized additive models techniques, developed and popularized by Hastie and Tibshirani (1990), are a generalization of multiple regression (see Chapters 19 and 26), where the additive nature of the model is maintained but the simple lines of the linear regression are replaced by nonparametric functions with multiple parameters but can model highly nonlinear, multivariate relationships between variables. A detailed description of these and related techniques, the algorithms used to fit these models, and discussions of recent research in this area of statistical modeling can also be found in Schimek (2000).

Overview

The methods described in this chapter represent a generalization of multiple regression (which is a special case of general linear models). Specifically, in linear regression, a linear least-squares fit is computed for a set of predictor or X variables, to predict a dependent Y variable. The well known linear regression equation with m predictors, to predict a dependent variable Y, can be stated as:

$$Y = b_0 + b_1 {}^* X_1 + ... + b_m {}^* X_m$$

Where Y stands for the (predicted values of the) dependent variable, X_1 through X_m represent the m values for the predictor variables, and b_0, and b_1 through b_m are the regression coefficients estimated by multiple regression. A generalization of the multiple regression model would be to maintain the additive nature of the model, but to replace the simple terms of the linear equation $b_i {}^* X_i$ with $f_i(X_i)$ where f_i is a non-parametric function of the predictor X_i. In other words, instead of a single coefficient for each variable (additive term) in the model, in additive models an unspecified (non-parametric) function is estimated for each predictor, to achieve the best prediction of the dependent variable values.

Generalized linear/nonlinear models. To summarize the basic idea, the generalized linear model differs from the general linear model (of which multiple regression is a special case) in two major respects: First, the distribution of the dependent or response variable can be (explicitly) non-normal, and does not have to be continuous, e.g., it can be binomial; second, the dependent variable values are predicted from a linear combination of predictor variables, which are connected to the dependent variable via a link function. The general linear model for a single dependent variable can be considered a special case of the generalized linear model: In the general linear model the dependent variable values are expected to follow the normal distribution, and the link function is a simple identity function (i.e., the linear combination of values for the predictor variables is not transformed).

To illustrate, in the general linear model a response variable Y is linearly associated with values on the X variables while the relationship in the generalized linear model is assumed to be:

$$Y = g(b_0 + b_1 {}^* X_1 + ... + b_m {}^* X_m)$$

where $g(...)$ is a function. Formally, the inverse function of $g(...)$, say gi$(...)$, is called the link function; so that:

$$gi(muY) = b_0 + b_1 {}^* X_1 + ... + b_m {}^* X_m$$

where $mu\text{-}Y$ stands for the expected value of Y.

Distributions and link functions. With generalized additive models, you can choose from a wide variety of distributions for the dependent variable, and link functions for the effects of the predictor variables on the dependent variable (see McCullagh and Nelder, 1989; Hastie and Tibshirani, 1990; see also, *Computational Approach*, page 300, in Chapter 21 – *Generalized Linear/Nonlinear Models (GLZ)* for a discussion of link functions and distributions):

Normal, gamma, and Poisson distributions:

Log link: $f(z) = \log(z)$

Inverse link: $f(z) = 1/z$

Identity link: $f(z) = z$

Binomial distributions:

Logit link: $f(z) = \log(z/(1-z))$

Generalized Additive Models

We can combine the notion of additive models with generalized linear models, to derive the notion of generalized additive models, as:

$$gi(muY) = \Sigma_i(f_i(X_i))$$

In other words, the purpose of generalized additive models is to maximize the quality of prediction of a dependent variable Y from various distributions, by estimating unspecific (non-parametric) functions of the predictor variables that are connected to the dependent variable via a link function.

Estimating the non-parametric function of predictors via scatterplot smoothers. A unique aspect of generalized additive models is the non-parametric functions f_i of the predictor variables X_i. Specifically, instead of some kind of simple or complex parametric functions, Hastie and Tibshirani (1990) discuss various general scatterplot smoothers that can be applied to the X variable values, with the target criterion to maximize the quality of prediction of the (transformed) Y variable values. One such scatterplot smoother is the cubic smoothing splines smoother, which generally produces a smooth generalization of the relationship between the two variables in the scatterplot. Computational details regarding this smoother can be found in Hastie and Tibshirani (1990; see also, Schimek, 2000).

To summarize, instead of estimating single parameters (such as the regression weights in multiple regression), in generalized additive models, we find a general unspecific (non-parametric) function that relates the predicted (transformed) Y values to the predictor values.

A specific example: The generalized additive logistic model. Let's consider a specific example of the generalized additive models – a

generalization of the logistic (logit) model for binary dependent variable values. As also described in detail in Chapter 28 – *Nonlinear Estimation* and Chapter 21 – *Generalized Linear/Nonlinear Models*, the logistic regression model for binary responses can be written as follows:

$$y = e^{(b_0 + b_1 x_1 \ldots + b_m x_m)} / \{1 + e^{(b_0 + b_1 x_1 \ldots + b_m x_m)}\}$$

Note that the distribution of the dependent variable is assumed to be binomial, i.e., the response variable can only assume the values 0 or 1 (e.g., in a market research study, the purchasing decision would be binomial: The customer either did or did not make a particular purchase). We can apply the logistic link function to the probability p (ranging between 0 and 1) so that:

$$p' = \log \{p/(1-p)\}$$

By applying the logistic link function, we can now rewrite the model as:

$$p' = b_0 + b_1 {}^* X_1 + \ldots + b_m {}^* X_m$$

Finally, we substitute the simple single-parameter additive terms to derive the generalized additive logistic model:

$$p' = b_0 + f_1(X_1) + \ldots + f_m(X_m)$$

An example application of the this model can be found in Hastie and Tibshirani (1990).

Fitting generalized additive models. Detailed descriptions of how generalized additive models are fit to data can be found in Hastie and Tibshirani (1990), as well as Schimek (2000, p. 300). In general there are two separate iterative operations involved in the algorithm, which are usually labeled the outer and inner loop. The purpose of the outer loop is to maximize the overall fit of the model, by minimizing the overall likelihood of the data given the model (similar to the maximum likelihood estimation procedures as described in, for example, the context of nonlinear estimation). The purpose of the inner loop is to refine the scatterplot smoother, which is the cubic splines smoother. The smoothing is performed with respect to the partial residuals; i.e., for every predictor k, the weighted cubic spline fit is found that best represents the relationship between variable k and the (partial) residuals computed by removing the effect of all other j predictors (j ≠ k). The iterative estimation procedure will terminate, when the likelihood of the data given the model cannot be improved.

Interpreting the results. Many of the standard results statistics computed by generalized additive models are similar to those customarily reported by linear or nonlinear model fitting procedures. For example, predicted and residual values for the final model can be computed, and various graphs of the residuals can be displayed to help you identify possible outliers, etc. Refer also to the description of the residual statistics computed by generalized linear/nonlinear models for details.

The main result of interest, of course, is how the predictors are related to the dependent variable. Scatterplots can be computed showing the smoothed predictor variable values plotted against the partial residuals, i.e., the residuals after removing the effect of all other predictor variables.

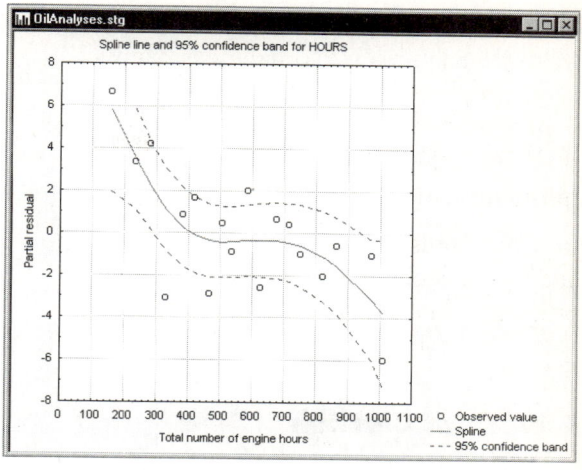

With this plot, you can evaluate the nature of the relationship between the predictor with the residualized (adjusted) dependent variable values (see Hastie & Tibshirani, 1990; in particular formula 6.3), and hence the nature of the influence of the respective predictor in the overall model.

Degrees of freedom. To reiterate, the generalized additive models approach replaces the simple products of (estimated) parameter values times the predictor values with a cubic spline smoother for each predictor. When estimating a single parameter value, we lose one degree of freedom, i.e., we add one degree of freedom to the overall model. It is not clear how many degrees of freedom are lost due to estimating the cubic spline smoother for each variable. Intuitively, a smoother can either be very smooth, not following the pattern of data in the scatterplot very closely, or it can be less smooth, following the pattern of the data more closely. In the most extreme case, a simple line would be very smooth, and require us to estimate a single slope parameter, i.e., we would use one degree of freedom to fit the smoother (simple straight line); on the other hand, we could force a very "non-smooth" line to connect each actual data point, in which case we could use-up approximately as many degrees of freedom as there are points in the plot. With generalized additive models, you can specify the degrees of freedom for the cubic spline smoother; the fewer degrees of freedom you specify, the smoother is the cubic spline fit to the partial residuals, and typically, the worse is the overall fit of the model. The issue of degrees of freedom for smoothers is discussed in detail in Hastie and Tibshirani (1990).

A Word of Caution

Generalized additive models are very flexible, and can provide an excellent fit in the presence of nonlinear relationships and significant noise in the predictor variables. However, note that because of this flexibility, you must be extra cautious not to over-fit the data, i.e., apply an overly complex model (with many degrees of freedom) to data so as to produce a good fit that likely will not replicate in subsequent validation studies. Also, compare the quality of the fit obtained from generalized additive models to the fit obtained via generalized linear/nonlinear models. In other words, evaluate whether the added complexity (generality) of generalized additive models (regression smoothers) is necessary in order to obtain a satisfactory fit to the data. Often, this is not the case, and given a comparable fit of the models, the simpler generalized linear model is preferable to the more complex generalized additive model. These issues are discussed in greater detail in Hastie and Tibshirani (1990).

Another issue to keep in mind pertains to the interpretability of results obtained from (generalized) linear models vs. generalized additive models. Linear models are easily understood, summarized, and communicated to others (e.g., in technical reports). Moreover, parameter estimates can be used to predict or classify new cases in a simple and straightforward manner. Generalized additive models are not easily interpreted, in particular when they involve complex nonlinear effects of some or all of the predictor variables (and, of course, it is in those instances where generalized additive models may yield a better fit than generalized linear models). To reiterate, it is usually preferable to rely on a simple, well understood model for predicting future cases, than on a complex model that is difficult to interpret and summarize.

GENERALIZED LINEAR/NONLINEAR MODELS (GLZ)

This chapter describes the use of the generalized linear/nonlinear model for analyzing linear and nonlinear effects of continuous and categorical predictor variables on a discrete or continuous dependent variable. If you are unfamiliar with the basic methods of regression in linear models, it may be useful to first review the information on these topics in Chapter 1 – *Elementary Concepts in Statistics*. Discussion of the ways in which the linear regression model is extended by the general linear model can be found in Chapter 18 – *General Linear Models*.

For additional information about generalized linear models, see also, Dobson (1990), Green and Silverman (1994), or McCullagh and Nelder (1989).

Overview

The generalized linear/nonlinear model (GLZ) is a generalization of the general linear model (see, e.g., Chapter 3 – *ANOVA/MANOVA*, Chapter 18 – *General Linear Models*, and Chapter 26 – *Multiple Linear Regression*). In its simplest form, a linear model specifies the (linear) relationship between a dependent (or response) variable Y, and a set of predictor variables, the X's, so that:

$$Y = b_0 + b_1X_1 + b_2X_2 + ... + b_kX_k$$

In this equation b_0 is the regression coefficient for the intercept and the b_i values are the regression coefficients (for variables 1 through k) computed from the data.

So for example, you could estimate (i.e., predict) a person's weight as a function of the person's height and gender. You could use linear regression to estimate the respective regression coefficients from a sample of data, measuring height, weight, and observing the subjects' gender. For many data analysis problems, estimates of the linear relationships between variables are adequate to describe the observed data, and to make reasonable predictions for new observations (see Chapter 26 – *Multiple Linear Regression* for additional details).

However, there are many relationships that cannot adequately be summarized by a simple linear equation, for two major reasons:

Distribution of dependent variable. First, the dependent variable of interest may have a non-continuous distribution, and thus, the predicted values should also follow the respective distribution; any other predicted values are not logically possible. For example, a researcher may be interested in predicting one of three possible discrete outcomes (e.g., a consumer's choice of one of three alternative products). In that case, the dependent variable can only take on 3 distinct values, and the distribution of the dependent variable is said to be multinomial. Or suppose you are trying to predict people's family planning choices, specifically, how many children families will have, as a function of income and various other socioeconomic indicators. The dependent variable – number of children – is discrete (i.e., a family may have 1, 2, or 3 children and so on, but cannot have 2.4 children), and most likely the distribution of that variable is highly skewed (i.e., most families have 1, 2, or 3 children, fewer will have 4 or 5, very few will have 6 or 7, and so on). In this case it would be reasonable to assume that the dependent variable follows a Poisson distribution.

Link function. A second reason why the linear (multiple regression) model might be inadequate to describe a particular relationship is that the effect of the predictors on the dependent variable may not be linear in nature. For example, the relationship between a person's age and various indicators of health is most likely not linear in nature: During early adulthood, the (average) health status of people who are 30 years old as compared to the (average) health status of people who are 40 years old is not markedly different. However, the difference in health status of 60-year-old people and 70-year-old people is probably greater. Thus, the relationship between age and health status is likely nonlinear in nature. Probably some kind of a power function would be adequate to describe the relationship between a person's age and health, so that each increment in years of age at older ages will have greater impact on health status, as compared to each increment in years of age during early adulthood. Put in other words, the link between age and health status is best described as nonlinear, or as a power relationship in this particular example.

The generalized linear/nonlinear model can be used to predict responses both for dependent variables with discrete distributions and for dependent variables that are nonlinearly related to the predictors.

Computational Approach

To summarize the basic ideas, the generalized linear/nonlinear model differs from the general linear model (of which, for example, multiple regression is a special case) in two major respects. First, the distribution of the dependent or response variable can be (explicitly) non-normal, and does not have to be continuous, i.e., it can be binomial, multinomial, or ordinal multinomial (i.e., contain information on ranks only); second, the dependent variable values are predicted from a linear combination of predictor variables, which are connected to the dependent variable via a link function. The general linear model for a single dependent variable can be considered a special case of the generalized linear model: In the general linear model the dependent variable values are expected to follow the normal distribution, and the link function is a simple identity function (i.e., the linear combination of values for the predictor variables is not transformed).

To illustrate, in the general linear model a response variable Y is linearly associated with values on the X variables by

$$Y = b_0 + b_1X_1 + b_2X_2 + ... + b_kX_k) + e$$

(where e stands for the error variability that cannot be accounted for by the predictors; note that the expected value of e is assumed to be 0), while the relationship in the generalized linear model is assumed to be:

$$Y = g (b_0 + b_1X_1 + b_2X_2 + ... + b_kX_k + e$$

where e is the error, and $g(...)$ is a function. Formally, the inverse function of $g(...)$, say $f(...)$, is called the link function; so that:

$$f(mu_y) = b_0 + b_1X_1 + b_2X_2 + ... + b_kX_k$$

where mu_y stands for the expected value of y.

Link functions and distributions. Various link functions (see McCullagh and Nelder, 1989) can be chosen, depending on the assumed distribution of the y variable values:

Normal, gamma, inverse normal, and Poisson distributions:

Identity link: $f(z) = z$

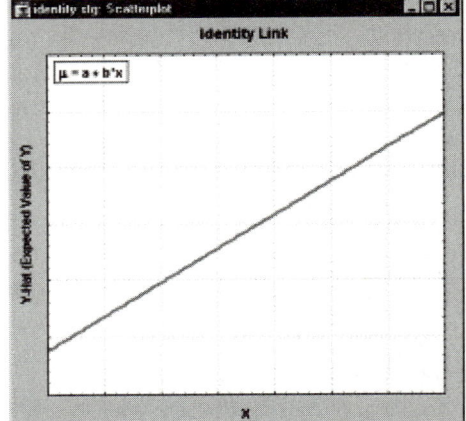

Log link: $f(z) = \log(z)$

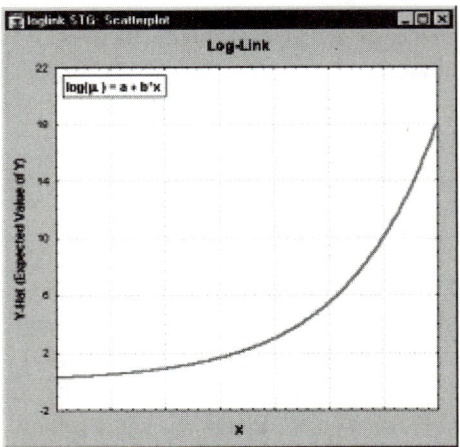

Power link: $f(z) = z^a$, for a given a

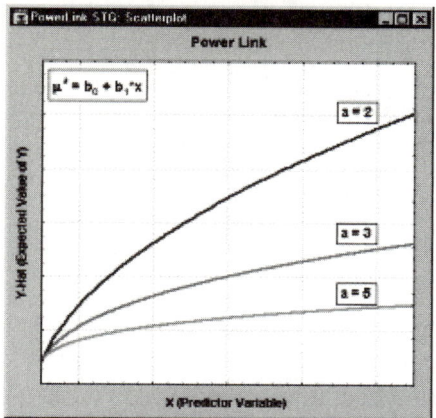

Binomial and ordinal multinomial distributions:

Logit link: $f(z)=\log(z/(1-z))$

Probit link: $f(z)=\text{invnorm}(z)$

where *invnorm* is the inverse of the standard normal cumulative distribution function.

Complementary log-log link:

$f(z)=\log(-\log(1-z))$

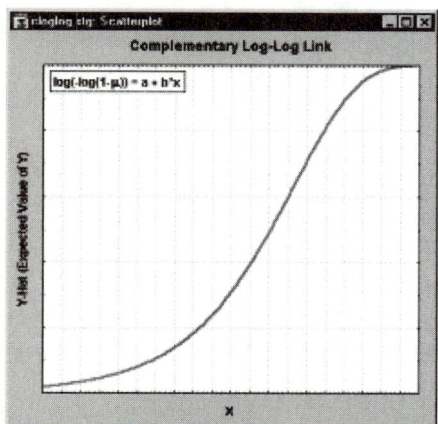

Log-log link: $f(z)=-\log(-\log(z))$

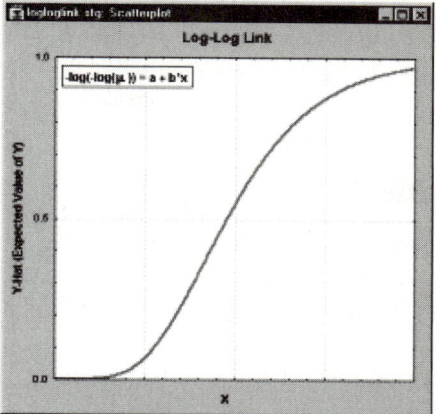

Multinomial distribution:

Generalized logit link:

$f(z1|z2,\ldots,zc)=\log(x1/(1-z1-\ldots-zc))$

where the model has c+1 categories.

Estimation in the generalized linear model.
The values of the parameters (b_0 through b_k and the scale parameter) in the generalized linear model are obtained by maximum likelihood (ML) estimation, which requires iterative computational procedures. There are many iterative methods for ML estimation in the generalized linear model, of which the Newton-Raphson and Fisher-Scoring methods are among the most efficient and widely used (see Dobson, 1990). The Fisher-scoring (or iterative re-weighted least squares) method in particular provides a unified algorithm for all generalized linear models, as well as providing the expected variance-covariance matrix of parameter estimates as a byproduct of its computations.

Statistical significance testing.
Tests for the significance of the effects in the model can be performed via the Wald statistic, the likelihood ratio (LR), or score statistic. Detailed descriptions of these tests can be found in

McCullagh and Nelder (1989). The Wald statistic (e.g., see Dobson, 1990), which is computed as the generalized inner product of the parameter estimates with the respective variance-covariance matrix, is an easily computed, efficient statistic for testing the significance of effects. The score statistic is obtained from the generalized inner product of the score vector with the Hessian matrix (the matrix of the second-order partial derivatives of the maximum likelihood parameter estimates). The likelihood ratio (LR) test requires the greatest computational effort (another iterative estimation procedure) and is thus not as fast as the first two methods; however, the LR test provides the most asymptotically efficient test known. For details concerning these different test statistics, see Agresti(1996), McCullagh and Nelder(1989), and Dobson(1990).

Diagnostics in the generalized linear/nonlinear model.

The two basic types of residuals are the so-called Pearson residuals and deviance residuals. Pearson residuals are based on the difference between observed responses and the predicted values; deviance residuals are based on the contribution of the observed responses to the log-likelihood statistic. In addition, leverage scores, studentized residuals, generalized Cook's D, and other observational statistics (statistics based on individual observations) can be computed. For a description and discussion of these statistics, see Hosmer and Lemeshow (1989).

Types of Analyses

The design for an analysis can include effects for continuous as well as categorical predictor variables. Designs can include polynomials for continuous predictors (e.g., squared or cubic terms) as well as interaction effects (i.e., product terms) for continuous predictors. For categorical predictor variables, you can fit ANOVA-like designs, including full factorial, nested, and fractional factorial designs, etc. Designs can be incomplete (i.e., involve missing cells), and effects for categorical predictor variables can be represented using either the *sigma*-restricted parameterization or the overparameterized (i.e., indicator variable) representation of effects.

The following topics contain descriptions of the types of designs that can be analyzed using the generalized linear/nonlinear model, as well as types of designs that can be analyzed using the general linear model.

Signal detection theory. The following list of designs is by no means comprehensive, i.e., it does not describe all possible research problems to which the generalized linear/nonlinear model can be applied. For example, an important application of the generalized linear model is the estimation of parameters for signal detection theory (SDT) models. SDT is an application of statistical decision theory used to detect a signal embedded in noise. SDT is used in psychophysical studies of detection, recognition, and discrimination, and in other areas such as medical research, weather forecasting, survey research, and marketing research. For example, DeCarlo (1998) shows how signal detection models based on different underlying distributions can easily be considered by using the generalized linear model with different link functions.

For discussion of the generalized linear/nonlinear model and the link functions that it uses, see *Computational Approach*, page 300.

Between-Subject Designs

Overview. The levels or values of the predictor variables in an analysis describe the differences between the n subjects or the n valid cases that are analyzed. Thus, when we speak of the between subject design (or simply the between design) for an analysis, we are referring to the nature, number, and arrangement of the predictor variables.

Concerning the nature or type of predictor variables, between designs that contain only categorical predictor variables can be called ANOVA (analysis of variance) designs, between designs that contain only continuous predictor variables can be called regression designs, and between designs that contain both categorical and continuous predictor variables can be called ANCOVA (analysis of covariance) designs. Further, continuous predictors are always considered to have fixed values, but the levels of categorical predictors can be considered to be fixed or to vary randomly. Designs that contain random categorical factors are called mixed-model designs (see Chapter 39 – *Variance Components and Mixed Model ANOVA/ANCOVA*).

Between designs can involve only a single predictor variable and therefore be described as simple (e.g., simple regression) or can employ numerous predictor variables (e.g., multiple regression).

Concerning the arrangement of predictor variables, some between designs employ only main effect or first-order terms for predictors, that is, the values for different predictor variables are independent and raised only to the first power. Other between designs can employ higher-order terms for predictors by raising the values for the original predictor variables to a power greater than

1 (e.g., in polynomial regression designs), or by forming products of different predictor variables (i.e., interaction terms). A common arrangement for ANOVA designs is the full-factorial design, in which every combination of levels for each of the categorical predictor variables is represented in the design. Designs with some but not all combinations of levels for each of the categorical predictor variables are aptly called fractional factorial designs. Designs with a hierarchy of combinations of levels for the different categorical predictor variables are called nested designs.

These basic distinctions about the nature, number, and arrangement of predictor variables can be used in describing a variety of different types of between designs. Some of the more common between designs can now be described.

One-way ANOVA. A design with a single categorical predictor variable is called a one-way ANOVA design. For example, a study of 4 different fertilizers used on different individual plants could be analyzed via one-way ANOVA, with four levels for the factor *Fertilizer*.

In general, consider a single categorical predictor variable A with 1 case in each of its 3 categories. Using the *sigma*-restricted coding of A into 2 quantitative contrast variables, the matrix **X** defining the between design is:

$$\mathbf{X} = \begin{array}{c} \\ A_1 \\ A_2 \\ A_3 \end{array} \begin{array}{ccc} X_0 & X_1 & X_2 \\ \left[\begin{array}{ccc} 1 & 1 & 0 \\ 1 & 0 & 1 \\ 1 & -1 & -1 \end{array} \right. & & \left] \right. \end{array}$$

That is, cases in groups A_1, A_2, and A_3 are all assigned values of 1 on X_0 (the intercept), the case in group A_1 is assigned a value of 1 on X_1 and a value 0 on X_2, the case in group A_2 is assigned a value of 0 on X_1 and a value 1 on X_2, and the case in group A_3 is assigned a value of -1 on X_1 and a value -1 on X_2. Of course, any additional cases in any of the 3 groups would be coded similarly. If there were 1 case in group A_1, 2 cases in group A_2, and 1 case in group A_3, the **X** matrix would be:

$$\mathbf{X} = \begin{array}{c} \\ A_{11} \\ A_{12} \\ A_{22} \\ A_{13} \end{array} \begin{array}{ccc} X_0 & X_1 & X_2 \\ \left[\begin{array}{ccc} 1 & 1 & 0 \\ 1 & 0 & 1 \\ 1 & 0 & 1 \\ 1 & -1 & -1 \end{array}\right] \end{array}$$

where the first subscript for A gives the replicate number for the cases in each group. For brevity, replicates usually are not shown when describing ANOVA design matrices.

Note that in one-way designs with an equal number of cases in each group, *sigma*-restricted coding yields $X_1 \dots X_k$ variables all of which have means of 0.

Using the overparameterized model to represent A, the **X** matrix defining the between design is simply:

$$\mathbf{X} = \begin{array}{c} \\ A_1 \\ A_2 \\ A_3 \end{array} \begin{array}{cccc} X_0 & X_1 & X_2 & X_3 \\ \left[\begin{array}{cccc} 1 & 1 & 0 & 0 \\ 1 & 0 & 1 & 0 \\ 1 & 0 & 0 & 1 \end{array}\right] \end{array}$$

These simple examples show that the **X** matrix actually serves two purposes. It specifies 1) the coding for the levels of the original predictor variables on the X variables used in the analysis as well as 2) the nature, number, and arrangement of the X variables, that is, the between design.

Main effect ANOVA. Main effect ANOVA designs contain separate one-way ANOVA designs for 2 or more categorical predictors. A good example of main effect ANOVA would be the typical analysis performed on screening designs as described in Chapter 15 – *Experimental Design (DOE)*.

Consider 2 categorical predictor variables A and B each with 2 categories. Using the *sigma*-

restricted coding, the **X** matrix defining the between design is:

$$\mathbf{X} = \begin{array}{c} \\ A_1B_1 \\ A_1B_2 \\ A_2B_1 \\ A_2B_2 \end{array} \begin{array}{ccc} X_0 & X_1 & X_2 \\ \left[\begin{array}{ccc} 1 & 1 & 1 \\ 1 & 1 & -1 \\ 1 & -1 & 1 \\ 1 & -1 & -1 \end{array}\right] \end{array}$$

Note that if there are equal numbers of cases in each group, the sum of the cross-products of values for the X_1 and X_2 columns is 0, for example, with 1 case in each group $(1*1)+(1*-1)+(-1*1)+(-1*-1)=0$. Using the overparameterized model, the matrix **X** defining the between design is:

$$\mathbf{X} = \begin{array}{c} \\ A_1B_1 \\ A_1B_2 \\ A_2B_1 \\ A_2B_2 \end{array} \begin{array}{ccccc} X_0 & X_1 & X_2 & X_3 & X_4 \\ \left[\begin{array}{ccccc} 1 & 1 & 0 & 1 & 0 \\ 1 & 1 & 0 & 0 & 1 \\ 1 & 0 & 1 & 1 & 0 \\ 1 & 0 & 1 & 0 & 1 \end{array}\right] \end{array}$$

Comparing the two types of coding, it can be seen that the overparameterized coding takes almost twice as many values as the *sigma*-restricted coding to convey the same information.

Factorial ANOVA. Factorial ANOVA designs contain X variables representing combinations of the levels of 2 or more categorical predictors (e.g., a study of boys and girls in four age groups, resulting in a *2 (Gender) x 4 (Age Group)* design). In particular, full-factorial designs represent all possible combinations of the levels of the categorical predictors. A full-factorial design with 2 categorical predictor variables A and B each with 2 levels each would be called a 2x2 full-factorial design. Using the *sigma*-restricted coding, the **X** matrix for this design would be:

$$
\mathbf{X} = \begin{array}{c} \\ A_1B_1 \\ A_1B_2 \\ A_2B_1 \\ A_2B_2 \end{array}
\begin{array}{cccc}
X_0 & X_1 & X_2 & X_3 \\
\begin{bmatrix} 1 & 1 & 1 & 1 \\ 1 & 1 & -1 & -1 \\ 1 & -1 & 1 & -1 \\ 1 & -1 & -1 & 1 \end{bmatrix}
\end{array}
$$

Several features of this **X** matrix deserve comment. Note that the X_1 and X_2 columns represent main effect contrasts for one variable, (i.e., *A* and *B*, respectively) collapsing across the levels of the other variable. The X_3 column instead represents a contrast between different combinations of the levels of *A* and *B*. Note also that the values for X_3 are products of the corresponding values for X_1 and X_2. Product variables such as X_3 represent the multiplicative or interaction effects of their factors, so X_3 would be said to represent the 2-way interaction of *A* and *B*. The relationship of such product variables to the dependent variables indicate the interactive influences of the factors on responses above and beyond their independent (i.e., main effect) influences on responses. Thus, factorial designs provide more information about the relationships between categorical predictor variables and responses on the dependent variables than is provided by corresponding one-way or main effect designs.

When many factors are being investigated, however, full-factorial designs sometimes require more data than reasonably can be collected to represent all possible combinations of levels of the factors, and high-order interactions between many factors can become difficult to interpret. With many factors, a useful alternative to the full-factorial design is the fractional factorial design. As an example, consider a 2x2 x 2 fractional factorial design to degree 2 with 3 categorical predictor variables each with 2 levels. The design would include the main effects for each variable, and all 2-way interactions between the three variables, but

would not include the 3-way interaction between all three variables. Using the overparameterized model, the **X** matrix for this design is:

$$
\mathbf{X} = \begin{array}{c}
A_1B_1C_1 \\ A_1B_1C_2 \\ A_1B_2C_1 \\ A_1B_2C_2 \\ A_2B_1C_1 \\ A_2B_1C_2 \\ A_2B_2C_1 \\ A_2B_2C_2
\end{array}
\begin{bmatrix}
1 & 1 & 0 & 1 & 0 & 1 & 0 & 1 & 0 & 0 & 0 & 1 & 0 & 0 & 0 & 1 & 0 & 0 & 0 \\
1 & 1 & 0 & 1 & 0 & 0 & 1 & 1 & 0 & 0 & 0 & 1 & 0 & 0 & 1 & 0 & 0 & 0 & 0 \\
1 & 1 & 0 & 0 & 1 & 1 & 0 & 0 & 1 & 0 & 0 & 1 & 0 & 0 & 0 & 0 & 1 & 0 \\
1 & 1 & 0 & 0 & 1 & 0 & 1 & 0 & 1 & 0 & 0 & 1 & 0 & 0 & 0 & 0 & 0 & 1 \\
1 & 0 & 1 & 1 & 0 & 1 & 0 & 0 & 0 & 1 & 0 & 0 & 1 & 0 & 1 & 0 & 1 & 0 & 0 & 0 \\
1 & 0 & 1 & 1 & 0 & 0 & 1 & 0 & 0 & 1 & 0 & 0 & 0 & 1 & 0 & 1 & 0 & 0 & 1 & 0 & 0 \\
1 & 0 & 1 & 0 & 1 & 1 & 0 & 0 & 0 & 1 & 0 & 0 & 1 & 0 & 0 & 0 & 1 & 0 \\
1 & 0 & 1 & 0 & 1 & 0 & 1 & 0 & 0 & 1 & 0 & 0 & 0 & 1 & 0 & 0 & 0 & 1
\end{bmatrix}
$$

The 2-way interactions are the highest degree effects included in the design. These types of designs are discussed in detail in $2^{(k-p)}$ *Fractional Factorial Designs* in Chapter 15 – *Experimental Design (DOE)*.

Nested ANOVA designs. Nested designs are similar to fractional factorial designs in that all possible combinations of the levels of the categorical predictor variables are not represented in the design. In nested designs, however, the omitted effects are lower-order effects. Nested effects are effects in which the nested variables never appear as main effects. Suppose that for 2 variables *A* and *B* with 3 and 2 levels, respectively, the design includes the main effect for *A* and the effect of *B* nested within the levels of *A*. The **X** matrix for this design using the overparameterized model is:

$$
\mathbf{X} = \begin{array}{c}
A_1B_1 \\ A_1B_2 \\ A_2B_1 \\ A_2B_2 \\ A_3B_1 \\ A_3B_2
\end{array}
\begin{array}{cccccccccc}
X_0 & X_1 & X_2 & X_3 & X_4 & X_5 & X_6 & X_7 & X_8 & X_9 \\
\begin{bmatrix}
1 & 1 & 0 & 0 & 1 & 0 & 0 & 0 & 0 & 0 \\
1 & 1 & 0 & 0 & 0 & 1 & 0 & 0 & 0 & 0 \\
1 & 0 & 1 & 0 & 0 & 0 & 1 & 0 & 0 & 0 \\
1 & 0 & 1 & 0 & 0 & 0 & 0 & 1 & 0 & 0 \\
1 & 0 & 0 & 1 & 0 & 0 & 0 & 0 & 1 & 0 \\
1 & 0 & 0 & 1 & 0 & 0 & 0 & 0 & 0 & 1
\end{bmatrix}
\end{array}
$$

Note that if the *sigma*-restricted coding were used, there would be only 2 columns in the **X** matrix for the *B* nested within *A* effect instead of the 6 columns in the **X** matrix for this effect when the overparameterized model coding is used (i.e., columns X_4 through X_9). The *sigma*-restricted coding method is overly restrictive for nested designs, so only the overparameterized model is used to represent nested designs.

Simple regression. Simple regression designs involve a single continuous predictor variable. If there were 3 cases with values on a predictor variable P of, say, 7, 4, and 9, and the design is for the first-order effect of P, the **X** matrix would be:

$$\mathbf{X} = \begin{matrix} X_0 & X_1 \\ \begin{bmatrix} 1 & 7 \\ 1 & 4 \\ 1 & 9 \end{bmatrix} \end{matrix}$$

and using P for X_1 the regression equation would be:

$$Y = b_0 + b_1 P$$

If the simple regression design is for a higher-order effect of P, say the quadratic effect, the values in the X_1 column of the design matrix would be raised to the 2nd power, that is, squared:

$$\mathbf{X} = \begin{matrix} X_0 & X_1 \\ \begin{bmatrix} 1 & 49 \\ 1 & 16 \\ 1 & 81 \end{bmatrix} \end{matrix}$$

and using P^2 for X_1 the regression equation would be:

$$Y = b_0 + b_1 P^2$$

The *sigma*-restricted and overparameterized coding methods do not apply to simple regression designs and any other design containing only continuous predictors (since there are no categorical predictors to code). Regardless of which coding method is chosen, values on the continuous predictor variables are raised to the desired power and used as the values for the X variables. No recoding is performed. It is therefore sufficient, in describing regression designs, to simply describe the regression equation without explicitly describing the design matrix **X**.

Multiple regression. Multiple regression designs are to continuous predictor variables as main effect ANOVA designs are to categorical predictor variables, that is, multiple regression designs contain the separate simple regression designs for 2 or more continuous predictor variables. The regression equation for a multiple regression design for the first-order effects of 3 continuous predictor variables P, Q, and R would be:

$$Y = b_0 + b_1 P + b_2 Q + b_3 R$$

Factorial regression. Factorial regression designs are similar to factorial ANOVA designs, in which combinations of the levels of the factors are represented in the design. In factorial regression designs, however, there may be many more such possible combinations of distinct levels for the continuous predictor variables than there are cases in the data set. To simplify matters, full-factorial regression designs are defined as designs in which all possible products of the continuous predictor variables are represented in the design. For example, the full-factorial regression design for two continuous predictor variables P and Q would include the main effects (i.e., the first-order effects) of P and Q and their 2-way P by Q interaction effect, which is represented by the product of P and Q scores for each case. The regression equation is:

$$Y = b_0 + b_1 P + b_2 Q + b_3 P^* Q$$

Factorial regression designs can also be fractional, that is, higher-order effects can be omitted from the design. A fractional factorial design to degree 2 for 3 continuous predictor variables P, Q, and R would include the main effects and all 2-way interactions between the predictor variables:

$$Y = b_0 + b_1P + b_2Q + b_3R + b_4P*Q + b_5P*R + b_6Q*R$$

Polynomial regression. Polynomial regression designs are designs that contain main effects and higher-order effects for the continuous predictor variables but do not include interaction effects between predictor variables. For example, the polynomial regression design to degree 2 for three continuous predictor variables P, Q, and R would include the main effects (i.e., the first-order effects) of P, Q, and R and their quadratic (i.e., second-order) effects, but not the 2-way interaction effects or the P by Q by R 3-way interaction effect:

$$Y = b_0 + b_1P + b_2P^2 + b_3Q + b_4Q^2 + b_5R + b_6R^2$$

Polynomial regression designs do not have to contain all effects up to the same degree for every predictor variable. For example, main, quadratic, and cubic effects could be included in the design for some predictor variables, and effects up the fourth degree could be included in the design for other predictor variables.

Response surface regression. Quadratic response surface regression designs are a hybrid type of design with characteristics of both polynomial regression designs and fractional factorial regression designs. Quadratic response surface regression designs contain all the same effects of polynomial regression designs to degree 2 and additionally the 2-way interaction effects of the predictor variables. The regression equation for a quadratic response surface regression design for 3 continuous predictor variables P, Q, and R is:

$$Y = b_0 + b_1P + b_2P^2 + b_3Q + b_4Q^2 + b_5R + b_6R^2 + b_7P*Q + b_8P*R + b_9Q*R$$

These types of designs are commonly employed in applied research (e.g., in industrial experimentation), and a detailed discussion of these types of designs is also presented in Chapter 15 – *Experimental Design* (see *Central composite designs*).

Mixture surface regression. Mixture surface regression designs are identical to factorial regression designs to degree 2 except for the omission of the intercept. Mixtures, as the name implies, add up to a constant value; the sum of the proportions of ingredients in different recipes for some material all must add up 100%. Thus, the proportion of one ingredient in a material is redundant with the remaining ingredients. Mixture surface regression designs deal with this redundancy by omitting the intercept from the design. The design matrix for a mixture surface regression design for 3 continuous predictor variables P, Q, and R is:

$$Y = b_1P + b_2Q + b_3R + b_4P*Q + b_5P*R + b_6Q*R$$

These types of designs are commonly employed in applied research (e.g., in industrial experimentation), and a detailed discussion of these types of designs is also presented in Chapter 15 – *Experimental Design* (see *Mixture Designs and Triangular Surfaces.*

Analysis of covariance. In general, between designs which contain both categorical and continuous predictor variables can be called ANCOVA designs. Traditionally, however, ANCOVA designs have referred more specifically to designs in which the first-order effects of one or more continuous predictor variables are taken into account when assessing the effects of one or more categorical predictor variables. A basic introduction to analysis of covariance can also be found in Chapter 3 – *ANOVA/MANOVA.*

To illustrate, suppose a researcher wants to assess the influences of a categorical predictor variable A with 3 levels on some outcome, and that measurements on a continuous predictor variable P, known to covary with the outcome, are available. If the data for the analysis are:

$$\begin{array}{cc} P & Group \\ \begin{bmatrix} 7 \\ 4 \\ 9 \\ 3 \\ 6 \\ 8 \end{bmatrix} & \begin{bmatrix} A_1 \\ A_1 \\ A_2 \\ A_2 \\ A_3 \\ A_3 \end{bmatrix} \end{array}$$

then the *sigma*-restricted $\mathbf{X}$ matrix for the design that includes the separate first-order effects of P and A would be:

$$\mathbf{X} = \begin{array}{cccc} X_0 & X_1 & X_2 & X_3 \\ \begin{bmatrix} 1 & 7 & 1 & 0 \\ 1 & 4 & 1 & 0 \\ 1 & 9 & 0 & 1 \\ 1 & 3 & 0 & 1 \\ 1 & 6 & -1 & -1 \\ 1 & 8 & -1 & -1 \end{bmatrix} \end{array}$$

The b_2 and b_3 coefficients in the regression equation:

$$Y = b_0 + b_1 X_1 + b_2 X_2 + b_3 X_3$$

represent the influences of group membership on the A categorical predictor variable, controlling for the influence of scores on the P continuous predictor variable. Similarly, the b_1 coefficient represents the influence of scores on P controlling for the influences of group membership on A. This traditional ANCOVA analysis gives a more sensitive test of the influence of A to the extent that P reduces the prediction error, that is, the residuals for the outcome variable.

The $\mathbf{X}$ matrix for the same design using the overparameterized model is:

$$\mathbf{X} = \begin{array}{ccccc} X_0 & X_1 & X_2 & X_3 & X_4 \\ \begin{bmatrix} 1 & 7 & 1 & 0 & 0 \\ 1 & 4 & 1 & 0 & 0 \\ 1 & 9 & 0 & 1 & 0 \\ 1 & 3 & 0 & 1 & 0 \\ 1 & 6 & 0 & 0 & 1 \\ 1 & 8 & 0 & 0 & 1 \end{bmatrix} \end{array}$$

The interpretation is unchanged except that the influences of group membership on the A categorical predictor variables are represented by the b_2, b_3 and b_4 coefficients in the regression equation:

$$Y = b_0 + b_1 X_1 + b_2 X_2 + b_3 X_3 + b_4 X_4$$

Separate slope designs. The traditional analysis of covariance (ANCOVA) design for categorical and continuous predictor variables is inappropriate when the categorical and continuous predictors interact in influencing responses on the outcome. The appropriate design for modeling the influences of the predictors in this situation is called the separate slope design. For the same example data used to illustrate traditional ANCOVA, the overparameterized $\mathbf{X}$ matrix for the design that includes the main effect of the three-level categorical predictor A and the 2-way interaction of P by A is:

$$\mathbf{X} = \begin{array}{ccccccc} X_0 & X_1 & X_2 & X_3 & X_4 & X_5 & X_6 \\ \begin{bmatrix} 1 & 1 & 0 & 0 & 7 & 0 & 0 \\ 1 & 1 & 0 & 0 & 4 & 0 & 0 \\ 1 & 0 & 1 & 0 & 0 & 9 & 0 \\ 1 & 0 & 1 & 0 & 0 & 3 & 0 \\ 1 & 0 & 0 & 1 & 0 & 0 & 6 \\ 1 & 0 & 0 & 1 & 0 & 0 & 8 \end{bmatrix} \end{array}$$

The b_4, b_5, and b_6 coefficients in the regression equation:

$$Y = b_0 + b_1 X_1 + b_2 X_2 + b_3 X_3 + b_4 X_4 + b_5 X_5 + b_6 X_6$$

give the separate slopes for the regression of the outcome on P within each group on A, controlling for the main effect of A.

As with nested ANOVA designs, the *sigma*-restricted coding of effects for separate slope designs is overly restrictive, so only the overparameterized model is used to represent separate slope designs. In fact, separate slope designs are identical in form to nested ANOVA designs, since the main effects for continuous predictors are omitted in separate slope designs.

Homogeneity of slopes. The appropriate design for modeling the influences of continuous and categorical predictor variables depends on whether the continuous and categorical predictors interact in influencing the outcome. The traditional analysis of covariance (ANCOVA) design for continuous and categorical predictor variables is appropriate when the continuous and categorical predictors do not interact in influencing responses on the outcome, and the separate slope design is appropriate when the continuous and categorical predictors do interact in influencing responses. The homogeneity of slopes designs can be used to test whether the continuous and categorical predictors interact in influencing responses, and thus, whether the traditional ANCOVA design or the separate slope design is appropriate for modeling the effects of the predictors. For the same example data used to illustrate the traditional ANCOVA and separate slope designs, the overparameterized $\mathbf{X}$ matrix for the design that includes the main effect of P, the main effect of the three-level categorical predictor A, and the 2-way interaction of P by A would be:

$$\mathbf{X} = \begin{matrix} X_0 & X_1 & X_2 & X_3 & X_4 & X_5 & X_6 & X_7 \\ \begin{bmatrix} 1 & 7 & 1 & 0 & 0 & 7 & 0 & 0 \\ 1 & 4 & 1 & 0 & 0 & 4 & 0 & 0 \\ 1 & 9 & 0 & 1 & 0 & 0 & 9 & 0 \\ 1 & 3 & 0 & 1 & 0 & 0 & 3 & 0 \\ 1 & 6 & 0 & 0 & 1 & 0 & 0 & 6 \\ 1 & 8 & 0 & 0 & 1 & 0 & 0 & 8 \end{bmatrix} \end{matrix}$$

If the b_5, b_6, or b_7 coefficient in the regression equation:

$$Y = b_0 + b_1X_1 + b_2X_2 + b_3X_3 + b_4X_4 + b_5X_5 + b_6X_6 + b_7X_7$$

is non-zero, the separate slope model should be used. If instead all 3 of these regression coefficients are zero the traditional ANCOVA design should be used.

The *sigma*-restricted $\mathbf{X}$ matrix for the homogeneity of slopes design would be:

$$\mathbf{X} = \begin{matrix} X_0 & X_1 & X_2 & X_3 & X_4 & X_5 \\ \begin{bmatrix} 1 & 7 & 1 & 0 & 7 & 0 \\ 1 & 4 & 1 & 0 & 4 & 0 \\ 1 & 9 & 0 & 1 & 0 & 9 \\ 1 & 3 & 0 & 1 & 0 & 3 \\ 1 & 6 & -1 & -1 & -6 & -6 \\ 1 & 8 & -1 & -1 & -8 & -8 \end{bmatrix} \end{matrix}$$

Using this $\mathbf{X}$ matrix, if the b_4, or b_5 coefficient in the regression equation:

$$Y = b_0 + b_1X_1 + b_2X_2 + b_3X_3 + b_4X_4 + b_5X_5$$

is non-zero, the separate slope model should be used. If instead both of these regression coefficients are zero the traditional ANCOVA design should be used.

Model Building

In addition to fitting the whole model for the specified type of analysis, different methods for automatic model building can be employed in analyses using the generalized linear model. Specifically, forward entry, backward removal, forward stepwise, and backward stepwise procedures can be performed, as well as best-subset search procedures. In forward methods of selection of effects to include in the model (i.e., forward entry and forward stepwise methods), score statistics are compared to select new (significant) effects. The Wald statistic can be used for backward removal methods (i.e., backward removal and backward stepwise, when effects are selected for removal from the model).

The best subsets search method can be based on three different test statistics: the score statistic, the model likelihood, and the AIC (Akaike Information Criterion, see Akaike, 1973). Note that, since the score statistic does not require iterative computations, best subset selection based on the score statistic is computationally fastest, while selection based on the other two statistics usually provides more accurate results; see McCullagh and Nelder(1989), for additional details.

Interpretation of Results and Diagnostics

Simple estimation and test statistics may not be sufficient for adequate interpretation of the effects in an analysis. Especially for higher order (e.g., interaction) effects, inspection of the observed and predicted means can be invaluable for understanding the nature of an effect. Plots of these means (with error bars) can be useful for quickly grasping the role of the effects in the model. Inspection of the distributions of variables is critically important when using the generalized linear model. Histograms and probability plots for variables, and scatterplots showing the relationships between observed values, predicted values, and residuals (e.g., Pearson residuals, deviance residuals, studentized residuals, differential *chi*-square statistics, differential deviance statistics, and generalized Cook's D) provide invaluable model-checking tools.

LOG LINEAR ANALYSIS OF FREQUENCY TABLES

One basic and straightforward method for analyzing data is via crosstabulation. For example, a medical researcher may tabulate the frequency of different symptoms by patients' age and gender; an educational researcher may tabulate the number of high school dropouts by age, gender, and ethnic background; an economist may tabulate the number of business failures by industry, region, and initial capitalization; a market researcher may tabulate consumer preferences by product, age, and gender; etc. In all of these cases, the major results of interest can be summarized in a multi-way frequency table, that is, in a crosstabulation table with two or more factors.

Log linear provides a more sophisticated way of looking at crosstabulation tables. Specifically, you can test the different factors that are used in the crosstabulation (e.g., gender, region, etc.) and their interactions for statistical significance (see Chapter 1 – *Elementary Concepts in Statistics*, for a discussion of statistical significance testing). This chapter presents an introduction to these methods and the logic and interpretation.

Correspondence analysis (see Chapter 11) is a descriptive/exploratory technique designed to analyze two-way and multi-way tables containing some measure of correspondence between the rows and columns. The results provide information similar in nature to those produced by factor analysis techniques, and they enable you to explore the structure of the categorical variables included in the table.

Two-Way Frequency Tables

Let's begin with the simplest possible crosstabulation, the 2x2 table. Suppose we are interested in the relationship between age and the graying of people's hair. We take a sample of 100 subjects, and determine who does and does not have gray hair. We also record the approximate age of the subjects. The results of this study can be summarized as follows:

Data: Log Linear Analysis Example Spreadsheet 1 (3v by 3c)			
Gray Hair	Age Below 40	Age 40 or older	Total
No	40	5	45
Yes	20	35	55
Total	60	40	100

While interpreting the results of our study, let's introduce the terminology that will enable us to generalize to complex tables more easily.

Design variables and response variables. In multiple regression (see Chapter 26) or analysis of variance (see Chapter 3 – ANOVA/ MANOVA) you customarily distinguish between independent and dependent variables. Dependent variables are those that we are trying to explain, that is, that we hypothesize to depend on the independent variables. We could classify the factors in the 2x2 table accordingly: we can think of hair color (gray, not gray) as the dependent variable, and age as the independent variable. Alternative terms that are often used in the context of frequency tables are response variables and design variables, respectively. Response variables are those that vary in response to the design variables. Thus, in the example table above, hair color can be considered to be the response variable, and age the design variable.

Fitting marginal frequencies. Let's now turn to the analysis of our example table. We could ask ourselves what the frequencies would look

like if there were no relationship between variables (the null hypothesis). Without going into details, intuitively you could expect that the frequencies in each cell would proportionately reflect the marginal frequencies (*Totals*). For example, consider the following table:

Data: Log Linear Analysis Example Spreadsheet 2 (3v by 3c)			
Gray Hair	Age Below 40	Age 40 or older	Total
No	27	18	45
Yes	33	22	55
Total	60	40	100

In this table, the proportions of the marginal frequencies are reflected in the individual cells. Thus, $27/33=18/22=45/55$ and $27/18=33/22=60/40$. Given the marginal frequencies, these are the cell frequencies that we would expect if there were no relationship between age and graying. If you compare this table with the previous one you will see that the previous table does reflect a relationship between the two variables: There are more than expected (under the null hypothesis) cases below age 40 without gray hair, and more cases above age 40 with gray hair.

This example illustrates the general principle on which the log-linear analysis is based: Given the marginal totals for two (or more) factors, we can compute the cell frequencies that would be expected if the two (or more) factors are unrelated. Significant deviations of the observed frequencies from those expected frequencies reflect a relationship between the two (or more) variables.

Model fitting approach. Let's now rephrase our discussion of the 2x2 table so far. We can say that fitting the model of two variables that are not related (age and hair color) amounts to computing the cell frequencies in the table based on the respective marginal frequencies (totals). Significant deviations of the observed table from those fitted frequencies reflect the lack of fit of the independence (between two variables) model. In that case we would reject that model for our data, and instead accept the model that allows for a relationship or association between age and hair color.

Multi-Way Frequency Tables

The reasoning presented for the analysis of the 2x2 table can be generalized to more complex tables. For example, suppose we had a third variable in our study, namely whether the individuals in our sample experience stress at work. Because we are interested in the effect of stress on graying, we will consider Stress as another design variable. (Note that, if our study were concerned with the effect of gray hair on subsequent stress, variable stress would be the response variable, and hair color would be the design variable.). The resultant table is a three-way frequency table.

Fitting models. We can apply our previous reasoning to analyze this table. Specifically, we could fit different models that reflect different hypotheses about the data. For example, we could begin with a model that hypothesizes independence between all factors. As before, the expected frequencies in that case would reflect the respective marginal frequencies. If any significant deviations occur, we would reject this model.

Interaction effects. Another conceivable model would be that age is related to hair color, and stress is related to hair color, but the two (age and stress) factors do not interact in their effect. In that case, we would need to simultaneously fit the marginal totals for the two-way table of age by hair color collapsed across levels of stress, and the two-way table of stress by hair color collapsed across the levels of age. If this model does not fit the data, we would have to

conclude that age, stress, and hair color all are interrelated. Put another way, we would conclude that age and stress interact in their effect on graying.

The concept of interaction here is analogous to that used in analysis of variance (ANOVA /MANOVA). For example, the age by stress interaction could be interpreted such that the relationship of age to hair color is modified by stress. While age brings about only little graying in the absence of stress, age is highly related when stress is present. Put another way, the effects of age and stress on graying are not additive, but interactive.

If you are not familiar with the concept of interaction, it is recommended that you read *Basic Ideas* in Chapter 3 – *ANOVA/MANOVA*. Many aspects of the interpretation of results from a log-linear analysis of a multi-way frequency table are very similar to ANOVA.

Iterative proportional fitting. The computation of expected frequencies becomes increasingly complex when there are more than two factors in the table. However, they can be computed, and, therefore, we can easily apply the reasoning developed for the 2x2 table to complex tables. The commonly used method for computing the expected frequencies is the so-called iterative proportional fitting procedure.

Log Linear Model

The term log linear derives from the fact that you can, through logarithmic transformations, restate the problem of analyzing multi-way frequency tables in terms that are very similar to ANOVA. Specifically, you can think of the multi-way frequency table to reflect various main effects and interaction effects that add together in a linear fashion to bring about the observed table of frequencies. Bishop, Fienberg, and Holland (1974) provide details on how to derive log-linear equations to express the relationship between factors in a multi-way frequency table.

Goodness of Fit

In the previous discussion we have repeatedly made reference to the "significance" of deviations of the observed frequencies from the expected frequencies. You can evaluate the statistical significance of the goodness-of-fit of a particular model via a *chi*-square test. You can compute two types of *chi*-squares, the traditional Pearson *chi*-square statistic and the maximum likelihood ratio *chi*-square statistic (the term likelihood ratio was first introduced by Neyman and Pearson, 1931; the term maximum likelihood was first used by Fisher, 1922a). In practice, the interpretation and magnitude of those two *chi*-square statistics are essentially identical. Both tests evaluate whether the expected cell frequencies under the respective model are significantly different from the observed cell frequencies. If so, the respective model for the table is rejected.

Reviewing and plotting residual frequencies. After you have chosen a model for the observed table, it is always a good idea to inspect the residual frequencies, that is, the observed minus the expected frequencies. If the model is appropriate for the table, all residual frequencies should be "random noise," that is, consist of positive and negative values of approximately equal magnitudes that are distributed evenly across the cells of the table.

Statistical significance of effects. The *chi*-squares of models that are hierarchically related to each other can be directly compared. For example, if we first fit a model with the age by

hair color interaction and the stress by hair color interaction, and then fit a model with the age by stress by hair color (three-way) interaction, then the second model is a superset of the previous model. We could evaluate the difference in the *chi*-square statistics, based on the difference in the degrees of freedom; if the differential *chi*-square statistic is significant, we would conclude that the three-way interaction model provides a significantly better fit to the observed table than the model without this interaction. Therefore, the three-way interaction is statistically significant.

In general, two models are hierarchically related to each other if one can be produced from the other by either adding terms (variables or interactions) or deleting terms (but not both at the same time).

Automatic Model Fitting

When analyzing four- or higher-way tables, finding the best fitting model can become increasingly difficult. You can use automatic model fitting options to facilitate the search for a good model that fits the data. The general logic of this algorithm is as follows. First, fit a model with no relationships between factors; if that model does not fit (i.e., the respective *chi*-square statistic is significant), then fit a model with all two-way interactions. If that model does not fit either, fit all three-way interactions, and so on. Let's assume that this process found the model with all two-way interactions to fit the data. Then proceed to eliminate all two-way interactions that are not statistically significant. The resulting model will be the one that includes the least number of interactions necessary to fit the observed table.

MACHINE LEARNING

Written by Cazhaow Qazaz, Ph.D.

Machine Learning includes a number of advanced statistical methods for handling regression and classification tasks with multiple dependent and independent variables. These methods, useful in data mining applications, include Support Vector Machines (SVM) for regression and classification, Naive Bayes for classification, and k-Nearest Neighbours (KNN) for regression and classification. Detailed discussions of these techniques can be found in Hastie, Tibshirani, & Freedman (2001); a specialized comprehensive introduction to support vector machines can also be found in Cristianini and Shawe-Taylor (2000).

Support Vector Machines (SVM). This method performs regression and classification tasks by constructing nonlinear decision boundaries. Because of the nature of the feature space in which these boundaries are found, Support Vector Machines can exhibit a large degree of flexibility in handling classification and regression tasks of varied complexities. There are several types of Support Vector models including linear, polynomial, RBF, and sigmoid.

Naive Bayes. This is a well-established Bayesian method primarily formulated for performing classification tasks. Given its simplicity, i.e., the assumption that the independent variables are statistically independent, Naive Bayes models are effective classification tools that are easy to use and interpret. Naive Bayes is particularly appropriate when the dimensionality of the independent space (i.e., number of input variables) is high (a problem known as the curse of dimensionality). For the reasons given above, Naive Bayes can often outperform other more sophisticated classification methods. A variety of methods exist for modeling the conditional distributions of the inputs including normal, lognormal, *gamma*, and Poisson.

***k*-Nearest Neighbors (KNN).** *k*-Nearest Neighbors is a memory-based method that, in contrast to other statistical methods, requires no training (i.e., no model to fit). It falls into the category of Prototype Methods. It functions on the intuitive idea that close objects are more likely to be in the same category. Thus, in KNN, predictions are based on a set of prototype examples that are used to predict new (i.e., unseen) data based on the majority vote (for classification tasks) and averaging (for regression) over a set of k-nearest prototypes (hence the name k-nearest neighbors).

Support Vector Machines Overview

Support Vector Machines are based on the concept of decision planes that define decision boundaries. A decision plane is one that separates a set of objects having different class memberships. A schematic example is shown in the illustration below. In this example, the objects belong either to class *GREEN* or *RED*. The separating line defines a boundary on the right side of which all objects are *GREEN* and to the left of which all objects are *RED*. Any new object (white circle) falling to the right is labeled, i.e., classified, as *GREEN* (or classified as *RED* should it fall to the left of the separating line).

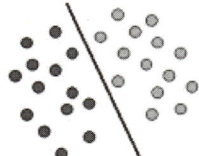

The above is a classic example of a linear classifier, i.e., a classifier that separates a set of objects into their respective groups (*GREEN* and *RED* in this case) by a line. Most classification tasks, however, are not that simple, and often more complex structures are needed in order to make an optimal separation, i.e., correctly classify new objects (test cases) on the basis of the examples that are available (train cases). This situation is depicted in the next illustration.

Compared to the previous schematic, it is clear that a full separation of the *GREEN* and *RED* objects require a curve (which is more complex than a line). Classification tasks based on drawing separating lines to distinguish between objects of different class memberships are known as hyperplane classifiers. Support Vector Machines are particularly suited to handle such tasks.

The next illustration shows the basic idea behind Support Vector Machines. Here we see the original objects (left side of the schematic) mapped, i.e., rearranged, using a set of mathematical functions known as kernels. The process of rearranging the objects is known as mapping (transformation). Note that in this new setting, the mapped objects (right side of the schematic) is linearly separable and, thus, instead of constructing the complex curve (left schematic), all we have to do is to find an optimal line to separate the *GREEN* and the *RED* objects.

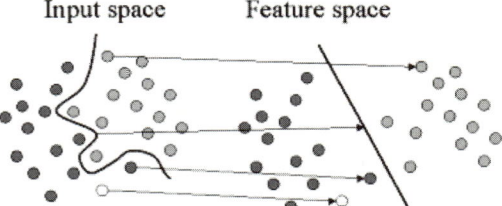

Input space   Feature space

Technical Notes

Support Vector Machine (SVM) is primarily a classier method that performs classification tasks by constructing hyperplanes in a multidimensional space that separates cases of different class labels. SVM supports both regression and classification tasks and can handle multiple continuous and categorical variables. For categorical variables, a dummy variable is created with case values as either *0* or *1*. Thus, a categorical dependent variable consisting of three levels, say (*A, B, C*), is represented by a set of three dummy variables:

A: (1 0 0), B: (0 1 0), C: (0 0 1)

To construct an optimal hyperplane, SVM employees an iterative training algorithm, which is used to minimize an error function. According to the form of the error function, SVM models can be classified into four distinct groups:

- Classification SVM Type 1 (also known as C-SVM classification)
- Classification SVM Type 2 (also known as nu-SVM classification)
- Regression SVM Type 1 (also known as epsilon-SVM regression)
- Regression SVM Type 2 (also known as nu-SVM regression)

Following is a brief summary of each model.

Classification SVM

Classification SVM Type 1. For this type of SVM, training involves the minimization of the error function:

$$\frac{1}{2} w^T w + C \sum_{i=1}^{N} \xi_i$$

subject to the constraints:

$$y_i \left(w^T \phi(x_i) + b \right) \geq 1 - \xi_i \ \text{ and } \ \xi_i \geq 0, \ \ i = 1, ..., N$$

where C is the capacity constant, $\mathbf{w}$ is the vector of coefficients, b is a constant, and ξ_i are parameters for handling nonseparable data (inputs). The index i labels the N training cases. Note that $y \in \pm 1$ is the class labels and x_i is the independent variables. The kernel ϕ is used to transform data from the input (independent) to the feature space. It should be noted that the larger the C, the more the error is penalized. Thus, C should be chosen with care to avoid over fitting.

Classification SVM Type 2. In contrast to Classification SVM Type 1, the Classification SVM Type 2 model minimizes the error function:

$$\frac{1}{2} w^T w - v\rho + \frac{1}{N} \sum_{i=1}^{N} \xi_i$$

subject to the constraints:

$$y_i \left(w^T \phi(x_i) + b \right) \geq \rho - \xi_i, \ \xi_i \geq 0, \ \ i = 1, ..., N \text{ and } \rho \geq 0$$

Regression SVM

In a regression SVM, you have to estimate the functional dependence of the dependent variable y on a set of independent variables x. It assumes, like other regression problems, that the relationship between the independent and dependent variables is given by a deterministic function f plus the addition of some additive noise:

$$y = f(x) + \text{noise}$$

The task is then to find a functional form for f that can correctly predict new cases that the SVM has not been presented with before. This can be achieved by training the *SVM* model on a sample set, i.e., training set, a process that involves, like classification (see above), the sequential optimization of an error function. Depending on the definition of this error function, two types of SVM models can be recognized.

Regression SVM Type 1. For this type of SVM the error function is:

$$\frac{1}{2} w^T w + C \sum_{i=1}^{N} \xi_i + C \sum_{i=1}^{N} \xi_i^*$$

which we minimize subject to:

$$w^T \phi(x_i) + b - y_i \leq \varepsilon + \xi_i^*$$
$$y_i - w^T \phi(x_i) - b_i \leq \varepsilon + \xi_i$$
$$\xi_i, \xi_i^* \geq 0, i = 1, ..., N$$

Regression SVM Type 2. For this SVM model, the error function is given by:

$$\frac{1}{2} w^T w - C \left(v\varepsilon + \frac{1}{N} \sum_{i=1}^{N} \left(\xi_i + \xi_i^* \right) \right)$$

which we minimize subject to:

$$\left(w^T \phi(x_i) + b\right) - y_i \le \varepsilon + \xi_i$$

$$y_i - \left(w^T \phi(x_i) + b_i\right) \le \varepsilon + \xi^*_i$$

$$\xi_i, \xi^*_i \ge 0, i = 1, ..., N, \varepsilon \ge 0$$

Kernel Functions

There are a number of kernels that can be used in Support Vector Machines models. These include linear, polynomial, radial basis function (RBF) and sigmoid:

$$\phi = \begin{cases} x_i * x_i & \text{Linear} \\ \left(\gamma x_i x_j + \text{coefficient}\right)^{\text{degree}} & \text{Polynomial} \\ \exp\left(-\gamma \mid x_i - x_j \mid^2\right) & \text{RBF} \\ \tanh\left(\gamma x_i x_j + \text{coefficient}\right) & \text{Sigmoid} \end{cases}$$

The RBF is by far the most popular choice of kernel types used in Support Vector Machines. This is mainly because of their localized and finite responses across the entire range of the real x-axis.

Naive Bayes Classifier Overview

The Naïve Bayes classifier technique is based on the so-called Bayesian theorem and is particularly suited when the dimensionality of the inputs is high. Despite its simplicity, Naïve Bayes can often outperform more sophisticated classification methods.

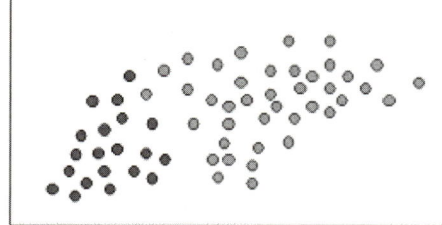

To demonstrate the concept of Naïve Bayes Classification, consider the example displayed in the illustration above. As indicated, the objects can be classified as either *GREEN* or *RED*. Our task is to classify new cases as they arrive, i.e., decide to which class label they belong, based on the currently exiting objects.

Since there are twice as many *GREEN* objects as *RED*, it is reasonable to believe that a new case (which hasn't been observed yet) is twice as likely to have membership *GREEN* rather than *RED*. In the Bayesian analysis, this belief is known as the prior probability. Prior probabilities are based on previous experience, in this case the percentage of *GREEN* and *RED* objects, and often used to predict outcomes before they actually happen.

Thus, we can write:

Prior probability for GREEN $\propto \dfrac{\text{Number of GREEN objects}}{\text{Total number of objects}}$

Prior probability for RED $\propto \dfrac{\text{Number of RED objects}}{\text{Total number of objects}}$

Since there is a total of 60 objects, 40 of which are *GREEN* and 20 *RED*, our prior probabilities for class membership are:

Prior probability for GREEN $\propto \dfrac{40}{60}$

Prior probability for RED $\propto \dfrac{20}{60}$

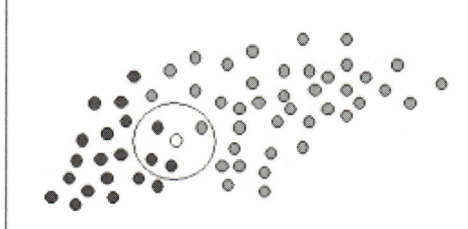

Having formulated our prior probability, we are now ready to classify a new object (*WHITE* circle). Since the objects are well clustered, it is reasonable to assume that the more *GREEN* (or *RED*) objects in the vicinity of *X*, the more likely that the new cases belong to that particular color. To measure this likelihood, we draw a circle around X which encompasses a number (to be chosen *a priori*) of points irrespective of their class labels. Then we calculate the number of points in the circle belonging to each class label. From this we calculate the likelihood:

$Likelihood\ of\ X\ given\ GREEN \propto \dfrac{Number\ of\ GREEN\ in\ the\ vicinity\ of\ X}{Total\ number\ of\ GREEN\ cases}$

$Likelihood\ of\ X\ given\ RED \propto \dfrac{Number\ of\ RED\ in\ the\ vicinity\ of\ X}{Total\ number\ of\ RED\ cases}$

From the illustration above, it is clear that *Likelihood of X given GREEN* is larger than *Likelihood of X given RED*, since the circle encompasses 1 *GREEN* object and 3 *RED* ones. Thus:

$\Pr obability\ of\ X\ given\ GREEN \propto \dfrac{1}{40}$

$\Pr obability\ of\ X\ given\ RED \propto \dfrac{3}{40}$

Although the prior probabilities indicate that X may belong to *GREEN* (given that there are twice as many *GREEN* compared to *RED*) the likelihood indicates otherwise; that the class membership of X is *RED* (given that there are more *RED* objects in the vicinity of X than *GREEN*). In the Bayesian analysis, the final classification is produced by combining both sources of information, i.e., the prior and the likelihood, to form a posterior probability using the so-called Bayes' rule (named after Rev. Thomas Bayes 1702-1761).

$Posterior\ probability\ of\ X\ being\ GREEN \propto$
$\quad Prior\ probability\ of\ GREEN \times Likelihood\ of\ X\ given\ GREEN$

$= \dfrac{4}{6} \times \dfrac{1}{40} = \dfrac{1}{60}$

$Posterior\ probability\ of\ X\ being\ RED \propto$
$\quad Prior\ probability\ of\ RED \times Likelihood\ of\ X\ given\ RED$

$= \dfrac{2}{6} \times \dfrac{3}{40} = \dfrac{1}{40}$

Finally, we classify X as *RED* since its class membership achieves the largest posterior probability.

Note. The above probabilities are not normalized. However, this does not affect the classification outcome since their normalizing constants are the same.

Technical Notes

In the previous section, we provided an intuitive example for understanding classification using Naive Bayes. In this section are further details of the technical issues involved. Naive Bayes classifiers can handle an arbitrary number of independent variables whether continuous or categorical. Given a set of variables, $X = \{x_1, x_2, x..., x_d\}$, we want to construct the posterior probability for the event C_j among a set of possible outcomes $C = \{c_1, c_2, c..., c_d\}$. In a more familiar language, X is the set of predictors and C is the set of categorical levels present in the dependent variable. Using Bayes' rule:

$$p\left(C_j \mid x_1, x_2, ..., x_d\right) \propto p\left(x_1, x_2, ..., x_d \mid C_j\right) p\left(C_j\right)$$

where $p(C_j \mid x_1, x_2, x..., x_d)$ is the posterior probability of class membership, i.e., the probability that X belongs to C_j. Since Naive Bayes assumes that the conditional probabilities of the independent variables are statistically independent we can decompose the likelihood to a product of terms:

$$p\left(X \mid C_j\right) \propto \prod_{k=1}^{d} p\left(x_k \mid C_j\right)$$

and rewrite the posterior as:

$$p\left(C_j \mid X\right) \propto p\left(C_j\right) \prod_{k=1}^{d} p\left(x_k \mid C_j\right)$$

Using Bayes' rule above, we label a new case X with a class level C_j that achieves the highest posterior probability.

Although the assumption that the predictor (independent) variables are independent is not always accurate, it does simplify the classification task dramatically, since it allows the class conditional densities $p(x_k \mid C_j)$ to be calculated separately for each variable, i.e., it reduces a multidimensional task to a number of one-dimensional ones. In effect, Naive Bayes reduces a high-dimensional density estimation task to a one-dimensional kernel density estimation. Furthermore, the assumption does not seem to greatly affect the posterior probabilities, especially in regions near decision boundaries, thus, leaving the classification task unaffected.

Naive Bayes can be modeled in several different ways including normal, lognormal, *gamma* and Poisson density functions:

Note that the indices k and j should be read as

$$
p(x_k \mid C_j) = \begin{cases}
\dfrac{1}{\sigma_{kj}\sqrt{2\pi}} \exp\left(\dfrac{-\left(x-\mu_{kj}\right)^2}{2\sigma_{kj}}\right), & -\infty < x < \infty, -\infty < \mu_{kj} <, \sigma_{kj} > 0 \quad \text{Normal} \\[4pt]
\quad \mu_{kj} : \text{mean}, \sigma_{kj} : \text{standard deviation} \\[6pt]
\dfrac{1}{x\sigma_{kj}(2\pi)^{1/2}} \exp\left\{\dfrac{-\left[\log\left(x/m_{kj}\right)\right]^2}{2\sigma_{kj}^2}\right\}, & 0 < x < \infty, m_{kj} > 0, \sigma_{kj} > 0 \quad \text{Lognormal} \\[4pt]
\quad m_{kj} : \text{scale parameter}, \sigma_{kj} : \text{shape parameter} \\[6pt]
\dfrac{\left(\dfrac{x}{b_{kj}}\right)^{c_{kj}-1}}{b_{kj}\Gamma\left(c_{kj}\right)} \exp\left(\dfrac{-x}{b_{kj}}\right), & 0 \le x < \infty, b_{kj} > 0, c_{kj} > 0 \quad \text{Gamma} \\[4pt]
\quad b_{kj} : \text{scale parameter}, c_{kj} : \text{shape parameter} \\[6pt]
\dfrac{\lambda_{kj} \exp\left(-\lambda_{kj}\right)}{x!}, & 0 \le x < \infty, \lambda_{kj} > 0, x = 0,1,2,\ldots \quad \text{Poisson} \\[4pt]
\quad \lambda_{kj} : \text{mean}
\end{cases}
$$

follows: take μ_{kj} of the normal distribution, for example. For $k=1$ and $j=2$, μ_{12} is simply the mean of the normal distribution of the *1st* independent variable conditioned on the *2nd* categorical level C_l of the dependent variable. This simply means μ_{12} is the average of the *1st* independent variable for which the dependent variable entries belonged to the *2nd* categorical level C_l. Similarly, σ_{12} is the standard deviation of the normal distribution of the *1st* independent variable conditioned on the *2nd* categorical level C_2 of the dependent variable. Thus $p(x_k \mid C_j)$ is the distribution of the *1st* independent variable conditioned on the *2nd* categorical level C_l of the dependent variable.

Note. Poisson variables are regarded here as continuous since they are ordinal rather than truly categorical. For categorical variables, a discrete probability is used with values of the categorical level being proportional to their conditional frequency in the training data.

k-Nearest Neighbors Overview

Classification

To demonstrate a *k*-nearest neighbor analysis, let's consider the task of classifying a new object (query point) among a number of known examples. This is shown in the figure below, which depicts the examples (instances) with the plus and minus signs and the query point with a red circle. Our task is to estimate (classify) the outcome of the query point based on a selected number of its nearest neighbors. In other words, we want to know whether the query point can be classified as a plus or a minus sign.

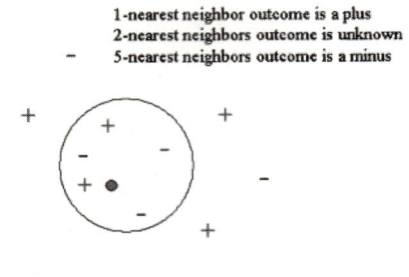

1-nearest neighbor outcome is a plus
2-nearest neighbors outcome is unknown
5-nearest neighbors outcome is a minus

To proceed, let's consider the outcome of KNN based on 1-nearest neighbor. It is clear that in this case KNN will predict the outcome of the query point with a plus (since the closest point carries a plus sign). Now let's increase the number of nearest neighbors to 2, i.e., 2-nearest neighbors. This time KNN will not be able to classify the outcome of the query point since the second closest point is a minus, and so both the plus and the minus signs achieve the same score (i.e., win the same number of votes). For the next step, let's increase the number of nearest neighbors to 5 (5-nearest neighbors). This will define a nearest neighbor region, which is indicated by the circle shown in the illustration above. Since there are 2 and 3 plus and minus signs, respectively, in this circle KNN will assign a minus sign to the outcome of the query point.

Regression

In this section we will generalize the concept of k-nearest neighbors to include regression problems. Regression problems are concerned with predicting the outcome of a dependent variable given a set of independent variables. To start with, we consider the schematic shown above, where a set of points (green squares) are drawn from the relationship between the independent variable x and the dependent

variable y (red curve). Given the set of green objects (known as examples) we use the k-nearest neighbors method to predict the outcome of X (also known as query point) given the example set (green squares).

To begin with, let's consider the 1-nearest neighbor method as an example. In this case we search the example set (green squares) and locate the one closest to the query point X. For this particular case, this happens to be x_4. The outcome of x_4 (i.e., y_4) is thus then taken to be the answer for the outcome of X (i.e., Y). Thus for 1-nearest neighbor we can write

$$Y = y_4$$

For the next step, let's consider the 2-nearest neighbor method. In this case, we locate the first two closest points to X, which happen to be y_3 and y_4. Taking the average of their outcome, the solution for Y is then given by:

$$Y = \frac{y_3 + y_4}{2}$$

The above discussion can be extended to an arbitrary number of nearest neighbors k. To summarize, in a k-nearest neighbor method, the outcome Y of the query point X is taken to be the average of the outcomes of its k-nearest neighbors.

Technical Details

k-Nearest Neighbors (KNN) is a memory-based model defined by a set of objects known as examples (also known as instances) for which the outcome are known (i.e., the examples are labeled). Each example consists of a data case having a set of independent values labeled by a set of dependent outcomes. The independent and dependent variables can be either continuous or categorical. For continuous dependent variables, the task is regression;

otherwise it is a classification. Thus, KNN can handle both regression and classification tasks.

Given a new case of dependent values (query point), we would like to estimate the outcome based on the KNN examples. KNN achieves this by finding k examples that are closest in distance to the query point, hence, the name k-Nearest Neighbors. For regression problems, KNN predictions are based on averaging the outcomes of the k nearest neighbors; for classification problems, a majority of voting is used.

The choice of k is essential in building the KNN model. In fact, k can be regarded as one of the most important factors of the model that can strongly influence the quality of predictions. One appropriate way to look at the number of nearest neighbors k is to think of it as a smoothing parameter. For any given problem, a small value of k will lead to a large variance in predictions. Alternatively, setting k to a large value may lead to a large model bias. Thus, k should be set to a value large enough to minimize the probability of misclassification and small enough (with respect to the number of cases in the example sample) so that the k nearest points are close enough to the query point. Thus, like any smoothing parameter, there is an optimal value for k that achieves the right trade off between the bias and the variance of the model. KNN can provide an estimate of k using an algorithm known as cross-validation (Bishop, 1995).

Cross-Validation. Cross-validation is a well-established technique that can be used to obtain estimates of model parameters that are unknown. Here we discuss the applicability of this technique to estimating k.

The general idea of this method is to divide the data sample into a number of v folds (randomly drawn, disjointed sub-samples or segments). For a fixed value of k, we apply the KNN model to make predictions on the vth segment (i.e., use the v-1 segments as the examples) and evaluate the error. The most common choice for this error for regression is sum-of-squared and for classification it is most conveniently defined as the accuracy (the percentage of correctly classified cases). This process is then successively applied to all possible choices of v. At the end of the v folds (cycles), the computed errors are averaged to yield a measure of the stability of the model (how well the model predicts query points). The above steps are then repeated for various k and the value achieving the lowest error (or the highest classification accuracy) is then selected as the optimal value for k (optimal in a cross-validation sense). Note that cross-validation is computationally. Alternatively, you can specify k. This may be a reasonable course of action should you have an idea of which value k may take (i.e., from previous KNN analyses you may have conducted on similar data).

Distance metric. As mentioned before, given a query point, *KNN* makes predictions based on the outcome of the k neighbors closest to that point. Therefore, to make predictions with KNN, we need to define a metric for measuring the distance between the query point and cases from the examples sample. One of the most popular choices to measure this distance is known as Euclidean. Other measures include Euclidean squared, City-block, and Chebychev:

$$D(x, p) = \begin{cases} \sqrt{(x-p)^2} & Euclidean \\ (x-p)^2 & Euclidean\ squared \\ Abs(x-p) & Cityblock \\ Max(|x-p|) & Chebychev \end{cases}$$

where x and p are the query point and a case from the examples sample, respectively.

k-Nearest Neighbor Predictions. After selecting the value of k, you can make predictions based on the KNN examples. For regression, KNN predictions is the average of the k nearest neighbors outcome.

$$y = \frac{1}{K}\sum_{i=1}^{K} y_i$$

where y_i is the ith case of the examples sample and y is the prediction (outcome) of the query point. In contrast to regression, in classification problems, KNN predictions are based on a voting scheme in which the winner is used to label the query.

Note. For binary classification tasks, odd values of $y = 1,3,5$ are used to avoid ties, i.e., two classes labels achieving the same score.

So far we have discussed KNN analysis without paying any attention to the relative distance of the k-nearest examples to the query point. In other words, we let the k neighbors have equal influence on predictions irrespective of their relative distance from the query point. An alternative approach (Shepard 1968) is to use arbitrarily large values of k (if not the entire prototype sample) with more importance given to cases closest to the query point. This is achieved using so-called distance weighting.

Distance weighting. Since KNN predictions are based on the intuitive assumption that objects close in distance are potentially similar, it makes good sense to discriminate between the k-nearest neighbors when making predictions, i.e., let the closest points among the k-nearest neighbors have more say in affecting the outcome of the query point. This can be achieved by introducing a set of weights W, one for each nearest neighbor, defined by the relative closeness of each neighbor with respect to the query point. Thus:

$$W(x, p_i) = \frac{\exp(-D(x, p_i))}{\sum_{i=1}^{K} \exp(-D(x, p_i))}$$

where $D(x, p_i)$ is the distance between the query point x and the ith case p_i of the example sample. It is clear that the weights defined in this manner above will satisfy:

$$\sum_{i=1}^{K} W(x_o, x_i) = 1$$

Thus, for regression problems, we have:

$$y = \sum_{i=1}^{K} W(x_o, x_i) y_i$$

For classification problems, the maximum of the above equation is taken for each class variables.

It is clear from the above discussion that when $k>1$, one can naturally define the standard deviation for predictions in regression tasks using:

$$error\ bar = \mp\sqrt{\frac{1}{K-1}\sum_{i=1}^{K}(y - y_i)^2}$$

MULTIVARIATE ADAPTIVE REGRESSION SPLINES (MARSPLINES)

Multivariate adaptive regression splines (MARSplines) is an implementation of techniques popularized by Friedman (1991) for solving regression-type problems (see also, Chapter 26 – *Multiple Linear Regression*), with the main purpose to predict the values of a continuous dependent or outcome variable from a set of independent or predictor variables. There are a large number of methods available for fitting models to continuous variables, such as a linear regression [e.g., multiple regression, general linear models (GLM)], nonlinear regression (generalized linear/nonlinear models), regression trees (see Chapter 8 – *Classification and Regression Trees*), CHAID, neural networks, etc. (see also, Hastie, Tishirani, and Friedman, 2001, for an overview).

Overview

Multivariate adaptive regression splines (MARSplines) is a nonparametric regression procedure that makes no assumption about the underlying functional relationship between the dependent and independent variables. Instead, MARSplines constructs this relation from a set of coefficients and basis functions that are entirely "driven" from the regression data. In a sense, the method is based on the divide-and-conquer strategy, which partitions the input space into regions, each with its own regression equation. This makes MARSplines particularly suitable for problems with higher input dimensions (i.e., with more than 2 variables), where the curse of dimensionality would likely create problems for other techniques.

The MARSplines technique has become particularly popular in the area of data mining because it does not assume or impose any particular type or class of relationship (e.g., linear, logistic, etc.) between the predictor variables and the dependent (outcome) variable of interest. Instead, useful models (i.e., models that yield accurate predictions) can be derived even in situations where the relationship between the predictors and the dependent variables is non-monotone and difficult to approximate with parametric models. For more information about this technique and how it compares to other methods for nonlinear regression (or regression trees), see Hastie, Tishirani, and Friedman (2001).

Regression Problems

Regression problems are used to determine the relationship between a set of dependent variables (also called output, outcome, or response variables) and one or more independent variables (also known as input or predictor variables). The dependent variable is the one whose values you want to predict, based on the values of the independent (predictor) variables. For instance, you might be interested in the number of car accidents on the roads, which can be caused by 1) bad weather and 2) drunk driving. In this case you might write, for example,

Number_of_Accidents = Some Constant + 0.5*Bad_Weather + 2.0*Drunk_Driving

The variable *Number of Accidents* is the dependent variable that is thought to be caused by (among other variables) *Bad Weather* and *Drunk Driving* (hence the name dependent variable). Note that the independent variables are multiplied by factors, i.e., *0.5* and *2.0*. These are known as regression coefficients. The larger these coefficients, the stronger the influence of the independent variables on the dependent variable. If the two predictors in this simple (fictitious) example were measured on the same scale (e.g., if the variables were standardized to a mean of *0.0* and standard deviation *1.0*), *Drunk Driving* could be inferred to contribute 4 times more to car accidents than *Bad Weather*. (If the variables are not measured on the same scale, direct comparisons between these coefficients are not meaningful, and, usually, some other standardized measure of predictor importance is included in the results.)

For additional details regarding these types of statistical models, refer to Chapter 18 – *General Linear Models (GLM)*, Chapter 19 – *General Regression Models (GRM)*, or Chapter 26 – *Multiple Linear Regression*. In general, in the social and natural sciences, regression procedures are widely used in research. Regression enables the researcher to ask (and hopefully answer) the general question "what is the best predictor of ..." For example, educational researchers might want to learn what the best predictors of success in high-school are. Psychologists may want to determine which personality variable best predicts social adjustment. Sociologists may want to find out which of the multiple social indicators best predict whether a new immigrant group will adapt and be absorbed into society.

Multivariate Adaptive Regression Splines

The car accident example we considered previously is a typical application for linear regression, where the response variable is hypothesized to depend linearly on the predictor variables. Linear regression also falls into the category of so-called parametric regression, which assumes that the nature of the relationships (but not the specific parameters) between the dependent and independent variables is known *a priori* (e.g., is linear). By contrast, nonparametric regression (see Chapter 29 – *Nonparametric Statistics*) does not make any such assumption as to how the dependent variables are related to the predictors. Instead it allows the regression function to be "driven" directly from data.

Multivariate adaptive regression splines (MARSplines) is a nonparametric regression procedure that makes no assumption about the underlying functional relationship between the dependent and independent variables. Instead, MARSplines constructs this relation from a set of coefficients and so-called basis functions that are entirely determined from the regression data. You can think of the general "mechanism by which the MARSplines algorithm operates as multiple piecewise linear regression (see Chapter 28 – *Nonlinear Estimation*), where each breakpoint (estimated from the data)

defines the region of application for a particular (very simple) linear regression equation.

Basis functions. Specifically, MARSplines uses two-sided truncated functions of the form (as shown in the next illustration) as basis functions for linear or nonlinear expansion, which approximates the relationships between the response and predictor variables.

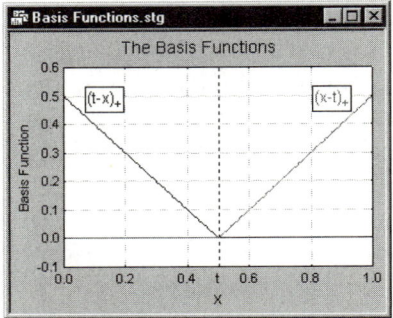

Shown above is a simple example of two basis functions (t-x)+ and (x-t)+ (adapted from Hastie, et al., 2001, Figure 9.9). Parameter *t* is the knot of the basis functions (defining the pieces of the piecewise linear regression); these knots (parameters) are also determined from the data. The "+" signs next to the terms *(t-x)* and *(x-t)* simply denote that only positive results of the respective equations are considered; otherwise the respective functions evaluate to zero. This can also be seen in the illustration.

The MARSplines model. The basis functions together with the model parameters (estimated via least squares estimation) are combined to produce the predictions given the inputs. The general MARSplines model equation (see Hastie et al., 2001, equation 9.19) is given as:

$$y = f(X) = \beta_0 + \sum_{m=1}^{M} \beta_m h_m(X)$$

where the summation is over the M nonconstant terms in the model (further details regarding the model are also provided in *Technical Notes*, page 332). To summarize, *y* is predicted as a function of the predictor variables *X* (and their interactions); this function consists of an intercept parameter (β_o) and the weighted (by β_m) sum of one or more basis functions $h_m(X)$, of the kind illustrated earlier. You can also think of this model as selecting a weighted sum of basis functions from the set of (a large number of) basis functions that span all values of each predictor (i.e., that set would consist of one basis function, and parameter *t*, for each distinct value for each predictor variable). The MARSplines algorithm then searches over the space of all inputs and predictor values (knot locations t) as well as interactions between variables. During this search, an increasingly larger number of basis functions are added to the model (selected from the set of possible basis functions), to maximize an overall least squares goodness-of-fit criterion. As a result of these operations, MARSplines automatically determines the most important independent variables as well as the most significant interactions among them. The details of this algorithm are further described in *Technical Notes*, as well as in Hastie et al., 2001).

Categorical predictors. In practice, both continuous and categorical predictors could be used, and will often yield useful results. However, the basic MARSplines algorithm assumes that the predictor variables are continuous in nature, and, for example, the computed knots program will usually not coincide with actual class codes found in the categorical predictors. For a detailed discussion of categorical predictor variables in MARSplines, see Friedman (1993).

Multiple dependent (outcome) variables. The MARSplines algorithm can be applied to multiple dependent (outcome) variables. In this case, the algorithm will determine a common set of basis functions in the predictors, but estimate different coefficients for each dependent variable. This method of treating multiple outcome variables is not unlike some neural networks architectures, where multiple outcome variables can be predicted from common neurons and hidden layers; in the case of MARSplines, multiple outcome variables are predicted from common basis functions, with different coefficients.

MARSplines and classification problems. Because MARSplines can handle multiple dependent variables, it is easy to apply the algorithm to classification problems as well. First, code the classes in the categorical response variable into multiple indicator variables (e.g., 1 = observation belongs to class k, 0 = observation does not belong to class k); then apply the MARSplines algorithm to fit a model, and compute predicted (continuous) values or scores; finally, for prediction, assign each case to the class for which the highest score is predicted (see also, Hastie, Tibshirani, and Freedman, 2001, for a description of this procedure). Note that this type of application will yield heuristic classifications that may work very well in practice, but is not based on a statistical model for deriving classification probabilities.

Model Selection and Pruning

In general, nonparametric models are adaptive and can exhibit a high degree of flexibility that can ultimately result in overfitting if no measures are taken to counteract it. Although such models can achieve zero error on training data, they have the tendency to perform poorly when presented with new observations or instances (i.e., they do not generalize well to the prediction of new cases). MARSplines, as most methods of this kind, tend to overfit the data as well. To combat this problem, MARSplines uses a pruning technique (similar to pruning in classification trees) to limit the complexity of the model by reducing the number of its basis functions.

MARSplines as a predictor (feature) selection method. This feature – the selection of and pruning of basis functions – makes this method a very powerful tool for predictor selection. The MARSplines algorithm will pick up only those basis functions (and those predictor variables) that make a sizeable contribution to the prediction (refer to *Technical Notes* for details).

Applications

Multivariate adaptive regression splines (MARSplines) have become very popular recently for finding predictive models for difficult data mining problems, i.e., when the predictor variables do not exhibit simple and/or monotone relationships to the dependent variable of interest. Alternative models or approaches that you can consider for such cases are CHAID, classification and regression trees, or any of the many neural networks architectures available. Because of the specific manner in which MARSplines selects predictors (basis functions) for the model, it does generally well in situations where regression tree models are also appropriate, i.e., where hierarchically organized successive splits on the predictor variables yield good (accurate) predictions. In fact, instead of considering this technique as a generalization of multiple regression (as it was presented in this introduction), you can consider MARSplines as a generalization of regression trees, where the "hard" binary splits are replaced by "smooth" basis functions. Refer to

Hastie, Tibshirani, and Friedman (2001) for additional details.

The MARSplines Algorithm

Implementing MARSplines involves a two-step procedure that is applied successively until a desired model is found. In the first step, we build the model, i.e. increase its complexity by adding basis functions until a preset (user-defined) maximum level of complexity has been reached. Then we begin a backward procedure to remove the least significant basis functions from the model, i.e. those whose removal will lead to the least reduction in the (least-squares) goodness of fit. This algorithm is implemented as follows:

Start with the simplest model involving only the constant basis function.

Search the space of basis functions, for each variable and for all possible knots, and add those that maximize a certain measure of goodness of fit (minimize prediction error).

Step 2 is recursively applied until a model of pre-determined maximum complexity is derived.

Finally, in the last stage, a pruning procedure is applied where those basis functions are removed that contribute least to the overall (least squares) goodness of fit.

The MARSplines Model

The MARSplines algorithm builds models from two sided truncated functions of the predictors (x) of the form:

$$(x - t)_+ = \begin{cases} x - t & x > t \\ 0 & otherwise \end{cases}$$

These serve as basis functions for linear or nonlinear expansion that approximates some true underlying function $f(x)$.

The MARSplines model for a dependent (outcome) variable y, and M terms, can be summarized in the following equation:

$$y = f(x) = \beta_0 + \sum_{m=1}^{M} \beta_m H_{km}(x_{v(k,m)})$$

where the summation is over the M terms in the model, and β_o and β_m are parameters of the model (along with the knots t for each basis function, which are also estimated from the data). Function H is defined as:

$$H_{km}(x_{v(k,m)}) = \prod_{k=1}^{K} h_{km}$$

where $x_{v(k,m)}$ is the predictor in the k'th of the m'th product. For order of interactions $K=1$, the model is additive and for $K=2$ the model pairwise interactive.

During forward stepwise, a number of basis functions are added to the model according to a pre-determined maximum that should be considerably larger (twice as much at least) than the optimal (best least-squares fit).

After implementing the forward stepwise selection of basis functions, a backward procedure is applied in which the model is pruned by removing those basis functions that are associated with the smallest increase in the (least squares) goodness of fit. A least squares error function (inverse of goodness-of-fit) is computed. The so-called generalized cross validation error is a measure of the goodness of fit that takes into account not only the residual error but also the model complexity as well. It is given by:

$$GCV = \frac{\sum_{i=1}^{N}(y_i - f(x_i))^2}{\left(1 - \frac{C}{N}\right)^2}$$

with

C = 1 + cd

where N is the number of cases in the data set, d is the effective degrees of freedom, which is equal to the number of independent basis functions. The quantity c is the penalty for adding a basis function. Experiments have shown that the best value for C can be found somewhere in the range $2 < d < 3$ (see Hastie et al., 2001).

MULTIDIMENSIONAL SCALING (MDS)

Multidimensional scaling (MDS) can be considered to be an alternative to factor analysis (see Chapter 16 – *Factor Analysis*). In general, the goal of the analysis is to detect meaningful underlying dimensions that enable the researcher to explain observed similarities or dissimilarities (distances) between the investigated objects. In factor analysis, the similarities between objects (e.g., variables) are expressed in the correlation matrix. With MDS, you can analyze any kind of similarity or dissimilarity matrix, in addition to correlation matrices.

Logic of Multidimensional Scaling

This simple example may demonstrate the logic of a multidimensional scaling (MDS) analysis. Suppose we take a matrix of distances between major US cities from a map. We then analyze this matrix, specifying that we want to reproduce the distances based on two dimensions. As a result of the MDS analysis, we would most likely obtain a two-dimensional representation of the locations of the cities, that is, we would basically obtain a two-dimensional map.

In general then, MDS attempts to arrange objects (major cities in this example) in a space with a particular number of dimensions (two-dimensional in this example) so as to reproduce the observed distances. As a result, we can explain the distances in terms of underlying dimensions; in our example, we could explain the distances in terms of the two geographical dimensions: north/south and east/west.

Orientation of axes. As in factor analysis, the actual orientation of axes in the final solution is arbitrary. To return to our example, we could rotate the map in any way we want; the distances between cities remain the same. Thus, the final orientation of axes in the plane or space is mostly the result of a subjective decision by the researcher, who will choose an orientation that can be most easily explained. To return to our example, we could have chosen an orientation of axes other than north/south and east/west; however, that orientation is most convenient because it makes the most sense (i.e., it is easily interpretable).

Computational Approach

MDS is not so much an exact procedure as rather a way to rearrange objects in an efficient manner, so as to arrive at a configuration that best approximates the observed distances. It actually moves objects around in the space defined by the specified number of dimensions, and checks how well the distances between objects can be reproduced by the new configuration. In more technical terms, it uses a function minimization algorithm that evaluates different configurations with the goal of maximizing the goodness-of-fit (or minimizing lack of fit).

Measures of goodness-of-fit: Stress. The most common measure that is used to evaluate how well (or poorly) a particular configuration reproduces the observed distance matrix is the stress measure. The raw stress value *phi* of a configuration is defined by:

$$\text{Phi} = \Sigma\,[d_{ij} - f\,(\delta_{ij})]^2$$

In this formula, d_{ij} stands for the reproduced distances, given the respective number of dimensions, and δ_{ij} (*delta*$_{ij}$) stands for the input data (i.e., observed distances). The expression $f\,(\delta_{ij})$ indicates a nonmetric, monotone transformation of the observed input data (distances). Thus, it will attempt to reproduce the general rank ordering of distances between the objects in the analysis.

There are several similar related measures that are commonly used; however, most of them amount to the computation of the sum of squared deviations of observed distances (or some monotone transformation of those distances) from the reproduced distances. Thus, the smaller the stress value, the better is the fit of the reproduced distance matrix to the observed distance matrix.

Shepard diagram. You can plot the reproduced distances for a particular number of dimensions against the observed input data (distances). This

scatterplot is referred to as a Shepard diagram. This plot shows the reproduced distances plotted on the vertical (Y) axis versus the original similarities plotted on the horizontal (X) axis (hence, the generally negative slope). This plot also shows a step-function. This line represents the so-called *D-hat* values, that is, the result of the monotone transformation $f(\delta_{ij})$ of the input data. If all reproduced distances fall onto the step-line, the rank ordering of distances (or similarities) would be perfectly reproduced by the respective solution (dimensional model). Deviations from the step-line indicate lack of fit.

How Many Dimensions to Specify?

If you are familiar with factor analysis, you will be quite aware of this issue. If you are not familiar with factor analysis, you may want to read Chapter 16 – *Factor Analysis*; however, this is not necessary in order to understand the following discussion. In general, the more dimensions we use in order to reproduce the distance matrix, the better is the fit of the reproduced matrix to the observed matrix (i.e., the smaller is the stress). In fact, if we use as many dimensions as there are variables, we can perfectly reproduce the observed distance matrix. Of course, our goal is to reduce the observed complexity of nature, that is, to explain the distance matrix in terms of fewer underlying dimensions. To return to the example of distances between cities, once we have a two-dimensional map it is much easier to visualize the location of and navigate between cities, as compared to relying on the distance matrix only.

Sources of misfit. Let's consider for a moment why fewer factors may produce a worse representation of a distance matrix than would

more factors. Imagine the three cities *A, B,* and *C,* and the three cities *D, E,* and *F;* shown in the next illustrations are their distances from each other.

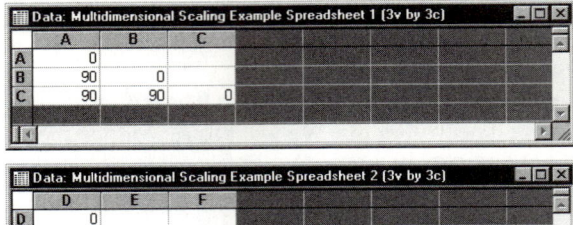

In the first matrix, all cities are exactly 90 miles apart from each other; in the second matrix, cities *D* and *F* are 180 miles apart. Now, can we arrange the three cities (objects) on one dimension (line)? Indeed, we can arrange cities *D, E,* and *F* on one dimension:

D---90 miles---E---90 miles---F

D is 90 miles away from *E,* and *E* is 90 miles away form *F;* thus, *D* is 90+90=180 miles away from *F.* If you try to do the same thing with cities *A, B,* and *C* you will see that there is no way to arrange the three cities on one line so that the distances can be reproduced. However, we can arrange those cities in two dimensions, in the shape of a triangle:

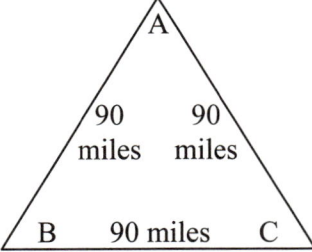

Arranging the three cities in this manner, we can perfectly reproduce the distances between them. Without going into much detail, this small example illustrates how a particular distance matrix implies a particular number of dimensions. Of course, real data are never this clean, and contain a lot of noise, that is, random variability that contributes to the differences between the reproduced and observed matrix.

Scree test. A common way to decide how many dimensions to use is to plot the stress value against different numbers of dimensions. This test was first proposed by Cattell (1966) in the context of the number-of-factors problem in factor analysis (see Chapter 16 – *Factor Analysis*); Kruskal and Wish (1978; pp. 53-60) discuss the application of this plot to MDS.

Cattell suggests finding the place where the smooth decrease of stress values (eigenvalues in factor analysis) appears to level off to the right of the plot. To the right of this point you find, presumably, only factorial scree – scree is the geological term referring to the debris that collects on the lower part of a rocky slope.

Interpretability of configuration. A second criterion for deciding how many dimensions to interpret is the clarity of the final configuration. Sometimes, as in our example of distances between cities, the resultant dimensions are easily interpreted. At other times, the points in the plot form a sort of "random cloud," and there is no straightforward and easy way to interpret the dimensions. In the latter case, you should try to include more or fewer dimensions and examine the resultant final configurations. Often, more interpretable solutions emerge. However, if the data points in the plot do not follow any pattern, and if the stress plot does not show any clear "elbow," the data are most likely random noise.

Interpreting the Dimensions

The interpretation of dimensions usually represents the final step of the analysis. As mentioned earlier, the actual orientations of the axes from the MDS analysis are arbitrary, and can be rotated in any direction. A first step is to produce scatterplots of the objects in the different two-dimensional planes.

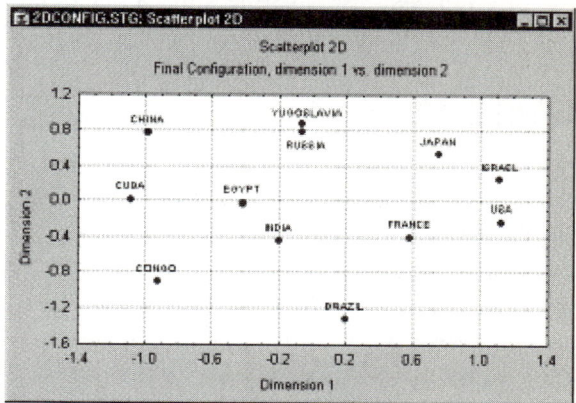

Three-dimensional solutions can also be illustrated graphically, however, their interpretation is somewhat more complex.

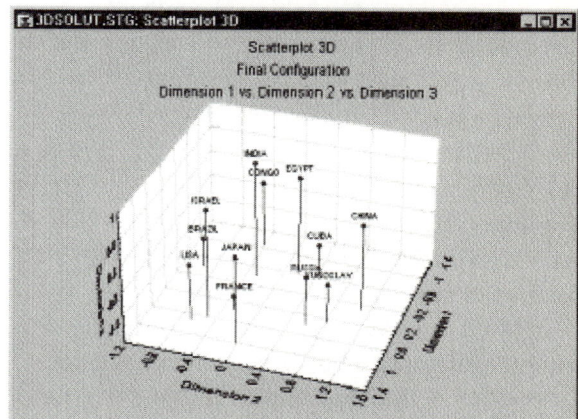

In addition to meaningful dimensions, you should also look for clusters of points or particular patterns and configurations (such as circles, manifolds, etc.). For a detailed discussion of how to interpret final configurations, see Borg and Lingoes (1987), Borg and Shye (in press), or Guttman (1968).

Use of multiple regression techniques. An analytical way of interpreting dimensions (described in Kruskal & Wish, 1978) is to use multiple regression techniques to regress some meaningful variables on the coordinates for the different dimensions. Note that this can be done easily via multiple regression.

Applications

The beauty of MDS is that we can analyze any kind of distance or similarity matrix. These similarities can represent people's ratings of similarities between objects, the percent agreement between judges, the number of times a subjects fails to discriminate between stimuli, etc. For example, MDS methods used to be very popular in psychological research on person perception where similarities between trait descriptors were analyzed to uncover the underlying dimensionality of people's perceptions of traits (see, for example Rosenberg, 1977). They are also very popular in marketing research, in order to detect the number and nature of dimensions underlying the perceptions of different brands or products & Carmone, 1970).

In general, MDS methods enable the researcher to ask relatively unobtrusive questions ("how similar is brand A to brand B") and to derive from those questions underlying dimensions without the respondents ever knowing what is the researcher's real interest.

MDS and Factor Analysis

Even though there are similarities in the type of research questions to which these two procedures can be applied, MDS and factor analysis are fundamentally different methods. Factor analysis requires that the underlying data are distributed as multivariate normal, and that the relationships are linear. MDS imposes no such restrictions. As long as the rank ordering of distances (or similarities) in the matrix is meaningful, MDS can be used. In terms of resultant differences, factor analysis tends to extract more factors (dimensions) than MDS; as a result, MDS often yields more readily, interpretable solutions. Most importantly, however, MDS can be applied to any kind of distances or similarities, while factor analysis requires us to first compute a correlation matrix. MDS can be based on subjects' direct assessment of similarities between stimuli, while factor analysis requires subjects to rate those stimuli on some list of attributes (for which the factor analysis is performed).

In summary, MDS methods are applicable to a wide variety of research designs because distance measures can be obtained in any number of ways (for different examples, refer to the references provided at the beginning of this chapter).

MULTIPLE LINEAR REGRESSION

The general purpose of multiple regression (the term was first used by Pearson, 1908) is to analyze the relationship between several independent or predictor variables and a dependent or criterion variable. The computational problem that needs to be solved in multiple regression analysis is to fit a straight line (or plane in an n-dimensional space, where n is the number of independent variables) to a number of points. In the simplest case - one dependent and one independent variable - one can visualize this in a scatterplot (scatterplots are two-dimensional plots of the scores on a pair of variables). It is used as either a hypothesis testing or exploratory method.

Overview

The general purpose of multiple regression (the term was first used by Pearson, 1908) is to learn more about the relationship between several independent or predictor variables and a dependent or criterion variable. For example, a real estate agent might record for each listing the size of the house, the number of bedrooms, the average income in the neighborhood according to census data, and a subjective rating of appeal of the house. Once this information has been compiled, it would be interesting to see whether and how these measures relate to the price for which a house is sold. For example, you might learn that the number of bedrooms is a better predictor of the price in a particular neighborhood than how pretty the house is (subjective rating). You can also detect outliers, that is, houses that should really sell for more, given their location and characteristics.

Personnel professionals customarily use multiple regression procedures to determine equitable compensation. You can determine a number of factors or dimensions such as "amount of responsibility" (*Resp*) or "number of people to supervise" (*No_Super*) that you believe to contribute to the value of a job. The personnel analyst then usually conducts a salary survey among comparable companies in the market, recording the salaries and respective characteristics (i.e., values on dimensions) for different positions. This information can be used in a multiple regression analysis to build a regression equation of the form:

Salary = .5*Resp + .8*No_Super

Once this regression line has been determined, the analyst can easily construct a graph of the expected (predicted) salaries and the actual salaries of job incumbents in the company. Thus, the analyst is able to determine which position is underpaid (below the regression line) or overpaid (above the regression line), or paid equitably.

In the social and natural sciences, multiple regression procedures are widely used in research. In general, multiple regression enables the researcher to ask (and hopefully answer) the general question "what is the best predictor of ...". For example, educational researchers might want to learn what the best predictors of success in high school are. Psychologists may want to determine which personality variable best predicts social adjustment. Sociologists may want to find out which of the multiple social indicators best predict whether a new immigrant group will adapt and be absorbed into society.

See also, *Exploratory Data Analysis (EDA) and Data Mining Techniques* in Chapter 12 – *Data Mining Techniques*, Chapter 18 – *General Linear Models*, and Chapter 19 – *General Regression Models*.

Computational Approach

The general computational problem that needs to be solved in multiple regression analysis is to fit a straight line to a number of points.

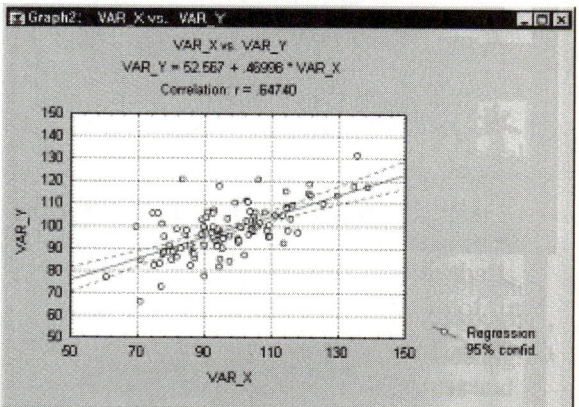

In the simplest case – one dependent and one independent variable – you can visualize this in a scatterplot.

Least Squares

In the scatterplot, we have an independent or X variable, and a dependent or Y variable. These variables may, for example, represent IQ (intelligence as measured by a test) and school achievement (grade point average; GPA). Each point in the plot represents one student, that is, the student's IQ and GPA. The goal of linear regression procedures is to fit a line through the points. Specifically, a line is computed so that the squared deviations of the observed points from that line are minimized. Thus, this general procedure is sometimes also referred to as least squares estimation.

Regression Equation

A line in a two dimensional or two-variable space is defined by the equation $Y=a+b*X$; in full text: the Y variable can be expressed in terms of a constant (a) and a slope (b) times the X variable. The constant is also referred to as the intercept, and the slope as the regression coefficient or B coefficient. For example, GPA may best be predicted as $1+.02*IQ$. Thus, knowing that a student has an IQ of 130 would lead us to predict that her GPA would be 3.6 (since, $1+.02*130=3.6$).

For example, the following illustrations show a 2-dimensional regression equation plotted with three different confidence intervals (90%, 95% and 99%).

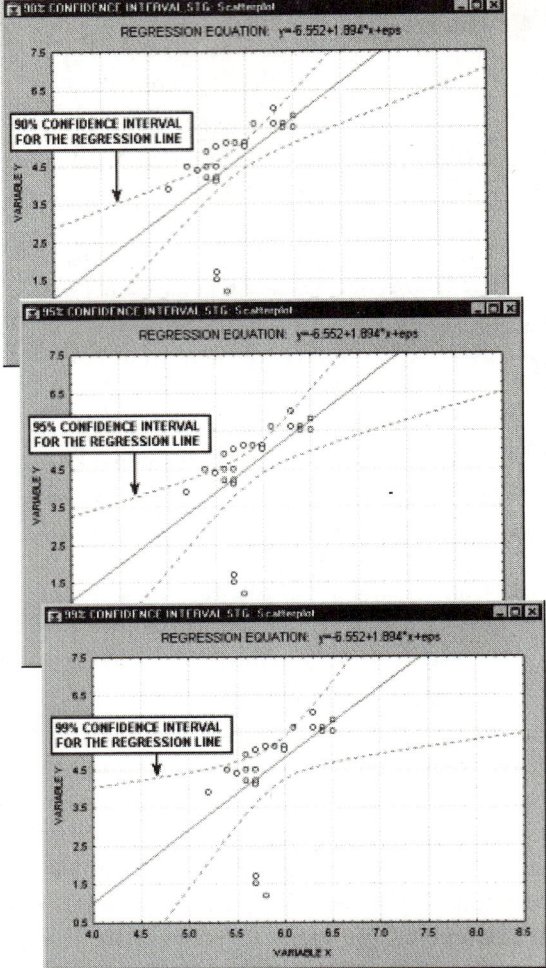

In the multivariate case, when there is more than one independent variable, the regression line cannot be visualized in the two-dimensional space, but can be computed just as easily. For example, if in addition to IQ we had additional predictors of achievement (e.g., *Motivation, Self-discipline*), we could construct a linear equation containing all those variables. In general then, multiple regression procedures will estimate a linear equation of the form:

$$Y = a + b1*X1 + b2*X2 + ... + bp*Xp$$

Unique Prediction and Partial Correlation

Note that in this equation, the regression coefficients (or B coefficients) represent the independent contributions of each independent variable to the prediction of the dependent variable. Another way to express this is to say that, for example, variable X_1 is correlated with the Y variable, after controlling for all other independent variables. This type of correlation is also referred to as a partial correlation (term first used by Yule, 1907). Perhaps an example will clarify this issue. You would probably find a significant negative correlation between hair length and height in the population (i.e., short people have longer hair). At first this may seem odd; however, if we add the variable *Gender* into the multiple regression equation, this correlation would probably disappear. This is because women, on the average, have longer hair than men and, on the average, are shorter than men. Thus, after we remove this gender difference by entering *Gender* into the equation, the relationship between hair length and height disappears because hair length does not make any unique contribution to the prediction of height beyond what it shares in the prediction with variable *Gender*. Put another way, after controlling for the variable *Gender*, the partial correlation between hair length and height is zero.

Predicted and Residual Scores

The regression line expresses the best prediction of the dependent variable (Y), given the independent variables (X). However, nature is rarely (if ever) perfectly predictable, and usually there is substantial variation of the observed points around the fitted regression line (as in the scatterplot shown earlier). The deviation of a particular point from the regression line (its predicted value) is called the residual value.

Residual Variance and R-Square

The smaller the variability of the residual values around the regression line relative to the overall variability, the better is our prediction. For example, if there is no relationship between the X and Y variables, the ratio of the residual variability of the Y variable to the original variance is equal to 1.0. If X and Y are perfectly related, there is no residual variance and the ratio of variance would be 0.0. In most cases, the ratio would fall somewhere between these extremes, that is, between 0.0 and 1.0. 1.0 minus this ratio is referred to as R-square or the coefficient of determination. This value is immediately interpretable in the following manner. If we have an R-square of 0.4, we know that the variability of the Y values around the regression line is 1-0.4 times the original variance; in other words we have explained 40% of the original variability, and are left with 60% residual variability. Ideally, we want to explain most if not all of the original variability. The R-square value is an indicator of how well the model fits the data (e.g., an R-square close to 1.0 indicates that we have accounted for almost all of the variability with the variables specified in the model).

Interpreting the Correlation Coefficient R

Customarily, the degree to which two or more predictors (independent or X variables) are related to the dependent (Y) variable is expressed in the correlation coefficient R, which is the square root of *R-square*. In multiple regression, R can assume values between 0 and 1. To interpret the direction of the relationship between variables, look at the signs (plus or

minus) of the regression or B coefficients. If a B coefficient is positive, the relationship of this variable with the dependent variable is positive (e.g., the greater the IQ the better the grade point average); if the B coefficient is negative, the relationship is negative (e.g., the lower the class size the better the average test scores). Of course, if the B coefficient is equal to 0, there is no relationship between the variables.

Assumptions, Limitations, Practical Considerations

Assumption of Linearity

As evident in the name multiple linear regression, it is assumed that the relationship between variables is linear. In practice this assumption can virtually never be confirmed. Fortunately, multiple regression procedures are not greatly affected by minor deviations from this assumption. However, as a rule it is prudent to always look at a bivariate scatterplot of the variables of interest. If curvature in the relationships is evident, you can consider either transforming the variables, or explicitly allowing for nonlinear components.

Normality Assumption

It is assumed in multiple regression that the residuals (predicted minus observed values) are distributed normally (i.e., follow the normal distribution). Again, even though most tests (specifically the F-test) are quite robust with regard to violations of this assumption, it is *always* a good idea, before drawing final conclusions, to review the distributions of the major variables of interest. You can produce histograms for the residuals as well as normal probability plots in order to inspect the distribution of the residual values.

Limitations

The major conceptual limitation of all regression techniques is that you can only ascertain relationships, but never be sure about the underlying causal mechanism. For example, you would find a strong positive relationship (correlation) between the damage that a fire does and the number of firemen involved in fighting the blaze. Do we conclude that the firemen cause the damage? Of course, the most likely explanation of this correlation is that the size of the fire (an external variable that we forgot to include in our study) caused the damage as well as the involvement of a certain number of firemen (i.e., the bigger the fire, the more firemen are called to fight the blaze). Even though this example is fairly obvious, in real correlation research, alternative causal explanations are often not considered.

Choice of the Number of Variables

Include as many predictor variables as you can think of and usually at least a few of them will come out significant. This is because you are capitalizing on chance when simply including as many variables as you can think of as predictors of some other variable of interest. This problem is compounded when, in addition, the number of observations is relatively low. Intuitively, it is clear that you can hardly draw conclusions from an analysis of 100 questionnaire items based on 10 respondents. Most authors recommend that you should have at least 10 to 20 times as many observations (cases, respondents) as you have variables, otherwise the estimates of the regression line are probably very unstable and unlikely to replicate if you were to do the study again.

Multicollinearity and Matrix III-Conditioning

This is a common problem in many correlation analyses. Imagine that you have two predictors (X variables) of a person's height: 1) weight in pounds and 2) weight in ounces. Obviously, our two predictors are redundant; weight is the same variable regardless of whether it is measured in pounds or ounces. Trying to decide which one of the two measures is a better predictor of height would be rather silly; however, this is exactly what you would try to do if you were to perform a multiple regression analysis with height as the dependent (Y) variable and the two measures of weight as the independent (X) variables. When there are many variables involved, it is often not immediately apparent that this problem exists, and it may only manifest itself after several variables have already been entered into the regression equation. Nevertheless, when this problem occurs it means that at least one of the predictor variables is (practically) completely redundant with other predictors. There are many statistical indicators of this type of redundancy (tolerances, semi-partial R, etc.), as well as some remedies (e.g., Ridge regression).

Fitting Centered Polynomial Models

The fitting of higher-order polynomials of an independent variable with a mean not equal to zero can create difficult multicollinearity problems. Specifically, the polynomials will be highly correlated due to the mean of the primary independent variable. With large numbers (e.g., Julian dates), this problem is serious, and if proper protections are not put in place, can cause wrong results. The solution is to center the independent variable (sometimes, this procedures is referred to as centered

polynomials), i.e., to subtract the mean, and then to compute the polynomials. See, for example, the classic text by Neter, Wasserman, & Kutner (1985, Chapter 9), for a detailed discussion of this issue (and analyses with polynomial models in general).

Importance of Residual Analysis

Even though most assumptions of multiple regression cannot be tested explicitly, gross violations can be detected and should be dealt with appropriately. In particular, outliers (i.e., extreme cases) can seriously bias the results by "pulling" or "pushing" the regression line in a particular direction (see the following illustrations), thereby leading to biased regression coefficients. Often, excluding just a single extreme case can yield a completely different set of results.

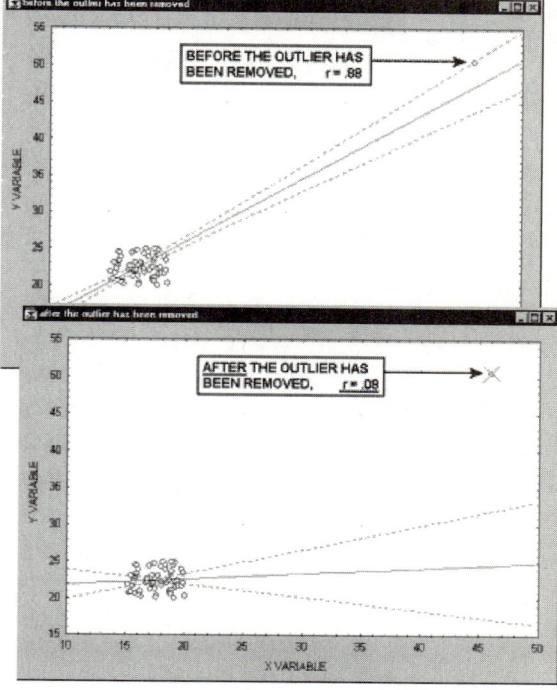

NEURAL NETWORKS

Written by Cazhaow Qazaz, Ph.D.

Neural Networks are analytic techniques modeled after the (hypothesized) processes of learning in the cognitive system and are capable of predicting new observations (on specific variables) from other observations (on the same or other variables) after executing a process of so-called *learning* from existing data. Neural Networks is one of the Data Mining (see Chapter 12) techniques.

Many concepts related to the neural networks methodology are best explained if they are illustrated with applications of a specific neural network program. Therefore, this chapter contains many references to *STATISTICA Neural Networks* (*SNN*) – a neural networks application available from StatSoft – a particularly comprehensive neural network tool.

Artificial Neural Networks

Artificial neural networks have emerged as a reliable tool in modern applied statistics for tackling many of the real world problems in which conventional methods may prove not just as useful. The success of artificial neural networks is due mainly to its ability to describe and model data sets irrespective of the nature of the relationship between the variables of the data set and no matter how strong the intrinsic rules of nonlinearity are that govern the inter-variables relationships. This point of advantage is gained by artificial neural networks through its ability to model and approximate any function no matter how complicated (see *Universal Approximator*, page 355). For this reason, neural networks is capable of performing non-parametric tasks that do not assume a *priori* known relationship between the variables of the data set [see *Nonparametric (Black Box) Models*, page 352]. Instead, neural networks discovers this relationship from the data and through a process known as *training* (see *Learning By Example – Training*, page 359).

Biological Inspiration

The human brain contains on the order of a 100 billion neurons, with each neuron having no less than 1,000 synapses connecting one neuron to its neighbors in an immensely complex communication system. A typical neuron collects electro chemical signals from a host of fine structures called *dendrites*. Each neuron is capable of sending out spikes of electro chemical activity through a long, thin cable like projection known as an *axon*, which splits into thousands of branches. Located at the end of each branch is a structure called *synapse* whose function is to convert the activity from the axon

into electrical effects that inhibit or excite activity from the axon into electrical effects. When a neuron receives an input signal that is sufficiently large compared with its inhibitory input, it sends a spike-like pulse of electrical activity down its axon. It is believed that the learning occurs by changing the effectiveness of the synapses so that the influence of one neuron on another changes. Thus, an electrical signal fired by a neuron is vastly propagated through the brain's complex communication system.

Although a neuron's firing time is much slower than that of a modern computer, it is this massive connectivity that makes our brain distinct and capable of amazing tasks such as speech, vision, thinking, experiencing emotions, and other forms of neural information processing.

Artificial neural networks are statistical tools whose architectures are modeled after the brain. A neural network consists of many neurons, also known as nodes or hidden units, which fires if it receives a sufficiently strong signal from the other nodes to which it is connected. The nature of the signal fired by a neuron depends on its type of activation function. The neurons can be considered as feeble parallel processing units, which individually represent nothing but an unspecialized mathematical function that cannot achieve any meaningful task on its own, but when combined in a suitable way and given a sufficient number of them, the neurons in principal can collectively achieve any task.

Why Use Neural Networks?

Neural networks has a remarkable ability to derive and extract meaning, rules, and trends from complicated, noisy, and imprecise data. It can be used to extract patterns and detect trends that are governed by complicated mathematical functions, which are too difficult, if not impossible, to model using analytic or parametric techniques [see *Nonparametric (Black Box) Models*, page 352]. One of the abilities of neural networks is to accurately predict patterns that are not part of the training data set (see *Learning By Example – Training*, page 359), a process known as generalization (see *Validating Neural Networks and Generalization*, page 362). Given these characteristics and their broad applicability, neural networks is suitable for applications of real-world problems in research and science, business and industry. Listed below are some areas where neural networks has been successfully applied:

- Signal processing
- Process control
- Robotics
- Classification
- Data preprocessing
- Pattern recognition
- Image and speech analysis
- Medical diagnostics and monitoring
- Stock market and forecasting
- Loan or credit solicitations
- Task oriented marketing

Nonparametric (Black Box) Models

One of the most straightforward and perhaps the simplest approach to statistical inference is to assume that the data can be modeled using a closed functional form that can contain a number of adjustable parameters that can be estimated so the model can provide us with the best explanation of the data in hand. As an example, consider a regression problem in

which we aim to model or approximate a single output variable y as a linear function of an input variable x. The mathematical function used to model such a relationship is given by a linear transformation f with two parameters, namely the *intercept a* and *slope b*,

$$y = f(x) = a + bx$$

Our task is to find suitable values for a and b that relate an input x to the outcome variable y. This type of problem is known as *linear regression*.

Another example of parametric regression is the quadratic problem where the input output relationship is described by the quadratic form

$$y = f(x) = a + bx^2$$

input → f=a+bx → output

Parametric model

input → [black box] → output

Non-parametric model

The examples above belong to the category of the so-called *parametric* methods. They strictly rely on the assumption that y is related to x in a *priori* known way, or can be sufficiently approximated by a closed mathematical from, e.g., a line or a quadratic function. Once the mathematical function is chosen, all we have to do is adjust the parameters of the assumed model so they best approximate y given an instance of x.

By contrast, *nonparametric* models generally make no assumptions regarding the relationship of x and y. In other words, they assume that the true underlying function governing the relationship between x and y is not known *a priori*, hence the term "black box." Instead, they attempt to discover a mathematical function (which often does not have a closed form) that can approximate the representation of x and y sufficiently well. The most popular examples of nonparametric models are polynomial functions with adaptable parameters and, indeed, neural networks. Since no closed form for the relationship between x and y is assumed, the nonparametric method must be sufficiently flexible to be able to model a wide spectrum of functional relationships. The higher the order of a polynomial, for example, the more flexible the model is. Similarly, the more neurons a neural network has, the stronger the model becomes.

Parametric models enjoy the advantage of being easy to use and having outputs easy to interpret. Alternately, they suffer from the disadvantage of limited flexibility. Consequently, their usefulness strictly depends on how well the assumed input-output relationship survives the test of reality. Unfortunately many real-world problems simply do not lend themselves to a closed form, and the parametric representation may often prove too restrictive. No wonder then that statisticians and engineers often consider using nonparametric models, especially neural networks, as alternatives to parametric methods.

Just as with parametric methods, nonparametric models too, including neural networks, have their own advantages and disadvantages. While they have enough flexibility to discover or model the data adaptively, their outputs may not be easily subjected to a straightforward interpretation, an issue that has a profound significance in critical and safety applications of these models. Nonetheless, nonparametric

models remain as acceptable alternatives when parametric models fail to be adequate.

Single Neuron Model

Like the cells in our brain, artificial neural networks consists of a network of neurons. A single neuron model contains just one. The body of the cell is represented by a mathematical function, known as the *activation* or *transfer* function, that combines and transforms the incoming signals from the inputs to form an output.

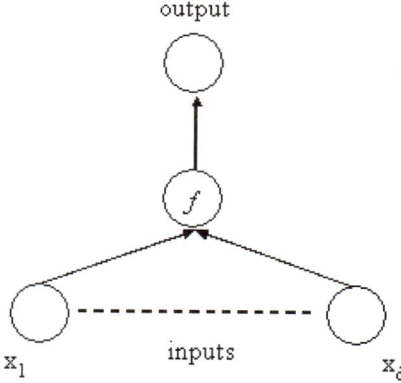

Single Neuron Model

The incoming signals are received through connections to the neuron, *inputs*, with their strength controlled by a set of *weights*, one per connection. After processing these signals the neuron fires a new signal, *output*. In mathematical notation:

$$\text{output} = f\left(w_o + w_1 x_1 + w_2 x_2 + \cdots\right)$$

where $f(...)$ is the neurons activation function, w_i is a typical weight. The nature of the activation function and value of the weights (which control the strength of the inputs into the neuron) completely and uniquely determine the behavior of the single neuron model. In other words, by specifying the weights and the activation function we can completely determine the behavior of the single neural model. The activation function $f(...)$ is typically chosen to be *linear*, *Gaussian*, *logistic sigmoid*, or *tanh* functions, although $f(...)$ is by no means restricted only to these categories:

$$f = \text{Gaussian} = \exp\left(-\frac{(a-\mu)^2}{2\sigma^2}\right)$$

$$f = \text{sigmoid} = \frac{1}{1 + \exp(-a)}$$

$$f = \tanh = \frac{\exp(a) - \exp(-a)}{\exp(a) + \exp(-a)}$$

For linear activation functions, the output is simply the linear weighted sum of the inputs.

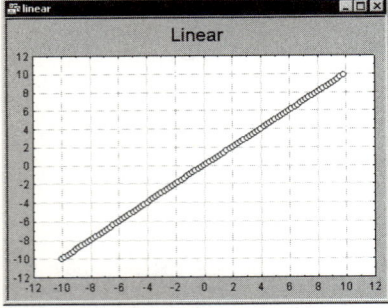

For Gaussian, sigmoid, and tanh functions, the output varies as a continuous and bounded function of the inputs.

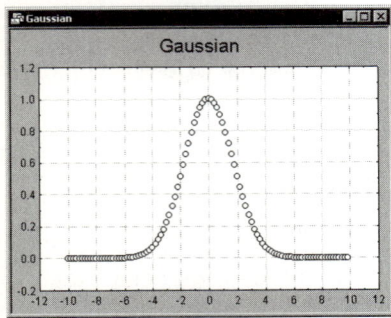

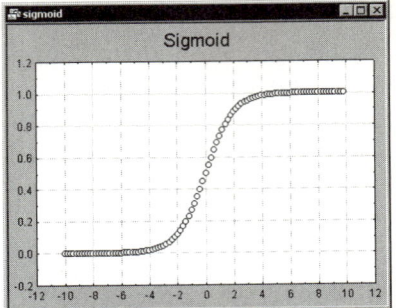

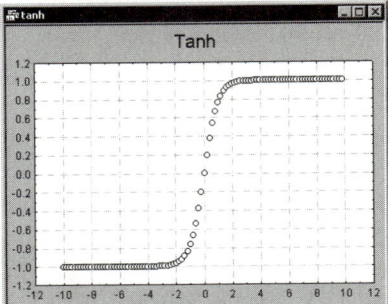

Gaussian activation functions also have the property of being local implying that the neuron would fire only if the input was close to the center of the Gaussian function. In contrast, logistic sigmoid and tanh functions have no local responses.

Multi-Neuron Model

In the previous section, it is mentioned that the nature of the activation function and the strength of the weights completely determine the behavior of the single neuron model. This statement is not entirely correct, however, for a multi-neuron model, a true artificial neural network. Since a neural network typically involves many connecting neurons, it is not only the nature of the activation functions and the strength of the weights that entirely define the model. The manner in which the neurons are arranged and their signals combined also play an important role in defining the model. If we think of a neural network as a parallel processing system, then it becomes clear that there is a strict relationship between its topology and the range of mathematical functions that it can represent.

Artificial neural networks typically consist of layers in which the neurons are arranged in a manner specific to the particular architecture in mind. Invariably, all neural networks have an *input layer* that represents the entry to the complex neural system. The input layer is connected to the *first hidden* layer via connections with strength values determined by a set of weights. After the first layer, there is a succession of hidden layers each consisting of an array of neurons that receive, transform, and transmit the incoming signals from the layers below all the way up through the network until received by the final layer known as the *outputs*. The output layer will finally receive and transform the signals to form the neural network response.

It is this connectivity and parallelism together with the nonlinear nature of the activation functions of the neurons that give neural networks versatility, flexibility, and adaptability to perform any task, provided that they have a sufficient number of neurons.

It should be noted that the neural network architecture discussed in this section is known as *feedforward* neural networks. The term feedforward arises from the very nature of the network in which the signals are propagated, beginning from the input layer, then up all the way through the hidden layers until the signals are finally received and processed by the output layer. Thus, feedforward networks are uni-directional models and, hence, the name *feed forward neural networks*. In addition, such networks are also known as *fully connected*

networks in which every processing unit connects to the units above without exception. In general, a feedforward network has units that receive connections only from the immediate units below.

Neural Network Complexity

In *Nonparametric (Black Box) Models*, page 352, it was mentioned that neural networks, as black box tools, are flexible enough to be capable of performing any task so long as they contain enough neurons. Roughly speaking, the number of neurons in a neural network defines the complexity of the network itself. Thus, the more neurons we have in a network, the more complex the network becomes and, hence, the more capable the model will be to perform complex tasks. By throwing more neurons in a single layer and by adding more layers into a network, we can create increasingly complex models to meet the demands of the task in hand.

However, it should be noted that a network with a single hidden layer containing enough neurons is typically sufficient to carry out any task. Thus, in practice, there is seldom a need for more complex architectures with two or more hidden layers.

Universal Approximator

Neural networks are *universal approximators* particularly suitable for nonparametric modeling of data since they do not restrict the underlying variable relationships to a particular form. Given a sufficient complexity, they can approximate any mathematical function, so they are particularly adaptable to generalization, a term that is used for assessing the performance of neural networks on unseen future data (see

Validating Neural Networks and Generalization, page 362).

Multilayer Perceptrons (MLP)

Multilayer Perceptrons (MLP) are feedforward neural networks with successive layers of neurons and weights and are one of the most commonly used types of networks in practical applications. They provide a general framework for representing nonlinear mappings between the inputs and the outputs. A typical multilayer perceptron network consists of a set of nodes forming the input layer, one or more hidden layers of computing *perceptrons* (neurons), and finally a set of output nodes.

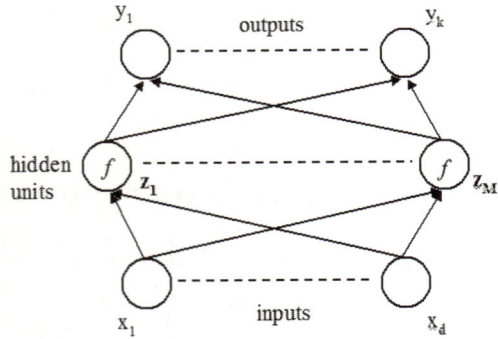

Two Layer Perceptron (MLP2)

The above schematic shows an example of a multilayer perceptron with one hidden layer, i.e,. three layers in total. It is called a *two-layer perceptron (MLP2)*. The output of a neuron in the hidden layer is formed by combining a weighted linear combination of the incoming signals from the inputs,

$$a_j = \sum_{i=1}^{d} w_{ij} \times x_i$$

where w_{ij} is the weight of the connection between input i and neuron j. The new signal,

i.e., neuron output, is then formed using the activation function,

$$z_j = f(a_j)$$

As mentioned before, the nature of the activation function plays an important role in determining the behavior and nonlinearity of the neural network. The most commonly used ones are *logistic sigmoid* and t*anh*.

The input to an output unit is given by the linear weighted sum of the neurons' activation functions in the hidden layer,

$$a_k = \sum_{j=1}^{M} w_{jk} \times z_j$$

where w_{ij} is the weight of the connection between hidden unit j and output k. The output of the network itself is then obtained by transforming the combined signals using the output activation function,

$$\text{output}_k = f(a_j)$$

For multilayer perceptrons, it is common practice, but not always, to choose f as linear.

A neuron on its own is not very useful because of the limited mapping that it can perform. When combined, however, the perceptrons can be used as building blocks of a larger and much more meaningful structure.

Radial Basis Function (RBF)

In this section we consider another major class of neural network architecture that has a single hidden layer with values of neuron activation functions determined by the Euclidian distance between the inputs and the location of its activation functions. Such networks are known as *Radial Basis Function* (RBF) networks, and they form the second, after MLP, most commonly used type of networks. As the name implies, this network makes use of *radial* functions.

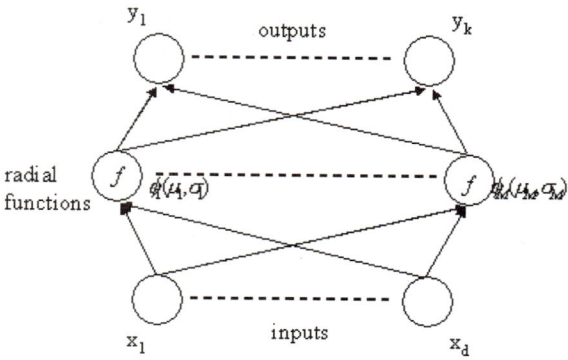

radial functions

Radial Basis Function

An RBF is a network consisting of three layers: one for the inputs, one for the outputs, and a single hidden layer located in between. The activation of the neurons in an RBF network are known as *radial centers*. A typical architecture of this network is shown in the illustration above. Each basis center acts in effect as a neuron with activation defined by a radial function (hence the name radial basis function network). Using the language of MLP2 networks, the inputs connect to the neurons via weights with vales set to unity. In other words, RBF networks have no input-hidden unit weights. Nonetheless, the strength of the neuron's response to an input signal is determined by the Euclidian distance of the inputs to the center of the radial basis function. Thu, upon receiving an input x, the response from a typical neuron (basis function) is given by:

$$\varphi_j = f\left(\|x - \mu_j\|\right)$$

Typically the activation function φ_j is chosen to be Gaussian in which case we can write

$$\varphi_j(x) = \exp\left(\frac{\|x - \mu_j\|^2}{2\sigma^2}\right)$$

where μ_j is the location of the basis center of the jth neuron and σ_j is its radial spread. After processing the incoming signals each neuron sends its response to the output layer where the signals are linearly combined to form the final output of the RBF network,

$$\text{output}_k = \sum_{j=1}^{M} w_{jk}\varphi_j(x)$$

Again the weight w_{jk} pertains to the connection between the jth radial center and the kth output. Note that the output of an RBF is linear in the hidden-output weights. Thus, the only source of nonlinearity in the networks comes from the radial centers.

Labeled and Unlabeled Data

In most statistical applications, data is in the form of input-target pairs (x, t). The inputs x record variables that are believed to influence the targets t and the targets are thought to be dependent on the inputs via a mathematical function, which is often unknown [see *Nonparametric (Black Box) Models*, page 352]. It is by modeling this function that we can represent the relationship between x and t. Data sets falling into this category are known as *labeled* data. In other words, for every input x there is, a hopefully unique, answer t. For example, in a regression problems the pair (x, t) shows what numeric value t will take should the input be x. Similarly, in classification problems t takes on discrete values indicating the class membership of x. Neural networks can be equally used to perform regression or classification tasks.

Neural Networks for Regression

Regression is concerned with predicting the outcome of a number of continuous dependent variables *t* given a set of inputs *x*. We have already discussed the linear regression problem in *Nonparametric (Black Box) Models*, page 352, in relation to parametric models and showed that, when the functional relationship between *x* and *t* does not conform to a closed form, using nonparametric models becomes a more appropriate choice. This is exactly why neural networks is used for tackling regression problems, since it doesn't make any assumptions regarding the functional form of the input-target relationship.

As stated before, it is the task of regression analysis to model the relationship between *x* and *t*, such that for any instance of *x* one can predict *t* with a sufficient degree of accuracy. Since neural networks are universal approximators, they can perform any regression task provided they have enough number of neurons. Some examples of the most commonly used neural architectures in regression are MLP2 and RBF [see *Multilayer Perceptrons (MLP)*, page 356, and *Radial Basis Function (RBF)*, page 356]. Both types of networks discover the structure of the underlying relationship between *x* and *y* by adjusting their adaptive parameters through a process known as *training* (see *Learning By Example – Training*, page 359).

Neural Networks for Classification

In classification problems, the purpose of the network is to label each input pattern *x* with one category (class) among many. Thus, for an input value *x* we want to decide whether *x* belongs to class, say, A, B, or C... etc.

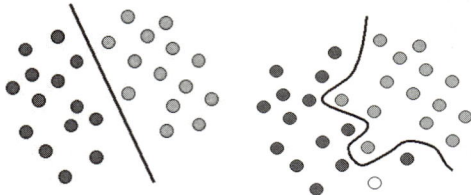

The above schematic shows a two-class (red and green) data *x* in two dimensions. A parametric approach toward solving the problem would, for example, involve assuming that the class categories can be totally separated by a line $a + bx$. Thus, any *x* falling on the right side of the line can be labeled as green; otherwise it is red. Since the intercept and slope completely determine a line, by adjusting these parameters you can make a complete separation between the greens and the reds.

However, in most classification problems the boundary between class memberships isn't simple enough to be modeled with a line or even a curve with a closed form. This is why we often have to resort to nonparametric models, especially neural networks, for tackling real-world classification problems. Here, instead of assuming a closed form for the separating boundary between greens and reds, you let the neural network discover this boundary from the data itself without making any restricting assumptions.

Other Tasks Performed by Neural Networks

So far we have discussed neural networks in the context of regression and classification. Neural networks, however, are by no means restricted or limited to performing such tasks. In fact there

are many inference and predictive tasks for which neural networks is a likely candidate if not one of the best. The most immediate examples are *time series analysis*, *self-organizing feature maps*, *dimensionality reduction*, and *density estimation*.

Learning by Example – Training

One of the capabilities of neural networks is its adaptability to learn any mathematical function provided that it has a sufficient number of neurons. The following illustrations show how a single neural network with one layer and 10 neurons can learn the mapping of the *sine* and the *cosine* functions. The following graph shows the outputs of the network where the relationship between the inputs and the targets is governed by the well known and elementary *sine* function.

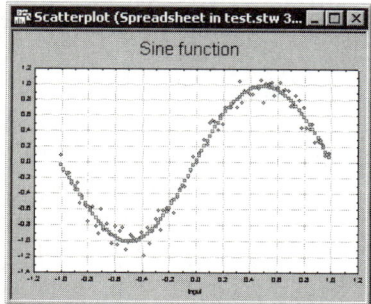

In the following graph, the network has learned the *cosine* relationship as well.

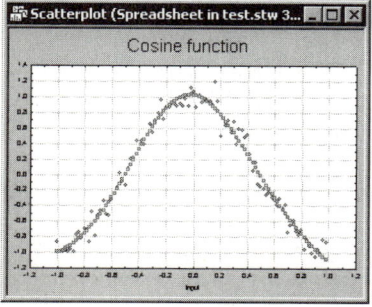

So how can the same network adapt to two problems? The answer lies, as previously mentioned, in the adaptive weights. The weights determine the strength with which a unit, whether an input or a neuron, can influence the rest the processing units in the model. By tuning these adaptive parameters to suitable values, we in effect can adapt the network for the task in hand, i.e., to approximate a *sine* or a *cosine* function in this case. In short, the weights of a neural network are adaptable (adjustable) parameters. That is why artificial neural networks are sometimes called *adaptive neural networks*. The process in which the adaptation takes place is known as *training* or *learning*.

Neural Network Training

The types of neural networks discussed in this chapter have been limited to a brand of models that are also known as the family of *supervised* neural networks. These types of models are capable of adjusting their adaptive weights to approximate the underlying mathematical function that represents the relationship between an input-target set. In other words, they require the data set to consist of *training* pairs *(x, t)* recording the values of the inputs *x* and the answers *t* known as targets. Each pair is known as a *pattern* or *example* that guides the neural network in learning the problem (see *Labeled and Unlabeled Data*, page 357).

The methodology with which a neural network is trained depends on its architecture. For example, MLP2 networks can learn from examples in an iterative procedure. Each time a pattern is presented to the network, the outputs of the network are calculated and an *error function E(y)* is evaluated to measure how good a job the network is doing in predicting the data. If the error is large, the network is penalized and, as a result, the weights are further adjusted (updated) to reduce the error in predictions. This process is then repeated a finite number of times until the error reaches an acceptable level. At that point we have a *fully trained* network.

Alternately, Radial Basis Function (RBF) networks have Gaussian functions for the processing elements of the hidden layer. This function has a local response (i.e., responds only to a narrow region of the input space where the Gaussian is centered). The training procedure for these networks is to find suitable centers and radial spreads for the Gaussian functions that approximate the clustering structure of the inputs data x. This can be done with supervised learning, but it is reported that an *unsupervised* approach, such as clustering algorithms (*Gaussian mixtures* for example), may produce better results. In unsupervised learning, only the inputs are used to adjust the centers and the spreads of the basis functions. By contrast, supervised learning involves the use of input-target pairs to train the network.

After fixing the radial basis centers, the next step is to adjust the hidden-output weights. Since the outputs of an RBF are linear in these parameters, this procedure requires a single matrix inversion.

Unlike the training algorithms discussed previously, the mathematical from of the error function depends on the nature of the statistical task in hand. For regression problems, the most commonly used error is known as *sum-of-squares* E_{SOS},

$$E_{SOS} = \sum_{i=1}^{k} \sum_{j=1}^{N} \left(y_{ij} - t_{ij} \right)^2$$

where the indices i and j run over the outputs of the network and the patterns in the data set. Although the sum-of-squares error function are

most suited for regression problems (see *Neural Networks For Regression*, page 358), this error function is also used for classification problems where the task of the neural network is to classify (label) the input variables x as one of C categories (see *Neural Networks For Classification*, page 358). However, a more principled approach is to use the so-called *cross-entropy* error,

$$E_{Entropy} = \sum_{i=1}^{C} \sum_{j=1}^{N} t_{ij} \log y_{ij}$$

One of the advantages of the cross-entropy error is that it allows the outputs of the network to be interpreted directly as probabilities.

The Problem of Overfitting

In general, the term *overfitting* refers to the condition where a predictive model (e.g., for predictive data mining) is so "specific" that it reproduces various idiosyncrasies (random "noise" variation) of the particular data from which the parameters of the model were estimated. As a result, such models often may not yield accurate predictions for new observations.

Overfitting a model with one predictor. To demonstrate the concept of overfitting in statistics, consider the case of a data set with a single predictor variable. This is shown in the following illustration.

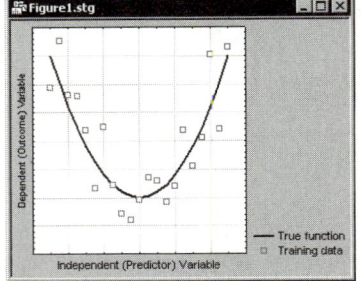

Although the relationship between the predictor and dependent variable is a smooth function (blue circles), in reality what we see in a statistical experiment is the original data "corrupted" with noise (red squares). Thus, our task is to discover the true relationship (the U-shaped curve) between the predictor and the dependent variable given the noisy data set.

Underfitting the data. First, we try to fit the red square points using a function that has too few parameters (a weak model), a straight line in this case. This is shown in the next illustration.

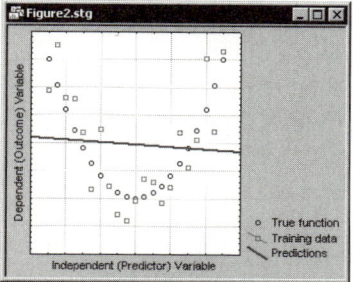

Note that, being too simple (inflexible), the model predictions (green line) cannot fit the data well (i.e., join the data cases together), thus missing the underlying structure in the data. Such a model would not perform well on new data since it missed capturing the relationship between the dependent and predictor variables given its insufficient complexity. A model that is too simple for a given task is known to underfit. Such models will usually yield poor generalization ability, i.e., cannot predict well when presented with new data.

Overfitting the data. Let's now greatly increase the complexity of our model such that it will include as many parameters as we have observations in the data, i.e., fit a model with too many parameters. The result of fitting such a model is shown in the following illustration.

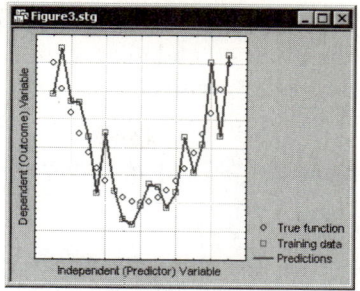

Note that the line depicting the model predictions passes through each data point, thus "discovering" the idiosyncratic (random noise) patterns in the training data that are not part of the true model in the population to which we want to generalize. This is known as overfitting, which is typical of overly complex models. Although such models can achieve very low (even zero) error on training data (since they are flexible enough to pass through each data point), they will perform poorly when presented with unseen (test) data (i.e., when we attempt to apply the model to a new sample of observations drawn from the same population).

Fitting a model of the "right" complexity. At this point, it is clear that if we are to discover the true function (a U-shaped relationship) from what is presented to us by the sample data (red squares), we need a model in between, i.e., a model that is more complex (with more parameters) than a straight line, but not so complex that it will not detect patterns in the data that are due to noise.

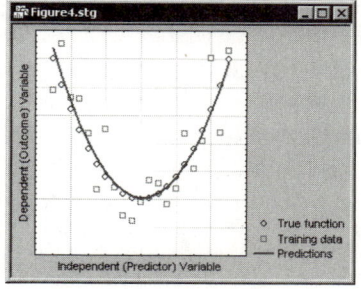

Such a model, as shown above, will capture the true relationship in the data best and, thus, can achieve the best accuracy for prediction.

Network Complexity Revisited

In the previous section, the problem of underfitting and overfitting data was discissed without referring to any particular statistical model. In this example we will reconsider these important issues in the context of neural networks.

In particular, it will be demonstrated how crucial the complexity of neural networks is for tackling statistical problems. To this end we will consider an example based, once again, on regression in which we train an MLP2 with one and three neurons (hidden units) on a data set with the input-output relationship represented by a *sine* function.

Given the insufficiency of the model with a single neuron (see the next illustration) we can clearly see how bad the network is in fitting the data set.

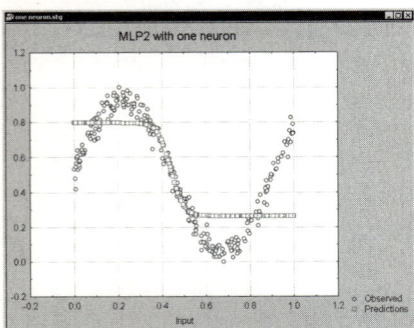

This can be compared to the second model (see the next illustration) with three neurons, where we can observe a substantial increase in performance.

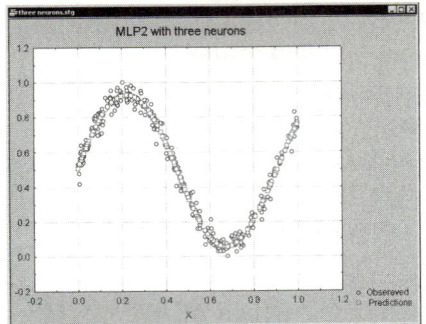

This example highlights the need for using neural networks with complexity large enough to capture the underlying relationship between the inputs and outputs. However, this does not mean we should always use maximally complex networks. The reason is, while we can make arbitrarily complex models and, hence, fit the training data more and more accurately, such models usually perform poorly in predicting future data, i.e., data that was not used to train the model. This problem is known as *generalization*. This argument further highlights the need for achieving a trade off between model complexity and goodness of fit - an issue that is dealt with by *validation* techniques.

Validating Neural Networks and Generalization

But how can we decide on the complexity of a neural network model? In the previous section, the answer was known beforehand, that the input-target relationship was governed by a *sine* function. In real-world problems, we seldom have this knowledge. Besides, it is the task of the network to discover the functional relationship in the first place. But how do we know if the model gets it wrong? The answer lies in neural network validation and verification methods that are a matter of paramount importance especially in safety and critical applications of this technology.

There are several methods for validating trained neural networks, most of which go beyond the scope of this book. Here we will discuss the most commonly used and perhaps the simplest one, namely using a *test* data set.

Test data is a portion of the original data set that was not used to train the network but rather held out and was not presented to the network until training was over. As a metaphor, think of the training data as the teacher who helps the neural network to learn the data, while the test data is the examiner assessing how well the network has learned the underlying relationship in the data set so it can generalize and predict new observations well. Thus, by assessing the performance of neural networks on a test data set, we can somehow get a measure of how well the network can generalize on unseen examples.

Recommended Textbooks

Bishop, C. (1995). *Neural Networks for Pattern Recognition.* Oxford: University Press. Extremely well written, up-to-date. Requires a good mathematical background, but rewards careful reading, putting neural networks firmly into a statistical context.

Carling, A. (1992). *Introducing Neural Networks.* Wilmslow, UK: Sigma Press. A relatively gentle introduction. Starting to show its age a little, but still a good starting point.

Fausett, L. (1994). *Fundamentals of Neural Networks.* New York: Prentice Hall. A well written book with very detailed worked examples to explain how the algorithms function.

Haykin, S. (1994). *Neural Networks: A Comprehensive Foundation.* New York: Macmillan Publishing. A comprehensive book with an engineering perspective. Requires a good mathematical background, and contains a great deal of background theory.

Patterson, D. (1996). *Artificial Neural Networks.* Singapore: Prentice Hall. Good wide-ranging coverage of topics, although less detailed than some other books.

Ripley, B.D. (1996). *Pattern Recognition and Neural Networks.* Cambridge University Press. A very good advanced discussion of neural networks, firmly putting them in the wider context of statistical modeling.

NONLINEAR ESTIMATION

In the most general terms, nonlinear estimation involves finding the best fitting relationship between the values of a dependent variable and the values of a set of one or more independent variables (it is used as either a hypothesis testing or exploratory method). For example, we may want to compute the relationship between the dose of a drug and its effectiveness, the relationship between training and subsequent performance on a task, the relationship between the price of a house and the time it takes to sell it, etc. Research issues in these examples are commonly addressed by such techniques as multiple regression or analysis of variance. In fact, you can think of nonlinear estimation as a generalization of those methods. Specifically, multiple regression (and ANOVA) assumes that the relationship between the independent variable(s) and the dependent variable is linear in nature. Nonlinear estimation leaves it up to you to specify the nature of the relationship; for example, you can specify the dependent variable to be a logarithmic function of the independent variable(s), an exponential function, a function of some complex ratio of independent measures, etc. (However, if all variables of interest are categorical in nature, or can be converted into categorical variables, you can also consider correspondence analysis as an alternative analysis technique.)

Overview

In the most general terms, nonlinear estimation will compute the relationship between a set of independent variables and a dependent variable. For example, we may want to compute the relationship between the dose of a drug and its effectiveness, the relationship between training and subsequent performance on a task, the relationship between the price of a house and the time it takes to sell it, etc. You may recognize research issues in these examples that are commonly addressed by such techniques as multiple regression (see Chapter 26 – *Multiple Linear Regression*) or analysis of variance (see Chapter 3 – *ANOVA/MANOVA*). In fact, you can think of nonlinear estimation as a generalization of those methods. Specifically, multiple regression (and ANOVA) assumes that the relationship between the independent variable(s) and the dependent variable is linear in nature. Nonlinear estimation leaves it up to you to specify the nature of the relationship; for example, you can specify the dependent variable to be a logarithmic function of the independent variable(s), an exponential function, a function of some complex ratio of independent measures, etc. (However, if all variables of interest are categorical in nature or can be converted into categorical variables, you can also consider correspondence analysis.)

When allowing for any type of relationship between the independent variables and the dependent variable, two issues are raised. First, what types of relationships make sense, that is, are interpretable in a meaningful manner? Note that the simple linear relationship is very convenient in that it enables us to make such straightforward interpretations as "the more of x (e.g., the higher the price of a house), the more there is of y (the longer it takes to sell it); and

given a particular increase in x, a proportional increase in y can be expected." Nonlinear relationships cannot usually be interpreted and verbalized in such a simple manner. The second issue that needs to be addressed is how to exactly compute the relationship, that is, how to arrive at results that allow us to say whether or not there is a nonlinear relationship as predicted.

Let's now discuss the nonlinear regression problem in a somewhat more formal manner, that is, introduce the common terminology that will enable us to examine the nature of these techniques more closely, and how they are used to address important questions in various research domains (medicine, social sciences, physics, chemistry, pharmacology, engineering, etc.).

Estimating Linear and Nonlinear Models

Technically speaking, nonlinear estimation is a general fitting procedure that will estimate any kind of relationship between a dependent (or response variable), and a list of independent variables. In general, all regression models can be stated as:

$$y = F(x_1, x_2, \ldots, x_n)$$

In most general terms, we are interested in whether and how a dependent variable is related to a list of independent variables; the term $F(x\ldots)$ in the expression above means that y, the dependent or response variable, is a function of the x's, that is, the independent variables.

An example of this type of model would be the linear multiple regression model as described in *Linear Multiple Regression*. For this model, we assume the dependent variable to be a linear function of the independent variables, that is:

$$y = a + b_1{}^*x_1 + b_2{}^*x_2 + \ldots + b_n{}^*x_n$$

If you are not familiar with multiple linear regression, you may want to read the introductory section to *Multiple Linear Regression* (however, it is not necessary to understand all of the nuances of multiple linear regression techniques in order to understand the methods discussed here).

With nonlinear estimation, you can specify essentially any type of continuous or discontinuous regression model. Some of the most common nonlinear models are probit, logit, exponential growth, and breakpoint regression. However, you can also define any type of regression equation to fit to your data. Moreover, you can specify either standard least squares estimation, maximum likelihood estimation (where appropriate), or, again, define your own loss function by defining the respective equation.

In general, whenever the simple linear regression model does not appear to adequately represent the relationships between variables, the nonlinear regression model approach is appropriate. See the following topics for overviews of the common nonlinear regression models, nonlinear estimation procedures, and evaluation of the fit of the data to the nonlinear model.

Common Nonlinear Regression Models

Intrinsically Linear Regression Models

Polynomial regression. A common "nonlinear" model is polynomial regression. We place the term *nonlinear* in quotes here because the nature of this model is actually linear. For example, suppose we measure in a learning experiment subjects' physiological arousal and their performance on a complex tracking task. Based on the well-known Yerkes-Dodson law we could expect a curvilinear relationship between arousal and performance; this expectation can be expressed in the regression equation:

Performance = $a + b_1$*Arousal + b_2*Arousal^2

In this equation, a represents the intercept, and b_1 and b_2 are regression coefficients. The nonlinearity of this model is expressed in the term *Arousal^2*. However, the nature of the model is still linear, except that when estimating it, we would square the measure of arousal. These types of models, where we include some transformation of the independent variables in a linear equation, are also referred to as models that are nonlinear in the variables.

Models that are nonlinear in the parameters. To contrast the example above, consider the relationship between a human's age from birth (the x variable) and his or her growth rate (the y variable). Clearly, the relationship between these two variables in the first year of a person's life (when most growth occurs) is very different than during adulthood (when almost no growth occurs). Thus, the relationship could probably best be expressed in terms of some negative exponential function:

Growth = exp($-b_1$*Age)

If you plotted this relationship for a particular estimate of the regression coefficient you would obtain a curve that looks something like the following illustration. Note that the nature of this model is no longer linear, that is, the expression shown above does not simply represent a linear regression model, with some transformation of the independent variable.

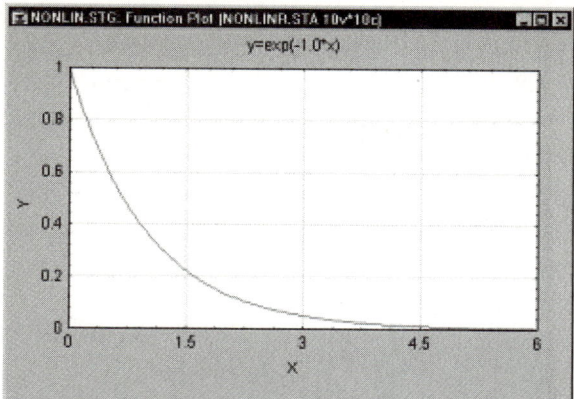

This type of model is said to be nonlinear in the parameters.

Making nonlinear models linear. In general, whenever a regression model can be "made" into a linear model, this is the preferred route to pursue (for estimating the respective model). The multiple linear regression model (see also, Chapter 26 – *Multiple Linear Regression*) is very well understood mathematically, and, from a pragmatic standpoint, is most easily interpreted. Therefore, returning to the simple exponential regression model of *Growth* as a function of *Age* shown above, we could convert this nonlinear regression equation into a linear one by simply taking the logarithm of both sides of the equations, so that:

log(Growth) = -b$_1$*Age

If we now substitute *log(Growth)* with *y*, we have the standard linear regression model as shown earlier (without the intercept which was ignored here to simplify matters). Thus, we could log-transform the *Growth* rate data and then use multiple regression to estimate the relationship between *Age* and *Growth*, that is, compute the regression coefficient b_1.

Model adequacy. Of course, by using the wrong transformation, you could end up with an inadequate model. Therefore, after linearizing a model such as the one shown above, it is particularly important to use extensive residual statistics in multiple regression.

Intrinsically Nonlinear Regression Models

Some regression models that cannot be transformed into linear ones can only be estimated via nonlinear estimation. In the growth rate example above, we purposely forgot about the random error in the dependent variable. Of course, the growth rate is affected by very many other variables (other than time), and we can expect a considerable amount of random (residual) fluctuation around the fitted line. If we add this error or residual variability to the model, we could rewrite it as follows:

Growth = exp(-b$_1$*Age) + error

Additive error. In this model we assume that the error variability is independent of age, that is, that the amount of residual error variability is the same at any age. Because the error term in this model is additive, you can no longer linearize this model by taking the logarithm of both sides. If, for a given data set, you were to log-transform variable *Growth* anyway and fit the simple linear model, you would find that the residuals from the analysis would no longer be evenly distributed over the range of variable Age; and thus, the standard linear regression analysis (via multiple regression) would no longer be appropriate. Therefore, the only way to estimate the parameters for this model is via nonlinear estimation.

Multiplicative error. To defend our previous example, in this particular instance it is not likely that the error variability is constant at all ages, that is, that the error is additive. Most likely, there is more random and unpredictable

fluctuation of the growth rate at the earlier ages than the later ages, when growth comes to a virtual standstill anyway. Thus, a more realistic model including the error would be:

$$Growth = e^{(-b_1 Age)} * error$$

Put in words, the greater the age, the smaller the term $exp(-b_1 * Age)$, and, consequently, the smaller the resultant error variability. If we now take the log of both sides of the equation, the residual error term will become an additive factor in a linear equation, and we can go ahead and estimate b1 via standard multiple regression.

$$Log\ (Growth) = -b_1 * Age + error$$

Let's now consider some regression models (that are nonlinear in their parameters) that cannot be "made into" linear models through simple transformations of the raw data.

General growth model. The general growth model, is similar to the example that we previously considered:

$$y = b_0 + b_1 e^{b_2 x} + error$$

This model is commonly used in studies of any kind of growth (y), when the rate of growth at any given point in time (x) is proportional to the amount of growth remaining. The parameter b_0 in this model represents the maximum growth value. A typical example where this model would be adequate is when you want to describe the concentration of a substance (e.g., in water) as a function of elapsed time.

Models for binary responses: probit and logit. It is not uncommon that a dependent or response variable is binary in nature, that is, that it can have only two possible values. For example, patients either do or do not recover from an injury; job applicants either succeed or fail at an employment test, subscribers to a journal either do or do not renew a subscription,

coupons may or may not be returned, etc. In all of these cases, you may be interested in estimating a model that describes the relationship between one or more continuous independent variable(s) to the binary dependent variable.

Using linear regression. Of course, you could use standard multiple regression procedures to compute standard regression coefficients. For example, if you studied the renewal of journal subscriptions, you could create a y variable with *1*'s and *0*'s, where *1* indicates that the respective subscriber renewed, and *0* indicates that the subscriber did not renew. However, there is a problem: multiple regression does not know that the response variable is binary in nature. Therefore, it will inevitably fit a model that leads to predicted values that are greater than *1* or less than *0*. However, predicted values that are greater than *1* or less than *0* are not valid; thus, the restriction in the range of the binary variable (e.g., between *0* and *1*) is ignored if you use the standard multiple regression procedure.

Continuous response functions. We could rephrase the regression problem so that, rather than predicting a binary variable, we are predicting a continuous variable that naturally stays within the 0-1 bounds. The two most common regression models that accomplish exactly this are the logit and the probit regression models.

Logit regression. In the logit regression model, the predicted values for the dependent variable will never be less than (or equal to) *0*, or greater than (or equal to) *1*, regardless of the values of the independent variables. This is accomplished by applying the following regression equation, which actually has some deeper meaning, as we will see shortly (the term logit was first used by Berkson, 1944):

$$y = e^{(b_0 + b_1 x_1 \ldots + b_n x_n)} / \{1 + e^{(b_0 + b_1 x_1 \ldots + b_n x_n)}\}$$

You can easily recognize that, regardless of the regression coefficients or the magnitude of the x values, this model will always produce predicted values (predicted y's) in the range of 0 to 1.

The name logit stems from the fact that you can easily linearize this model via the logit transformation. Suppose we think of the binary dependent variable y in terms of an underlying continuous probability p, ranging from 0 to 1. We can then transform that probability p as:

$$p' = log_e\{p/(1-p)\}$$

This transformation is referred to as the logit or logistic transformation. Note that p' can theoretically assume any value between minus and plus infinity. Since the logit transform solves the issue of the 0/1 boundaries for the original dependent variable (probability), we could use those (logit transformed) values in an ordinary linear regression equation. In fact, if we perform the logit transform on both sides of the logit regression equation stated earlier, we obtain the standard linear regression model:

$$p' = b_0 + b_1{}^*x_1 + b_2{}^*x_2 + \ldots + b_n{}^*x_n$$

Probit regression. You can consider the binary response variable to be the result of a normally distributed underlying variable that actually ranges from minus infinity to positive infinity. For example, a subscriber to a journal can feel very strongly about not renewing a subscription, be almost undecided, tend toward renewing the subscription, or feel very much in favor of renewing the subscription. In any event, all that we (the publisher of the journal) will see is the binary response of renewal or failure to renew the subscription. However, if we set up the standard linear regression equation based on the underlying feeling or attitude we could write:

$$feeling\ldots = b_0 + b_1{}^*x_1 + \ldots$$

which is, of course, the standard regression model. It is reasonable to assume that these feelings are normally distributed, and that the probability p of renewing the subscription is about equal to the relative *space* under the normal curve. Therefore, if we transform each side of the equation so as to reflect normal probabilities, we obtain:

$$NP(feeling\ldots) = NP(b_0 + b_1{}^*x_1 + \ldots)$$

where *NP* stands for *normal probability* (space under the normal curve), as tabulated in practically all statistics texts. The equation shown above is also referred to as the probit regression model. (The term probit was first used by Bliss, 1934.)

General logistic regression model. The general logistic model can be stated as:

$$y = b_0 / (1 + b_1 e^{b_2 x})$$

You can think of this model as an extension of the logit or logistic model for binary responses. However, while the logit model restricts the dependent response variable to only two values, this model allows the response to vary within a particular lower and upper limit. For example, suppose we are interested in the population growth of a species that is introduced to a new habitat, as a function of time. The dependent variable would be the number of individuals of that species in the respective habitat. Obviously, there is a lower limit on the dependent variable, since fewer than 0 individuals cannot exist in the habitat; however, there also is most likely an upper limit that will be reached at some point in time.

Drug responsiveness and half-maximal response. In pharmacology, the following model is often used to describe the effects of different dose levels of a drug:

$$y = b_0 - b_0/\{1 + (x/b_2)b_1\}$$

In this model, x is the dose level (usually in some coded form, so that $x \geq 1$) and y is the responsiveness, in terms of the percent of maximum possible responsiveness. The parameter b_0 then denotes the expected response at the level of dose saturation and b_2 is the concentration that produces a half-maximal response; the parameter b_1 determines the slope of the function.

Discontinuous Regression Models

Piecewise linear regression. It is not uncommon that the nature of the relationship between one or more independent variables and a dependent variable changes over the range of the independent variables. For example, suppose we monitor the per-unit manufacturing cost of a particular product as a function of the number of units manufactured (output) per month. In general, the more units per month we produce, the lower is our per-unit cost, and this linear relationship may hold over a wide range of different levels of production output. However, it is conceivable that above a certain point, there is a discontinuity in the relationship between these two variables. For example, the per-unit cost may decrease relatively less quickly when older (less efficient) machines have to be put on-line in order to cope with the larger volume. Suppose that the older machines go on-line when the production output rises above 500 units per month; we can specify a regression model for cost-per-unit as:

$$y = b_0 + b_1*x*(x \leq 500) + b_2*x*(x > 500)$$

In this formula, y stands for the estimated per-unit cost; x is the output per month. The expressions $(x \leq 500)$ and $(x > 500)$ denote

logical conditions that evaluate to 0 if false, and to 1 if true. Thus, this model specifies a common intercept (b_0), and a slope that is either equal to b_1 (if $x \leq 500$ is true, that is, equal to 1) or b_2 (if $x > 500$ is true, that is, equal to 1).

Instead of specifying the point where the discontinuity in the regression line occurs (at 500 units per months in the example above), you could also estimate that point. For example, you might have noticed or suspected that there is a discontinuity in the cost-per-unit at one particular point; however, you may not know where that point is. In that case, simply replace the 500 in the equation above with an additional parameter (e.g., b_3).

Breakpoint regression. You could also adjust the equation above to reflect a jump in the regression line. For example, imagine that, after the older machines are put on-line, the per-unit-cost jumps to a higher level, and then slowly goes down as volume continues to increase. In that case, simply specify an additional intercept (b_3), so that:

$$y = (b_0 + b_1*x)*(x \leq 500) + (b_3 + b_2*x)*(x > 500)$$

Comparing groups. The method described here to estimate different regression equations in different domains of the independent variable can also be used to distinguish between groups. For example, suppose in the example above, there are three different plants; to simplify the example, let's ignore the breakpoint for now. If we coded the three plants in a grouping variable by using the values $1, 2$, and 3, we could simultaneously estimate three different regression equations by specifying:

$$y = (x_p=1)*(b_{10} + b_{11}*x) + (x_p=2)*(b_{20} + b_{21}*x) + (x_p=3)*(b_{30} + b_{31}*x)$$

In this equation, x_p denotes the grouping variable containing the codes that identify each plant, b_{10}, b_{20}, and b_{30} are the three different intercepts, and b_{11}, b_{21}, and b_{31} refer to the slope parameters (regression coefficients) for each plant. You could compare the fit of the common regression model without considering the different groups (plants) with this model in order to determine which model is more appropriate.

Nonlinear Estimation Procedures

Least squares estimation. Some of the more common nonlinear regression models are reviewed in *Common Nonlinear Regression Models*, page 368. Now, the question arises as to how these models are estimated. If you are familiar with linear regression techniques (as described in Chapter 26 – *Multiple Linear Regression*) or analysis of variance (ANOVA) techniques (as described in Chapter 3 – *ANOVA/MANOVA*), you may be aware of the fact that all of those methods use so-called least squares estimation procedures. In the most general terms, least squares estimation is aimed at minimizing the sum of squared deviations of the observed values for the dependent variable from those predicted by the model. (The term least squares was first used by Legendre, 1805.)

Loss functions. In standard multiple regression, we estimate the regression coefficients by finding those coefficients that minimize the residual variance (sum of squared residuals) around the regression line. Any deviation of an observed score from a predicted score signifies some loss in the accuracy of our prediction, for example, due to random noise (error). Therefore, we can say that the goal of least squares estimation is to minimize a loss function; specifically, this loss function is defined as the sum of the squared deviation about the predicted values (the term *loss* was first used by Wald, 1939). When this function is at its minimum, we get the same parameter estimates (intercept, regression coefficients) as we would in multiple regression; because of the particular loss functions that yielded those estimates, we can call the estimates least squares estimates.

Phrased in this manner, there is no reason why you cannot consider other loss functions. For example, rather than minimizing the sum of squared deviations, why not minimize the sum of absolute deviations? Indeed, this is sometimes useful in order to de-emphasize outliers. Relative to all other residuals, a large residual will become much larger when squared. However, if you only take the absolute value of the deviations, outliers will most likely less affect the resulting regression line.

There are several function minimization methods that can be used to minimize any kind of loss function. For more information, see:

- Weighted Least Squares, page 374
- Maximum Likelihood, page 374
- Maximum Likelihood and Probit/Logit Models, page 375
- Function Minimization Algorithms, page 375
- Start Values, Step Sizes, Convergence Criteria, page 376
- Penalty Functions, Constraining Parameters, page 376
- Local Minima, page 376
- Quasi-Newton Method, page 376
- Simplex Procedure, page 377
- Hooke-Jeeves Pattern Moves, page 377
- Rosenbrock Pattern Search, page 377

● Hessian Matrix and Standard Errors, page 377

Weighted least squares. In addition to least squares and absolute deviation regression, weighted least squares estimation is probably the most commonly used technique. Ordinary least squares techniques assume that the residual variance around the regression line is the same across all values of the independent variable(s). Put another way, it is assumed that the error variance in the measurement of each case is identical. Often, this is not a realistic assumption; in particular, violations frequently occur in business, economic, or biological applications.

For example, suppose we wanted to study the relationship between the projected cost of construction projects, and the actual cost. This may be useful in order to gage the expected cost overruns. In this case it is reasonable to assume that the absolute magnitude (dollar amount) by which the estimates are off, is proportional to the size of the project. Thus, we would use a weighted least squares loss function to fit a linear regression model. Specifically, the loss function would be (see, for example, Neter, Wasserman, & Kutner, 1985, p. 168):

Loss = $(Obs\text{-}Pred)^2 * (1/x^2)$

In this equation, the loss function first specifies the standard least squares loss function (*Observed minus Predicted squared*; i.e., the squared residual), and then weighs this loss by the inverse of the squared value of the independent variable (x) for each case. In the actual estimation, you sum up the value of the loss function for each case (e.g., construction project), as specified above, and estimate the parameters that minimize that sum. To return to our example, the larger the project (x) the less weight is placed on the deviation from the predicted value (cost). This method will yield more stable estimates of the regression

parameters (for more details, see Neter, Wasserman, & Kutner, 1985).

Maximum likelihood. An alternative to the least squares loss function (see above) is to maximize the likelihood or log-likelihood function (or to minimize the negative log-likelihood function; the term maximum likelihood was first used by Fisher, 1922a). In most general terms, the likelihood function is defined as:

$$L = F(Y, Model) = \prod_{i=1}^{n} \{p[y_i, Model\ Parameters(x_i)]\}$$

In theory, we can compute the probability (now called L, the *likelihood*) of the specific dependent variable values to occur in our sample, given the respective regression model. Provided that all observations are independent of each other, this likelihood is the geometric sum ($\prod$, across $i = 1$ to n cases) of probabilities for each individual observation (i) to occur, given the respective model and parameters for the x values. (The geometric sum means that we would multiply out the individual probabilities across cases.) It is also customary to express this function as a natural logarithm, in which case the geometric sum becomes a regular arithmetic sum ($\sum$, across $i = 1$ to n cases).

Given the respective model, the larger the likelihood of the model, the larger is the probability of the dependent variable values to occur in the sample. Therefore, the greater the likelihood, the better is the fit of the model to the data. The actual computations for particular models here can become quite complicated because we need to track (compute) the probabilities of the y-values to occur (given the model and the respective x-values). As it turns out, if all assumptions for standard multiple regression are met (as described in Chapter 26 – *Multiple Linear Regression*), the standard least squares estimation method will yield results identical to the maximum likelihood method. If the assumption of equal error variances across the range of the x variable(s) is violated, the weighted least squares method described earlier will yield maximum likelihood estimates.

Maximum likelihood and probit/logit models. The maximum likelihood function has been worked out for probit and logit regression models. Specifically, the loss function for these models is computed as the sum of the natural log of the logit or probit likelihood L_1 so that:

$$\log(L_1) = \sum_{i=1}^{n} [y_i \log(p_i) + (1 - y_i)\log(1 - p_i)]$$

where $log(L_1)$ is the natural log of the (logit or probit) likelihood (log-likelihood) for the current model, y_i is the observed value for case I, and p_i is the expected (predicted or fitted) probability (between 0 and 1)

The log-likelihood of the null model (L_0), that is, the model containing the intercept only (and no regression coefficients) is computed as:

$$\log(L_0) = n_0 \log(n_0 / n) + n_1 \log(n_1 / n)$$

where $log(L_0)$ is the natural log of the (logit or probit) likelihood of the null model (intercept only), n_0 is the number of observations with a value of 0 (zero), n_1 is the number of observations with a value of 1, and n is the total number of observations

Function minimization algorithms. Now that we have discussed different regression models, and the loss functions that can be used to estimate them, the only mystery that is left is how to minimize the loss functions (to find the best fitting set of parameters), and how to estimate the standard errors of the parameter estimates. There is one very efficient algorithm (quasi-Newton) that approximates the second-

order derivatives of the loss function to guide the search for the minimum (i.e., for the best parameter estimates, given the respective loss function). In addition, there are several more general function minimization algorithms that follow different search strategies (which do not depend on the second-order derivatives). These strategies are sometimes more effective for estimating loss functions with local minima; therefore, these methods are often particularly useful to find appropriate start values for the estimation via the quasi-Newton method.

In all cases, you can compute the standard errors of the parameter estimates. These standard errors are based on the second-order partial derivatives for the parameters, which are computed via finite difference approximation.

If you are not interested in how the minimization of the loss function is done, only that it can be done, you can skip the following paragraphs. However, you may find it useful to know a little about these procedures in case your regression model refuses to be fit to the data. In that case, the iterative estimation procedure will fail to converge, producing ever stranger (e.g., very large or very small) parameter estimates.

In the following paragraphs, we will first discuss some general issues involved in unconstrained optimization, and then briefly review the methods used. For more detailed discussions of these procedures, refer to Brent (1973), Gill and Murray (1974), Peressini, Sullivan, and Uhl (1988), and Wilde and Beightler (1967). For specific algorithms, see Dennis and Schnabel (1983), Eason and Fenton (1974), Fletcher (1969), Fletcher and Powell (1963), Fletcher and Reeves (1964), Hooke and Jeeves (1961), Jacoby, Kowalik, and Pizzo (1972), and Nelder and Mead (1964).

Start values, step sizes, convergence criteria. A common aspect of all estimation procedures is that they require you to specify some start values, initial step sizes, and a criterion for convergence. All methods will begin with a particular set of initial estimates (start values), which will be changed in some systematic manner from iteration to iteration; in the first iteration, the step size determines by how much the parameters will be moved. Finally, the convergence criterion determines when the iteration process will stop. For example, the process may stop when the improvements in the loss function from iteration to iteration are less than a specific amount.

Penalty functions, constraining parameters. These estimation procedures are unconstrained in nature. That means that it will move parameters around without any regard for whether or not permissible values result. For example, in the course of logit regression we may get estimated values that are equal to 0.0, in which case the logarithm cannot be computed (since the log of 0 is undefined). When this happens, it will assign a penalty to the loss function, that is, a very large value. As a result, the various estimation procedures usually move away from the regions that produce those functions. However, in some circumstances, the estimation will get stuck, and as a result, you would see a very large value of the loss function. This could happen, if, for example, the regression equation involves taking the logarithm of an independent variable that has a value of zero for some cases (in which case the logarithm cannot be computed).

If you want to constrain a procedure, this constraint must be specified in the loss function as a penalty function (assessment). By doing this, you can control what permissible values of the parameters to be estimated may be

manipulated. For example, if two parameters (*a* and *b*) are to be constrained to be greater than or equal to zero, you must assess a large penalty to these parameters if this condition is not met. Below is an example of a user-specified regression and loss function, including a penalty assessment designed to penalize the parameters *a* and/or *b* if either one is not greater than or equal to zero:

Estimated function: v3 = a + b*v1 + (c*v2)
Loss function: L = (obs − pred)**2 + (a<0)*100000 + (b<0)*100000

Local minima. The most treacherous threat to unconstrained function minimization are local minima. For example, a particular loss function may become slightly larger, regardless of how a particular parameter is moved. However, if the parameter were to be moved into a completely different place, the loss function may actually become smaller. You can think of such local minima as local "valleys" or minor "dents" in the loss function. However, in most practical applications, local minima will produce outrageous and extremely large or small parameter estimates with very large standard errors. In those cases, specify different start values and try again. Also note, that the simplex method (see page 377) is particularly smart in avoiding such minima; therefore, this method may be particularly suited in order to find appropriate start values for complex functions.

Quasi-Newton method. As you may remember, the slope of a function at a particular point can be computed as the first-order derivative of the function (at that point). The "slope of the slope" is the second-order derivative, which tells us how fast the slope is changing at the respective point, and in which direction. The quasi-Newton method will, at each step, evaluate the function at different points in order to estimate the first-order derivatives and second-order derivatives. It will then use this information to follow a path toward the minimum of the loss function.

Simplex procedure. This algorithm does not rely on the computation or estimation of the derivatives of the loss function. Instead, at each iteration the function will be evaluated at $m+1$ points in the m dimensional parameter space. For example, in two dimensions (i.e., when there are two parameters to be estimated), it will evaluate the function at three points around the current optimum. These three points would define a triangle; in more than two dimensions, the figure produced by these points is called a simplex. Intuitively, in two dimensions, three points will enable us to determine which way to go, that is, in which direction in the two dimensional space to proceed in order to minimize the function. The same principle can be applied to the multidimensional parameter space, that is, the simplex will move downhill; when the current step sizes become too crude to detect a clear downhill direction, (i.e., the simplex is too large), the simplex will contract and try again.

An additional strength of this method is that when a minimum appears to have been found, the simplex will again be expanded to a larger size to see whether the respective minimum is a local minimum. Thus, in a way, the simplex moves like a smooth single cell organism down the loss function, contracting and expanding as local minima or significant ridges are encountered.

Hooke-Jeeves pattern moves. In a sense this is the simplest of all algorithms. At each iteration, this method first defines a pattern of points by moving each parameter one by one, so as to optimize the current loss function. The entire pattern of points is then shifted or moved to a new location; this new location is determined by extrapolating the line from the old base point in the m dimensional parameter space to the new base point. The step sizes in this process are constantly adjusted to zero in on the respective optimum. This method is usually quite effective, and should be tried if both the quasi-Newton and simplex methods fail to produce reasonable estimates.

Rosenbrock pattern search. Where all other methods fail, the Rosenbrock pattern search method often succeeds. This method will rotate the parameter space and align one axis with a ridge (this method is also called the method of rotating coordinates); all other axes will remain orthogonal to this axis. If the loss function is unimodal and has detectable ridges pointing toward the minimum of the function, this method will proceed with sure-footed accuracy toward the minimum of the function. However, note that this search algorithm may terminate early when there are several constraint boundaries (resulting in the penalty value; see above) that intersect, leading to a discontinuity in the ridges.

Hessian matrix and standard errors. The matrix of second-order (partial) derivatives is also called the Hessian matrix. It turns out that the inverse of the Hessian matrix approximates the variance/ covariance matrix of parameter estimates. Intuitively, there should be an inverse relationship between the second-order derivative for a parameter and its standard error: If the change of the slope around the minimum of the function is very sharp, the second-order derivative will be large; however, the parameter estimate will be quite stable in the sense that the minimum with respect to the parameter is clearly identifiable. If the second-order derivative is nearly zero, the change in the slope around the minimum is zero, meaning that we can practically move the parameter in any direction without greatly affecting the loss function. Thus, the standard error of the parameter will be very large.

The Hessian matrix (and asymptotic standard errors for the parameters) can be computed via finite difference approximation. This procedure yields very precise asymptotic standard errors for all estimation methods.

Evaluating the Fit of the Model

After estimating the regression parameters, an essential aspect of the analysis is to test the appropriateness of the overall model. For example, if you specified a linear regression model, but the relationship is intrinsically nonlinear, the parameter estimates (regression coefficients) and the estimated standard errors of those estimates may be significantly off. Let's review some of the ways to evaluate the appropriateness of a model.

Proportion of variance explained. Regardless of the model, you can always compute the total variance of the dependent variable (total sum of squares, SST), the proportion of variance due to the residuals (error sum of squares, SSE), and the proportion of variance due to the regression model (regression sum of squares, SSR=SST-SSE). The ratio of the regression sum of squares to the total sum of squares (SSR/SST) explains the proportion of variance accounted for in the dependent variable (y) by the model; thus, this ratio is equivalent to the *R-square* ($0 \leq R\text{-}square \leq 1$, the *coefficient of determination*). Even when the dependent variable is not normally distributed across cases, this measure may help evaluate how well the model fits the data.

Goodness-of-fit chi-square. For probit and logit regression models, you can use maximum likelihood estimation (i.e., maximize the likelihood function). As it turns out, you can

directly compare the likelihood L_0 for the null model where all slope parameters are zero, with the likelihood L_1 of the fitted model. Specifically, you can compute the *chi*-square statistic for this comparison as:

Chi-square = -2 * (log(L_0) – log(L_1))

The degrees of freedom for this *chi*-square value are equal to the difference in the number of parameters for the null and the fitted model; thus, the degrees of freedom will be equal to the number of independent variables in the logit or probit regression. If the *p*-level associated with this *chi*-square is significant, we can say that the estimated model yields a significantly better fit to the data than the null model, that is, that the regression parameters are statistically significant.

Plot of observed vs. predicted values. It is always a good idea to inspect a scatterplot of predicted vs. observed values. If the model is appropriate for the data, we would expect the points to roughly follow a straight line; if the model is incorrectly specified, this plot will indicate a nonlinear pattern.

Normal and half-normal probability plots. The normal probability plot of residual will give us an indication of whether or not the residuals (i.e., errors) are normally distributed.

Plot of the fitted function. For models involving two or three variables (one or two predictors) it is useful to plot the fitted function using the final parameter estimates. Following is an example of a 3D plot with two predictor variables.

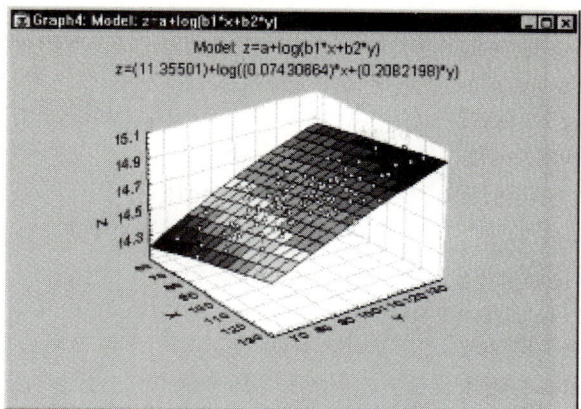

This type of plot represents the most direct visual check of whether or not a model fits the data, and whether there are apparent outliers.

Variance/covariance matrix for parameters.

When a model is grossly misspecified, or the estimation procedure gets hung up in a local minimum, the standard errors for the parameter estimates can become very large. This means that regardless of how the parameters were moved around the final values, the resulting loss function did not change much. Also, the correlations between parameters may become very large, indicating that parameters are very redundant; put another way, when the estimation algorithm moved one parameter away from the final value, then the increase in the loss function could be almost entirely compensated for by moving another parameter. Thus, the effect of those two parameters on the loss function was very redundant.

NONPARAMETRIC STATISTICS

Nonparametric methods were developed to be used in cases when the researcher does not know the parameters of the distribution of the variable of interest in the population (hence the name nonparametric). In more technical terms, nonparametric methods do not rely on the estimation of parameters (such as the mean or the standard deviation) describing the distribution of the variable of interest in the population. Therefore, these methods are also sometimes (and more appropriately) called parameter-free methods or distribution-free methods.

Overview

Brief review of the idea of significance testing.

To understand the idea of nonparametric statistics (the term nonparametric was first used by Wolfowitz, 1942) first requires a basic understanding of parametric statistics. Chapter 1 – *Elementary Concepts in Statistics* introduces the concept of statistical significance testing based on the sampling distribution of a particular statistic (you may want to review that chapter before reading on). In short, if we have a basic knowledge of the underlying distribution of a variable, we can make predictions about how, in repeated samples of equal size, this particular statistic will behave, that is, how it is distributed. For example, if we draw 100 random samples of 100 adults each from the general population, and compute the mean height in each sample, the distribution of the standardized means across samples will likely approximate the normal distribution (to be precise, student's *t* distribution with 99 degrees of freedom). Now imagine that we take an additional sample in a particular city ("Tallburg") where we suspect that people are taller than the average population. If the mean height in that sample falls outside the upper 95% tail area of the *t* distribution, we conclude that, indeed, the people of Tallburg are taller than the average population.

Are most variables normally distributed?

In the example described above, we rely on our knowledge that, in repeated samples of equal size, the standardized means (for height) will be distributed following the *t* distribution (with a particular mean and variance). However, this will only be true if in the population the variable of interest (height in this example) is normally distributed, that is, if the distribution of people of particular height follows the normal distribution (the bell-shape distribution).

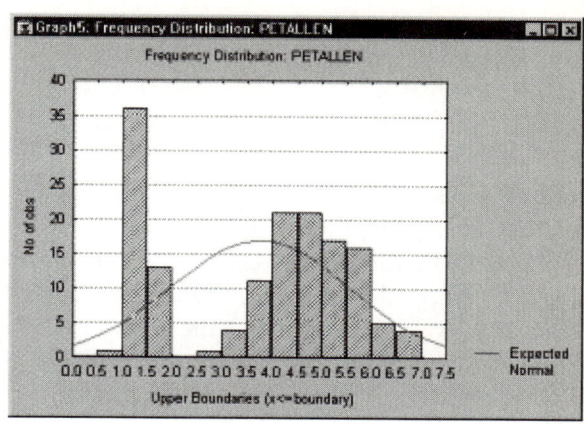

For many variables of interest, we simply do not know for sure that this is the case. For example, is income distributed normally in the population? Probably not. The incidence rates of rare diseases are not normally distributed in the population, the number of car accidents is also not normally distributed, and neither are very many other variables in which a researcher might be interested.

For more information on the normal distribution, see Chapter 1 – *Elementary Concepts in Statistics*; for information on *Normality Tests*, see also, the *Statistical Glossary*.

Sample size.

Another factor that often limits the applicability of tests based on the assumption that the sampling distribution is normal is the size of the sample of data available for the analysis (sample size; *n*). We can assume that the sampling distribution is normal even if we are not sure that the distribution of the variable in the population is normal, as long as our sample is large enough (e.g., 100 or more observations). However, if our sample is very small, those tests can be used only if we are sure that the variable is normally distributed, and there is no way to test this assumption if the sample is small.

Problems in measurement. Applications of tests that are based on the normality assumptions are further limited by a lack of precise measurement. For example, let's consider a study where grade point average (GPA) is measured as the major variable of interest. Is an A average twice as good as a C average? Is the difference between a B and an A average comparable to the difference between a D and a C average? Somehow, the GPA is a crude measure of scholastic accomplishments that only allows us to establish a rank ordering of students from good students to poor students. This general measurement issue is usually discussed in statistics textbooks in terms of types of measurement or scale of measurement. Without going into too much detail, most common statistical techniques such as analysis of variance (and t-tests), regression, etc. assume that the underlying measurements are at least of interval, meaning that equally spaced intervals on the scale can be compared in a meaningful manner (e.g., B minus A is equal to D minus C). However, as in our example, this assumption is very often not tenable, and the data rather represent a rank ordering of observations (ordinal) rather than precise measurements.

Parametric and nonparametric methods.

Hopefully, after this somewhat lengthy introduction, the need is evident for statistical procedures that enable us to process data of low quality from small samples on variables about which nothing is known (concerning their distribution). Specifically, nonparametric methods were developed to be used in cases when the researcher knows nothing about the parameters of the variable of interest in the population (hence the name nonparametric). In more technical terms, nonparametric methods do not rely on the estimation of parameters (such as the mean or the standard deviation)

describing the distribution of the variable of interest in the population. Therefore, these methods are also sometimes (and more appropriately) called parameter-free methods or distribution-free methods.

Brief Overview of Nonparametric Methods

Basically, there is at least one nonparametric equivalent for each parametric general type of test. In general, these tests fall into the following categories:

Differences between independent groups. Usually, when we have two samples that we want to compare concerning their mean value for some variable of interest, we would use the t-test for independent samples; nonparametric alternatives for this test are the Wald-Wolfowitz runs test, the Mann-Whitney U test, and the Kolmogorov-Smirnov two-sample test. If we have multiple groups, we would use analysis of variance (see Chapter 3 – *ANOVA/ MANOVA*; the nonparametric equivalents to this method are the Kruskal-Wallis analysis of ranks and the median test.

Differences between dependent groups. If we want to compare two variables measured in the same sample we would customarily use the t-test for dependent samples. Nonparametric alternatives to this test are the sign test and Wilcoxon matched pairs test. If the variables of interest are dichotomous in nature (i.e., pass vs. no pass), McNemar *chi*-square test is appropriate. If there are more than two variables that were measured in the same sample, we would customarily use repeated measures ANOVA. Nonparametric alternatives to this method are Friedman's two-way analysis of variance and Cochran Q test (if the variable was measured in terms of categories, e.g., passed vs.

failed). Cochran Q is particularly useful for measuring changes in frequencies (proportions) across time.

Relationships between variables. To express a relationship between two variables, you usually compute the correlation coefficient. Nonparametric equivalents to the standard correlation coefficient are Spearman R, Kendall *tau*, and coefficient *gamma* (see *Nonparametric Correlations*, page 386). If the two variables of interest are categorical in nature (e.g., passed vs. failed by male vs. female) appropriate nonparametric statistics for testing the relationship between the two variables are the *chi*-square test, the *phi* coefficient, and the Fisher exact test. In addition, a simultaneous test for relationships between multiple cases is available: Kendall coefficient of concordance. This test is often used for expressing inter-rater agreement among independent judges who are rating (ranking) the same stimuli.

Descriptive statistics. When your data is not normally distributed, and the measurements at best contain rank order information, computing the standard descriptive statistics (e.g., mean and standard deviation) is sometimes not the most informative way to summarize the data. For example, in the area of psychometrics it is well known that the rated intensity of a stimulus (e.g., perceived brightness of a light) is often a logarithmic function of the actual intensity of the stimulus (brightness as measured in objective units of Lux). In this example, the simple mean rating (sum of ratings divided by the number of stimuli) is not an adequate summary of the average actual intensity of the stimuli. (In this example, you would probably rather compute the geometric mean.) Nonparametrics and distributions will compute a wide variety of measures of location (mean, median, mode, etc.) and dispersion (variance,

average deviation, quartile range, etc.) to provide the complete picture of your data.

When to Use Which Method

It is not easy to give simple advice concerning the use of nonparametric procedures. Each nonparametric procedure has its peculiar sensitivities and blind spots. For example, the Kolmogorov-Smirnov two-sample test is not only sensitive to differences in the location of distributions (for example, differences in means) but is also greatly affected by differences in their shapes. The Wilcoxon matched pairs test assumes that you can rank order the magnitude of differences in matched observations in a meaningful manner. If this is not the case, you should instead use the sign test. In general, if the result of a study is important (e.g., does a very expensive and painful drug therapy help people get better?), it is always advisable to run different nonparametric tests; should discrepancies in the results occur contingent upon which test is used, you should try to understand why some tests give different results. On the other hand, nonparametric statistics are less statistically powerful (sensitive) than their parametric counterparts, and if it is important to detect even small effects (e.g., is this food additive harmful to people?) you should be very careful in the choice of a test statistic.

Large data sets and nonparametric methods. Nonparametric methods are most appropriate when the sample sizes are small. When the data set is large (e.g., $n > 100$), it often makes little sense to use nonparametric statistics at all. Chapter 1 – *Elementary Concepts in Statistics* briefly discusses the idea of the central limit theorem. In a nutshell, when the samples become very large, the sample means will

follow the normal distribution even if the respective variable is not normally distributed in the population, or is not measured very well. Thus, parametric methods, which are usually much more sensitive (i.e., have more statistical power) are in most cases appropriate for large samples. However, the tests of significance of many of the nonparametric statistics described here are based on asymptotic (large sample) theory; therefore, meaningful tests can often not be performed if the sample sizes become too small. Refer to the descriptions of the specific tests to learn more about their power and efficiency.

Nonparametric Correlations

The following paragraphs describe three types of commonly used nonparametric correlation coefficients – Spearman R, Kendall *tau*, and *gamma* coefficients. Note that the *chi*-square statistic computed for two-way frequency tables also provides a careful measure of a relation between the two (tabulated) variables, and unlike the correlation measures listed below, it can be used for variables that are measured on a simple nominal scale.

Spearman R. Spearman R (Siegel & Castellan, 1988) assumes that the variables under consideration were measured on at least an ordinal (rank order) scale, that is, that the individual observations can be ranked into two ordered series. Spearman R can be thought of as the regular Pearson product moment correlation coefficient, that is, in terms of proportion of variability accounted for, except that Spearman R is computed from ranks.

Kendall tau. Kendall *tau* is equivalent to Spearman R with regard to the underlying assumptions. It is also comparable in terms of its statistical power. However, Spearman R and

Kendall *tau* are usually not identical in magnitude because their underlying logic, as well as their computational formulas, is very different. Siegel and Castellan (1988) express the relationship of the two measures in terms of the inequality:

$-1 \leq 3 * \text{Kendall tau} - 2 * \text{Spearman R} \leq 1$

More importantly, Kendall *tau* and Spearman R imply different interpretations: Spearman R can be thought of as the regular Pearson product moment correlation coefficient, that is, in terms of proportion of variability accounted for, except that Spearman R is computed from ranks. Kendall *tau*, on the other hand, represents a probability, that is, it is the difference between the probability that in the observed data the two variables are in the same order versus the probability that the two variables are in different orders.

Gamma. The *gamma* statistic (Siegel & Castellan, 1988) is preferable to Spearman R or Kendall *tau* when the data contain many tied observations. In terms of the underlying assumptions, *gamma* is equivalent to Spearman R or Kendall *tau*; in terms of its interpretation and computation it is more similar to Kendall *tau* than Spearman R. In short, *gamma* is also a probability; specifically, it is computed as the difference between the probability that the rank ordering of the two variables agree minus the probability that they disagree, divided by 1 minus the probability of ties. Thus, *gamma* is basically equivalent to Kendall *tau*, except that ties are explicitly taken into account.

2x2 Tables, Chi/V/Phi Square, McNemar, Fisher Exact

By entering frequencies into a 2x2 table, you can calculate various statistics to evaluate the relationship between two dichotomous variables. Thus, the 2x2 option can be used as an alternative to correlation when the two variables of interest are dichotomous.

Yates corrected chi-square. The approximation of the *chi*-square statistic in small 2x2 tables can be improved by reducing the absolute value of differences between expected and observed frequencies by 0.5 before squaring (Yates' correction). This correction, which makes the estimation more conservative, is usually applied when the table contains only small observed frequencies, so that some expected frequencies become less than 10 (for further discussion of this correction, see Conover, 1974; Everitt, 1977; Hays, 1988; Kendall & Stuart, 1979; and Mantel, 1974).

Phi-square. The *phi*-square is a measure of correlation between the two categorical variables in the table.

McNemar chi-square (A/D, B/C). This test is applicable in situations where the frequencies in the 2x2 table represent dependent samples. For example, in a before-after design study, we may count the number of students who fail a test of minimal math skills at the beginning of the semester and at the end of the semester. Two *chi*-square values are reported: *A/D* and *B/C*. The *chi*-square *A/D* tests the hypothesis that the frequencies in cells *A* and *D* (upper left, lower right) are identical. The *Chi*-square *B/C* tests the hypothesis that the frequencies in cells *B* and *C* (upper right, lower left) are identical.

Fisher exact test. Given the marginal frequencies in the table, and assuming that in the population the two factors in the table are not related, how likely is it to obtain cell frequencies as uneven or worse than the ones that were observed? For small n, this probability can be computed exactly by counting all possible tables that can be constructed based on the marginal frequencies. This is the underlying rationale for the Fisher exact test. It computes the exact probability under the null hypothesis of obtaining the current distribution of frequencies across cells, or one that is more uneven.

Correlations (Spearman, Kendall Tau, Gamma)

You can compute three different alternatives to the parametric Pearson product-moment correlation coefficient: Spearman rank R, Kendall *tau,* and *gamma.*

Spearman rank R. Spearman rank R can be thought of as the regular Pearson product-moment correlation coefficient (Pearson r); that is, in terms of the proportion of variability accounted for, except that Spearman R is computed from ranks. Spearman R assumes that the variables under consideration were measured on at least an ordinal (rank order) scale; that is, the individual observations (cases) can be ranked into two ordered series. Detailed discussions of the Spearman R statistic and its power and efficiency can be found in Gibbons (1985), Hays (1981), McNemar (1969), Siegel and Castellan (1988), Kendall (1948), Olds (1949), or Hotelling and Pabst (1936).

Kendall tau. Kendall *tau* is equivalent to Spearman R with regard to the underlying assumptions. It is also comparable in terms of its statistical power. However, Spearman R and Kendall *tau* are usually not identical in magnitude because their underlying logic as well as their computational formulas are very different. Siegel and Castellan (1988) express the relationship of the two measures in terms of the inequality:

-1 ≤ 3 * Kendall *tau* - 2 * Spearman *R* ≤ 1

More importantly, Kendall *tau* and Spearman R imply different interpretations: Spearman R can be thought of as the regular Pearson product-moment correlation coefficient; that is, in terms of proportion of variability accounted for, except that Spearman R is computed from ranks. Kendall *tau*, on the other hand, represents a probability; that is, it is the difference between the probability that the two variables are in the same order in the observed data versus the probability that the two variables are in different orders.

Gamma. The *gamma* statistic is preferable to Spearman R or Kendall *tau* when the data contain many tied observations. In terms of the underlying assumptions, *Gamma* is equivalent to Spearman R or Kendall *tau*; in terms of its interpretation and computation, it is more similar to Kendall *tau* than Spearman R. In short, *gamma* is also a probability; specifically, it is computed as the difference between the probability that the rank ordering of the two variables agree minus the probability that they disagree, divided by 1 minus the probability of ties. Thus, *gamma* is basically equivalent to Kendall *tau*, except that ties are explicitly taken into account.

Wald-Wolfowitz Runs Test

The Wald-Wolfowitz runs test is a nonparametric alternative to the *t*-test for independent samples.

Assumptions and interpretation. The Wald-Wolfowitz runs test works as follows: Imagine you want to compare male and female subjects on some variable. You can sort the data by that variable and look for cases when, in the sorted data, same-gender subjects are adjacent to each other. If there are no differences between male and female subjects, then the number and "lengths" of such adjacent "runs" of subjects of the same gender will be more or less random. If not, the two groups (genders in our example) are somehow different from each other. This test assumes that the variable under consideration is continuous, and that it was measured on at least an ordinal scale (i.e., rank order). The Wald-Wolfowitz runs test assesses the hypothesis that two independent samples were drawn from two populations that differ in some respect, i.e., not just with respect to the mean, but also with respect to the general shape of the distribution. The null hypothesis is that the two samples were drawn from the same population. In this respect, this test is different from the parametric *t*-test, which strictly tests for differences in locations (means) of two samples.

Kolmogorov-Smirnov Two-Sample Test

The Kolmogorov-Smirnov test is another nonparametric alternative to the *t*-test for independent samples.

Assumptions and interpretation. The Kolmogorov-Smirnov test assesses the hypothesis that two samples were drawn from different populations. Unlike the parametric *t*-test for independent samples or the Mann-Whitney *U* test, which test for differences in the location of two samples (differences in means, differences in average ranks, respectively), the Kolmogorov-Smirnov test is also sensitive to differences in the general shapes of the distributions in the two samples (i.e., to differences in dispersion, skewness, etc.). Thus, its interpretation is similar to that of the Wald-Wolfowitz runs test.

Mann-Whitney U Test

The Mann-Whitney U test is a nonparametric alternative to the *t*-test for independent samples.

Assumptions and interpretation. The Mann-Whitney U test assumes that the variable under consideration was measured on at least an ordinal (rank order) scale. The interpretation of the test is essentially identical to the interpretation of the result of a *t*-test for independent samples, except that the U test is computed based on rank sums rather than means. The U test is the most powerful (or sensitive) nonparametric alternative to the *t*-test for independent samples; in fact, in some instances it may offer even greater power to reject the null hypothesis than the *t*-test. With samples larger than 20, the sampling distribution of the U statistic rapidly approaches the normal distribution (see Siegel, 1956). Hence, the U statistic (adjusted for ties) will be accompanied by a z value (normal distribution variate value), and the respective *p*-value.

Kruskal-Wallis ANOVA by Ranks and Median Test

These two tests are nonparametric alternatives to between-groups one-way analysis of variance.

Assumptions and interpretation. The Kruskal-Wallis ANOVA by Ranks test assumes that the variable under consideration is continuous and that it was measured on at least an ordinal (rank order) scale. The test assesses the hypothesis that the different samples in the comparison were drawn from the same distribution or from distributions with the same median. Thus, the interpretation of the Kruskal-Wallis test is basically identical to that of the parametric one-way ANOVA, except that it is based on ranks rather than means.

The Median test is a "crude" version of the Kruskal-Wallis ANOVA in that it frames the computation in terms of a contingency table. Under the null hypothesis (all samples come from populations with identical medians), we expect approximately 50% of all cases in each sample to fall above (or below) the common median. The Median test is particularly useful when the scale contains artificial limits, and many cases fall at either extreme of the scale ("off the scale"). In this case, the Median test is in fact the only appropriate method for comparing samples.

Sign Test

The *Sign test* is a nonparametric alternative to the *t*-test for dependent samples. The test is applicable to situations when the researcher has two measures (e.g., under two conditions) for each subject and wants to establish that the two measurements (or conditions) are different.

Assumptions and interpretation. The only assumption required by this test is that the underlying distribution of the variable of interest is continuous; no assumptions about the nature or shape of the underlying distribution are required. The test simply computes the number of times (across subjects) that the value of the first variable (A) is larger than that of the second variable (B). Under the null hypothesis (stating that the two variables are not different from each other) we expect this to be the case about 50% of the time. Based on the binomial distribution we can compute a z value for the observed number of cases where $A > B$, and compute the associated tail probability for that z value. For small n's (less than 20) you may prefer to use the tabulated values found in Siegel and Castellan (1988) to evaluate statistical significance.

Wilcoxon Matched Pairs Test

The Wilcoxon matched pairs test is a nonparametric alternative to the *t*-test for dependent (correlated) samples.

Assumptions and interpretation. The procedure assumes that the variables under consideration were measured on a scale that allows the rank ordering of observations based on each variable (i.e., ordinal scale) and that allows rank ordering of the differences between variables (this type of scale is sometimes referred to as an ordered metric scale, see Coombs, 1950). Thus, the required assumptions for this test are more stringent than those for the Sign test. However, if they are met, that is, if the magnitudes of differences (e.g., different ratings by the same individual) contain meaningful information, then this test is more powerful than the Sign test. In fact, if the assumptions for the parametric *t*-test for dependent samples are met, then this test is almost as powerful as the *t*-test.

Friedman ANOVA and Kendall Concordance

These two tests are somewhat different in nature, however, they require similar user input. Friedman ANOVA is a nonparametric alternative to one-way repeated measures analysis of variance. The Kendall concordance statistic is similar to Spearman R (nonparametric correlation between two variables) except that it expresses the relationship between multiple cases.

Assumptions and interpretation: Friedman ANOVA. The Friedman ANOVA by ranks test assumes that the variables (levels) under consideration were measured on at least an ordinal (rank order) scale. The null hypothesis

for the procedure is that the different columns of data contain samples drawn from the same population, or specifically, populations with identical medians. Thus, the interpretation of results from this procedure is similar to that of a repeated measures ANOVA.

Assumptions and interpretation: Kendall concordance. The Kendall concordance coefficient expresses the simultaneous association (relatedness) between k sets of rankings (i.e., cases; correlated samples). For example, this statistic is commonly used to assess inter-judge reliability. Basically, the concordance coefficient is the average of all Spearman Rs between cases; specifically:

average Spearman $R = (k * \text{concordance} - 1) / (k-1)$

Thus the general assumptions of this test are identical to that of the Spearman rank order correlation.

The range of Kendall concordance is from 0 to +1. Values close to zero represent lack of agreement in the rankings of the variables (e.g., objects) among cases (e.g., judges), while values close to 1 represent perfect agreement in the rankings of the variables (objects) among cases (judges).

Cochran Q Test

The Cochran Q test is an extension of McNemar's *Chi*-square test for changes in frequencies or proportions to k (more than two) dependent samples. Specifically, it tests whether several matched frequencies or proportions differ significantly among themselves.

Assumptions and interpretation. The Cochran Q test only requires a nominal scale, or that the data have been artificially dichotomized. A typical example where the Q test is useful is when you want to compare the difficulty of dichotomous questionnaire items that can either be answered right or wrong. Here, each variable in the data file would represent one item, and contain 0's (wrong) and 1's (right). If the Q test is significant, then we conclude that the items are of different difficulty since different items were answered correctly by more or fewer respondents.

Ordinal Descriptive Statistics

Ordinal descriptive statistics (median, mode, ...) in addition to the standard descriptive statistics (minimum value, maximum value, mean, valid n) can be computed for each variable.

Median. The median value is the value that "splits the sample in half," given the respective variable. Fifty percent of the cases will fall below the median, and fifty percent will fall above the median. If the median value is very different from the mean, then the distribution of data is skewed.

Mode. The mode is the value that occurs with the greatest frequency. The frequency with which the mode occurs is also displayed; if there is a tie (i.e., more than one value occurs with equal frequency) then the respective frequency column will contain the label "multiple" to indicate that more than one mode was found.

Geometric mean. The geometric mean is the product of all scores to the power of 1/N (one over the valid number of cases). The geometric mean is useful in instances when we know that the measurement scale is not linear. For example, in the area of psychometrics it is well known that the rated intensity of a stimulus (e.g., brightness of a light) is often a logarithmic

function of the actual intensity of the stimulus (brightness measured in units of Lux). In this instance, the geometric mean is a better "summary" of ratings than the simple mean. Note that if a variable contains negative values or a zero (0), then the geometric mean cannot be calculated.

Harmonic mean. The harmonic mean is sometimes used to average frequencies (sample sizes). The harmonic mean is calculated as:

$$HM = \frac{n}{\sum_{i=1}^{n} \frac{1}{x_i}}$$

where: HM is the harmonic mean, n is the number of valid cases, and x_i is the score for the i'th valid case. If a variable contains a zero (0) as a valid score, then the harmonic mean cannot be calculated (since it implies division by zero).

Variance and standard deviation. The variance and standard deviation are standard measures of variability (see *Basic Statistics*).

Average deviation. The average deviation is another measure of variability. It is calculated as the sum of absolute deviations (mean for respective variable minus raw score) divided by n (number of valid cases).

Range. The range of a variable is also an indicator of variability. It is calculated as the largest valid score minus the smallest valid score.

Quartile range. The quartile range of a variable is calculated as the value of the 75th percentile minus the value of the 25th percentile. Thus, it is the width of the range about the median that includes 50% of the cases.

Skewness. As implied by the term, the skewness is a measure of the extent to which the distribution of the respective variable is skewed to the left (negative value) or right (positive value), relative to the standard normal distribution (for which the skewness is 0). The measure skewness is related to the third moment of the distribution. The skewness is defined as:

Skewness = $n*M_3/[(n-1)*(n-2)*\sigma^3]$

where: M_3 is equal to: $\Sigma(xi-Meanx)^3$, n is the valid number of cases, and σ^3 is the standard deviation (sigma) raised to the third power.

Kurtosis. The kurtosis is a measure of how "wide" or "skinny" ("flat" or "peaked") the distribution is for the respective variable, relative to the standard normal distribution (for which the kurtosis is equal to 0). It is also sometimes referred to as the fourth moment of the distribution. The kurtosis is defined as:

Kurtosis = $[n*(n+1)*M_4 - 3*M_2*M_2*(n-1)] / [(n-1)*(n-2)*(n-3)*\sigma^4]$

where: M_j is equal to: $\Sigma(x_i-Meanx)^j$, n is the valid number of cases, and σ^4 is the standard deviation (sigma) raised to the fourth power.

PARTIAL LEAST SQUARES (PLS)

This chapter describes the use of partial least squares regression analysis, an extension of general multiple regression (see Chapter 19) that removes restrictions of the traditional regression models and is particularly suited for exploratory data analysis. If you are unfamiliar with the basic methods of regression in linear models, it may be useful to first review the information on these topics in Chapter 1 – *Elementary Concepts in Statistics*. The different designs discussed in this chapter are also described in Chapter 18 – *General Linear Models*, Chapter 19 – *General Regression Models*, and Chapter 21 – *Generalized Linear/Nonlinear Models*.

Overview

Partial least squares regression is an extension of the multiple linear regression model (see, e.g., Chapter 19 – *General Regression Models* or Chapter 26 – *Multiple Linear Regression*). In its simplest form, a linear model specifies the (linear) relationship between a dependent (response) variable Y, and a set of predictor variables, the X's, so that:

$$Y = b_0 + b_1X_1 + b_2X_2 + ... + b_pX_p$$

In this equation b_0 is the regression coefficient for the intercept and the b_i values are the regression coefficients (for variables 1 through p) computed from the data.

So for example, you could estimate (i.e., predict) a person's weight as a function of the person's height and gender. You could use linear regression to estimate the respective regression coefficients from a sample of data, measuring height, weight, and observing the subjects' gender. For many data analysis problems, estimates of the linear relationships between variables are adequate to describe the observed data, and to make reasonable predictions for new observations.

The multiple linear regression model has been extended in a number of ways to address more sophisticated data analysis problems. The multiple linear regression model serves as the basis for a number of multivariate methods such as discriminant analysis (i.e., the prediction of group membership from the levels of continuous predictor variables), principal components regression (i.e., the prediction of responses on the dependent variables from factors underlying the levels of the predictor variables), and canonical correlation (i.e., the prediction of factors underlying responses on the dependent variables from factors underlying the levels of the predictor variables). These multivariate methods all have two important properties in common. These methods impose restrictions such that 1) factors underlying the Y and X variables are extracted from the Y'Y and X'X matrices, respectively, and never from cross-product matrices involving both the Y and X variables, and 2) the number of prediction functions can never exceed the minimum of the number of Y variables and X variables.

Partial least squares regression extends multiple linear regression without imposing the restrictions employed by discriminant analysis, principal components regression, and canonical correlation. In partial least squares regression, prediction functions are represented by factors extracted from the **Y'XX'Y** matrix. The number of such prediction functions that can be extracted typically will exceed the maximum of the number of Y and X variables.

In short, partial least squares regression is probably the least restrictive of the various multivariate extensions of the multiple linear regression model. This flexibility enables it to be used in situations where the use of traditional multivariate methods is severely limited, such as when there are fewer observations than predictor variables. Furthermore, partial least squares regression can be used as an exploratory analysis tool to select suitable predictor variables and to identify outliers before classical linear regression.

Partial least squares regression has been used in various disciplines such as chemistry, economics, medicine, psychology, and pharmaceutical science where predictive linear modeling, especially with a large number of predictors, is necessary. Especially in chemometrics, partial least squares regression has become a standard tool for modeling linear relations between multivariate measurements (de Jong, 1993).

Computational Approach

Basic Model

As in multiple linear regression, the main purpose of partial least squares regression is to build a linear model, $Y=XB+E$, where Y is an n cases by m variables response matrix, X is an n cases by p variables predictor (design) matrix, B is a p by m regression coefficient matrix, and E is a noise term for the model which has the same dimensions as Y. Usually, the variables in X and Y are centered by subtracting their means and scaled by dividing by their standard deviations. For more information about centering and scaling in partial least squares regression, you can refer to Geladi and Kowalsky (1986).

Both principal components regression and partial least squares regression produce factor scores as linear combinations of the original predictor variables, so that there is no correlation between the factor score variables used in the predictive regression model. For example, suppose we have a data set with response variables Y (in matrix form) and a large number of predictor variables X (in matrix form), some of which are highly correlated. A regression using factor extraction for this type of data computes the factor score matrix $T=XW$ for an appropriate weight matrix W, and then considers the linear regression model $Y=TQ+E$, where $\mathbf{Q}$ is a matrix of regression coefficients (loadings) for T, and E is an error (noise) term. Once the loadings Q are computed, the above regression model is equivalent to $Y=XB+E$, where $B=WQ$, which can be used as a predictive regression model.

Principal components regression and partial least squares regression differ in the methods used in extracting factor scores. In short, principal components regression produces the weight matrix $\mathbf{W}$ reflecting the covariance structure between the predictor variables, while partial least squares regression produces the weight matrix $\mathbf{W}$ reflecting the covariance structure between the predictor and response variables.

For establishing the model, partial least squares regression produces a p by c weight matrix $\mathbf{W}$ for X such that $T=XW$, i.e., the columns of W are weight vectors for the X columns producing the corresponding n by c factor score matrix $\mathbf{T}$. These weights are computed so that each of them maximizes the covariance between responses and the corresponding factor scores. Ordinary least squares procedures for the regression of Y on T are then performed to produce Q, the loadings for Y (or weights for Y) such that $Y=TQ+E$. Once Q is computed, we have $Y=XB+E$, where $B=WQ$, and the prediction model is complete.

One additional matrix that is necessary for a complete description of partial least squares regression procedures is the p by c factor loading matrix $\mathbf{P}$, which gives a factor model $X=TP+F$, where F is the unexplained part of the X scores. We now can describe the algorithms for computing partial least squares regression.

NIPALS Algorithm

The standard algorithm for computing partial least squares regression components (i.e., factors) is nonlinear iterative partial least squares (NIPALS). There are many variants of the NIPALS algorithm that normalize or do not normalize certain vectors. The following algorithm, which assumes that the X and Y variables have been transformed to have means of zero, is considered to be one of most efficient NIPALS algorithms.

For each $h=1,\ldots,c$, where $A_0=X'Y$, $M_0=X'X$, $C_0=I$, and c given,

1. compute q_h, the dominant eigenvector of $A_h'A_h$

2. $w_h=G_hA_hq_h$, $w_h=w_h/||w_h||$, and store w_h into W as a column

3. $p_h=M_hw_h$, $c_h=w_h'M_hw_h$, $p_h=p_h/c_h$, and store p_h into P as a column

4. $q_h=A_h'w_h/c_h$, and store q_h into Q as a column

5. $A_{h+1}=A_h - c_hp_hq_h'$ and $B_{h+1}=M_h - c_hp_hp_h'$

6. $C_{h+1}=C_h - w_hp_h'$

The factor scores matrix **T** is then computed as $T=XW$ and the partial least squares regression coefficients B of Y on X are computed as $B=WQ$.

SIMPLS Algorithm

An alternative estimation method for partial least squares regression components is the SIMPLS algorithm (de Jong, 1993), which can be described as follows.

For each $h=1,\ldots,c$, where $A_0=X'Y$, $M_0=X'X$, $C_0=I$, and c given,

1. compute q_h, the dominant eigenvector of $A_h'A_h$

2. $w_h=A_hq_h$, $c_h=w_h'M_hw_h$, $w_h=w_h/sqrt(c_h)$, and store w_h into W as a column

3. $p_h=M_hw_h$, and store p_h into P as a column

4. $q_h=A_h'w_h$, and store q_h into Q as a column

5. $v_h=C_hp_h$, and $v_h=v_h/||v_h||$

6. $C_{h+1}=C_h - v_hv_h'$ and $M_{h+1}=M_h - p_hp_h'$

7. $A_{h+1}=C_hA_h$

Similarly to NIPALS, the T of SIMPLS is computed as $T=XW$ and B for the regression of Y on X is computed as $B=WQ'$.

Training (Analysis) and Verification (Cross-Validation) Samples

A very important step when fitting models to be used for prediction of future observation is to verify (cross-validate) the results, i.e., to apply the current results to a new set of observations that was not used to compute those results (estimate the parameters). Some software programs offer very flexible methods for computing detailed predicted value and residual statistics for observations that 1) were not used in the computations for fitting the current model and have observed values for the dependent variables (the so-called cross-validation sample), and 2) were not used in the computations for fitting the current model, and have missing data for the dependent variables (prediction sample).

Types of Analyses

The design for an analysis can include effects for continuous as well as categorical predictor variables. Designs can include polynomials for continuous predictors (e.g., squared or cubic terms) as well as interaction effects (i.e., product terms) for continuous predictors. For categorical predictor, you can fit ANOVA-like designs, including full factorial, nested, and fractional factorial designs, etc. Designs can be incomplete (i.e., involve missing cells), and effects for categorical predictor variables can be represented using either the *sigma*-restricted parameterization or the overparameterized (i.e., indicator variable) representation of effects.

The following topics contain complete descriptions of the types of designs that can be analyzed using partial least squares regression, as well as types of designs that can be analyzed using the general linear model.

Between-Subject Designs

Overview. The levels or values of the predictor variables in an analysis describe the differences between the n subjects or the n valid cases that are analyzed. Thus, when we speak of the between subject design (or simply the between design) for an analysis, we are referring to the nature, number, and arrangement of the predictor variables.

Concerning the nature or type of predictor variables, between designs which contain only categorical predictor variables can be called ANOVA (analysis of variance) designs, between designs which contain only continuous predictor variables can be called regression designs, and between designs which contain both categorical and continuous predictor variables can be called ANCOVA (analysis of covariance) designs. Further, continuous predictors are always considered to have fixed values, but the levels of categorical predictors can be considered to be fixed or to vary randomly. Designs that contain random categorical factors are called mixed-model designs (see Chapter 39 – *Variance Components and Mixed Model ANOVA/ANCOVA*).

Between designs can involve only a single predictor variable and therefore be described as simple (e.g., simple regression) or can employ numerous predictor variables (e.g., multiple regression).

Concerning the arrangement of predictor variables, some between designs employ only "main effect" or first-order terms for predictors, that is, the values for different predictor variables are independent and raised only to the first power. Other between designs can employ higher-order terms for predictors by raising the values for the original predictor variables to a power greater than 1 (e.g., in polynomial regression designs), or by forming products of different predictor variables (i.e., interaction terms). A common arrangement for ANOVA designs is the full-factorial design, in which every combination of levels for each of the categorical predictor variables is represented in the design. Designs with some but not all combinations of levels for each of the categorical predictor variables are aptly called fractional factorial designs. Designs with a hierarchy of combinations of levels for the different categorical predictor variables are called nested designs.

These basic distinctions about the nature, number, and arrangement of predictor variables can be used in describing a variety of different types of between designs. Some of the more common between designs can now be described.

One-way ANOVA. A design with a single categorical predictor variable is called a one-way ANOVA design. For example, a study of 4 different fertilizers used on different individual plants could be analyzed via one-way ANOVA, with four levels for the factor *Fertilizer*.

In general, consider a single categorical predictor variable A with 1 case in each of its 3 categories. Using the *sigma*-restricted coding of A into 2 quantitative contrast variables, the matrix $\mathbf{X}$ defining the between design is:

$$\mathbf{X} = \begin{array}{c} \\ A_1 \\ A_2 \\ A_3 \end{array} \begin{array}{ccc} X_0 & X_1 & X_2 \\ \left[\begin{array}{ccc} 1 & 1 & 0 \\ 1 & 0 & 1 \\ 1 & -1 & -1 \end{array}\right] \end{array}$$

That is, cases in groups A_1, A_2, and A_3 are all assigned values of 1 on X_0 (the intercept), the case in group A_1 is assigned a value of 1 on X_1 and a value 0 on X_2, the case in group A_2 is assigned a value of 0 on X_1 and a value 1 on X_2, and the case in group A_3 is assigned a value of -

1 on X_1 and a value -1 on X_2. Of course, any additional cases in any of the 3 groups would be coded similarly. If there were 1 case in group A_1, 2 cases in group A_2, and 1 case in group A_3, the **X** matrix would be:

$$
\mathbf{X} = \begin{array}{c} \\ A_{11} \\ A_{12} \\ A_{22} \\ A_{13} \end{array} \begin{array}{ccc} X_0 & X_1 & X_2 \\ \begin{bmatrix} 1 & 1 & 0 \\ 1 & 0 & 1 \\ 1 & 0 & 1 \\ 1 & -1 & -1 \end{bmatrix} \end{array}
$$

where the first subscript for A gives the replicate number for the cases in each group. For brevity, replicates usually are not shown when describing ANOVA design matrices.

Note that in one-way designs with an equal number of cases in each group, *sigma*-restricted coding yields $X_1 \ldots X_k$ variables all of which have means of 0.

Using the overparameterized model to represent A, the **X** matrix defining the between design is simply:

$$
\mathbf{X} = \begin{array}{c} \\ A_1 \\ A_2 \\ A_3 \end{array} \begin{array}{cccc} X_0 & X_1 & X_2 & X_3 \\ \begin{bmatrix} 1 & 1 & 0 & 0 \\ 1 & 0 & 1 & 0 \\ 1 & 0 & 0 & 1 \end{bmatrix} \end{array}
$$

These simple examples show that the **X** matrix actually serves two purposes. It specifies 1) the coding for the levels of the original predictor variables on the X variables used in the analysis as well as 2) the nature, number, and arrangement of the X variables, that is, the between design.

Main effect ANOVA. Main effect ANOVA designs contain separate one-way ANOVA designs for 2 or more categorical predictors. A good example of main effect ANOVA is the typical analysis performed on screening designs as described in Chapter 15 – *Experimental Design*.

Consider 2 categorical predictor variables A and B each with 2 categories. Using the *sigma*-restricted coding, the **X** matrix defining the between design is:

$$
\mathbf{X} = \begin{array}{c} \\ A_1B_1 \\ A_1B_2 \\ A_2B_1 \\ A_2B_2 \end{array} \begin{array}{ccc} X_0 & X_1 & X_2 \\ \begin{bmatrix} 1 & 1 & 1 \\ 1 & 1 & -1 \\ 1 & -1 & 1 \\ 1 & -1 & -1 \end{bmatrix} \end{array}
$$

Note that if there are equal numbers of cases in each group, the sum of the cross-products of values for the X_1 and X_2 columns is 0, for example, with 1 case in each group (1*1)+(1*-1)+(-1*1)+(-1*-1)=0. Using the overparameterized model, the matrix **X** defining the between design is:

$$
\mathbf{X} = \begin{array}{c} \\ A_1B_1 \\ A_1B_2 \\ A_2B_1 \\ A_2B_2 \end{array} \begin{array}{ccccc} X_0 & X_1 & X_2 & X_3 & X_4 \\ \begin{bmatrix} 1 & 1 & 0 & 1 & 0 \\ 1 & 1 & 0 & 0 & 1 \\ 1 & 0 & 1 & 1 & 0 \\ 1 & 0 & 1 & 0 & 1 \end{bmatrix} \end{array}
$$

Comparing the two types of coding, it can be seen that the overparameterized coding takes almost twice as many values as the *sigma*-restricted coding to convey the same information.

Factorial ANOVA. Factorial ANOVA designs contain X variables representing combinations of the levels of 2 or more categorical predictors (e.g., a study of boys and girls in four age groups, resulting in a *2 (Gender) x 4 (Age Group)* design). In particular, full-factorial designs represent all possible combinations of the levels of the categorical predictors. A full-factorial design with 2 categorical predictor variables A and B each with 2 levels each would be called a 2x2 full-factorial design. Using the *sigma*-restricted coding, the **X** matrix for this design would be:

$$
\mathbf{X} = \begin{array}{c} \\ A_1B_1 \\ A_1B_2 \\ A_2B_1 \\ A_2B_2 \end{array}
\begin{array}{cccc} X_0 & X_1 & X_2 & X_3 \\ \left[\begin{array}{cccc} 1 & 1 & 1 & 1 \\ 1 & 1 & -1 & -1 \\ 1 & -1 & 1 & -1 \\ 1 & -1 & -1 & 1 \end{array}\right. \end{array}
$$

Several features of this **X** matrix deserve comment. Note that the X_1 and X_2 columns represent main effect contrasts for one variable, (i.e., A and B, respectively) collapsing across the levels of the other variable. The X_3 column instead represents a contrast between different combinations of the levels of A and B. Note also that the values for X_3 are products of the corresponding values for X_1 and X_2. Product variables such as X_3 represent the multiplicative or interaction effects of their factors, so X_3 would be said to represent the 2-way interaction of A and B. The relationship of such product variables to the dependent variables indicates the interactive influences of the factors on responses above and beyond their independent (i.e., main effect) influences on responses. Thus, factorial designs provide more information about the relationships between categorical predictor variables and responses on the dependent variables than are provided by corresponding one-way or main effect designs.

When many factors are being investigated, however, full-factorial designs sometimes require more data than reasonably can be collected to represent all possible combinations of levels of the factors, and high-order interactions between many factors can become difficult to interpret. With many factors, a useful alternative to the full-factorial design is the fractional factorial design. As an example, consider a 2x2 x 2 fractional factorial design to degree 2 with 3 categorical predictor variables each with 2 levels. The design would include the main effects for each variable, and all 2-way interactions between the three variables, but

would not include the 3-way interaction between all three variables. Using the overparameterized model, the **X** matrix for this design is:

$$
\mathbf{X} = \begin{array}{c} \\ A_1B_1C_1 \\ A_1B_1C_2 \\ A_1B_2C_1 \\ A_1B_2C_2 \\ A_2B_1C_1 \\ A_2B_1C_2 \\ A_2B_2C_1 \\ A_2B_2C_2 \end{array}
\left[\begin{array}{ccccccccccccccccccc}
1 & 1 & 0 & 1 & 0 & 1 & 0 & 1 & 0 & 0 & 0 & 1 & 0 & 0 & 0 & 1 & 0 & 0 & 0 \\
1 & 1 & 0 & 1 & 0 & 0 & 1 & 1 & 0 & 0 & 0 & 0 & 1 & 0 & 0 & 0 & 1 & 0 & 0 \\
1 & 1 & 0 & 0 & 1 & 1 & 0 & 0 & 1 & 0 & 0 & 1 & 0 & 0 & 0 & 0 & 0 & 1 & 0 \\
1 & 1 & 0 & 0 & 1 & 0 & 1 & 0 & 1 & 0 & 0 & 0 & 1 & 0 & 0 & 0 & 0 & 0 & 1 \\
1 & 0 & 1 & 1 & 0 & 1 & 0 & 0 & 0 & 1 & 0 & 0 & 0 & 1 & 0 & 1 & 0 & 0 & 0 \\
1 & 0 & 1 & 1 & 0 & 0 & 1 & 0 & 0 & 1 & 0 & 0 & 0 & 0 & 1 & 0 & 1 & 0 & 0 \\
1 & 0 & 1 & 0 & 1 & 1 & 0 & 0 & 0 & 0 & 1 & 0 & 0 & 1 & 0 & 0 & 0 & 1 & 0 \\
1 & 0 & 1 & 0 & 1 & 0 & 1 & 0 & 0 & 0 & 1 & 0 & 0 & 0 & 1 & 0 & 0 & 0 & 1
\end{array}\right]
$$

with column groupings labeled "......main effects......" and "...............2 - way interactions...........".

The 2-way interactions are the highest degree effects included in the design. These types of designs are discussed in detail in $2^{(k-p)}$ *Fractional Factorial Designs* in Chapter 15 – *Experimental Design (Industrial DOE)*.

Nested ANOVA designs. Nested designs are similar to fractional factorial designs in that all possible combinations of the levels of the categorical predictor variables are not represented in the design. In nested designs, however, the omitted effects are lower-order effects. Nested effects are effects in which the nested variables never appear as main effects. Suppose that for 2 variables A and B with 3 and 2 levels, respectively, the design includes the main effect for A and the effect of B nested within the levels of A. The **X** matrix for this design using the overparameterized model is:

$$
\mathbf{X} = \begin{array}{c} \\ A_1B_1 \\ A_1B_2 \\ A_2B_1 \\ A_2B_2 \\ A_3B_1 \\ A_3B_2 \end{array}
\begin{array}{cccccccccc} X_0 & X_1 & X_2 & X_3 & X_4 & X_5 & X_6 & X_7 & X_8 & X_9 \\
\left[\begin{array}{cccccccccc}
1 & 1 & 0 & 0 & 1 & 0 & 0 & 0 & 0 & 0 \\
1 & 1 & 0 & 0 & 0 & 1 & 0 & 0 & 0 & 0 \\
1 & 0 & 1 & 0 & 0 & 0 & 1 & 0 & 0 & 0 \\
1 & 0 & 1 & 0 & 0 & 0 & 0 & 1 & 0 & 0 \\
1 & 0 & 0 & 1 & 0 & 0 & 0 & 0 & 1 & 0 \\
1 & 0 & 0 & 1 & 0 & 0 & 0 & 0 & 0 & 1
\end{array}\right] \end{array}
$$

Note that if the *sigma*-restricted coding were used, there would be only 2 columns in the **X** matrix for the B nested within A effect instead of the 6 columns in the **X** matrix for this effect when the overparameterized model coding is used (i.e., columns X_4 through X_9). The *sigma-*

restricted coding method is overly restrictive for nested designs, so only the overparameterized model is used to represent nested designs.

Simple regression. Simple regression designs involve a single continuous predictor variable. If there were 3 cases with values on a predictor variable P of, say, 7, 4, and 9, and the design is for the first-order effect of P, the **X** matrix would be:

$$X = \begin{matrix} X_0 & X_1 \\ \begin{bmatrix} 1 & 7 \\ 1 & 4 \\ 1 & 9 \end{bmatrix} \end{matrix}$$

and using P for X_1 the regression equation would be:

$$Y = b_0 + b_1 P$$

If the simple regression design is for a higher-order effect of P, say the quadratic effect, the values in the X_1 column of the design matrix would be raised to the 2nd power, that is, squared:

$$X = \begin{matrix} X_0 & X_1 \\ \begin{bmatrix} 1 & 49 \\ 1 & 16 \\ 1 & 81 \end{bmatrix} \end{matrix}$$

and using P^2 for X_1 the regression equation would be:

$$Y = b_0 + b_1 P^2$$

The *sigma*-restricted and overparameterized coding methods do not apply to simple regression designs and any other design containing only continuous predictors (since there are no categorical predictors to code). Regardless of which coding method is chosen, values on the continuous predictor variables are raised to the desired power and used as the values for the X variables. No recoding is performed. It is therefore sufficient, in describing regression designs, to simply describe the regression equation without explicitly describing the design matrix **X**.

Multiple regression. Multiple regression designs are to continuous predictor variables as main effect ANOVA designs are to categorical predictor variables, that is, multiple regression designs contain the separate simple regression designs for 2 or more continuous predictor variables. The regression equation for a multiple regression design for the first-order effects of 3 continuous predictor variables P, Q, and R would be:

$$Y = b_0 + b_1 P + b_2 Q + b_3 R$$

Factorial regression. Factorial regression designs are similar to factorial ANOVA designs, in which combinations of the levels of the factors are represented in the design. In factorial regression designs, however, there may be many more such possible combinations of distinct levels for the continuous predictor variables than there are cases in the data set. To simplify matters, full-factorial regression designs are defined as designs in which all possible products of the continuous predictor variables are represented in the design. For example, the full-factorial regression design for two continuous predictor variables P and Q would include the main effects (i.e., the first-order effects) of P and Q and their 2-way P by Q interaction effect, which is represented by the product of P and Q scores for each case. The regression equation would be:

$$Y = b_0 + b_1 P + b_2 Q + b_3 P*Q$$

Factorial regression designs can also be fractional, that is, higher-order effects can be omitted from the design. A fractional factorial design to degree 2 for 3 continuous predictor variables P, Q, and R would include the main effects and all 2-way interactions between the predictor variables:

$Y = b_0 + b_1P + b_2Q + b_3R + b_4P*Q + b_5P*R + b_6Q*R$

Polynomial regression. Polynomial regression designs are designs that contain main effects and higher-order effects for the continuous predictor variables but do not include interaction effects between predictor variables. For example, the polynomial regression design to degree 2 for three continuous predictor variables P, Q, and R would include the main effects (i.e., the first-order effects) of P, Q, and R and their quadratic (i.e., second-order) effects, but not the 2-way interaction effects or the P by Q by R 3-way interaction effect.

$Y = b_0 + b_1P + b_2P^2 + b_3Q + b_4Q^2 + b_5R + b_6R^2$

Polynomial regression designs do not have to contain all effects up to the same degree for every predictor variable. For example, main, quadratic, and cubic effects could be included in the design for some predictor variables, and effects up the fourth degree could be included in the design for other predictor variables.

Response surface regression. Quadratic response surface regression designs are a hybrid type of design with characteristics of both polynomial regression designs and fractional factorial regression designs. Quadratic response surface regression designs contain all the same effects of polynomial regression designs to degree 2 and additionally the 2-way interaction effects of the predictor variables. The regression equation for a quadratic response surface regression design for 3 continuous predictor variables P, Q, and R would be:

$Y = b_0 + b_1P + b_2P^2 + b_3Q + b_4Q^2 + b_5R + b_6R^2 + b_7P*Q + b_8P*R + b_9Q*R$

These types of designs are commonly employed in applied research (e.g., in industrial experimentation), and a detailed discussion of these types of designs is also presented in Chapter 15 – *Experimental Design* (see *Central Composite Designs*).

Analysis of covariance. In general, between designs that contain both categorical and continuous predictor variables can be called ANCOVA designs. Traditionally, however, ANCOVA designs have referred more specifically to designs in which the first-order effects of one or more continuous predictor variables are taken into account when assessing the effects of one or more categorical predictor variables. A basic introduction to analysis of covariance can also be found in *Analysis of Covariance (ANCOVA)* in Chapter 3 – *ANOVA/MANOVA*.

To illustrate, suppose a researcher wants to assess the influences of a categorical predictor variable A with 3 levels on some outcome, and that measurements on a continuous predictor variable P, known to covary with the outcome, are available. If the data for the analysis are:

$$\begin{array}{cc} P & Group \\ \begin{bmatrix} 7 \\ 4 \\ 9 \\ 3 \\ 6 \\ 8 \end{bmatrix} & \begin{bmatrix} A_1 \\ A_1 \\ A_2 \\ A_2 \\ A_3 \\ A_3 \end{bmatrix} \end{array}$$

the *sigma*-restricted $\mathbf{X}$ matrix for the design that includes the separate first-order effects of P and A would be:

$$\mathbf{X} = \begin{array}{c} \begin{array}{cccc} X_0 & X_1 & X_2 & X_3 \end{array} \\ \begin{bmatrix} 1 & 7 & 1 & 0 \\ 1 & 4 & 1 & 0 \\ 1 & 9 & 0 & 1 \\ 1 & 3 & 0 & 1 \\ 1 & 6 & -1 & -1 \\ 1 & 8 & -1 & -1 \end{bmatrix} \end{array}$$

The b_2 and b_3 coefficients in the regression equation:

$$Y = b_0 + b_1X_1 + b_2X_2 + b_3X_3$$

represent the influences of group membership on the A categorical predictor variable, controlling for the influence of scores on the P continuous predictor variable. Similarly, the b_1 coefficient represents the influence of scores on P controlling for the influences of group membership on A. This traditional ANCOVA analysis gives a more sensitive test of the influence of A to the extent that P reduces the prediction error, that is, the residuals for the outcome variable.

The **X** matrix for the same design using the overparameterized model would be:

$$X = \begin{array}{cccccc} X_0 & X_1 & X_2 & X_3 & X_4 \\ \left[\begin{array}{ccccc} 1 & 7 & 1 & 0 & 0 \\ 1 & 4 & 1 & 0 & 0 \\ 1 & 9 & 0 & 1 & 0 \\ 1 & 3 & 0 & 1 & 0 \\ 1 & 6 & 0 & 0 & 1 \\ 1 & 8 & 0 & 0 & 1 \end{array}\right] \end{array}$$

The interpretation is unchanged except that the influences of group membership on the A categorical predictor variables are represented by the b_2, b_3 and b_4 coefficients in the regression equation:

$$Y = b_0 + b_1X_1 + b_2X_2 + b_3X_3 + b_4X_4$$

Separate slope designs. The traditional analysis of covariance (ANCOVA) design for categorical and continuous predictor variables is inappropriate when the categorical and continuous predictors interact in influencing responses on the outcome. The appropriate design for modeling the influences of the predictors in this situation is called the separate slope design. For the same example data used to illustrate traditional ANCOVA, the

overparameterized **X** matrix for the design that includes the main effect of the three-level categorical predictor A and the 2-way interaction of P by A would be:

$$X = \begin{array}{ccccccc} X_0 & X_1 & X_2 & X_3 & X_4 & X_5 & X_6 \\ \left[\begin{array}{ccccccc} 1 & 1 & 0 & 0 & 7 & 0 & 0 \\ 1 & 1 & 0 & 0 & 4 & 0 & 0 \\ 1 & 0 & 1 & 0 & 0 & 9 & 0 \\ 1 & 0 & 1 & 0 & 0 & 3 & 0 \\ 1 & 0 & 0 & 1 & 0 & 0 & 6 \\ 1 & 0 & 0 & 1 & 0 & 0 & 8 \end{array}\right] \end{array}$$

The b_4, b_5, and b_6 coefficients in the regression equation:

$$Y = b_0 + b_1X_1 + b_2X_2 + b_3X_3 + b_4X_4 + b_5X_5 + b_6X_6$$

give the separate slopes for the regression of the outcome on P within each group on A, controlling for the main effect of A.

As with nested ANOVA designs, the *sigma*-restricted coding of effects for separate slope designs is overly restrictive, so only the overparameterized model is used to represent separate slope designs. In fact, separate slope designs are identical in form to nested ANOVA designs, since the main effects for continuous predictors are omitted in separate slope designs.

Homogeneity of slopes. The appropriate design for modeling the influences of continuous and categorical predictor variables depends on whether the continuous and categorical predictors interact in influencing the outcome. The traditional analysis of covariance (ANCOVA) design for continuous and categorical predictor variables is appropriate when the continuous and categorical predictors do not interact in influencing responses on the outcome, and the separate slope design is appropriate when the continuous and categorical predictors do interact in influencing responses. The homogeneity of slopes designs can be used to test whether the continuous and categorical predictors interact in influencing responses, and thus, whether the traditional ANCOVA design or the separate slope design is appropriate for modeling the effects of the predictors. For the same example data used to illustrate the traditional ANCOVA and separate slope designs, the overparameterized $\mathbf{X}$ matrix for the design that includes the main effect of P, the main effect of the three-level categorical predictor A, and the 2-way interaction of P by A would be:

$$\mathbf{X} = \begin{matrix} X_0 & X_1 & X_2 & X_3 & X_4 & X_5 & X_6 & X_7 \\ \begin{bmatrix} 1 & 7 & 1 & 0 & 0 & 7 & 0 & 0 \\ 1 & 4 & 1 & 0 & 0 & 4 & 0 & 0 \\ 1 & 9 & 0 & 1 & 0 & 0 & 9 & 0 \\ 1 & 3 & 0 & 1 & 0 & 0 & 3 & 0 \\ 1 & 6 & 0 & 0 & 1 & 0 & 0 & 6 \\ 1 & 8 & 0 & 0 & 1 & 0 & 0 & 8 \end{bmatrix} \end{matrix}$$

If the b_5, b_6, or b_7 coefficient in the regression equation

$$Y = b_0 + b_1 X_1 + b_2 X_2 + b_3 X_3 + b_4 X_4 + b_5 X_5 + b_6 X_6 + b_7 X_7$$

is non-zero, the separate slope model should be used. If instead all 3 of these regression coefficients are zero the traditional ANCOVA design should be used.

The *sigma*-restricted $\mathbf{X}$ matrix for the homogeneity of slopes design would be:

$$\mathbf{X} = \begin{matrix} X_0 & X_1 & X_2 & X_3 & X_4 & X_5 \\ \begin{bmatrix} 1 & 7 & 1 & 0 & 7 & 0 \\ 1 & 4 & 1 & 0 & 4 & 0 \\ 1 & 9 & 0 & 1 & 0 & 9 \\ 1 & 3 & 0 & 1 & 0 & 3 \\ 1 & 6 & -1 & -1 & -6 & -6 \\ 1 & 8 & -1 & -1 & -8 & -8 \end{bmatrix} \end{matrix}$$

Using this $\mathbf{X}$ matrix, if the b_4, or b_5 coefficient in the regression equation:

$$Y = b_0 + b_1 X_1 + b_2 X_2 + b_3 X_3 + b_4 X_4 + b_5 X_5$$

is non-zero, the separate slope model should be used. If instead both of these regression coefficients are zero the traditional ANCOVA design should be used.

Distance Graphs

A graphic technique that is useful in analyzing partial least squares designs is a distance graph. With these graphs, you can compute and plot distances from the origin (zero for all dimensions) for the predicted and residual statistics, loadings, and weights for the respective number of components.

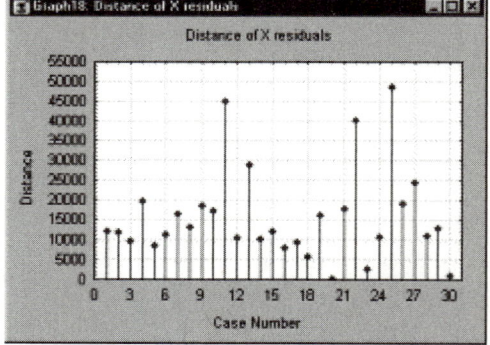

Based on Euclidean distances, these observation plots can be helpful in determining major contributors to the prediction of the conceptual variable(s) (plotting weights) as well as outliers that have a disproportionate influence (relative to the other observation) on the results (plotting residual values).

POWER ANALYSIS

Written by James Steiger, Ph.D.

The techniques of statistical power analysis, sample size estimation, and advanced techniques for confidence interval estimation are discussed in this chapter. The main goal of first two techniques is to enable you to decide, while in the process of designing an experiment, a) how large a sample is needed to enable statistical judgments that are accurate and reliable and b) how likely your statistical test will be to detect effects of a given size in a particular situation. The third technique is useful in implementing objectives a and b and in evaluating the size of experimental effects in practice.

Overview

Performing power analysis and sample size estimation is an important aspect of experimental design, because without these calculations, sample size may be too high or too low. If sample size is too low, the experiment will lack the precision to provide reliable answers to the questions it is investigating. If sample size is too large, time and resources will be wasted, often for minimal gain.

In some power analysis software programs, a number of graphical and analytical tools are available to enable precise evaluation of the factors affecting power and sample size in many of the most commonly encountered statistical analyses. This information can be crucial to the design of a study that is cost-effective and scientifically useful.

Noncentrality interval estimation procedures and other sophisticated confidence interval procedures provide some sophisticated confidence interval methods for analyzing the importance of an observed experimental result. An increasing number of influential statisticians are suggesting that confidence interval estimation should augment or replace traditional hypothesis testing approaches in the analysis of experimental data.

Power Analysis and Sample Size Calculation in Experimental Design

There is a growing recognition of the importance of power analysis and sample size calculation in the proper design of experiments. See the following topics for a discussion of the fundamental ideas behind these methods.

Sampling Theory

In most situations in statistical analysis, we do not have access to an entire statistical population of interest, either because the population is too large, is not willing to be measured, or the measurement process is too expensive or time-consuming to allow more than a small segment of the population to be observed. As a result, we often make important decisions about a statistical population on the basis of a relatively small amount of sample data.

Typically, we take a sample and compute a quantity called a statistic in order to estimate some characteristic of a population called a parameter.

For example, suppose a politician is interested in the proportion of people who currently favor her position on a particular issue. Her constituency is a large city with a population of about 1,500,000 potential voters. In this case, the parameter of interest, which we might call π, is the proportion of people in the entire population who favor the politician's position. The politician is going to commission an opinion poll, in which a (hopefully) random sample of people will be asked whether or not they favor her position. The number (call it N) of people to be polled will be quite small, relative to the size of the population. Once these people have been polled, the proportion of them favoring the politician's position will be computed. This proportion, which is a statistic, can be called p.

One thing is virtually certain before the study is ever performed: p will not be equal to π. Because p involves "the luck of the draw," it will deviate from π. The amount by which p is wrong, i.e., the amount by which it deviates from π, is called sampling error.

In any one sample, it is virtually certain there will be some sampling error (except in some highly unusual circumstances), and that we will never be

certain exactly how large this error is. If we knew the amount of the sampling error, this would imply that we also knew the exact value of the parameter, in which case we would not need to be doing the opinion poll in the first place.

In general, the larger the sample size N, the smaller sampling error tends to be. (One can never be sure what will happen in a particular experiment, of course.) If we are to make accurate decisions about a parameter such as π, we need to have an N large enough so that sampling error will tend to be reasonably small. If N is too small, there is not much point in gathering the data, because the results will tend to be too imprecise to be of much use.

On the other hand, there is also a point of diminishing returns beyond which increasing N provides little benefit. Once N is large enough to produce a reasonable level of accuracy, making it larger simply wastes time and money.

So some key decisions in planning any experiment are, "How precise will my parameter estimates tend to be if I select a particular sample size?" and "How big a sample do I need to attain a desirable level of precision?"

The purpose of power analysis and sample size estimation is to provide you with the statistical methods to answer these questions quickly, easily, and accurately. A good statistical software program will provide simple dialogs for performing power calculations and sample size estimation for many of the classic statistical procedures as well as special noncentral distribution routines to enable the advanced user to perform a variety of additional calculations.

Hypothesis Testing Logic

Suppose that the politician was interested in showing that more than the majority of people supported her position. Her question, in statistical terms: "Is $\pi > .50$?" Being an optimist, she believes that it is.

In statistics, the following strategy is quite common. State as a "statistical null hypothesis" something that is the logical opposite of what you believe. Call this hypothesis *H0*. Gather data. Then, using statistical theory, show from the data that it is likely *H0* is false, and should be rejected.

By rejecting *H0*, you support what you actually believe. This kind of situation, which is typical in many fields of research, for example, is called Reject-Support testing (RS testing) because rejecting the null hypothesis supports the experimenter's theory.

The null hypothesis is either true or false, and the statistical decision process is set up so that there are no ties. The null hypothesis is either rejected or not rejected. Consequently, before undertaking the experiment, we can be certain that only 4 possible things can happen. These are summarized in the following table.

		State of the World	
		H0	H1
Decision	H0	Correct Acceptance	Type II Error β
	H1	Type I Error α	Correct Rejection

Note that there are two kinds of errors represented in the table. Many statistics textbooks present a point of view that is common in the social sciences, i.e., that α, the Type I error rate, must be kept at or below .05, and that, if at all possible, β, the Type II error rate, must be kept low as well. Statistical power, which is equal to $1 - \beta$, must be kept correspondingly high. Ideally, power should be at least .80 to detect a reasonable departure from the null hypothesis.

The conventions are, of course, much more rigid with respect to α than with respect to β. For example, in the social sciences seldom, if ever, is α allowed to stray above the magical .05 mark.

Significance testing (RS/AS). In the context of significance testing, we can define two basic kinds of situations, reject-support (RS) (discussed above) and accept-support (AS). In RS testing, the null hypothesis is the opposite of what the researcher actually believes, and rejecting it supports the researcher's theory. In a two-group RS experiment involving comparison of the means of an experimental and control group, the experimenter believes the treatment has an effect, and seeks to confirm it through a significance test that rejects the null hypothesis.

In the RS situation, a Type I error represents, in a sense, a false positive for the researcher's theory. From society's standpoint, such false positives are particularly undesirable. They result in much wasted effort, especially when the false positive is interesting from a theoretical or political standpoint (or both), and as a result stimulates a substantial amount of research. Such follow-up research will usually not replicate the (incorrect) original work, and much confusion and frustration will result.

In RS testing, a Type II error is a tragedy from the researcher's standpoint, because a theory that is true is, by mistake, not confirmed. So, for example, if a drug designed to improve a medical condition is found (incorrectly) not to produce an improvement relative to a control group, a worthwhile therapy will be lost, at least temporarily, and an experimenter's worthwhile idea will be discounted.

As a consequence, in RS testing, society (in the person of journal editors and reviewers) insists on keeping α low. The statistically well-informed researcher makes it a top priority to keep β low

as well. Ultimately, of course, everyone benefits if both error probabilities are kept low, but unfortunately there is often, in practice, a trade-off between the two types of error.

The RS situation is by far the more common one, and the conventions relevant to it have come to dominate popular views on statistical testing. As a result, the prevailing views on error rates are that relaxing α beyond a certain level is unthinkable, and that it is up to the researcher to make sure *statistical power* is adequate. You might argue how appropriate these views are in the context of RS testing, but they are not altogether unreasonable.

In AS testing, the common view on error rates we described above is clearly inappropriate. In AS testing, *H0* is what the researcher actually believes, so accepting it supports the researcher's theory. In this case, a Type I error is a false negative for the researcher's theory, and a Type II error constitutes a false positive. Consequently, acting in a way that might be construed as highly virtuous in the RS situation, for example, maintaining a very low Type I error rate such as .001, is actually "stacking the deck" in favor of the researcher's theory in AS testing.

In both AS and RS situations, it is easy to find examples where significance testing seems strained and unrealistic. Consider first the RS situation. In some such situations, it is simply not possible to have very large samples. An example that comes to mind is social or clinical psychological field research. Researchers in these fields sometimes spend several days interviewing a single subject. A year's research may only yield valid data from 50 subjects. Correlational tests, in particular, have very low power when samples are that small. In such a case, it probably makes sense to relax α beyond .05, if it means that reasonable power can be achieved.

On the other hand, it is possible, in an important sense, to have power that is too high. For example, you might be testing the hypothesis that two population means are equal (i.e., $Mu1 = Mu2$) with sample sizes of a million in each group. In this case, even with trivial differences between groups, the null hypothesis would virtually always be rejected.

The situation becomes even more unnatural in AS testing. Here, if N is too high, the researcher almost inevitably decides against the theory, even when it turns out, in an important sense, to be an excellent approximation to the data. It seems paradoxical indeed that in this context experimental precision seems to work against the researcher.

To summarize, in reject-support research:

1. The researcher wants to reject $H0$.

2. Society wants to control Type I error.

3. The researcher must be very concerned about Type II error.

4. High sample size works for the researcher.

5. If there is too much power, trivial effects become highly significant.

In accept-support research:

1. The researcher wants to accept $H0$.

2. "Society" should be worrying about controlling Type II error, although it sometimes gets confused and retains the conventions applicable to RS testing.

3. The researcher must be very careful to control Type I error.

4. High sample size works against the researcher.

5. If there is too much power, the researcher's theory can be rejected by a significance test even though it fits the data almost perfectly.

Calculating Power

Properly designed experiments must ensure that power will be reasonably high to detect reasonable departures from the null hypothesis. Otherwise, an experiment is hardly worth doing. Elementary textbooks contain detailed discussions of the factors influencing power in a statistical test. These include

1. What kind of statistical test is being performed? Some statistical tests are inherently more powerful than others.

2. Sample size. In general, the larger the sample size, the larger the power. However, generally increasing sample size involves tangible costs, both in time, money, and effort. Consequently, it is important to make sample size large enough, but not wastefully large.

3. The size of experimental effects. If the null hypothesis is wrong by a substantial amount, power will be higher than if it is wrong by a small amount.

4. The level of error in experimental measurements. Measurement error acts like "noise" that can bury the "signal" of real experimental effects. Consequently, anything that enhances the accuracy and consistency of measurement can increase statistical power.

Calculating Required Sample Size

To ensure a statistical test will have adequate power, you usually must perform special analyses prior to running the experiment, to calculate how large an N is required.

Let's briefly examine the kind of statistical theory that lies at the foundation of the calculations used to estimate power and sample size. Return to the original example of the politician, contemplating how large an opinion poll should be taken to suit her purposes.

Statistical theory, of course, cannot tell us what will happen with any particular opinion poll. However, through the concept of a sampling distribution, it can tell us what will tend to happen in the long run, over many opinion polls of a particular size.

A sampling distribution is the distribution of a statistic over repeated samples. Consider the sample proportion p resulting from an opinion poll of size N, in the situation where the population proportion π is exactly .50. Sampling distribution theory tells us that p will have a distribution that can be calculated from the binomial theorem. This distribution, for reasonably large N, and for values of p not too close to 0 or 1, looks very much like a normal distribution with a mean of π and a standard deviation (called the standard error of the proportion) of:

$$\sigma_p = \sqrt{\frac{\pi(1-\pi)}{N}}$$

Suppose, for example, the politician takes an opinion poll based on an N of 100. Then the distribution of p, over repeated samples, will look like this if $\pi = .5$.

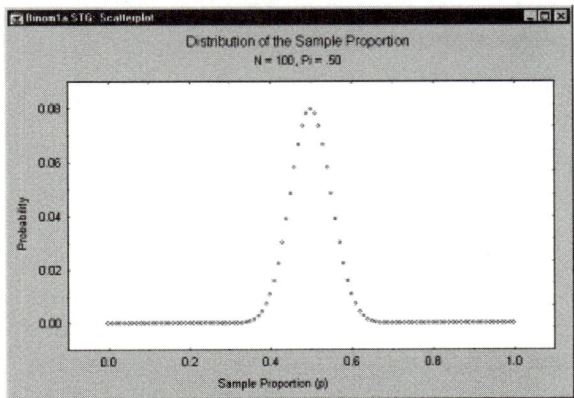

The values are centered around .5, but a small percentage of values are greater than .6 or less than .4. This distribution of values reflects the

fact that an opinion poll based on a sample of 100 is an imperfect indicator of the population proportion π.

If p were a perfect estimate of π, the standard error of the proportion would be zero, and the sampling distribution would be a spike located at 0.5. The spread of the sampling distribution indicates how much "noise" is mixed in with the signal generated by the parameter.

Notice from the equation for the standard error of the proportion that, as N increases, the standard error of the proportion gets smaller. If N becomes large enough, we can be very certain that our estimate p will be a very accurate one.

Suppose the politician uses a decision criterion as follows. If the observed value of p is greater than .58, she will decide that the null hypothesis that π is less than or equal to .50 is false. This rejection rule is diagrammed in the next illustration.

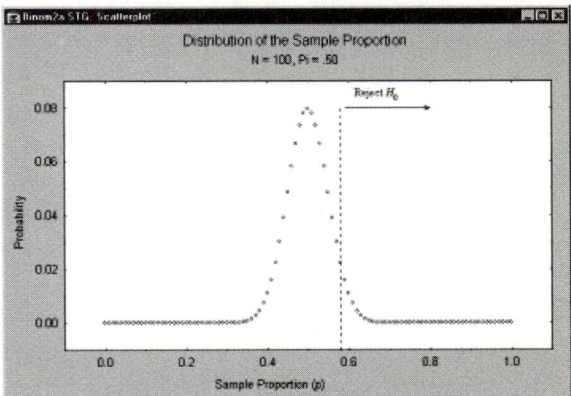

You can determine, by adding up all the probabilities (computable from the binomial distribution) that the probability of rejecting the null hypothesis when $p = .50$ is .044. Hence, this decision rule controls the Type I Error rate, α, at or below .044. It turns out, this is the lowest decision criterion that maintains α at or below .05.

However, the politician is also concerned about power in this situation, because it is by rejecting the null hypothesis that she is able to support the notion that she has public opinion on her side.

Suppose that 55% of the people support the politician, that is, that $\pi = .55$ and the null hypothesis is actually false. In this case, the correct decision is to reject the null hypothesis. What is the probability that she will obtain a sample proportion greater than the cut-off value of .58 required to reject the null hypothesis?

In the next illustration, we have superimposed the sampling distribution for p when $\pi = .55$. Clearly, only a small percentage of the time will the politician reach the correct decision that she has majority support. The probability of obtaining a p greater than .58 is only .241.

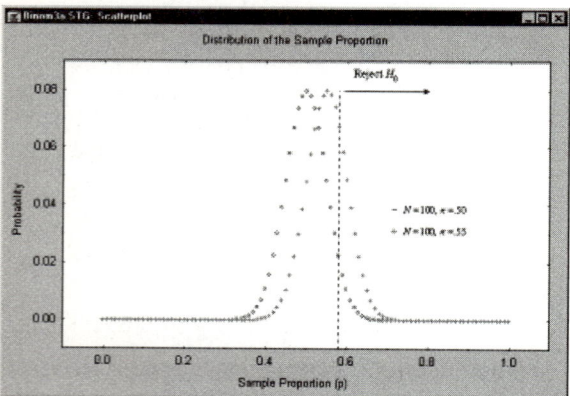

Needless to say, there is no point in conducting an experiment in which, if your position is correct, it will only be verified 24.1% of the time. In this case a statistician would say that the significance test has "inadequate power to detect a departure of 5 percentage points from the null hypothesized value."

The crux of the problem lies in the width of the two distributions in the preceding figure. If the sample size were larger, the standard error of the proportion would be smaller, and there would be

little overlap between the distributions. Then it would be possible to find a decision criterion that provides a low α and high power.

The question is, "How large an N is necessary to produce a power that is reasonably high" in this situation, while maintaining α at a reasonably low value.

You could, of course, go through laborious, repetitive calculations in order to arrive at such a sample size. However, a good software program will perform them automatically, with just a few clicks of the mouse. Moreover, for each analytic situation that it handles, it will provide extensive capabilities for analyzing and graphing the theoretical relationships between power, sample size, and the variables that affect them. Assuming that you will be employing the well-known *chi*-square test, rather than the exact binomial test, suppose that the politician decides that she requires a power of .80 to detect a p of .80. It turns out, a sample size of 607 will yield a power of exactly .8009. (The actual *alpha* of this test, which has a nominal level of .05, is .0522 in this situation.)

Graphical Approaches to Power Analysis

In the preceding discussion, we arrived at a necessary sample size of 607 under the assumption that p is precisely .80. In practice, of course, we would be foolish to perform only one power calculation, based on one hypothetical value. For example, suppose the function relating required sample size to p is particularly steep in this case. It might then be that the sample size required for a p of .70 is much different than that required to reliably detect a p of .80.

Intelligent analysis of power and sample size requires the construction, and careful evaluation, of graphs relating power, sample size, the amount by which the null hypothesis is wrong (i.e., the experimental effect), and other factors such as Type I error rate.

In the example discussed in the preceding section, the goal, from the standpoint of the politician, is to plan a study that can decide, with a low probability of error, whether the support level is greater than .50. Graphical analysis can shed a considerable amount of light on the capabilities of a statistical test to provide the desired information under such circumstances.

For example, researchers could plot power against sample size, under the assumption that the true level is .55, i.e., 55%. They might start with a graph that covers a very wide range of sample sizes, to get a general idea of how the statistical test behaves. The following graph shows power as a function of sample sizes ranging from 20 to 2000, using a normal approximation to the exact binomial distribution.

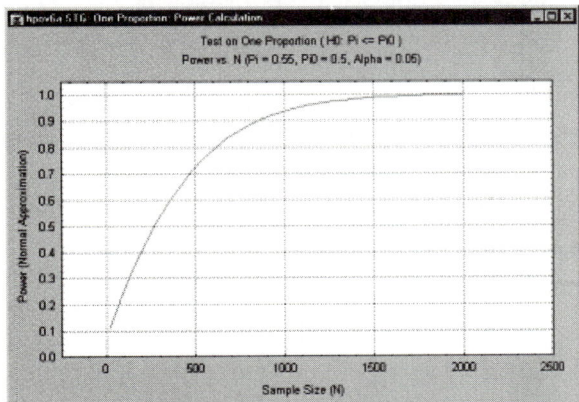

The previous graph demonstrates that power reaches an acceptable level (often considered to be between .80 and .90) at a sample size of approximately 600.

Remember, however, that this calculation is based on the supposition that the true value of p is .55. It may be that the shape of the curve relating power and sample size is very sensitive

to this value. The question immediately arises, "how sensitive is the slope of this graph to changes in the actual value of p?

There are a number of ways to address this question. You can plot power vs. sample size for other values of p, for example. In the next illustration is a graph of power vs. sample size for $p = .6$.

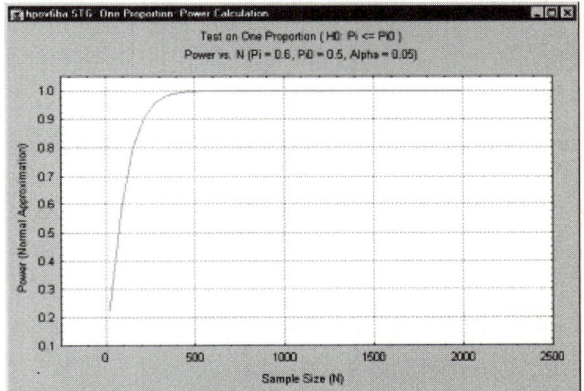

You can see immediately in the preceding graph that the improvement in power for increases in N occurs much more rapidly for $p = .6$ than for $p = .55$. The difference is striking if you merge the two graphs into one, as shown in the next illustration.

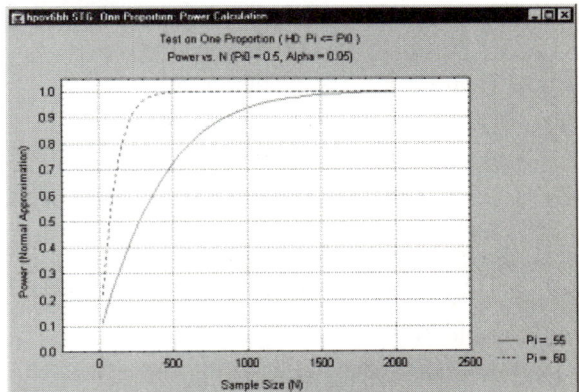

In planning a study, particularly when a grant proposal must be submitted with a proposed

sample size, you must estimate what constitutes a reasonable minimum effect that you want to detect, a minimum power to detect that effect, and the sample size that will achieve that desired level of power. This sample size can be obtained by analyzing the previous graphs (additionally, some software packages can calculate it directly). For example, if you specify the minimum sample size required to achieve a power of .90 when $p = .55$, some programs can calculate this directly. The result is reported in a spreadsheet, as in the next illustration:

Data: Power Analysis Example Spreadsheet 1	
	Sample Size Calculation (Factor.sta) One Proportion, Z, Chi-Square Test H0: Pi <= Pi0
	Value
Null Proportion (Pi0)	0.5000
Population Proportion (Pi)	0.5500
Alpha (Nominal)	0.0500
Actual Alpha (Exact)	0.0501
Power Goal	0.9000
Actual Power (Normal Approx.)	0.9001
Actual Power (Exact)	0.9002
Required Sample Size (N)	853.0000

For a given level of power, a graph of sample size vs. p can show how sensitive the required sample size is to the actual value of p. This can be important in gauging how sensitive the estimate of a required sample size is. For example, the following graph shows values of N needed to achieve a power of .90 for various values of p, when the null hypothesis is that $p = .50$.

The following graph demonstrates how the required N drops off rapidly as p varies from .55 to .60. To be able to reliably detect a difference of .05 (from the null hypothesized value of .50) requires an N greater than 800, but reliable detection of a difference of .10 requires an N of only around 200. Obviously, then, required sample size is somewhat difficult to pinpoint in this situation. It is much better to be aware of the overall performance of the statistical test against a range of possibilities before beginning

an experiment, than to be informed of an unpleasant reality after the fact.

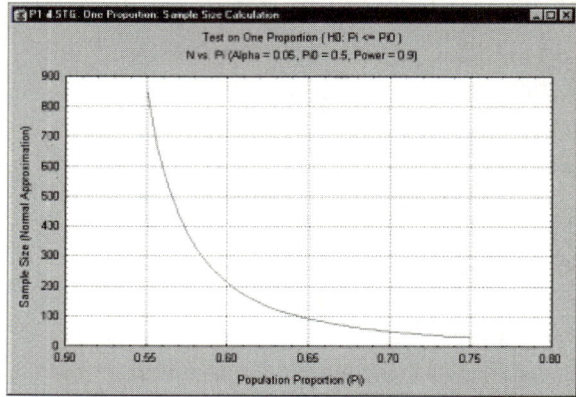

For example, imagine that the experimenter had estimated the required sample size on the basis of reliably (with power of .90) detecting a p of .6. The experimenter budgets for a sample size of, say, 220, and imagines that minor departures of p from .6 will not require substantial differences in N. Only later does the experimenter realize that a small change in requires a huge increase in N, and that the planning for the experiment was optimistic. In some such situations, a window of opportunity may close before the sample size can be adjusted upward.

Across a wide variety of analytic situations, power analysis and sample size estimation involve steps that are fundamentally the same.

1. The type of analysis and null hypothesis are specified

2. Power and required sample size for a reasonable range of effects is investigated.

3. The sample size required to detect a reasonable experimental effect (i.e., departure from the null hypothesis), with a reasonable level of power, is calculated, while allowing for a reasonable margin of error.

Noncentrality Interval Estimation and the Evaluation of Statistical Models

Power analysis and interval estimation includes a number of confidence intervals that are not widely available in general purpose statistics packages. Several of these are discussed within a common theoretical framework, called noncentrality interval estimation by Steiger and Fouladi (1997). In this section, we briefly review some of the basic rationale behind the emerging popularity of confidence intervals.

Inadequacies of the Hypothesis Testing Approach

Strictly speaking, the outcome of a significance test is the dichotomous decision whether or not to reject the null hypothesis. This dichotomy is inherently dissatisfying to many scientists who use the null hypothesis as a statement of no effect, and are more interested in knowing how big an effect is than whether it is (precisely) zero. This has led to behavior such as putting one, two, or three asterisks next to results in tables, or listing p levels next to results, when, in fact, such numbers across (or sometimes even within) studies need not be monotonically related to the best estimates of strength of experimental effects, and hence can be extremely misleading. Some writers (e.g., Guttman, 1977) view asterisk-placing behavior as inconsistent with the foundations of significance testing logic.

Probability levels can deceive about the strength of a result, especially when presented without supporting information. For example, if, in an ANOVA table, one effect had a p level of .019, and the other a p level of .048, it might be an error to conclude that the statistical evidence supported the view that the first effect was stronger than the second. A meaningful interpretation would require additional information. To see why, suppose someone reports a p level of .001. This could be representative of a trivial population effect combined with a huge sample size, or a powerful population effect combined with a moderate sample size, or a huge population effect with a small sample. Similarly a p level of .075 could represent a powerful effect operating with a small sample, or a tiny effect with a huge sample. Clearly then, we need to be careful when comparing p levels.

In accept-support testing, which occurs frequently in the context of model fitting in factor analysis or "causal modeling," significance testing logic is basically inappropriate. Rejection of an "almost true" null hypothesis in such situations frequently has been followed by vague statements that the rejection shouldn't be taken too seriously. Failure to reject a null hypothesis usually results in a demand by a vigilant journal editor for cumbersome power calculations. Such problems can be avoided to some extent by using confidence intervals.

Advantages of Interval Estimation

Much research is exploratory. The fundamental questions in exploratory research are "What is our best guess for the size of the population effect?" and "How precisely have we determined the population effect size from our sample data?" Significance testing fails to answer these questions directly. Many a researcher, faced with an "overwhelming rejection" of a null hypothesis, cannot resist the temptation to report that it was "significant well beyond the .001 level." Yet it is widely agreed that a p level following a significance test can be a poor vehicle for conveying what we have learned about the strength of population effects.

Confidence interval estimation provides a convenient alternative to significance testing in most situations. Consider the 2-tailed hypothesis of no difference between means. Recall first that the significance test rejects at the α significance level if and only if the $1 - \alpha$ confidence interval for the mean difference excludes the value zero. Thus the significance test can be performed with the confidence interval. Most undergraduate texts in behavioral statistics show how to compute such a confidence interval. The interval is exact under the assumptions of the standard t test. However, the confidence interval contains information about experimental precision that is not available from the result of a significance test. Assuming we are reasonably confident about the metric of the data, it is much more informative to state a confidence interval on Mu1 – Mu2 than it is to give the p level for the t test of the hypothesis that Mu1 – Mu2 = 0 In summary, we might say that, in general, a confidence interval conveys more information, in a more naturally usable form, than a significance test.

This is seen most clearly when confidence intervals from several studies are graphed alongside one another, as in the next illustration.

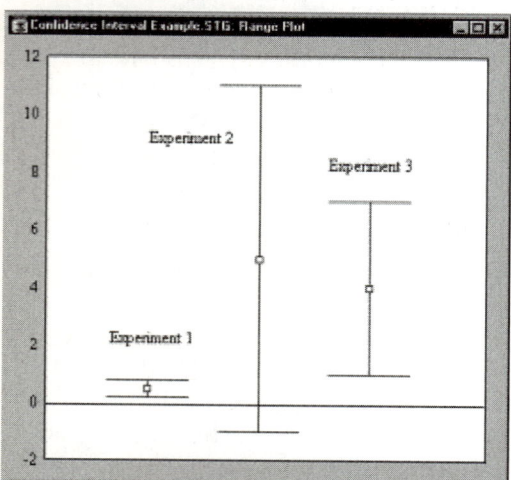

The figure shows confidence intervals for the difference between means for 3 experiments, all performed in the same domain, using measures with approximately the same variability. Experiments 1 and 3 yield a confidence interval that fails to include zero. For these experiments, the null hypothesis was rejected. The second experiment yields a confidence interval that includes zero, so the null hypothesis of no difference is not rejected. A significance testing approach would yield the impression that the second experiment did not agree with the first and the third.

The confidence intervals suggest a different interpretation, however. The first experiment had a very large sample size, and very high precision of measurement, reflected in a very narrow confidence interval. In this experiment, a small effect was found, and determined with such high precision that the null hypothesis of no difference could be rejected at a stringent significance level.

The second experiment clearly lacked precision, and this is reflected in the very wide confidence interval. Evidently, the sample size was too small. It may well be that the actual effect in conditions assessed in the second experiment was larger than that in the first experiment, but the experimental precision was simply inadequate to detect it.

The third experiment found an effect that was statistically significant, and perhaps substantially higher than the first experiment, although this is partly masked by the lower level of precision, reflected in a confidence interval that, though narrower than Experiment 2, is substantially wider than Experiment 1.

Suppose the 3 experiments involved testing groups for differences in IQ. In the final analysis, we may have had too much power in

Experiment 1, as we are declaring highly significant a rather miniscule effect substantially less than a single IQ point. We had far too little power in Experiment 2. Experiment 3 seems about right.

Many of the arguments we have made on behalf of confidence intervals have been made by others as cogently as we have made them here. Yet, confidence intervals are seldom reported in the literature. Most important, as we demonstrate in the succeeding sections, there are several extremely useful confidence intervals that virtually never are reported. In what follows, we discuss why the intervals are seldom reported.

Why Interval Estimates are Seldom Reported

In spite of the obvious advantages of interval estimates, they are seldom employed in published articles in many areas of science. On those infrequent occasions when interval estimates are reported, they are often not the optimal ones. There are several reasons for this status quo:

Tradition. Traditional approaches to statistics emphasize significance testing much more than interval estimation.

Pragmatism. In RS situations, interval estimates are sometimes embarrassing. When they are narrow but close to zero, they suggest that a highly significant result may be statistically significant but trivial. When they are wide, they betray a lack of experimental precision.

Ignorance. Many people are simply unaware of some of the very valuable interval estimation procedures that are available. For example, many textbooks on multivariate analysis never mention that it is possible to compute a confidence interval on the squared multiple correlation coefficient.

Lack of availability. Some of the most desirable interval estimation procedures are computer intensive, and have not been implemented in major statistical packages. This has made it less likely that anyone will try the procedure.

Replacing Traditional Hypothesis Tests with Interval Estimates

There is a number of confidence interval procedures that can replace and/or augment the traditional hypothesis tests used in classical testing situations. For a review of these techniques, see Steiger & Fouladi (1997).

Analysis of variance. One area where confidence intervals have seldom been employed is in assessing strength of effects in the analysis of variance (ANOVA).

For example, suppose you are reading a paper, which reports that, in a 1-Way ANOVA, with 4 groups, and $N = 60$ per group, an F statistic was found that is significant at the .05 level ("$F = 2.70$, $p = .0464$"). This result is statistically significant, but how meaningful is it in a practical sense? What have we learned about the size of the experimental effects?

Fleischman (1980) discusses a technique for setting a confidence interval on the overall effect size in the analysis of variance. This technique enables you to set a confidence interval on the RMSSE, the root-mean-square standardized effect. Standardized effects are reported in standard deviation units, and are hence remain constant when the unit of measurement changes. So, for example, an experimental effect reported in pounds would be different from the same effect reported in kilograms, whereas the standardized effect would be the same in each case. In the case of the data mentioned above, the F statistic that is significant at the .05 level yields a 90% confidence interval for the RMSSE that ranges from .0190 to .3139. The lower limit of this interval stands for a truly mediocre effect, less than 1/50th of a standard deviation. The upper limit of the interval represents effects on the order of 1/3 of a standard deviation, moderate but not overwhelming. It seems, then, that the results from this study need not imply really strong experimental effects, even though the effects are statistically significant.

Multiple regression. The squared multiple correlation is reported frequently as an index of the overall strength of a prediction equation. After fitting a regression equation, the most natural questions to ask are a) "How effective is the regression equation at predicting the criterion" and b) "How precisely has this effectiveness been determined."

Hence, one very common statistical application that practically cries out for a confidence interval is multiple regression analysis. Publishing an observed squared multiple R together with the result of a hypothesis test that the population squared multiple correlation is zero, conveys little of the available statistical information. A confidence interval on the populations squared multiple correlation is much more informative.

One software package computes exact confidence intervals for the population squared multiple correlation, following the approach of Steiger and Fouladi (1992). As an example,

suppose a criterion is predicted from 45 independent observations on 5 variables and the observed squared multiple correlation is .40. In this case a 95% confidence interval for the population squared multiple correlation ranges from .095 to .562! A 95% lower confidence limit is at .129. On the other hand the sample multiple correlation value is significant "beyond the .001 level," because the p level is .0009, and the shrunken estimator is .327. Clearly, it is far more impressive to state "the squared multiple R value is significant at the .001 level" than it is to state "we are 95% confident that the population squared multiple correlation is between .095 and .562." But we believe the latter statement conveys the quality and meaning of the statistical result more accurately than the former.

Some writers, such as Lee (1972), prefer a lower confidence limit, or "statistical lower bound" on the squared multiple correlation to a confidence interval. The rationale, apparently, is that one is primarily interested in assuring that the percentage of variance "accounted for" in the regression equation exceeds some value. Although we understand the motivation behind this view, we hesitate to accept it. The confidence interval, in fact, contains a lower bound, but also includes an upper bound and, in the interval width, a measure of precision of estimation. It seems to us that adoption of a lower confidence limit can lead to a false sense of security, and reduces that amount of information available in the model assessment process.

PROCESS ANALYSIS

Sampling plans are discussed in detail in Duncan (1974) and Montgomery (1985); most process capability procedures (and indices) were only recently introduced to the United States from Japan (Kane, 1986), however, they are discussed in three excellent recent hands-on books by Bohte (1988), Hart and Hart (1989), and Pyzdek (1989); detailed discussions of these methods can also be found in Montgomery (1991).

Step-by-step instructions for the computation and interpretation of capability indices are also provided in the *Fundamental Statistical Process Control Reference Manual* published by the ASQC (American Society for Quality Control) and AIAG (Automotive Industry Action Group, 1991; referenced as ASQC/AIAG, 1991). Repeatability and reproducibility (R & R) methods are discussed in Grant and Leavenworth (1980), Pyzdek (1989) and Montgomery (1991); a more detailed discussion of the subject (of variance estimation) is also provided in Duncan (1974).

Step-by-step instructions on how to conduct and analyze R & R experiments are presented in the *Measurement Systems Analysis Reference Manual* published by ASQC/AIAG (1990). In this chapter, we will briefly introduce the purpose and logic of each of these procedures. For more information on analyzing designs with random effects and for estimating components of variance, see Chapter 39 – *Variance Components and Mixed Model ANOVA/ANCOVA*.

StatSoft®
Copyright © StatSoft, 2006

www.statsoft.com

Sampling Plans

Overview

A common question that quality control engineers face is how many items from a batch (e.g., a shipment from a supplier) to inspect in order to ensure that the items (products) in that batch are of acceptable quality. For example, suppose we have a supplier of piston rings for small automotive engines that our company produces, and our goal is to establish a sampling procedure (of piston rings from the delivered batches) that ensures a specified quality. In principle, this problem is similar to that of on-line quality control discussed in Chapter 33 – *Quality Control Charts*. In fact, you may want to read that chapter at this point to familiarize yourself with the issues involved in industrial statistical quality control.

Acceptance sampling. The procedures described here are useful whenever we need to decide whether a batch or lot of items complies with specifications without having to inspect 100% of the items in the batch. Because of the nature of the problem – whether to accept a batch – these methods are also sometimes discussed under the heading of acceptance sampling.

Advantages over 100% inspection. An obvious advantage of acceptance sampling over 100% inspection of the batch or lot is that reviewing only a sample requires less time, effort, and money. In some cases, inspection of an item is destructive (e.g., stress testing of steel), and testing 100% would destroy the entire batch. From a managerial standpoint, rejecting an entire batch or shipment (based on acceptance sampling) from a supplier, rather than just a certain percent of defective items (based on 100% inspection) often provides a stronger incentive to the supplier to adhere to quality standards.

Computational Approach

In principle, the computational approach to the question of how large a sample to take is straightforward. Chapter 1 – *Elementary Concepts in Statistics* discusses the concept of the sampling distribution. Briefly, if we were to take repeated samples of a particular size from a population of, for example, piston rings and compute their average diameters, the distribution of those averages (means) would approach the normal distribution with a particular mean and standard deviation (or standard error; in sampling distributions the term standard error is preferred, in order to distinguish the variability of the means from the variability of the items in the population). Fortunately, we do not need to take repeated samples from the population in order to estimate the location (mean) and variability (standard error) of the sampling distribution. If we have a good idea (estimate) of what the variability (standard deviation or *sigma*) is in the population, we can infer the sampling distribution of the mean. In principle, this information is sufficient to estimate the sample size that is needed in order to detect a certain change in quality (from target specifications). Without going into the details about the computational procedures involved, let's next review the particular information that the engineer must supply in order to estimate required sample sizes.

Means for H₀ and H₁

To formalize the inspection process of, for example, a shipment of piston rings, we can formulate two alternative hypotheses: First, we can hypothesize that the average piston ring diameters comply with specifications. This hypothesis is called the null hypothesis (H_0). The second and alternative hypothesis (H_1) is that the diameters of the piston rings delivered to us deviate from specifications by more than a certain amount. Note that we can specify these types of hypotheses not just for measurable variables such as diameters of piston rings, but also for attributes. For example, we can hypothesize (H_1) that the number of defective parts in the batch exceeds a certain percentage. Intuitively, it should be clear that the larger the difference between H_0 and H_1, the smaller the sample necessary to detect this difference (see Chapter 1 – *Elementary Concepts in Statistics*).

Alpha and Beta Error Probabilities

To return to the piston rings example, there are two types of mistakes that we can make when inspecting a batch of piston rings that has just arrived at our plant. First, we may erroneously reject H_0, that is, reject the batch because we erroneously conclude that the piston ring diameters deviate from target specifications. The probability of committing this mistake is usually called the *alpha* error probability. The second mistake that we can make is to erroneously not reject H_0 (accept the shipment of piston rings), when, in fact, the mean piston ring diameter deviates from the target specification by a certain amount. The probability of committing this mistake is usually called the *beta* error probability. Intuitively, the more certain we want to be, that is, the lower we set the *alpha* and *beta* error probabilities, the larger the sample will have to be; in fact, in order to be 100% certain, we would have to measure every single piston ring delivered to our company.

Fixed Sampling Plans

To construct a simple sampling plan, we would first decide on a sample size, based on the means under H_0/H_1 and the particular *alpha* and

beta error probabilities. Then, we would take a single sample of this fixed size and, based on the mean in this sample, decide whether to accept or reject the batch. This procedure is referred to as a fixed sampling plan.

Operating characteristic (OC) curve. The power of the fixed sampling plan can be summarized via the operating characteristic curve. In that plot, the probability of rejecting H_0 (and accepting H_1) is plotted on the Y axis, as a function of an actual shift from the target (nominal) specification to the respective values shown on the X axis of the plot (see the next illustration). This probability is, of course, one minus the *beta* error probability of erroneously rejecting H_1 and accepting H_0; this value is referred to as the power of the fixed sampling plan to detect deviations. Also indicated in this plot are the power functions for smaller sample sizes.

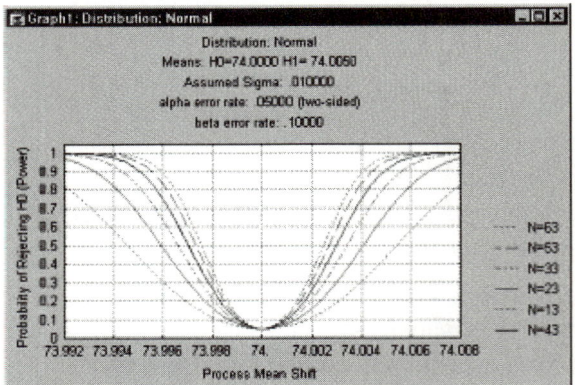

Sequential Sampling Plans

As an alternative to the fixed sampling plan, we could randomly choose individual piston rings and record their deviations from specification. As we continue to measure each piston ring, we could keep a running total of the sum of deviations from specification. Intuitively, if H_1 is true, that is, if the average piston ring diameter in the batch is not on target, we would expect to observe a slowly increasing or decreasing cumulative sum of deviations, depending on whether the average diameter in the batch is larger or smaller than the specification, respectively. It turns out that this kind of sequential sampling of individual items from the batch is a more sensitive procedure than taking a fixed sample. In practice, we continue sampling until we either accept or reject the batch.

Using a sequential sampling plan. Typically, we would produce a graph in which the cumulative deviations from specification (plotted on the *Y*-axis) are shown for successively sampled items (e.g., piston rings, plotted on the *X*-axis). Then two sets of lines are drawn in this graph to denote the "corridor" along which we will continue to draw samples, that is, as long as the cumulative sum of deviations from specifications stays within this corridor, we continue sampling.

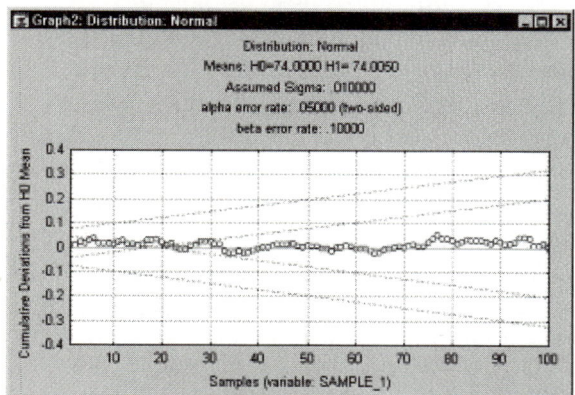

If the cumulative sum of deviations steps outside the corridor we stop sampling. If the cumulative sum moves above the upper line or below the lowest line, we reject the batch. If the cumulative sum steps out of the corridor to the inside, that is, if it moves closer to the center line, we accept the batch (since this indicates zero deviation from specification). Note that the

inside area starts only at a certain sample number; this indicates the minimum number of samples necessary to accept the batch (with the current error probability).

Summary

To summarize, the idea of (acceptance) sampling is to use statistical "inference" to accept or reject an entire batch of items, based on the inspection of only relatively few items from that batch. The advantage of applying statistical reasoning to this decision is that we can be explicit about the probabilities of making a wrong decision.

Whenever possible, sequential sampling plans are preferable to fixed sampling plans because they are more powerful. In most cases, relative to the fixed sampling plan, using sequential plans requires fewer items to be inspected in order to arrive at a decision with the same degree of certainty.

Process (Machine) Capability Analysis

Overview

(See also, *Non-Normal Distributions*, page 430.) Quality control describes numerous methods for monitoring the quality of a production process. However, once a process is under control the question arises, "to what extent does the long-term performance of the process comply with engineering requirements or managerial goals?" For example, to return to our piston ring example, how many of the piston rings that we are using fall within the design specification limits? In more general terms, the question is, "how capable is our process (or supplier) in terms of producing items within the specification limits?" Most of the procedures and indices described here were only recently introduced to the US by Ford Motor Company (Kane, 1986). They enable us to summarize the process capability in terms of meaningful percentages and indices.

In this topic, the computation and interpretation of process capability indices will first be discussed for the normal distribution case. If the distribution of the quality characteristic of interest does not follow the normal distribution, modified capability indices can be computed based on the percentiles of a fitted non-normal distribution.

Order of business. Note that it makes little sense to examine the process capability if the process is not in control. If the means of successively taken samples fluctuate widely or are clearly off the target specification, those quality problems should be addressed first. Therefore, the first step toward a high-quality process is to bring the process under control using the charting techniques available in quality control.

Computational Approach

Once a process is in control, we can ask the question concerning the process capability. Again, the approach to answering this question is based on statistical reasoning, and is actually quite similar to that presented earlier in the context of sampling plans. To return to the piston ring example, given a sample of a particular size, we can estimate the standard deviation of the process, that is, the resultant ring diameters. We can then draw a histogram of the distribution of the piston ring diameters. As we discussed earlier, if the distribution of the diameters is normal, we can make inferences concerning the proportion of piston rings within specification limits. (For non-

normal distributions, see *Percentile Method*, page 433.)

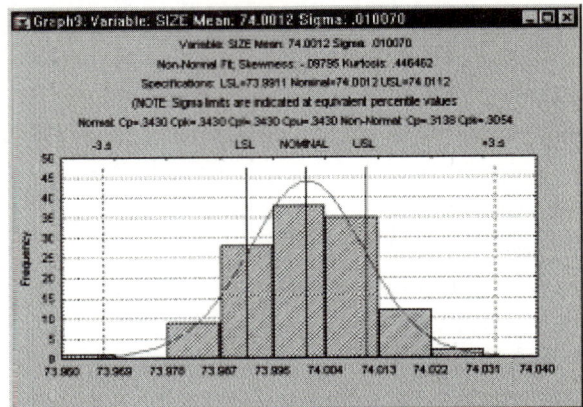

Let's now review some of the major indices that are commonly used to describe process capability.

Capability Analysis - Process Capability Indices

Process range. First, it is customary to establish the ± 3 *sigma* limits around the nominal specifications. Actually, the *sigma* limits should be the same as the ones used to bring the process under control using Shewhart control charts (see Chapter 33 – *Quality Control Charts*). These limits denote the range of the process (i.e., process range). If we use the ± 3 *sigma* limits then, based on the normal distribution, we can estimate that approximately 99% of all piston rings fall within these limits.

Specification limits LSL, USL. Usually, engineering requirements dictate a range of acceptable values. In our example, it may have been determined that acceptable values for the piston ring diameters would be 74.0 ± .02 millimeters. Thus, the lower specification limit (LSL) for our process is 74.0 − 0.02 = 73.98; the upper specification limit (USL) is 74.0 +

0.02 = 74.02. The difference between USL and LSL is called the specification range.

Potential capability (Cp). This is the simplest and most straightforward indicator of process capability. It is defined as the ratio of the specification range to the process range; using ± 3 *sigma* limits we can express this index as:

$$C_p = (USL-LSL)/(6*Sigma)$$

Put into words, this ratio expresses the proportion of the range of the normal curve that falls within the engineering specification limits (provided that the mean is on target, that is, that the process is centered).

Bhote (1988) reports that prior to the widespread use of statistical quality control techniques (prior to 1980), the normal quality of US manufacturing processes was approximately C_p = .67. This means that the two 33/2 percent tail areas of the normal curve fall outside specification limits. As of 1988, only about 30% of US processes are at or below this level of quality (see Bhote, 1988, p. 51). Ideally, of course, we want this index to be greater than 1; that is, we want to achieve a process capability so that no (or almost no) items fall outside specification limits. Interestingly, in the early 1980's the Japanese manufacturing industry adopted as their standard C_p = 1.33! The process capability required to manufacture high-tech products is usually even higher than this; Minolta has established a C_p index of 2.0 as their minimum standard (Bhote, 1988, p. 53), and as the standard for its suppliers. Note that high process capability usually implies lower, not higher costs, taking into account the costs due to poor quality. We will return to this point shortly.

Capability ratio (Cr). This index is equivalent to C_p; specifically, it is computed as $1/C_p$ (the inverse of C_p).

Lower/upper potential capability: Cpl, Cpu. A major shortcoming of the C_p (and C_r) index is that it may yield erroneous information if the process is not on target, that is, if it is not centered. We can express non-centering via the following quantities. First, upper and lower potential capability indices can be computed to reflect the deviation of the observed process mean from the LSL and USL. Assuming ± 3 *sigma* limits as the process range, we compute:

$$C_{pl} = (Mean - LSL)/3*Sigma$$

and

$$C_{pu} = (USL - Mean)/3*Sigma$$

Obviously, if these values are not identical to each other, the process is not centered.

Non-centering correction (K). We can correct C_p for the effects of non-centering. Specifically, we can compute:

$$K=abs(D - Mean)/(1/2*(USL - LSL))$$

where D = (USL+LSL)/2.

This correction factor expresses the non-centering (target specification minus mean) relative to the specification range.

Demonstrated excellence (Cpk). Finally, we can adjust C_p for the effect of non-centering by computing:

$$C_{pk} = (1-k)*C_p$$

If the process is perfectly centered, *k* is equal to zero, and C_{pk} is equal to C_p. However, as the process drifts from the target specification, *k* increases and C_{pk} becomes smaller than C_p.

Potential capability II: Cpm. A recent modification (Chan, Cheng, & Spiring, 1988) to C_p is directed at adjusting the estimate of *sigma* for the effect of (random) non-centering. Specifically, we can compute the alternative *sigma (Sigma_2)* as:

$$Sigma_2 = \{\Sigma(x_i - TS)^2/(n-1)\}^{\frac{1}{2}}$$

Where *Sigma₂* is the alternative estimate of *sigma*, x_i is the value of the *i*'th observation in the sample, *TS* is the target or nominal specification, and *n* is the number of observations in the sample

We can then use this alternative estimate of *sigma* to compute C_p as before; however, we will refer to the resultant index as C_{pm}.

Process Performance vs. Process Capability

When monitoring a process via a quality control chart (e.g., the X-bar and R-chart) it is often useful to compute the capability indices for the process. Specifically, when the data set consists of multiple samples, such as data collected for the quality control chart, you can compute two different indices of variability in the data. One is the regular standard deviation for all observations, ignoring the fact that the data consist of multiple samples; the other is to estimate the process's inherent variation from the within-sample variability. For example, when plotting X-bar and R-charts, you can use the common estimator R-bar/d_2 for the process *sigma* (e.g., see Duncan, 1974; Montgomery, 1985, 1991). Note however, that this estimator is only valid if the process is statistically stable. For a detailed discussion of the difference between the total process variation and the inherent variation refer to ASQC/AIAG reference manual (ASQC/AIAG, 1991, page 80).

When the total process variability is used in the standard capability computations, the resulting indices are usually referred to as process performance indices (as they describe the actual performance of the process), while indices computed from the inherent variation (within-sample *sigma*) are referred to as capability indices (since they describe the inherent capability of the process).

Using Experiments to Improve Process Capability

As mentioned before, the higher the C_p index, the better the process, and there is virtually no upper limit to this relationship. The issue of quality costs, that is, the losses due to poor quality, is discussed in detail in the context of Taguchi robust design methods [see Chapter 15 – *Experimental Design (DOE)*]. In general, higher quality usually results in lower costs overall; even though the costs of production may increase, the losses due to poor quality, for example, due to customer complaints, loss of market share, etc. are usually much greater. In practice, two or three well-designed experiments carried out over a few weeks can often achieve a C_p of 5 or higher. If you are not familiar with the use of designed experiments, but are concerned with the quality of a process, we strongly recommend that you review the methods detailed in *Experimental Design*.

Testing the Normality Assumption

The indices we have just reviewed are only meaningful if, in fact, the quality characteristic that is being measured is normally distributed. A specific test of the normality assumption (Kolmogorov-Smirnov and *chi*-square test of goodness-of-fit) is available; these tests are described in most statistics textbooks, and they are also discussed in greater detail in Chapter 14 – *Distribution Fitting* and Chapter 29 – *Nonparametric Statistics*.

A visual check for normality is to examine the probability-probability and quantile-quantile plots for the normal distribution. For more information, see *Non-Normal Distributions*, page 430.

Tolerance Limits

Before the introduction of process capability indices in the early 1980's, the common method for estimating the characteristics of a production process was to estimate and examine the tolerance limits of the process (see, for example, Hald, 1952). The logic of this procedure is as follows. Let's assume that the respective quality characteristic is normally distributed in the population of items produced; we can then estimate the lower and upper interval limits that will ensure with a certain level of confidence (probability) that a certain percent of the population is included in those limits. Put another way, given:

1. a specific sample size (*n*),

2. the process mean,

3. the process standard deviation (*sigma*),

4. a confidence level, and

5. the percent of the population that we want to be included in the interval,

we can compute the corresponding tolerance limits that will satisfy all these parameters. You can also compute parameter-free tolerance limits that are not based on the assumption of normality (Scheffe & Tukey, 1944, p. 217; Wilks, 1946, p. 93; see also, Duncan, 1974, or Montgomery, 1985, 1991).

Gage Repeatability and Reproducibility

Overview

Gage repeatability and reproducibility analysis addresses the issue of precision of measurement. The purpose of repeatability and reproducibility experiments is to determine the proportion of measurement variability that is due to 1) the items or parts being measured (part-to-part variation), 2) the operator or appraiser of the gages (reproducibility), and 3) errors (unreliabilities) in the measurements over several trials by the same operators of the same parts (repeatability). In the ideal case, all variability in measurements will be due to the part-to-part variation, and only a negligible proportion of the variability will be due to operator reproducibility and trial-to-trial repeatability.

To return to the piston ring example, if we require detection of deviations from target specifications of the magnitude of .01 millimeters, we obviously need to use gages of sufficient precision. The procedures described here enable the engineer to evaluate the precision of gages and different operators (users) of those gages, relative to the variability of the items in the population.

You can compute the standard indices of repeatability, reproducibility, and part-to-part variation, based either on ranges (as is still common in these types of experiments) or from the analysis of variance (ANOVA) table (as, for example, recommended in ASQC/AIAG, 1990, page 65). The ANOVA table will also contain an *F* test (statistical significance test) for the operator-by-part interaction, and report the estimated variances, standard deviations, and confidence intervals for the components of the ANOVA model.

Finally, you can compute the respective percentages of total variation, and report so-called percent-of-tolerance statistics. These measures are briefly discussed in the following sections of this introduction. Additional information can be found in Duncan (1974), Montgomery (1991), or the *DataMyte Handbook* (1992); step-by-step instructions and examples are also presented in the ASQC/AIAG

Measurement Systems Analysis Reference Manual (1990) and the ASQC/AIAG *Fundamental Statistical Process Control Reference Manual* (1991).

Note that there are several other statistical procedures that can be used to analyze these types of designs; for details, see *Methods for Analysis of Variance* in Chapter 3 – *ANOVA/MANOVA*. In particular, the methods discussed in Chapter 39 – *Variance Components and Mixed Model ANOVA/ ANCOVA* are very efficient for analyzing very large nested designs (e.g., with more than 200 levels overall), or hierarchically nested designs (with or without random factors).

Computational Approach

You can think of each measurement as consisting of the following components:

1. a component due to the characteristics of the part or item being measured,

2. a component due to the reliability of the gage, and

3. a component due to the characteristics of the operator (user) of the gage.

The method of measurement (measurement system) is reproducible if different users of the gage come up with identical or very similar measurements. A measurement method is repeatable if a repeated measurement of the same part produces identical results. Both of these characteristics – repeatability and reproducibility – will affect the precision of the measurement system.

We can design an experiment to estimate the magnitudes of each component, that is, the repeatability, reproducibility, and the variability between parts, and thus assess the precision of the measurement system. In essence, this

procedure amounts to an analysis of variance (ANOVA) on an experimental design that includes as factors different parts, operators, and repeated measurements (trials). We can then estimate the corresponding variance components (the term was first used by Daniels, 1939) to assess the repeatability (variance due to differences across trials), reproducibility (variance due to differences across operators), and variability between parts (variance due to differences across parts). If you are not familiar with the general idea of ANOVA, you may want to refer to Chapter 3 – *ANOVA/MANOVA*. In fact, the extensive features provided there can also be used to analyze repeatability and reproducibility studies.

Plots of Repeatability and Reproducibility

There are several ways to summarize via graphs the findings from a repeatability and reproducibility experiment. For example, suppose we are manufacturing small kilns that are used for drying materials for other industrial production processes. The kilns should operate at a target temperature of around 100 degrees Celsius. In this study, 5 different engineers (operators) measured the same sample of 8 kilns (parts), three times each (three trials). We can plot the mean ratings of the 8 parts by operator. If the measurement system is reproducible, the pattern of means across parts should be quite consistent across the 5 engineers who participated in the study.

R and S charts. Quality control discusses in detail the idea of R (range) and S (*sigma*) plots for controlling process variability. We can apply these ideas here and produce a plot of ranges (or *sigmas*) by operators or by parts; these plots will enable us to identify outliers among operators or parts. If one operator produced particularly wide ranges of measurements, we may want to find out why that particular person had problems producing reliable measurements (e.g., perhaps he or she failed to understand the instructions for using the measurement gage).

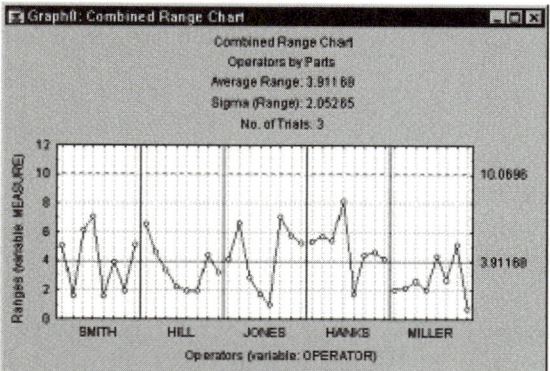

Analogously, producing an *R* chart by parts may enable us to identify parts that are particularly difficult to measure reliably; again, inspecting that particular part may give us some insights into the weaknesses in our measurement system.

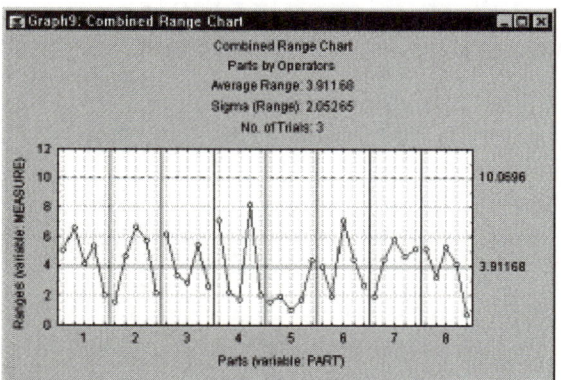

Repeatability and reproducibility summary plot. The summary plot shows the individual measurements by each operator; specifically, the measurements are shown in terms of deviations from the respective average rating for the respective part. Each trial is represented

by a point, and the different measurement trials for each operator for each part are connected by a vertical line. Boxes drawn around the measurements give us a general idea of a particular operator's bias (see the graph in the next illustration).

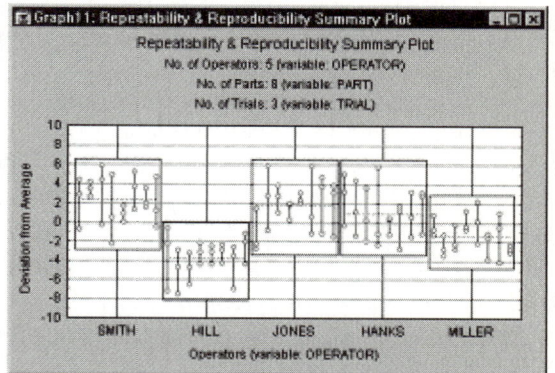

Components of Variance

Percent of process variation and tolerance. Use percent tolerance to evaluate the performance of the measurement system with regard to the overall process variation, and the respective tolerance range. You can specify the tolerance range (total tolerance for parts) and the number of *sigma* intervals. The latter value is used in the computations to define the range (spread) of the respective (repeatability, reproducibility, part-to-part, etc.) variability. Specifically, the default value (*5.15*) defines *5.15* times the respective *sigma* estimate as the respective range of values; if the data are normally distributed, this range defines 99% of the space under the normal curve, that is, the range that will include 99% of all values (or reproducibility/repeatability errors) due to the respective source of variation.

Percent of process variation. This value reports the variability due to different sources relative to the total variability (range) in the measurements.

Analysis of variance. Rather than computing variance components estimates based on ranges, an accurate method for computing these estimates is based on the ANOVA mean squares (see Duncan, 1974, ASQC/AIAG, 1990).

You can treat the three factors in the R & R experiment (operator, parts, trials) as random factors in a three-way ANOVA model (see also, Chapter 3 – *ANOVA/MANOVA*). For details concerning the different models that are typically considered, refer to ASQC/AIAG (1990, pages 92-95), or to Duncan (1974, pages 716-734). Customarily, it is assumed that all interaction effects by the trial factor are non-significant. This assumption seems reasonable, since, for example, it is difficult to imagine how the measurement of some parts will be systematically different in successive trials, in particular when parts and trials are randomized.

However, the operator by parts interaction may be important. For example, it is conceivable that certain less experienced operators will be more prone to particular biases, and hence will arrive at systematically different measurements for particular parts. If so, you would expect a significant two-way interaction (again, refer to Chapter 3 – *ANOVA/MANOVA* if you are not familiar with ANOVA terminology).

In the case when the two-way interaction is statistically significant, you can separately estimate the variance components due to operator variability, and due to the operator by parts variability

In the case of significant interactions, the combined repeatability and reproducibility variability is defined as the sum of three

components: repeatability (gage error), operator variability, and the operator-by-part variability.

If the operator by part interaction is not statistically significant a simpler additive model can be used without interactions.

Summary

To summarize, the purpose of the repeatability and reproducibility procedures is to enable the quality control engineer to assess the precision of the measurement system (gages) used in the quality control process. Obviously, if the measurement system is not repeatable (large variability across trials) or reproducible (large variability across operators) relative to the variability between parts, the measurement system is not sufficiently precise to be used in the quality control efforts. For example, it should not be used in charts produced via quality control, or product capability analyses and acceptance sampling procedures via process analysis.

Non-Normal Distributions

Overview

General purpose. The concept of process capability is described in detail in the *Process Capability Overview*, page 423. To reiterate, when judging the quality of a (e.g., production) process it is useful to estimate the proportion of items produced that fall outside a predefined acceptable specification range. For example, the so-called C_p index is computed as:

$C_p - (USL-LSL)/(6*sigma)$

where *sigma* is the estimated process standard deviation, and *USL* and *LSL* are the upper and lower specification limits, respectively. If the distribution of the respective quality characteristic or variable (e.g., size of piston rings) is normal, and the process is perfectly centered (i.e., the mean is equal to the design center), this index can be interpreted as the proportion of the range of the standard normal curve (the process width) that falls within the engineering specification limits. If the process is not centered, an adjusted index C_{pk} is used instead.

Non-normal distributions. You can fit non-normal distributions to the observed histogram, and compute capability indices based on the respective fitted non-normal distribution (via the percentile method). In addition, instead of computing capability indices by fitting specific distributions, you can compute capability indices based on two different general families of distributions – the Johnson distributions (Johnson, 1965; see also, Hahn and Shapiro, 1967) and Pearson distributions (Johnson, Nixon, Amos, and Pearson, 1963; Gruska, Mirkhani, and Lamberson, 1989; Pearson and Hartley, 1972) – which enable you to approximate a wide variety of continuous distributions. For all distributions, you can also compute the table of expected frequencies, the expected number of observations beyond specifications, and quantile-quantile and probability-probability plots. The specific method for computing process capability indices from these distributions is described in Clements (1989).

Quantile-quantile plots and probability-probability plots. There are various methods for assessing the quality of respective fit to the observed data. In addition to the table of observed and expected frequencies for different intervals, and the Kolmogorov-Smirnov and *chi*-square goodness-of-fit tests, you can compute quantile and probability plots for all distributions. These scatterplots are constructed so that if the observed values follow the respective distribution, the points will form a

straight line in the plot. These plots are described further on page 432.

Fitting Distributions by Moments

In addition to the specific continuous distributions described above, you can fit general "families" of distributions – the so-called Johnson and Pearson curves – with the goal to match the first four moments of the observed distribution.

General approach. The shapes of most continuous distributions can be sufficiently summarized in the first four moments. Put another way, if you fit to a histogram of observed data a distribution that has the same mean (first moment), variance (second moment), skewness (third moment) and kurtosis (fourth moment) as the observed data, you can usually approximate the overall shape of the distribution very well. Once a distribution has been fitted, you can then calculate the expected percentile values under the (standardized) fitted curve, and estimate the proportion of items produced by the process that fall within the specification limits.

Johnson curves. Johnson (1949) described a system of frequency curves that represents transformations of the standard normal curve (see Hahn and Shapiro, 1967, for details). By applying these transformations to a standard normal variable, a wide variety of non-normal distributions can be approximated, including distributions that are bounded on either one or both sides (e.g., *U*-shaped distributions). The advantage of this approach is that once a particular Johnson curve has been fit, the normal integral can be used to compute the expected percentage points under the respective curve. Methods for fitting Johnson curves, so as to approximate the first four moments of an empirical distribution, are described in detail in Hahn and Shapiro, 1967, pages 199-220; and Hill, Hill, and Holder, 1976.

Pearson curves. Another system of distributions was proposed by Karl Pearson (e.g., see Hahn and Shapiro, 1967, pages 220-224). The system consists of seven solutions (of 12 originally enumerated by Pearson) to a differential equation that also approximate a wide range of distributions of different shapes. Gruska, Mirkhani, and Lamberson (1989) describe in detail how the different Pearson curves can be fit to an empirical distribution. A method for computing specific Pearson percentiles is also described in Davis and Stephens (1983).

Assessing the Fit: Quantile and Probability Plots

For each distribution, you can compute the table of expected and observed frequencies and the respective *chi*-square goodness-of-fit test, as well as the Kolmogorov-Smirnov *d* test. However, the best way to assess the quality of the fit of a theoretical distribution to an observed distribution is to review the plot of the observed distribution against the theoretical fitted distribution. There are two standard types of plots used for this purpose: Quantile-quantile plots and probability-probability plots.

Quantile-quantile plots. In quantile-quantile plots (Q-Q plots), the observed values of a variable are plotted against the theoretical quantiles. To produce a Q-Q plot, you first sort the *n* observed data points into ascending order, so that:

$$x_1 \le x_2 \le \ldots \le x_n$$

These observed values are plotted against one axis of the graph; on the other axis the plot will show:

$$F^{-1}((i - r_{adj})/(n + n_{adj}))$$

where *i* is the rank of the respective observation, r_{adj} and n_{adj} are adjustment factors (≤ 0.5) and F^{-1} denotes the inverse of the probability integral for the respective standardized distribution. The resulting plot (see the next illustration) is a scatterplot of the observed values against the (standardized) expected values, given the respective distribution. Note that, in addition to the inverse probability integral value, you can also show the respective cumulative probability values on the opposite axis, that is, the plot will show not only the standardized values for the theoretical distribution, but also the respective *p*-values.

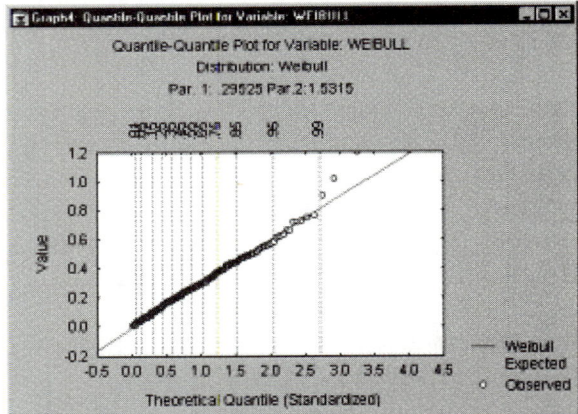

A good fit of the theoretical distribution to the observed values would be indicated by this plot if the plotted values fall onto a straight line. Note that the adjustment factors r_{adj} and n_{adj} ensure that the *p*-value for the inverse probability integral will fall between 0 and 1, but not including 0 and 1 (see Chambers, Cleveland, Kleiner, and Tukey, 1983).

Probability-probability plots. In probability-probability plots (P-P plots) the observed cumulative distribution function is plotted against the theoretical cumulative distribution function. As in the Q-Q plot, the values of the respective variable are first sorted into ascending order. The *i*'th observation is plotted against one axis as *i/n* (i.e., the observed cumulative distribution function), and against the other axis as $F(x_{(i)})$, where $F(x_{(i)})$ stands for the value of the theoretical cumulative distribution function for the respective observation $x_{(i)}$. If the theoretical cumulative distribution approximates the observed distribution well, all points in this plot should fall onto the diagonal line (as in the graph in the next illustration).

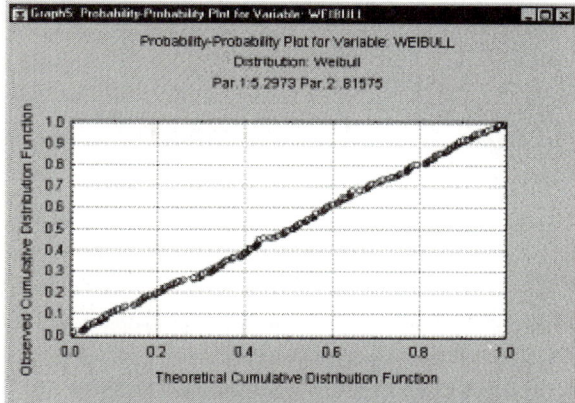

Non-Normal Process Capability Indices (Percentile Method)

Process capability indices are generally computed to evaluate the quality of a process, that is, to estimate the relative range of the items manufactured by the process (process width) with regard to the engineering specifications. For the standard, normal distribution-based, process capability indices, the process width is typically defined as 6 times *sigma*, that is, as plus/minus 3 times the estimated process standard deviation. For the standard normal curve, these limits ($z_l = -3$ and $z_u = +3$) translate into the *0.135* percentile and *99.865* percentile, respectively. In the non-normal case, the 3 times *sigma* limits as well as the mean ($z_M = 0.0$) can be replaced by the corresponding

standard values, given the same percentiles, under the non-normal curve. This procedure is described in detail by Clements (1989).

Process capability indices. The formulas for the non-normal process capability indices:

$C_p = (USL-LSL)/(U_p-L_p)$

$C_{pL} = (M-LSL)/(M-L_p)$

$C_{pU} = (USL-M)/(U_p-M)$

$C_{pk} = Min(C_{pU}, C_{pL})$

In these equations, *M* represents the 50'th percentile value for the respective fitted distribution, and U_p and L_p are the *99.865* and *.135* percentile values, respectively, if the computations are based on a process width of ±3 times *sigma*. Note that the values for U_p and L_p may be different, if the process width is defined by different *sigma* limits (e.g., ±2 times *sigma*).

Weibull and Reliability/ Failure Time Analysis

A key aspect of product quality is product reliability. A number of specialized techniques have been developed to quantify reliability and to estimate the "life expectancy" of a product. Standard references and textbooks describing these techniques include Lawless (1982), Nelson (1990), Lee (1980, 1992), and Dodson (1994); the relevant functions of the Weibull distribution (hazard, CDF, reliability) are also described in *Weibull CDF, reliability, and hazard function s*, page 440. Note that very similar statistical procedures are used in the analysis of survival data (see also, the description of *Survival Analysis*), and, for example, the descriptions in Lee's book (Lee, 1992) are primarily addressed to biomedical research applications. An excellent overview with many examples of engineering applications is provided by Dodson (1994).

General Purpose

The reliability of a product or component constitutes an important aspect of product quality. Of particular interest is the quantification of a product's reliability so that you can derive estimates of the product's expected useful life. For example, suppose you are flying a small single engine aircraft. It would be very useful (in fact vital) information to know what the probability of engine failure is at different stages of the engine's life (e.g., after 500 hours of operation, 1000 hours of operation, etc.). Given a good estimate of the engine's reliability, and the confidence limits of this estimate, you can then make a rational decision about when to swap or overhaul the engine.

Weibull Distribution

A useful general distribution for describing failure time data is the Weibull distribution (see also, *Weibull CDF, reliability, and hazard function s*, page 440), named after the Swedish professor Waloddi Weibull, who demonstrated the appropriateness of this distribution for modeling a wide variety of different data sets (see also, Hahn and Shapiro, 1967; for example, the Weibull distribution has been used to model the life times of electronic components, relays, ball bearings, or even some businesses).

Hazard function and the bathtub curve. It is often meaningful to consider the function that describes the probability of failure during a very small time increment (assuming that no failures have occurred prior to that time). This function is called the hazard function (or, sometimes, also the conditional failure, intensity, or force of mortality function), and is generally defined as:

h(t) = f(t)/(1-F(t))

where *h(t)* stands for the hazard function (of time *t*), and *f(t)* and *F(t)* are the probability density and cumulative distribution functions, respectively. The hazard (conditional failure) function for most machines (components, devices) can best be described in terms of the "bathtub" curve: Very early during the life of a machine, the rate of failure is relatively high (so-called Infant Mortality Failures); after all components settle, and the electronic parts are burned in, the failure rate is relatively constant and low. Then, after some time of operation, the failure rate again begins to increase (so-called Wear-out Failures), until all components or devices will have failed.

For example, new automobiles often suffer several small failures right after they were purchased. Once these have been "ironed out," a (hopefully) long relatively trouble-free period of operation will follow. Then, as the car reaches a particular age, it becomes more prone to breakdowns, until finally, after 20 years and 250,000 miles, practically all cars will have failed. A typical bathtub hazard function is shown in the next illustration.

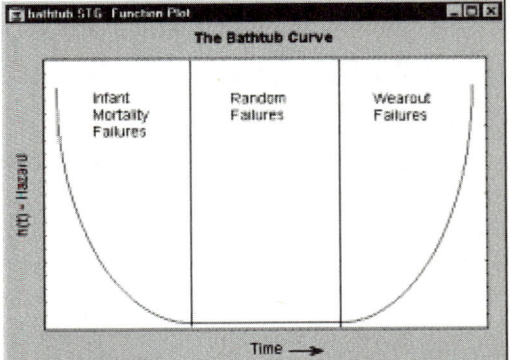

The Weibull distribution is flexible enough for modeling the key stages of this typical bathtub-shaped hazard function. Shown in the next illustration are the hazard functions for shape parameters *c=.5*, *c=1*, *c=2*, and *c=5*.

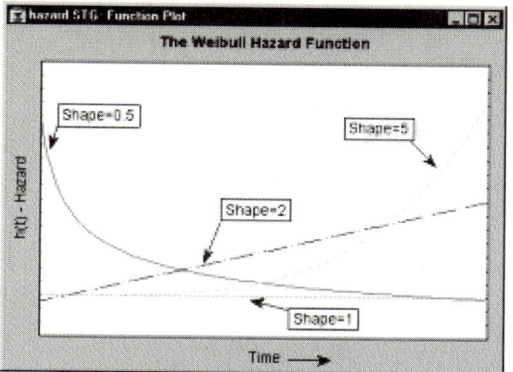

Clearly, the early (infant mortality) phase of the bathtub can be approximated by a Weibull hazard function with shape parameter *c<1*; the constant hazard phase of the bathtub can be modeled with a shape parameter *c=1*, and the final (wear-out) stage of the bathtub with *c>1*.

Cumulative distribution and reliability functions. Once a Weibull distribution (with a particular set of parameters) has been fit to the data, a number of additional important indices and measures can be estimated. For example, you can compute the cumulative distribution function (commonly denoted as *F(t)*) for the fitted distribution, along with the standard errors for this function. Thus, you can determine the percentiles of the cumulative survival (and failure) distribution, and, for example, predict the time at which a predetermined percentage of components can be expected to have failed.

The reliability function (commonly denoted as *R(t)*) is the complement to the cumulative distribution function (i.e., *R(t)=1-F(t)*); the reliability function is also sometimes referred to as the survivorship or survival function (since it describes the probability of not failing or of surviving until a certain time *t*; e.g., see Lee, 1992). Shown in the next illustration is the

reliability function for the Weibull distribution, for different shape parameters.

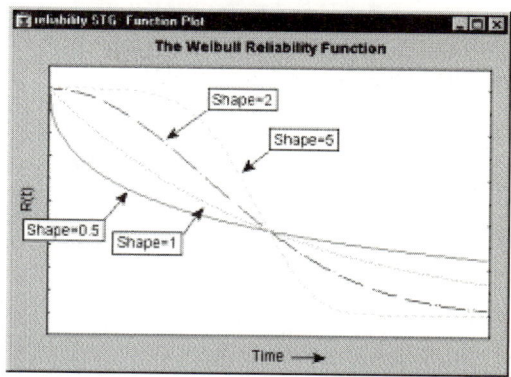

For shape parameters less than 1, the reliability decreases sharply very early in the respective product's life, and then slowly thereafter. For shape parameters greater than 1, the initial drop in reliability is small, and then the reliability drops relatively sharply at some point later in time. The point where all curves intersect is called the *characteristic life*: regardless of the shape parameter, 63.2 percent of the population will have failed at or before this point (i.e., *R(t)* = 1-0.632 = .368). This point in time is also equal to the respective scale parameter *b* of the two-parameter Weibull distribution (with $\theta = 0$; otherwise it is equal to b+ θ).

The formulas for the Weibull cumulative distribution, reliability, and hazard functions are shown in *Weibull CDF, Reliability, and Hazard Functions*, page 440.

Censored Observations

In most studies of product reliability, not all items in the study will fail. In other words, by the end of the study the researcher only knows that a certain number of items have not failed for a particular amount of time, but has no knowledge of the exact failure times (i.e., when the items would have failed). Those types of data are called censored

observations. The issue of censoring, and several methods for analyzing censored data sets, are also described in great detail in Chapter 36 – *Survival/Failure Time Analysis*. Censoring can occur in many different ways.

Type I and II censoring. So-called Type I censoring describes the situation when a test is terminated at a particular point in time, so that the remaining items are only known not to have failed up to that time (e.g., we start with 100 light bulbs, and terminate the experiment after a certain amount of time). In this case, the censoring time is often fixed, and the number of items failing is a random variable. In Type II censoring the experiment would be continued until a fixed proportion of items have failed (e.g., we stop the experiment after exactly 50 light bulbs have failed). In this case, the number of items failing is fixed, and time is the random variable.

Left and right censoring. An additional distinction can be made to reflect the "side" of the time dimension at which censoring occurs. In the examples described above, the censoring always occurred on the right side (right censoring), because the researcher knows when exactly the experiment started, and the censoring always occurs on the right side of the time continuum. Alternatively, it is conceivable that the censoring occurs on the left side (left censoring). For example, in biomedical research you may know that a patient entered the hospital at a particular date, and that s/he survived for a certain amount of time thereafter; however, the researcher does not know when exactly the symptoms of the disease first occurred or were diagnosed.

Single and multiple censoring. Finally, there are situations in which censoring can occur at different times (multiple censoring), or only at a particular point in time (single censoring). To return to the light bulb example, if the experiment is terminated at a particular point in time, a single point of censoring exists and the data set is said to be single-censored. However, in biomedical research multiple censoring often exists, for example, when patients are discharged from a hospital after different amounts (times) of treatment, and the researcher knows that the patient survived up to those (differential) points of censoring.

The methods described in this section are applicable primarily to right censoring, and single- as well as multiple-censored data.

Two- and Three-Parameter Weibull Distribution

The Weibull distribution is bounded on the left side. If you look at the probability density function, you can see that that the term $x - \theta$ must be greater than 0. In most cases, the location parameter θ (*theta*) is known (usually 0): it identifies the smallest possible failure time. However, sometimes the probability of failure of an item is 0 (zero) for some time after a study begins, and in that case it may be necessary to estimate a location parameter that is greater than 0. There are several methods for estimating the location parameter of the three-parameter Weibull distribution. To identify situations where the location parameter is greater than 0, Dodson (1994) recommends looking for downward of upward sloping tails on a probability plot, as well as large (>6) values for the shape parameter after fitting the two-parameter Weibull distribution, which may indicate a non-zero location parameter.

Parameter Estimation

Maximum likelihood estimation. Standard iterative function minimization methods can be used to compute maximum likelihood parameter estimates for the two- and three-

parameter Weibull distribution. The specific methods for estimating the parameters are described in Dodson (1994); a detailed description of a Newton-Raphson iterative method for estimating the maximum likelihood parameters for the two-parameter distribution is provided in Keats and Lawrence (1997).

The estimation of the location parameter for the three-parameter Weibull distribution poses a number of special problems, which are detailed in Lawless (1982). Specifically, when the shape parameter is less than 1, a maximum likelihood solution does not exist for the parameters. In other instances, the likelihood function may contain more than one maximum (i.e., multiple local maxima). In the latter case, Lawless basically recommends using the smallest failure time (or a value that is a little bit less) as the estimate of the location parameter.

Nonparametric (rank-based) probability plots. You can derive a descriptive estimate of the cumulative distribution function (regardless of distribution) by first rank-ordering the observations, and then computing any of the following expressions:

Median rank:
$$F(t) = (j-0.3)/(n+0.4)$$

Mean rank:
$$F(t) = j/(n+1)$$

White's plotting position:
$$F(t) = (j-3/8)/(n+1/4)$$

where j denotes the failure order (rank; for multiple-censored data a weighted average ordered failure is computed; see Dodson, p. 21), and n is the total number of observations. You can then construct the following plot.

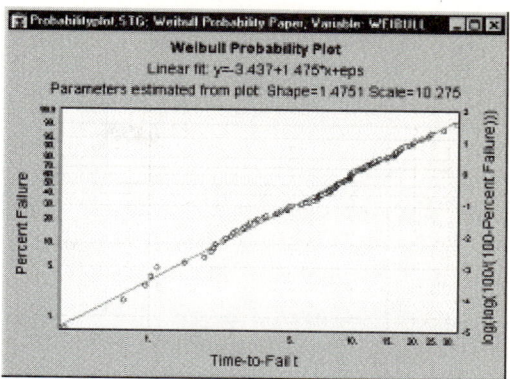

Note that the horizontal *Time* axis is scaled logarithmically; on the vertical axis the quantity *log(log(100/(100-F(t)))* is plotted (a probability scale is shown on the left-y axis). From this plot the parameters of the two-parameter Weibull distribution can be estimated; specifically, the shape parameter is equal to the slope of the linear fit-line, and the scale parameter can be estimated as *exp(-intercept/slope)*.

Estimating the location parameter from probability plots. It is apparent in the plot shown above that the regression line provides a good fit to the data. When the location parameter is misspecified (e.g., not equal to zero), the linear fit is worse as compared to the case when it is appropriately specified. Therefore, you can compute the probability plot for several values of the location parameter, and observe the quality of the fit. These computations are summarized in the following plot.

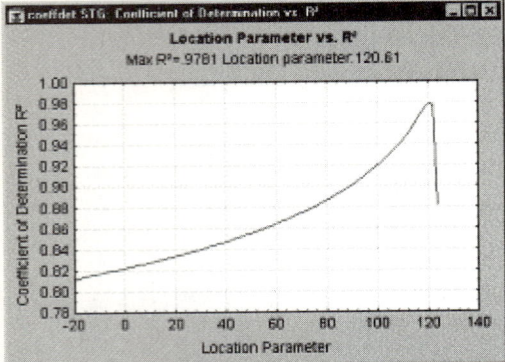

Here the common *R-square* measure (correlation squared) is used to express the quality of the linear fit in the probability plot, for different values of the location parameter shown on the horizontal *x* axis (this plot is based on the example data set in Dodson, 1994, Table 2.9). This plot is often very useful when the maximum likelihood estimation procedure for the three-parameter Weibull distribution fails, because it shows whether or not a unique (single) optimum value for the location parameter exists (as in the plot shown above).

Hazard plotting. Another method for estimating the parameters for the two-parameter Weibull distribution is via hazard plotting (as discussed earlier, the hazard function describes the probability of failure during a very small time increment, assuming that no failures have occurred prior to that time). This method is very similar to the probability plotting method. First plot the cumulative hazard function against the logarithm of the survival times; then fit a linear regression line and compute the slope and intercept of that line. As in probability plotting, the shape parameter can then be estimated as the slope of the regression line, and the scale parameter as exp (-intercept/slope). See Dodson (1994) for additional details; see also, *Weibull CDF, reliability, and hazard functions*, page 440.

Method of moments. This method – to approximate the moments of the observed distribution by choosing the appropriate parameters for the Weibull distribution – is also widely described in the literature. In fact, this general method is used for fitting the Johnson curves general non-normal distribution to the data to compute non-normal process capability indices (see *Fitting Distributions by Moments*, page 431). However, the method is not suited for censored data sets, and is therefore not very useful for the analysis of failure time data.

Comparing the estimation methods. Dodson (1994) reports the result of a Monte Carlo simulation study, comparing the different methods of estimation. In general, the maximum likelihood estimates proved to be best for large sample sizes (e.g., $n>15$), while probability plotting and hazard plotting appeared to produce better (more accurate) estimates for smaller samples.

A note of caution regarding maximum likelihood based confidence limits. Many software programs will compute confidence intervals for maximum likelihood estimates, and for the reliability function based on the standard errors of the maximum likelihood estimates. Dodson (1994) cautions against the interpretation of confidence limits computed from maximum likelihood estimates, or more precisely, estimates that involve the information matrix for the estimated parameters. When the shape parameter is less than 2, the variance estimates computed for maximum likelihood estimates lack accuracy, and it is advisable to compute the various results graphs based on nonparametric confidence limits as well.

Goodness of Fit Indices

A number of different tests have been proposed for evaluating the quality of the fit of the

Weibull distribution to the observed data. These tests are discussed and compared in detail in Lawless (1982).

Hollander-Proschan. This test compares the theoretical reliability function to the Kaplan-Meier estimate. The actual computations for this test are somewhat complex, and you may refer to Dodson (1994, Chapter 4) for a detailed description of the computational formulas. The Hollander-Proschan test is applicable to complete, single-censored, and multiple-censored data sets; however, Dodson (1994) cautions that the test may sometimes indicate a poor fit when the data are heavily single-censored. The Hollander-Proschan C statistic can be tested against the normal distribution (z).

Mann-Scheuer-Fertig. This test, proposed by Mann, Scheuer, and Fertig (1973), is described in detail in, for example, Dodson (1994) or Lawless (1982). The null hypothesis for this test is that the population follows the Weibull distribution with the estimated parameters. Nelson (1982) reports this test to have reasonably good power, and this test can be applied to Type II censored data. For computational details refer to Dodson (1994) or Lawless (1982); the critical values for the test statistic have been computed based on Monte Carlo studies, and have been tabulated for n (sample sizes) between 3 and 25.

Anderson-Darling. The Anderson-Darling procedure is a general test to compare the fit of an observed cumulative distribution function to an expected cumulative distribution function. However, this test is only applicable to complete data sets (without censored observations). The critical values for the Anderson-Darling statistic have been tabulated (see, for example, Dodson, 1994, Table 4.4) for

sample sizes between 10 and 40; this test is not computed for n less than 10 and greater than 40.

Interpreting Results

Once a satisfactory fit of the Weibull distribution to the observed failure time data has been obtained, there are a number of different plots and tables that are of interest to understand the reliability of the item under investigation. If a good fit for the Weibull cannot be established, distribution-free reliability estimates (and graphs) should be reviewed to determine the shape of the reliability function.

Reliability plots. This plot will show the estimated reliability function along with the confidence limits.

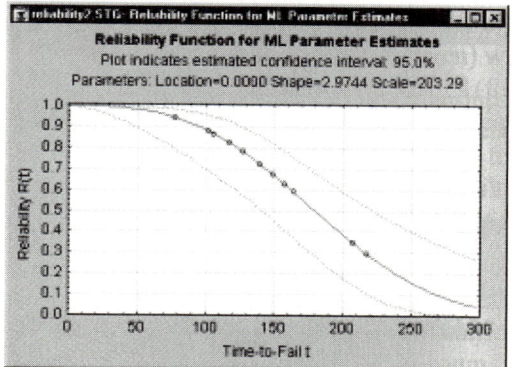

Note that nonparametric (distribution-free) estimates and their standard errors can also be computed and plotted.

Hazard plots. As mentioned earlier, the hazard function describes the probability of failure during a very small time increment (assuming that no failures have occurred prior to that time). The plot of hazard as a function of time gives valuable information about the conditional failure probability.

Percentiles of the reliability function. Based on the fitted Weibull distribution, you can compute the percentiles of the reliability (survival) function, along with the confidence limits for these estimates (for maximum likelihood parameter estimates). These estimates are particularly valuable for determining the percentages of items that can be expected to have failed at particular points in time.

Grouped Data

In some cases, failure time data are presented in grouped form. Specifically, instead of having available the precise failure time for each observation, only aggregate information is available about the number of items that failed or were censored in a particular time interval. Such life-table data input is also described in Chapter 36 – *Survival/Failure Time Analysis*. There are two general approaches for fitting the Weibull distribution to grouped data.

First, you can treat the tabulated data as if they were continuous. In other words, you can expand the tabulated values into continuous data by assuming 1) that each observation in a given time interval failed exactly at the interval mid-point (interpolating out "half a step" for the last interval), and 2) that censoring occurred after the failures in each interval (in other words, censored observations are sorted after the observed failures). Lawless (1982) advises that this method is usually satisfactory if the class intervals are relatively narrow.

Alternatively, you can treat the data explicitly as a tabulated life table, and use a weighted least squares methods algorithm (based on Gehan and Siddiqui, 1973; see also, Lee, 1992) to fit the Weibull distribution (Lawless, 1982, also describes methods for computing maximum likelihood parameter estimates from grouped data).

Modified Failure Order for Multiple-Censored Data

For multiple-censored data a weighted average ordered failure is calculated for each failure after the first censored data point. These failure orders are then used to compute the median rank, to estimate the cumulative distribution function.

The modified failure order j is computed as (see Dodson 1994):

$$I_j = ((n+1)-O_p)/(1+c)$$

Where I_j is the increment for the j'th failure, n is the total number of data points, O_p is the failure order of the previous observation (and $O_j = O_p + I_j$), and c is the number of data points remaining in the data set, including the current data point

The median rank is then computed as:

$$F(t) = (I_j - 0.3)/(n+0.4)$$

where I_j denotes the modified failure order, and n is the total number of observations.

Weibull CDF, Reliability, and Hazard Functions

Density function. The Weibull distribution (Weibull, 1939, 1951; see also, Lieblein, 1955) has density function (for positive parameters b, c, and θ):

$$f(x) = \frac{c}{b}\left(\frac{x-\theta}{b}\right)^{c-1} e^{-\left[\frac{(x-\theta)}{b}\right]^c}$$

where b is the scale parameter of the distribution, c is the shape parameter of the distribution, θ is the location parameter of the distribution, and e is the base of the natural logarithm, sometimes called Euler's e (2.71...) .

Cumulative distribution function (CDF). The Weibull distribution has the cumulative distribution function (for positive parameters b, c, and θ):

$$F(x) = 1 - e^{-\left[\frac{(x-\theta)}{b}\right]^c}$$

using the same notation and symbols as described above for the density function.

Reliability function. The Weibull reliability function is the complement of the cumulative distribution function:

$R(x) = 1 - F(x)$

Hazard function. The hazard function describes the probability of failure during a very small time increment, assuming that no failures have occurred prior to that time. The Weibull distribution has the hazard function (for positive parameters b, c, and θ):

$h(t) = f(t)/R(t) = [c^*(x-\theta)^{(c-1)}] / b^c$

using the same notation and symbols as described above for the density and reliability functions.

Cumulative hazard function. The Weibull distribution has the cumulative hazard function (for positive parameters b, c, and θ):

$H(t) = (x-\theta) / b^c$

using the same notation and symbols as described above for the density and reliability functions.

QUALITY CONTROL CHARTS

In all production processes, we need to monitor the extent to which our products meet specifications. In the most general terms, there are two enemies of product quality: 1) deviations from target specifications and 2) excessive variability around target specifications. During the earlier stages of developing the production process, designed experiments are often used to optimize these two quality characteristics [see Chapter 15 – *Experimental Design (DOE)*]; the methods provided in quality control are on-line or in-process quality control procedures to monitor an on-going production process. For detailed descriptions of these charts and extensive annotated examples, see Buffa (1972), Duncan (1974) Grant and Leavenworth (1980), Juran (1962), Juran and Gryna (1970), Montgomery (1985, 1991), Shirland (1993), or Vaughn (1974). Two recent excellent introductory texts with a how-to approach are Hart & Hart (1989) and Pyzdek (1989); two recent German language texts on this subject are Rinne and Mittag (1995) and Mittag (1993).

Overview

The general approach to on-line quality control is straightforward: We simply extract samples of a certain size from the ongoing production process. We then produce line charts of the variability in those samples, and consider their closeness to target specifications. If a trend emerges in those lines, or if samples fall outside pre-specified limits, we declare the process to be out of control and take action to find the cause of the problem. These types of charts are sometimes also referred to as Shewhart control charts (named after W. A. Shewhart who is generally credited as being the first to introduce these methods; see Shewhart, 1931).

Interpreting the chart. The most standard display actually contains two charts (and two histograms); one is called an X-bar chart, the other is called an R chart.

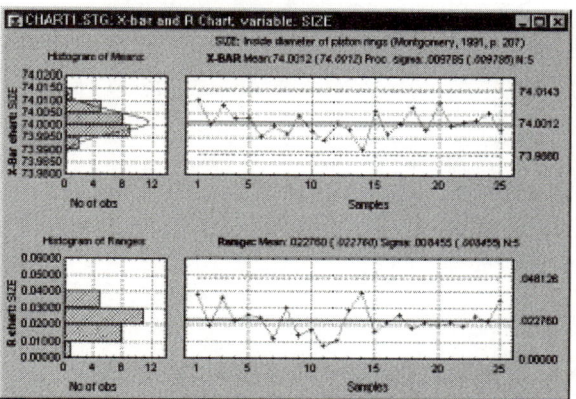

In both line charts, the horizontal axis represents the different samples; the vertical axis for the X-bar chart represents the means for the characteristic of interest; the vertical axis for the R chart represents the ranges. For example, suppose we wanted to control the diameter of piston rings that we are producing. The center line in the X-bar chart would represent the desired standard size (e.g., diameter in

millimeters) of the rings, while the center line in the R chart would represent the acceptable (within-specification) range of the rings within samples; thus, this latter chart is a chart of the variability of the process (the larger the variability, the larger the range). In addition to the center line, a typical chart includes two additional horizontal lines to represent the upper and lower control limits (*UCL, LCL,* respectively); we will return to those lines shortly. Typically, the individual points in the chart, representing the samples, are connected by a line. If this line moves outside the upper or lower control limits or exhibits systematic patterns across consecutive samples (see Runs Tests), a quality problem may potentially exist.

Establishing Control Limits

Even though you could arbitrarily determine when to declare a process out of control (that is, outside the UCL-LCL range), it is common practice to apply statistical principles to do so. Chapter 1 -*Elementary Concepts in Statistics* discusses the concept of the sampling distribution, and the characteristics of the normal distribution. The method for constructing the upper and lower control limits is a straightforward application of the principles described there.

Example. Suppose we want to control the mean of a variable, such as the size of piston rings. Under the assumption that the mean (and variance) of the process does *not* change, the successive sample means will be distributed normally around the actual mean. Moreover, without going into details regarding the derivation of this formula, we also know (because of the central limit theorem, and thus approximate normal distribution of the means; see, for example, Hoyer and Ellis, 1996) that

the distribution of sample means will have a standard deviation of *sigma* (the standard deviation of individual data points or measurements) over the square root of *n* (the sample size). It follows that approximately 95% of the sample means will fall within the limits μ ± *1.96 * Sigma/Square Root(n)* (refer to *Elementary Concepts* for a discussion of the characteristics of the normal distribution and the central limit theorem). In practice, it is common to replace the 1.96 with 3 (so that the interval will include approximately 99% of the sample means), and to define the upper and lower control limits as plus and minus 3 *sigma* limits, respectively.

General case. The general principle for establishing control limits just described applies to all control charts. After deciding on the characteristic we want to control, for example, the standard deviation, we estimate the expected variability of the respective characteristic in samples of the size we are about to take. Those estimates are then used to establish the control limits on the chart.

Common Types of Charts

The types of charts are often classified according to the type of quality characteristic that they are supposed to monitor: there are quality control charts for variables and control charts for attributes. Specifically, the following charts are commonly constructed for controlling variables:

- **X-bar chart.** In this chart the sample means are plotted in order to control the mean value of a variable (e.g., size of piston rings, strength of materials, etc.).

- **R chart.** In this chart, the sample ranges are plotted in order to control the variability of a variable.

- **S chart.** In this chart, the sample standard deviations are plotted in order to control the variability of a variable.

- **S² chart.** In this chart, the sample variances are plotted in order to control the variability of a variable.

For controlling quality characteristics that represent attributes of the product, the following charts are commonly constructed:

- **C chart.** In this chart (see the next illustration), we plot the number of defectives (per batch, per day, per machine, per 100 feet of pipe, etc.). This chart assumes that defects of the quality attribute are rare, and the control limits in this chart are computed based on the Poisson distribution (distribution of rare events).

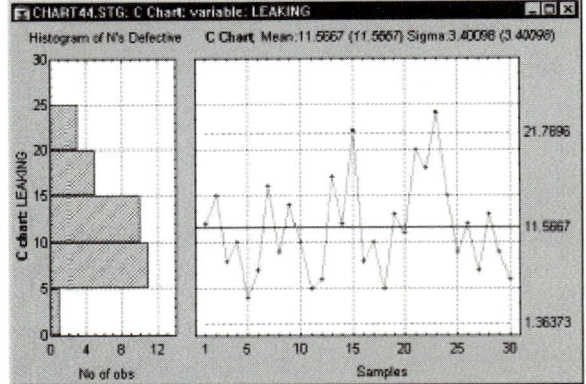

- **U chart.** In this chart we plot the rate of defectives, that is, the number of defectives divided by the number of units inspected (the *n*; e.g., feet of pipe, number of batches). Unlike the C chart, this chart does not require a constant number of units, and it can be used, for example, when the batches (samples) are of different sizes.

- **Np chart.** In this chart, we plot the number of defectives (per batch, per day, per machine) as in the C chart. However, the control limits in this chart are not based on the distribution of rare events, but rather on the binomial distribution. Therefore, this chart should be used if the occurrence of defectives is not rare (e.g., they occur in more than 5% of the units inspected). For example, we may use this chart to control the number of units produced with minor flaws.

- **P chart.** In this chart, we plot the percent of defectives (per batch, per day, per machine, etc.) as in the U chart. However, the control limits in this chart are not based on the distribution of rare events but rather on the binomial distribution (of proportions). Therefore, this chart is most applicable to situations where the occurrence of defectives is not rare (e.g., we expect the percent of defectives to be more than 5% of the total number of units produced).

All of these charts can be adapted for short production runs (short run charts), and for multiple process streams.

Short Run Charts

The short run control chart, or control chart for short production runs, plots observations of variables or attributes for multiple parts on the same chart. Short run control charts were developed to address the requirement that several dozen measurements of a process must be collected before control limits are calculated. Meeting this requirement is often difficult for operations that produce a limited number of a particular part during a production run.

For example, a paper mill may produce only three or four (huge) rolls of a particular kind of paper (i.e., part) and then shift production to another kind of paper. But if variables, such as

paper thickness, or attributes, such as blemishes, are monitored for several dozen rolls of paper of, say, a dozen different kinds, control limits for thickness and blemishes could be calculated for the transformed (within the short production run) variable values of interest. Specifically, these transformations will rescale the variable values of interest such that they are of compatible magnitudes across the different short production runs (or parts). The control limits computed for those transformed values could then be applied in monitoring thickness, and blemishes, regardless of the types of paper (parts) being produced. Statistical process control procedures could be used to determine if the production process is in control, to monitor continuing production, and to establish procedures for continuous quality improvement.

For additional discussions of short run charts refer to Bothe (1988), Johnson (1987), or Montgomery (1991).

Short Run Charts for Variables

Nominal chart, target chart. There are several different types of short run charts. The most basic are the nominal short run chart, and the target short run chart. In these charts, the measurements for each part are transformed by subtracting a part-specific constant. These constants can either be the nominal values for the respective parts (*nominal* short run chart), or they can be target values computed from the (historical) means for each part (Target X-bar and R chart). For example, the diameters of piston bores for different engine blocks produced in a factory can only be meaningfully compared (for determining the consistency of bore sizes) if the mean differences between bore diameters for different sized engines are first removed. The nominal or target short run chart makes such comparisons possible. Note that for

the nominal or target chart it is assumed that the variability across parts is identical, so that control limits based on a common estimate of the process *sigma* are applicable.

Standardized short run chart. If the variability of the process for different parts cannot be assumed to be identical, a further transformation is necessary before the sample means for different parts can be plotted in the same chart. Specifically, in the standardized short run chart the plot points are further transformed by dividing the deviations of sample means from part means (or nominal or target values for parts) by part-specific constants that are proportional to the variability for the respective parts. For example, for the short run X-bar and R chart, the plot points (that are shown in the X-bar chart) are computed by first subtracting from each sample mean a part specific constant (e.g., the respective part mean, or nominal value for the respective part), and then dividing the difference by another constant, for example, by the average range for the respective chart. These transformations will result in comparable scales for the sample means for different parts.

Short Run Charts for Attributes

For attribute control charts (C, U, Np, or P charts), the estimate of the variability of the process (proportion, rate, etc.) is a function of the process average (average proportion, rate, etc.; for example, the standard deviation of a proportion p is equal to the square root of $p*(1-p)/n)$. Hence, only standardized short run charts are available for attributes. For example, in the short run P chart, the plot points are computed by first subtracting from the respective sample p values the average part p's, and then dividing by the standard deviation of the average p's.

Unequal Sample Sizes

When the samples plotted in the control chart are not of equal size, the control limits around the center line (target specification) cannot be represented by a straight line. For example, to return to the formula *Sigma/Square Root(n)* presented earlier for computing control limits for the X-bar chart, it is obvious that unequal n's will lead to different control limits for different sample sizes. There are three ways of dealing with this situation.

Average sample size. If you want to maintain the straight-line control limits (e.g., to make the chart easier to read and easier to use in presentations), you can compute the average n per sample across all samples, and establish the control limits based on the average sample size. This procedure is not exact; however, as long as the sample sizes are reasonably similar to each other, this procedure is quite adequate.

Variable control limits. Alternatively, you can compute different control limits for each sample, based on the respective sample sizes. This procedure will lead to variable control limits, and result in step-chart like control lines in the plot. This procedure ensures that the correct control limits are computed for each sample. However, you lose the simplicity of straight-line control limits.

Stabilized (normalized) chart. The best of two worlds (straight line control limits that are accurate) can be accomplished by standardizing the quantity to be controlled (mean, proportion, etc.) according to units of *sigma*. The control limits can then be expressed in straight lines, while the location of the sample points in the plot depend not only on the characteristic to be controlled, but also on the respective sample n's. The disadvantage of this procedure is that the values on the vertical (Y) axis in the control chart are in terms of *sigma* rather than the original units of measurement, and therefore, those numbers cannot be taken at face value (e.g., a sample with a value of 3 is 3 times *sigma* away from specifications; in order to express the value of this sample in terms of the original units of measurement, we need to perform some computations to convert this number back).

Control Charts for Variables vs. Charts for Attributes

Sometimes, the quality control engineer has a choice between variable control charts and attribute control charts.

Advantages of attribute control charts.
Attribute control charts have the advantage of allowing for quick summaries of various aspects of the quality of a product, that is, the engineer may simply classify products as acceptable or unacceptable, based on various quality criteria. Thus, attribute charts sometimes bypass the need for expensive, precise devices and time-consuming measurement procedures. Also, this type of chart tends to be more easily understood by managers unfamiliar with quality control procedures; therefore, it may provide more persuasive (to management) evidence of quality problems.

Advantages of variable control charts.
Variable control charts are more sensitive than attribute control charts (see Montgomery, 1985, p. 203). Therefore, variable control charts can alert us to quality problems before any actual "unacceptables" (as detected by the attribute chart) will occur. Montgomery (1985) calls the variable control charts leading indicators of trouble that will sound an alarm before the number of rejects (scrap) increases in the production process.

Control Charts for Individual Observations

Variable control charts can by constructed for individual observations taken from the production line, rather than samples of observations. This is sometimes necessary when testing samples of multiple observations would be too expensive, inconvenient, or impossible. For example, the number of customer complaints or product returns may only be available on a monthly basis; yet, you want to chart those numbers to detect quality problems. Another common application of these charts occurs in cases when automated testing devices inspect every single unit that is produced. In this case, you are often primarily interested in detecting small shifts in the product quality (for example, gradual deterioration of quality due to machine wear). The *CUSUM, MA*, and *EWMA* charts of cumulative sums and weighted averages discussed below may be most applicable in those situations.

Out-of-Control Process: Runs Tests

As mentioned earlier, when a sample point (e.g., mean in an X-bar chart) falls outside the control lines, you have reason to believe that the process may no longer be in control. In addition, you should look for systematic patterns of points (e.g., means) across samples, because such patterns may indicate that the process average has shifted. These tests are also sometimes referred to as AT&T runs rules (see AT&T, 1959) or tests for special causes (e.g., see Nelson, 1984, 1985; Grant and Leavenworth, 1980; Shirland, 1993). The term special or assignable causes as opposed to chance or common causes was used by Shewhart to distinguish between a process that

is in control, with variation due to random (chance) causes only, from a process that is out of control, with variation that is due to some non-chance or special (assignable) factors (cf. Montgomery, 1991, p. 102).

As the *sigma* control limits discussed earlier, the runs rules are based on "statistical" reasoning. For example, the probability of any sample mean in an X-bar control chart falling above the center line is equal to 0.5, provided 1) that the process is in control (i.e., that the center line value is equal to the population mean), 2) that consecutive sample means are independent (i.e., not auto-correlated), and 3) that the distribution of means follows the normal distribution. Simply stated, under those conditions there is a 50-50 chance that a mean will fall above or below the center line. Thus, the probability that two consecutive means will fall above the center line is equal to 0.5 times 0.5 = 0.25.

Accordingly, the probability that 9 consecutive samples (or a *run* of 9 samples) will fall on the same side of the center line is equal to 0.5^9 = .00195. Note that this is approximately the probability with which a sample mean can be expected to fall outside the 3-times *sigma* limits (given the normal distribution, and a process in control). Therefore, you could look for 9 consecutive sample means on the same side of the center line as another indication of an out-of-control condition. Refer to Duncan (1974) for details concerning the "statistical" interpretation of the other (more complex) tests.

Zone A, B, C. Customarily, to define the runs tests, the area above and below the chart center line is divided into three zones.

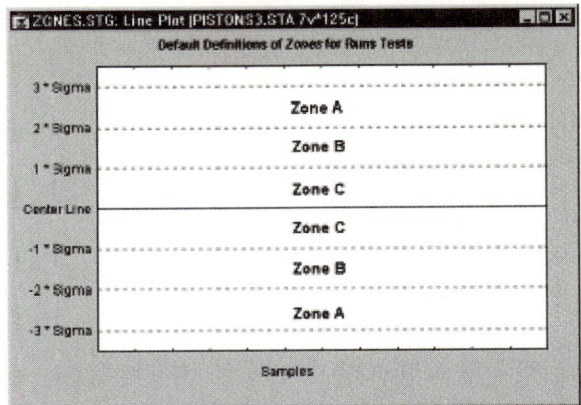

By default, *Zone A* is defined as the area between 2 and 3 times *sigma* above and below the center line; *Zone B* is defined as the area between 1 and 2 times *sigma*, and *Zone C* is defined as the area between the center line and 1 times *sigma*.

Nine points in zone C or beyond (on one side of central line). If this test is positive (i.e., if this pattern is detected), the process average has probably changed. Note that it is assumed that the distribution of the respective quality characteristic in the plot is symmetrical around the mean. This is, for example, not the case for R charts, S charts, or most attribute charts. However, this is still a useful test to alert the quality control engineer to potential shifts in the process. For example, successive samples with less-than-average variability may be worth investigating, since they may provide hints on how to decrease the variation in the process.

Six points in a row steadily increasing or decreasing. This test signals a drift in the process average. Often, such drift can be the result of tool wear, deteriorating maintenance, improvement in skill, etc. (Nelson, 1985).

Fourteen points in a row alternating up and down. If this test is positive, it indicates that two systematically alternating causes are producing different results. For example, use two alternating suppliers or monitor the quality for two different (alternating) shifts.

Two out of three points in a row in Zone A or beyond. This test provides an early warning of a process shift. Note that the probability of a false-positive (test is positive but process is in control) for this test in X-bar charts is approximately 2%.

Four out of five points in a row in Zone B or beyond. Like the previous test, this test can be considered to be an early warning indicator of a potential process shift. The false-positive error rate for this test is also about 2%.

Fifteen points in a row in Zone C (above and below the center line). This test indicates a smaller variability than is expected (based on the current control limits).

Eight points in a row in Zone B, A, or beyond, on either side of the center line (without points in Zone C). This test indicates that different samples are affected by different factors, resulting in a bimodal distribution of means. This may happen, for example, if different samples in an X-bar chart where produced by one of two different machines, where one produces above average parts, and the other below average parts.

Operating Characteristic (OC) Curves

A common supplementary plot to standard quality control charts is the so-called operating characteristic or OC curve (see the next illustration). One question that comes to mind when using standard variable or attribute charts is how sensitive is the current quality control procedure? More specifically, how likely is it that you will not find a sample (e.g., mean in an X-bar chart) outside the control limits (i.e., accept the production process as in control), when, in fact, it has shifted by a certain

amount? This probability is usually referred to as the β (*beta*) error probability, i.e., the probability of erroneously accepting a process (mean, mean proportion, mean rate defectives, etc.) as being in control. Note that operating characteristic curves pertain to the false-acceptance probability using the sample-outside-of-control-limits criterion only, and not the runs tests described earlier.

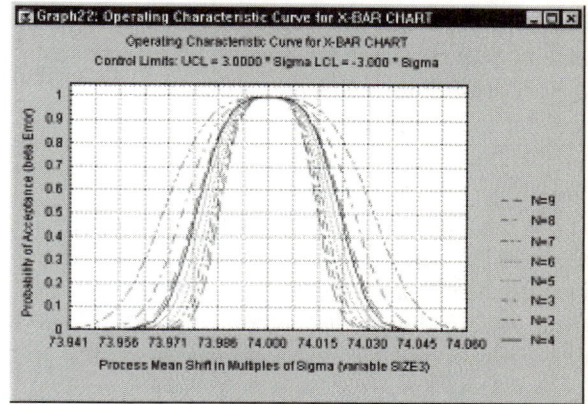

Operating characteristic curves are extremely useful for exploring the power of our quality control procedure. The actual decision concerning sample sizes should depend not only on the cost of implementing the plan (e.g., cost per item sampled), but also on the costs resulting from not detecting quality problems. The OC curve enables the engineer to estimate the probabilities of not detecting shifts of certain sizes in the production quality.

Process Capability Indices

For variable control charts, it is often desired to include so-called process capability indices in the summary graph. In short, process capability indices express (as a ratio) the proportion of parts or items produced by the current process that fall within user-specified limits (e.g., engineering tolerances).

For example, the so-called Cp index is computed as:

$$C_p = (USL-LSL)/(6*sigma)$$

where *sigma* is the estimated process standard deviation, and *USL* and *LSL* are the upper and lower specification (engineering) limits, respectively. If the distribution of the respective quality characteristic or variable (e.g., size of piston rings) is normal, and the process is perfectly centered (i.e., the mean is equal to the design center), this index can be interpreted as the proportion of the range of the standard normal curve (the process width) that falls within the engineering specification limits. If the process is not centered, an adjusted index C_{pk} is used instead. For a capable process, the C_p index should be greater than 1, that is, the specification limits would be larger than 6 times the *sigma* limits, so that over 99% of all items or parts produced could be expected to fall inside the acceptable engineering specifications. For a detailed discussion of this and other indices, refer to Chapter 32 – *Process Analysis*.

Other Specialized Control Charts

The types of control charts mentioned so far are the "workhorses" of quality control, and are probably the most widely used methods. However, with the advent of inexpensive desktop computing, procedures requiring more computational effort have become increasingly popular.

X-bar charts for non-normal data. The control limits for standard X-bar charts are constructed based on the assumption that the sample means are approximately normally distributed. Thus, the underlying individual observations do not have to be normally distributed, since, as the sample size increases, the distribution of the means will become approximately normal (see

the discussion of the central limit theorem in the *Are All Test Statistics Normally Distributed?* section of Chapter 1 – *Elementary Concepts in Statistics*; however, note that for *R, S, and S^2* charts, it is assumed that the individual observations are normally distributed). Shewhart (1931) in his original work experimented with various non-normal distributions for individual observations, and evaluated the resulting distributions of means for samples of size four. He concluded that, indeed, the standard normal distribution-based control limits for the means are appropriate, as long as the underlying distribution of observations are approximately normal. (See also, Hoyer and Ellis, 1996, for an introduction and discussion of the distributional assumptions for quality control charting.)

However, as Ryan (1989) points out, when the distribution of observations is highly skewed and the sample sizes are small, the resulting standard control limits may produce a large number of false alarms (increased *alpha* error rate), as well as a larger number of false negative (process-is-in-control) readings (increased *beta* error rate). You can compute control limits (as well as process capability indices) for X-bar charts based on so-called Johnson curves (Johnson, 1949), which enable you to approximate the skewness and kurtosis for a large range of non-normal distributions (see also, *Fitting Distributions by Moments*, page 431). These non-normal X-bar charts are useful when the distribution of means across the samples is clearly skewed, or otherwise non-normal.

Hotelling T^2 chart. When there are multiple related quality characteristics (recorded in several variables), we can produce a simultaneous plot (see the next illustration) for all means based on Hotelling multivariate T^2 statistic (first proposed by Hotelling, 1947).

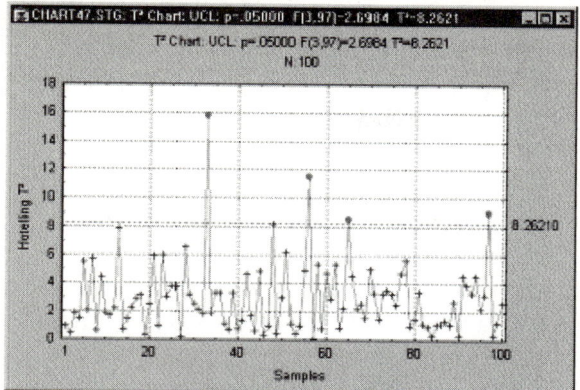

Cumulative Sum (CUSUM) chart. The CUSUM chart was first introduced by Page (1954); the mathematical principles involved in its construction are discussed in Ewan (1963), Johnson (1961), and Johnson and Leone (1962).

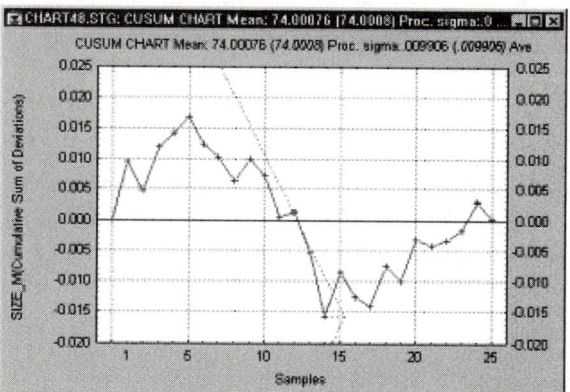

If you plot the cumulative sum of deviations of successive sample means from a target specification, even minor, permanent shifts in the process mean will eventually lead to a sizable cumulative sum of deviations. Thus, this chart is particularly well suited for detecting such small permanent shifts that may go undetected when using the X-bar chart. For example, if, due to machine wear, a process slowly slides out of control to produce results above target specifications, this plot would show a steadily

increasing (or decreasing) cumulative sum of deviations from specification.

To establish control limits in such plots, Barnhard (1959) proposed the so-called V-mask, which is plotted after the last sample (on the right). The V-mask can be thought of as the upper and lower control limits for the cumulative sums. However, rather than being parallel to the center line; these lines converge at a particular angle to the right, producing the appearance of a *V* rotated on its side. If the line representing the cumulative sum crosses either one of the two lines, the process is out of control.

Moving average (MA) chart. To return to the piston ring example, suppose we are mostly interested in detecting small trends across successive sample means. For example, we may be particularly concerned about machine wear, leading to a slow but constant deterioration of quality (i.e., deviation from specification). The CUSUM chart described above is one way to monitor such trends and to detect small permanent shifts in the process average. Another way is to use a weighting scheme that summarizes the means of several successive samples; moving such a weighted mean across the samples will produce a moving average chart (as shown in the next illustration).

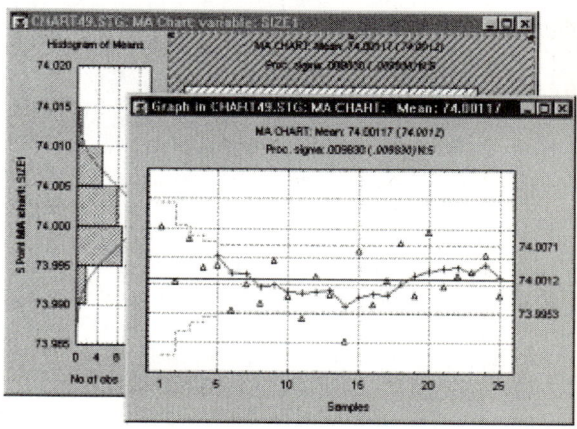

Exponentially-weighted moving average (EWMA) chart. The idea of moving averages of successive (adjacent) samples can be generalized. In principle, in order to detect a trend we need to weight successive samples to form a moving average; however, instead of a simple arithmetic moving average, we could compute a geometric moving average (this chart (see the next illustration) is also called Geometric Moving Average chart, see Montgomery, 1985, 1991).

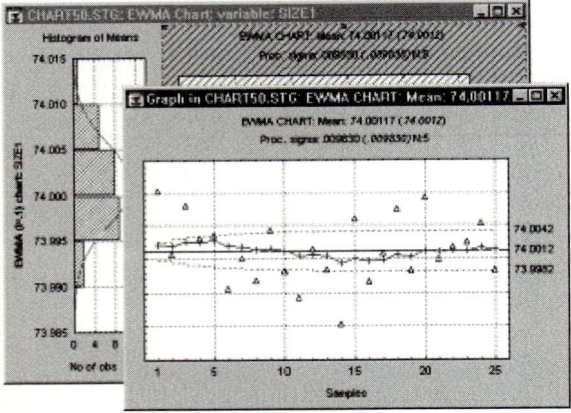

Specifically, we could compute each data point for the plot as:

$$z_t = \lambda * x\text{-}bar_t + (1 - \lambda) * z_{t-1}$$

In this formula, each point z_t is computed as λ (lambda) times the respective mean $x\text{-}bar_t$, plus one minus λ times the previous (computed) point in the plot. The parameter λ (*lambda*) here should assume values greater than 0 and less than 1. Without going into detail (see Montgomery, 1985, p. 239), this method of averaging specifies that the weight of historically old sample means decreases geometrically as you continue to draw samples. The interpretation of this chart is much like that of the moving average chart, and it enables us to detect small shifts in the means and, therefore, in the quality of the production process.

Regression control charts. Sometimes we want to monitor the relationship between two aspects of our production process. For example, a post office may want to monitor the number of worker-hours that are spent to process a certain amount of mail. These two variables should roughly be linearly correlated with each other, and the relationship can probably be described in terms of the well-known Pearson product-moment correlation coefficient r. This statistic is also described in Chapter 2 – *Basic Statistics and Tables*. The regression control chart contains a regression line that summarizes the linear relationship between the two variables of interest. The individual data points are also shown in the same graph. Around the regression line we establish a confidence interval within which we would expect a certain proportion (e.g., 95%) of samples to fall. Outliers in this plot may indicate samples where, for some reason, the common relationship between the two variables of interest does not hold.

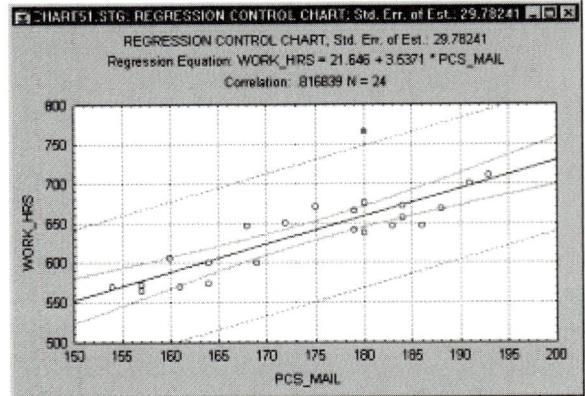

Applications. There are many useful applications for the regression control chart. For example, professional auditors may use this chart to identify retail outlets with a greater than expected number of cash transactions given the overall volume of sales, or grocery stores with a greater than expected number of coupons redeemed, given the total sales. In both instances, outliers in the regression control charts (e.g., too many cash transactions; too many coupons redeemed) may deserve closer scrutiny.

Pareto chart analysis. Quality problems are rarely spread evenly across the different aspects of the production process or different plants. Rather, a few "bad apples" often account for the majority of problems. This principle has come to be known as the Pareto principle, which basically states that quality losses are mal-distributed in such a way that a small percentage of possible causes are responsible for the majority of the quality problems. For example, a relatively small number of "dirty" cars are probably responsible for the majority of air pollution; the majority of losses in most companies result from the failure of only one or two products. To illustrate the bad apples, plot the Pareto chart:

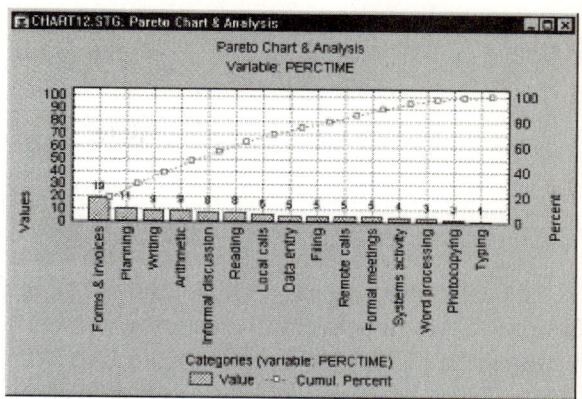

which simply amounts to a histogram showing the distribution of the quality loss (e.g., dollar loss) across some meaningful categories; usually, the categories are sorted into descending order of importance (frequency, dollar amounts, etc.). Very often, this chart provides useful guidance as to where to direct quality improvement efforts.

RELIABILITY/ITEM ANALYSIS

This chapter discusses the concept of reliability of measurement as used in social sciences (but not in industrial statistics or biomedical research). The term reliability used in industrial statistics denotes a function describing the probability of failure (as a function of time). For a discussion of the concept of reliability as applied to product quality (e.g., in industrial statistics), refer to Chapter 32 – *Process Analysis*: *Reliability/Failure Time Analysis*, page 433 (see also, *Repeatability and Reproducibility* in the same chapter, page 427, and Chapter 36 – *Survival/Failure Time Analysis*). For a comparison between these two (very different) concepts of reliability, see *Reliability* in Chapter 41 – *Statistical Glossary*.

Overview

In many areas of research, the precise measurement of hypothesized processes or variables (theoretical *constructs*) poses a challenge by itself. For example, in psychology, the precise measurement of personality variables or attitudes is usually a necessary first step before any theories of personality or attitudes can be considered. In general, in all social sciences, unreliable measurements of people's beliefs or intentions will obviously hamper efforts to predict their behavior. The issue of precision of measurement will also come up in applied research, whenever variables are difficult to observe. For example, reliable measurement of employee performance is usually a difficult task; yet, it is obviously a necessary precursor to any performance-based compensation system.

In all of these cases, reliability and item analysis can be used to construct reliable measurement scales, to improve existing scales, and to evaluate the reliability of scales already in use. Specifically, reliability and item analysis will aid in the design and evaluation of sum scales, that is, scales that are made up of multiple individual measurements (e.g., different items, repeated measurements, different measurement devices, etc.). You can compute numerous statistics that enables you to build and evaluate scales following the so-called classical testing theory model.

The assessment of scale reliability is based on the correlations between the individual items or measurements that make up the scale, relative to the variances of the items. If you are not familiar with the correlation coefficient or the variance statistic, we recommend that you review the respective discussions provided in Chapter 2 – *Basic Statistics and Tables*.

The classical testing theory model of scale construction has a long history, and there are many textbooks available on the subject. For additional detailed discussions, refer to, for example, Carmines and Zeller (1980), De Gruitjer and Van Der Kamp (1976), Kline (1979, 1986), or Thorndyke and Hagen (1977). A widely acclaimed classic in this area, with an emphasis on psychological and educational testing, is Nunally (1970).

Testing hypotheses about relationships between items and tests. Using structural equation modeling and path analysis, you can test specific hypotheses about the relationship between sets of items or different tests (e.g., test whether two sets of items measure the same construct, analyze multi-trait, multi-method matrices, etc.).

Basic Ideas

Suppose we want to construct a questionnaire to measure people's prejudices against foreign-made cars. We could start out by generating a number of items such as "Foreign cars lack personality," "Foreign cars all look the same," etc. We could then submit those questionnaire items to a group of subjects (for example, people who have never owned a foreign-made car). We could ask subjects to indicate their agreement with these statements on 9-point scales, anchored at *1=disagree* and *9=agree*.

True scores and error. Let's now consider more closely what we mean by precise measurement in this case. We hypothesize that there is such a thing (theoretical construct) as "prejudice against foreign cars," and that each item taps into this concept to some extent. Therefore, we can say that a subject's response to a particular item reflects two aspects: first, the response reflects the prejudice against foreign cars, and second, it will reflect some esoteric aspect of the respective question. For example, consider the item "Foreign cars all look the same." A subject's agreement or disagreement with that statement will partially depend on his or her general prejudices, and partially on some other aspects of the question or person. For example, the subject may have a friend who just bought a very different looking foreign car.

Testing hypotheses about relationships between items and tests. To test specific hypotheses about the relationship between sets of items or different tests (e.g., whether two sets of items measure the same construct, analyze multi-trait, multi-method matrices, etc.) use *Structural Equation Modeling*.

Classical Testing Model

To summarize, each measurement (response to an item) reflects to some extent the true score for the intended concept (prejudice against foreign cars), and to some extent esoteric, random error. We can express this in an equation as:

X = tau + error

In this equation, X refers to the respective actual measurement, that is, subject's response to a particular item; tau is commonly used to refer to the true score, and error refers to the random error component in the measurement.

Reliability

In this context, the definition of reliability is straightforward: a measurement is reliable if it reflects mostly true score, relative to the error. For example, an item such as "Red foreign cars are particularly ugly" would likely provide an unreliable measurement of prejudices against foreign-made cars. This is because there

probably are ample individual differences concerning the likes and dislikes of colors. Thus, this item would capture not only a person's prejudice but also his or her color preference. Therefore, the proportion of true score (for prejudice) in subjects' response to that item would be relatively small.

Measures of reliability. From the above discussion, we can easily infer a measure or statistic to describe the reliability of an item or scale. Specifically, we can define an index of reliability in terms of the proportion of true score variability that is captured across subjects or respondents, relative to the total observed variability. In equation form, we can say:

$$\text{Reliability} = \sigma^2_{\text{(true score)}} / \sigma^2_{\text{(total observed)}}$$

Sum Scales

What will happen when we sum up several more or less reliable items designed to measure prejudice against foreign-made cars? Suppose the items were written so as to cover a wide range of possible prejudices against foreign-made cars. If the error component in subjects' responses to each question is truly random, we can expect that the different components will cancel each other out across items. In slightly more technical terms, the expected value or mean of the error component across items will be zero. The true score component remains the same when summing across items. Therefore, the more items are added, the more true score (relative to the error score) will be reflected in the sum scale.

Number of items and reliability. This conclusion describes a basic principle of test design. Namely, the more items there are in a scale designed to measure a particular concept, the more reliable will the measurement (sum scale) be. Perhaps a somewhat more practical example will further clarify this point. Suppose you want to measure the height of 10 persons, using only a crude stick as the measurement device. Note that we are not interested in this example in the absolute correctness of measurement (i.e., in inches or centimeters), but rather in the ability to distinguish reliably between the 10 individuals in terms of their height. If you measure each person only once in terms of multiples of lengths of your crude measurement stick, the resultant measurement may not be very reliable. However, if you measure each person 100 times, and then take the average of those 100 measurements as the summary of the respective person's height, you will be able to make very precise and reliable distinctions between people (based solely on the crude measurement stick).

Let's now look at some of the common statistics that are used to estimate the reliability of a sum scale.

Cronbach's Alpha

To return to the prejudice example, if there are several subjects who respond to our items, we can compute the variance for each item, and the variance for the sum scale. The variance of the sum scale will be smaller than the sum of item variances if the items measure the same variability between subjects, that is, if they measure some true score. Technically, the variance of the sum of two items is equal to the sum of the two variances minus (two times) the covariance, that is, the amount of true score variance common to the two items.

We can estimate the proportion of true score variance that is captured by the items by comparing the sum of item variances with the variance of the sum scale. Specifically, we can compute:

$$\alpha = (k/(k-1)) * [1- \sum (s_i^2)/s_{sum}^2]$$

This is the formula for the most common index of reliability, namely, Cronbach's coefficient *alpha* (α). In this formula, the s_i^2 denotes the variances for the k individual items; s_{sum}^2 denotes the variance for the sum of all items. If there is no true score but only error in the items (which is esoteric and unique, and, therefore, uncorrelated across subjects), the variance of the sum will be the same as the sum of variances of the individual items. Therefore, coefficient *alpha* will be equal to zero. If all items are perfectly reliable and measure the same thing (true score), coefficient *alpha* is equal to 1. (Specifically, $1-\sum (s_i^2)/s_{sum}^2$ will become equal to $(k-1)/k$; if we multiply this by $k/(k-1)$ we obtain 1.)

Alternative terminology. Cronbach's *alpha*, when computed for binary (e.g., true/false) items, is identical to the so-called Kuder-Richardson-20 formula of reliability for sum scales. In either case, because the reliability is actually estimated from the consistency of all items in the sum scales, the reliability coefficient computed in this manner is also referred to as the internal-consistency reliability.

Split-Half Reliability

An alternative way of computing the reliability of a sum scale is to divide it in some random manner into two halves. If the sum scale is perfectly reliable, we would expect that the two halves are perfectly correlated (i.e., $r = 1.0$). Less than perfect reliability will lead to less than perfect correlations. We can estimate the reliability of the sum scale via the Spearman-Brown split half coefficient:

$$r_{sb} = 2r_{xy} /(1+r_{xy})$$

In this formula, r_{sb} is the split-half reliability coefficient, and r_{xy} represents the correlation between the two halves of the scale.

Correction for Attenuation

Let's now consider some of the consequences of less than perfect reliability. Suppose we use our scale of prejudice against foreign-made cars to predict some other criterion, such as subsequent actual purchase of a car. If our scale correlates with such a criterion, it would raise our confidence in the validity of the scale, that is, that it really measures prejudices against foreign-made cars, and not something completely different. In actual test design, the validation of a scale is a lengthy process that requires the researcher to correlate the scale with various external criteria that, in theory, should be related to the concept that is supposedly being measured by the scale.

How will validity be affected by less than perfect scale reliability? The random error portion of the scale is unlikely to correlate with some external criterion. Therefore, if the proportion of true score in a scale is only 60% (that is, the reliability is only .60), the correlation between the scale and the criterion variable will be attenuated, that is, it will be smaller than the actual correlation of true scores. In fact, the validity of a scale is always limited by its reliability.

Given the reliability of the two measures in a correlation (i.e., the scale and the criterion variable), we can estimate the actual correlation of true scores in both measures. Put another way, we can correct the correlation for attenuation:

$$r_{xy,corrected} = r_{xy} / (r_{xx} * r_{yy})^{1/2}$$

In this formula, $r_{xy,corrected}$ stands for the corrected correlation coefficient, that is, it is the estimate of the correlation between the true scores in the two measures x and y. The term r_{xy} denotes the uncorrected correlation, and r_{xx} and r_{yy} denote the reliability of measures (scales) x and y. You can compute the attenuation correction based on specific values, or based on

actual raw data (in which case the reliabilities of the two measures are estimated from the data).

Designing a Reliable Scale

After the discussion so far, it should be clear that the more reliable a scale, the better (e.g., more valid) the scale. As mentioned earlier, one way to make a sum scale more valid is by adding items. You can compute how many items would have to be added in order to achieve a particular reliability, or how reliable the scale would be if a certain number of items were added. However, in practice, the number of items on a questionnaire is usually limited by various other factors (e.g., respondents get tired, overall space is limited, etc.). Let's return to our prejudice example, and outline the steps that you would generally follow in order to design the scale so that it will be reliable:

Step 1: Generating items. The first step is to write the items. This is essentially a creative process where the researcher makes up as many items as possible that seem to relate to prejudices against foreign-made cars. In theory, you should "sample items" from the domain defined by the concept. In practice, for example in marketing research, focus groups are often utilized to illuminate as many aspects of the concept as possible. For example, we could ask a small group of highly committed American car buyers to express their general thoughts and feelings about foreign-made cars. In educational and psychological testing, you commonly look at other similar questionnaires at this stage of the scale design, again, in order to gain as wide a perspective on the concept as possible.

Step 2: Choosing items of optimum difficulty. In the first draft of our prejudice questionnaire, we will include as many items as possible. We then administer this questionnaire to an initial sample of typical respondents, and examine the results for each item. First, we would look at various characteristics of the items, for example, in order to identify *floor* or *ceiling* effects. If all respondents agree or disagree with an item, it obviously does not help us discriminate between respondents, and thus, it is useless for the design of a reliable scale. In test construction, the proportion of respondents who agree or disagree with an item, or who answer a test item correctly, is often referred to as the item difficulty. In essence, we would look at the item means and standard deviations and eliminate those items that show extreme means, and zero or nearly zero variances.

Step 3: Choosing internally consistent items.

Remember that a reliable scale is made up of items that proportionately measure mostly true score; in our example, we want to select items that measure mostly prejudice against foreign-made cars, and few esoteric aspects we consider random error. To do so, we would look at the following:

variable	Mean if deleted	Var. if deleted	StDv. if deleted	Itm-Totl Correl.	Squared Multp. R	Alpha if deleted
ITEM1	41.61000	51.93790	7.206795	0.656298	0.507160	0.752243
ITEM2	41.37000	53.79310	7.334378	0.666111	0.533015	0.754691
ITEM3	41.41000	54.86190	7.406882	0.549226	0.363895	0.766778
ITEM4	41.63000	56.57310	7.521509	0.470852	0.305573	0.776015
ITEM5	41.52000	64.16961	8.010593	0.054609	0.057399	0.824907
ITEM6	41.56000	62.68640	7.917474	0.118561	0.045653	0.817907
ITEM7	41.46000	54.02840	7.350401	0.587637	0.443563	0.762033
ITEM8	41.33000	53.32110	7.302130	0.609204	0.446298	0.758992
ITEM9	41.44000	55.06640	7.420674	0.502529	0.328149	0.772012
ITEM10	41.66000	53.78440	7.333785	0.572875	0.410561	0.763314

Data: Reliability Item Analysis Example Spreadsheet 1

Summary for scale: Mean=46.1100 Std.Dv.=8.26444 Valid N:100 (10Items Cronbach alpha: .794313 Standardized alpha: .800491 Average inter-item corr.: .297818

Shown in the previous illustration are the results for 10 items. Of most interest to us are the three right-most columns. They show us the correlation between the respective item and the total sum score (without the respective item),

the squared multiple correlation between the respective item and all others, and the internal consistency of the scale (coefficient *alpha*) if the respective item would be deleted. Clearly, items *5* and *6* stick out, in that they are not consistent with the rest of the scale. Their correlations with the sum scale are *.05* and *.12*, respectively, while all other items correlate at *.45* or better. In the right-most column, we can see that the reliability of the scale would be about *.82* if either of the two items were to be deleted. Thus, we would probably delete the two items from this scale.

Step 4: Returning to Step 1. After deleting all items that are not consistent with the scale, we may not be left with enough items to make up an overall reliable scale (remember that, the fewer items, the less reliable the scale). In practice, you often go through several rounds of generating items and eliminating items until you arrive at a final set that makes up a reliable scale.

Tetrachoric correlations. In educational and psychological testing, it is common to use yes/no type items, that is, to prompt the respondent to answer either yes or no to a question. An alternative to the regular correlation coefficient in that case is the so-called tetrachoric correlation coefficient. Usually, the tetrachoric correlation coefficient is larger than the standard correlation coefficient, therefore, Nunally (1970, p. 102) discourages the use of this coefficient for estimating reliabilities. However, it is a widely used statistic (e.g., in mathematical modeling).

STRUCTURAL EQUATION MODELING

Structural equation modeling is a very general, very powerful multivariate analysis technique that includes specialized versions of a number of other analysis methods as special cases. We will assume that you are familiar with the basic logic of statistical reasoning as described in Chapter 1 – *Elementary Concepts in Statistics*. Moreover, we will also assume that you are familiar with the concepts of variance, covariance, and correlation; if not, we advise that you read Chapter 2 – *Basic Statistics and Tables*. Although it is not absolutely necessary, it is highly desirable that you have some background in factor analysis (Chapter 16) before attempting to use structural modeling.

Overview

Major applications of structural equation modeling include:

1. **Causal modeling, or path analysis**, which hypothesizes causal relationships among variables and tests the causal models with a linear equation system. Causal models can involve either manifest variables, latent variables, or both;

2. **Confirmatory factor analysis**, an extension of factor analysis in which specific hypotheses about the structure of the factor loadings and intercorrelations are tested;

3. **Second order factor analysis**, a variation of factor analysis in which the correlation matrix of the common factors is itself factor analyzed to provide second order factors;

4. **Regression models**, an extension of linear regression analysis in which regression weights can be constrained to be equal to each other or to specified numerical values;

5. **Covariance structure models**, which hypothesize that a covariance matrix has a particular form. For example, you can test the hypothesis that a set of variables all have equal variances with this procedure;

6. **Correlation structure models**, which hypothesize that a correlation matrix has a particular form. A classic example is the hypothesis that the correlation matrix has the structure of a circumplex (Guttman, 1954; Wiggins, Steiger, & Gaelick, 1981).

Many different kinds of models fall into each of the above categories, so structural modeling as an enterprise is very difficult to characterize.

Most structural equation models can be expressed as path diagrams. Consequently even beginners to structural modeling can perform complicated analyses with a minimum of training.

Basic Idea behind Structural Modeling

One of the fundamental ideas taught in intermediate applied statistics courses is the effect of additive and multiplicative transformations on a list of numbers. Students are taught that, if you multiply every number in a list by some constant K, you multiply the mean of the numbers by K. Similarly, you multiply the standard deviation by the absolute value of K.

For example, suppose you have the list of numbers 1,2,3. These numbers have a mean of 2 and a standard deviation of 1. Now, suppose you were to take these 3 numbers and multiply them by 4. Then the mean would become 8, and the standard deviation would become 4, the variance thus 16.

The point is, if you have a set of numbers X related to another set of numbers Y by the equation Y = 4X, then the variance of Y must be 16 times that of X, so you can test the hypothesis that Y and X are related by the equation Y = 4X indirectly by comparing the variances of the Y and X variables.

This idea generalizes, in various ways, to several variables inter-related by a group of linear equations. The rules become more complex, the calculations more difficult, but the basic message remains the same – you can test whether variables are interrelated through a set of linear relationships by examining the variances and covariances of the variables.

Statisticians have developed procedures for testing whether a set of variances and covariances in a covariance matrix fits a specified structure. The way structural modeling works is as follows:

1. State the way that you believe the variables are interrelated, often with the use of a path diagram.

2. Work out, via some complex internal rules, what the implications of this are for the variances and covariances of the variables.

3. Test whether the variances and covariances fit this model of them.

4. Results of the statistical testing, and also parameter estimates and standard errors for the numerical coefficients in the linear equations are reported.

5. On the basis of this information, decide whether the model seems like a good fit to your data.

There are some important, and very basic logical points to remember about this process. First, although the mathematical machinery required to perform structural equations modeling is extremely complicated, the basic logic is embodied in the five steps. In the next illustration, we diagram the process.

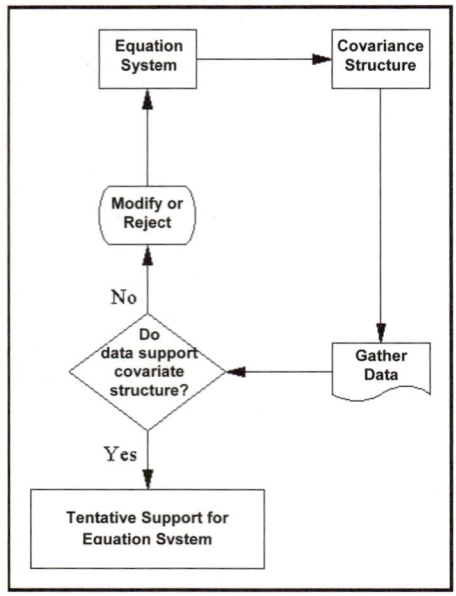

Second, we must remember that for a number of reasons it is unreasonable to expect a structural model to fit perfectly. A structural model with linear relations is only an approximation. The world is unlikely to be linear. Indeed, the true relations between variables are probably nonlinear. Moreover, many of the statistical assumptions are somewhat questionable as well. The real question is not so much, "Does the model fit perfectly?" but rather, "Does it fit well enough to be a useful approximation to reality, and a reasonable explanation of the trends in our data?"

Third, we must remember that simply because a model fits the data well does not mean that the model is necessarily correct. You cannot prove that a model is true; to assert this is the fallacy of affirming the consequent. For example, we could say, "If Joe is a cat, Joe has hair." However, "Joe has hair" does not imply "Joe is a cat." Similarly, we can say, "If a certain causal model is true, it will fit the data." However, the model fitting the data does not necessarily imply the model is the correct one. There may be another model that fits the data equally well.

Structural Equation Modeling and the Path Diagram

Path diagrams play a fundamental role in structural modeling. Path diagrams are like flowcharts. They show variables interconnected with lines that are used to indicate causal flow.

You can think of a path diagram as a device for showing which variables cause changes in other variables. However, path diagrams need not be thought of strictly in this way. They can also be given a narrower, more specific interpretation.

Consider the classic linear regression equation:

Y = aX + e

Any such equation can be represented in a path diagram as in the following illustration:

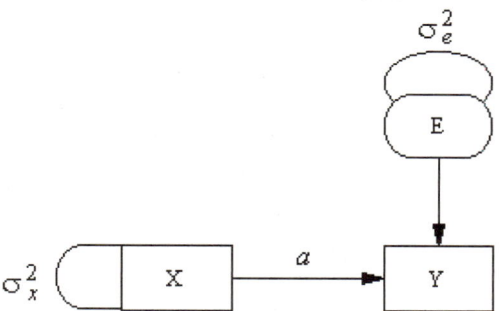

Such diagrams establish a simple isomorphism. All variables in the equation system are placed in the diagram, either in boxes or ovals. Each equation is represented on the diagram as follows: All independent variables (the variables on the right side of an equation) have arrows pointing to the dependent variable. The weighting coefficient is placed above the arrow. The above diagram shows a simple linear equation system and its path diagram representation.

Notice that, besides representing the linear equation relationships with arrows, the diagrams also contain some additional aspects. First, the variances of the independent variables, which we must know in order to test the structural relations model, are shown on the diagrams using curved lines without arrowheads attached. We refer to such lines as wires. Second, some variables are represented in ovals, others in rectangular boxes. Manifest variables are placed in boxes in the path diagram. Latent variables are placed in an oval or circle. For example, the variable *E* in the above diagram can be thought of as a linear regression residual when Y is predicted from X. Such a residual is not observed directly, but calculated from Y and

X, so we treat it as a latent variable and place it in an oval.

The example discussed above is an extremely simple one. Generally, we are interested in testing models that are much more complicated than these. As the equation systems we examine become increasingly complicated, so do the covariance structures they imply. Ultimately, the complexity can become so bewildering that we lose sight of some very basic principles. For one thing, the train of reasoning that supports testing causal models with linear structural equations testing has several weak links. The variables may be nonlinear. They may be linearly related for reasons unrelated to what we commonly view as causality. The ancient adage "correlation is not causation" remains true, even if the correlation is complex and multivariate. What causal modeling does enable us to do is examine the extent to which data fail to agree with one reasonably viable consequence of a model of causality. If the linear equations system isomorphic to the path diagram does fit the data well, it is encouraging, but hardly proof of the truth of the causal model.

Although path diagrams can be used to represent causal flow in a system of variables, they need not imply such a causal flow. Such diagrams can be viewed as simply an isomorphic representation of a linear equations system. As such, they can convey linear relationships when no causal relations are assumed. Hence, although you might interpret the diagram in the above figure to mean that "X causes Y," the diagram can also be interpreted as a visual representation of the linear regression relationship between X and Y.

SURVIVAL/FAILURE TIME ANALYSIS

Survival analysis (exploratory and hypothesis testing) techniques include descriptive methods for estimating the distribution of survival times from a sample, methods for comparing survival in two or more groups, and techniques for fitting linear or nonlinear regression models to survival data. A defining characteristic of survival time data is that they usually include so-called censored observations, e.g., observations that "survived" to a certain point in time, and then dropped out from the study (e.g., patients who are discharged from a hospital). Instead of discarding such observations from the data analysis altogether (i.e., unnecessarily lose potentially useful information) survival analysis techniques can accommodate censored observations, and "use" them in statistical significance testing and model fitting.

Typical survival analysis methods include life table, survival distribution, and Kaplan-Meier survival function estimation, and additional techniques for comparing the survival in two or more groups. Finally, survival analysis includes the use of regression models for estimating the relationship of (multiple) continuous variables to survival times.

Overview

These techniques were primarily developed in the medical and biological sciences, but they are also widely used in the social and economic sciences, as well as in engineering (reliability and failure time analysis).

Imagine that you are a researcher in a hospital who is studying the effectiveness of a new treatment for a generally terminal disease. The major variable of interest is the number of days that the respective patients survive. In principle, you could use the standard parametric and nonparametric statistics for describing the average survival, and for comparing the new treatment with traditional methods (see Chapter 2 – *Basic Statistics and Tables* and Chapter 29 – *Nonparametric Statistics*). However, at the end of the study there will be patients who survived over the entire study period, in particular among those patients who entered the hospital (and the research project) late in the study; there will be other patients with whom you will have lost contact. Surely you would not want to exclude all of those patients from the study by declaring them to be missing data (since most of them are survivors and, therefore, they reflect on the success of the new treatment method). Those observations, which contain only partial information, are called censored observations (e.g., "patient A survived at least 4 months before he moved away and we lost contact"; the term censoring was first used by Hald, 1949).

Censored Observations

In general, censored observations arise whenever the dependent variable of interest represents the time to a terminal event, and the duration of the study is limited in time. Censored observations can occur in a number of different areas of research. For example, in the social sciences, we may study the survival of marriages, high school dropout rates (time to drop-out), turnover in organizations, etc. In each case, by the end of the study period, some subjects will still be married, will not have dropped out, or are still working at the same company; thus, those subjects represent censored observations.

In economics, we may study the survival of new businesses or the survival times of products such as automobiles. In quality control research, it is common practice to study the survival of parts under stress (failure time analysis).

Analytic Techniques

Essentially, the methods offered in survival analysis address the same research questions as many of the other procedures; however, all methods in survival analysis will handle censored data. The life table, survival distribution, and Kaplan-Meier survival function estimation are all descriptive methods for estimating the distribution of survival times from a sample. Several techniques are available for comparing the survival in two or more groups. Finally, survival analysis offers several regression models for estimating the relationship of (multiple) continuous variables to survival times.

Life Table Analysis

The most straightforward way to describe the survival in a sample is to compute the life table. The life table technique is one of the oldest methods for analyzing survival (failure time) data (e.g., see Berkson & Gage, 1950; Cutler & Ederer, 1958; Gehan, 1969). This table can be thought of as an enhanced frequency

distribution table. The distribution of survival times is divided into a certain number of intervals. For each interval we can then compute the number and proportion of cases or objects that entered the respective interval alive, the number and proportion of cases that failed in the respective interval (i.e., number of terminal events, or number of cases that died), and the number of cases that were lost or censored in the respective interval.

Based on those numbers and proportions, several additional statistics can be computed:

- Number of cases at risk
- Proportion failing
- Proportion surviving
- Cumulative proportion surviving (survival function)
- Probability density
- Hazard rate
- Median survival time
- Required sample sizes

Number of cases at risk. This is the number of cases that entered the respective interval alive, minus half of the number of cases lost or censored in the respective interval.

Proportion failing. This proportion is computed as the ratio of the number of cases failing in the respective interval, divided by the number of cases at risk in the interval.

Proportion surviving. This proportion is computed as 1 minus the proportion failing.

Cumulative proportion surviving (survival function). This is the cumulative proportion of cases surviving up to the respective interval. Since the probabilities of survival are assumed to be independent across the intervals, this probability is computed by multiplying out the probabilities of survival across all previous intervals. The resulting function is also called the survivorship or survival function.

Probability Density. This is the estimated probability of failure in the respective interval, computed per unit of time, that is:

$$F_i = (P_i - P_{i+1}) / h_i$$

In this formula, F_i is the respective probability density in the i'th interval, P_i is the estimated cumulative proportion surviving at the beginning of the i'th interval (at the end of interval i-1), P_{i+1} is the cumulative proportion surviving at the end of the i'th interval, and h_i is the width of the respective interval.

Hazard rate. The hazard rate (the term was first used by Barlow, 1963) is defined as the probability per time unit that a case that has survived to the beginning of the respective interval will fail in that interval. Specifically, it is computed as the number of failures per time units in the respective interval, divided by the average number of surviving cases at the mid-point of the interval.

Median survival time. This is the survival time at which the cumulative survival function is equal to *0.5*. Other percentiles (25th and 75th percentile) of the cumulative survival function can be computed accordingly. Note that the 50th percentile (median) for the cumulative survival function is usually not the same as the point in time up to which 50% of the sample survived. (This would only be the case if there were no censored observations prior to this time).

Required sample sizes. In order to arrive at reliable estimates of the three major functions (survival, probability density, and hazard) and their standard errors at each time interval the minimum recommended sample size is 30.

Distribution Fitting

General introduction. In summary, the life table gives us a good indication of the distribution of failures over time. However, for predictive purposes it is often desirable to understand the shape of the underlying survival function in the population. The major distributions that have been proposed for modeling survival or failure times are the exponential (and linear exponential) distribution, the Weibull distribution of extreme events, and the Gompertz distribution.

Estimation. The parameter estimation procedure (for estimating the parameters of the theoretical survival functions) is essentially a least squares linear regression algorithm (see Gehan & Siddiqui, 1973). A linear regression algorithm can be used because all four theoretical distributions can be made linear by appropriate transformations. Such transformations sometimes produce different variances for the residuals at different times, leading to biased estimates.

Goodness of fit. Given the parameters for the different distribution functions and the respective model, you can compute the likelihood of the data. You can also compute the likelihood of the data under the null model, that is, a model that allows for different hazard rates in each interval. Without going into details, these two likelihoods can be compared via an incremental *chi*-square test statistic. If this *chi*-square is statistically significant, we conclude that the respective theoretical distribution fits the data significantly worse than the null model; that is, we reject the respective distribution as a model for our data.

Graphs. You can produce plots of the survival function, hazard, and probability density for the observed data and the respective theoretical distributions. These plots provide a quick visual check of the goodness-of-fit of the theoretical distribution. The next illustration shows an observed survivorship function and the fitted Weibull distribution.

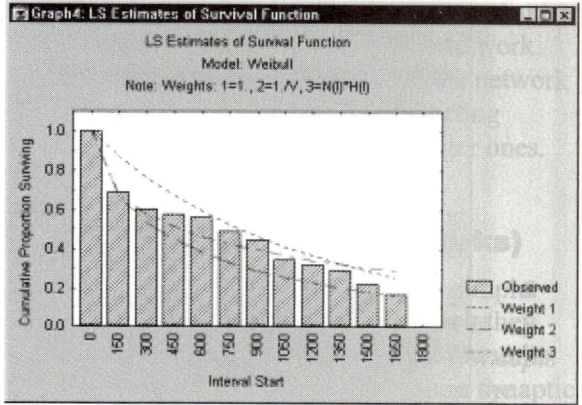

Specifically, the three lines in this plot denote the theoretical distributions that resulted from three different estimation procedures (least squares and two methods of weighted least squares).

Kaplan-Meier Product-Limit Estimator

Rather than classifying the observed survival times into a life table, we can estimate the survival function directly from the continuous survival or failure times. Intuitively, imagine that we create a life table so that each time interval contains exactly one case. Multiplying out the survival probabilities across the intervals (i.e., for each single observation) we would get for the survival function:

$$S(t) = \prod_{j=1}^{t} [(n-j)/(n-j+1)]^{\delta(j)}$$

In this equation, $S(t)$ is the estimated survival function, n is the total number of cases, and $\prod$ denotes the multiplication (geometric sum) across all cases less than or equal to t; $\delta(j)$ is a

constant that is either *1* if the *j*'th case is uncensored (complete), and *0* if it is censored. This estimate of the survival function is also called the product-limit estimator, and was first proposed by Kaplan and Meier (1958). An example plot of this function is shown in the next illustration.

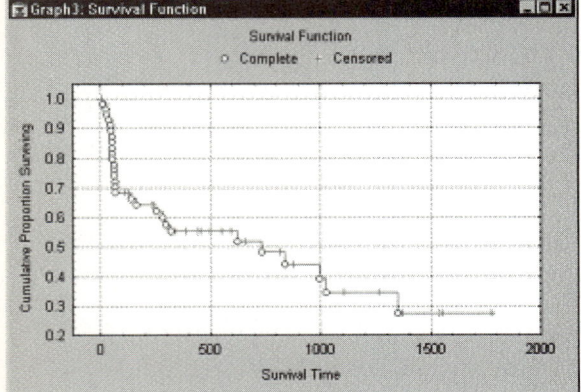

The advantage of the Kaplan-Meier Product-Limit method over the life table method for analyzing survival and failure time data is that the resulting estimates do not depend on the grouping of the data (into a certain number of time intervals). Actually, the Product-Limit method and the life table method are identical if the intervals of the life table contain at most one observation.

Comparing Samples

General introduction. You can compare the survival or failure times in two or more samples. In principle, because survival times are not normally distributed, nonparametric tests that are based on the rank ordering of survival times should be applied. A wide range of nonparametric tests can be used in order to compare survival times; however, the tests cannot handle censored observations.

Available tests. The following five different (mostly nonparametric) tests for censored data are available: Gehan's generalized Wilcoxon test, the Cox-Mantel test, Cox's *F* test, the log-rank test, and Peto and Peto's generalized Wilcoxon test. A nonparametric test for the comparison of multiple groups is also available. Most of these tests are accompanied by appropriate *z*-values (values of the standard normal distribution); these *z*-values can be used to test for the statistical significance of any differences between groups. However, note that most of these tests will only yield reliable results with fairly large samples sizes; the small sample "behavior" is less well understood.

Choosing a two-sample test. There are no widely accepted guidelines concerning which test to use in a particular situation. Cox's *F* test tends to be more powerful than Gehan's generalized Wilcoxon test when:

1. Sample sizes are small (i.e., *n* per group less than 50);

2. If samples are from an exponential or Weibull;

3. If there are no censored observations (see Gehan & Thomas, 1969).

Lee, Desu, and Gehan (1975) compared Gehan's test to several alternatives and showed that the Cox-Mantel test and the log-rank test are more powerful (regardless of censoring) when the samples are drawn from a population that follows an exponential or Weibull distribution; under those conditions there is little difference between the Cox-Mantel test and the log-rank test. Lee (1980) discusses the power of different tests in greater detail.

Multiple sample test. There is a multiple-sample test that is an extension (or generalization) of Gehan's generalized Wilcoxon test, Peto and Peto's generalized Wilcoxon test, and the log-rank test. First, a score is assigned to each survival time using Mantel's procedure (Mantel, 1967); next a *chi*-square value is computed based on the sums (for each group) of this score. If only two groups are specified, this test is equivalent to Gehan's generalized Wilcoxon test, and the computations will default to that test in this case.

Unequal proportions of censored data. When comparing two or more groups it is very important to examine the number of censored observations in each group. Particularly in medical research, censoring can be the result of, for example, the application of different treatments: patients who get better faster or get worse as the result of a treatment may be more likely to drop out of the study, resulting in different numbers of censored observations in each group. Such systematic censoring may greatly bias the results of comparisons.

Regression Models

Introduction

A common research question in medical, biological, or engineering (failure time) research is to determine whether or not certain continuous (independent) variables are correlated with the survival or failure times. There are two major reasons why this research issue cannot be addressed via straightforward multiple regression techniques (as available in multiple regression): First, the dependent variable of interest (survival/failure time) is most likely not normally distributed – a serious violation of an assumption for ordinary least squares multiple regression. Survival times usually follow an exponential or Weibull distribution. Second, there is the problem of

censoring, that is, some observations will be incomplete.

Cox's Proportional Hazard Model

The proportional hazard model is the most general of the regression models because it is not based on any assumptions concerning the nature or shape of the underlying survival distribution. The model assumes that the underlying hazard rate (rather than survival time) is a function of the independent variables (covariates); no assumptions are made about the nature or shape of the hazard function. Thus, in a sense, Cox's regression model may be considered to be a nonparametric method. The model can be written as:

$$h\{(t), (z_1, z_2, ..., z_m)\} = h_0(t) * \exp(b_1 * z_1 + ... + b_m * z_m)$$

where $h(t,...)$ denotes the resultant hazard, given the values of the m covariates for the respective case $(z_1, z_2, ..., z_m)$ and the respective survival time (t). The term $h_0(t)$ is called the baseline hazard; it is the hazard for the respective individual when all independent variable values are equal to zero. We can linearize this model by dividing both sides of the equation by $h_0(t)$ and then taking the natural logarithm of both sides:

$$\log[h\{(t), (z...)\}/h_0(t)] = b_1 * z_1 + ... + b_m * z_m$$

We now have a fairly simple linear model that can be readily estimated.

Assumptions. While no assumptions are made about the shape of the underlying hazard function, the model equations shown above do imply two assumptions. First, they specify a multiplicative relationship between the underlying hazard function and the log-linear function of the covariates. This assumption is also called the proportionality assumption. In practical terms, it is assumed that, given two observations with different values for the independent variables, the ratio of the hazard functions for those two observations does not depend on time. The second assumption of course, is that there is a log-linear relationship between the independent variables and the underlying hazard function.

Cox's Proportional Hazard Model with Time-Dependent Covariates

An assumption of the proportional hazard model is that the hazard function for an individual (i.e., observation in the analysis) depends on the values of the covariates and the value of the baseline hazard. Given two individuals with particular values for the covariates, the ratio of the estimated hazards over time will be constant – hence the name of the method: the proportional hazard model. The validity of this assumption may often be questionable. For example, age is often included in studies of physical health. Suppose you studied survival after surgery. It is likely, that age is a more important predictor of risk immediately after surgery, than some time after the surgery (after initial recovery). In accelerated life testing, you sometimes use a stress covariate (e.g., amount of voltage) that is slowly increased over time until failure occurs (e.g., until the electrical insulation fails; see Lawless, 1982, page 393). In this case, the impact of the covariate is clearly dependent on time. You can specify arithmetic expressions to define covariates as functions of several variables and survival time.

Testing the proportionality assumption. As indicated by the previous examples, there are many applications where it is likely that the proportionality assumption does not hold. In this case, you can explicitly define covariates as functions of time. For example, the analysis of a

data set presented by Pike (1966) consists of survival times for two groups of rats that had been exposed to a carcinogen (see also, Lawless, 1982, page 393, for a similar example). Suppose that z is a grouping variable with codes 1 and 0 to denote whether or not the respective rat was exposed. You could then fit the proportional hazard model:

$$h(t, z) = h_0(t)e^{b_1 z + b_2[z \log(t) - 5.4]}$$

Thus, in this model the conditional hazard at time t is a function of 1) the baseline hazard h_0, 2) the covariate z, and 3) of z times the logarithm of time. Note that the constant 5.4 is used here for scaling purposes only: the mean of the logarithm of the survival times in this data set is equal to 5.4. In other words, the conditional hazard at each point in time is a function of the covariate and time; thus, the effect of the covariate on survival is dependent on time; hence the name time-dependent covariate. This model enables you to specifically test the proportionality assumption. If parameter b_2 is statistically significant (e.g., if it is at least twice as large as its standard error), you can conclude that, indeed, the effect of the covariate z on survival is dependent on time, and, therefore, that the proportionality assumption does not hold.

Exponential Regression

Basically, this model assumes that the survival time distribution is exponential, and contingent on the values of a set of independent variables (z_i). The rate parameter of the exponential distribution can then be expressed as:

$$S(z) = e^{a + b_1 z_1 + b_2 z_2 + \ldots + b_m z_m}$$

$S(z)$ denotes the survival times, a is a constant, and the b_i's are the regression parameters.

Goodness-of-fit. The *chi*-square goodness-of-fit value is computed as a function of the log-likelihood for the model with all parameter estimates (*L1*), and the log-likelihood of the model in which all covariates are forced to 0 (zero; *L0*). If this *chi*-square value is significant, we reject the null hypothesis and assume that the independent variables are significantly related to survival times.

Standard exponential order statistic. One way to check the exponentiality assumption of this model is to plot the residual survival times against the standard exponential order statistic *theta*. If the exponentiality assumption is met, all points in this plot will be arranged roughly in a straight line.

Normal and Lognormal Regression

In this model, it is assumed that the survival times (or log survival times) come from a normal distribution; the resulting model is basically identical to the ordinary multiple regression model, and can be stated as:

$$t = a + b_1{}^*z_1 + b_2{}^*z_2 + \ldots + b_m{}^*z_m$$

where *t* denotes the survival times. For lognormal regression, *t* is replaced by its natural logarithm. The normal regression model is particularly useful because many data sets can be transformed to yield approximations of the normal distribution. Thus, in a sense this is the most general fully parametric model (as opposed to Cox's proportional hazard model which is non-parametric), and estimates can be obtained for a variety of different underlying survival distributions.

Goodness-of-fit. The *chi*-square value is computed as a function of the log-likelihood for the model with all independent variables (L1), and the log-likelihood of the model in which all independent variables are forced to 0 (zero, L0).

Stratified Analyses

The purpose of a stratified analysis is to test the hypothesis whether identical regression models are appropriate for different groups, that is, whether the relationships between the independent variables and survival are identical in different groups. To perform a stratified analysis, you must first fit the respective regression model separately within each group. The sum of the log-likelihoods from these analyses represents the log-likelihood of the model with different regression coefficients (and intercepts where appropriate) in different groups. The next step is to fit the requested regression model to all data in the usual manner (i.e., ignoring group membership), and compute the log-likelihood for the overall fit. The difference between the log-likelihoods can then be tested for statistical significance (via the *chi*-square statistic).

TEXT MINING

The purpose of text mining is to process unstructured (textual) information, extract meaningful numeric indices from the text and, thus, make the information contained in the text accessible to the various data mining (statistical and machine learning) algorithms. Information can be extracted to derive summaries for the words contained in the documents or to compute summaries for the documents based on the words contained in them. Hence, you can analyze words, clusters of words used in documents, etc., or you could analyze documents and determine similarities between them or how they are related to other variables of interest in the data mining project. In the most general terms, text mining will "turn text into numbers" (meaningful indices), which can then be incorporated in other analyses such as predictive data mining projects, the application of unsupervised learning methods (clustering), etc. These methods are described and discussed in great detail in the comprehensive overview work by Manning and Schütze (2002), and for an in-depth treatment of these and related topics as well as the history of this approach to text mining, we highly recommend that source.

Overview

The purpose of Text Mining is to process unstructured (textual) information, extract meaningful numeric indices from the text, and, thus, make the information contained in the text accessible to the various data mining (statistical and machine learning) algorithms. Information can be extracted to derive summaries for the words contained in the documents or to compute summaries for the documents based on the words contained in them. Hence, you can analyze words, clusters of words used in documents, etc., or you could analyze documents and determine similarities between them or how they are related to other variables of interest in the data mining project. In the most general terms, text mining will "turn text into numbers" (meaningful indices), which can then be incorporated in other analyses such as predictive data mining projects, the application of unsupervised learning methods (clustering), etc. These methods are described and discussed in great detail in the comprehensive overview work by Manning and Schütze (2002), and for an in-depth treatment of these and related topics as well as the history of this approach to text mining, we highly recommend that source.

Some Typical Applications for Text Mining

Unstructured text is very common, and in fact may represent the majority of information available to a particular research or data mining project.

Analyzing open-ended survey responses. In survey research (e.g., marketing), it is not uncommon to include various open-ended questions pertaining to the topic under investigation. The idea is to permit respondents to express their "views" or opinions without constraining them to particular dimensions or a particular response format. This may yield insights into customers' views and opinions that might otherwise not be discovered when relying solely on structured questionnaires designed by "experts." For example, you may discover a certain set of words or terms that are commonly used by respondents to describe the pro's and con's of a product or service (under investigation), suggesting common misconceptions or confusion regarding the items in the study.

Automatic processing of messages, emails, etc. Another common application for text mining is to aid in the automatic classification of texts. For example, it is possible to "filter" out automatically most undesirable "junk email" based on certain terms or words that are not likely to appear in legitimate messages, but instead identify undesirable electronic mail. In this manner, such messages can automatically be discarded. Such automatic systems for classifying electronic messages can also be useful in applications where messages need to be routed (automatically) to the most appropriate department or agency; e.g., email messages with complaints or petitions to a municipal authority are automatically routed to the appropriate departments; at the same time, the emails are screened for inappropriate or obscene messages, which are automatically returned to the sender with a request to remove the offending words or content.

Analyzing warranty or insurance claims, diagnostic interviews, etc. In some business domains, the majority of information is collected in open-ended, textual form. For example, warranty claims or initial medical (patient) interviews can be summarized in brief narratives, or when you take your automobile to a service station for repairs, typically, the attendant will write some notes about the problems that you report and what you believe needs to be fixed. Increasingly, those notes are collected electronically, so those types of narratives are readily available for input into text mining algorithms. This information can then be usefully exploited to, for example, identify common clusters of problems and complaints on certain automobiles, etc. Likewise, in the medical field, open-ended descriptions by patients of their own symptoms might yield useful clues for the actual medical diagnosis.

Investigating competitors by crawling their web sites. Another type of potentially very useful application is to automatically process the contents of Web pages in a particular domain. For example, you could go to a Web page, and begin "crawling" the links you find there to process all Web pages that are referenced. In this manner, you could automatically derive a list of terms and documents available at that site, and hence quickly determine the most important terms and features that are described. It is easy to see how these capabilities could efficiently deliver valuable business intelligence about the activities of competitors.

Approaches to Text Mining

To reiterate, text mining can be summarized as a process of "numericizing" text. At the simplest level, all words found in the input documents will be indexed and counted in order to compute a table of documents and words, i.e., a matrix of frequencies that enumerates the number of times that each word occurs in each document. This basic process can be further refined to exclude certain common words such

as "the" and "a" (stop word lists) and to combine different grammatical forms of the same words such as "traveling," "traveled," "travel," etc. (stemming). However, once a table of (unique) words (terms) by documents has been derived, all standard statistical and data mining techniques can be applied to derive dimensions or clusters of words or documents, or to identify "important" words or terms that best predict another outcome variable of interest.

Using well-tested methods and understanding the results of text mining. Once a data matrix has been computed from the input documents and words found in those documents, various well-known analytic techniques can be used for further processing those data including methods for clustering, factoring, or predictive data mining (see, for example, Manning and Schütze, 2002).

"Black-box" approaches to text mining and extraction of concepts. There are text mining applications which offer "black-box" methods to extract "deep meaning" from documents with little human effort (to first read and understand those documents). These text mining applications rely on proprietary algorithms for presumably extracting "concepts" from text, and may even claim to be able to summarize large numbers of text documents automatically, retaining the core and most important meaning of those documents. While there are numerous algorithmic approaches to extracting "meaning from documents," this type of technology is very much still in its infancy, and the aspiration to provide meaningful automated summaries of large numbers of documents may forever remain elusive. We urge skepticism when using such algorithms because 1) if it is not clear to the user how those algorithms work, it cannot possibly be clear how to interpret the results of those algorithms, and 2) the methods used in those programs are not open to scrutiny, for example by the academic community and peer review and, hence, one simply doesn't know how well they might perform in different domains. As a final thought on this subject, you may consider this concrete example: Try the various automated translation services available via the Web that can translate entire paragraphs of text from one language into another. Then translate some text, even simple text, from your native language to some other language and back, and review the results. Almost every time, the attempt to translate even short sentences to other languages and back while retaining the original meaning of the sentence produces humorous rather than accurate results. This illustrates the difficulty of automatically interpreting the meaning of text.

Text mining as document search. There is another type of application that is often described and referred to as "text mining" - the automatic search of large numbers of documents based on key words or key phrases. This is the domain of, for example, the popular internet search engines that have been developed over the last decade to provide efficient access to Web pages with certain content. While this is obviously an important type of application with many uses in any organization that needs to search very large document repositories based on varying criteria, it is very different from what has been described here.

Issues and Considerations for "Numericizing" Text

Large numbers of small documents vs. small numbers of large documents.
Examples of scenarios using large numbers of small or moderate sized documents were given earlier (e.g., analyzing warranty or insurance claims, diagnostic interviews, etc.). On the other hand, if your intent is to extract "concepts" from only a few documents that are very large (e.g., two lengthy books), then statistical analyses are generally less powerful because the "number of cases" (documents) in this case is very small while the "number of variables" (extracted words) is very large.

Excluding certain characters, short words, numbers, etc. Excluding numbers, certain characters, or sequences of characters, or words that are shorter or longer than a certain number of letters can be done before the indexing of the input documents starts. You may also want to exclude "rare words," defined as those that only occur in a small percentage of the processed documents.

Include lists, exclude lists (stop-words). Specific list of words to be indexed can be defined; this is useful when you want to search explicitly for particular words, and classify the input documents based on the frequencies with which those words occur. Also, "stop-words," i.e., terms that are to be excluded from the indexing can be defined. Typically, a default list of English stop words includes "the", "a", "of", "since," etc, i.e., words that are used in the respective language very frequently, but communicate very little unique information about the contents of the document.

Synonyms and phrases. Synonyms, such as "sick" or "ill," or words that are used in particular phrases where they denote unique meaning can be combined for indexing. For example, "Microsoft Windows" might be such a phrase, which is a specific reference to the computer operating system, but has nothing to do with the common use of the term "Windows" as it might, for example, be used in descriptions of home improvement projects.

Stemming algorithms. An important pre-processing step before indexing of input documents begins is the stemming of words. The term "stemming" refers to the reduction of words to their roots so that, for example, different grammatical forms or declinations of verbs are identified and indexed (counted) as the same word. For example, stemming will ensure that both "traveling" and "traveled" will be recognized by the text mining program as the same word.

Support for different languages. Stemming, synonyms, the letters that are permitted in words, etc. are highly language dependent operations. Therefore, support for different languages is important.

Transforming Word Frequencies

Once the input documents have been indexed and the initial word frequencies (by document) computed, a number of additional transformations can be performed to summarize and aggregate the information that was extracted.

Log-frequencies. First, various transformations of the frequency counts can be performed. The raw word or term frequencies generally reflect on how salient or important a word is in each document. Specifically, words that occur with greater frequency in a document are better descriptors of the contents of that document. However, it is not reasonable to assume that the word counts themselves are proportional to their importance as descriptors of the documents. For example, if a word occurs 1 time in document A, but 3 times in document B, then it is not necessarily reasonable to conclude that this word is 3 times as important a descriptor of document B as compared to document A. Thus,

a common transformation of the raw word frequency counts (wf) is to compute:

$$f(wf) = 1 + \log(wf), \text{ for } wf > 0$$

This transformation will "dampen" the raw frequencies and how they will affect the results of subsequent computations.

Binary frequencies. Likewise, an even simpler transformation can be used that enumerates whether a term is used in a document; i.e.:

$$f(wf) = 1, \text{ for } wf > 0$$

The resulting documents-by-words matrix will contain only 1s and 0s to indicate the presence or absence of the respective words. Again, this transformation will dampen the effect of the raw frequency counts on subsequent computations and analyses.

Inverse document frequencies. Another issue that you may want to consider more carefully and reflect in the indices used in further analyses are the relative document frequencies (df) of different words. For example, a term such as "guess" may occur frequently in all documents, while another term such as "software" may only occur in a few. The reason is that one might make "guesses" in various contexts, regardless of the specific topic, while "software" is a more semantically focused term that is only likely to occur in documents that deal with computer software. A common and very useful transformation that reflects both the specificity of words (document frequencies) as well as the overall frequencies of their occurrences (word frequencies) is the so-called inverse document frequency (for the i'th word and j'th document):

$$idf(i,j) = \begin{cases} 0 & \text{if } wf_{ij} = 0 \\ (1 + \log(wf_{ij}))\log\dfrac{N}{df_i} & \text{if } wf_{ij} \geq 1 \end{cases}$$

In this formula (see also formula 15.5 in Manning and Schütze, 2002), N is the total number of documents, and df_i is the document frequency for the i'th word (the number of documents that include this word). Hence, it can be seen that this formula includes both the dampening of the simple word frequencies via the log function (described above), and also includes a weighting factor that evaluates to 0 if the word occurs in all documents $(log(N/N=1)=0)$, and to the maximum value when a word only occurs in a single document $(log(N/1)=log(N))$. It can easily be seen how this transformation will create indices that both reflect the relative frequencies of occurrences of words, as well as their semantic specificities over the documents included in the analysis.

Latent Semantic Indexing via Singular Value Decomposition

As described above, the most basic result of the initial indexing of words found in the input documents is a frequency table with simple counts, i.e., the number of times that different words occur in each input document. Usually, one would transform those raw counts to indices that better reflect the (relative) "importance" of words and/or their semantic specificity in the context of the set of input documents (see the discussion of inverse document frequencies, above).

A common analytic tool for interpreting the "meaning" or "semantic space" described by the words that were extracted, and hence by the documents that were analyzed, is to create a mapping of the word and documents into a common space, computed from the word frequencies or transformed word frequencies (e.g., inverse document frequencies). In general, here is how it works:

Suppose you indexed a collection of customer reviews of their new automobiles (e.g., for different makes and models). You may find that every time a review includes the word "gas-mileage," it also includes the term "economy." Further, when reports include the word "reliability" they also include the term "defects" (e.g., make reference to "no defects"). However, there is no consistent pattern regarding the use of the terms "economy" and "reliability," i.e., some documents include either one or both. In other words, these four words "gas-mileage" and "economy," and "reliability" and "defects," describe two independent dimensions - the first having to do with the overall operating cost of the vehicle, the other with the quality and workmanship. The idea of latent semantic indexing is to identify such underlying dimensions (of "meaning"), into which the words and documents can be mapped. As a result, we may identify the underlying (latent) themes described or discussed in the input documents, and also identify the documents that mostly deal with economy, reliability, or both. Hence, we want to map the extracted words or terms and input documents into a common latent semantic space.

Singular value decomposition. The use of singular value decomposition in order to extract a common space for the variables and cases (observations) is used in various statistical techniques, most notably in Correspondence Analysis. The technique is also closely related to Principal Components Analysis and Factor Analysis. In general, the purpose of this technique is to reduce the overall dimensionality of the input matrix (number of input documents by number of extracted words) to a lower-dimensional space, where each consecutive dimension represents the largest degree of variability (between words and documents) possible. Ideally, you might identify the two or three most salient dimensions, accounting for most of the variability (differences) between the words and documents and, hence, identify the latent semantic space that organizes the words and documents in the analysis. In some way, once such dimensions can be identified, you have extracted the underlying "meaning" of what is contained (discussed, described) in the documents.

Incorporating Text Mining Results in Data Mining Projects

After significant (e.g., frequent) words have been extracted from a set of input documents, and/or after singular value decomposition has been applied to extract salient semantic dimensions, typically the next and most important step is to use the extracted information in a data mining project.

Graphics (visual data mining methods). Depending on the purpose of the analyses, in some instances the extraction of semantic dimensions alone can be a useful outcome if it clarifies the underlying structure of what is contained in the input documents. For example, a study of new car owners' comments about their vehicles may uncover the salient dimensions in the minds of those drivers when they think about or consider their automobile (or how they "feel" about it). For marketing research purposes, that in itself can be a useful and significant result. You can use the graphics (e.g., 2D scatterplots or 3D scatterplots) to help you visualize and identify the semantic space extracted from the input documents.

Clustering and factoring. You can use cluster analysis methods to identify groups of documents (e.g., vehicle owners who described their new cars), to identify groups of similar input texts. This type of analysis also could be extremely useful in the context of market research studies, for example of new car owners. You can also use Factor Analysis and Principal Components and Classification Analysis (to factor analyze words or documents).

Predictive data mining. Another possibility is to use the raw or transformed word counts as predictor variables in predictive data mining projects.

TIME SERIES/FORECASTING

In this chapter, we will first review techniques used to identify patterns in time series data (such as smoothing and curve fitting techniques and autocorrelations). Then we will introduce a general class of models that can be used to represent time series data and generate predictions (autoregressive and moving average models). Finally, we will review some simple but commonly used modeling and forecasting techniques based on linear regression.

Overview

In the following topics, we will review techniques that are useful for analyzing time series data, that is, sequences of measurements that follow non-random orders. Unlike the analyses of random samples of observations that are discussed in the context of most other statistics, the analysis of time series is based on the assumption that successive values in the data file represent consecutive measurements taken at equally spaced time intervals.

Detailed discussions of the methods described in this section can be found in Anderson (1976), Box and Jenkins (1976), Kendall (1984), Kendall and Ord (1990), Montgomery, Johnson, and Gardiner (1990), Pankratz (1983), Shumway (1988), Vandaele (1983), Walker (1991), and Wei (1989).

Two Main Goals

There are two main goals of time series analysis: a) identifying the nature of the phenomenon represented by the sequence of observations, and b) forecasting (predicting future values of the time series variable). Both of these goals require that the pattern of observed time series data is identified and more or less formally described. Once the pattern is established, we can interpret and integrate it with other data (i.e., use it in our theory of the investigated phenomenon, e.g., seasonal commodity prices). Regardless of the depth of our understanding and the validity of our interpretation (theory) of the phenomenon, we can extrapolate the identified pattern to predict future events.

Identifying Patterns in Time Series Data

For more information on simple auto-correlations (introduced in this section) and other auto correlations, see Anderson (1976), Box and Jenkins (1976), Kendall (1984), Pankratz (1983), and Vandaele (1983).

Systematic Pattern and Random Noise

As in most other analyses, in time series analysis it is assumed that the data consist of a systematic pattern (usually a set of identifiable components) and random noise (error) that usually makes the pattern difficult to identify. Most time series analysis techniques involve some form of filtering out noise in order to make the pattern more salient.

Two General Aspects of Time Series Patterns

Most time series patterns can be described in terms of two basic classes of components: trend and seasonality. The former represents a general systematic linear or (most often) nonlinear component that changes over time and does not repeat or at least does not repeat within the time range captured by our data (e.g., a plateau followed by a period of exponential growth). The latter may have a formally similar nature (e.g., a plateau followed by a period of exponential growth), however, it repeats itself in systematic intervals over time. Those two general classes of time series components may coexist in real-life data. For example, sales of a company can rapidly grow over years but they still follow consistent seasonal patterns (e.g., as much as 25% of yearly sales each year are made in December, whereas only 4% in August).

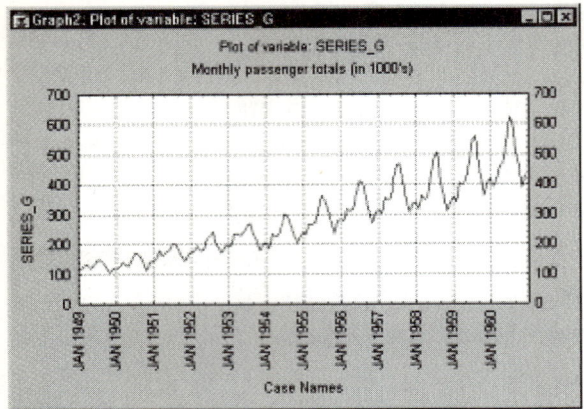

This general pattern is well illustrated in a classic *Series G* data set (Box and Jenkins, 1976, p. 531) representing monthly international airline passenger totals (measured in thousands) in twelve consecutive years from 1949 to 1960. If you plot the successive observations (months) of airline passenger totals, a clear, almost linear trend emerges, indicating that the airline industry enjoyed a steady growth over the years (approximately 4 times more passengers traveled in 1960 than in 1949). At the same time, the monthly figures will follow an almost identical pattern each year (e.g., more people travel during holidays then during any other time of the year). This example data file also illustrates a very common general type of pattern in time series data, where the amplitude of the seasonal changes increases with the overall trend (i.e., the variance is correlated with the mean over the segments of the series). This pattern, which is called multiplicative seasonality, indicates that the relative amplitude of seasonal changes is constant over time, thus it is related to the trend.

Trend Analysis

There are no proven automatic techniques to identify trend components in the time series data; however, as long as the trend is monotonous (consistently increasing or decreasing) that part of data analysis is typically not very difficult. If the time series data contain considerable error, then the first step in the process of trend identification is smoothing.

Smoothing. Smoothing always involves some form of local averaging of data such that the nonsystematic components of individual observations cancel each other out. The most common technique is moving average smoothing, which replaces each element of the series by either the simple or weighted average of n surrounding elements, where n is the width of the smoothing window (see Box & Jenkins, 1976; Velleman & Hoaglin, 1981). Medians can be used instead of means. The main advantage of median as compared to moving average smoothing is that its results are less biased by outliers (within the smoothing window). Thus, if there are outliers in the data (e.g., due to measurement errors), median smoothing typically produces smoother or at least more reliable curves than moving average based on the same window width. The main disadvantage of median smoothing is that in the absence of clear outliers it may produce more "jagged" curves than moving average and it does not allow for weighting.

In the relatively less common cases (in time series data), when the measurement error is very large, the distance weighted least squares smoothing or negative exponentially weighted smoothing techniques can be used. All those methods will filter out the noise and convert the data into a smooth curve that is relatively unbiased by outliers (see the respective sections on each of those methods for more details). Series with relatively few and systematically distributed points can be smoothed with bicubic splines.

Fitting a function. Many monotonous time series data can be adequately approximated by a linear function; if there is a clear monotonous nonlinear component, the data first need to be transformed to remove the nonlinearity. Usually a logarithmic, exponential, or (less often) polynomial function can be used.

Analysis of Seasonality

Seasonal dependency (seasonality) is another general component of the time series pattern. The concept was illustrated in the example of the airline passengers data above. It is formally defined as correlational dependency of order k between each i'th element of the series and the $(i-k)$'th element (Kendall, 1976) and measured by autocorrelation (i.e., a correlation between the two terms); k is usually called the lag. If the measurement error is not too large, seasonality can be visually identified in the series as a pattern that repeats every k elements.

Autocorrelation correlogram. Seasonal patterns of time series can be examined via correlograms. The correlogram (autocorrelogram) displays graphically and numerically the autocorrelation function (ACF), that is, serial correlation coefficients (and their standard errors) for consecutive lags in a specified range of lags (e.g., 1 through 30). Ranges of two standard errors for each lag are usually marked in correlograms but typically the size of auto correlation is of more interest than its reliability (see Chapter 1 – *Elementary Concepts in Statistics*) because we are usually interested only in very strong (and thus highly significant) autocorrelations.

Examining correlograms. While examining correlograms, you should keep in mind that autocorrelations for consecutive lags are formally dependent. Consider this example. If the first

element is closely related to the second, and the second to the third, then the first element must also be somewhat related to the third one, etc. This implies that the pattern of serial dependencies can change considerably after removing the first order auto correlation (i.e., after differencing the series with a lag of 1).

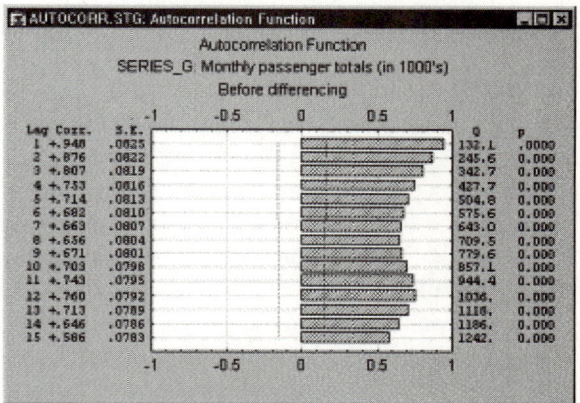

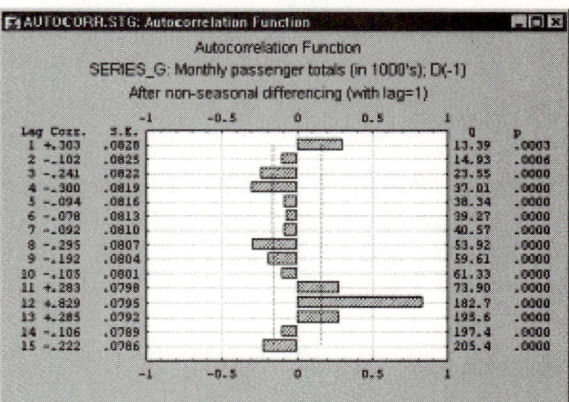

Partial autocorrelations. Another useful method to examine serial dependencies is to examine the partial autocorrelation function ($PACF$), an extension of autocorrelation, where the dependence on the intermediate elements (those within the lag) is removed. In other words the partial autocorrelation is similar to autocorrelation, except that when calculating it, the (auto) correlations with all the elements within the lag are partialled out (Box & Jenkins, 1976; see also, McDowall, McCleary, Meidinger, & Hay, 1980). If a lag of 1 is specified (i.e., there are no intermediate elements within the lag), then the partial autocorrelation is equivalent to auto correlation. In a sense, the partial autocorrelation provides a cleaner picture of serial dependencies for individual lags (not confounded by other serial dependencies).

Removing serial dependency. Serial dependency for a particular lag of k can be removed by differencing the series, that is converting each i'th element of the series into its difference from the $(i-k)$''th element. There are two major reasons for such transformations. First, you can identify the hidden nature of seasonal dependencies in the series. Remember that, as mentioned in the previous paragraph, autocorrelations for consecutive lags are interdependent. Therefore, removing some of the autocorrelations will change other auto correlations, that is, it may eliminate them or it may make some other seasonalities more apparent.

The other reason for removing seasonal dependencies is to make the series stationary, which is necessary for ARIMA and other techniques.

ARIMA

Overview

The modeling and forecasting procedures discussed in *Identifying Patterns in Time Series Data*, page 491, involved knowledge about the mathematical model of the process. However, in real-life research and practice, patterns of the data are unclear, individual observations involve considerable error, and we still need not only to uncover the hidden patterns in the data

but also generate forecasts. The ARIMA methodology developed by Box and Jenkins (1976) allows us to do just that; it has gained enormous popularity in many areas and research practice confirms its power and flexibility (Hoff, 1983; Pankratz, 1983; Vandaele, 1983). However, because of its power and flexibility, ARIMA is a complex technique; it is not easy to use, it requires a great deal of experience, and although it often produces satisfactory results, those results depend on the researcher's level of expertise (Bails & Peppers, 1982). The following sections will introduce the basic ideas of this methodology. For those interested in a brief, applications-oriented (non-mathematical), introduction to ARIMA methods, we recommend McDowall, McCleary, Meidinger, and Hay (1980).

Two Common Processes

Autoregressive process. Most time series consist of elements that are serially dependent in the sense that you can estimate a coefficient or a set of coefficients that describe consecutive elements of the series from specific, time-lagged (previous) elements. This can be summarized in the equation:

$$x_t = \zeta + \phi_1{}^*x_{(t-1)} + \phi_2{}^*x_{(t-2)} + \phi_3{}^*x_{(t-3)} + \dots + \epsilon$$

Where ζ is a constant (intercept), and ϕ_1, ϕ_2, ϕ_3 are the autoregressive model parameters.

Put in words, each observation is made up of a random error component (random shock, ϵ) and a linear combination of prior observations.

Stationarity requirement. Note that an autoregressive process will only be stable if the parameters are within a certain range; for example, if there is only one autoregressive parameter then is must fall within the interval of $-1 < \phi < 1$. Otherwise, past effects would accumulate and the values of successive x_t' s

would move toward infinity, that is, the series would not be stationary. If there is more than one autoregressive parameter, similar (general) restrictions on the parameter values can be defined (e.g., see Box & Jenkins, 1976; Montgomery, 1990).

Moving average process. Independent from the autoregressive process, each element in the series can also be affected by the past error (or random shock) that cannot be accounted for by the autoregressive component, that is:

$$x_t = \mu + \epsilon_t - \theta_1 {}^* \epsilon_{(t-1)} - \theta_2 {}^* \epsilon_{(t-2)} - \theta_3 {}^* \epsilon_{(t-3)} - \dots$$

Where μ is a constant, and θ_1, θ_2, θ_3 are the moving average model parameters.

Put in words, each observation is made up of a random error component (random shock, ϵ) and a linear combination of prior random shocks.

Invertibility requirement. Without going into too much detail, there is a duality between the moving average process and the autoregressive process (e.g., see Box & Jenkins, 1976; Montgomery, Johnson, & Gardiner, 1990), that is, the moving average equation above can be rewritten (*inverted*) into an autoregressive form (of infinite order). However, analogous to the stationarity condition described above, this can only be done if the moving average parameters follow certain conditions, that is, if the model is invertible. Otherwise, the series will not be stationary.

ARIMA Methodology

Autoregressive moving average model. The general model introduced by Box and Jenkins (1976) includes autoregressive as well as moving average parameters, and explicitly includes differencing in the formulation of the model. Specifically, the three types of parameters in the model are: the autoregressive

parameters (p), the number of differencing passes (d), and moving average parameters (q). In the notation introduced by Box and Jenkins, models are summarized as ARIMA (p, d, q); so, for example, a model described as (0, 1, 2) means that it contains 0 (zero) autoregressive (p) parameters and 2 moving average (q) parameters which were computed for the series after it was differenced once.

Identification. As mentioned earlier, the input series for ARIMA needs to be stationary, that is, it should have a constant mean, variance, and autocorrelation through time. Therefore, usually the series first needs to be differenced until it is stationary (this also often requires log transforming the data to stabilize the variance). The number of times the series needs to be differenced to achieve stationarity is reflected in the d parameter (see the previous paragraph). In order to determine the necessary level of differencing, you should examine the plot of the data and autocorrelogram. Significant changes in level (strong upward or downward changes) usually require first order non-seasonal (lag=1) differencing; strong changes of slope usually require second order non-seasonal differencing. Seasonal patterns require respective seasonal differencing. If the estimated autocorrelation coefficients decline slowly at longer lags, first order differencing is usually needed. However, you should keep in mind that some time series may require little or no differencing, and that over differenced series produce less stable coefficient estimates.

At this stage (which is usually called the identification phase) we also need to decide how many autoregressive (p) and moving average (q) parameters are necessary to yield an effective but still parsimonious model of the process (parsimonious means that it has the fewest parameters and greatest number of

degrees of freedom among all models that fit the data). In practice, the numbers of the p or q parameters very rarely need to be greater than 2 (see below for more specific recommendations).

Estimation and forecasting. At the next step (estimation), the parameters are estimated (using function minimization procedures, see below for more information on minimization procedures see also, Chapter 28 – *Nonlinear Estimation*), so that the sum of squared residuals is minimized. The estimates of the parameters are used in the last stage (forecasting) to calculate new values of the series (beyond those included in the input data set) and confidence intervals for those predicted values. The estimation process is performed on transformed (differenced) data; before the forecasts are generated, the series needs to be integrated (integration is the inverse of differencing) so that the forecasts are expressed in values compatible with the input data. This automatic integration feature is represented by the letter I in the name of the methodology (ARIMA = Auto-Regressive Integrated Moving Average).

The constant in ARIMA models. In addition to the standard autoregressive and moving average parameters, ARIMA models can also include a constant, as described above. The interpretation of a (statistically significant) constant depends on the model that is fit. Specifically, 1) if there are no autoregressive parameters in the model, then the expected value of the constant is μ, the mean of the series; 2) if there are autoregressive parameters in the series, then the constant represents the intercept. If the series is differenced, then the constant represents the mean or intercept of the differenced series; For example, if the series is differenced once, and there are no autoregressive parameters in the model, then the constant represents the mean of

the differenced series, and therefore the linear trend slope of the un-differenced series.

Identification

Number of parameters to be estimated.
Before the estimation can begin, we need to decide on (identify) the specific number and type of ARIMA parameters to be estimated. The major tools used in the identification phase are plots of the series, correlograms of auto correlation (ACF), and partial autocorrelation (PACF). The decision is not straightforward and in less typical cases requires not only experience but also a good deal of experimentation with alternative models (as well as the technical parameters of ARIMA). However, a majority of empirical time series patterns can be sufficiently approximated using one of the 5 basic models that can be identified based on the shape of the autocorrelogram (ACF) and partial auto correlogram (PACF). The following brief summary is based on practical recommendations of Pankratz (1983); for additional practical advice, see also, Hoff (1983), McCleary and Hay (1980), McDowall, McCleary, Meidinger, and Hay (1980), and Vandaele (1983). Also, note that since the number of parameters (to be estimated) of each kind is almost never greater than 2, it is often practical to try alternative models on the same data.

1. *One autoregressive (p) parameter*: ACF – exponential decay; PACF – spike at lag 1, no correlation for other lags.

2. *Two autoregressive (p) parameters*: ACF – a sine-wave shape pattern or a set of exponential decays; PACF – spikes at lags 1 and 2, no correlation for other lags.

3. *One moving average (q) parameter*: ACF – spike at lag 1, no correlation for other lags; PACF – damps out exponentially.

4. *Two moving average (q) parameters*: ACF – spikes at lags 1 and 2, no correlation for other lags; PACF – a sine-wave shape pattern or a set of exponential decays.

5. *One autoregressive (p) and one moving average (q) parameter*: ACF – exponential decay starting at lag 1; PACF – exponential decay starting at lag 1.

Seasonal models. Multiplicative seasonal ARIMA is a generalization and extension of the method introduced in the previous paragraphs to series in which a pattern repeats seasonally over time. In addition to the non-seasonal parameters, seasonal parameters for a specified lag (established in the identification phase) need to be estimated. Analogous to the simple ARIMA parameters, these are: seasonal autoregressive (*ps*), seasonal differencing (*ds*), and seasonal moving average parameters (*qs*). For example, the model *(0,1,2)(0,1,1)* describes a model that includes no autoregressive parameters, 2 regular moving average parameters and 1 seasonal moving average parameter, and these parameters were computed for the series after it was differenced once with lag 1, and once seasonally differenced. The seasonal lag used for the seasonal parameters is usually determined during the identification phase and must be explicitly specified.
The general recommendations concerning the selection of parameters to be estimated (based on ACF and PACF) also apply to seasonal models. The main difference is that in seasonal series, ACF and PACF will show sizable coefficients at multiples of the seasonal lag (in addition to their overall patterns reflecting the non seasonal components of the series).

Parameter Estimation

There are several different methods for estimating the parameters. All of them should produce very similar estimates, but may be more or less efficient for any given model. In general, during the parameter estimation phase a function minimization algorithm is used (the quasi-Newton method; refer to Chapter 28 – *Nonlinear Estimation*) to maximize the likelihood (probability) of the observed series, given the parameter values. In practice, this requires the calculation of the (conditional) sums of squares (SS) of the residuals, given the respective parameters. Different methods have been proposed to compute the SS for the residuals: 1) the approximate maximum likelihood method according to McLeod and Sales (1983), 2) the approximate maximum likelihood method with backcasting, and 3) the exact maximum likelihood method according to Melard (1984).

Comparison of methods. In general, all methods should yield very similar parameter estimates. Also, all methods are about equally efficient in most real-world time series applications. However, method *1* above, (approximate maximum likelihood, no backcasts) is the fastest, and should be used in particular for very long time series (e.g., with more than 30,000 observations). Melard's exact maximum likelihood method (number *3* above) may also become inefficient when used to estimate parameters for seasonal models with long seasonal lags (e.g., with yearly lags of 365 days). On the other hand, you should always use the approximate maximum likelihood method first in order to establish initial parameter estimates that are very close to the actual final values; thus, usually only a few iterations with the exact maximum likelihood method (*3*, above) are necessary to finalize the parameter estimates.

Parameter standard errors. For all parameter estimates, you will compute asymptotic standard errors. These are computed from the matrix of second-order partial derivatives that is approximated via finite differencing (see also, *Large data sets and nonparametric methods* in Chapter 29 – *Nonparametric Statistics*).

Penalty value. The estimation procedure requires that the (conditional) sums of squares of the ARIMA residuals be minimized. If the model is inappropriate, it may happen during the iterative estimation process that the parameter estimates become very large and, in fact, invalid. In that case, it will assign a very large value (a so-called penalty value) to the SS. This usually entices the iteration process to move the parameters away from invalid ranges. However, in some cases even this strategy fails, and you may see on the screen (during the estimation procedure) very large values for the SS in consecutive iterations. In that case, carefully evaluate the appropriateness of your model. If your model contains many parameters and perhaps an intervention component, you can try again with different parameter start values.

Evaluation of the Model

Parameter estimates. You will report approximate *t* values, computed from the parameter standard errors (see above). If not significant, the respective parameter can in most cases be dropped from the model without affecting substantially the overall fit of the model.

Other quality criteria. Another straightforward and common measure of the reliability of the model is the accuracy of its forecasts generated based on partial data so that the forecasts can be compared with known (original) observations.

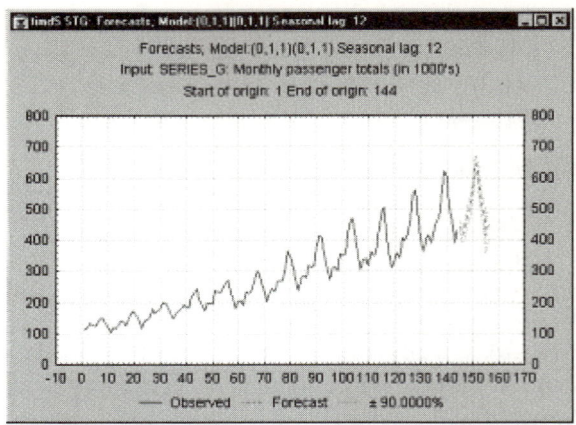

However, a good model should not only provide sufficiently accurate forecasts, it should also be parsimonious and produce statistically independent residuals that contain only noise and no systematic components (e.g., the correlogram of residuals should not reveal any serial dependencies). A good test of the model is a) to plot the residuals and inspect them for any systematic trends, and b) to examine the autocorrelogram of residuals (there should be no serial dependency between residuals).

Analysis of residuals. The major concern here is that the residuals are systematically distributed across the series (e.g., they could be negative in the first part of the series and approach zero in the second part) or that they contain some serial dependency, which may suggest that the ARIMA model is inadequate. The analysis of ARIMA residuals constitutes an important test of the model. The estimation procedure assumes that the residual are not (auto-) correlated and that they are normally distributed.

Limitations. The ARIMA method is appropriate only for a time series that is stationary (i.e., its mean, variance, and autocorrelation should be approximately constant through time) and it is recommended that there are at least 50 observations in the input data. It is also assumed that the values of the estimated parameters are constant throughout the series.

Interrupted Time Series

A common research questions in time series analysis is whether an outside event affected subsequent observations. For example, did the implementation of a new economic policy improve economic performance; did a new anti-crime law affect subsequent crime rates; and so on. In general, we want to evaluate the impact of one or more discrete events on the values in the time series. This type of interrupted time series analysis is described in detail in McDowall, McCleary, Meidinger, & Hay (1980). McDowall, et. al., distinguish between three major types of impacts that are possible: 1) permanent abrupt, 2) permanent gradual, and 3) abrupt temporary.

Exponential Smoothing

Overview

Exponential smoothing has become very popular as a forecasting method for a wide variety of time series data. Historically, the method was independently developed by Brown and Holt. Brown worked for the US Navy during World War II, where his assignment was to design a tracking system for fire-control information to compute the location of submarines. Later, he applied this technique to the forecasting of demand for spare parts (an inventory control problem). He described those ideas in his 1959 book on inventory control. Holt's research was sponsored by the Office of Naval Research; independently, he developed exponential smoothing models for constant

processes, processes with linear trends, and for seasonal data.

Gardner (1985) proposed a unified classification of exponential smoothing methods. Excellent introductions can also be found in Makridakis, Wheelwright, and McGee (1983), Makridakis and Wheelwright (1989), Montgomery, Johnson, & Gardiner (1990).

Simple Exponential Smoothing

A simple and pragmatic model for a time series would be to consider each observation as consisting of a constant (b) and an error component ε (epsilon), that is: $X_t = b + \varepsilon_t$. The constant b is relatively stable in each segment of the series, but may change slowly over time. If appropriate, then one way to isolate the true value of b, and thus the systematic or predictable part of the series, is to compute a kind of moving average, where the current and immediately preceding (younger) observations are assigned greater weight than the respective older observations. Simple exponential smoothing accomplishes exactly such weighting, where exponentially smaller weights are assigned to older observations. The specific formula for simple exponential smoothing is:

$$S_t = \alpha * X_t + (1 - \alpha) * S_{t-1}$$

When applied recursively to each successive observation in the series, each new smoothed value (forecast) is computed as the weighted average of the current observation and the previous smoothed observation; the previous smoothed observation was computed in turn from the previous observed value and the smoothed value before the previous observation, and so on. Thus, in effect, each smoothed value is the weighted average of the previous observations, where the weights decrease exponentially depending on the value of parameter α (alpha). If α is equal to 1 (one) then the previous observations are ignored entirely; if α is equal to 0 (zero), then the current observation is ignored entirely, and the smoothed value consists entirely of the previous smoothed value (which in turn is computed from the smoothed observation before it, and so on; thus all smoothed values will be equal to the initial smoothed value S_0). Values of α in-between will produce intermediate results.

Even though significant work has been done to study the theoretical properties of (simple and complex) exponential smoothing (e.g., see Gardner, 1985; Muth, 1960; see also, McKenzie, 1984, 1985), the method has gained popularity mostly because of its usefulness as a forecasting tool. For example, empirical research by Makridakis *et al.* (1982, Makridakis, 1983), has shown simple exponential smoothing to be the best choice for one-period-ahead forecasting, from among 24 other time series methods and using a variety of accuracy measures (see also, Gross and Craig, 1974, for additional empirical evidence). Thus, regardless of the theoretical model for the process underlying the observed time series, simple exponential smoothing will often produce quite accurate forecasts.

Choosing the Best Value for Parameter α (alpha)

Gardner (1985) discusses various theoretical and empirical arguments for selecting an appropriate smoothing parameter. Obviously, looking at the formula presented above, α should fall into the interval between 0 (zero) and 1 (although, see Brenner et al., 1968, for an ARIMA perspective, implying 0< α <2). Gardner (1985) reports that among practitioners, an α smaller than .30 is usually recommended. However, in the study by

Makridakis et al. (1982), α values above .30 frequently yielded the best forecasts. After reviewing the literature on this topic, Gardner (1985) concludes that it is best to estimate an optimum α from the data, rather than to guess and set an artificially low value.

Estimating the best α value from the data. In practice, the smoothing parameter is often chosen by a *grid search* of the parameter space; that is, different solutions for α are tried starting, for example, with α = 0.1 to α = 0.9, with increments of 0.1. Then α is chosen so as to produce the smallest sums of squares (or mean squares) for the residuals (i.e., observed values minus one-step-ahead forecasts; this mean squared error is also referred to as *ex post* mean squared error, *ex post* MSE for short).

Indices of Lack of Fit (Error)

The most straightforward way of evaluating the accuracy of the forecasts based on a particular α value is to simply plot the observed values and the one-step-ahead forecasts. This plot can also include the residuals (scaled against the right *Y*-axis), so that regions of better or worst fit can also easily be identified.

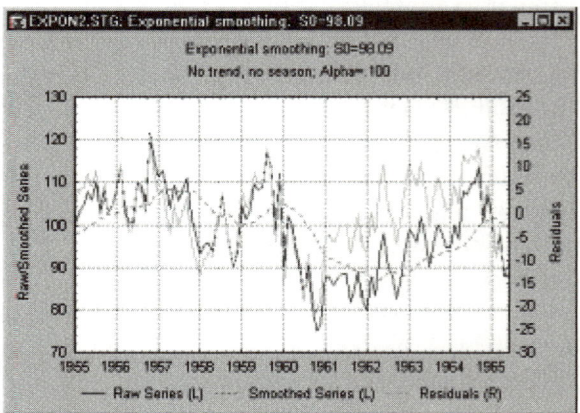

This visual check of the accuracy of forecasts is often the most powerful method for determining

whether or not the current exponential smoothing model fits the data. In addition, besides the *ex post* MSE criterion (see previous paragraph), there are other statistical measures of error that can be used to determine the optimum α parameter (see Makridakis, Wheelwright, and McGee, 1983):

Mean error. The mean error (ME) value is simply computed as the average error value (average of observed minus one-step-ahead forecast). Obviously, a drawback of this measure is that positive and negative error values can cancel each other out, so this measure is not a very good indicator of overall fit.

Mean absolute error. The mean absolute error (MAE) value is computed as the average *absolute* error value. If this value is 0 (zero), the fit (forecast) is perfect. As compared to the mean squared error value, this measure of fit will de-emphasize outliers, that is, unique or rare large error values will affect the MAE less than the MSE value.

Sum of squared error (SSE), Mean squared error. These values are computed as the sum (or average) of the squared error values. This is the most commonly used lack-of-fit indicator in statistical fitting procedures.

Percentage error (PE). All the above measures rely on the actual error value. It may seem reasonable to rather express the lack of fit in terms of the relative deviation of the one-step-ahead forecasts from the observed values, that is, relative to the magnitude of the observed values. For example, when trying to predict monthly sales that may fluctuate widely (e.g., seasonally) from month to month, we may be satisfied if our prediction "hits the target" with about ±10% accuracy. In other words, the absolute errors may be not so much of interest as are the relative errors in the forecasts. To

assess the relative error, various indices have been proposed (see Makridakis, Wheelwright, and McGee, 1983). The first one, the percentage error value, is computed as:

$$PE_t = 100 * (X_t - F_t)/X_t$$

where X_t is the observed value at time t, and F_t is the forecasts (smoothed values).

Mean percentage error (MPE). This value is computed as the average of the PE values.

Mean absolute percentage error (MAPE). As is the case with the mean error value (ME, see above), a mean percentage error near 0 (zero) can be produced by large positive and negative percentage errors that cancel each other out. Thus, a better measure of relative overall fit is the mean absolute percentage error. Also, this measure is usually more meaningful than the mean squared error. For example, knowing that the average forecast is off by ±5% is a useful result in and of itself, whereas a mean squared error of 30.8 is not immediately interpretable.

Automatic search for best parameter. A quasi-Newton function minimization procedure (the same as in ARIMA is used to minimize either the mean squared error, mean absolute error, or mean absolute percentage error. In most cases, this procedure is more efficient than the grid search (particularly when more than one parameter must be determined), and the optimum α parameter can quickly be identified.

The first smoothed value S_0. A final issue that we have neglected up to this point is the problem of the initial value, or how to start the smoothing process. If you look back at the formula above, it is evident that you need an S_0 value in order to compute the smoothed value (forecast) for the first observation in the series. Depending on the choice of the α parameter (i.e., when α is close to zero), the initial value for the smoothing process can affect the quality of the forecasts for many observations. As with most other aspects of exponential smoothing it is recommended to choose the initial value that produces the best forecasts. On the other hand, in practice, when there are many leading observations prior to a crucial actual forecast, the initial value will not affect that forecast by much, since its effect will have long faded from the smoothed series (due to the exponentially decreasing weights, the older an observation the less it will influence the forecast).

Seasonal and Non-Seasonal Models with or without Trend

The discussion above concerning simple exponential smoothing introduced the basic procedure for identifying a smoothing parameter, and for evaluating the goodness-of-fit of a model. In addition to simple exponential smoothing, more complex models have been developed to accommodate time series with seasonal and trend components. The general idea here is that forecasts are not only computed from consecutive previous observations (as in simple exponential smoothing), but an independent (smoothed) trend and seasonal component can be added. Gardner (1985) discusses the different models in terms of seasonality (none, additive, or multiplicative) and trend (none, linear, exponential, or damped).

Additive and multiplicative seasonality. Many time series data follow recurring seasonal patterns. For example, annual sales of toys will probably peak in the months of November and December, and perhaps during the summer (with a much smaller peak) when children are on their summer break. This pattern will likely repeat every year, however, the relative amount of increase in sales during December may slowly change from year to year. Thus, it may be useful to smooth the seasonal component independently with an extra parameter, usually denoted as δ (*delta*). Seasonal components can be additive in nature or multiplicative. For example, during the month of December the sales for a particular toy may increase by 1 million dollars every year. Thus, we could add to our forecasts for every December the amount of 1 million dollars (over the respective annual average) to account for this seasonal fluctuation. In this case, the seasonality is additive. Alternatively, during the month of December the sales for a particular toy may increase by 40%, that is, increase by a factor of 1.4. Thus, when the sales for the toy are generally weak, than the absolute (dollar) increase in sales during December will be relatively weak (but the percentage will be constant); if the sales of the toy are strong, than the absolute (dollar) increase in sales will be proportionately greater. Again, in this case the sales increase by a certain factor, and the seasonal component is thus multiplicative in nature (i.e., the multiplicative seasonal component in this case would be 1.4). In plots of the series, the distinguishing characteristic between these two types of seasonal components is that in the additive case, the series shows steady seasonal fluctuations, regardless of the overall level of the series; in the multiplicative case, the size of the seasonal fluctuations vary, depending on the overall level of the series.

The seasonal smoothing parameter δ. In general the one-step-ahead forecasts are computed as (for no trend models, for linear and exponential trend models a trend component is added to the model; see below):

Additive model: $\text{Forecast}_t = S_t + I_{t-p}$

Multiplicative model: $\text{Forecast}_t = S_t {}^* I_{t-p}$

In this formula, S_t stands for the (simple) exponentially smoothed value of the series at time t, and I_{t-p} stands for the smoothed seasonal factor at time t minus p (the length of the season). Thus, compared to simple exponential smoothing, the forecast is "enhanced" by adding or multiplying the simple smoothed value by the predicted seasonal component. This seasonal component is derived analogous to the S_t value from simple exponential smoothing as:

Additive model: $I_t = I_{t-p} + \delta*(1- \alpha)*e_t$

Multiplicative model: $I_t = I_{t-p} + \delta *(1- \alpha)*e_t/S_t$

Put in words, the predicted seasonal component at time t is computed as the respective seasonal component in the last seasonal cycle plus a portion of the error (e_t; the observed minus the forecast value at time t). Considering the formulas above, it is clear that parameter δ can assume values between 0 and 1. If it is zero, then the seasonal component for a particular point in time is predicted to be identical to the predicted seasonal component for the respective time during the previous seasonal cycle, which in turn is predicted to be identical to that from the previous cycle, and so on. Thus, if δ is zero, a constant unchanging seasonal component is used to generate the one-step-ahead forecasts. If the δ parameter is equal to 1, then the seasonal component is modified maximally at every step by the respective forecast error (times $(1-\alpha)$, which we will ignore for the purpose of this brief introduction). In most cases, when seasonality is present in the time series, the optimum δ parameter will fall somewhere between 0 (zero) and 1(one).

Linear, exponential, and damped trend. To remain with the toy example above, the sales for a toy can show a linear upward trend (e.g., each year, sales increase by 1 million dollars),

exponential growth (e.g., each year, sales increase by a factor of 1.3), or a damped trend (during the first year sales increase by 1 million dollars; during the second year the increase is only 80% over the previous year, i.e., $800,000; during the next year it is again 80% less than the previous year, i.e., $800,000 * .8 = $640,000; etc.). Each type of trend leaves a clear signature that can usually be identified in the series; shown below in the brief discussion of the different models are icons that illustrate the general patterns. In general, the trend factor may change slowly over time, and, again, it may make sense to smooth the trend component with a separate parameter [denoted γ (*gamma*) for linear and exponential trend models, and ϕ (*phi*) for damped trend models].

The trend smoothing parameters γ (linear and exponential trend) and Φ (damped trend). Analogous to the seasonal component, when a trend component is included in the exponential smoothing process, an independent trend component is computed for each time, and modified as a function of the forecast error and the respective parameter. If the γ parameter is 0 (zero), than the trend component is constant across all values of the time series (and for all forecasts). If the parameter is 1, then the trend component is modified "maximally" from observation to observation by the respective forecast error. Parameter values that fall in-between represent mixtures of those two extremes. Parameter Φ is a trend modification parameter, and affects how strongly changes in the trend will affect estimates of the trend for subsequent forecasts, that is, how quickly the trend will be damped or increased.

Classical Seasonal Decomposition (Census Method I)

Overview

Suppose you recorded the monthly passenger load on international flights for a period of 12 years (see Box & Jenkins, 1976). If you plot those data, it is apparent that 1) there appears to be a linear upward trend in the passenger loads over the years, and 2) there is a recurring pattern or seasonality within each year (i.e., most travel occurs during the summer months, and a minor peak occurs during the December holidays). The purpose of the seasonal decomposition method is to isolate those components, that is, to decompose the series into the trend effect, seasonal effects, and remaining variability. The classic technique designed to accomplish this decomposition is known as the Census I method. This technique is described and discussed in detail in Makridakis, Wheelwright, and McGee (1983), and Makridakis and Wheelwright (1989).

General model. The general idea of seasonal decomposition is straightforward. In general, a time series such as the one described above can be thought of as consisting of four different components: 1) A seasonal component (denoted as S_t, where t stands for the particular point in time) 2) a trend component (T_t), 3) a cyclical component (C_t), and 4) a random, error, or irregular component (I_t). The difference between a cyclical and a seasonal component is that the latter occurs at regular (seasonal) intervals, while cyclical factors have usually a longer duration that varies from cycle to cycle. In the Census I method, the trend and cyclical components are customarily combined into a trend-cycle component (TC_t). The specific functional relationship between these components can assume different forms. However, two straightforward possibilities are that they combine in an additive or a multiplicative fashion:

Additive model:

$$X_t = TC_t + S_t + I_t$$

Multiplicative model:

$$X_t = T_t {}^* C_t {}^* S_t {}^* I_t$$

Here X_t stands for the observed value of the time series at time t. Given some *a priori* knowledge about the cyclical factors affecting the series (e.g., business cycles), the estimates for the different components can be used to compute forecasts for future observations. (However, the exponential smoothing method, which can also incorporate seasonality and trend components, is the preferred technique for forecasting purposes.)

Additive and multiplicative seasonality. Let's consider the difference between an additive and multiplicative seasonal component in an example: The annual sales of toys will probably peak in the months of November and December, and perhaps during the summer (with a much smaller peak) when children are on their summer break. This seasonal pattern will likely repeat every year. Seasonal components can be additive or multiplicative in nature. For example, during the month of December the sales for a particular toy may increase by 3 million dollars every year. Thus, we could add to our forecasts for every December the amount of 3 million to account for this seasonal fluctuation. In this case, the seasonality is additive. Alternatively, during the month of December the sales for a particular toy may increase by 40%, that is, increase by a factor of 1.4. Thus, when the sales for the toy are generally weak, then the absolute (dollar) increase in sales during December will be

relatively weak (but the percentage will be constant); if the sales of the toy are strong, then the absolute (dollar) increase in sales will be proportionately greater. Again, in this case the sales increase by a certain factor, and the seasonal component is thus multiplicative in nature (i.e., the multiplicative seasonal component in this case would be 1.4). In plots of series, the distinguishing characteristic between these two types of seasonal components is that in the additive case, the series shows steady seasonal fluctuations, regardless of the overall level of the series; in the multiplicative case, the size of the seasonal fluctuations vary, depending on the overall level of the series.

Additive and multiplicative trend-cycle.

We can extend the previous example to illustrate the additive and multiplicative trend-cycle components. In terms of our toy example, a fashion trend may produce a steady increase in sales (e.g., a trend toward more educational toys in general); as with the seasonal component, this trend may be additive (sales increase by 3 million dollars per year) or multiplicative (sales increase by 30%, or by a factor of 1.3, annually) in nature. In addition, cyclical components may impact sales; to reiterate, a cyclical component is different from a seasonal component in that it usually is of longer duration, and that it occurs at irregular intervals. For example, a particular toy may be particularly "hot" during a summer season (e.g., a particular doll tied to the release of a major children's movie and promoted with extensive advertising). Again such a cyclical component can effect sales in an additive manner or multiplicative manner.

Computations

The Seasonal Decomposition (Census I) standard formulas are shown in Makridakis,

Wheelwright, and McGee (1983), and Makridakis and Wheelwright (1989).

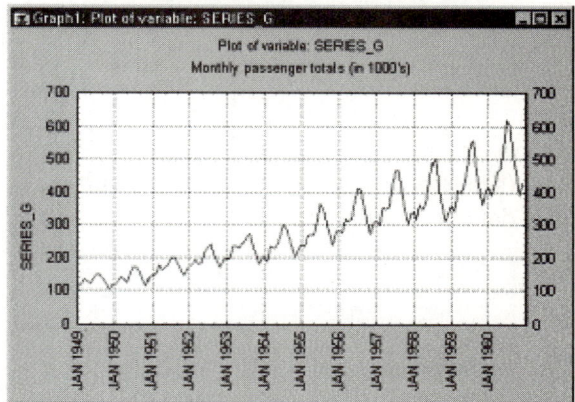

Moving average. First a moving average is computed for the series, with the moving average window width equal to the length of one season. If the length of the season is even, then you can choose to use either equal weights for the moving average or unequal weights can be used, where the first and last observation in the moving average window are averaged.

Ratios or differences. In the moving average series, all seasonal (within-season) variability will be eliminated; thus, the differences (in additive models) or ratios (in multiplicative models) of the observed and smoothed series will isolate the seasonal component (plus irregular component). Specifically, the moving average is subtracted from the observed series (for additive models) or the observed series is divided by the moving average values (for multiplicative models).

Seasonal components. The seasonal component is then computed as the average (for additive models) or medial average (for multiplicative models) for each point in the season.

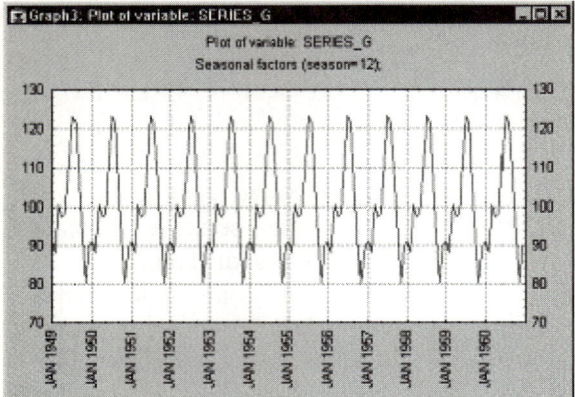

(The medial average of a set of values is the mean after the smallest and largest values are excluded.) The resulting values represent the (average) seasonal component of the series.

Seasonally adjusted series. The original series can be adjusted by subtracting from it (additive models) or dividing it by (multiplicative models) the seasonal component.

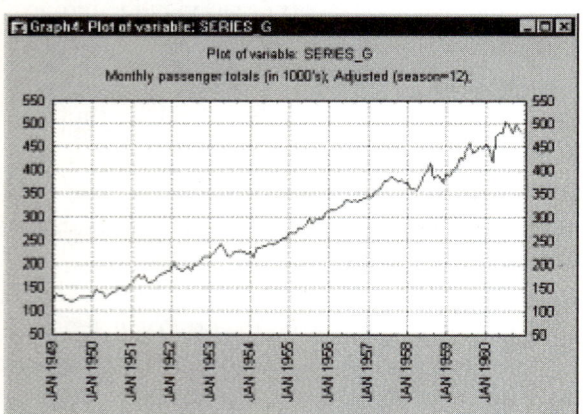

The resulting series is the seasonally adjusted series (i.e., the seasonal component will be removed).

Trend-cycle component. Remember that the cyclical component is different from the seasonal component in that it is usually longer than one season, and different cycles can be of different lengths. The combined trend and cyclical component can be approximated by applying to the seasonally adjusted series a 5 point (centered) weighed moving average smoothing transformation with the weights of 1, 2, 3, 2, 1.

Random or irregular component. Finally, the random or irregular (error) component can be isolated by subtracting from the seasonally adjusted series (additive models) or dividing the adjusted series by (multiplicative models) the trend-cycle component.

X-11 Census Method II Seasonal Adjustment

The general idea of seasonal decomposition and adjustment are discussed in the context of the Census I seasonal adjustment method [Seasonal Decomposition (Census I)]. The Census method II (2) is an extension and refinement of the simple adjustment method. Over the years, different versions of the Census method II evolved at the Census Bureau; the method that has become most popular and is used most widely in government and business is the so-called X-11 variant of the Census method II (see Hiskin, Young, & Musgrave, 1967). Subsequently, the term X-11 has become synonymous with this refined version of the Census method II. In addition to the documentation that can be obtained from the Census Bureau, a detailed summary of this method is also provided in Makridakis, Wheelwright, and McGee (1983) and Makridakis and Wheelwright (1989).

Seasonal Adjustment: Basic Ideas and Terms

Suppose you recorded the monthly passenger load on international flights for a period of 12 years (see Box & Jenkins, 1976). If you plot those data, it is apparent that 1) there appears to be an upward linear trend in the passenger loads over the years, and 2) there is a recurring pattern or seasonality within each year (i.e., most travel occurs during the summer months, and a minor peak occurs during the December holidays). The purpose of seasonal decomposition and adjustment is to isolate those components, that is, to de-compose the series into the trend effect, seasonal effects, and remaining variability. The classic technique designed to accomplish this decomposition was developed in the 1920's and is also known as the Census I method (see *Census Method I*, page 504). This technique is also described and discussed in detail in Makridakis, Wheelwright, and McGee (1983), and Makridakis and Wheelwright (1989).

General model. The general idea of seasonal decomposition is straightforward. In general, a time series such as the one described above can be thought of as consisting of four different components: 1) A seasonal component (denoted as S_t, where t stands for the particular point in time) 2) a trend component (T_t), 3) a cyclical component (C_t), and 4) a random, error, or irregular component (I_t). The difference between a cyclical and a seasonal component is that the latter occurs at regular (seasonal) intervals, while cyclical factors usually have a longer duration that varies from cycle to cycle. The trend and cyclical components are customarily combined into a trend-cycle component (TC_t). The specific functional relationship between these components can assume different forms. However, two straightforward possibilities are that they combine in an additive or a multiplicative fashion:

Additive Model: $X_t = TC_t + S_t + I_t$

Multiplicative Model: $X_t = T_t * C_t * S_t * I_t$

Where:

X_t represents the observed value of the time series at time t.

Given some *a priori* knowledge about the cyclical factors affecting the series (e.g., business cycles), the estimates for the different components can be used to compute forecasts for future observations. (However, the exponential smoothing method, which can also incorporate seasonality and trend components, is the preferred technique for forecasting purposes.)

Additive and multiplicative seasonality. Consider the difference between an additive and multiplicative seasonal component in an example: The annual sales of toys will probably peak in the months of November and December, and perhaps during the summer (with a much smaller peak) when children are on their summer break. This seasonal pattern will likely repeat every year. Seasonal components can be additive or multiplicative in nature. For example, during the month of December the sales for a particular toy may increase by 3 million dollars every year. Thus, you could add to your forecasts for every December the amount of 3 million to account for this seasonal fluctuation. In this case, the seasonality is additive. Alternatively, during the month of December the sales for a particular toy may increase by 40%, that is, increase by a factor of 1.4. Thus, when the sales for the toy are generally weak, then the absolute (dollar) increase in sales during December will be relatively weak (but the percentage will be constant); if the sales of the toy are strong, then the absolute (dollar) increase in sales will be proportionately greater. Again, in this case the sales increase by a certain factor, and the seasonal component is thus multiplicative in

nature (i.e., the multiplicative seasonal component in this case would be 1.4). In plots of series, the distinguishing characteristic between these two types of seasonal components is that in the additive case, the series shows steady seasonal fluctuations, regardless of the overall level of the series; in the multiplicative case, the size of the seasonal fluctuations vary, depending on the overall level of the series.

Additive and multiplicative trend-cycle. The previous example can be extended to illustrate the additive and multiplicative trend-cycle components. In terms of the toy example, a "fashion" *trend* may produce a steady increase in sales (e.g., a trend toward more educational toys in general); as with the seasonal component, this trend may be additive (sales increase by 3 million dollars per year) or multiplicative (sales increase by 30%, or by a factor of 1.3, annually) in nature. In addition, cyclical components may impact sales. To reiterate, a cyclical component is different from a seasonal component in that it usually is of longer duration, and that it occurs at irregular intervals. For example, a particular toy may be particularly "hot" during a summer season (e.g., a particular doll tied to the release of a major children's movie promoted with extensive advertising). Again such a cyclical component can affect sales in an additive manner or multiplicative manner.

Census II Method

The basic method for seasonal decomposition and adjustment outlined in *Basic Ideas and Terms*, page 506, can be refined in several ways. In fact, unlike many other time-series modeling techniques (e.g., ARIMA) which are grounded in some theoretical model of an underlying process, the *X-11* variant of the

Census II method simply contains many *ad hoc* features and refinements, that over the years have proven to provide excellent estimates for many real-world applications (see Burman, 1979, Kendal & Ord, 1990, Makridakis & Wheelwright, 1989; Wallis, 1974). Some of the major refinements are listed below.

Trading-day adjustment. Different months have different numbers of days, and different numbers of trading-days (i.e., Mondays, Tuesdays, etc.). When analyzing, for example, monthly revenue figures for an amusement park, the fluctuation in the different numbers of Saturdays and Sundays (peak days) in the different months will surely contribute significantly to the variability in monthly revenues. The *X-11* variant of the Census II method allows you to test whether such trading-day variability exists in the series, and, if so, to adjust the series accordingly.

Extreme values. Most real-world time series contain outliers, that is, extreme fluctuations due to rare events. For example, a strike may affect production in a particular month of one year. Such extreme outliers may bias the estimates of the seasonal and trend components. The *X-11* procedure includes provisions to deal with extreme values through the use of statistical control principles, that is, values that are above or below a certain range (expressed in terms of multiples of *sigma*, the standard deviation) can be modified or dropped before final estimates for the seasonality are computed.

Multiple refinements. The refinement for outliers, extreme values, and different numbers of trading-days can be applied more than once, in order to obtain successively improved estimates of the components. The X-11 method applies a series of successive refinements of the estimates to arrive at the final trend-cycle, seasonal, and irregular components, and the seasonally adjusted series.

Tests and summary statistics. In addition to estimating the major components of the series, various summary statistics can be computed. For example, analysis of variance tables can be prepared to test the significance of seasonal variability and trading-day variability (see above) in the series; the X-11 procedure will also compute the percentage change from month to month in the random and trend-cycle components. As the duration or span in terms of months (or quarters for quarterly X-11) increases, the change in the trend-cycle component will likely also increase, while the change in the random component should remain about the same. The width of the average span at which the changes in the random component are about equal to the changes in the trend-cycle component is called the month (quarter) for cyclical dominance, or MCD (QCD) for short. For example, if the MCD is equal to 2, you can infer that over a 2 month span the trend-cycle will dominate the fluctuations of the irregular (random) component. These and various other results are discussed in greater detail below.

Results Tables Computed by the X-11 Method

The computations performed by the X-11 procedure are best discussed in the context of the results tables that are reported. The adjustment process is divided into seven major steps, which are customarily labeled with consecutive letters A through G.

A. Prior adjustment (monthly seasonal adjustment only). Before any seasonal adjustment is performed on the monthly time series, various prior user-defined adjustments can be incorporated. You can specify a second series that contains prior adjustment factors; the values in that series will either be subtracted (additive model) from the original series, or the original series will be divided by these values (multiplicative model). For multiplicative models, user-specified trading-day adjustment weights can also be specified. These weights will be used to adjust the monthly observations depending on the number of respective trading-days represented by the observation.

B. Preliminary estimation of trading-day variation (monthly X-11) and weights. Next, preliminary trading-day adjustment factors (monthly X-11 only) and weights for reducing the effect of extreme observations are computed.

C. Final estimation of trading-day variation and irregular weights (monthly X- 11). The adjustments and weights computed in B above are then used to derive improved trend-cycle and seasonal estimates. These improved estimates are used to compute the final trading-day factors (monthly X-11 only) and weights.

D. Final estimation of seasonal factors, trend-cycle, irregular, and seasonally adjusted series. The final trading-day factors and weights computed in *C* above are used to compute the final estimates of the components.

E. Modified original, seasonally adjusted, and irregular series. The original and final seasonally adjusted series and the irregular component are modified for extremes. The resulting modified series enable you to examine the stability of the seasonal adjustment.

F. Month (quarter) for cyclical dominance (MCD, QCD), moving average, and summary measures. In this part of the

computations, various summary measures are computed to enable you to examine the relative importance of the different components, the average fluctuation from month-to-month (quarter-to-quarter), the average number of consecutive changes in the same direction (average number of runs), etc.

G. Charts. Finally, you will compute various charts (graphs) to summarize the results. For example, the final seasonally adjusted series will be plotted, in chronological order, or by month (see below).

Distributed Lags Analysis Overview

General Purpose

Distributed lags analysis is a specialized technique for examining the relationships between variables that involve some delay. For example, suppose that you are a manufacturer of computer software, and you want to determine the relationship between the number of inquiries that are received, and the number of orders that are placed by your customers. You could record those numbers monthly for a one-year period, and then correlate the two variables. However, obviously inquiries will precede actual orders, and you can expect that the number of orders will follow the number of inquiries with some delay. Put another way, there will be a (time) lagged correlation between the number of inquiries and the number of orders that are received.

Time-lagged correlations are particularly common in econometrics. For example, the benefits of investments in new machinery usually only become evident after some time. Higher income will change people's choice of rental apartments, however, this relationship will be lagged because it will take some time

for people to terminate their current leases, find new apartments, and move. In general, the relationship between capital appropriations and capital expenditures will be lagged, because it will require some time before investment decisions are actually acted upon.

In all of these cases, we have an independent or explanatory variable that affects the dependent variables with some lag. The distributed lags method allows you to investigate those lags. Detailed discussions of distributed lags correlation can be found in most econometrics textbooks, for example, in Judge, Griffith, Hill, Luetkepohl, and Lee (1985), Maddala (1977), and Fomby, Hill, and Johnson (1984). In the following paragraphs we will present a brief description of these methods. We will assume that you are familiar with the concept of correlation (see Chapter 2 – *Basic Statistics and Tables*), and the basic ideas of multiple regression (see Chapter 26 – *Multiple Linear Regression*).

General Model

Suppose we have a dependent variable *y* and an independent or explanatory variable *x* that are both measured repeatedly over time. In some textbooks, the dependent variable is also referred to as the endogenous variable, and the independent or explanatory variable the exogenous variable. The simplest way to describe the relationship between the two would be in a simple linear relationship:

$$Y_t = \Sigma \; \beta_i {}^* x_{t-i}$$

In this equation, the value of the dependent variable at time *t* is expressed as a linear function of *x* measured at times *t, t-1, t-2*, etc. Thus, the dependent variable is a linear function of *x*, and *x* is lagged by *1, 2*, etc. time periods. The *beta* weights (β_i) can be considered slope parameters in this equation. You may recognize this equation as a special case of the general linear regression equation (see Chapter 26 – *Multiple Linear Regression*). If the weights for the lagged time periods are statistically significant, we can conclude that the y variable is predicted (or explained) with the respective lag.

Almon Distributed Lag

A common problem that often arises when computing the weights for the multiple linear regression model shown above is that the values of adjacent (in time) values in the *x* variable are highly correlated. In extreme cases, their independent contributions to the prediction of y may become so redundant that the correlation matrix of measures can no longer be inverted, and thus, the *beta* weights cannot be computed. In less extreme cases, the computation of the *beta* weights and their standard errors can become very imprecise, due to round-off error. In the context of multiple regression, this general computational problem is discussed as the multicollinearity or matrix ill-conditioning issue.

Almon (1965) proposed a procedure that will reduce the multicollinearity in this case. Specifically, suppose we express each weight in the linear regression equation in the following manner:

$$\beta_i = \alpha_0 + \alpha_1 * i + \ldots + \alpha_q * i^q$$

Almon could show that in many cases it is easier (i.e., it avoids the multicollinearity problem) to estimate the *alpha* values than the *beta* weights directly. Note that with this method, the precision of the beta weight estimates is dependent on the degree or order of the polynomial approximation.

Misspecifications. A general problem with this technique is that, of course, the lag length and correct polynomial degree are not known *a priori*. The effects of misspecifications of

these parameters are potentially serious (in terms of biased estimation). This issue is discussed in greater detail in Frost (1975), Schmidt and Waud (1973), Schmidt and Sickles (1975), and Trivedi and Pagan (1979).

Single Spectrum (Fourier) Analysis

Spectrum analysis is concerned with the exploration of cyclical patterns of data. The purpose of the analysis is to decompose a complex time series with cyclical components into a few underlying sinusoidal (sine and cosine) functions of particular wavelengths. The term spectrum provides an appropriate metaphor for the nature of this analysis: Suppose you study a beam of white sun light, which at first looks like a random (white noise) accumulation of light of different wavelengths. However, when put through a prism, we can separate the different wave lengths or cyclical components that make up white sun light. In fact, via this technique we can now identify and distinguish between different sources of light. Thus, by identifying the important underlying cyclical components, we have learned something about the phenomenon of interest. In essence, performing spectrum analysis on a time series is like putting the series through a prism in order to identify the wave lengths and importance of underlying cyclical components. As a result of a successful analysis, you might uncover just a few recurring cycles of different lengths in the time series of interest, which at first looked more or less like random noise.

A much cited example for spectrum analysis is the cyclical nature of sun spot activity (e.g., see Bloomfield, 1976, or Shumway, 1988). It turns out that sun spot activity varies over 11 year cycles. Other examples of celestial phenomena, weather patterns, fluctuations in commodity

prices, economic activity, etc. are also often used in the literature to demonstrate this technique. To contrast this technique with ARIMA or exponential smoothing, the purpose of spectrum analysis is to identify the seasonal fluctuations of different lengths, while in the former types of analysis, the length of the seasonal component is usually known (or guessed) *a priori* and then included in some theoretical model of moving averages or autocorrelations.

The classic text on spectrum analysis is Bloomfield (1976); however, other detailed discussions can be found in Jenkins and Watts (1968), Brillinger (1975), Brigham (1974), Elliott and Rao (1982), Priestley (1981), Shumway (1988), or Wei (1989).

Cross-Spectrum Analysis

Overview

Cross-spectrum analysis is an extension of single spectrum (Fourier) analysis to the simultaneous analysis of two series. In the following paragraphs, we will assume that you have already read *Single Spectrum (Fourier) Analysis*, page 511. Detailed discussions of this technique can be found in Bloomfield (1976), Jenkins and Watts (1968), Brillinger (1975), Brigham (1974), Elliott and Rao (1982), Priestley (1981), Shumway (1988), or Wei (1989).

Strong periodicity in the series at the respective frequency.
A much cited example for spectrum analysis is the cyclical nature of sun spot activity (e.g., see Bloomfield, 1976, or Shumway, 1988). It turns out that sun spot activity varies over 11 year cycles. Other examples of celestial phenomena, weather patterns, fluctuations in commodity prices, economic activity, etc. are also often used in the literature to demonstrate this technique.

The purpose of cross-spectrum analysis is to uncover the correlations between two series at different frequencies. For example, sun spot activity may be related to weather phenomena here on earth. If so, then if we were to record those phenomena (e.g., yearly average temperature) and submit the resulting series to a cross-spectrum analysis together with the sun spot data, we may find that the weather indeed correlates with the sunspot activity at the 11 year cycle. That is, we may find a periodicity in the weather data that is in-sync with the sun spot cycles. You can easily think of other areas of research where such knowledge could be very useful; for example, various economic indicators may show similar (correlated) cyclical behavior; various physiological measures likely will also display "coordinated" (i.e., correlated) cyclical behavior, and so on.

Basic Notation and Principles

A simple example. Consider the following two series with 16 cases:

	Var1	Var2
1	1.000	-0.058
2	1.664	-0.713
3	1.148	-0.383
4	-0.058	0.006
5	-0.713	-0.483
6	-0.383	-1.441
7	0.006	-1.637
8	-0.483	-0.707
9	-1.441	0.331
10	-1.637	0.441
11	-0.707	-0.058
12	0.331	-0.006
13	0.441	0.924
14	-0.058	1.713
15	-0.006	1.365
16	0.924	0.266

Data: Time Series Example Spreadsheet 1 (2v by 16c)

At first sight, it is not easy to see the relationship between the two series. However, as shown in the next illustration, the series were created so that they would contain two strong correlated periodicities. Parts of the summary from the cross-spectrum analysis are shown in the next illustration (the spectral estimates were smoothed with a Parzen window of width 3).

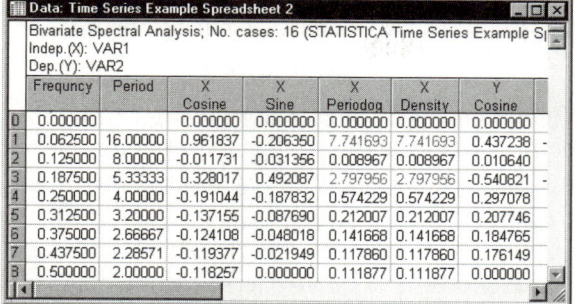

Results for Each Variable

The complete summary contains all spectrum statistics computed for each variable, as described in *Single Spectrum (Fourier) Analysis*, page 511. Looking at the results shown in the previous spreadsheet; it is clear that both variables show strong periodicities at the frequencies .0625 and .1875.

Cross-Periodogram, Cross-Density, Quadrature-Density, Cross-Amplitude

Analogous to the results for the single variables, the complete summary will also display periodogram values for the cross periodogram. However, the cross-spectrum consists of complex numbers that can be divided into a real and an imaginary part. These can be smoothed to obtain the cross-density and quadrature density (quad density for short) estimates, respectively. [The reasons for smoothing, and the different common weight functions for smoothing are discussed in *Single Spectrum (Fourier) Analysis*, page 511.] The square root of the sum of the squared cross-density and quad-density values is called the cross-amplitude. The cross-amplitude can be interpreted as a measure of covariance between the respective frequency components in the two

series. Thus, we can conclude from the results shown in the previous illustration that the .0625 and .1875 frequency components in the two series covary.

Squared Coherency, Gain, and Phase Shift

There are additional statistics that can be displayed in the complete summary.

Squared coherency. You can standardize the cross-amplitude values by squaring them and dividing by the product of the spectrum density estimates for each series. The result is called the squared coherency, which can be interpreted similar to the squared correlation coefficient (see *Correlations* in Chapter 2 – *Basic Statistics and Tables*), that is, the coherency value is the squared correlation between the cyclical components in the two series at the respective frequency. However, the coherency values should not be interpreted by themselves; for example, when the spectral density estimates in both series are very small, large coherency values may result (the divisor in the computation of the coherency values will be very small), even though there are no strong cyclical components in either series at the respective frequencies.

Gain. The gain value is computed by dividing the cross-amplitude value by the spectrum density estimates for one of the two series in the analysis. Consequently, two gain values are computed, which can be interpreted as the standard least squares regression coefficients for the respective frequencies.

Phase shift. Finally, the phase shift estimates are computed as $\tan^{-1}$ of the ratio of the quad density estimates over the cross-density estimate. The phase shift estimates (usually denoted by the Greek letter) are measures of the extent to which each frequency component of one series leads the other.

How the Example Data Were Created

Now, let's return to the example data set presented above. The large spectral density estimates for both series, and the cross-amplitude values at frequencies $v = 0.0625$ and $v = .1875$ suggest two strong synchronized periodicities in both series at those frequencies. In fact, the two series were created as:

v1 = cos(2*π*.0625*(v0-1)) + .75*sin(2* π *.2*(v0-1))

v2 = cos(2* π *.0625*(v0+2)) + .75*sin(2* π *.2*(v0+2))

(where *v0* is the case number). Indeed, the analysis presented in this overview reproduced the periodicity inserted into the data very well.

Spectrum Analysis – Basic Notation and Principles

Frequency and Period

The "wave length" of a sine or cosine function is typically expressed in terms of the number of cycles per unit time (frequency), often denoted by the Greek letter *nu* (v; some textbooks also use *f*). For example, the number of letters handled in a post office may show 12 cycles per year. On the first of every month a large amount of mail is sent (many bills come due on the first of the month), then the amount of mail decreases in the middle of the month, then it increases again toward the end of the month. Therefore, every month the fluctuation in the amount of mail handled by the post office will go through a full cycle. Thus, if the unit of analysis is one year, then n would be equal to

12, as there would be 12 cycles per year. Of course, there will likely be other cycles with different frequencies. For example, there might be annual cycles ($v =1$), and perhaps weekly cycles <($v =52$ weeks per year).

The *period T* of a sine or cosine function is defined as the length of time required for one full cycle. Thus, it is the reciprocal of the frequency, or: $T = 1/ v$. To return to the mail example in the previous paragraph, the monthly cycle, expressed in yearly terms, would be equal to $1/12 = 0.0833$. Put into words, there is a period in the series of length 0.0833 years.

The General Structural Model

As mentioned before, the purpose of spectrum analysis is to decompose the original series into underlying sine and cosine functions of different frequencies, in order to determine those that appear particularly strong or important. One way to do so would be to cast the issue as a linear multiple regression problem, where the dependent variable is the observed time series, and the independent variables are the sine functions of all possible (discrete) frequencies. Such a linear multiple regression model may be written as:

$x_t = a_0 + \Sigma [a_k*cos(\lambda_k*t) + b_k*sin(\lambda_k*t)]$ (for k = 1 to q)

Following the common notation from classical harmonic analysis, in this equation λ (*lambda*) is the frequency expressed in terms of radians per unit time, that is: $\lambda = 2* \pi *v_k$, where π is the constant *pi*=3.14... and $v_k = k/q$. What is important here is to recognize that the computational problem of fitting sine and cosine functions of different lengths to the data can be considered in terms of multiple linear regression. Note that the cosine parameters a_k and sine parameters b_k are regression coefficients that tell us the degree to which the respective functions are correlated with the data. Overall there are q different sine and cosine functions; intuitively (as also discussed in Chapter 26 – *Multiple Linear Regression*), it should be clear that we cannot have more sine and cosine functions than there are data points in the series. Without going into detail, if there are N data points in the series, then there will be $N/2+1$ cosine functions and $N/2-1$ sine functions. In other words, there will be as many different sinusoidal waves as there are data points, and we will be able to completely reproduce the series from the underlying functions. (Note that if the number of cases in the series is odd, then the last data point will usually be ignored; in order for a sinusoidal function to be identified, you need at least two points: the high peak and the low peak.)

To summarize, spectrum analysis will identify the correlation of sine and cosine functions of different frequency with the observed data. If a large correlation (sine or cosine coefficient) is identified, you can conclude that there is a strong periodicity of the respective frequency (or period) in the data.

Complex numbers (real and imaginary numbers). In many textbooks on spectrum analysis, the structural model shown above is presented in terms of complex numbers, that is, the parameter estimation process is described in terms of the Fourier transform of a series into real and imaginary parts. Complex numbers are the superset that includes all real and imaginary numbers. Imaginary numbers, by definition, are numbers that are multiplied by the constant i, where i is defined as the square root of -1. Obviously, the square root of -1 does not exist, hence the term imaginary number; however, meaningful arithmetic operations on imaginary numbers can still be performed (e.g., $[i*2]**2=$

-4). It is useful to think of real and imaginary numbers as forming a two-dimensional plane, where the horizontal or X-axis represents all real numbers, and the vertical or Y-axis represents all imaginary numbers. Complex numbers can then be represented as points in the two-dimensional plane. For example, the complex number $3+i*2$ can be represented by a point with coordinates $\{3,2\}$ in this plane. You can also think of complex numbers as angles, for example, you can connect the point representing a complex number in the plane with the origin (complex number $0+i*0$), and measure the angle of that vector to the horizontal line. Thus, intuitively you can see how the spectrum decomposition formula shown above, consisting of sine and cosine functions, can be rewritten in terms of operations on complex numbers. In fact, in this manner the mathematical discussion and required computations are often more elegant and easier to perform; which is why many textbooks prefer the presentation of spectrum analysis in terms of complex numbers.

A Simple Example

Shumway (1988) presents a simple example to clarify the underlying mechanics of spectrum analysis. Let's create a series with 16 cases following the equation shown above, and then see how we may extract the information that was put in it. First, create a variable and define it as:

$$x = 1*\cos(2*\pi*.0625*(v0-1)) + .75*\sin(2*\pi*.2*(v0-1))$$

This variable is made up of two underlying periodicities: The first at the frequency of $v =.0625$ (or period $1/v=16$; one observation completes $1/16$'th of a full cycle, and a full cycle is completed every 16 observations) and

the second at the frequency of $v = .2$ (or period of 5). The cosine coefficient (1.0) is larger than the sine coefficient (.75). The spectrum analysis summary is shown in the next illustration.

Data: Time Series Example Spreadsheet 3

Spectral analysis: VAR1 (STATISTICA Time Series Example Spreadsheet 1.sta)
No. of cases: 16

	Frequncy	Period	Cosine Coeffs	Sine Coeffs	Periodog	Density	Parzen Weights
0	0.000000		0.000000	0.000000	0.000000	0.000000	0.000000
1	0.062500	16.00000	0.961837	-0.206350	7.741693	7.741693	1.000000
2	0.125000	8.00000	-0.011731	-0.031356	0.008967	0.008967	0.000000
3	0.187500	5.33333	0.328017	0.492087	2.797956	2.797956	
4	0.250000	4.00000	-0.191044	-0.187832	0.574229	0.574229	
5	0.312500	3.20000	-0.137155	-0.087690	0.212007	0.212007	
6	0.375000	2.66667	-0.124108	-0.048018	0.141668	0.141668	
7	0.437500	2.28571	-0.119377	-0.021949	0.117860	0.117860	
8	0.500000	2.00000	-0.118257	0.000000	0.111877	0.111877	

Let's now review the columns. Clearly, the largest cosine coefficient can be found for the .0625 frequency. A smaller sine coefficient can be found at frequency = .1875. Thus, clearly the two sine/cosine frequencies that were inserted into the example data file are reflected in the previous table.

Periodogram

The sine and cosine functions are mutually independent (or orthogonal); thus we may sum the squared coefficients for each frequency to obtain the periodogram. Specifically, the periodogram values above are computed as:

$$P_k = \text{sine coefficient}_k^2 + \text{cosine coefficient}_k^2 * N/2$$

where P_k is the periodogram value at frequency v_k and N is the overall length of the series. The periodogram values can be interpreted in terms of variance (sums of squares) of the data at the respective frequency or period. Customarily, the periodogram values are plotted against the frequencies or periods.

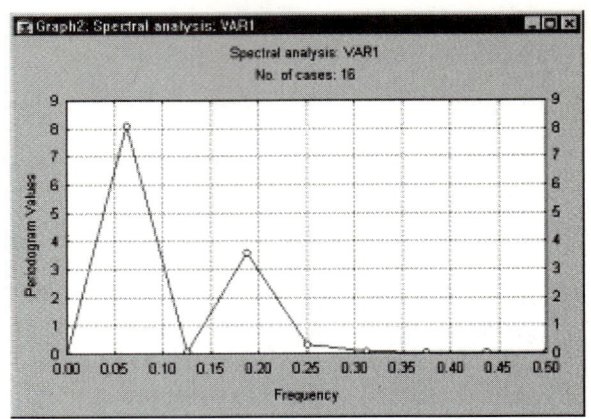

The Problem of Leakage

In the example above, a sine function with a frequency of 0.2 was inserted into the series. However, because of the length of the series (16), none of the frequencies reported exactly hits on that frequency. In practice, what often happens in those cases is that the respective frequency will leak into adjacent frequencies. For example, you may find large periodogram values for two adjacent frequencies, when, in fact, there is only one strong underlying sine or cosine function at a frequency that falls in-between those implied by the length of the series. There are three ways in which you can approach the problem of leakage:

- By padding the series, you can apply a finer frequency roster to the data,

- By tapering the series prior to the analysis, you can reduce leakage, or

- By smoothing the periodogram, you can identify the general frequency regions or (spectral densities) that significantly contribute to the cyclical behavior of the series.

See the following paragraphs for descriptions of each of these approaches.

Padding the Time Series

Because the frequency values are computed as N/t (the number of units of times) you can simply pad the series with a constant (e.g., zeros) and thereby introduce smaller increments in the frequency values. In a sense, padding enables you to apply a finer roster to the data. In fact, if we padded the example data file described in the example above with ten zeros, the results would not change, that is, the largest periodogram peaks would still occur at the frequency values closest to .0625 and .2. (Padding is also often desirable for computational efficiency reasons; see below.)

Tapering

The so-called process of split-cosine-bell tapering is a recommended transformation of the series prior to the spectrum analysis. It usually leads to a reduction of leakage in the periodogram. The rationale for this transformation is explained in detail in Bloomfield (1976, p. 80-94). In essence, a proportion (p) of the data at the beginning and at the end of the series is transformed via multiplication by the weights:

$w_t = 0.5*\{1-\cos[\pi*(t-0.5)/m]\}$ (for t=0 to m-1)

$w_t = 0.5*\{1-\cos[\pi*(N-t+0.5)/m]\}$ (for t=N-m to N-1)

where m is chosen so that $2*m/N$ is equal to the proportion of data to be tapered (p).

Data Windows and Spectral Density Estimates

In practice, when analyzing actual data, it is usually not of crucial importance to identify exactly the frequencies for particular underlying sine or cosine functions. Rather, because the periodogram values are subject to substantial random fluctuation, you are faced with the problem of very many chaotic periodogram spikes. In this case, you want to find the frequencies with the greatest spectral densities, that is, the frequency regions, consisting of many adjacent frequencies, that contribute most to the overall periodic behavior of the series. This can be accomplished by smoothing the periodogram values via a weighted moving average transformation. Suppose the moving average window is of width m (which must be an odd number); the following are the most commonly used smoothers (note: $p = (m-1)/2$).

Daniell (or equal weight) window. The Daniell window (Daniell 1946) amounts to a simple (equal weight) moving average transformation of the periodogram values, that is, each spectral density estimate is computed as the mean of the $m/2$ preceding and subsequent periodogram values.

Tukey window. In the Tukey (Blackman and Tukey, 1958) or Tukey-Hanning window (named after Julius Von Hann), for each frequency, the weights for the weighted moving average of the periodogram values are computed as:

$w_j = 0.5 + 0.5*\cos(\pi*j/p)$ (for j=0 to p)
$w_{-j} = w_j$ (for j $\neq$ 0)

Hamming window. In the Hamming (named after R. W. Hamming) window or Tukey-Hamming window (Blackman and Tukey, 1958), for each frequency, the weights for the weighted moving average of the periodogram values are computed as:

$w_j = 0.54 + 0.46*\cos(\pi*j/p)$ (for j=0 to p)
$w_{-j} = w_j$ (for j $\neq$ 0)

Parzen window. In the Parzen window (Parzen, 1961), for each frequency, the weights for the weighted moving average of the periodogram values are computed as:

$$w_j = 1-6*(j/p)^2 + 6*(j/p)^3 \quad \text{(for } j = 0 \text{ to } p/2)$$
$$w_j = 2*(1-j/p)^3 \quad \text{(for } j = p/2 + 1 \text{ to } p)$$
$$w_{-j} = w_j \quad \text{(for } j \neq 0)$$

Bartlett window. In the Bartlett window (Bartlett, 1950) the weights are computed as:

$$w_j = 1-(j/p) \text{ (for } j = 0 \text{ to } p)$$
$$w_{-j} = w_j \quad \text{(for } j \neq 0)$$

With the exception of the Daniell window, all weight functions will assign the greatest weight to the observation being smoothed in the center of the window, and increasingly smaller weights to values that are further away from the center. In many cases, all of these data windows will produce very similar results

Preparing the Data for Analysis

Let's now consider a few other practical points in spectrum analysis. Usually, you want to subtract the mean from the series, and detrend the series (so that it is stationary) prior to the analysis. Otherwise the periodogram and density spectrum will mostly be overwhelmed by a very large value for the first cosine coefficient (for frequency 0.0). In a sense, the mean is a cycle of frequency 0 (zero) per unit time; that is, it is a constant. Similarly, a trend is also of little interest when you want to uncover the periodicities in the series. In fact, both of those potentially strong effects may mask the more interesting periodicities in the data, and thus both the mean and the trend (linear) should be removed from the series prior to the analysis. Sometimes, it is also useful to smooth the data prior to the analysis, in order to tame the random noise that may obscure meaningful periodic cycles in the periodogram.

Results when No Periodicity in the Series Exists

Finally, what if there are no recurring cycles in the data, that is, if each observation is completely independent of all other observations? If the distribution of the observations follows the normal distribution, such a time series is also referred to as a white noise series (such as the white noise you hear on the radio when tuned in between stations). A white noise input series will result in periodogram values that follow an exponential distribution. Thus, by testing the distribution of periodogram values against the exponential distribution, you can test whether the input series is different from a white noise series. In addition, you can also specify to compute the Kolmogorov-Smirnov one-sample d statistic (see also, Chapter 29 – *Nonparametric Statistics* for more details).

Testing for white noise in certain frequency bands. Note that you can also plot the periodogram values for a particular frequency range only. Again, if the input is a white noise series with respect to those frequencies (i.e., it there are no significant periodic cycles of those frequencies), then the distribution of the periodogram values should again follow an exponential distribution.

Fast Fourier Transforms (FFT)

Overview

The interpretation of the results of spectrum analysis is discussed in *Basic Notation and Principles*, page 512, however, we have not described how it is done computationally. Up until the mid-1960s the standard way of performing the spectrum decomposition was to use explicit formulae to solve for the sine and cosine parameters. The computations involved required at least N^2 (complex) multiplications. Thus, even with today's high-speed computers, it would be very time consuming to analyze even small time series (e.g., 8,000 observations would result in at least 64 million multiplications).

The time requirements changed drastically with the development of the so-called Fast Fourier Transformation algorithm, or FFT. In the mid-1960s, J.W. Cooley and J.W. Tukey (1965) popularized this algorithm, which, in retrospect, had in fact been discovered independently by various individuals. Various refinements and improvements of this algorithm can be found in Monro (1975) and Monro and Branch (1976). Readers interested in the computational details of this algorithm may refer to any of the texts cited in the overview. Suffice it to say that via the FFT algorithm, the time to perform a spectral analysis is proportional to $N*\log2(N)$ – a huge improvement.

However, a draw-back of the standard FFT algorithm is that the number of cases in the series must be equal to a power of 2 (i.e., 16, 64, 128, 256, ...). Usually, this necessitated padding of the series, which, as described above, will in most cases not change the characteristic peaks of the periodogram or the spectral density estimates. In cases, however, where the time units are meaningful, such padding may make the interpretation of results more cumbersome.

Computation of FFT in Time Series

The implementation of the FFT algorithm allows you to take full advantage of the savings afforded by this algorithm. On most standard computers, series with over 100,000 cases can easily be analyzed. However, there are a few

things to remember when analyzing series of that size.

As mentioned above, the standard (and most efficient) FFT algorithm requires that the length of the input series is equal to a power of 2. If this is not the case, additional computations have to be performed. It will use the simple explicit computational formulas as long as the input series is relatively small, and the number of computations can be performed in a relatively short amount of time. For long time series, in order to still utilize the FFT algorithm, an implementation of the general approach described by Monro and Branch (1976) is used. This method requires significantly more storage space, however, series of considerable length can still be analyzed very quickly, even if the number of observations is not equal to a power of 2.

For time series of lengths not equal to a power of 2, we want to make the following recommendations: If the input series is small to moderately sized (e.g., only a few thousand cases), do not worry. The analysis will typically only take a few seconds anyway. In order to analyze moderately large and large series (e.g., more than 100,000 cases), pad the series to a power of 2 and then taper the series during the exploratory part of your data analysis.

VARIANCE COMPONENTS AND MIXED MODEL ANOVA/ANCOVA

This chapter describes a comprehensive set of techniques for analyzing research designs that include random effects; however, these techniques are also well suited for analyzing large main effect designs (e.g., designs with more than 200 levels per factor), designs with many factors where the higher order interactions are not of interest, and analyses involving case weights.

There are several chapters in this textbook that discuss analysis of variance for factorial or specialized designs. For a discussion of these chapters and the types of designs for which they are best suited refer to *Methods for Analysis of Variance* in Chapter 3 – *ANOVA/MANOVA*. Note, however, that Chapter 18 – *General Linear Models* describes how to analyze designs with any number and type of between effects and compute ANOVA-based variance component estimates for any effect in a mixed-model analysis.

Overview

Experimentation is sometimes mistakenly thought to involve only the manipulation of levels of the independent variables and the observation of subsequent responses on the dependent variables. Independent variables whose levels are determined or set by the experimenter are said to have fixed effects. There is a second class of effects, however, which is often of great interest to the researcher. Random effects are classification effects where the levels of the effects are assumed to be randomly selected from an infinite population of possible levels. Many independent variables of research interest are not fully amenable to experimental manipulation, but nevertheless can be studied by considering them to have random effects. For example, the genetic makeup of individual members of a species cannot at present be (fully) experimentally manipulated, yet it is of great interest to the geneticist to assess the genetic contribution to individual variation on outcomes such as health, behavioral characteristics, and the like. As another example, a manufacturer might want to estimate the components of variation in the characteristics of a product for a random sample of machines operated by a random sample of operators. The statistical analysis of random effects is accomplished by using the random effect model, if all of the independent variables are assumed to have random effects, or by using the mixed model, if some of the independent variables are assumed to have random effects and other independent variables are assumed to have fixed effects.

Properties of random effects. To illustrate some of the properties of random effects, suppose you collect data on the amount of insect damage done to different varieties of wheat. It is impractical to study insect damage

for every possible variety of wheat, so to conduct the experiment, you randomly select four varieties of wheat to study. Plant damage is rated for up to a maximum of four plots per variety. Ratings are on a 0 (no damage) to 10 (great damage) scale. The following data for this example are presented in Milliken and Johnson (1992, p. 237).

	VARIETY	PLOT	DAMAGE
1	105	1	3.90
2	105	2	4.05
3	105	3	4.25
4	106	4	3.60
5	106	5	4.20
6	106	6	4.05
7	106	7	3.85
8	107	8	4.15
9	107	9	4.60
10	107	10	4.15
11	107	11	4.40
12	108	12	3.35
13	108	13	3.80

Data: Wheat [3v by 13c] — Milliken & Johnson (1992, p.237)

To determine the components of variation in resistance to insect damage for *Variety* and *Plot*, an ANOVA can first be performed. Perhaps surprisingly, in the ANOVA, *Variety* can be treated as a fixed or as a random factor without influencing the results (provided that *Type I Sums of squares* are used and that *Variety* is always entered first in the model). The spreadsheet in the next illustration shows the ANOVA results of a mixed model analysis treating *Variety* as a fixed effect and ignoring *Plot*, i.e., treating the plot-to-plot variation as a measure of random error.

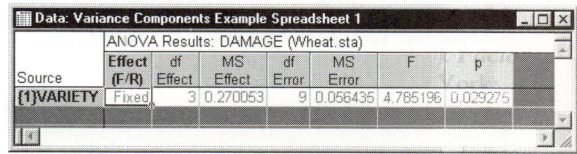

Source	Effect (F/R)	df Effect	MS Effect	df Error	MS Error	F	p
(1)VARIETY	Fixed	3	0.270053	9	0.056435	4.785196	0.029275

Data: Variance Components Example Spreadsheet 1 — ANOVA Results: DAMAGE (Wheat.sta)

Another way to perform the same mixed model analysis is to treat *Variety* as a fixed effect and *Plot* as a random effect. The spreadsheet in the next illustration shows the ANOVA results for this mixed model analysis.

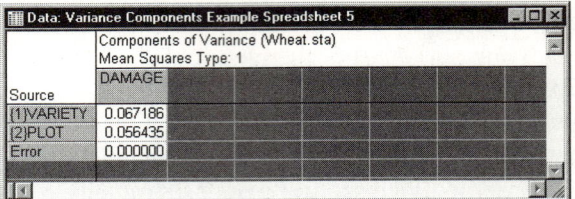

The spreadsheet in the next illustration shows the ANOVA results for a random effect model treating *Plot* as a random effect nested within *Variety*, which is also treated as a random effect.

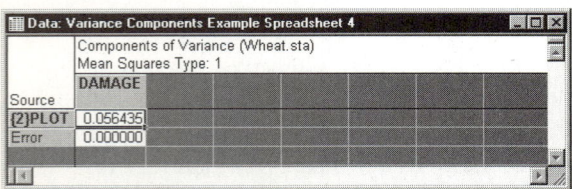

As can be seen, the tests of significance for the *Variety* effect are identical in all three analyses (and in fact, there are even more ways to produce the same result). When components of variance are estimated, however, the difference between the mixed model (treating *Variety* as fixed) and the random model (treating *Variety* as random) becomes apparent. The spreadsheet in the next illustration shows the variance component estimates for the mixed model treating *Variety* as a fixed effect.

The spreadsheet in the next illustration shows the variance component estimates for the random effects model treating *Variety* and *Plot* as random effects.

As can be seen, the difference in the two sets of estimates is that a variance component is estimated for *Variety* only when it is considered to be a random effect. This reflects the basic distinction between fixed and random effects. The variation in the levels of random factors is assumed to be representative of the variation of the whole population of possible levels. Thus, variation in the levels of a random factor can be used to estimate the population variation. Even more importantly, covariation between the levels of a random factor and responses on a dependent variable can be used to estimate the population component of variance in the dependent variable attributable to the random factor. The variation in the levels of fixed factors is instead considered to be arbitrarily determined by the experimenter (i.e., the experimenter can make the levels of a fixed factor vary as little or as much as desired). Thus, the variation of a fixed factor cannot be used to estimate its population variance, nor can the population covariance with the dependent variable be meaningfully estimated. With this basic distinction between fixed effects and random effects in mind, we now can look more closely at the properties of variance components.

Estimation of Variance Components (Technical Overview)

The basic goal of variance component estimation is to estimate the population covariation between random factors and the dependent variable. Depending on the method used to estimate variance components, the population variances of the random factors can also be estimated, and significance tests can be performed to test whether the population covariation between the random factors and the dependent variable are nonzero.

Estimating the variation of random factors.
The ANOVA method provides an integrative approach to estimating variance components, because ANOVA techniques can be used to estimate the variance of random factors, to estimate the components of variance in the dependent variable attributable to the random factors, and to test whether the variance components differ significantly from zero. The ANOVA method for estimating the variance of the random factors begins by constructing the Sums of squares and cross products (SSCP) matrix for the independent variables. The sums of squares and cross products for the random effects are then residualized on the fixed effects, leaving the random effects independent of the fixed effects, as required in the mixed model (see, for example, Searle, Casella, & McCulloch, 1992). The residualized Sums of squares and cross products for each random factor are then divided by their degrees of freedom to produce the coefficients in the Expected mean squares matrix. Nonzero off-diagonal coefficients for the random effects in this matrix indicate confounding, which must be taken into account when estimating the population variance for each factor. For the *wheat.sta* data, treating both *Variety* and *Plot* as random effects, the coefficients in the Expected mean squares matrix show that the two factors are at least somewhat confounded. The Expected Mean Squares spreadsheet is shown in the next illustration.

Source	Effect (F/R)	VARIETY	PLOT	Error		
{1} VARIETY	Random	3.179487	1.000000	1.000000		
{2} PLOT	Random		1.000000	1.000000		
Error				1.000000		

The coefficients in the Expected Mean Squares matrix are used to estimate the population variation of the random effects by equating their variances to their expected mean squares. For example, the estimated population variance for *Variety* using *Type I Sums of squares* would be 3.179487 times the *Mean square* for *Variety* plus 1 times the *Mean square* for *Plot* plus 1 times the *Mean square* for *Error*.

The ANOVA method provides an integrative approach to estimating variance components, but it is not without problems (i.e., ANOVA estimates of variance components are generally biased, and can be negative, even though variances, by definition, must be either zero or positive). An alternative to ANOVA estimation is provided by maximum likelihood estimation. Maximum likelihood methods for estimating variance components are based on quadratic forms, and typically, but not always, require iteration to find a solution. Perhaps the simplest form of maximum likelihood estimation is MIVQUE(0) estimation. MIVQUE(0) produces Minimum Variance Quadratic Unbiased Estimators (i.e., MIVQUE). In MIVQUE(0) estimation, there is no weighting of the random effects (thus the 0 [zero] after MIVQUE), so an iterative solution for estimating variance components is not required. MIVQUE(0) estimation begins by constructing the Quadratic sums of squares (SSQ) matrix. The elements for the random effects in the SSQ matrix can most simply be described as the sums of squares of the sums of squares and cross products for each random effect in the model (after residualization on the fixed effects). The elements of this matrix provide coefficients, similar to the elements of the Expected Mean Squares matrix, which are used to estimate the covariances among the random factors and the dependent variable. The SSQ matrix for the *wheat.sta* data is shown in the next illustration. Note that the nonzero off-diagonal element for *Variety* and *Plot* again shows that the two random factors are at least somewhat confounded.

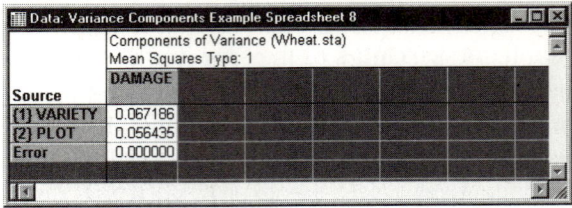

Restricted Maximum Likelihood (REML) and Maximum Likelihood (ML) variance component estimation methods are closely related to MIVQUE(0). In fact, in the program, REML and ML use MIVQUE(0) estimates as start values for an iterative solution for the variance components, so the elements of the SSQ matrix serve as initial estimates of the covariances among the random factors and the dependent variable for both REML and ML.

Estimating Components of Variation. For ANOVA methods for estimating variance components, a solution is found for the system of equations relating the estimated population variances and covariances among the random factors to the estimated population covariances between the random factors and the dependent variable. The solution then defines the variance components. The spreadsheet in the next illustration shows the Type I Sums of squares estimates of the variance components for the *wheat.sta* data.

MIVQUE(0) variance components are estimated by inverting the partition of the SSQ matrix that does not include the dependent variable (or finding the generalized inverse, for singular matrices), and postmultiplying the inverse by the dependent variable column vector. This amounts to solving the system of equations that relates the dependent variable to the random independent variables, taking into account the covariation among the independent variables. The MIVQUE(0) estimates for the *wheat.sta* data are listed in the spreadsheet in the next illustration.

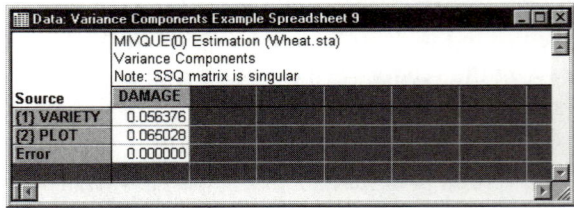

REML and ML variance components are estimated by iteratively optimizing the parameter estimates for the effects in the model. REML differs from ML in that the likelihood of the data is maximized only for the random effects, thus REML is a restricted solution. In both REML and ML estimation, an iterative solution is found for the weights for the random effects in the model that maximize the likelihood of the data. The program uses MIVQUE(0)) estimates as the start values for both REML and ML estimation, so the relation between these three techniques is close indeed. The statistical theory underlying maximum likelihood variance component estimation techniques is an advanced topic (Searle, Casella, & McCulloch, 1992, is recommended as an authoritative and comprehensive source). Implementation of maximum likelihood estimation algorithms, furthermore, is difficult (see, for example, Hemmerle & Hartley, 1973, and Jennrich & Sampson, 1976, for descriptions of these algorithms), and faulty implementation can lead to variance component estimates that lie outside the parameter space, converge prematurely to nonoptimal solutions, or give nonsensical results. Milliken and Johnson (1992) noted all of these problems with the commercial software packages they used to estimate variance components.

The basic idea behind both REML and ML estimation is to find the set of weights for the random effects in the model that minimize the negative of the natural logarithm times the likelihood of the data (the likelihood of the data can vary from zero to one, so minimizing the negative of the natural logarithm times the likelihood of the data amounts to maximizing the probability, or the likelihood, of the data). The logarithm of the REMLlikelihood and the REML variance component estimates for the *wheat.sta* data are listed in the last row of the Iteration History spreadsheet shown in the next illustration.

Data: Variance Components Example Spreadsheet 10
Iteration History (Wheat.sta)
Variable: DAMAGE

Iter.	Log LL	Error	VARIETY
0	-2.30618	0.065028	0.056376
1	-2.25253	0.057430	0.068746
2	-2.25082	0.057065	0.072813
3	-2.25081	0.057003	0.073152
4	-2.25081	0.057003	0.073155
5	-2.25081	0.057003	0.073155

The logarithm of the MLlikelihood and the ML estimates for the variance components for the *wheat.sta* data are listed in the last row of the Iteration History spreadsheet shown in the next illustration.

Data: Variance Components Example Spreadsheet 11
Iteration History (Wheat.sta)
Variable: DAMAGE

Iter.	Log LL	Error	VARIETY
0	-2.53585	0.065028	0.056376
1	-2.48382	0.057454	0.048799
2	-2.48381	0.057480	0.048586
3	-2.48381	0.057492	0.048552
4	-2.48381	0.057492	0.048552
5	-2.48381	0.057492	0.048552

As can be seen, the estimates of the variance components for the different methods are quite similar. In general, components of variance using different estimation methods tend to agree fairly well (see, for example, Swallow & Monahan, 1984).

Testing the significance of variance components. When maximum likelihood estimation techniques are used, standard linear model significance testing techniques may not be applicable. ANOVA techniques such as decomposing sums of squares and testing the significance of effects by taking ratios of mean squares are appropriate for linear methods of estimation, but generally are not appropriate for quadratic methods of estimation. When ANOVA methods are used for estimation, standard significance testing techniques can be employed, with the exception that any confounding among random effects must be taken into account.

To test the significance of effects in mixed or random models, error terms must be constructed that contain all the same sources of random variation except for the variation of the respective effect of interest. This is done using Satterthwaite's method of denominator synthesis (Satterthwaite, 1946), which finds the linear combinations of sources of random variation that serve as appropriate error terms for testing the significance of the respective effect of interest. The spreadsheet in the next illustration shows the coefficients used to construct these linear combinations for testing the *Variety* and *Plot* effects.

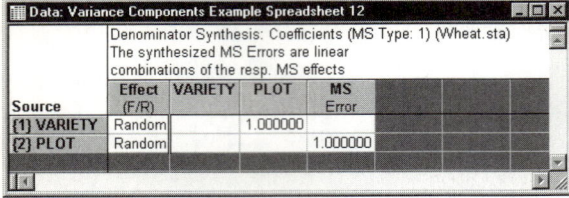

The coefficients show that the mean square for *Variety* should be tested against the mean square for *Plot*, and that the mean square for *Plot* should be tested against the mean square for *Error*. Referring back to the Expected mean squares spreadsheet, it is clear that the denominator synthesis has identified appropriate error terms for testing the *Variety* and *Plot* effects. Although this is a simple example, in more complex analyses with various degrees of confounding among the random effects, the denominator synthesis can identify appropriate error terms for testing the random effects that would not be readily apparent.

To perform the tests of significance of the random effects, ratios of appropriate Mean squares are formed to compute *F* statistics and *p*-levels for each effect. Note that in complex analyses the degrees of freedom for random effects can be fractional rather than integer values, indicating that fractions of sources of variation were used in synthesizing appropriate error terms for testing the random effects. The spreadsheet displaying the results of the ANOVA for the *Variety* and *Plot* random effects is shown in the next illustration. Note that for this simple design the results are identical to the results presented earlier in the spreadsheet for the ANOVA treating *Plot* as a random effect nested within *Variety*.

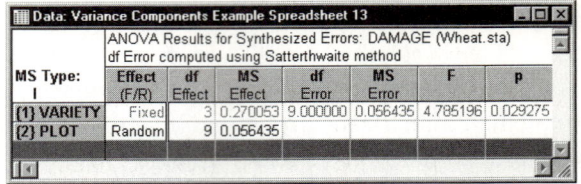

As shown in the spreadsheet, the *Variety* effect is found to be significant at $p < .05$, but as would be expected, the *Plot* effect cannot be tested for significance because plots served as the basic unit of analysis. If data on samples of plants taken within plots were available, a test of the significance of the *Plot* effect could be constructed.

Appropriate tests of significance for MIVQUE(0) variance component estimates generally cannot be constructed, except in special cases (see Searle, Casella, & McCulloch, 1992). Asymptotic (large sample) tests of significance of REML and ML variance component estimates, however, can be constructed for the parameter estimates from the final iteration of the solution. The spreadsheet in the next illustration shows the asymptotic (large sample) tests of significance for the REML estimates for the *wheat.sta* data.

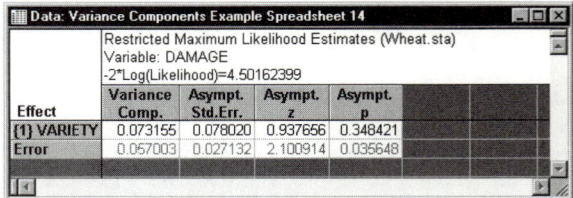

The spreadsheet in the next illustration shows the asymptotic (large sample) tests of significance for the *ML* estimates for the *wheat.sta* data.

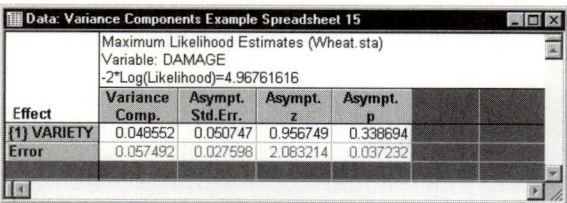

Maximum Likelihood Estimates (Wheat.sta)
Variable: DAMAGE
-2*Log(Likelihood)=4.96761616

Effect	Variance Comp.	Asympt. Std.Err.	Asympt. z	Asympt. p		
{1} VARIETY	0.048552	0.050747	0.956749	0.338694		
Error	0.057492	0.027598	2.083214	0.037232		

It should be emphasized that the asymptotic tests of significance for REML and ML variance component estimates are based on large sample sizes, which certainly is not the case for the *wheat.sta* data. For this data set, the tests of significance from both analyses agree in suggesting that the *Variety* variance component does not differ significantly from zero. For basic information on ANOVA in linear models, see Chapter 1 – *Elementary Concepts in Statistics*.

Estimating the population intraclass correlation. Note that if the variance component estimates for the random effects in the model are divided by the sum of all components (including the error component), the resulting percentages are population intraclass correlation coefficients for the respective effects.

GRAPHICAL TECHNIQUES

Graphical data visualization methods can identify relations, trends, and biases "hidden" in unstructured data sets. Brushing, perhaps the most common and historically first widely used technique explicitly identified as graphical exploratory data analysis, is an interactive method allowing you to select on-screen specific data points or subsets of data and identify their (e.g., common) characteristics, or to examine their effects on relations between relevant variables. Those relations between variables can be visualized by fitted functions (e.g., 2D lines or 3D surfaces) and their confidence intervals, e.g., to examine changes in those functions by interactively (temporarily) removing or adding specific subsets of data.

Other graphical exploratory analytic techniques include function fitting and plotting; data smoothing, overlaying and merging of multiple displays; categorizing data; splitting/merging subsets of data in graphs; aggregating data in graphs; identifying and marking subsets of data that meet specific conditions; shading; plotting confidence intervals and confidence areas (ellipses); generating tessellations, spectral planes, integrated layered compressions, and projected contours; data image reduction techniques; interactive (and continuous) rotation with animated stratification (cross-sections) of 3D displays; and selective highlighting of specific series and blocks of data.

For a brief overview of the following graph types, see the *Statistical Glossary*, page 555.

2D Graphs

- Bar/Column
- Bar Deviation
- Bar Left Y
- Bar Right Y
- Bar Top
- Bar X
- Box
- Histograms
- Line
- Pie Charts
- Probability
- Probability, Detrended
- Probability, Half-Normal
- Probability-Probability
- Quantile-Quantile
- Range
- Scatterplots
- Sequential/Stacked
- Voronoi Scatterplot

3D XYZ Graphs

- Contour
- Deviation
- Scatterplots
- Space
- Spectral
- Trace

3D Sequential Graphs

- Bivariate Histograms
- Box
- Contour
- Contour/Discrete Raw Data
- Range
- Surface
- Surface Raw Data

2D Categorized Graphs

- Probability Detrended
- Probability Half-Normal
- Probability Normal
- Probability-Probability
- Quantile-Quantile

3D Categorized Graphs

- Contour
- Deviation
- Scatterplots
- Space
- Spectral
- Surface
- Ternary

Icon Graphs

- Chernoff Faces
- Columns
- Lines
- Pies
- Polygons
- Profiles
- Stars
- Sun Rays

Matrix Graphs

- Columns
- Lines
- Scatterplot

Categorized Graphs

One of the most important, general, and also powerful analytic methods involves dividing (splitting) the data set into categories in order compare the patterns of data between the resulting subsets. This common technique is known under a variety of terms (such as breaking down, grouping, categorizing, splitting, slicing, drilling down, or conditioning) and it is used both in exploratory data analyses and hypothesis testing. For example: A positive relation between the age and the risk of a heart attack may be different in males and females (it may be stronger in males). A promising relation between taking a drug and a decrease of the cholesterol level may be present only in women with a low blood pressure and only in their thirties and forties. The process capability indices or capability histograms can be different for periods of time supervised by different operators. The regression slopes can be different in different experimental groups.

There are many computational techniques that capitalize on grouping and that are designed to quantify the differences that the grouping will reveal (e.g., ANOVA/MANOVA). However, graphical techniques (such as categorized graphs discussed in this section) offer unique advantages that cannot be substituted by any computational method alone: they can reveal patterns that cannot be easily quantified (e.g., complex interactions, exceptions, anomalies) and they provide unique, multidimensional, global analytic perspectives to explore or mine the data.

What are Categorized Graphs?

Categorized graphs (the term first used in *STATISTICA* software by StatSoft in 1990; also recently called Trellis graphs, by Becker,

Cleveland, and Clark, at Bell Labs) produce a series of 2D, 3D, ternary, or nD graphs (such as histograms, scatterplots, line plots, surface plots, ternary scatterplots, etc.), one for each selected category of cases (i.e., subset of cases), for example, respondents from New York, Chicago, Dallas, etc. These component graphs are placed sequentially in one display, allowing for comparisons between the patterns of data shown in graphs for each of the specified groups (e.g., cities).

A variety of methods can be used to select the subsets; the simplest of them is using a categorical variable (e.g., a variable *City*, with three values *New York*, *Chicago*, and *Dallas*). For example, the following graph shows histograms of a variable representing self-reported stress levels in each of the three cities.

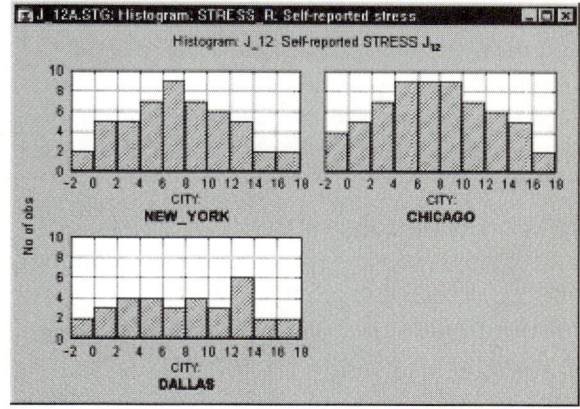

You could conclude that the data suggest that people who live in Dallas are less likely to report being stressed, while the patterns (distributions) of stress reporting in New York and Chicago are quite similar.

Categorized graphs in some software systems also support two-way or multi-way categorizations, where not one criterion (e.g., *City*) but two or more criteria (e.g., *City* and *Time* of the day) are used to create the subsets.

Two-way categorized graphs can be thought of as crosstabulations of graphs, where each component graph represents a cross-section of one level of one grouping variable (e.g., *City*) and one level of the other grouping variable (e.g., *Time*).

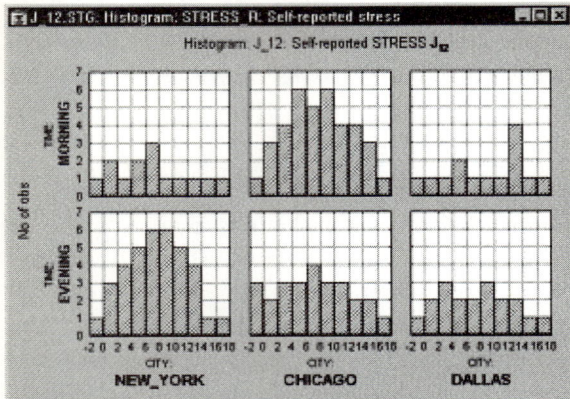

Adding this second factor reveals that the patterns of stress reporting in *New York* and *Chicago* are actually quite different when the *Time* of questioning is taken into consideration, whereas the *Time* factor makes little difference in *Dallas*.

Categorized graphs vs. matrix graphs.
Matrix graphs also produce displays containing multiple component graphs; however, each of those component graphs are (or can be) based on the same set of cases and the graphs are generated for all combinations of variables from one or two lists. Categorized graphs require a selection of variables that normally would be selected for noncategorized graphs of the respective type (e.g., two variables for a scatterplot). However, in categorized plots, you also need to specify at least one grouping variable (or some criteria to be used for sorting the observations into the categories) that contains information on group membership of each case (e.g., *Chicago*, *Dallas*). That grouping variable will not be included in the graph directly (i.e., it will not be plotted) but it will serve as a criterion for dividing all analyzed cases into separate graphs. As shown in the previous illustration, one graph will be created for each group (category) identified by the grouping variable.

Common vs. independent scaling. Each individual category graph may be scaled according to its own range of values (independent scaling),

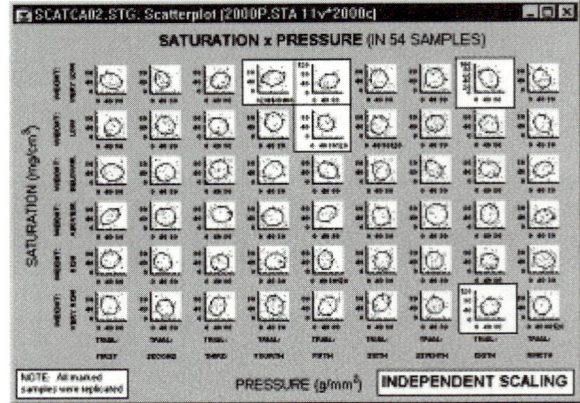

or all graphs may be scaled to a common scale wide enough to accommodate all values in all of the category graphs.

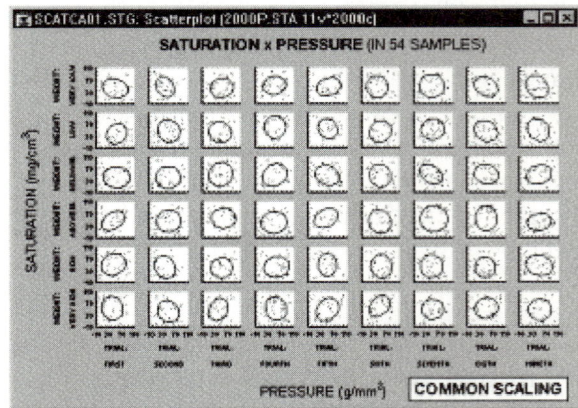

Common scaling enables the analyst to make comparisons of ranges and distributions of values among categories. However, if the

ranges of values in graph categories are considerably different (causing a very wide common scale), some of the graphs may be difficult to examine. The use of independent scaling may make it easier to spot trends and specific patterns within categories, but it may be more difficult to make comparisons of ranges of values among categories.

Categorization Methods

There are five general methods of categorization of values, and they will be reviewed briefly in this section: integer mode, categories, boundaries, codes, and multiple subsets. Note that the same methods of categorization can be used to categorize cases into component graphs and to categorize cases within component graphs (e.g., in histograms or box plots).

Integer mode. When you use integer mode, integer values of the selected grouping variable will be used to define the categories, and one graph will be created for all cases that belong each category (defined by those integer values). If the selected grouping variable contains non-integer values, the software will usually truncate each encountered value of the selected grouping variable to an integer value.

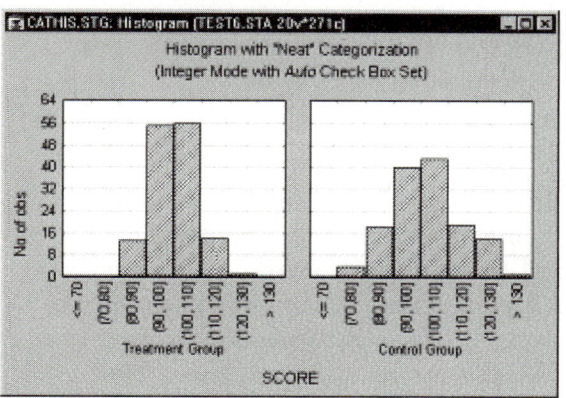

Categories. With this mode of categorization, you will specify the number of categories that

you want to use. The entire range of values of the selected grouping variable (from minimum to maximum) will be divided into the specified number of equal length intervals.

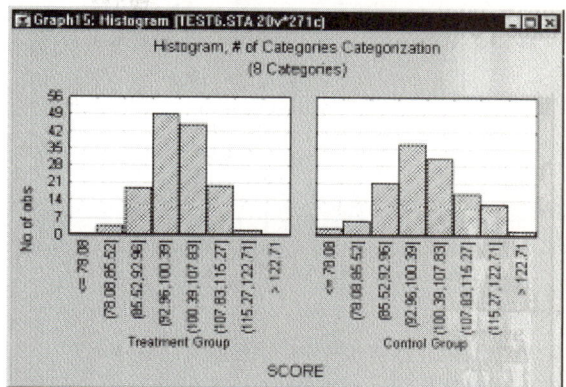

Boundaries. The boundaries method will also create interval categorization, however, the intervals can be of arbitrary (e.g., uneven) width as defined by custom interval boundaries (for example, "less than -10," "greater than or equal to -10 but less than 0," "greater than or equal to 0 but less than 10," and "equal to or greater than 10").

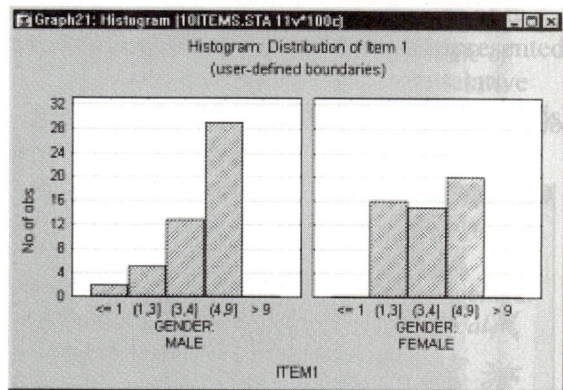

Codes. Use this method if the selected grouping variable contains codes (i.e., specific, meaningful values such as *Male*, *Female*) from which you want to specify the categories.

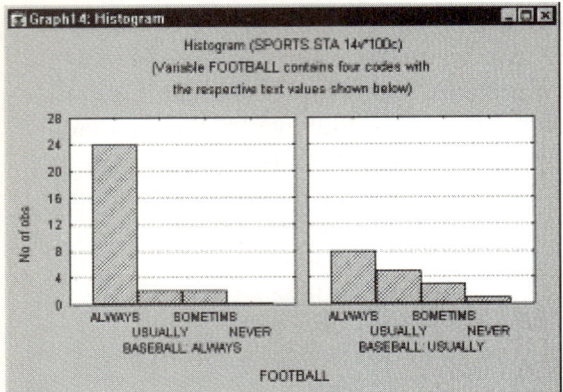

Multiple subsets. With this method, you can custom define the categories and use more than one variable to define the category. In other words, categorizations based on multiple subset definitions of categories may not represent distributions of specific (individual) variables but distributions of frequencies of specific events defined by particular combinations of values of several variables (and defined by conditions that may involve any number of variables from the current data set). For example, you might specify six categories based on combinations of three variables *Gender*, *Age*, and *Employment*.

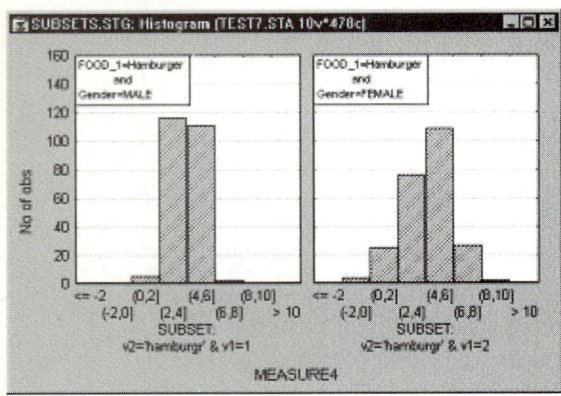

Histograms

In general, histograms are used to examine frequency distributions of values of variables.

For example, the frequency distribution plot shows which specific values or ranges of values of the examined variable are most frequent, how differentiated the values are, whether most observations are concentrated around the mean, whether the distribution is symmetrical or skewed, whether it is multimodal (i.e., has two or more peaks) or unimodal, etc. Histograms are also useful for evaluating the similarity of an observed distribution with theoretical or expected distributions.

With categorized histograms, you can produce histograms broken down by one or more categorical variables, or by any other one or more sets of logical categorization rules (see *Categorization Methods*, page 535).

There are two major reasons why frequency distributions are of interest.

- You can learn from the shape of the distribution about the nature of the examined variable (e.g., a bimodal distribution may suggest that the sample is not homogeneous and consists of observations that belong to two populations that are more or less normally distributed).

- Many statistics are based on assumptions about the distributions of analyzed variables; histograms help you to test whether those assumptions are met.

Often, the first step in the analysis of a new data set is to run histograms on all variables.

Histograms vs. breakdown. Categorized histograms provide information similar to breakdowns (e.g., mean, median, minimum, maximum, differentiation of values, etc.; see Chapter 2 – *Basic Statistics and Tables*). Although specific (numerical) descriptive statistics are easier to read in a table, the overall shape and global descriptive characteristics of a distribution are much easier to examine in a

graph. Moreover, the graph provides qualitative information about the distribution that cannot be fully represented by any single index. For example, the overall skewed distribution of income may indicate that the majority of people have an income that is much closer to the minimum than maximum of the range of income. Moreover, when broken down by gender and ethnic background, this characteristic of the income distribution may be found to be more pronounced in certain subgroups. Although this information will be contained in the index of skewness (for each sub-group), when presented in the graphical form of a histogram, the information is usually more easily recognized and remembered. The histogram may also reveal 'bumps' that may represent important facts about the specific social stratification of the investigated population or anomalies in the distribution of income in a particular group caused by a recent tax reform.

Categorized histograms and scatterplots. A useful application of the categorization methods for continuous variables is to represent the simultaneous relationships between three variables. Shown in the next illustration is a scatterplot for two variables *Load 1* and *Load 2*.

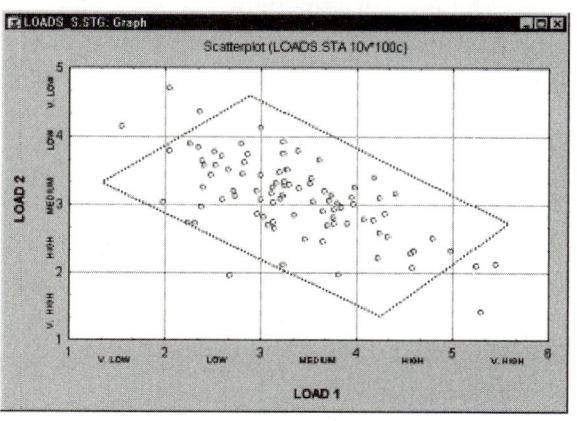

Now suppose you want to add a third variable (Output) and examine how it is distributed at different levels of the joint distribution of Load 1 and Load 2. The following graph could be produced:

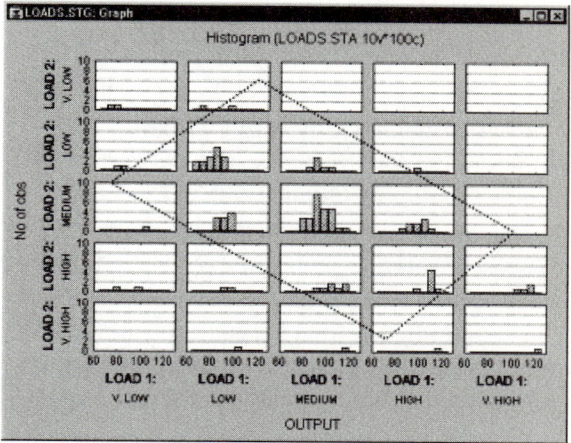

In this graph, *Load 1* and *Load 2* are both categorized into 5 intervals, and within each combination of intervals the distribution for variable *Output* is computed. Note that the box (parallelogram) encloses approximately the same observations (cases) in both graphs in the two previous illustrations.

Scatterplots

In general, two-dimensional scatterplots are used to visualize relations between two variables X and Y (e.g., weight and height). In scatterplots, individual data points are represented by point markers in two-dimensional space, where axes represent the variables. The two coordinates (X and Y) that determine the location of each point correspond to its specific values on the two variables. If the two variables are strongly related, the data points form a systematic shape (e.g., a straight line or a clear curve). If the variables are not related, the points form a round "cloud."

With a categorized scatterplot option, you can produce scatterplots categorized by one or more variables. With the multiple subsets method (see *Categorization Methods*, page 535), you can also categorize the scatterplot based on logical selection conditions that define each category or group of observations.

Categorized scatterplots offer a powerful exploratory and analytic technique for investigating relationships between two or more variables within different sub-groups.

Homogeneity of bivariate distributions (shapes of relations). Scatterplots are typically used to identify the nature of relations between two variables (e.g., blood pressure and cholesterol level), because they can provide much more information than a correlation coefficient.

For example, a lack of homogeneity in the sample from which a correlation was calculated can bias the value of the correlation. Imagine a case where a correlation coefficient is calculated from data points that came from two different experimental groups, but this fact was ignored when the correlation was calculated. Suppose the experimental manipulation in one of the groups increased the values of both correlated variables and, thus, the data from each group form a distinctive cloud in the scatterplot (as shown in the next illustration).

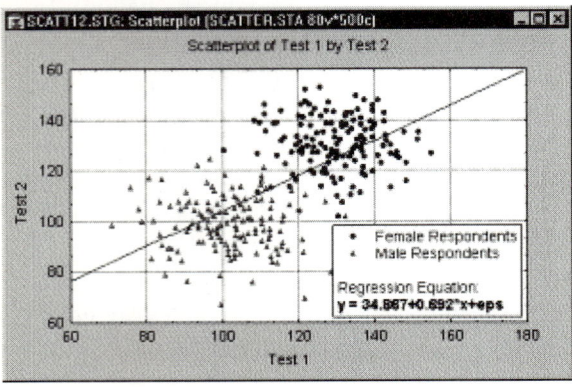

In this example, the high correlation is due entirely to the arrangement of the two groups, and it does not represent the true relation between the two variables, which is practically equal to 0 (as could be seen if you looked at each group separately).

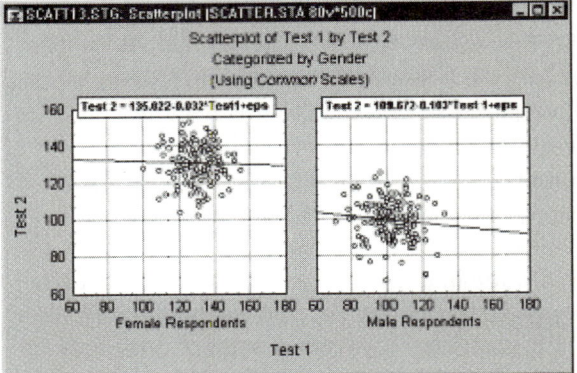

If you suspect that such a pattern may exist in your data, and you know how to identify the possible subsets of data, then producing a categorized scatterplot may yield a more accurate picture of the strength of the relationship between the X and Y variable, within each group (i.e., after controlling for group membership).

Curvilinear relations. Curvilinearity is another aspect of the relationships between variables that can be examined in scatterplots. There are no automatic or easy-to-use tests to measure curvilinear relationships between variables. The standard Pearson r coefficient measures only linear relations. Some nonparametric correlations such as the Spearman R can measure curvilinear relations, but not non-monotonous relations. Examining scatterplots enables you to identify the shape of relations, so that later an appropriate data transformation can be chosen to straighten the data or choose an appropriate nonlinear estimation equation to be fit.

For more information, refer to Chapter 2 – *Basic Statistics and Tables*, Chapter 26 – *Multiple Linear Regression*, Chapter 28 – *Nonlinear Estimation*, and Chapter 29 – *Nonparametric Statistics*.

Probability Plots

Three types of categorized probability plots are normal, half normal, and detrended. Normal probability plots provide a quick way to visually inspect to what extent the pattern of data follows a normal distribution. Via categorized probability plots, you can examine how closely the distribution of a variable follows the normal distribution in different sub-groups.

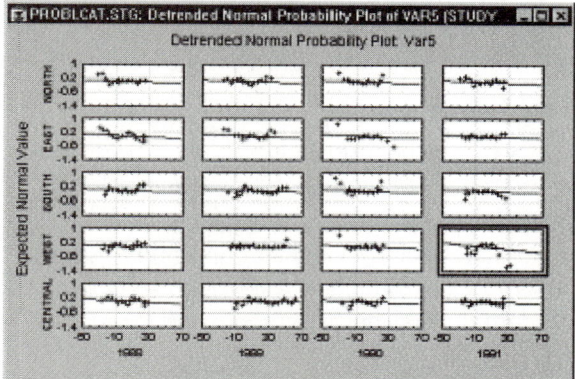

Categorized normal probability plots provide an efficient tool to examine the normality aspect of group homogeneity.

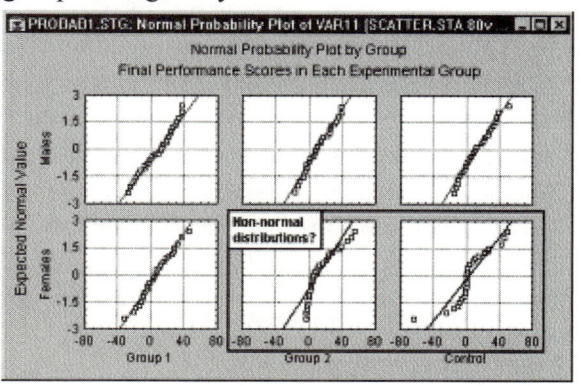

Quantile-Quantile Plots

The categorized quantile-quantile (Q-Q) plot is useful for finding the best fitting distribution within a family of distributions.

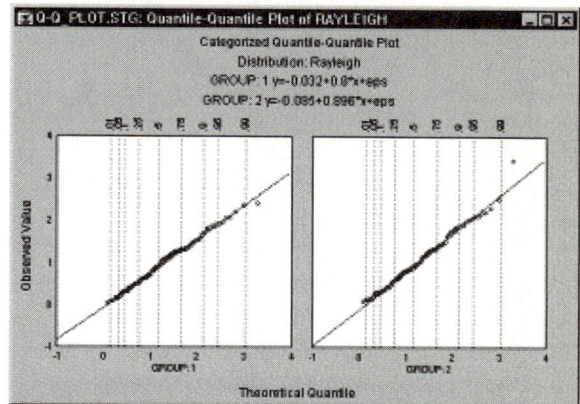

With categorized Q-Q plots, a series of Q-Q plots, one for each category of cases identified by the X or X and Y category variables (or identified by the Multiple Subset criteria, see *Categorization Methods*, page 535) are produced. Examples of distributions that are used for Q-Q plots are the exponential distribution, extreme distribution, normal, Rayleigh, *beta*, *gamma*, lognormal, and Weibull distributions.

Probability-Probability Plots

The categorized probability-probability (P-P) plot is useful for determining how well a specific theoretical distribution fits the observed data. This type of graph includes a series of P-P plots, one for each category of cases identified by the X or X and Y category variables (or identified by the multiple subset criteria, see *Categorization Methods*, page 535).

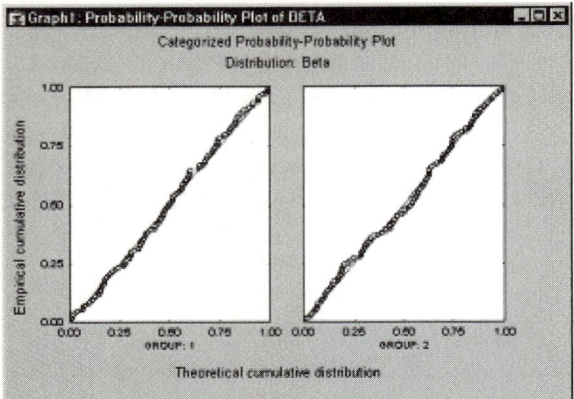

In the P-P plot, the observed cumulative distribution function (the proportion of non-missing values ≤ x) is plotted against a theoretical cumulative distribution function in order to assess the fit of the theoretical distribution to the observed data. If all points in this plot fall onto a diagonal line (with intercept 0 and slope 1), you can conclude that the theoretical cumulative distribution adequately approximates the observed distribution.

If the data points do not all fall on the diagonal line, you can use this plot to visually assess where the data do and do not follow the distribution (e.g., if the points form an S shape along the diagonal line, the data may need to be transformed in order to bring them to the desired distribution pattern).

Line Plots

In line plots, individual data points are connected by a line. Line plots provide a simple way to visually present a sequence of many values (e.g., stock market quotes over a number of days). The categorized line plots graph is useful when you want to view such data broken down (categorized) by a grouping variable (e.g., closing stock quotes on Mondays, Tuesdays, etc.) or some other logical criteria involving one

or more other variables (e.g., closing quotes only for those days when two other stocks and the Dow Jones index went up, versus all other closing quotes; see *Categorization Methods*, page 535).

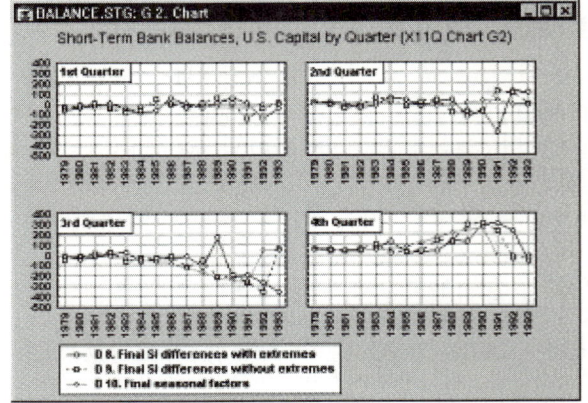

Box Plots

In box plots (the term first used by Tukey, 1970), ranges of values of a selected variable (or variables) are plotted separately for groups of cases defined by values of up to three categorical (grouping) variables, or as defined by multiple subsets categories.

The central tendency (e.g., median or mean), and range or variation statistics (e.g., quartiles, standard errors, or standard deviations) are computed for each group of cases, and the selected values are presented in one of five styles (box whiskers, whiskers, boxes, columns, or high-low close). Outlier data points can also be plotted.

For example, in the following graph, outliers (in this case, points greater or less than 1.5 times the inter-quartile range) indicate a particularly unfortunate flaw in an otherwise nearly perfect combination of factors:

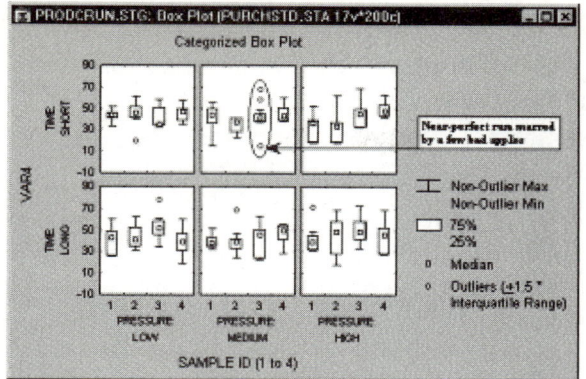

However, in the following graph, no outliers or extreme values are evident.

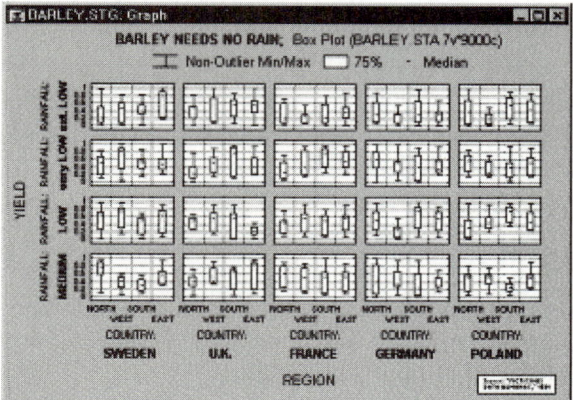

There are two typical applications for box plots: (a) showing ranges of values for individual items, cases or samples (e.g., a typical MIN-MAX plot for stocks or commodities or aggregated sequence data plots with ranges), and (b) showing variation of scores in individual groups or samples (e.g., box and whisker plots presenting the mean for each sample as a point inside the box, standard errors as the box, and standard deviations around the mean as a narrower box or a pair of whiskers).

Box plots showing variation of scores enable you to quickly evaluate and intuitively envision the strength of the relation between the grouping and dependent variable. Specifically,

assuming that the dependent variable is normally distributed, and knowing what proportion of observations fall, for example, within ±1 or ±2 standard deviations from the mean (see Chapter 1 – *Elementary Concepts in Statistics*), you can easily evaluate the results of an experiment and say that, for example, the scores in about 95% of cases in experimental group 1 belong to a different range than scores in about 95% of cases in group 2.

In addition, so-called trimmed means (this term was first used by Tukey, 1962) may be plotted by excluding a user-specified percentage of cases from the extremes (i.e., tails) of the distribution of cases.

Pie Charts

The pie chart is one of the most common graph formats used for representing proportions or values of variables. This graph enables you to produce pie charts broken down by one or more other variables (e.g., grouping variables such as gender) or categorized according to some logical selection conditions that identify Multiple Subsets (see *Categorization Methods*, page 535).

For purposes of this discussion, categorized pie charts will always be interpreted as frequency pie charts (as opposed to data pie charts). This type of pie chart (sometimes called a frequency pie chart) interprets data just as a histogram. It categorizes all values of the selected variable following the selected categorization technique and then displays the relative frequencies as pie slices of proportional sizes. Thus, these pie charts offer an alternative method to display frequency histogram data (see *Categorization Methods – Histograms*, page 536).

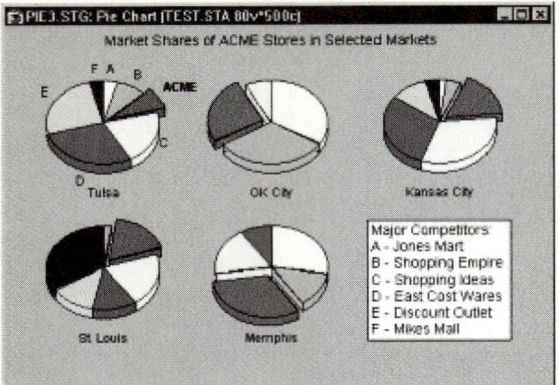

Pie scatterplots. Another useful application of categorized pie charts is to represent the relative frequency distribution of a variable at each location of the joint distribution of two other variables. Following is an example:

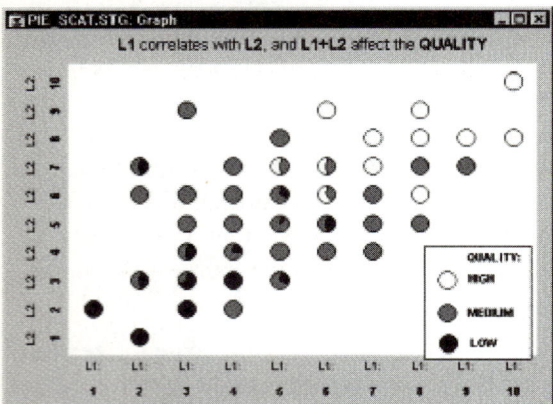

Note that pies are only drawn in places where there are data. Thus, the graph shown above takes on the appearance of a scatterplot (of variables L1 and L2), with the individual pies as point markers. However, in addition to the information contained in a simple scatterplot, each pie shows the relative distribution of a third variable at the respective location (i.e., low, medium, and high quality).

Missing/Range Data Points Plot

This plot displays the out of range and/or missing observations within a data set.

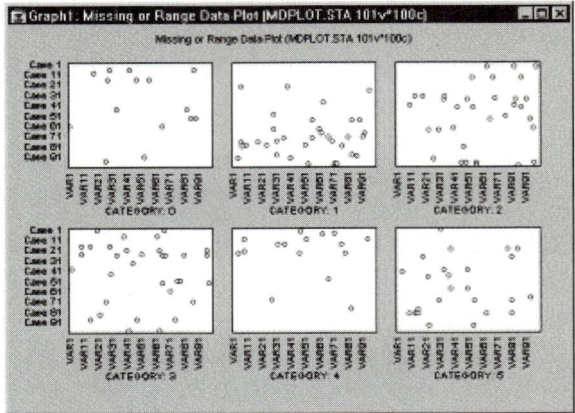

The graph is useful in exploratory data analysis to determine the extent of missing (and/or out of range) data and whether the patterns of those data occur randomly.

3D Plots

Graphical displays of data in three dimensions. Common types of 3D plots include scatterplots, surface plots, and contour plots.

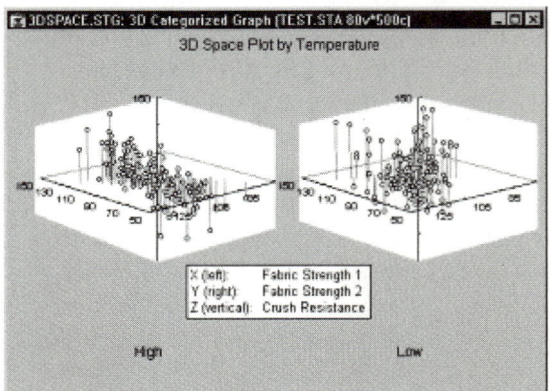

Applications. In general, 3D XYZ graphs summarize the interactive relationships between three variables. The different ways in which data can be categorized (in a categorized graph)

enable you to review those relationships contingent on some other criterion (e.g., group membership).

For example, from the categorized surface plot shown in the next illustration, you can conclude that the setting of the tolerance level in an apparatus does not affect the investigated relationship between the measurements (*Depend1*, *Depend2*, and *Height*) unless the setting is ≤ 3.

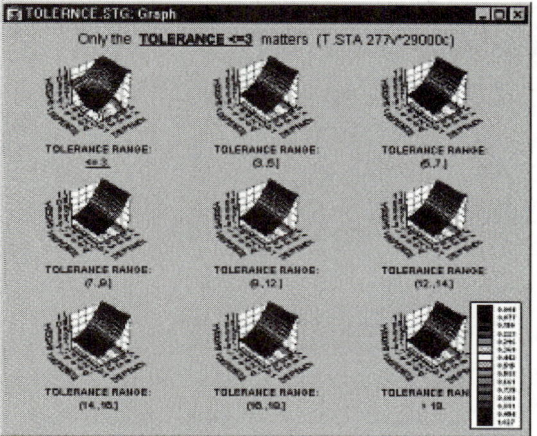

The effect is more salient when you switch to the contour plot representation.

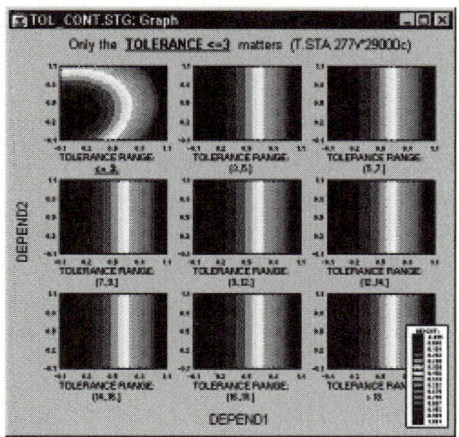

Ternary Plots

A categorized ternary plot can be used to examine relations between three or more dimensions where three of those dimensions represent components of a mixture (i.e., the relations between them is constrained such that the values of the three variables add up to the same constant for each case) for each level of a grouping variable.

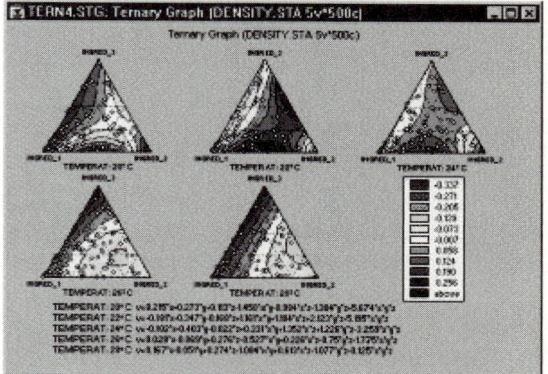

In ternary plots, the triangular coordinate systems are used to plot four (or more) variables (the components X, Y, and Z, and the responses V1, V2, etc.) in two dimensions (ternary scatterplots or contours) or three dimensions (ternary surface plots). In order to produce ternary graphs, the relative proportions of each component within each case are constrained to add up to the same value (e.g., 1).

In a categorized ternary plot, one component graph is produced for each level of the grouping variable (or user-defined subset of data) and all the component graphs are arranged in one display to allow for comparisons between the subsets of data (categories).

Applications. A typical application of this graph is when the measured response(s) from an experiment depends on the relative proportions of three components (e.g., three different chemicals) which are varied in order to determine an optimal combination of those components (e.g., in mixture designs). This type of graph can also be used for other applications where relations between constrained variables need to be compared across categories or subsets of data.

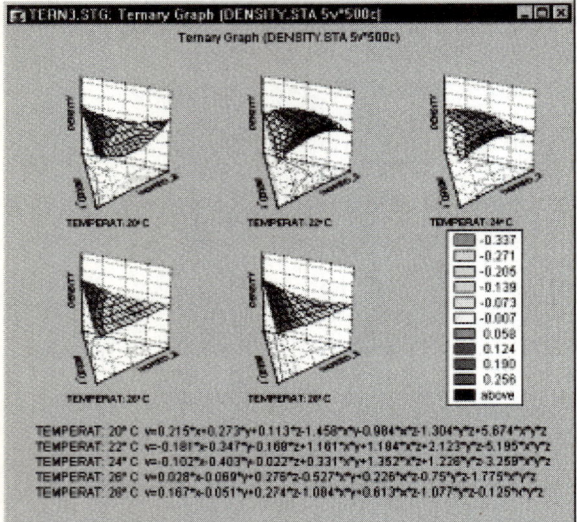

Brushing

Perhaps the most common and historically first widely used technique explicitly identified as graphical exploratory data analysis is brushing, an interactive method allowing you to select specific data points or subsets of data on screen and identify their (e.g., common) characteristics or examine their effects on relations between relevant variables (e.g., in scatterplot matrices) or identify (e.g., label) outliers.

Those relations between variables can be visualized by fitted functions (e.g., 2D lines or 3D surfaces) and their confidence intervals, thus, for example, you can examine changes in those functions by interactively (temporarily) removing or adding specific subsets of data. For example, one of many applications of the brushing technique is to select (i.e., highlight) in a matrix scatterplot all data points that belong to a certain category (e.g., a medium income level, see the highlighted subset in the upper-right component graph in the next illustration) in order to examine how those specific observations contribute to relations between other variables in the same data set (e.g., the correlation between the *debt* and *assets* in the current example).

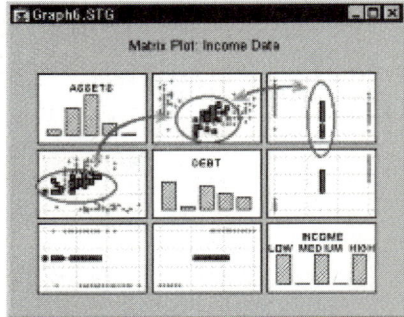

If the brushing facility supports features such as animated brushing (see an example in the next illustration) or automatic function re-fitting, you can define a dynamic brush that will move over the consecutive ranges of a criterion variable (e.g., "income" measured on a continuous scale and not a discrete scale as in the illustration to the above) and examine the dynamics of the contribution of the criterion variable to the relations between other relevant variables in the same data set.

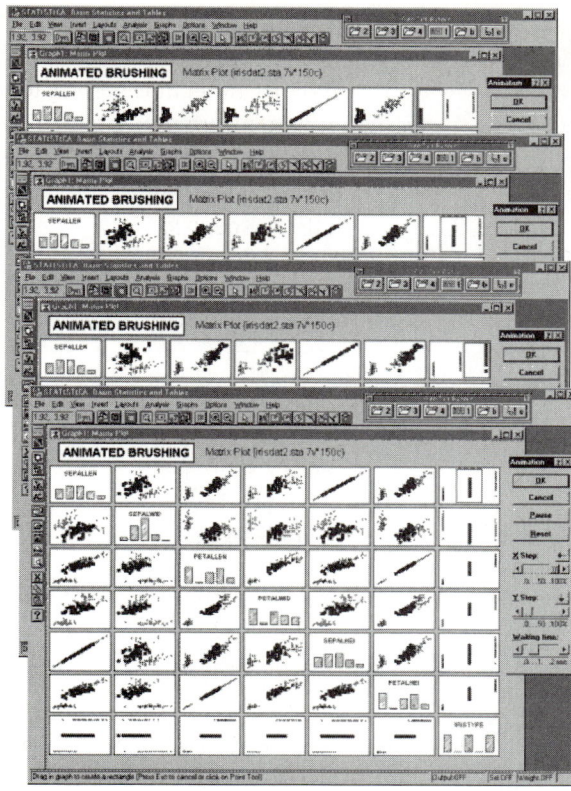

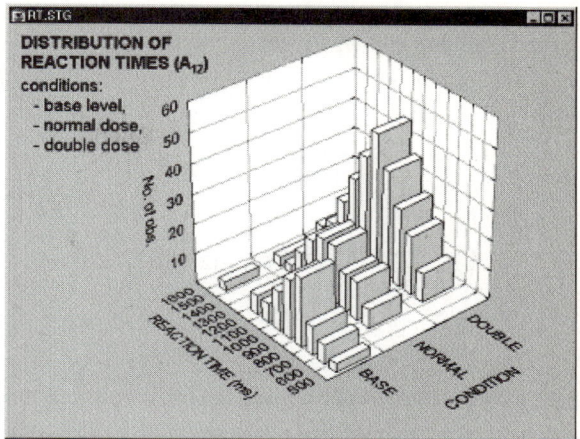

Smoothing Bivariate Distributions

Three-dimensional histograms are used to visualize crosstabulations of values in two variables. They can be considered to be a conjunction of two simple (i.e., univariate) histograms, combined such that the frequencies of co-occurrences of values on the two analyzed variables can be examined. In a most common format of this graph, a 3D bar is drawn for each cell of the crosstabulation table and the height of the bar represents the frequency of values for the respective cell of the table. Different methods of categorization can be used for each of the two variables for which the bivariate distribution is visualized.

If the software provides smoothing facilities, you can fit surfaces to 3D representations of bivariate frequency data. Thus, every 3D histogram can be turned into a smoothed surface. This technique is of relatively little help if applied to a simple pattern of categorized data (such as the histogram shown above).

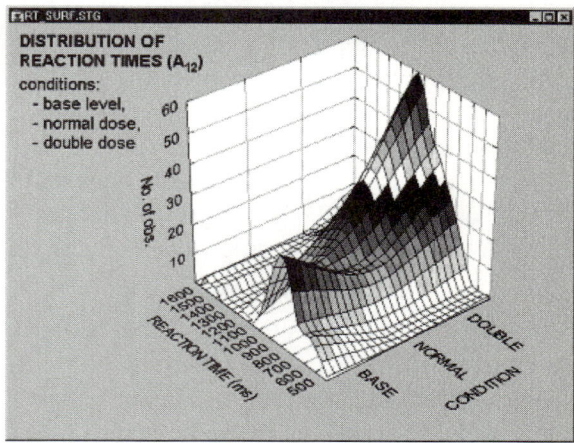

However, if applied to more complex patterns of frequencies, it may provide a valuable exploratory technique,

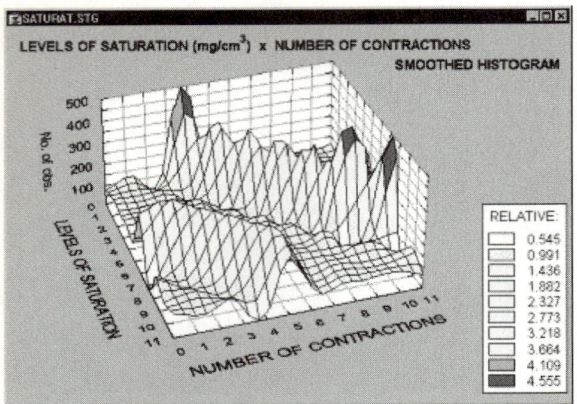

allowing identification of regularities that are less salient when examining the standard 3D histogram representations (e.g., see the systematic surface "wave-patterns" shown on the smoothed histogram in the previous illustration).

Layered Compression

When layered compression is used, the main graph plotting area is reduced in size to leave space for margin graphs in the upper and right side of the display (and a miniature graph in the corner). These smaller margin graphs represent vertically and horizontally compressed images (respectively) of the main graph.

In 2D graphs, layered compression is an exploratory data analysis technique that may facilitate the identification of otherwise obscured trends and patterns in 2-dimensional data sets. For example, in the following illustration (based on an example discussed by Cleveland, 1993), it can be seen that the number of sunspots in each cycle decays more slowly than it rises at the onset of each cycle.

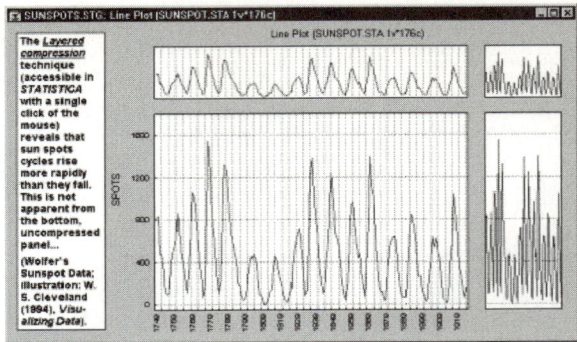

This tendency is not readily apparent when examining the standard line plot; however, the compressed graph uncovers the hidden pattern.

Projections of 3D Data Sets

Contour plots generated by projecting surfaces (created from multivariate, typically three-variable data sets) offer a useful method to explore and analytically examine the shapes of surfaces.

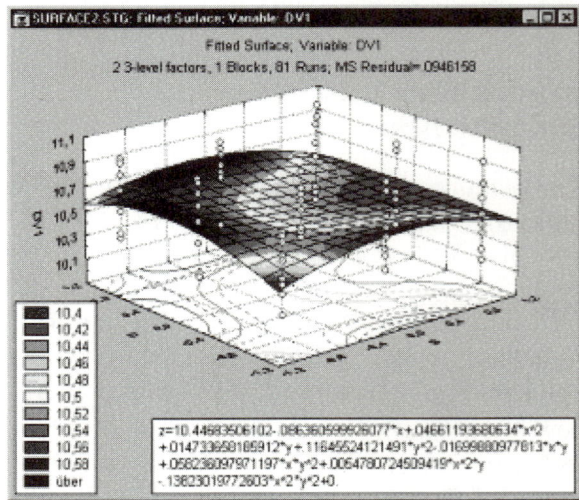

Compared to surface plots, they may be less effective to quickly visualize the overall shape of 3D data structures.

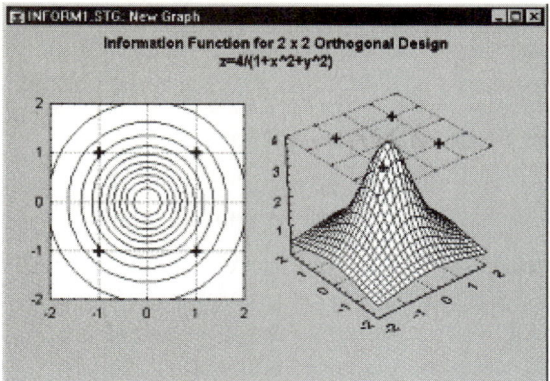

However, their main advantage is that they allow for precise examination and analysis of the shape of the surface.

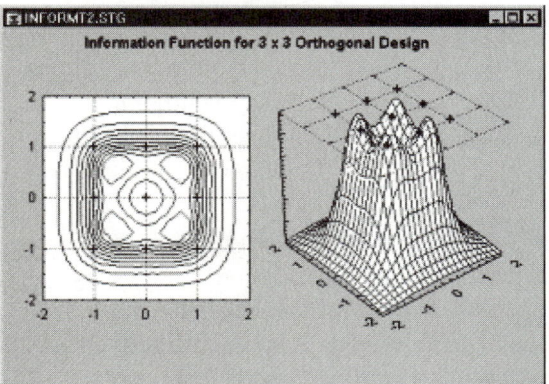

Contour plots display a series of undistorted horizontal "cross sections" of the surface.

Icon Plots

Icon graphs represent cases or units of observation as multidimensional symbols and they offer a powerful although not easy to use exploratory technique. The general idea behind this method capitalizes on the human ability to automatically spot complex (sometimes interactive) relations between multiple variables if those relations are consistent across a set of instances (in this case, icons). Sometimes the observation (or a "feeling")

that certain instances are somehow similar to each other comes before the observer (in this case an analyst) can articulate which specific variables are responsible for the observed consistency (Lewicki, Hill, & Czyzewska, 1992). However, further analysis that focuses on such intuitively spotted consistencies can reveal the specific nature of the relevant relations between variables.

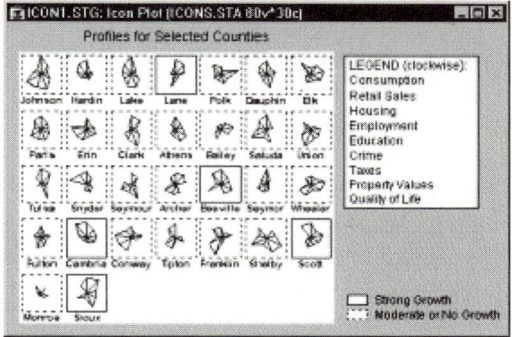

The basic idea of icon plots is to represent individual units of observation as particular graphical objects where values of variables are assigned to specific features or dimensions of the objects (usually one case = one object). The assignment is such that the overall appearance of the object changes as a function of the configuration of values.

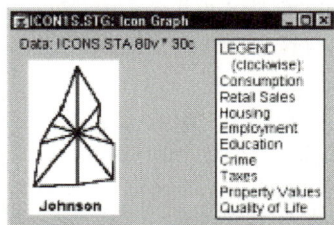

Thus, the objects are given visual identities that are unique for configurations of values and that can be identified by the observer. Examining such icons may help to discover specific clusters of both simple relations and interactions between variables.

Analyzing Icon Plots

The ideal design of the analysis of icon plots consists of five phases:

1. Select the order of variables to be analyzed. In many cases a random starting sequence is the best solution. You may also try to enter variables based on the order in a multiple regression equation, factor loadings on an interpretable factor (see Chapter 16 – *Factor Analysis*), or a similar multivariate technique. That method may simplify and "homogenize" the general appearance of the icons that may facilitate the identification of non-salient patterns. It may also, however, make some interactive patterns more difficult to find. No universal recommendations can be given at this point, other than to try the quicker (random order) method before getting involved in the more time-consuming method.

2. Look for any potential regularities, such as similarities between groups of icons, outliers, or specific relations between aspects of icons (e.g., "if the first two rays of the star icon are long, then one or two rays on the other side of the icon are usually short"). The circular type of icon plots is recommended for this phase.

3. If any regularities are found, try to identify them in terms of the specific variables involved.

4. Reassign variables to features of icons (or switch to one of the sequential icon plots) to verify the identified structure of relations (e.g., try to move the related aspects of the icon closer together to facilitate further comparisons). In some cases, at the end of this phase it is recommended to drop the variables that appear not to contribute to the identified pattern.

5. Finally, use a quantitative method (such as a regression method, nonlinear estimation,

discriminant function analysis, or cluster analysis) to test and quantify the identified pattern or at least some aspects of the pattern.

Taxonomy of Icon Plots

Most icon plots can be assigned to one of two categories: circular and sequential.

Circular icons. Circular icon plots (star plots, sun ray plots, polygon icons) follow a "spoked wheel" format where values of variables are represented by distances between the center (hub) of the icon and its edges.

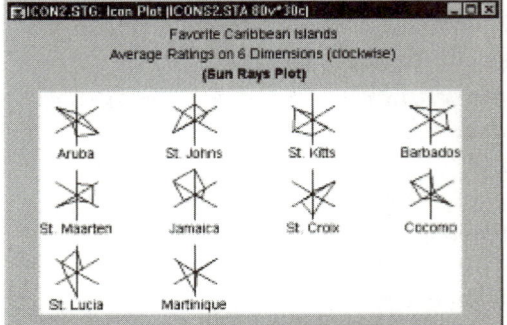

Those icons may help to identify interactive relations between variables because the overall shape of the icon may assume distinctive and identifiable overall patterns depending on multivariate configurations of values of input variables.

In order to translate such overall patterns into specific models (in terms of relations between variables) or verify specific observations about the pattern, it is helpful to switch to one of the sequential icon plots, which may prove more efficient when you already know what to look for.

Sequential icons. Sequential icon plots (column icons, profile icons, line icons) follow a simpler format where individual symbols are represented by small sequence plots (of different types).

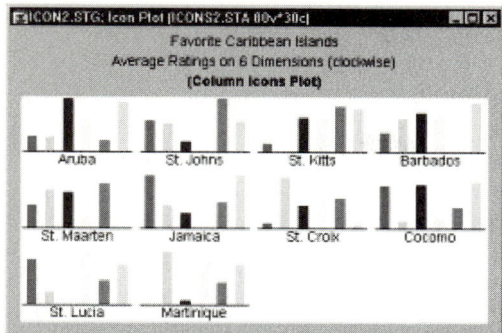

The values of consecutive variables are represented in those plots by distances between the base of the icon and the consecutive break points of the sequence (e.g., the height of the columns shown above). Those plots may be less efficient as a tool for the initial exploratory phase of icon analysis because the icons may look alike. However, as mentioned before, they may be helpful in the phase when some hypothetical pattern has already been revealed and you need to verify it or articulate it in terms of relations between individual variables.

Pie icons. Pie icon plots fall somewhere in-between the previous two categories; all icons have the same shape (pie) but are sequentially divided in a different way according to the values of consecutive variables.

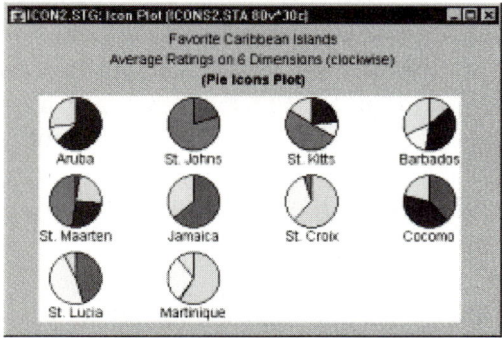

From a functional point of view, they belong rather to the sequential than circular category,

although they can be used for both types of applications.

Chernoff faces. This type of icon is a category by itself. Cases are visualized by schematic faces such that relative values of variables selected for the graph are represented by variations of specific facial features.

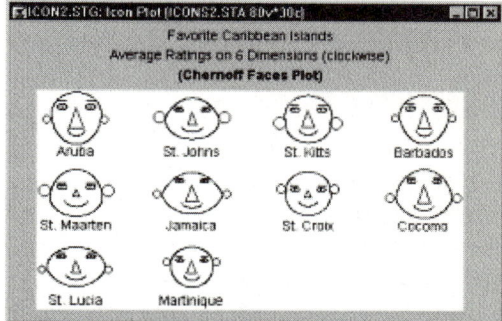

Due to its unique features, some researchers consider it an ultimate exploratory multivariate technique that is capable of revealing hidden patterns of interrelations between variables that cannot be uncovered by any other technique. This statement may be an exaggeration, however. Also, it must be admitted that Chernoff faces is a method that is difficult to use, and it requires a great deal of experimentation with the assignment of variables to facial features. See also, Chapter 12 – *Data Mining Techniques*.

Standardization of Values

Except for unusual cases when you intend for the icons to reflect the global differences in ranges of values between the selected variables, the values of the variables should be standardized once to assure within-icon compatibility of value ranges. For example, because the largest value sets the global scaling reference point for the icons, if there are variables that are in a range of much smaller order, they may not appear in the icon at all, e.g., in a star plot, the rays that represent them will be too short to be visible.

Applications

Icon plots are generally applicable 1) to situations where you want to find systematic patterns or clusters of observations, and 2) when you want to explore possible complex relationships between several variables. The first type of application is similar to cluster analysis; that is, it can be used to classify observations.

For example, suppose you studied the personalities of artists, and you recorded the scores for several artists on a number of personality questionnaires. The icon plot may help you determine whether there are natural clusters of artists distinguished by particular patterns of scores on different questionnaires (e.g., you may find that some artists are very creative, undisciplined, and independent, while a second group is particularly intelligent, disciplined, and concerned with publicly-acknowledged success).

The second type of application – the exploration of relationships between several variables – is more similar to factor analysis; that is, it can be used to detect which variables tend to go together. For example, suppose you were studying the structure of people's perception of cars. Several subjects completed detailed questionnaires rating different cars on numerous dimensions. In the data file, the average ratings on each dimension (entered as the variables) for each car (entered as cases or observations) are recorded.

When you now study the Chernoff faces (each face representing the perceptions for one car), it may occur to you that smiling faces tend to have big ears; if price was assigned to the amount of smile and acceleration to the size of ears, this discovery means that fast cars are more expensive. This, of course, is only a simple example; in real-life exploratory data analyses, non-obvious complex relationships between variables may become apparent.

Related Graphs

Matrix plots visualize relations between variables from one or two lists. If the software allows you to mark selected subsets, matrix plots may provide information similar to that in icon plots.

If the software allows you to create and identify user-defined subsets in scatterplots, simple 2D scatterplots can be used to explore the relationships between two variables; likewise, when exploring the relationships between three variables, 3D scatterplots provide an alternative to icon plots.

Graph Types

There are various types of icon plots.

Chernoff faces. A separate face icon is drawn for each case; relative values of the selected variables for each case are assigned to shapes and sizes of individual facial features (e.g., length of nose, angle of eyebrows, width of face).

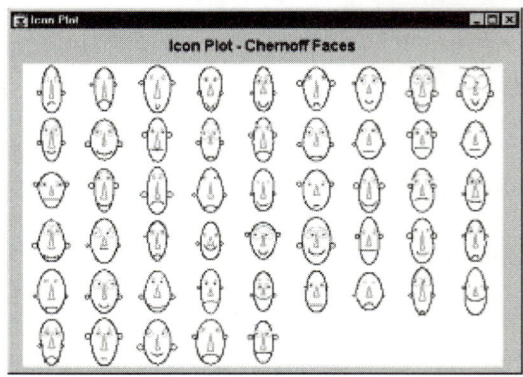

For more information, see Chernoff faces in *Taxonomy of Icon Plots*, page 548.

Stars. Star icons is a circular type of icon plot. A separate star-like icon is plotted for each case; relative values of the selected variables for each case are represented (clockwise, starting at 12:00) by the length of individual rays in each star. The ends of the rays are connected by a line.

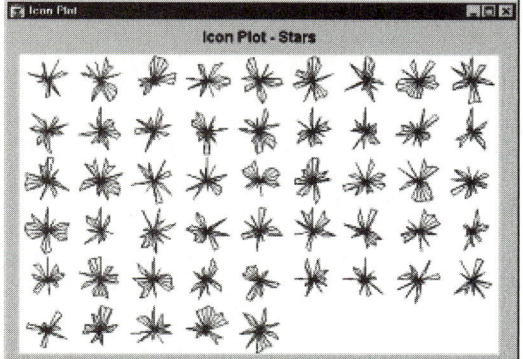

Sun rays. Sun ray icons is a circular type of icon plot. A separate sun-like icon is plotted for each case; each ray represents one of the selected variables (clockwise, starting at 12:00), and the length of the ray represents the relative value of the respective variable. Data values of the variables for each case are connected by a line.

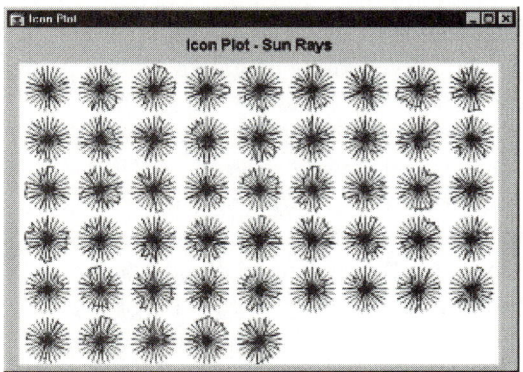

Polygons. Polygon icons is a circular type of icon plot. A separate polygon icon is plotted for each case; relative values of the selected variables for each case are represented by the distance from the center of the icon to

consecutive corners of the polygon (clockwise, starting at 12:00).

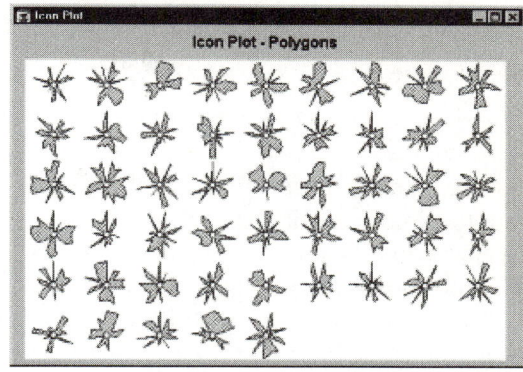

Pies. Pie icons is a circular type of icon plot. Data values for each case are plotted as a pie chart (clockwise, starting at 12:00); relative values of selected variables are represented by the size of the pie slices.

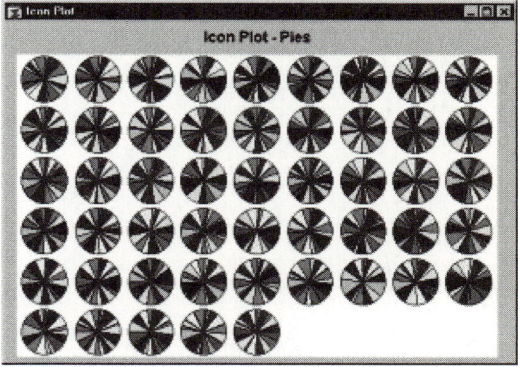

Columns. A sequential type of icon plot. An individual column graph is plotted for each case; relative values of the selected variables for each case are represented by the height of consecutive columns.

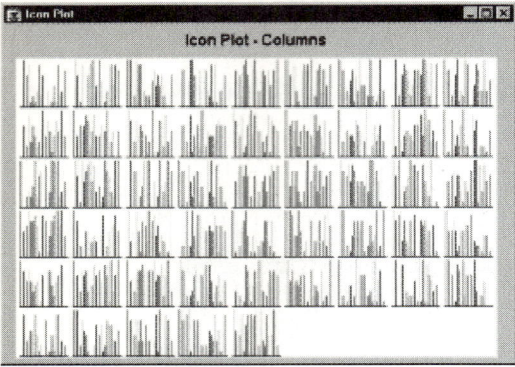

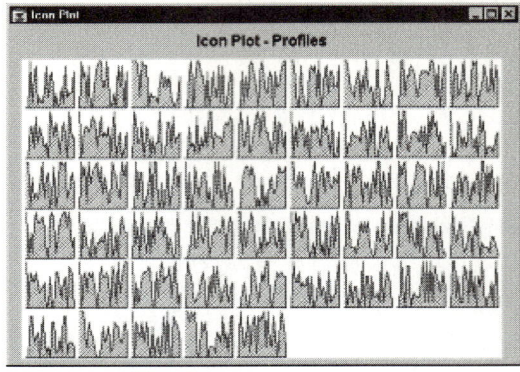

Lines. A sequential type of icon plot. An individual line graph is plotted for each case; relative values of the selected variables for each case are represented by the height of consecutive break points of the line above the baseline.

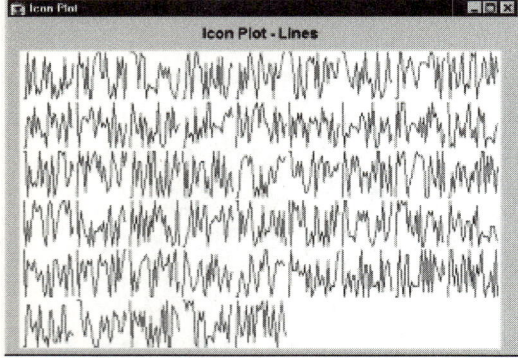

Profiles. A sequential type of icon plot. An individual area graph is plotted for each case; relative values of the selected variables for each case are represented by the height of consecutive peaks of the profile above the baseline.

Data Reduction

Sometimes plotting an extremely large data set can obscure an existing pattern. In this case, it can be useful to plot only a subset of the data so that the pattern is not hidden by the number of point markers. Some software products offer methods for data reduction (or optimizing) that can be useful in these instances. Ideally, a data reduction option enables you to specify an integer value n less than the number of cases in the data file. Then the software will randomly select approximately n cases from the available cases and create the plot based on these cases only.

Note that such data set (or sample size) reduction methods effectively draw a random sample from the current data set. Obviously, the nature of such data reduction is entirely different than when data are selectively reduced only to a specific subset or split into subgroups based on certain criteria (e.g., such as gender, region, or cholesterol level). The latter methods can be implemented interactively (e.g., using animated brushing facilities), or other techniques (e.g., categorized graphs or case selection conditions). All these methods can further aid in identifying patterns in large data sets. See the following illustrations.

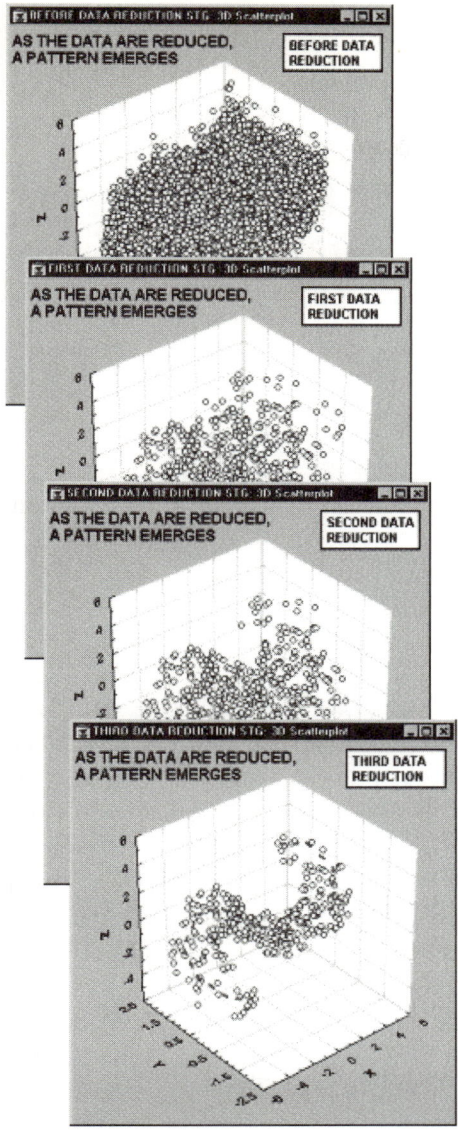

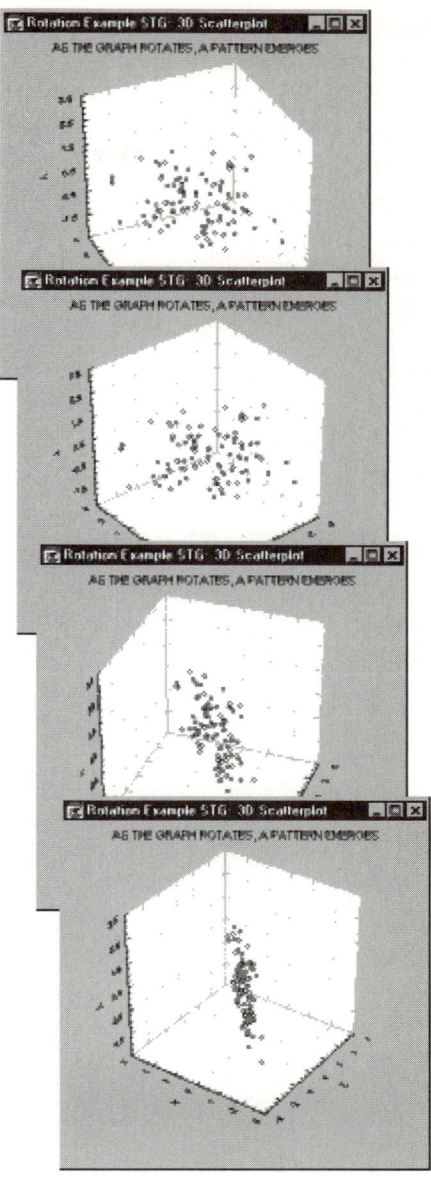

Data Rotation (in 3D Space)

Changing the viewpoint for 3D scatterplots (e.g., simple, spectral, or space plots) may prove to be an effective exploratory technique since it can reveal patterns that are easily obscured unless you look at the cloud of data points from an appropriate angle.

Some software products offer interactive perspective, rotation, and continuous spinning controls that can be useful in these instances. Ideally, these controls will enable you to adjust the graph's angle and perspective to find the most informative location of the viewpoint for the graph as well as allowing you to control the vertical and horizontal rotation of the graph.

While these facilities are useful for initial exploratory data analysis, they can also be quite beneficial in exploring the factorial space (see Chapter 16 – *Factor Analysis*) and exploring the dimensional space (see Chapter 25 – *Multidimensional Scaling*).

STATISTICAL GLOSSARY

557 – A

565 – B

582 – C

602 – D

610 – E

617 – F

620 – G

626 – H

630 – I

636 – J

638 – K

642 – L

655 – M

671 – N

677 – O

682 – P

697 – Q

703 – R

715 – S

740 – T

747 – U

749 – V

751 – W

755 – X

756 – Y

757 – Z

A

Aberration, Minimum

Minimum aberration is an experimental design criterion that is subsidiary to the criterion of design resolution. The minimum aberration design is defined as the design of maximum resolution "which minimizes the number of words in the defining relation that are of minimum length" (Fries & Hunter, 1980). Less technically, the criterion apparently operates by choosing design generators that produce the smallest number of pairs of confounded interactions of the crucial order. For example, the minimum aberration resolution IV design would have the minimum number of pairs of confounded 2-factor interactions. For discussions of the role of design criteria in experimental design see $2^{(k-p)}$ *Fractional Factorial Designs* and $2^{(k-p)}$ *Maximally Unconfounded and Minimum Aberration Designs* In Chapter 15 – *Experimental Design (Industrial DOE)*.

Abrupt Permanent Impact

In Time Series, a permanent abrupt impact pattern simply implies that the overall mean of the times series shifted after the intervention; the overall shift is denoted by ω (*omega*).

Abrupt Temporary Impact

In time series, the abrupt temporary impact pattern implies an initial abrupt increase or decrease due to the intervention, which then slowly decays without permanently changing the mean of the series. This type of intervention can be summarized by the expressions:

Prior to intervention: $Impact_t = 0$

At time of intervention: $Impact_t = \omega$

After intervention: $Impact_t = \delta * Impact_{t-1}$

Note that this impact pattern is again defined by the two parameters δ (*delta*) and ω (*omega*). As long as the δ parameter is greater than 0 and less than 1 (the bounds of system stability), the initial abrupt impact will gradually decay. If δ is near 0 (zero) than the decay will be very quick, and the impact will have entirely disappeared after only a few observations. If δ is close to 1, the decay will be slow and the intervention will affect the series over many observations. Note that, when evaluating a fitted model, it is again important that both parameters are statistically significant; otherwise you could reach paradoxical conclusions. For example, suppose the ω parameter is not statistically significant from 0 (zero) but the δ parameter is; this would mean that an intervention did not cause an initial abrupt change, which then showed significant decay.

Accept-Support (AS) Testing

In this type of statistical test, the statistical null hypothesis is the hypothesis that, if true, supports the experimenter's theoretical hypothesis. Consequently, in AS testing, the experimenter would prefer *not* to obtain "statistical significance." In AS testing, accepting the null hypothesis supports the experimenter's theoretical hypothesis. For more information see Chapter 31 – *Power Analysis*.

Accept Threshold (Confidence Limits)

Confidence limits are the same as confidence intervals. In Neural Networks, they represent the accept and reject thresholds used in classification tasks, to determine whether a pattern of outputs corresponds to a particular class or not. These are applied according to the conversion function of the output variable (*One-of-N*, *Two-state*, Kohonen, etc).

Activation Function (in Neural Networks)

A function used to transform the activation level of a unit (neuron) into an output signal. Typically, activation functions have a "squashing" effect. Together with the PSP function (which is applied first) this defines the unit type.

Neural networks supports a wide range of activation functions. Only a few of these are used by default; the others are available for customization.

Identity. The activation level is passed on directly as the output. Used in a variety of network types, including linear networks, and the output layer of radial basis function networks.

Logistic. This is an S-shaped (sigmoid) curve, with output in the range (0,1).

Hyperbolic. The hyperbolic tangent function (tanh): a sigmoid curve, like the logistic function, except that output lies in the range (-1,+1). Often performs better than the logistic function because of its symmetry. Ideal for customization of multilayer perceptrons, particularly the hidden layers.

Exponential. The negative exponential function. Ideal for use with radial units. The combination of radial synaptic function and negative exponential activation function produces units that model a Gaussian (bell-shaped) function centered at the weight vector. The standard deviation of the Gaussian is given by the formula below, where *d* is the deviation of the unit stored in the unit's threshold:

$$\sigma = \sqrt{1/d}$$

Softmax. Exponential function, with results normalized so that the sum of activations across the layer is 1.0. Can be used in the output layer of multilayer perceptrons for classification problems, so that the outputs can be interpreted as probabilities of class membership (Bishop, 1995; Bridle, 1990).

Unit sum. Normalizes the outputs to sum to 1.0. Used in PNNs to allow the outputs to be interpreted as probabilities.

Square root. Used to transform the squared distance activation in an SOFM network or Cluster network to the actual distance as an output.

Sine. Possibly useful if recognizing radially-distributed data; not used by default.

Ramp. A piece-wise linear version of the sigmoid function. Relatively poor training performance, but fast execution.

Step. Outputs either 1.0 or 0.0, depending on whether the Synaptic value is positive or negative. Can be used to model simple networks such as perceptrons.

See the mathematical definitions of the activation functions in the following table.

Activation Functions

Function	Definition	Range
Identity	x	(-inf,+inf)
Logistic	$\dfrac{1}{1 - e^{-1}}$	(0,+1)
Hyperbolic	$\dfrac{e^{x} - e^{-x}}{e^{x} + e^{-x}}$	(-1,+1)
-Exponential	e^{-x}	(0, +inf)

Softmax	$\dfrac{e^{x}}{\sum\limits_{i} e^{x_i}}$	(0,+1)
Unit sum	$\dfrac{x}{\sum\limits_{i} x_i}$	(0,+1)
Square root	$\sqrt{x}$	(0, +inf)
Sine	$\sin(x)$	[0,+1]
Ramp	$\begin{cases} -1 & x \le -1 \\ x & -1 < x < + \\ +1 & x \ge +1 \end{cases}$	[-1,+1]
Step	$\begin{cases} 0 & x < 0 \\ +1 & x \ge 0 \end{cases}$	[0,+1]

Additive Models

Additive models represent a generalization of multiple regression (which is a special case of general linear models). Specifically, in linear regression, a linear least-squares fit is computed for a set of predictor or X variables, to predict a dependent Y variable. The well know linear regression equation with m predictors, to predict a dependent variable Y, can be stated as:

$$Y = b_0 + b_1 * X_1 + .. b_m * X_m$$

where Y stands for the (predicted values of the) dependent variable, X_1 through X_m represent the m values for the predictor variables, and b_0, and b_1 through b_m are the regression coefficients estimated by multiple regression. A generalization of the multiple regression model would be to maintain the additive nature of the model, but to replace the simple terms of the linear equation $b_i * X_i$ with $f_i(X_i)$ where f_i is

nonparametric function of the predictor X_i. In other words, instead of a single coefficient for each variable (additive term) in the model, in additive models, an unspecified (non-parametric) function is estimated for each predictor, to achieve the best prediction of the dependent variable values.

For additional information, see Hastie and Tibshirani, 1990, or Schimek, 2000.

Additive Season, Damped Trend

In this time series model, the simple exponential smoothing forecasts are enhanced both by a damped trend component (independently smoothed with the single parameter ϕ, this model is an extension of Brown's one-parameter linear model, see Gardner, 1985, p. 12-13) and an additive seasonal component (smoothed with parameter δ).

For example, suppose we want to forecast from month to month the number of households that purchase a particular consumer electronics device (e.g., VCR). Every year, the number of households that purchase a VCR will increase, however, this trend will be damped (i.e., the upward trend will slowly disappear) over time as the market becomes saturated. In addition, there will be a seasonal component, reflecting the seasonal changes in consumer demand for VCRs from month to month (demand will likely be smaller in the summer and greater during the December holidays). This seasonal component may be additive, for example, a relatively stable number of additional households may purchase VCRs during the December holiday season. To compute the smoothed values for the first season, initial values for the seasonal components are necessary. Also, to compute the smoothed value (forecast) for the first observation in the series, both estimates of S_0

and T_0 (initial trend) are necessary. By default, these values are computed as:

$$T_0 = (1/\phi)*(M_k-M_1)/[(k-1)*p]$$

where ϕ is the smoothing parameter, k is the number of complete seasonal cycles, M_k is the mean for the last seasonal cycle, M_1 is the mean for the first seasonal cycle, p is the length of the seasonal cycle, and $S_0 = M_1 - p*T_0/2$.

Additive Season, Exponential Trend

In this time series model, the simple exponential smoothing forecasts are "enhanced" both by an exponential trend component (independently smoothed with parameter γ) and an additive seasonal component (smoothed with parameter δ). For example, suppose we wanted to forecast the monthly revenue for a resort area. Every year, revenue may increase by a certain percentage or factor, resulting in an exponential trend in overall revenue. In addition, there could be an additive seasonal component, for example a particular fixed (and slowly changing) amount of added revenue during the December holidays.

To compute the smoothed values for the first season, initial values for the seasonal components are necessary. Also, to compute the smoothed value (forecast) for the first observation in the series, both estimates of S_0 and T_0 (initial trend) are necessary. By default, these values are computed as:

$$T_0 = \exp((\log(M_2) - \log(M_1))/p)$$

where M_2 is the mean for the second seasonal cycle, M_1 is the mean for the first seasonal cycle, p is the length of the seasonal cycle, and $S_0 = \exp(\log(M_1) - p*\log(T_0)/2)$.

Additive Season, Linear Trend

In this time series model, the simple exponential smoothing forecasts are "enhanced" both by a linear trend component (independently smoothed with parameter γ) and an additive seasonal component (smoothed with parameter δ). For example, suppose we were to predict the monthly budget for snow-removal in a community. There may be a trend component (as the community grows, there is a steady upward trend for the cost of snow removal from year to year). At the same time, there is obviously a seasonal component, reflecting the differential likelihood of snow during different months of the year. This seasonal component could be additive, meaning that a particular fixed additional amount of money is necessary during the winter months, or multiplicative, that is, given the respective budget figure, it may increase by a factor of, for example, 1.4 during particular winter months.

To compute the smoothed values for the first season, initial values for the seasonal components are necessary. Also, to compute the smoothed value (forecast) for the first observation in the series, both estimates of S_0 and T_0 (initial trend) are necessary. By default, these values are computed as:

$$T_0 = (M_k-M_1)/((k-1)*p)$$

where k is the number of complete seasonal cycles, M_k is the mean for the last seasonal cycle, M_1 is the mean for the first seasonal cycle, p is the length of the seasonal cycle, and $S_0 = M_1 - T_0/2$.

Additive Season, No Trend

This time series model is partially equivalent to the simple exponential smoothing model; however, in addition, each forecast is enhanced by an additive seasonal component that is smoothed independently (see *Seasonal smoothing parameter δ* in Chapter 38 – *Time Series/Forecasting*). This model would, for example, be adequate when computing forecasts for the expected amount of rain monthly. The amount of rain will be stable from year to year, or change only slowly. At the same time, there will be seasonal changes (e.g., rainy seasons), which again may change slowly from year to year.

To compute the smoothed values for the first season, initial values for the seasonal components are necessary. The initial smoothed value S_0 will by default be computed as the mean for all values included in complete seasonal cycles.

Adjusted Means

These are the means that you would get after removing all differences that can be accounted for by the covariate in an analysis of variance design (see *ANOVA*). The general formula (see Kerlinger & Pedhazur, 1973, p. 272) is:

$$\text{Y-bar}_{j(adj)} = \text{Y-bar}_j - b(\text{X-bar}_j - \text{X-bar})$$

where $\text{Y-bar}_{j(adj)}$ is the adjusted mean of group j; Y-bar_j is the mean of group j before adjustment; b is the common regression coefficient; X-bar_j is the mean of the covariate for group j; and X-bar is the grand mean of the covariate.

See *Categorical Predictor Variable*, page 583; see also, Chapter 18 – *General Linear Models (GLM)* or Chapter 3 – *ANOVA/MANOVA*.

AID

AID (Automatic Interaction Detection) is a classification program developed by Morgan & Sonquist (1963) that led to the development of the THAID (Morgan & Messenger, 1973) and CHAID (Kass, 1980) classification tree programs. These programs perform multi-level

splits when computing classification trees. For discussion of the differences of AID from other classification tree programs, see *A Brief Comparison of Classification Tree Programs* in Chapter 9 – *Classification Trees*.

Akaike Information Criterion (AIC)

When a model involving q parameters is fitted to data, the criterion is defined as $-L_q + 2q$ where L_q is the maximized log likelihood. Akaike suggested maximizing the criterion to choose between models with different numbers of parameters. It was originally proposed for time-series models, but is also used in regression. Akaike Information Criterion (AIC) can be used in generalized linear/nonlinear models (GLZ) when comparing the subsets of effects during best subset regression. Since the evaluation of the score statistic does not require iterative computations, best subset selection based on the score statistic is computationally faster, while the selection based on the AIC statistic usually provides more accurate results.

Algorithm

As opposed to heuristics (which contain general recommendations based on statistical evidence or theoretical reasoning), algorithms are completely defined, finite sets of steps, operations, or procedures that will produce a particular outcome. For example, with a few exceptions, all computer programs, mathematical formulas, and (ideally) medical and food recipes are algorithms. See also, *Data Mining*, *Heuristic*, and *Neural Networks*.

Alpha (Type I Error Rate)

The probability of incorrectly rejecting a true statistical null hypothesis. For more information see Chapter 31 – *Power Analysis*.

Anderson-Darling Test

The Anderson-Darling procedure is a general test to compare the fit of an observed cumulative distribution function to an expected cumulative distribution function. This test is applicable to complete data sets (without censored observations). The critical values for the Anderson-Darling statistic have been tabulated (see, for example, Dodson, 1994, Table 4.4) for sample sizes between 10 and 40; this test is not computed for n less than 10 and greater than 40.

The Anderson-Darling test is used in Weibull and reliability/failure time analysis; see also, *Mann-Scheuer-Fertig Test* and *Hollander-Proschan Test*.

ANOVA (General ANOVA/MANOVA)

The purpose of analysis of variance (ANOVA) is to test for significant differences between means by comparing (i.e., analyzing) variances. More specifically, by partitioning the total variation into different sources (associated with the different effects in the design), we are able to compare the variance due to the between-groups (or treatments) variability with that due to the within-group (treatment) variability. Under the null hypothesis (that there are no mean differences between groups or treatments in the population), the variance estimated from the within-group (treatment) variability should be about the same as the variance estimated from between-groups (treatments) variability. For more information, see Chapter 3 – *ANOVA/MANOVA*.

Application Programming Interface (API)

Application Programming Interface is a set of functions that conform to conventions of a particular operating system (e.g., Windows), which enables you to programmatically access the functionality of another program. For example, the kernel of *STATISTICA Neural Networks* can be accessed by other program packages (e.g., Visual Basic, Delphi, C, C++) in a variety of ways.

Arrow

An element in a path diagram used to indicate causal flow from one variable to another, or, in narrower interpretation, to show which of two variables in a linear equation is the independent variable and which is the dependent variable.

Assignable Causes and Actions

In the context of monitoring quality characteristics, you have to distinguish between two different types of variability: Common cause variation describes random variability that is inherent in the process and affects all individual values. Ideally, when your process is in-control, only common cause variation will be present. In a quality control chart, it will show up as a random fluctuation of the individual samples around the center line with all samples falling between the upper and lower control limit and no non-random patterns (runs) of adjacent samples. Special cause or assignable cause variation is due to specific circumstances that can be accounted for. It will usually show up in the QC chart as outlier samples (i.e., exceeding the lower or upper control limit) or as a systematic pattern (run) of adjacent samples. It will also affect the calculation of the chart specifications (center line and control limits).

With some software programs, if you investigate the out-of-control conditions and you find an explanation for them, you can assign descriptive labels to those out-of-control samples

and explain the causes (e.g., valve defect) and actions that have been taken (e.g. valve fixed).

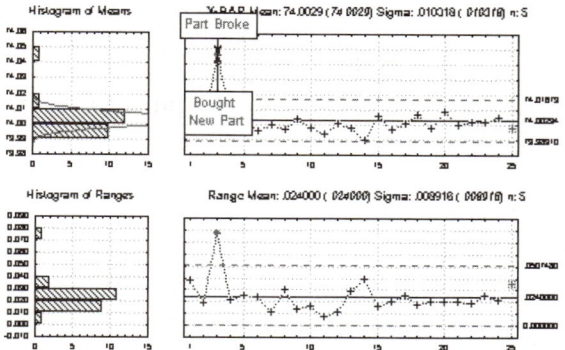

Having causes and actions displayed in the chart will document that the center line and the control limits of the chart are affected by special cause variation in the process. See the previous illustration.

Association Rules

Data mining for association rules is often the first and most useful method for analyzing data that describe transactions, lists of items, unique phrases (in text mining) etc. In general, association rules take the form If *Body* then *Head*, where *Body* and *Head* stand for simple codes, text values, items, consumer choices, phrases etc., or the conjunction of codes and text values etc. (e.g., *if (Car=Porsche and Age<20 and ThrillSeeking=High*) then (*Risk=High and Insurance=High*); here the logical conjunction before the then would be the *Body*, and the logical conjunction following the then would be the *Head* of the association rule). The *a priori* algorithm (see Agrawal and Swami, 1993; Agrawal and Srikant, 1994; Han and Lakshmanan, 2001; see also, Witten and Frank, 2000) is a popular and efficient algorithm for deriving such association rules from large data sets, based on some user-defined "threshold" values for rule.

Asymmetrical Distribution

If you split the distribution in half at its mean (or median), the distribution of values on the two sides of this central point would not be the same (i.e., not symmetrical) and the distribution would be considered skewed. See also, *Descriptive Statistics* in Chapter 2 – *Basic Statistics and Tables*.

AT&T Runs Rules – Runs Tests (in Quality Control)

These tests are designed to detect patterns measurement (e.g., sample means) that may indicate that the process is out of control. In quality control charting, when a sample point (e.g., a mean in an X-bar chart) falls outside the control lines, you have reason to believe that the process may no longer be in control. In addition, you should look for systematic patterns of points (e.g., means) across samples, because such patterns may indicate that the process average has shifted. Most quality control software packages will (optionally) perform the standard set of tests for such patterns; these tests are also sometimes referred to as AT&T runs rules (see AT&T, 1959) or tests for special causes (e.g., see Nelson, 1984, 1985; Grant and Leavenworth, 1980; Shirland, 1993). The term special or assignable causes as opposed to chance or common causes was used by Shewhart to distinguish between a process that is in control, with variation due to random (chance) causes only, from a process that is out of control, with variation that is due to some non-chance or special (assignable) factors (cf. Montgomery, 1991, p. 102).

As the *sigma* control limits for quality control charts, the runs rules are based on statistical reasoning. For example, the probability of any sample mean in an X-bar control chart falling above the center line is equal to 0.5, provided

1) that the process is in control (i.e., that the center line value is equal to the population mean), 2) that consecutive sample means are independent (i.e., not auto-correlated), and 3) that the distribution of means follows the normal distribution. Simply stated, under those conditions there is a 50-50 chance that a mean will fall above or below the center line. Thus, the probability that two consecutive means will fall above the center line is equal to 0.5 times $0.5 = 0.25$.

For additional information, see *Runs Tests*; see also, *Assignable Causes and Actions*, page 562.

Attribute (Attribute Variable)

An alternative name for a nominal variable.

Augmented Product Moment Matrix

For a set of p variables, this is a $(p + 1) \times (p + 1)$ square matrix. The first p rows and columns contain the matrix of moments about zero, while the last row and column contain the sample means for the p variables. The matrix is therefore of the form:

$$M_A = \begin{bmatrix} M & \bar{x} \\ \bar{x}' & 1 \end{bmatrix}$$

where **M** is a matrix with element:

$$M_{jk} = \frac{1}{N} \sum_{i=1}^{N} X_{ij} X_{ik}$$

N is the sample size, and $\bar{x}$ is a vector with the means of the variables, and $\bar{x}'$ is the transpose of $\bar{x}$. See also, Chapter 35 – *Structural Equation Modeling*.

Autoassociative Network

A neural network (usually a multilayer perceptron) designed to reproduce its inputs at its outputs, while "squeezing" the data through a lower-dimensionality middle layer. Used for compression or dimensionality reduction purposes (see Fausett, 1994; Bishop, 1995).

Automatic Interaction Detection (AID)

AID (Automatic Interaction Detection) is a classification program developed by Morgan & Sonquist (1963) that led to the development of the THAID (Morgan & Messenger, 1973) and CHAID (Kass, 1980) classification tree programs. These programs perform multi-level splits when computing classification trees. For discussion of the differences of these programs from Classification Trees, see the *A Brief Comparison of Classification Tree Programs* in Chapter 9.

B

B Coefficients

A line in a two dimensional or two-variable space is defined by the equation Y=a+b*X; in full text: the Y variable can be expressed in terms of a constant (a) and a slope (b) times the X variable. The constant is also referred to as the intercept, and the slope as the regression coefficient or *B coefficient*. In general then, multiple regression procedures will estimate a linear equation of the form:

$$Y = a + b_1*X_1 + b_2*X_2 + \dots + b_p*X_p$$

Note that in this equation, the regression coefficients (or *B coefficients*) represent the independent contributions of each independent variable to the prediction of the dependent variable. However, their values may not be comparable between variables because they depend on the units of measurement or ranges of the respective variables. Some software products will produce both the raw regression coefficients (B coefficients) and the Beta coefficients (note that the Beta coefficients are comparable across variables). See also, Chapter 26 – *Multiple Linear Regression*.

Back Propagation (in Neural Networks)

Back propagation is the best-known training algorithm for neural networks and still one of the most useful. Devised independently by Rumelhart et. al. (1986), Werbos (1974), and Parker (1985), it is thoroughly described in most neural network textbooks (e.g., Patterson, 1996; Fausett, 1994; Haykin, 1994). It has lower memory requirements than most algorithms, and usually reaches an acceptable

error level quite quickly, although it can then be very slow to converge properly on an error minimum. It can be used on most types of networks, although it is most appropriate for training multilayer perceptrons.

Back propagation includes:

- Time-dependent learning rate
- Time-dependent momentum rate
- Random shuffling of order of presentation.
- Additive noise during training
- Independent testing on selection set
- A variety of stopping conditions
- RMS error plotting: graph
- Selectable error function

The last five bulleted items are equally available in other iterative algorithms, including conjugate gradient descent, Quasi-Newton, Levenberg-Marquardt, quick propagation, Delta-bar-Delta, and Kohonen training (apart from noise in conjugate gradients, Kohonen and Levenberg-Marquardt, and selectable error function in Levenberg-Marquardt).

Technical details. The on-line version of back propagation calculates the local gradient of each weight with respect to each case during training. Weights are updated once per training case. The update formula is:

$$\Delta\omega_{ij}(t) = \eta\delta_j o_i + \alpha\Delta\omega_{ij}(t\text{-}1)$$

Where η is the learning rate, δ is the local error gradient, α is the momentum coefficient, and o_i is the output of the i'th unit. Thresholds are treated as weights with $o_i = -1$.

The local error gradient calculation depends on whether the unit into which the weights feed is in the output layer or the hidden layers.

Local gradients in output layers are the product of the derivatives of the network's error function and the units' activation functions.

Local gradients in hidden layers are the weighted sum of the unit's outgoing weights and the local gradients of the units to which these weights connect.

Bagging (Voting, Averaging)

The concept of bagging (voting for classification, averaging for regression-type problems with continuous dependent variables of interest) applies to the area of predictive data mining, to combine the predicted classifications (prediction) from multiple models, or from the same type of model for different learning data. It is also used to address the inherent instability of results when applying complex models to relatively small data sets.

Suppose your data mining task is to build a model for predictive classification, and the data set from which to train the model (learning data set, which contains observed classifications) is relatively small. You could repeatedly sub-sample (with replacement) from the data set, and apply, for example, a tree classifier (e.g., CART and CHAID) to the successive samples. In practice, very different trees will often be grown for the different samples, illustrating the instability of models often evident with small data sets. One method of deriving a single prediction (for new observations) is to use all trees found in the different samples, and to apply some simple voting: The final classification is the one most often predicted by the different trees. Note that some weighted combination of predictions (weighted vote, weighted average) is also possible, and commonly used. A sophisticated (machine learning) algorithm for generating weights for weighted prediction or voting is the boosting procedure.

Banner Tables (Stub and Banner Tables)

Stub-and-banner tables are essentially two-way tables, except that two lists of categorical variables (instead of just two individual variables) are crosstabulated. In the stub-and-banner table, one list will be tabulated in the columns (horizontally) and the second list will be tabulated in the rows (vertically) of the spreadsheet. For more information, see *Stub and Banner Tables* in Chapter 2 – *Basic Statistics and Tables*.

Bar/Column Plot, 2D

The bar/column plot represents sequences of values as bars or columns (one case is represented by one bar/column).

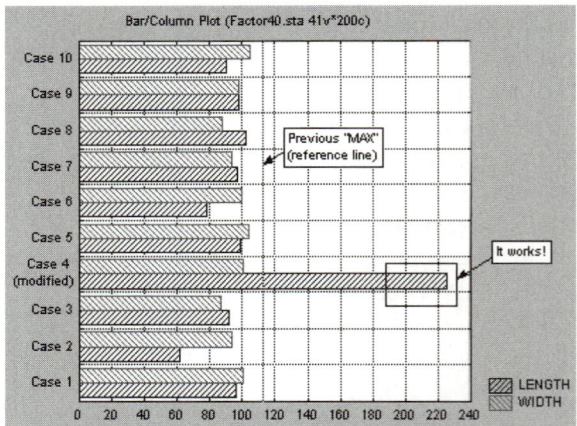

If more than one variable is selected, each plot can be represented in a separate graph or all of them can be combined in one display as multivariate clusters of bars/columns (one cluster per case, see the next illustration).

Bar/Column Plot, 2D - Multiple

Multiple bar/column plots present (in one 2D graph) multivariate clusters of bars/columns (one case = one cluster); one bar/column within each cluster represents one of the selected

variables. The values of all examined variables are plotted against a single y-axis (or x-axis if the horizontal orientation is specified), which facilitates comparisons between analyzed variables.

Bar/Column Plot, 2D – Regular

This graph is a simple bar/column plot for a single variable (if computed for more than one variable, a separate graph is produced for each variable.

Bar Deviation Plot

The bar deviation plot is similar to the Bar X plot, in that individual data points are represented by vertical bars, however, the bars connect the data points to a user-selectable baseline. If the baseline value is different than the plot's *Y-axis* minimum, individual bars will extend either up or down, depending on the direction of the deviation of individual data points from the baseline.

Bar Left Y Plot

In this plot, one horizontal bar is drawn for each data point (i.e., each pair of *XY* coordinates, see the next illustration), connecting the data point and the left *Y-axis*.

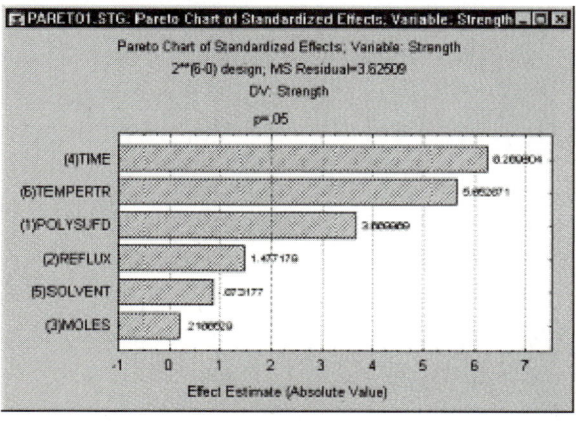

The vertical position of the bar is determined by the data point's *Y* value, and its length by the respective *X* value.

Bar Right Y Plot

In this plot, one horizontal bar is drawn for each data point (i.e., each pair of *XY* coordinates), connecting the data point and the right *Y-axis*. The vertical position of the bar is determined by the data point's *Y* value, and its length by the respective *X* value.

Bar Top Plot

(Also known as hanging column plots.) In this plot, one vertical bar is drawn for each data point (i.e., each pair of *XY* coordinates), connecting the data point and the upper *X-axis*. The horizontal position of the bar is determined by the data point's *X* value, and its length by the respective *Y* value.

Bar X Plot

In this plot, one vertical bar is drawn for each data point (i.e., each pair of *XY* coordinates), connecting the data point and the lower *X-axis*. The horizontal position of the bar is determined by the data point's *X* value, and its height by the respective *Y* value. See the following illustration.

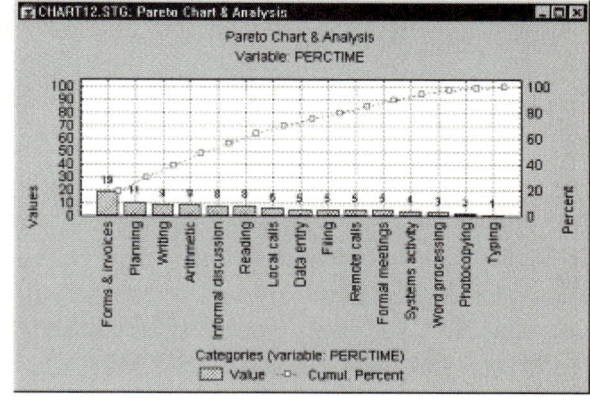

Bartlett Window

In time series, the Bartlett window is a weighted moving average transformation used to smooth the periodogram values. In the Bartlett window (Bartlett, 1950) the weights are computed as:

$$w_j = 1-(j/p) \quad \text{(for } j = 0 \text{ to } p)$$
$$w_{-j} = w_j \quad \text{(for } j \neq 0)$$

where $p = (m-1)/2$ and m is the width of the moving average window (which must be an odd number).

This weight function will assign the greatest weight to the observation being smoothed in the center of the window, and increasingly smaller weights to values that are further away from the center. See also, *Spectrum Analysis – Basic Notations and Principles*, page 514.

Basis Functions

Basis functions of predictor variables (X) play an important role in the estimation of Multivariate Adaptive Regression Splines (MARSplines). Specifically, MARSplines uses two-sided truncated functions of the form $\pm(x - t)_+$ (see the next illustration) as basis functions for linear or nonlinear expansion which approximates the relationships between the response and predictor variables.

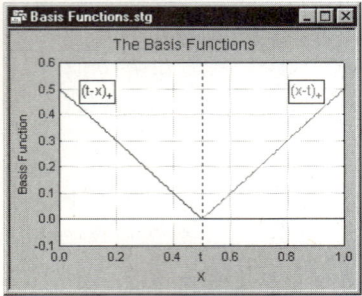

Shown above is a simple example of two basis functions $(t-x)_+$ and $(x-t)_+$. Parameter t is the knot of the basis functions (defining the "pieces" of the piecewise linear regression); these knots (parameters) are also determined from the data.

Bayesian Networks

Networks based on Bayes' theorem, on the inference of probability distributions from data sets. See also, *Probabilistic Neural Networks (PNN)* and *Generalized Regression Neural Networks (GRNN)*.

Bayesian Statistics (Analysis)

Bayesian analysis is an approach to statistical analysis that is based on the Bayes's law, which states that the posterior probability of a parameter p is proportional to the prior probability of parameter p multiplied by the likelihood of p derived from the data collected. This increasingly popular methodology represents an alternative to the traditional (or frequentist probability) approach: whereas the latter attempts to establish confidence intervals around parameters, and/or falsify *a priori* null-hypotheses, the Bayesian approach attempts to keep track of how *a priori* expectations about some phenomenon of interest can be refined, and how observed data can be integrated with such *a priori* beliefs, to arrive at updated posterior expectations about the phenomenon.

A good metaphor (and actual application) for the Bayesian approach is that of a physician who applies consecutive examinations to a patient so as to refine the certainty of a particular diagnosis: The results of each individual examination or test should be combined with the *a priori* knowledge about the patient, and expectation that the respective diagnosis is correct. The goal is to arrive at a final diagnosis that the physician believes to be correct with a known degree of certainty.

Bayes' Theorem

The Bayes' Theorem allows new information to be used to update the conditional probability (see page 594) of an event.

Bernoulli Distribution

The Bernoulli distribution best describes all situations where a trial is made resulting in either success or failure, such as when tossing a coin, or when modeling the success or failure of a surgical procedure. The Bernoulli distribution is defined as:

$$f(x) = p^x * (1-p)^{1-x} \qquad \text{for } x \in \{0,1\}$$

where p is the probability that a particular event (e.g., success) will occur. For a complete listing of all distribution functions, see *Types of Distributions* in Chapter 14 – *Distribution Fitting*.

Best Subset Regression

A model-building technique that finds subsets of predictor variables that best predict responses on a dependent variable by linear (or nonlinear) regression. For an overview of best subset regression, see Chapter 19 – *General Regression Models (GRM)*; for nonlinear stepwise and best subset regression, see Chapter 21 – *Generalized Linear/Nonlinear Models (GLZ)*.

Beta Coefficients

The *beta* coefficients are the regression coefficients you would have obtained had you first standardized all of your variables to a mean of *0* and a standard deviation of *1*. Thus, the advantage of *beta* coefficients (as compared to B coefficients which are not standardized) is that the magnitude of these Beta coefficients enables you to compare the relative contribution of each independent variable in the prediction of the dependent variable. See also, Chapter 26 – *Multiple Linear Regression*.

Beta Distribution

The *beta* distribution (the term first used by Gini, 1911) is defined as:

$$f(x) = \Gamma(v+\omega)/(\Gamma(v)\Gamma(\omega))*x^{v-1}*(1-x)^{\omega-1}$$
$$0 \le x \le 1$$
$$v > 0, \omega > 0$$

where Γ (*gamma*) is the *Gamma* function, and v and ω are the shape parameters.

The following illustrations show the *beta* distribution as the two shape parameters change.

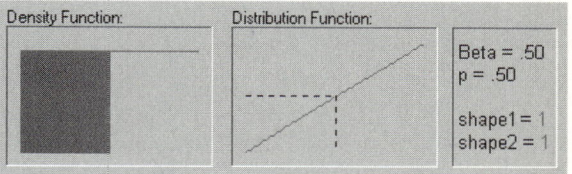

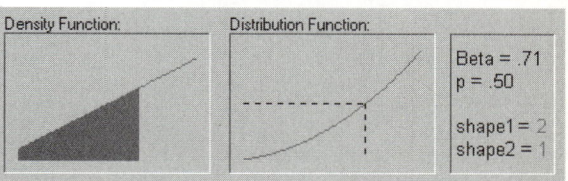

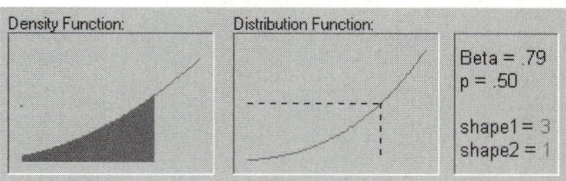

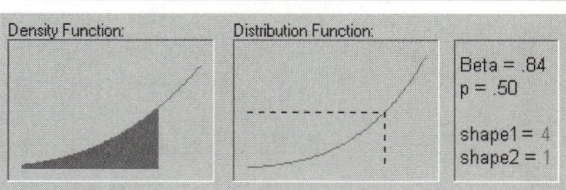

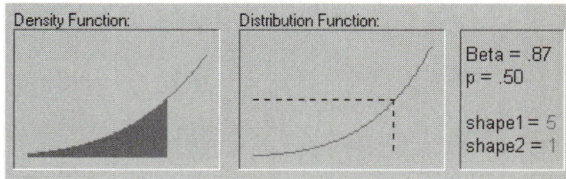

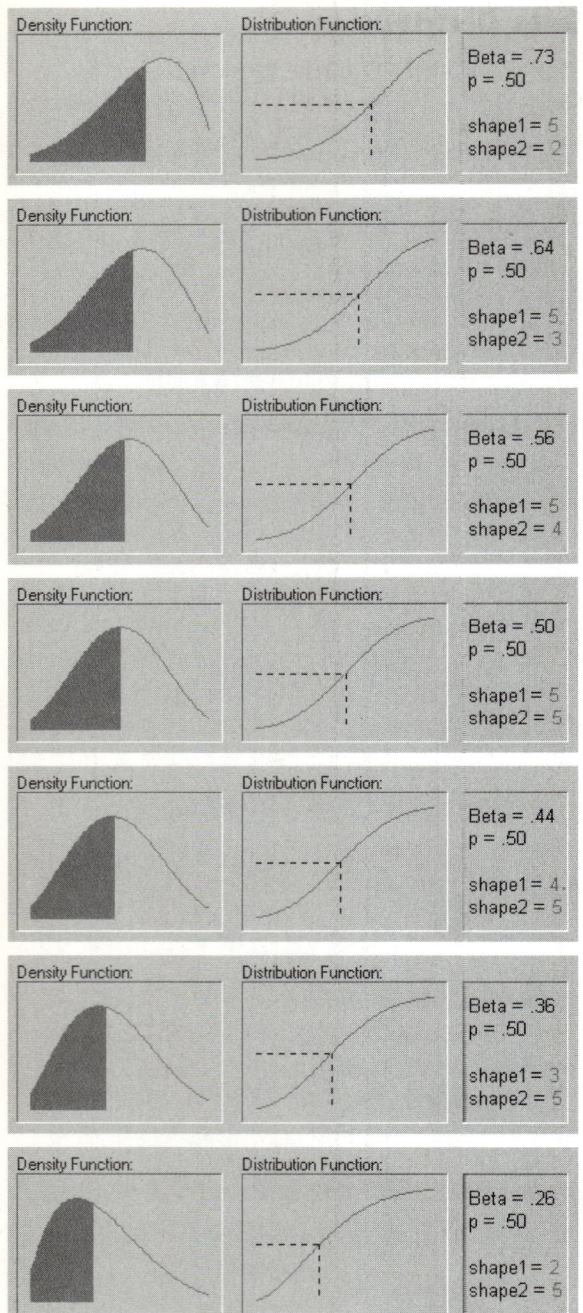

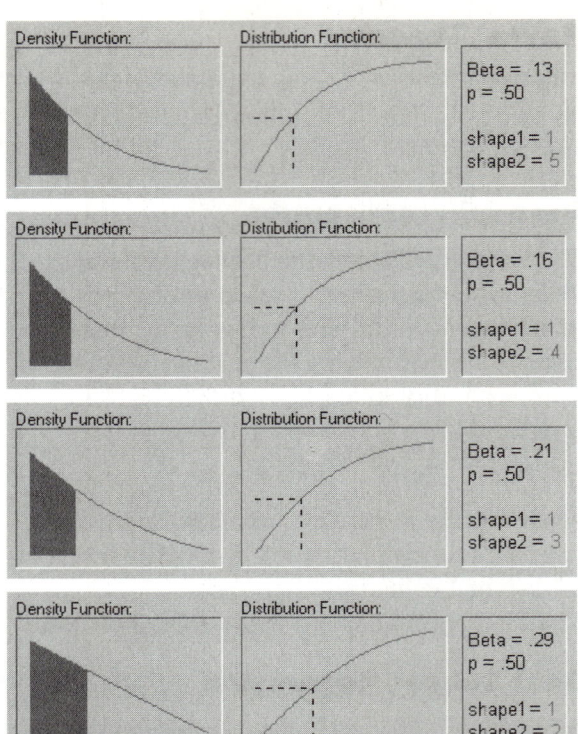

Bimodal Distribution

A distribution that has two modes (two peaks). Bimodality of the distribution in a sample is often a strong indication that the distribution of the variable in population is not normal.

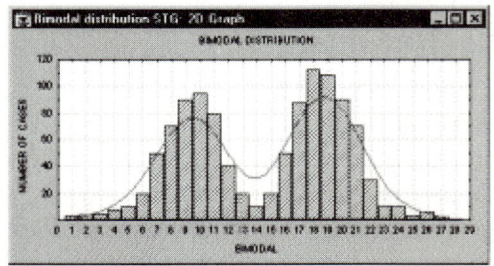

Bimodality of the distribution may provide important information about the nature of the investigated variable (i.e., the measured quality). For example, if the variable represents a reported preference or attitude, bimodality may indicate a polarization of opinions. Often however, the bimodality may indicate that the sample is not homogenous and the observations come in fact from two or more "overlapping" distributions.

Sometimes, bimodality of the distribution may indicate problems with the measurement instrument (e.g., gage calibration problems in natural sciences, or response biases in social sciences). See also, *Unimodal Distribution* and *Multimodal Distribution*.

Binomial Distribution

The binomial distribution (the term first used by Yule, 1911) is defined as:

f(x) = [n!/(x!*(n-x)!)] * p^x * q^{n-x}
for x = 0, 1, 2, ..., n

where *p* is the probability of success at each trial, *q* is equal to *1-p*, and *n* is the number of independent trials.

Bivariate Normal Distribution

Two variables follow the bivariate normal distribution if, for each value of one variable, the corresponding values of another variable are normally distributed. The bivariate normal probability distribution function for a pair of continuous random variables (*X* and *Y*) is given by:

$$f(x,y) = \frac{1}{2\pi\sigma_1\sigma_2\sqrt{1-\rho^2}}$$

$$\times \exp\left\{-\frac{1}{2(1-\rho^2)}\left[\left(\frac{x-\mu_1}{\sigma_1}\right)^2\right.\right.$$

$$\left.\left. -2\rho\left(\frac{x-\mu_1}{\sigma_1}\right)\left(\frac{y-\mu_2}{\sigma_2}\right)+\left(\frac{y-\mu_2}{\sigma_2}\right)^2\right]\right\}$$

$$-\infty < x < \infty, \ -\infty < y < \infty, \ -\infty < \mu_1 < \infty, -\infty < \mu_2 < \infty, \ \sigma_1 > 0, \sigma_2 > 0, \ -1 < p < 1$$

Where μ_1 and μ_2 are the respective means of the random variables *X* and *Y*, σ_1 and σ_2 are the respective standard deviations of the random variables *X* and *Y*, ρ is the correlation coefficient of *X* and *Y*, *e* is the base of the natural logarithm, sometimes called Euler's e (2.71...), and π is the constant *pi* (3.14...).

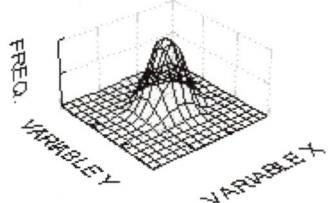

See also, *Why the Normal Distribution is Important* in Chapter 1 – *Elementary Concepts in Statistics* and *Normal Distribution* in the *Statistical Glossary*.

Blocking (in Experimental Design)

In some experiments, observations are organized in natural "chunks" or blocks. You want to make sure that these blocks do not bias your estimates of main effects or interactions. For example, consider an experiment to improve the quality of special ceramics produced in a kiln. The size of the kiln is limited so that you cannot produce all runs (observations) of your experiment at once. In this case, you need to break up the experiment into blocks. However, you do not want to run positive factor settings (for all factors in your experiment) in one block, and all negative settings in the other. Otherwise, any incidental differences between blocks would systematically affect all estimates of the main effects and interactions of the factors of interest. Rather, you want to distribute the runs over the blocks so that any differences between blocks (i.e., the blocking factor) do not bias your results for the factor effects of interest. This is accomplished by treating the blocking factor as another factor in the design. Blocked designs often have the advantage of being statistically more powerful because they enable you to estimate and control the variability in the production process that is due to differences between blocks.

For a detailed discussion of various blocked designs and for examples of how to analyze such designs, see Chapter 15 – *Experimental Design (DOE)* and Chapter 18 – *General Linear Models (GLM)*.

Bonferroni Adjustment

When performing multiple statistical significance tests on the same data, the Bonferroni adjustment can be applied to make it more difficult for any one test to be statistically significant. For example, when reviewing multiple correlation coefficients from a correlation matrix, accepting and interpreting the correlations that are statistically significant at the conventional .05 level may be inappropriate, given that multiple tests are performed. Specifically, the *alpha* error probability of erroneously accepting the observed correlation coefficient as not-equal-to-zero when in fact (in the population) it is equal to zero may be much larger than .05 in this case.

The Bonferroni adjustment usually is accomplished by dividing the alpha level (usually set to .05, .01, etc.) by the number of tests being performing. For instance, suppose you performed multiple tests of individual correlations from the same correlation matrix. The Bonferroni adjusted level of significance for any one correlation would be:

.05 / 5 = .01

Any test that results in a *p*-value of less than .01 is considered statistically significant;

correlations with a probability value greater than .01 (including those with *p*-values between .01 and .05) are considered non-significant.

Bonferroni Test

This post hoc test can be used to determine the significant differences between group means in an analysis of variance setting. The Bonferroni test is very conservative when a large number of group means are being compared (for a detailed discussion of different post hoc tests, see Winer, Michels, & Brown (1991).

For more details, see Chapter 18 – *General Linear Models (GLM)*. See also, *Post hoc Comparisons*. For a discussion of statistical significance, see Chapter 1 – *Elementary Concepts in Statistics*.

Boosting

The concept of boosting applies to the area of predictive data mining, to generate multiple models or classifiers (for prediction or classification), and to derive weights to combine the predictions from those models into a single prediction or predicted classification (see also, *Bagging*, page 566).

A simple algorithm for boosting works like this: Start by applying some method (e.g., a tree classifier such as CART or CHAID) to the learning data, where each observation is assigned an equal weight. Compute the predicted classifications, and apply weights to the observations in the learning sample that are inversely proportional to the accuracy of the classification. In other words, assign greater weight to those observations that were difficult to classify (where the misclassification rate was high), and lower weights to those that were easy to classify (where the misclassification rate was low). In the context of CART for example,

different misclassification costs (for the different classes) can be applied, inversely proportional to the accuracy of prediction in each class. Then apply the classifier again to the weighted data (or with different misclassification costs), and continue with the next iteration (application of the analysis method for classification to the re-weighted data).

Boosting will generate a sequence of classifiers, where each consecutive classifier in the sequence is an expert in classifying observations that were not well classified by those preceding it. During deployment (for prediction or classification of new cases), the predictions from the different classifiers can then be combined (e.g., via voting, or some weighted voting procedure) to derive a single best prediction or classification.

Note that boosting can also be applied to learning methods that do not explicitly support weights or misclassification costs. In that case, random sub-sampling can be applied to the learning data in the successive steps of the iterative boosting procedure, where the probability for selection of an observation into the subsample is inversely proportional to the accuracy of the prediction for that observation in the previous iteration (in the sequence of iterations of the boosting procedure).

Boundary Case

A boundary case occurs when a parameter iterates to the boundary of the permissible parameter space (see Chapter 35 – *Structural Equation Modeling*). For example, a variance can only take on values from 0 to infinity. If, during iteration, the program attempts to move an estimate of a variance below zero, the program will constrain it to be on the boundary value of 0.

For some problems (for example a Heywood Case in factor analysis), it may be possible to reduce the discrepancy function by estimating a variance to be a negative number. In that case, the program does "the best it can" within the permissible parameter space, but does not actually obtain the "global minimum" of the discrepancy function.

Box Plot, 2D

A box plot (the term first used Tukey, 1970) concisely displays the distribution of a variable. This type of graph will place a box around the midpoint (i.e., mean or median) which represents a selected range (i.e., standard error, standard deviation, min-max, or constant) and whiskers outside of the box which also represent a range. One of the most common configurations is shown below:

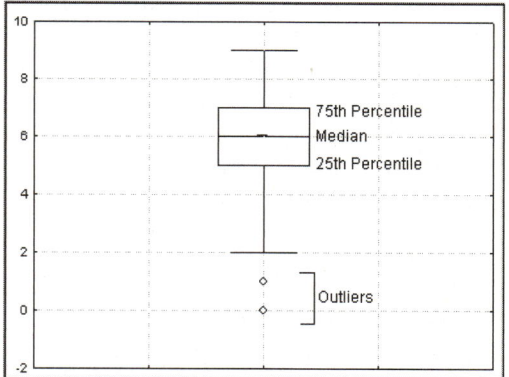

The central tendency (e.g., median or mean), and range or variation statistics (e.g., quartiles, standard errors, or standard deviations) are computed for each group of cases and the selected values are presented in the selected box plot style. Outlier data points can also be plotted.

Box Plot, 2D – Boxes

This type of box plot will place a box around the midpoint (i.e., mean or median) that represents the selected range (i.e., standard error, standard deviation, min-max, or constant).

Box Plot, 2D – Box Whiskers

This type of box plot will place a box around the midpoint (i.e., mean or median) that represents a selected range (i.e., standard error, standard deviation, min-max, or constant) and whiskers outside of the box that also represent a selected range (see the next illustration).

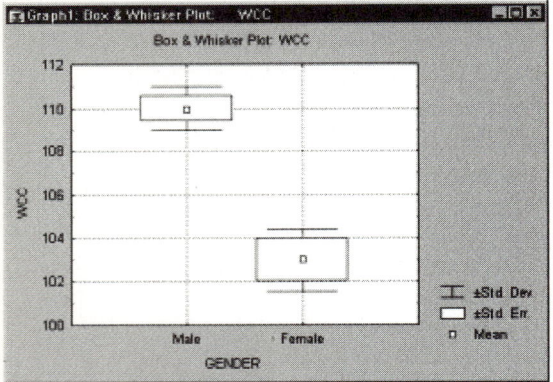

Box Plot, 2D – Columns

In this type of box plot, vertical columns are used to represents the variable's midpoint (i.e., mean or median). The whiskers superimposed on each column mark the selected range (i.e., standard error, standard deviation, min-max, or constant) around the midpoint.

Box Plot, 2D – Error Bars

In this style of 2D box plots, the ranges or error bars are calculated from the data. The central tendency (e.g., median or mean), and range or variation statistics (e.g., min-max values, quartiles, standard errors, or standard

deviations) are computed for each variable and the selected values are presented as error bars.

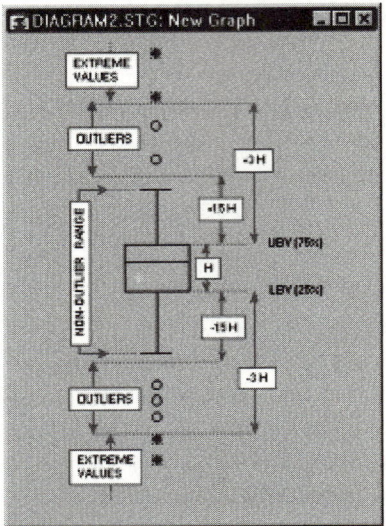

The diagram above illustrates the ranges of outliers and extremes in the "classic" box and whisker plot (for more information about box plots, see Tukey, 1977).

Box Plot, 2D – Mean/SD/1.96*SD

This box-whisker plot will show the mean (small box in the plot) of a category surrounded by a larger box (± 1 times the standard deviation). If the distribution is normal, the "whiskers" in this plot represent a "95% confidence interval" defined as the category mean ± 1.96 times the category standard deviation.

Box Plot, 2D – Mean/SE/1.96*SE

This box-whisker plot will show the mean (small box in the plot) of a category surrounded by a larger box (± 1 times the standard error). If the distribution is normal, then the "whiskers" in this plot represent a "95% confidence interval" defined as the category mean ± 1.96 times the category standard deviation.

Box Plot, 2D – Mean/SE/SD

In this type of box-whisker plot, the smallest box in the plot represents the mean (central tendency) of the category values, while the dispersion (variability) is represented by ± 1 times the standard error (large box) and ± 1 times the standard deviation about the mean ("whiskers").

Box Plots, 2D – Median/Quartiles/Range

This box plot describes the central tendency of each category of the variable in terms of the median of the values in the category (represented by the smallest box in the plot). The spread (variability) in each category's values are represented in this plot by the quartiles (the 25th and 75th percentiles, larger box in the plot) and the minimum and maximum values of the variable (the "whiskers" in the plot).

Box Plot, 2D – Whiskers

In this style of box plot, the range (i.e., standard error, standard deviation, min-max, or constant) is represented by whiskers (i.e., as a line with a serif on both ends, see the next illustration).

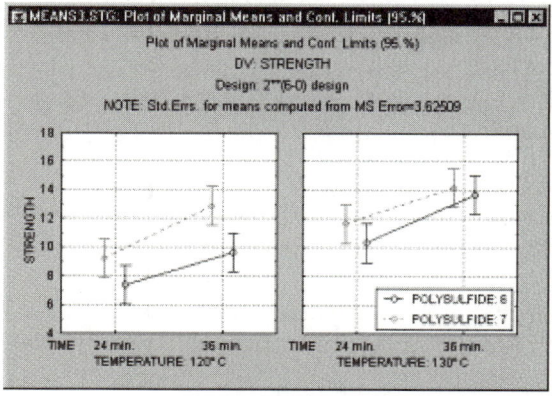

Box Plot, 3D

In box plots (this term was first used by Tukey, 1970), ranges or distribution characteristics of values of selected variables are plotted separately for groups of cases defined by values of a categorical (grouping) variable.

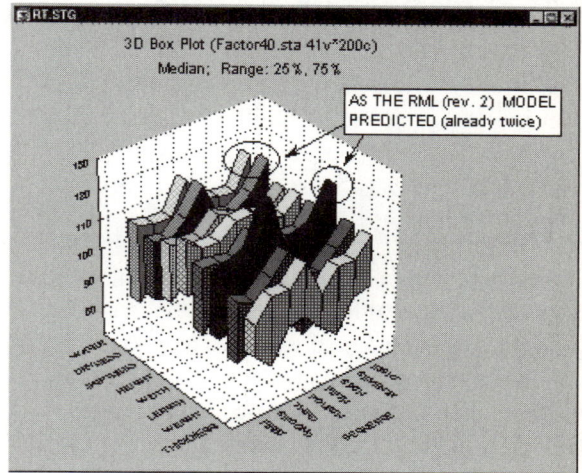

The central tendency (e.g., median or mean), and range or variation statistics (e.g., quartiles, standard errors, or standard deviations) are computed for each group of cases and the selected values are presented in the selected box plot style. Outlier data points can also be plotted.

Box Plot, 3D – Border-Style Ranges

In this style of 3D sequential box plot, the ranges of values of selected variables are plotted separately for groups of cases defined by values of a categorical (grouping) variable. The central tendency (e.g., median or mean), and range or variation statistics (e.g., quartiles, standard errors, or standard deviations) are computed for each variable and for each group of cases and the selected values are presented as points with whiskers, and the ranges marked by the whiskers are connected with lines (i.e., range borders) separately for each variable.

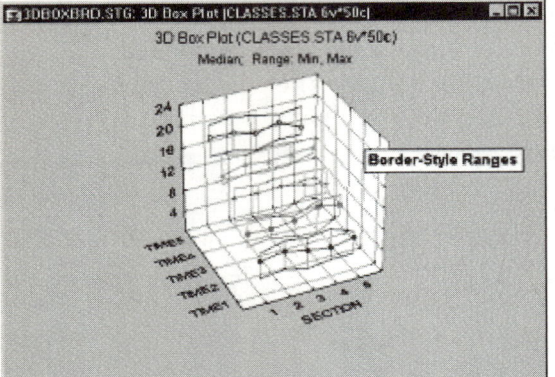

3D range plots (see the next illustration) differ from 3D box plots in that for range plots, the ranges are the values of the selected variables (e.g., one variable contains the minimum range values and another variable contains the maximum range values) while for box plots, the ranges are calculated from variable values (e.g., standard deviations, standard errors, or min-max value).

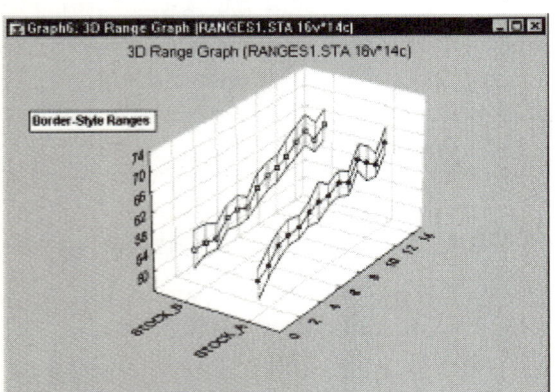

Box Plot, 3D – Double Ribbon Ranges

In this style of 3D sequential box plot, the ranges of values of selected variables are plotted separately for groups of cases defined by values of a categorical (grouping) variable.

The range or variation statistics (e.g., quartiles, standard errors, or standard deviations) are computed for each variable and for each group of cases and the selected values are presented as double ribbons.

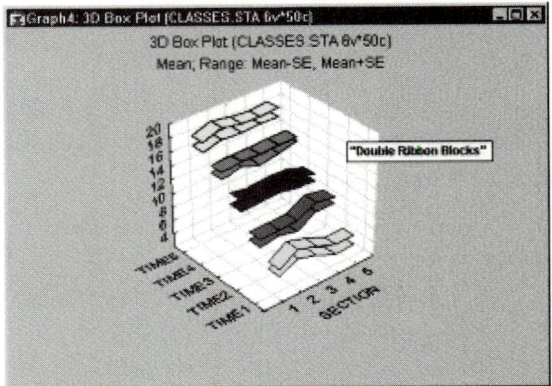

3D range plots (see the next illustration) differ from 3D box plots in that for range plots, the ranges are the values of the selected variables (e.g., one variable contains the minimum range values and another variable contains the maximum range values) while for box plots the ranges are calculated from variable values (e.g., standard deviations, standard errors, or min-max value).

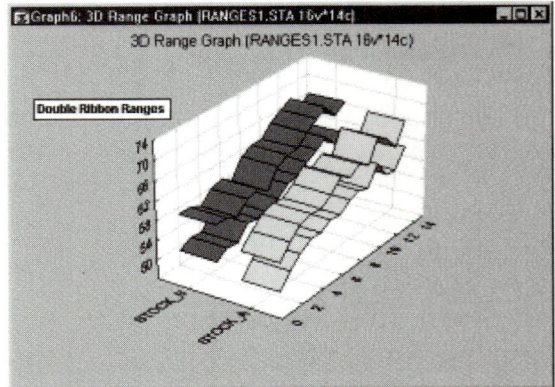

Box Plot, 3D – Error Bars

In this style of 3D sequential box plot, the ranges of values of selected variables are

plotted separately for groups of cases defined by values of a categorical (grouping) variable.

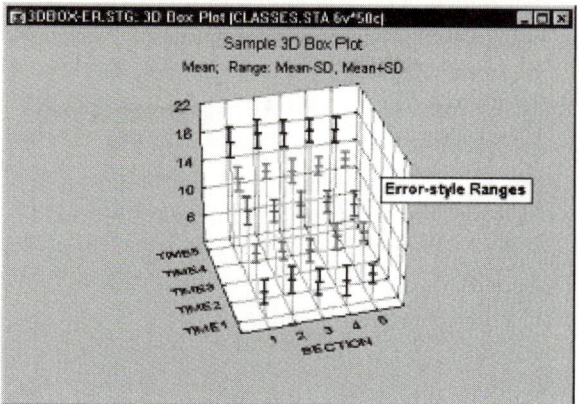

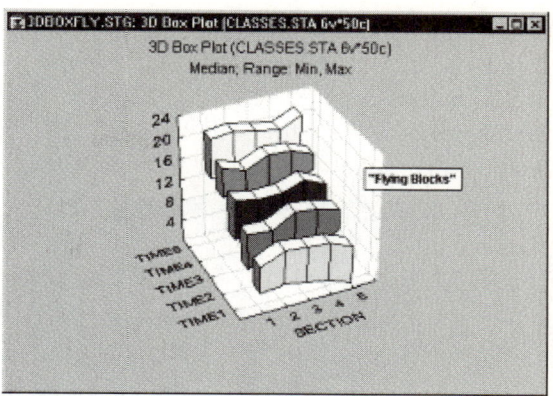

The central tendency (e.g., median or mean), and range or variation statistics (e.g., min-max values, quartiles, standard errors, or standard deviations) are computed for each variable and for each group of cases and the selected values are presented as error bars.

3D range plots differ from 3D box plots in that for range plots, the ranges are the values of the selected variables (e.g., one variable contains the minimum range values and another variable contains the maximum range values) while for box plots the ranges are calculated from variable values (e.g., standard deviations, standard errors, or min-max value).

Box Plot, 3D – Flying Blocks

In this style of 3D sequential box plot, the ranges of values of selected variables are plotted separately for groups of cases defined by values of a categorical (grouping) variable.

The central tendency (e.g., median or mean), and range or variation statistics (e.g., quartiles, standard errors, or standard deviations) are computed for each variable and for each group of cases and the selected values are presented as flying blocks.

3D range plots (see the next illustration) differ from 3D box plots in that for range plots, the ranges are the values of the selected variables (e.g., one variable contains the minimum range values and another variable contains the maximum range values) while for box plots the ranges are calculated from variable values (e.g., standard deviations, standard errors, or min-max value).

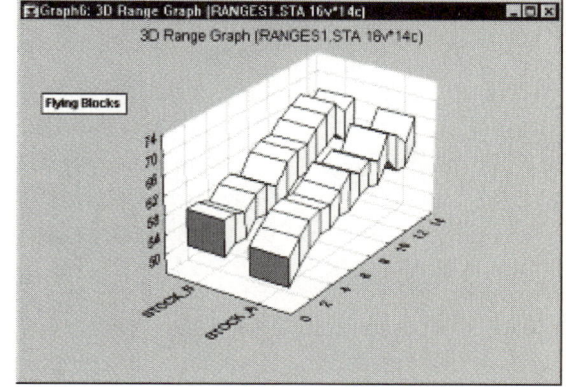

Box Plot, 3D – Flying Boxes

In this style of 3D sequential box plot, the ranges of values of selected variables are plotted separately for groups of cases defined by values of a categorical (grouping) variable.

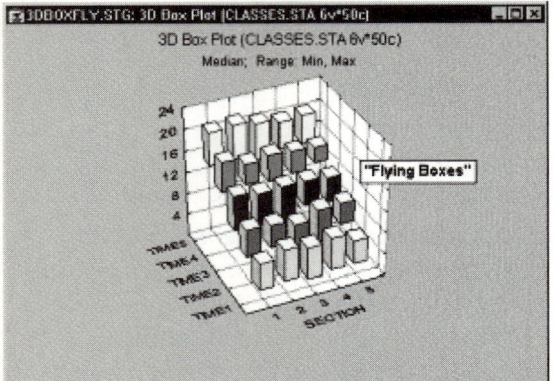

The central tendency (e.g., median or mean), and range or variation statistics (e.g., quartiles, standard errors, or standard deviations) are computed for each variable and for each group of cases and the selected values are presented as flying boxes.

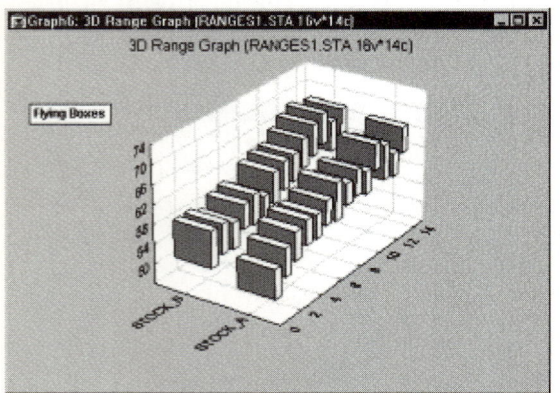

3D range plots (see the previous illustration) differ from 3D box plots in that for range plots, the ranges are the values of the selected variables (e.g., one variable contains the minimum range values and another variable

contains the maximum range values) while for box plots the ranges are calculated from variable values (e.g., standard deviations, standard errors, or min-max value).

Box Plot, 3D – Points

In this style of 3D sequential box plot, the ranges of values of selected variables are plotted separately for groups of cases defined by values of a categorical (grouping) variable. The central tendency (e.g., median or mean), and range or variation statistics (e.g., quartiles, standard errors, or standard deviations) are computed for each variable and for each group of cases and the selected values are presented as point markers connected by a line.

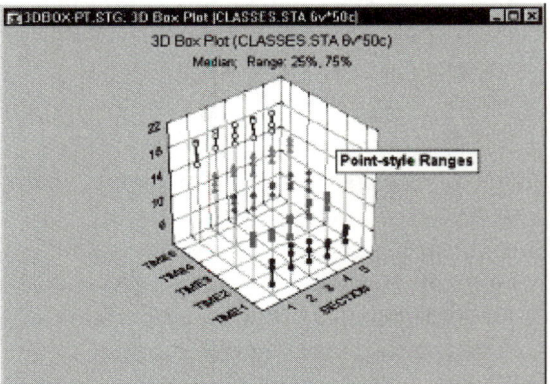

3D range plots (see the next illustration) differ from 3D box plots in that for range plots, the ranges are the values of the selected variables (e.g., one variable contains the minimum range values and another variable contains the maximum range values) while for box plots, the ranges are calculated from variable values (e.g., standard deviations, standard errors, or min-max value). See the next illustration.

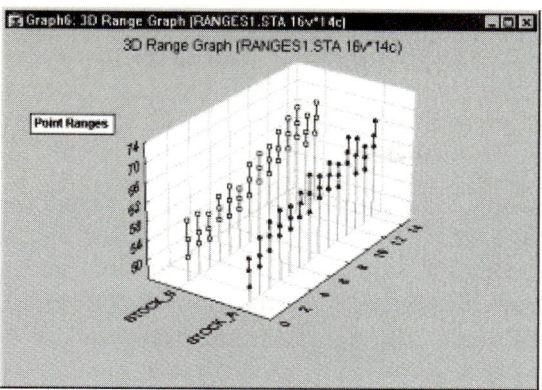

Box-Ljung Q Statistic

In Time Series analysis, you can shift a series by a given lag k. For that given lag, the Box-Ljung Q statistic is defined by:

$$Q_k = n*(n+2)*Sum(r_i^2/(n-1))$$
for i = 1 to k

When the number of observations is large, the Q statistic has a *chi*-square distribution with k-p-q degrees of freedom, where p and q are the number of autoregressive and moving average parameters, respectively.

Breakdowns

Breakdowns are procedures that enable us to calculate descriptive statistics and correlations for dependent variables in each of a number of groups defined by one or more grouping (independent) variables. It is used as either a hypothesis testing or exploratory method. For more information, see *Breakdowns* in Chapter 2 – *Basic Statistics and Tables*.

Breaking Down (Categorizing, Grouping, Slicing, Drilling-down)

One of the most important, general, and also powerful analytic methods involves dividing (splitting) the data set into categories in order

compare the patterns of data between the resulting subsets. This common technique is known under a variety of terms (such as breaking down, grouping, categorizing, splitting, slicing, drilling-down, or conditioning) and it is used both in exploratory data analyses and hypothesis testing. For example: A positive relation between the age and the risk of a heart attack may be different in males and females (it may be stronger in males). A promising relation between taking a drug and a decrease of the cholesterol level may be present only in women with a low blood pressure and only in their thirties and forties. The process capability indices or capability histograms can be different for periods of time supervised by different operators. The regression slopes can be different in different experimental groups.

There are many computational techniques that capitalize on grouping and that are designed to quantify the differences that the grouping will reveal (e.g., ANOVA/MANOVA). However, graphical techniques (such as categorized graphs) offer unique advantages that cannot be substituted by any computational method alone: they can reveal patterns that cannot be easily quantified (e.g., complex interactions, exceptions, anomalies) and they provide unique, multidimensional, global analytic perspectives to explore or mine the data.

Brown-Forsythe and Levene's Tests for Homogeneity of Variances (HOV)

An important assumption in analysis of variance (ANOVA and the *t*-test for mean differences) is that the variances in the different groups are equal (homogeneous). Two powerful and commonly used tests of this assumption are the Levene test and the Brown-Forsythe modification of this test. However, it is important to realize that 1) the homogeneity of variances assumption is usually not as crucial as other assumptions for ANOVA, in particular in the case of balanced (equal n) designs (see also, ANOVA Homogeneity of Variances and Covariances), and 2) that the tests described below are not necessarily very robust themselves (e.g., Glass and Hopkins, 1996, p. 436, call these tests "fatally flawed;" see also, the description of these tests below). If you are concerned about a violation of the HOV assumption, it is always advisable to repeat the key analyses using nonparametric methods.

Brown & Forsythe's test (homogeneity of variances). Recently, some authors (e.g., Glass and Hopkins, 1996) have called into question the power of the Levene test for unequal variances. Specifically, the absolute deviation (from the group means) scores can be expected to be highly skewed; thus, the normality assumption for the ANOVA of those absolute deviation scores is usually violated. This poses a particular problem when there is unequal n in the two (or more) groups that are to be compared. A more robust test that is very similar to the Levene test has been proposed by Brown and Forsythe (1974). Instead of performing the ANOVA on the deviations from the mean, you can perform the analysis on the deviations from the group medians. Olejnik and Algina (1987) have shown that this test will give quite accurate error rates even when the underlying distributions for the raw scores deviate significantly from the normal distribution. However, as Glass and Hopkins (1996, p. 436) have pointed out, both the Levene test as well as the Brown-Forsythe modification suffer from what those authors call a "fatal flaw," namely, that both tests themselves rely on the homogeneity of

variances assumption (of the absolute deviations from the means or medians); and hence, it is not clear how robust these tests are themselves in the presence of significant variance heterogeneity and unequal n.

Levene's test (homogeneity of variances). For each dependent variable, an analysis of variance is performed on the absolute deviations of values from the respective group means. If the Levene test is statistically significant, the hypothesis of homogeneous variances should be rejected.

Brushing

Perhaps the most common and historically first widely used technique explicitly identified as graphical exploratory data analysis is brushing, an interactive method enabling you to select specific data points or subsets of data on screen and identify their (e.g., common) characteristics, or to examine their effects on relations between relevant variables (e.g., in scatterplot matrices) or to identify (e.g., label) outliers. For more information, see *Brushing*, page 544.

Burt Table

Multiple correspondence analysis expects as input (i.e., the program will compute prior to the analysis) a so-called Burt table. The Burt table is the result of the inner product of a design or indicator matrix. If you denote the data (design or indicator matrix) as matrix **X**, then matrix product **X'X** is a *Burt* table); shown in the next illustration is an example of a *Burt* table that you might obtain in this manner.

Data: Burt Table* (8v by 8c)	Survival: No	Survival: Yes	Age: <50	Age: 50-69	Age: 69+	Location: Tokyo	Location: Boston	Location: Glamorgn
Survival: No	210	0	68	93	49	60	82	68
Survival: Yes	0	554	212	258	84	230	171	153
Age: Under 50	69	212	280	0	0	151	58	71
Age: 50-69	93	258	0	351	0	120	122	109
Age: Over 69	49	84	0	0	133	19	73	41
Location: Tokyo	60	230	151	120	19	290	0	0
Location: Boston	82	171	58	122	73	0	253	0
Location: Glamorgn	68	153	71	109	41	0	0	221

Overall, the data matrix is symmetrical. In the case of 3 categorical variables (as shown above), the data matrix consists 3 x 3 = 9 partitions, created by each variable being tabulated against itself, and against the categories of all other variables. Note that the sum of the diagonal elements in each diagonal partition (i.e., where the respective variables are tabulated against themselves) is constant (equal to 764 in this case). The off-diagonal elements in each partition in this example are all *0*. If the cases in the design or indicator matrix are assigned to categories via fuzzy coding, the off-diagonal elements of the diagonal partitions are not necessarily equal to 0.

C

Canonical Correlation

With canonical correlation, you can investigate the relationship between two sets of variables (this analysis type is used as either a hypothesis testing or exploratory method).

For example, an educational researcher may want to compute the (simultaneous) relationship between three measures of scholastic ability with five measures of success in school.

A sociologist may want to investigate the relationship between two predictors of social mobility based on interviews, with actual subsequent social mobility as measured by four different indicators.

A medical researcher may want to study the relationship of various risk factors to the development of a group of symptoms.

In all of these cases, the researcher is interested in the relationship between two *sets* of variables, and canonical correlation would be an appropriate method of analysis. See Chapter 6 – *Canonical Analysis*.

CART

See Chapter 8 – *Classification and Regression Trees (CART)* and *Classification and Regression Trees*, page 588.

Cartesian Coordinates

Cartesian coordinates (*x, y,* or *x, y, z*; also known as rectangular coordinates) are directed distances from two (or three) perpendicular axes.

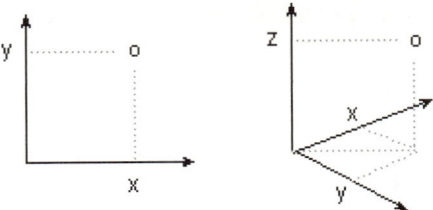

The location of a point in space is established by the corresponding coordinates on the *X*-and *Y*-axes (or *X-, Y-* , and *Z*-axes). See also, *Polar Coordinates*.

Casewise MD Deletion

When casewise deletion of missing data is selected, only cases that do not contain any missing data for any of the variables selected for the analysis will be included in the analysis.

In the case of correlations, all correlations are calculated by excluding cases that have missing data for any of the selected variables (all correlations are based on the same set of data). See also, *Casewise vs. Pairwise Deletion of Missing Data* in Chapter 2 – *Basic Statistics and Tables*.

Categorical Dependent Variable

A categorical dependent variable is a variable of interest (a researcher wants to predict), measured on a nominal scale, whose values identify class or group membership (e.g., *Gender*, with classes *Male* and *Female*; or *Education*, with classes *No High School Degree, High School Degree, Some College, College Degree, Some Graduate School, Graduate Degree*).

A researcher may be interested in predicting the group membership of observations based on the values of some independent or predictor variables. For example, *Credit Risk*, with values

Good and *Bad*, would be a categorical dependent variable that you might want to predict based on measured independent variables that are possibly related to credit risk.

Categorical Predictor Variable

A categorical predictor variable is a variable, measured on a nominal scale, whose categories identify class or group membership, which is used to predict responses on one or more dependent variables.

Gender is an example of a categorical predictor variable, with the two classes or groups *Male* and *Female*. See also, *Nominal Scale* of measurement.

Categorized Graphs (or Trellis Graphs)

Categorized graphs (the term first used in *STATISTICA* software by StatSoft in 1990; also recently called Trellis graphs, by Becker, Cleveland, and Clark, at Bell Labs) produce a series of 2D, 3D, ternary, or nD graphs (such as histograms, scatterplots, line plots, surface plots, ternary scatterplots, etc.), one for each selected category of cases (i.e., subset of cases); for example, respondents from New York, Chicago, Dallas, etc. These component graphs are placed sequentially in one display, allowing for comparisons between the patterns of data shown in graphs for each of the specified groups (e.g., cities).

A variety of methods can be used to select the subsets; the simplest of them is using a categorical variable (e.g., a variable *City*, with three values *New York*, *Chicago*, and *Dallas*). For example, the following graph shows histograms of a variable representing self-reported stress levels in each of the three cities.

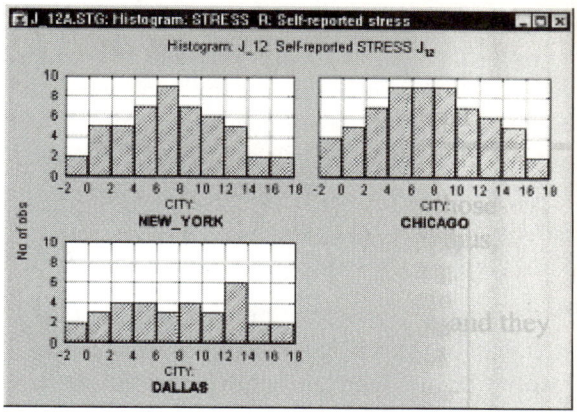

You could conclude that the data suggest that people who live in Dallas are less likely to report being stressed, while the patterns (distributions) of stress reporting in New York and Chicago are quite similar.

Categorized graphs in some software systems also support two-way or multi-way categorizations, where not one criterion (e.g., *City*) but two or more criteria (e.g., *City* and *Time* of the day) are used to create the subsets.

Categorizing, Grouping, Slicing, Drilling-Down

One of the most important, general, and powerful analytic methods involves dividing (splitting) the data set into categories in order to compare the patterns of data between the resulting subsets. This common technique is known under a variety of terms (such as breaking down, grouping, categorizing, splitting, slicing, drilling-down, or conditioning), and it is used both in exploratory data analyses and hypothesis testing.

For example: A positive relation between the age and the risk of a heart attack may be different in males and females (it may be stronger in males). A promising relation between taking a drug and a decrease of the cholesterol level may be present only in women with a low blood pressure and only in their thirties and forties. The process capability indices or capability histograms can be different for periods of time supervised by different operators. The regression slopes can be different in different experimental groups.

There are many computational techniques that capitalize on grouping and that are designed to quantify the differences that the grouping will reveal (e.g., ANOVA/MANOVA). However, graphical techniques (such as categorized graphs) offer unique advantages that cannot be substituted by any computational method alone: they can reveal patterns that cannot be easily quantified (e.g., complex interactions, exceptions, anomalies) and they provide unique, multidimensional, global analytic perspectives to explore or mine the data.

Cauchy Distribution

The Cauchy distribution (the term first used by Upensky, 1937) has density function:

$$f(x) = 1/(\theta\pi*\{1 + [(x-\eta)/\theta]^2\})$$
$$0 < \theta$$

where η is the location parameter (median), θ is the scale parameter, and π is the constant Pi (3.14...).

The following illustrations show the changing shape of the Cauchy distribution when the location parameter equals 0 and the scale parameter equals 1, 2, 3, and 4.

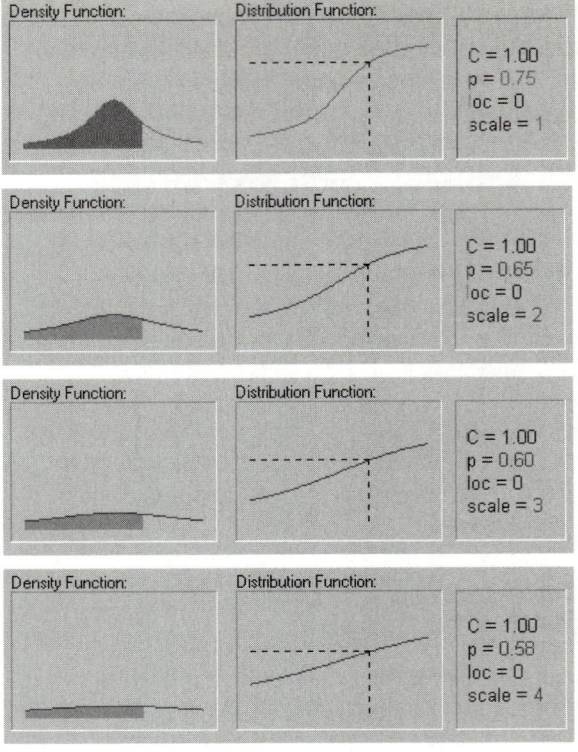

Cause-and-Effect Diagram

The cause-and-effect diagram provides an efficient summary of factors that impact a process and, hence, can be used as a map to guide the overall quality improvement efforts. Therefore, it is one of the important tools for the Define phase of Six Sigma quality control efforts.

The diagram is also sometimes referred to as a "fishbone chart" because of its appearance, or an Ishikawa chart. This name refers to the work of Professor Kaoru Ishikawa, Tokyo University, who developed this diagram to depict variables that are present in a process.

The general idea of the chart is straightforward. Suppose you want to turn on a reading light in your house one evening, and it won't light up.

Now consider the various variables or characteristics that make up the process (cause the light to come on), and which should be considered in order to fix this quality problem.

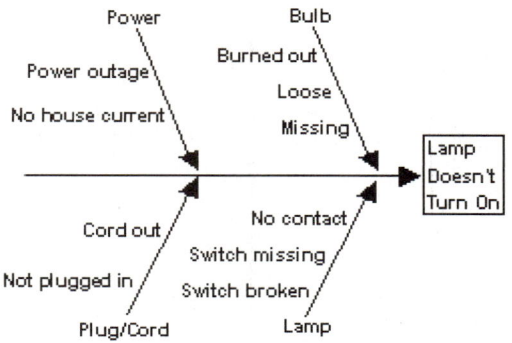

The cause-and-effect diagram shown above (adapted from Rath & Strong's Six Sigma pocket guide, 2000) spells out the various potential causes of the problem encountered.

Usually, the chart is constructed by identifying 1) the major categories of causes that affect the process (in this example *Power*, *Bulb*, *Plug/Cord* and *Lamp*), and 2) the individual factors or causes that can be classified into these major categories (e.g., *Power outage*, *No house current*, etc.). You could now use this map as a guide to troubleshooting the problem you encountered turning on your reading light. You can also further "augment" this chart by adding various sub-sub causes, causes that you ruled out, solutions you have tried, etc.

The cause-and-effect diagram plays a central role in Six Sigma quality programs. During the first stage of the Define-Measure-Analyze-Improve-Control (DMAIC) cycle, this diagram can be of great utility in order to identify the areas, departments, processes, and stakeholders that should be involved in the effort. See Harry and Schroeder (2000), Pyzdek (2001), or Rath and Strong (2000) for additional details; see also, *Six Sigma Process*.

Censoring (Censored Observations)

Observations are referred to as censored when the dependent variable of interest represents the time to a terminal event, and the duration of the study is limited in time. Although the concept was developed in the biomedical research, censored observations may occur in a number of different areas of research.

For example, in the social sciences we may study the "survival" of marriages, high school dropout rates (time to drop-out), turnover in organizations, etc. In each case, by the end of the study period, some subjects probably will still be married, will not have dropped out, or will still be working at the same company; thus, those subjects represent censored observations.

In economics, we may study the survival of new businesses or the survival times of products such as automobiles. In quality control research, it is common practice to study the survival of parts under stress (failure time analysis).

Data sets with censored observations can be analyzed via survival analysis or via Weibull and reliability/failure time analysis.

See also, *Type I and II Censoring*, *Single and Multiple Censoring*, and *Left and Right Censoring*.

Censoring (Left and Right)

When censoring, a distinction can be made to reflect the "side" of the time dimension at which censoring occurs. Consider an experiment where we start with 100 light bulbs, and terminate the experiment after a certain amount of time. In this experiment the censoring always occurs on the right side (right censoring), because the researcher knows exactly when the experiment started, and the censoring always occurs on the right side of the time continuum.

Alternatively, it is conceivable that the censoring occurs on the left side (left censoring). For example, in biomedical research you may know that a patient entered the hospital at a particular date, and that he or she survived for a certain amount of time thereafter; however, the researcher does not know exactly when the symptoms of the disease first occurred or were diagnosed.

Data sets with censored observations can be analyzed via survival analysis or via Weibull and reliability/failure time analysis. See also, *Type I and II Censoring* and *Single and Multiple Censoring.*

Central Limit Theorem

As the sample size of independent, identically distributed, random observations increases, the distribution of a variable representing a sum of these observations (i.e., the sampling distribution) approximates the normal distribution (regardless of the distribution of these observations in the population). This principle explains the importance and ubiquity of the normal distribution in statistical inference.

The following illustrations demonstrate the central limit theorem.

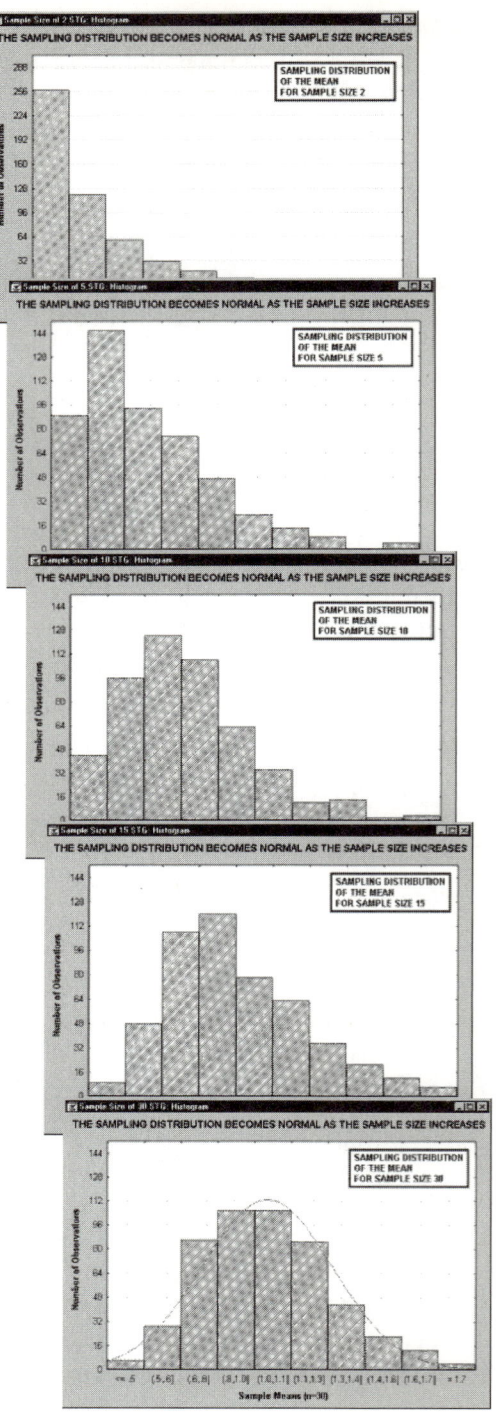

The term central limit theorem was first used by Polya, 1920; German, "Zentraler Grenzwertsatz."

CHAID

CHAID is a classification trees program developed by Kass (1980) that performs multi-level splits when computing classification trees. For discussion of the differences of *CHAID* from other classification tree programs, see *A Brief Comparison of Classification Tree Programs* in Chapter 9 – *Classification Trees*.

Characteristic Life

In Weibull and reliability/failure time analysis, the characteristic life is defined as the point in time where 63.2 percent of the population will have failed; this point is also equal to the respective scale parameter b of the 2-parameter Weibull distribution (with $\theta = 0$; otherwise it is equal to b+θ).

Chi-Square Distribution

The *chi*-square distribution is defined by:

$$f(x) = \{1/[2^{v/2} * \Gamma (v/2)]\} * [x^{(v/2)-1} * e^{-x/2}]$$
$$v = 1, 2, ..., 0 < x$$

where v is the degrees of freedom; e is the base of the natural logarithm, sometimes called Euler's e (2.71...), and Γ (*gamma*) is the *gamma* function.

The following illustrations show the shape of the *chi-square* distribution as the degrees of freedom increase (1, 2, 5, 10, 25 and 50).

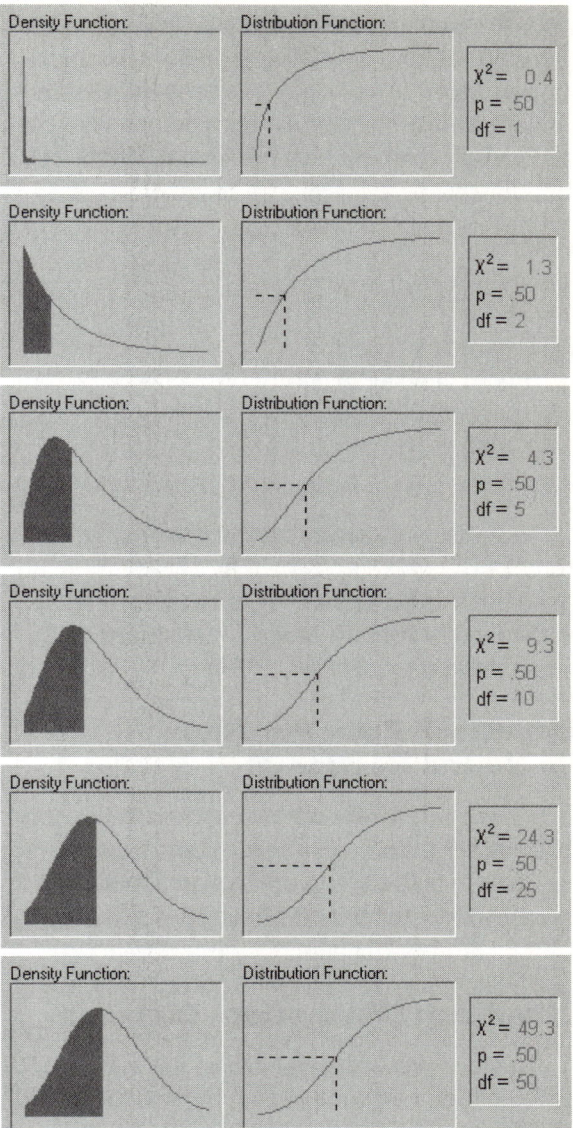

Circumplex

A circumplex is a set of variables that, when plotted as vectors in *N*-dimensional space, fall in a circular pattern. If a set of variables forms a circumplex, the correlation matrix for the variables will have an unusual pattern called circular structure. In this pattern, the correlations on diagonal strips below the main diagonal tend to be equal, or nearly so, first becoming smaller, then larger again as you move away from the main diagonal. Below is a circular structure for an 8x8 correlation matrix.

1.00
0.80 1.00
0.60 0.80 1.00
0.40 0.60 0.80 1.00
0.20 0.40 0.60 0.80 1.00
0.40 0.20 0.40 0.60 0.80 1.00
0.60 0.40 0.20 0.40 0.60 0.80 1.00
0.80 0.60 0.40 0.20 0.40 0.60 0.80 1.00

Circumplex is a special case of a more general concept of radex, developed by Louis Guttman (who contributed a number of innovative ideas to the theory of multidimensional scaling and factor analysis, Guttman, 1954).

City-Block Error Function (in Neural Networks)

Defines the error between two vectors as the sum of the differences in each component. Less sensitive to outliers than sum-squared, but usually causes poorer training performance (see Bishop, 1995). See also, Chapter 27 – *Neural Networks*.

City-Block (Manhattan) Distance

A distance measure computed as the average difference across dimensions. In most cases, this distance measure yields results similar to the simple Euclidean distance. However, note that in this measure, the effect of single large differences (outliers) is dampened (since they are not squared). See, Cluster Analysis.

Classification

Assigning data (i.e., cases or observations) cases to one of a fixed number of possible classes (represented by a nominal output variable).

Classification and Regression Trees

CART, or classification and regression trees, is a classification tree program developed by Breiman et. al. (1984).

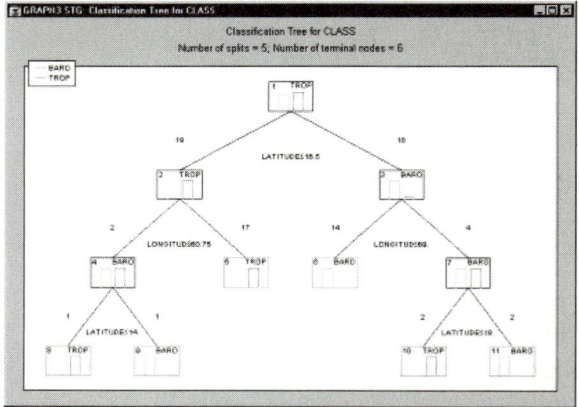

Classification and regression trees is an analytic procedure for predicting the values of a continuous response variable (e.g., *Height*) or categorical response variable (e.g., *Marital Status: Single, Married, Divorced*) from continuous or categorical predictors. When the dependent or response variable of interest is categorical in nature, the technique is referred to as classification trees; if the response variable of interest is continuous in nature, the method is referred to as regression trees.

The classic computational algorithms for classification and regression trees (*CART*) were popularized by Breiman, Friedman, Olshen, & Stone, 1984 (see also, Ripley, 1996; Hastie, Tibshirani, & Friedman, 2001). For classification problems, the goal is generally to find a tree where the terminal tree nodes are relatively pure, i.e., contain observations that (almost) all belong to the same category or class; for regression tree problems node purity is usually defined in terms of the sums-of-squares deviation within each node.

At each step, the program will find a logical split condition to assign observations to the two child nodes; for continuous predictors these logical conditions are usually of the type *If x > Value then NodeID=k*; for categorical predictors, the logical split conditions are usually of the type *If x=Category I then NodeID=k*. For more details, see Chapter 8 – *Classification and Regression Trees (CART)*.

Classification (in Neural Networks)

In classification, the aim is to assign input cases to one of a number of classes. Classification problems fall into two categories: two-class problems, and many-class problems. A two-class problem is usually encoded using a single output neuron. Many-class problems use one output neuron per class. It is also possible to encode two-class problems using this approach (i.e., using two output neurons), and this is in fact the approach taken by PNN networks.

Single output neuron. In these two-class networks, the target output is either 1.0 (indicating membership of one class) or 0.0 (representing membership of the other).

Multiple output neurons. In many-class problems, the target output is 1.0 in the correct class output and 0.0 in the others.

The output neuron activation levels provide confidence estimates for the output classes. It is desirable to be able to interpret these confidence levels as probabilities. If the correct choice of network error function is used during optimization, combined with the correct activation function, such as interpretation can be made. Specifically, a cross entropy error function is combined with the logistic activation function for a two-class problem encoded by a single output neuron or with softmax for a three or more class problem.

The entropic approach corresponds to maximum likelihood optimization, assuming that the data is drawn from the exponential family of distributions. An important feature is that the outputs may be interpreted as posterior estimates of class membership probability.

The alternative approach is to use a sum-squared error function with the logistic output activation function. This has less statistical justification; the network learns a discriminant function, and although the outputs can be treated as confidence measures, they are not probability estimates (and indeed may not even sum to 1.0). Alternatively, such networks sometimes train more quickly, the training process is more stable, and they may achieve higher classification rates.

Classification by Labeled Exemplars (in Neural Networks)

Classification neural networks must translate the numeric level on the output neuron(s) to a nominal output variable. There are two very different approaches to assigning classifications. In one of these, the activation level of the output layer units determines the class, usually by interpreting the activation level as a confidence measure and finding the highest confidence class. This approach is used in most neural network types.

This topic discusses the alternative approach, which is used in SOFM and cluster networks.

These types of networks store labeled exemplar vectors in their radial layer. When a new case is presented to the network, the network in essence calculates the distance between the (possibly normalized) new case and each exemplar vector; the activations of the neurons encode these distances. Each of these neurons has a class label. The class label of the winning (smallest distance from input case) neuron is typically used as the output of the network. The standard algorithm is extended slightly using the KL nearest neighbor algorithm; the class assigned by the network is the most common class among the K winning neurons, provided that at least L of them agree (otherwise, the class is unknown).

The input case might actually be very distant from any of the exemplar vectors, in which case it may be better to assign the case as unknown. You can optionally specify an accept threshold for this eventuality. If the normalized distance is greater than this threshold, the class is unknown.

Classification Statistics (in Neural Networks)

One of the major uses of neural networks is to perform classification tasks, i.e., to assign cases to one of a number of possible classes. The class of a case is indicated by the use of a nominal output variable. The classification statistics include, for each class:

Total. Number of cases of that class.

Correct. Number of cases correctly classified.

Wrong. Number of cases erroneously assigned to another class.

Unknown. Number of cases that could not be positively classified.

Correct (%). Percent of cases correctly classified.

Wrong (%). Percent of cases wrongly classified.

Unknown (%). Percent of cases classified as unknown.

Classification Thresholds (in Neural Networks)

Classification neural networks must translate the numeric activation level of the output neuron(s) to a nominal output variable. There are two very different approaches to assigning classifications. One of these, in which the neural network determines the "winning" neuron or neurons in the radial layer of the network and then uses the class labels on those neurons, is described in *Classification by Labeled Exemplars*, page 589. This approach is used in SOFM and cluster networks.

Here we discuss the alternative approach, where it is the activation level of the output layer units, which are not radial units, that determines the class. This approach is used in all other network types.

Two cases need to be distinguished: single output neuron versus multiple output neurons.

Single output neurons are typically used for two-class problems, with a high output neuron level indicating one class and a low activation the other class. This configuration uses the two-state conversion function.

Multiple output neurons are typically used for three or more class problems. One neuron is used for each class, and the highest activation neuron indicates the class. The neuron activation levels can be interpreted as confidence levels. This method is implemented by using the One-of-N conversion function for the output variable. You can optionally configure a multilayer perceptron to use two output neurons for a two-class output variable.

Single output neuron. Two thresholds are used: accept and reject. In the single output neuron case, the output is considered to be the first class if the output neuron's activation is below the reject threshold, and to be the second class if its activation is above the accept threshold. If the activation is between the two thresholds, the class is regarded as unknown (the so-called doubt option). If the two thresholds are equal, there is no doubt option. There are two common configurations: accept=reject=0.5 implies no doubt option, with the most likely class assigned; accept=0.95, reject=0.05 implies standard "95% confidence" in assignment of a class, with doubt expressed otherwise. Both of these cases assume the standard logistic output neuron activation function, which gives the output neuron a $(0,1)$ range; the thresholds should be adjusted accordingly for different activation functions [e.g., hyperbolic tangent uses output range $(-1, +1)$].

As an alternative to selecting the thresholds yourself, you can specify a loss coefficient that gives the relative cost of the two possible misclassifications (false-positive versus false-negative). A loss coefficient of 1.0 indicates that the two classes are equally important. A loss coefficient above 1.0 indicates that it is relatively more important to correctly recognize class two cases, even at the expense of misclassifying more class one cases. The thresholds (which are equal) are determined by calculating a ROC curve and determining the point on the curve where the ratio of false positives to false negatives equals the loss coefficient. This equalizes the weighted loss on each class, independent of the number of cases in each (i.e. with a loss coefficient of 1.0, it is the proportion of misclassifications in each class that is equalized, not the number of absolute number of misclassifications).

Multiple output neurons. If no thresholds are used, the network uses a "winner takes all" algorithm; the highest activation neuron gives the class. There is a "no doubt" option.

If you specify thresholds, the class is still assigned to the highest neuron, but there is a doubt option. The highest neuron's activation must be above the accept threshold and all other neuron's below the reject threshold in order for the class to be assigned; if this condition is not fulfilled, the class is unknown.

If your multiple output neuron classification network is using the softmax activation function, the output neuron activations are guaranteed to sum to 1.0 and can be interpreted as probabilities of class membership. However, in other cases, although the activations can be interpreted as confidence levels in some sense (i.e. a higher number indicates greater confidence), they are not probabilities, and should be interpreted with caution.

Ordinal classification. If you have a large number of classes, one-of-N encoding can become extremely unwieldy, as the number of neurons in the network proliferates. An alternative approach then is to use ordinal encoding. The output is mapped to a single neuron, with the output class represented by the ordinals 1, 2, 3, etc. The problem with this technique is that it falsely implies an ordering on the classes (i.e. class 1 is more like class 2 than class 3). However, in some circumstances it may be the only viable approach.

You can specify ordinal encoding by changing the conversion function of the output variable to minimax. The ordinals are then mapped into the output range of the neuron. With ordinal encoding, the output is determined as follows. The output neuron's activation is linearly scaled using the factors determined by minimax, then rounded to the nearest integer. This ordinal value gives the class. The only classification threshold used is the accept threshold. If the difference between the output and the selected ordinal is greater than the threshold, then the classification is instead unknown.

Example. The output is 3.8, which is rounded to the nearest ordinal 4. The difference is 0.2. If an accept threshold less than 0.2 has been selected, the classification is rejected, and unknown is generated. An accept threshold of 0.5 or above is equivalent to not using a threshold at all.

Classification Trees

Classification trees are used to predict membership of cases or objects in the classes of a categorical dependent variable from their measurements on one or more predictor variables. For a detailed description of classification trees, see Chapter 9 – *Classification Trees*.

Class Labeling (in Neural Networks)

A variety of clustering algorithms and networks are supported. All of these have networks where the second layer consists of Radial units, and these units contain exemplar vectors. In SOFM and cluster networks, these are combined in a two layer neural network with a single nominal output variable and the K nearest output conversion function to produce a classification based upon the nearest exemplar vector(s) to an input case.

The exemplar vectors can be positioned using a variety of cluster-center and sampling approaches. However, it is also necessary to apply class labels to the radial units (i.e. label each radial unit as representative of a particular class). Class labels can be applied using the class labeling algorithms described here.

KL nearest neighbor labeling. This algorithm assigns labels to units based upon the labels of the K nearest neighboring training cases. Provided that at least L of the K neighbors are of the same class, this class is used to label the unit. If not, a blank label is applied, signifying an unknown class.

Note that this is distinct from (although related to) the KL-Nearest algorithm used when executing cluster networks, which reports the class of at least L of the K nearest units to the input case.

Voronoi Labeling. This algorithm assigns labels to units based upon the labels of the training cases that are "assigned" to that unit. Assigned cases are those that are nearer to this unit than to any other (i.e. those that would be classified by this unit if using the 1-NN classification scheme). These are the cases in the Voronoi neighborhood of the unit. The class of the majority of the training cases is used to label the unit, provided that at least a given minimum proportion of the training cases belong to this majority. If not, a blank label is applied, signifying an unknown class.

Cluster Analysis

The term cluster analysis (first used by Tryon, 1939) actually encompasses a number of different classification algorithms that can be used to develop taxonomies (typically as part of exploratory data analysis). For example, biologists have to organize the different species of animals before a meaningful description of the differences between animals is possible. According to the modern system employed in biology, man belongs to the primates, the mammals, the amniotes, the vertebrates, and the

animals. Note how in this classification, the higher the level of aggregation the less similar are the members in the respective class. Man has more in common with all other primates (e.g., apes) than it does with the more "distant" members of the mammals (e.g., dogs), etc. For information on specific types of cluster analysis methods, see *Joining (Tree Clustering)*, *Two-Way Joining*, and *k-Means Clustering* in Chapter 10 – *Cluster Analysis*; see also, Chapter 9 – *Classification Trees*.

Cluster Diagram (in Neural Networks)

A scatter diagram plotting cases belonging to various classes in two dimensions. The dimensions are provided by the output levels of units in the neural network. See also, *Cluster Analysis*.

Cluster Networks (in Neural Networks)

Cluster networks are actually a non-neural model presented in a neural form for convenience. A cluster network consists of a number of class-labeled exemplar vectors (each represented by a radial neuron). The vectors are assigned centers by clustering algorithms such as *k*-means, and then labeled using nearby cases. After labeling, the centers positions can be fine-tuned using learned vector quantization.

Cluster networks are closely related to Kohonen networks, with a few differences. Cluster networks are intended for supervised learning situations where class labels are available in the training data and cluster networks do not have a topologically organized output layer.

Training consists of center assignment, followed by labeling. This can optionally be followed by LVQ training to improve center

location. After training, the network can be pruned, and classification factors set, including an acceptance threshold and K,L factors for KL nearest neighbor classification.

Codes

Codes are values of a grouping variable (e.g., *1, 2, 3, ...* or *MALE, FEMALE*) that identify the levels of the grouping variable in an analysis. Codes can either be text values or integer values.

Coefficient of Determination

This is the square of the product-moment correlation between two variables (r^2). It expresses the amount of common variation between the two variables. See also, Hays, 1988.

Columns (Box Plot)

In this type of box plot, vertical columns are used to represents the variable's midpoint (i.e., mean or median). The whiskers superimposed on each column mark the selected range (i.e., standard error, standard deviation, min-max, or constant) around the midpoint.

Common Causes

See *Assignable Causes and Actions*, page 562.

Communality

In Principal Components and Factor Analysis, communality is the proportion of variance that each item has in common with other items. The proportion of variance that is unique to each item is then the respective item's total variance minus the communality. A common starting point is to use the squared multiple correlation of an item with all other items as an estimate of the communality (refer to Chapter 26 – *Multiple Linear Regression* for details about multiple regression). Some authors have suggested various iterative post-solution improvements to the initial multiple regression communality estimate; for example, the so-called MINRES method (minimum residual factor method; Harman & Jones, 1966) will try various modifications to the factor loadings with the goal to minimize the residual (unexplained) sums of squares.

Complex Numbers

Complex numbers are the superset that includes all real and imaginary numbers. A complex number is usually represented by the expression $a + ib$ where a and b are real numbers and i is the imaginary part of the expression where i has the property that $i^2=-1$. See also, *Cross-Spectrum Analysis*, page 512, in Chapter 38 – *Time Series/Forecasting*.

Conditional Probability

In many situations, once more information becomes available, we are able to revise our estimates for the probability of further outcomes or events happening. The probability of the A event given that event B has occurred is known as the conditional probability of A given B and is often denoted by p(A|B).

Confidence Interval

The confidence intervals for specific statistics (e.g., means, or regression lines) give us a range of values around the statistic where the "true" (population) statistic can be expected to be located (with a given level of certainty, see also, Chapter 1 – *Elementary Concepts in Statistics*).

The following illustrations show a 90%, 95%, and 99% confidence interval for the regression line.

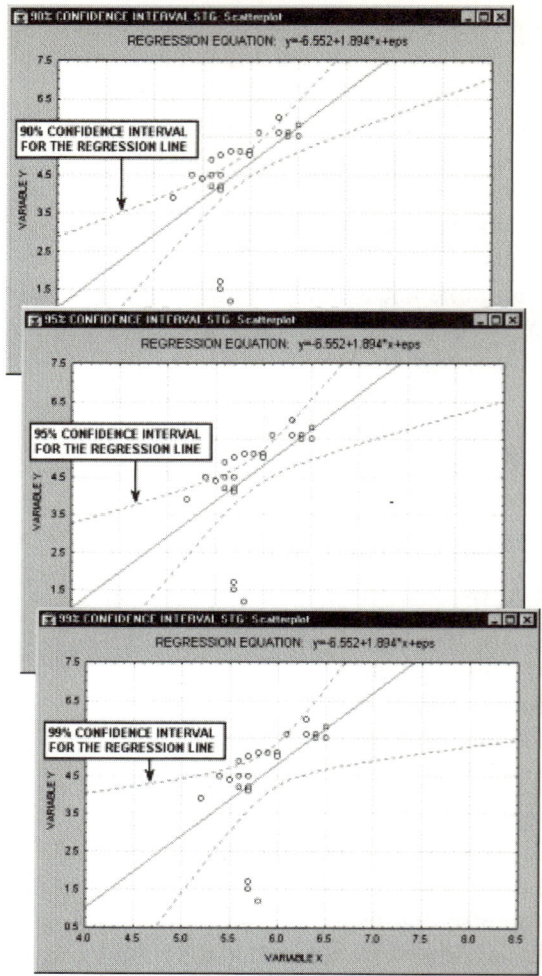

Confidence Interval for the Mean

The confidence intervals for the mean give us a range of values around the mean where we expect the "true" (population) mean is located (with a given level of certainty, see also, Chapter 1 – *Elementary Concepts in Statistics*). In some statistics or math software packages you can specify confidence intervals for any *p*-level; for example, if the mean in your sample is 23, and the lower and upper limits of the *p*=.05 confidence interval are 19 and 27 respectively, then you can conclude that there is a 95% probability that the population mean is greater than 19 and lower than 27. If you set the *p*-level to a smaller value, the interval would become wider thereby increasing the certainty of the estimate, and vice versa; as we all know from the weather forecast, the more vague the prediction (i.e., wider the confidence interval), the more likely it will materialize. Note that the width of the confidence interval depends on the sample size and on the variation of data values. The calculation of confidence intervals is based on the assumption that the variable is normally distributed in the population. This estimate may not be valid if this assumption is not met, unless the sample size is large, say $n = 100$ or more.

Confidence Interval vs. Prediction Interval

In regression, it is possible to predict the value of the dependent variable based on given values of the independent variables. When these values are predicted, it is also possible to calculate confidence intervals and/or prediction intervals for the dependent variable.

The confidence interval gives information on the expected value (mean) of the dependent variable. That is, a confidence interval for a predicted value of the dependent variable gives a range of values around which the "true" (population) mean (of the dependent variable for given levels of the independent variables) can be expected to be located (with a given level of certainty; see also, Chapter 1 – *Elementary Concepts in Statistics*).

The prediction interval gives information on individual predictions of the dependent variable. That is, a prediction interval for a predicted value of the dependent variable gives us a range of values around which an additional observation of the dependent variable can be expected to be located (with a given level of certainty; see also, Chapter 1 – *Elementary Concepts in Statistics*).

Note that the confidence interval will produce a smaller range of values, because it is an interval estimate for an average rather than an interval estimate for a single observation. See Neter, Wasserman, & Kutner, 1985.

Confidence Limits

The same as confidence intervals. In Neural Networks, they represent the accept and reject thresholds, used in classification tasks, to determine whether a pattern of outputs corresponds to a particular class or not. These are applied according to the conversion function of the output variable (One-of-N, Two-state, Kohonen, etc).

Confidence Value (Association Rules)

When applying (in data or text mining) algorithms for deriving association rules of the general form *If Body then Head* (e.g., *If (Car=Porsche and Age<20) then (Risk=High and Insurance=High)*), the *Confidence* value denotes the conditional probability of the *Head* of the association rule, given the *Body* of the association rule.

Confusion Matrix (in Neural Networks)

A name sometimes given to a matrix, in a classification problem, displaying the numbers of cases actually belonging to each class, and assigned by the neural network to that or other classes.

Conjugate Gradient Descent (in Neural Networks)

Conjugate gradient descent (Bishop, 1995; Shepherd, 1997) is an advanced method of training multilayer perceptrons. It usually performs significantly better than back propagation, and can be used wherever back propagation can be. It is the recommended technique for any network with a large number of weights (more than a few hundred) and/or multiple output units. For smaller networks, either Quasi-Newton or Levenberg-Marquardt may be better, the latter being preferred for low-residual regression problems.

Conjugate gradient descent is a batch update algorithm: whereas back propagation adjusts the network weights after each case, conjugate gradient descent works out the average gradient of the error surface across all cases before updating the weights once at the end of the epoch.

For this reason, there is no shuffle option available with conjugate gradient descent, since it would clearly serve no useful function. There is also no need to select learning or momentum rates for conjugate gradient descent, so it can be much easier to use than back propagation. Additive noise would destroy the assumptions made by conjugate gradient descent about the shape of search space, and so is also not available.

Conjugate gradient descent works by constructing a series of line searches across the error surface. It first works out the direction of steepest descent, just as back propagation would do. However, instead of taking a step proportional to a learning rate, conjugate gradient descent projects a straight line in that direction and then locates a minimum along this line, a process that is quite fast as it only involves searching in one dimension. Subsequently,

further line searches are conducted (one per epoch). The directions of the line searches (the conjugate directions) are chosen to try to ensure that the directions that have already been minimized stay minimized (contrary to intuition, this does not mean following the line of steepest descent each time).

The conjugate directions are actually calculated on the assumption that the error surface is quadratic, which is not generally the case. However, it is a fair working assumption, and if the algorithm discovers that the current line search direction isn't actually downhill, it simply calculates the line of steepest descent and restarts the search in that direction. Once a point close to a minimum is found, the quadratic assumption holds true and the minimum can be located very quickly.

Note: The line searches on each epoch of conjugate gradient descent actually involve one gradient calculation plus a variable number (perhaps as high as twenty) of error evaluations. Thus a conjugate gradient descent epoch is substantially more time-consuming (typically 3-10 times longer) than a back propagation epoch. If you want to compare performance of the two algorithms, you will need to record the time taken.

Technical details. Conjugate gradient descent is batch-based; it calculates the error gradient as the sum of the error gradients on each training case.

The initial search direction is given by:

$$d_0 = -g_0$$

Subsequently, the search direction is updated using the Polak-Rebiere formula:

$$d_{j+1} = -g_{j+1} + \beta_j d_j$$

$$\beta_{j+1} = \frac{g_{j+1}^T (g_{j+1} - g_j)}{g_j^T g_j}$$

If the search direction is not downhill, the algorithm restarts using the line of steepest descent. It restarts anyway after W directions (where W is the number of weights), as at that point the conjugacy has been exhausted.

Line searches are conducted using Brent's iterative line search procedure, which utilizes a parabolic interpolation to locate the line minima extremely quickly.

Continuous Dependent Variable

A continuous dependent variable is a variable of interest (a researcher wants to predict) measured on a continuous scale (such as *Weight in pounds*, or *Height in centimeters*). A researcher may be interested in predicting a continuous variable (e.g., *Annual Sales*) based on one or more predictors (e.g., based on *Marketing Expenditures*).

Contour Plot

A contour plot is the projection of a 3-dimensional surface onto a 2-dimensional plane.

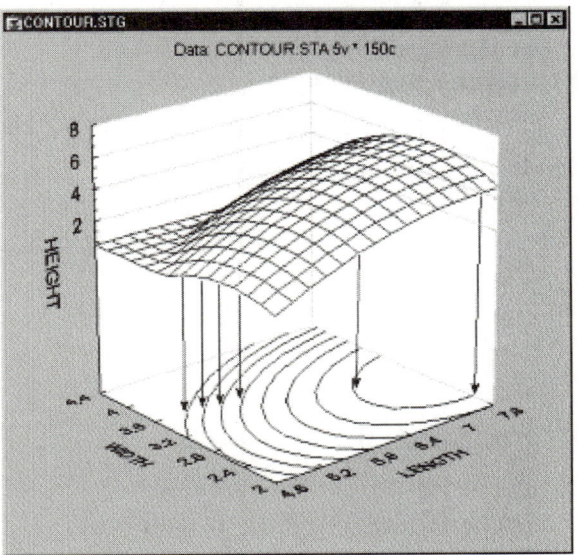

As compared to surface plots, they may be less effective to quickly visualize the overall shape of 3D data structures,

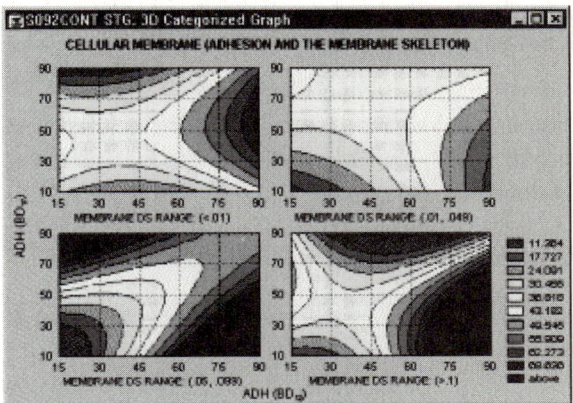

however, their main advantage is that they allow for precise examination and analysis of the shape of the surface (contour plots display a series of undistorted horizontal cross sections of the surface).

Contour/Discrete Raw Data Plot

This sequential plot can be considered to be a 2D projection of the 3D ribbons plot.

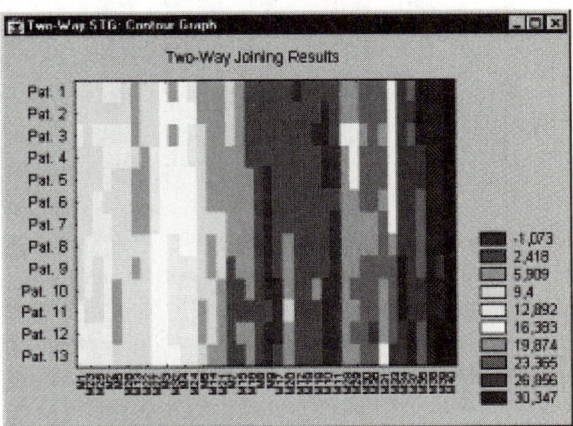

Each data point in this plot is represented as a rectangular region, with different colors and/or patterns corresponding to the values (or range of values of the data points). Values within each series are presented along the *X*-axis, with each series plotted along the *Y*-axis.

Contour Plot, 3D – Areas

A contour plot is the projection of a three-dimensional surface onto a two-dimensional plane. This type of contour plot is displayed as a series of areas.

Contour Plot, 3D (Categorized)

This type of graph projects a 3-dimensional surface onto a 2-dimensional plane as contour plots for each level of the grouping variable. The plots are arranged in one display to allow for comparisons between the subsets of data (categories).

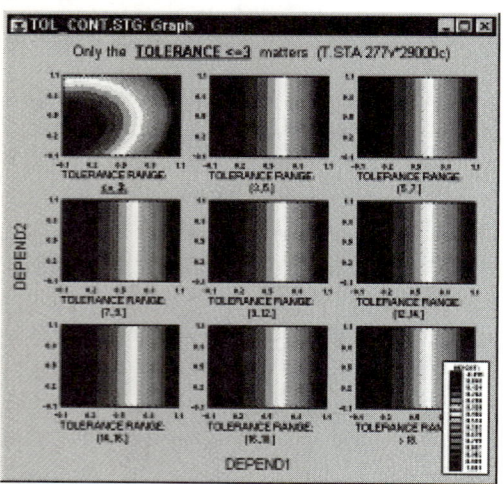

For a detailed discussion of categorized graphs, see *Categorized Graphs*, page 533.

Contour Plot, 3D – Lines

A contour plot is the projection of a three-dimensional surface onto a two-dimensional plane. This contour plot is displayed as a series of lines.

Cook's Distance

This is another measure of impact of the respective case on the regression equation. It indicates the difference between the computed B values and the values you would have obtained had the respective case been excluded. All distances should be of about equal magnitude; if not, there is reason to believe that the respective case(s) biased the estimation of the regression coefficients. See also, *Standard Residual Value*, *Mahalanobis Distance* and *Deleted Residual*.

Correlation

Correlation is a measure of the relation between two or more variables. Correlation coefficients can range from -1.00 to +1.00. The value of -1.00 represents a perfect *negative* correlation while a value of +1.00 represents a perfect *positive* correlation. A value of 0.00 represents a lack of correlation. See also, *Correlations* in Chapter 2 – *Basic Statistics and Tables*; *Partial Correlation*, *Pearson Correlation*, and *Spurious Correlations*.

Correlation Value (Association Rules)

When applying (in data or text mining) algorithms for deriving association rules of the general form *If Body then Head* (e.g., *If (Car=Porsche and Age<20) then (Risk=High and Insurance=High)*), a *Correlation* value can be computed as the support value for the rule, divided by the square root of the product of the support values for the *Body* and *Head* computed separately.

Correspondence Analysis

Correspondence analysis is a descriptive/ exploratory technique designed to analyze simple two-way and multi-way tables containing some measure of correspondence

between the rows and columns. The results provide information that is similar in nature to those produced by factor analysis techniques, and they enable you to explore the structure of categorical variables included in the table. The most common kind of table of this type is the two-way frequency crosstabulation table (see, for example, Chapter 2 – *Basic Statistics and Tables* or Chapter 22 – *Log-Linear Analysis of Frequency Tables*). See also, Chapter 11 – *Correspondence Analysis* for more information.

C_{pk}, C_p, C_r

Potential capability (C_p). This is the simplest and most straightforward indicator of process capability. It is defined as the ratio of the specification range to the process range; using $\pm$ 3 sigma limits we can express this index as:

$$C_p = (USL-LSL)/(6*Sigma)$$

Put into words, this ratio expresses the proportion of the range of the normal curve that falls within the engineering specification limits (provided that the mean is on target, that is, that the process is centered).

Non-centering correction (K). We can correct C_p for the effects of non-centering. Specifically, we can compute:

$$K = abs(Target\ Specification - Mean)/(1/2(USL-LSL))$$

This correction factor expresses the non-centering (target specification minus mean) relative to the specification range. Demonstrated excellence (C_{pk}). Finally, we can adjust C_p for the effect of non-centering by computing:

$$C_{pk} = (1-k)*C_p$$

If the process is perfectly centered, k is equal to zero, and C_{pk} is equal to C_p. However, as the process drifts from the target specification, k increases and C_{pk} becomes smaller than C_p.

Capability ratio (Cr). This index is equivalent to C_p; specifically, it is computed as $1/C_p$ (the inverse of C_p).

Estimate of sigma. When the data set consists of multiple samples, such as data collected for the quality control chart, you can compute two different indices of variability in the data. One is the regular standard deviation for all observations, ignoring the fact that the data consist of multiple samples; the other is to estimate the process's inherent variation from the within-sample variability. When the total process variability is used in the standard capability computations, the resulting indices are usually referred to as process performance indices (as they describe the actual performance of the process; common indices are P_p, P_r, and P_{pk}), while indices computed from the inherent variation (within-sample sigma) are referred to as capability indices (since they describe the inherent capability of the process; common indices are C_p, C_r, and C_{pk}). See Chapter 32 – *Process Analysis*, particularly the section on *Process (Machine) Capability Indices*.

CRISP

See *Models for Data Mining*. See also, Chapter 12 – *Data Mining Techniques*.

Cross Entropy (in Neural Networks)

Error functions based on information-theoretic measures, and particularly appropriate for classification networks. There are two versions, for single-output networks and multiple-output networks; these should be combined with the logistic and softmax activation functions respectively (Bishop, 1995). See also, Chapter 27 – *Neural Networks*.

Crossed Factors

Some experimental designs are completely crossed (factorial designs), that is, each level of each factor appears with each level of all others. For example, in a 2 (types of drug) x 2 (types of virus) design, each type of drug would be used with each type of virus. See also, Chapter 3 – *ANOVA/MANOVA*.

Crosstabulations (Tables, Multi-Way Tables)

A crosstabulation table is a combination of two (or more) frequency tables arranged such that each cell in the resulting table represents a unique combination of specific values of crosstabulated variables. Thus, crosstabulation enables us to examine frequencies of observations that belong to specific combinations of categories on more than one variable. For example, the following simple (two-way) table shows how many adults vs. children selected "cookie A" vs. "cookie B" in a taste preference test:

Data: Crosstabulation Table (3v by 3c)			
	Cookie: A	Cookie: B	Row Total
Age: Adult	50	0	50
Age: Child	0	50	50
All Grps	50	50	100

By examining these frequencies, we can identify relations between crosstabulated variables (e.g., children clearly prefer "cookie B"). Only categorical (nominal) variables or variables with a relatively small number of different meaningful values should be crosstabulated. Note that in the cases where we do want to include a continuous variable in a crosstabulation (e.g., income), we can first

recode it into a particular number of distinct ranges (e.g., low, medium, high). For more information, see *Crosstabulation and Stub-and-Banner Tables* in Chapter 2 – *Basic Statistics and Tables*.

Cross-Validation

Cross-validation refers to the process of assessing the predictive accuracy of a model in a test sample (sometimes also called a cross-validation sample) relative to its predictive accuracy in the learning sample from which the model was developed. Ideally, with a large sample size, a proportion of the cases (perhaps one-half or two-thirds) can be designated as belonging to the learning sample and the remaining cases can be designated as belonging to the test sample.

The model can be developed using the cases in the learning sample, and its predictive accuracy can be assessed using the cases in the test sample. If the model performs as well in the test sample as in the learning sample, it is said to cross-validate well, or simply to cross-validate.

For discussions of this type of test sample cross-validation, see *Computational Methods* in Chapter 9 – *Classification Trees*; Chapter 12 – *Data Mining Techniques*; and *Classification* in Chapter 13 – *Discriminant Function Analysis*.

A variety of techniques has been developed for performing cross-validation with small sample sizes by constructing test samples and learning samples that are partly but not wholly independent. For a discussion of some of these techniques, see *Computational Methods* in Chapter 9 – *Classification Trees*.

Cross Verification (in Neural Networks)

The same as cross-validation. In the context of neural networks, the use of an auxiliary set of data (the verification set) during iterative training. While the training set is used to adjust the network weights, the verification set maintains an independent check that the neural network is learning to generalize.

Cubic Spline Smoother

The cubic spline scatterplot smoother is a special smoothing technique for 2D scatterplots that generally produces a smooth generalization of the relationship between the two variables in the scatterplot. The cubic spline smoother is often used in generalized additive models to estimate the unspecific (non-parametric) function of the predictor variables that best predicts the (transformed) dependent variable values. Computational details regarding the cubic spline smoother, and comparisons to other smoothing algorithms, can be found in Hastie and Tibshirani, 1990, and Schimek, 2000.

Curse of Dimensionality

The term "curse of dimensionality" (Bellman, 1961, Bishop, 1995) generally refers to the difficulties involved in fitting models, estimating parameters, or optimizing a function in many dimensions, usually in the context of neural networks. As the dimensionality of the input data space (i.e., the number of predictors) increases, it becomes exponentially more difficult to find global optima for the parameter space, i.e., to fit models.

In practice, the complexity of neural networks becomes unmanageable when the number of inputs into the neural network exceeds a few hundreds or even less, depending on the complexity of the respective neural network architecture. Hence, it is simply a practical necessity to pre-screen and preselect from among a large set of input (predictor) variables those that are of likely utility for predicting the outputs (dependent variables) of interest.

D

Daniell (or Equal Weight) Window

In time series, the Daniell window (Daniell 1946) is a weighted moving average transformation used to smooth the periodogram values. This transformation amounts to a simple (equal weight) moving average transformation of the periodogram values, that is, each spectral density estimate is computed as the mean of the m/2 preceding and subsequent periodogram values. See also, *Basic Notations and Principles*, page 512, in Chapter 38 – *Time Series/Forecasting*.

Data Mining

Data mining is an analytic process designed to explore large amounts of (typically business or market related) data in search for consistent patterns and/or systematic relationships between variables, and then to validate the findings by applying the detected patterns to new subsets of data. Data mining uses many of the principles and techniques traditionally referred to as Exploratory Data Analysis (EDA).

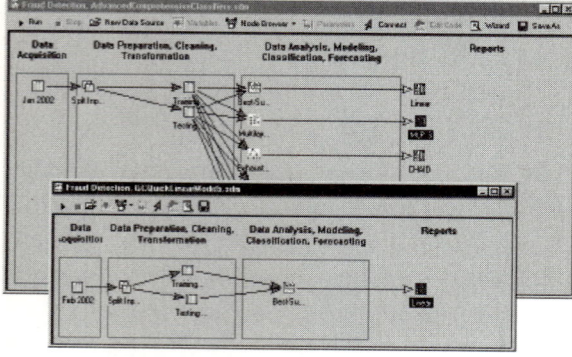

For more information, see Chapter 12 – *Data Mining Techniques*.

Data Preparation Phase

In data mining, the input data are often "noisy," containing many errors, and sometimes information in unstructured form (e.g., in text mining). For example, suppose you wanted to analyze a large database of information collected on-line via the web, based on voluntary responses of persons reviewing your web site (e.g., potential customers of a web-based retailer, who filled out suggestion forms). In those instances it is very important to first verify and "clean" the data in a data preparation phase, before applying any analytic procedures. For example, some individuals might enter clearly faulty information (e.g., age = 300), either by mistake or intentionally. If those types of data errors are not detected prior to the analysis phase of the data mining project, they can greatly bias the result, and potentially cause unjustified conclusions. Typically, during the data preparation phase, the data analyst applies filters to the data, to verify correct data ranges, and to delete impossible co-occurrences of values (e.g., Age=5; Retired=Yes).

Data Reduction

The term *Data Reduction* is used in two distinctively different meanings:

Data reduction by decreasing the dimensionality (exploratory multivariate statistics). This interpretation of the term data reduction pertains to analytic methods (typically multivariate exploratory techniques such as factor analysis, multidimensional scaling, cluster analysis, canonical correlation, or neural networks) that involve reducing the dimensionality of a data set by extracting a number of underlying factors, dimensions, clusters, etc., that can account for the variability in the (multidimensional) data set. For example,

in poorly designed questionnaires, all responses provided by the participants on a large number of variables (scales, questions, or dimensions) could be explained by a very limited number of trivial or artifactual factors. For example, two such underlying factors could be 1) the respondent's attitude toward the study (positive or negative) and 2) the social desirability factor (a response bias representing a tendency to respond in a socially desirable manner).

Data reduction by unbiased decreasing of the sample size (exploratory graphics). This type of data reduction is applied in exploratory graphical data analysis of extremely large data sets. The size of the data set can obscure an existing pattern (especially in large line graphs or scatterplots) due to the density of markers or lines. Then, it can be useful to plot only a representative subset of the data (so that the pattern is not hidden by the number of point markers) to reveal the otherwise obscured but still reliable pattern. For an animated illustration, see *Data Reduction*, page 552, in Chapter 40 – *Graphical Techniques*.

Data Rotation (in 3D Space)

Changing the viewpoint for 3D scatterplots (e.g., simple, spectral, or space plots) may prove to be an effective exploratory technique since it can reveal patterns that are easily obscured unless you look at the "cloud" of data points from an appropriate angle (see the next illustrations). Rotating or spinning a 3D graph will enable you to find the most informative location of the viewpoint for the graph.

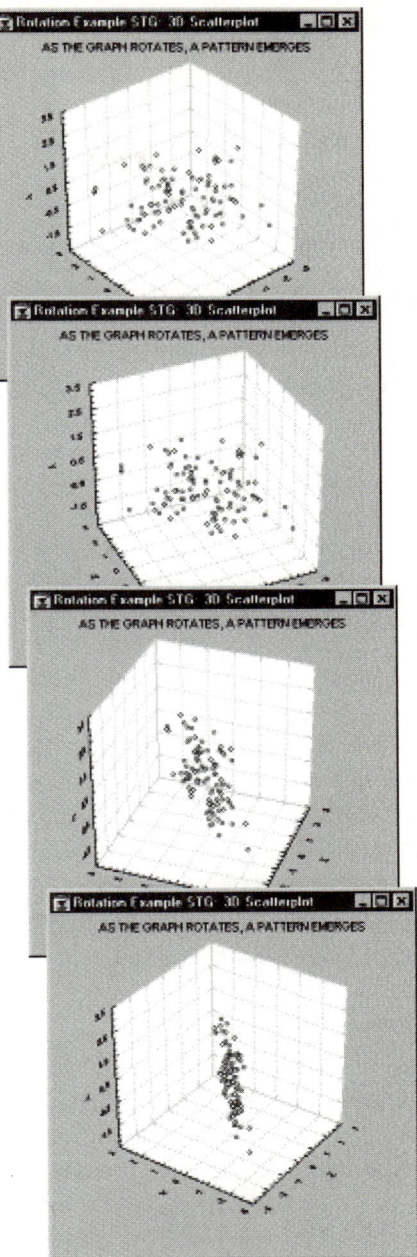

For more information, see *Data Rotation (in 3D Space)*, page 553.

Data Warehousing

Data warehousing is a process of organizing the storage of large, multivariate data sets in a way that facilitates the retrieval of information for specific analytic purposes. For more information, see *Data Warehousing* page in Chapter 12 – *Data Mining Techniques*.

Degrees of Freedom

Used in slightly different senses throughout the study of statistics, degrees of freedom were first introduced by Fisher based on the idea of degrees of freedom in a dynamical system (e.g., the number of independent coordinate values which are necessary to determine it). The degrees of freedom of a set of observations are the number of values which could be assigned arbitrarily within the specification of the system. For example, in a sample of size *n* grouped into *k* intervals, there are *k-1* degrees of freedom, because *k-1* frequencies are specified while the other one is specified by the total size *n*. Thus in a p by q contingency table with fixed marginal totals, there are (p-1)(q-1) degrees of freedom. In some circumstances the term degrees of freedom is used to denote the number of independent comparisons that can be made between the members of a sample.

Deleted Residual

The deleted residual is the residual value for a case had it not been included in the regression analysis, that is, if you would exclude this case from all computations. If the deleted residual differs greatly from the respective standardized residual value, this case is possibly an outlier because its exclusion changed the regression equation. See also, *Standard Residual Value*, *Mahalanobis Distance*, and *Cook's Distance*.

Delta-Bar-Delta (in Neural Networks)

Delta-bar-Delta (Jacobs, 1988; Patterson, 1996) is an alternative to back propagation that is sometimes more efficient, although it can be more inclined to stick in local minima than back propagation. Unlike quick propagation, it tends to be quite stable. Just as quick propagation, Delta-bar-Delta is a batch algorithm: the average error gradient across all the training cases is calculated on each epoch, and then the weights are updated once at the end of the epoch.

Delta-bar-Delta is inspired by the observation that the error surface may have a different gradient along each weight direction, and that consequently each weight should have its own learning rate (i.e. step size).

In Delta-bar-Delta, the individual learning rates for each weight are altered on each epoch to satisfy two important heuristics:

If the derivative has the same sign for several iterations, the learning rate is increased (the error surface has a low curvature, and so is likely to continue sloping the same way for some distance);

If the sign of the derivative alternates for several iterations, the learning rate is rapidly decreased (otherwise the algorithm may oscillate across points of high curvature).

To satisfy these heuristics, Delta-bar-Delta has an initial learning rate used for all weights on the first epoch, an increment factor added to learning rates when the derivative does not change sign, and a decay rate multiplied by the learning rates when the derivative does change sign. Using linear growth and exponential decay of learning rates contributes to stability.

The algorithm described above could still be prone to poor behavior on noisy error surfaces,

where the derivative changes sign rapidly even within an overall downward trend. Consequently, the increase or decrease of learning rate is actually based on a smoothed version of the derivative.

Technical details. Weights are updated using the same formula as in back propagation, except that momentum is not used, and each weight has its own time-dependent learning rate.

All learning rates are initially set to the same starting value; subsequently, they are adapted on each epoch using the formulae below.

The bar-Delta value is calculated as:

$$\bar{\delta}(t) = (1-\theta)\delta(t) + \bar{\delta}(t-1)$$

$\delta(t)$ is the derivative of the error surface,

θ is the smoothing constant.

The learning rate of each weight is updated using:

$$\Delta\eta(t) = \begin{cases} \kappa & \bar{\delta}(t-1)\delta(t) > 0 \\ -\phi\eta(t) & \bar{\delta}(t-1)\delta(t) < 0 \\ 0 & \bar{\delta}(t-1)\delta(t) = 0 \end{cases}$$

κ is the linear increment factor, ϕ is the exponential decay factor.

Denominator Synthesis

A method developed by Satterthwaite (1946) which finds the linear combinations of sources of random variation that serve as appropriate error terms for testing the significance of the respective effect of interest in mixed-model ANOVA/ANCOVA designs.

For descriptions of denominator synthesis, see Chapter 39 – *Variance Components and Mixed-Model ANOVA/ANCOVA* and Chapter 18 – *General Linear Models (GLM)*.

Deployment

The concept of deployment in predictive data mining refers to the application of a model for prediction or classification to new data. After a satisfactory model of set of models have been identified (trained) for a particular application, you usually want to deploy those models so that predictions or predicted classifications can quickly be obtained for new data. For example, a credit card company may want to deploy a trained model or set of models (e.g., neural networks, meta-learner) to quickly identify transactions that have a high probability of being fraudulent.

Derivative-Free Function Minimization Algorithms

Nonlinear estimation offers several general function minimization algorithms that follow different search strategies that do not depend on the second-order derivatives. These strategies are sometimes very effective for minimizing loss functions with local minima.

Design Matrix

In general linear models and generalized linear models, the design matrix is the matrix **X** for the predictor variables, which is used in solving the normal equations. **X** is a matrix, with 1 row for each case and 1 column for each coded predictor variable in the design, whose values identify the levels for each case on each coded predictor. See also, Chapter 18 – *General Linear Models (GLM)* and Chapter 21 – *Generalized Linear/Nonlinear Models (GLZ)*.

Desirability Profiles

The relationship between predicted responses on one or more dependent variables and the desirability of responses is called the desirability

function. Profiling the desirability of responses involves, first, specifying the desirability function for each dependent variable, by assigning predicted values a score ranging from 0 (very undesirable) to 1 (very desirable). The individual desirability scores for the predicted values for each dependent variable are then combined by computing their geometric mean. Desirability profiles consist of a series of graphs, one for each independent variable, of overall desirability scores at different levels of one independent variable, holding the levels of the other independent variables constant at specified values. Inspecting the desirability profiles can show which levels of the predictor variables produce the most desirable predicted responses on the dependent variables.

For a detailed description of response/ desirability profiling see *Profiling Predicted Responses and Response Desirability* in Chapter 15 – *Experimental Design (DOE)*.

Deviance

To evaluate the goodness of fit of a generalized linear model, a common statistic that is computed is the so-called deviance statistic. It is defined as:

Deviance = -2 * (Lm – Ls)

where *Lm* denotes the maximized log-likelihood value for the model of interest, and *Ls* is the log-likelihood for the saturated model, i.e., the most complex model given the current distribution and link function. For computational details, see Agresti (1996). See also, Chapter 21 – *Generalized Linear/ Nonlinear Models (GLZ)*.

Deviance Residuals

After fitting a generalized linear model to the data, to check the adequacy of the respective model, you usually compute various residual statistics. The deviance residual is computed as:

$r_D = sign(y-\mu)sqrt(d_i)$

Where $\Sigma d_i = D$, and D is the overall deviance measure of discrepancy of a generalized linear model (see McCullagh and Nelder, 1989, for details). Thus, the deviance statistic for an observation reflects its contribution to the overall goodness of fit (deviance) of the model. See also, Chapter 21 – *Generalized Linear/ Nonlinear Models (GLZ)*.

Deviation

In radial units, a figure multiplied by the radial exemplar's squared distance from the input pattern to generate the unit's activation level, before submission to the activation function. See also, *Neural Networks*.

Deviation Assignment Algorithms (in Neural Networks)

These algorithms assign deviations to the radial units in certain network types. The deviation is multiplied by the distance between the unit's exemplar vector and the input vector, to determine the unit's output. In essence, the deviation gives the size of the cluster represented by a radial unit.

Deviation assignment algorithms are used after radial centers have been set; see *Radial Sampling*, and *k-Means Algorithm*.

Explicit deviation assignment. The deviation is set to an explicit figure provided by the user.

Notes. The deviation assigned by this technique is not the standard deviation of the Gaussians; it is the value stored in the unit threshold, which is multiplied by the distance of the weight vector from the input vector. It is related to the standard deviation by:

$$d = \frac{1}{2\sigma^2}$$

Isotropic deviation assignment. This algorithm uses the isotropic deviation heuristic (Haykin, 1994) to assign the deviations to radial units. This heuristic attempts to determine a reasonable deviation (the same for all units), based upon the number of centers, and how spread out they are.

This isotropic deviation heuristic sets radial deviations to:

$$\frac{k}{d^2}$$

where d is the distance between the two most distant centers, and k is the number of centers.

K-nearest neighbor deviation. The K-nearest neighbor deviation assignment algorithm (Bishop, 1995) assigns deviations to radial units by using the RMS (Root Mean Squared) distance from the K units closest to (but not coincident with) each unit as the standard deviation (assuming the unit models a Gaussian). Each unit hence has its own independently calculated deviation, based upon the density of points close to itself.

If less than K non-coincident neighbors are available, the algorithm uses the neighbors that are available.

Deviation Plot, 3D

Data (representing the X, Y, and Z coordinates of each point) in this type of graph are represented in 3D space as deviations from a specified base level of the Z-axis.

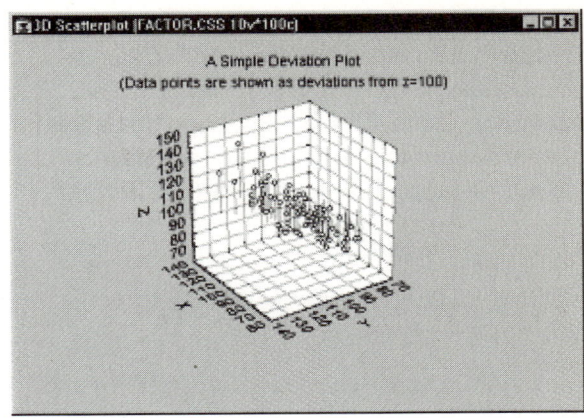

Deviation plots are similar to space plots, however, in deviation plots the deviations plane is invisible and not marked by the location of the X-Y axes (those axes are always fixed in the standard bottom position). Deviation plots can help explore the nature of 3D data sets by displaying them in the form of deviations from arbitrary (horizontal) levels. Such cutting methods can help identify interactive relations between variables. See also, *Data Rotation (in 3D Space)*, page 553.

Deviation Plot, 3D (Categorized)

Data points (representing the X, Y and Z coordinates of each point) in this graph are represented in 3D space as "deviations" from a specified base-level of the Z-axis. One component graph is produced for each level of the grouping variable (or user-defined subset of data) and all the component graphs are arranged in one display to allow for comparisons between the subsets of data (categories). For a detailed discussion of categorized graphs, see *Categorized Graphs*, page 533.

DFFITS

Several measures have been given for testing for leverage and influence of a specific case in

regression (including studentized residuals, studentized deleted residuals, DFFITS, and standardized DFFITS). Belsley et al. (1980) have suggested *DFFITS*, a measure that gives greater weight to outlying observations than Cook's distance. The formula for *DFFITS* is:

$$DFFITS_i = \tilde{h}_i e_i / (1 - \tilde{h}_i)$$

where e_i is the error for the *i*th case, h_i is the leverage for the *i*th case, and $\tilde{h}_i = 1/N + h_i$.

For more information see Hocking (1996) and Ryan (1997).

DIEHARD Suite of Tests and Random Number Generation

Many areas of statistical analysis, research, and simulation rely on the quality of random number generators. Most programs for statistical data analysis contain a function for generating uniform random numbers. A recent review of statistical packages (McCullough, 1998, 1999) that appeared in *The American Statistician* tested the random number generators of several programs using the so-called *DIEHARD suite of tests* (Marsaglia, 1998). DIEHARD applies various methods of assembling and combining uniform random numbers, and then performs statistical tests that are expected to be nonsignificant; this suite of tests has become a standard method of evaluating the quality of uniform random number generator routines.

Differencing (in Time Series)

In this time series transformation, the series will be transformed as: X=X-X(lag). After differencing, the resulting series will be of length *N-lag* (where *N* is the length of the original series).

Dimensionality Reduction

Data reduction by decreasing the dimensionality (exploratory multivariate statistics). This interpretation of the term Data Reduction pertains to analytic methods (typically multivariate exploratory techniques such as factor analysis, multidimensional scaling, cluster analysis, canonical correlation, or neural networks) that involve reducing the dimensionality of a data set by extracting a number of underlying factors, dimensions, clusters, etc., that can account for the variability in the (multidimensional) data set. For more information, see *Data Reduction*, page 602.

Discrepancy Function

A numerical value that expresses how badly a structural model reproduces the observed data. The larger the value of the discrepancy function, the worse (in some sense) the fit of model to data. In general, the parameter estimates for a given model are selected to make a discrepancy function as small as possible. The discrepancy functions employed in structural modeling all satisfy the following basic requirements:

1. They are non-negative, i.e., always greater than or equal to zero.

2. They are zero only if fit is perfect, i.e., if the model and parameter estimates perfectly reproduce the observed data.

3. The discrepancy function is a continuous function of the elements of **S**, the sample covariance matrix, and $\Sigma(\theta)$, the "reproduced" estimate of S obtained by using the parameter estimates and structural model.

Discriminant Function Analysis

Discriminant function analysis is used to determine which variables discriminate between two or more naturally occurring groups (it is used as either a hypothesis testing or exploratory method). For example, an educational researcher may want to investigate which variables discriminate between high school graduates who decide 1) to go to college, 2) to attend a trade or professional school, or 3) to seek no further training or education. For that purpose the researcher could collect data on numerous variables prior to students' graduation. After graduation, most students will naturally fall into one of the three categories. Discriminant analysis could then be used to determine which variable(s) are the best predictors of students' subsequent educational choice (e.g., IQ, GPA, SAT). For more information, see Chapter 9 – *Classification Trees* and Chapter 13 – *Discriminant Function Analysis*.

Drill-Down Analysis

The concept of drill-down analysis applies to the area of data mining, to denote the interactive exploration of data, in particular of large databases. The process of drill-down analyses begins by considering some simple breakdowns of the data by a few variables of interest (e.g., *Gender*, *geographic region*, etc.). Various statistics, tables, histograms, and other graphical summaries can be computed for each group. Next, you may want to drill-down to expose and further analyze the data underneath one of the categorizations, for example, one might want to further review the data for males from the mid-west. Again, various statistical and graphical summaries can be computed for those cases only, which might suggest further breakdowns by other variables (e.g., income, age, etc.). At the lowest (bottom) level are the raw data: For

example, you may want to review the addresses of male customers from one region, for a certain income group, etc., and to offer to those customers some particular services of particular utility to that group.

Duncan's Test

This post hoc test (or multiple comparison test) can be used to determine the significant differences between group means in an analysis of variance setting. Duncan's test, like the Newman-Keuls test, is based on the range statistic (for a detailed discussion of different post hoc tests, see Winer, Michels, & Brown (1991). For more details, see Chapter 18 – *General Linear Models*. See also, *Post hoc Comparisons*. For a discussion of statistical significance, see Chapter 1 – *Elementary Concepts in Statistics*.

Dunnett's test

This post hoc test (or multiple comparison test) can be used to determine the significant differences between a single control group mean and the remaining treatment group means in an analysis of variance setting. Dunnett's test is considered to be one of the least conservative post hoc tests (for a detailed discussion of different post hoc tests, see Winer, Michels, & Brown (1991). For more details, see Chapter 18 – *General Linear Models*. See also, *Post hoc Comparisons*. For a discussion of statistical significance, see Chapter 1 – *Elementary Concepts in Statistics*.

E

Effective Hypothesis Decomposition

When in a factorial ANOVA design there are missing cells, there is ambiguity regarding the specific comparisons between the (population, or least-squares) cell means that constitute the main effects and interactions of interest. Chapter 18 – *General Linear Models* discusses the methods commonly labeled *Type I, II, III,* and *IV* sums of squares, and a unique *Type V* sums of squares option.

In addition, for *sigma*-restricted models (e.g., in general regression models; some software offers you a choice between the *sigma*-restricted and overparameterized models), we propose a *Type VI sums of squares* option; this approach is identical to what is described as the *effective hypothesis* method in Hocking (1996). For details regarding these methods, refer to *Six Types of Sums of Squares* in Chapter 18 – *General Linear Models*.

Efficient Score Statistic

See *Score Statistic*.

Ellipse, Prediction Interval (Area) and Range

Prediction Interval (Area) Ellipse. This type of ellipse is useful for establishing confidence intervals for the prediction of single new observations (prediction intervals). Such bivariate confidence or control limits are, for example, often used in the context of multivariate control charts for industrial quality control (see, for example, Montgomery, 1996; see also, Hotelling T-square chart).

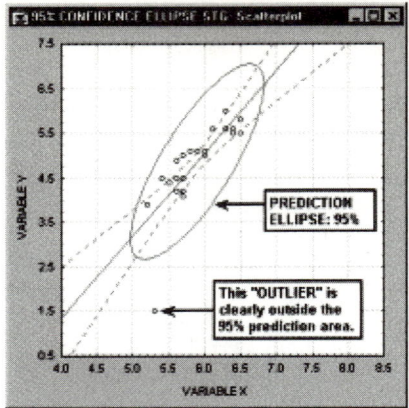

The ellipse is determined based on the assumption that the two variables follow the bivariate normal distribution. The orientation of this ellipse is determined by the sign of the linear correlation between two variables (the longer axis of the ellipse is superimposed on the regression line). The probability that the values will fall within the area marked by the ellipse is determined by the value of the coefficient that defines the ellipse (e.g., 95%). For additional information see, for example, Tracy, Young, and Mason (1992), or Montgomery 1996); see also, the description of the prediction interval ellipse.

Range Ellipse. This type of ellipse is a fixed size ellipse determined such that the length of its horizontal and vertical projection onto the X- and Y-axis (respectively) is equal to the mean (Range * I) where the mean and range refer to the X or Y variable, and I is the current value of the coefficient field.

EM Clustering

The EM clustering algorithm is a clustering technique similar in purpose to the *k*-means clustering method. The EM algorithm performs clustering by fitting a mixture of distributions to the data; for example:

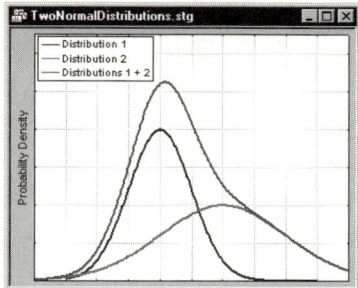

This illustration shows two normal distributions with different means and different standard deviations, and the sum of the two distributions. Only the mixture (sum) of the two normal distributions, with different means and standard deviations, would be observed (e.g., suppose clustering was performed based on a single continuous variable). The goal of EM clustering is to estimate the means and standard deviations for each cluster, so as to maximize the likelihood of the observed data (distribution). For additional details, see also, Witten and Frank (2001).

Endogenous Variable

An endogenous variable is a variable that appears as a dependent variable in at least one equation in a structural model. In a path diagram, endogenous variables can be recognized by the fact that they have at least one arrow pointing to them.

Ensembles (in Neural Networks)

Ensembles are collections of neural networks that cooperate in performing a prediction.

Output ensembles. Output ensembles are the most general form. Any combination of networks can be combined in an output ensemble. If the networks have different outputs, the resulting ensemble simply has multiple outputs. Thus, an output ensemble can be used to form a multiple output model where each output's prediction is formed separately.

If any networks in the ensemble have a shared output, the ensemble estimates a value for that output by combining the outputs from the individual networks. For classification (nominal outputs), the networks' predictions are combined in a winner-takes-all vote; the most common class among the combined networks is used. In the event of a tie, the unknown class is returned. For regression (numeric variables), the networks' predictions are averaged. In both cases, the vote or average is weighted using the networks' membership weights in the ensemble (usually all equal to 1.0).

Confidence ensembles. Confidence ensembles are much more restrictive than output ensembles. The network predictions are combined at the level of the output neurons. To make sense, the encoding of the output variables must therefore be the same for all the members. Given that restriction, there is no point in forming confidence ensembles for regression problems, as the effect is to produce the same output as an output ensemble, but with the averaging performed before scaling rather than after. Confidence ensembles are designed for use with classification problems.

The advantage of using a confidence ensemble for a classification problem is that it can estimate overall confidence levels for the various classes, rather than simply providing a final choice of class.

Why use ensembles? There are a number of uses for ensembles:

Ensembles can conveniently group together networks that provide predictions for related variables without requiring that all those variables be combined into a single network. Multiple output networks often suffer from cross talk in the hidden neurons, and make ineffective predictions. Using an ensemble, each output can be predicted separately.

Ensembles provide an important method to combat over-learning and improve generalization. Averaging predictions across models with different structures, and/or trained on different data subsets, can reduce model variance without increasing model bias. This is a relatively simple way to improve generalization. Ensembles therefore are particularly effective when combined with resampling. An important piece of theory shows that the expected performance of an ensemble is greater than or equal to the average performance of the members.

Ensembles report the average performance and error measures of their member networks. You can perform resampling experiments, and save the results to an ensemble. Then, these average measures give an unbiased estimate of an individual network's performance, if trained in the same fashion. It is standard practice to use resampling techniques such as cross validation to estimate network performance in this fashion.

Enterprise Resource Planning (ERP)

Enterprise Resource Planning (ERP) software offers an integrated set of applications that help manage various aspects of the business activity of a company (e.g., orders, production schedules, supply chain and inventories, accounting and bookkeeping, human resource management). Usually, an ERP application consists of a number of specialized modules that still offer a consistent user interface and integration of all relevant parts of the company data base system. Examples of widely used ERP applications are products of SAP and Oracle.

See also, *Enterprise-Wide Software Systems*, page 612.

Enterprise SPC

Enterprise SPC is a groupware based process control system (see SPC) designed to work in enterprise-wide environments and enable engineers and supervisors to share data, chart specifications (and other QC criteria), reports, and database queries. Enterprise SPC systems always include central QC data bases and if properly integrated, they enable managers to maintain quality standards for all products/ processes in a given corporation. See also, *Statistical Process Control*, *Quality Control*, and *Process Analysis*. For more information on process control systems, see the ASQC/AIAG's *Fundamental Statistical Process Control Reference Manual* (1991).

Enterprise-Wide Software Systems

Software applications designed to work in enterprise computer environments (e.g., such as large corporate computer systems). Such applications typically feature extensive groupware functionality, and they are usually well integrated with large repositories of data stored in corporate data warehouses. See also, *Data Warehousing*, page 604.

Epoch (in Neural Networks)

During iterative training of a neural network, an epoch is a single pass through the entire training set, followed by testing of the verification set. For more information, see Chapter 27 – *Neural Networks*.

EPSEM Samples

EPSEM samples are probability samples where each observation in the population has the same known probability of being selected into the

sample (EPSEM is an acronym for equal probability of selection method sampling; see Kish, 1965, for a comprehensive discussion of sampling techniques). EPSEM samples have certain desirable properties; for example, the simple formulas for computing means, standard deviations, and so on can be applied to estimate the respective parameters in the population.

Error Function (in Neural Networks)

The error function is used in training the network and in reporting the error. The error function used can have a profound effect on the performance of training algorithms (Bishop, 1995). The following four error functions are available.

Sum-squared. The error is the sum of the squared differences between the target and actual output values on each output unit. This is the standard error function used in regression problems. It can also be used for classification problems, giving robust performance in estimating discriminant functions, although arguably entropy functions are more appropriate for classification, as they correspond to maximum likelihood decision making (on the assumption that the generating distribution is drawn from the exponential family), and allow outputs to be interpreted as probabilities.

City-block. The error is the sum of the differences between the target and actual output values on each output unit; differences are always taken to be positive. The city-block error function is less sensitive to outlying points than the sum-squared error function (where a disproportionate amount of the error can be accounted for by the worst-behaved cases). Consequently, networks trained with this metric may perform better on regression problems if there are a few wide-flung outliers (either because the data naturally has such a structure, or because some cases may be mislabeled).

Cross-entropy (single & multiple). This error is the sum of the products of the target value and the logarithm of the error value on each output unit. There are two versions: one for single-output (two-class) networks, the other for multiple-output networks. The cross-entropy error function is specially designed for classification problems, where it is used in combination with the logistic (single output) or softmax (multiple output) activation functions in the output layer of the network. This is equivalent to maximum likelihood estimation of the network weights. An MLP with no hidden layers, a single output unit, and cross entropy error function is equivalent to a standard logistic regression function (logit or probit classification).

Kohonen. The Kohonen error assumes that the second layer of the network consists of radial units representing cluster centers. The error is the distance from the input case to the nearest of these. The Kohonen error function is intended for use with Kohonen networks and cluster networks only.

Estimable Functions

In general linear models and generalized linear/nonlinear models, if the $\mathbf{X'X}$ matrix (where $\mathbf{X}$ is the design matrix) is less than full *rank*, the regression coefficients depend on the particular generalized inverse used for solving the normal equations, and the regression coefficients will not be unique. When the regression coefficients are not unique, linear functions (*f*) of the regression coefficients having the form:

f=Lb

where $\mathbf{L}$ is a vector of coefficients, will also in general not be unique. However, *Lb* for an *L* which satisfies

L=L(X'X)`X'X

is invariant for all possible generalized inverses, and is therefore called an estimable function.

See also, *General Linear Model*, *Generalized Linear/Nonlinear Model*, *Design Matrix*, *Matrix Rank*, and *Generalized Inverse*; for additional details, see also, Chapter 18 – *General Linear Models (GLM)*.

Euclidean Distance

You can think of the independent variables (in a regression equation) as defining a multidimensional space in which each observation can be plotted. The Euclidean distance is the geometric distance in that multidimensional space. It is computed as:

$$\text{distance}(x,y) = \{\Sigma_i \, (x_i - y_i)^2\}^{1/2}$$

Note that Euclidean (and squared Euclidean) distances are computed from raw data, and not from standardized data. For more information on Euclidean distances and other distance measures, see *Distance Measures* in Chapter 10 – *Cluster Analysis*.

Euler's e

The base of the natural logarithm (numerical value: 2.71828182834905...), named after the Swiss mathematician Leonhard Euler (1707-1783).

Exogenous Variable

An exogenous variable is a variable that never appears as a dependent variable in any equation in a structural model. In a path diagram, exogenous variables can be recognized by the fact that they have no arrows pointing to them.

Experimental Design (DOE, Industrial Experimental Design)

In industrial settings, *Experimental design (DOE)* techniques apply analysis of variance principles to product development. The primary goal is usually to extract the maximum amount of unbiased information regarding the factors affecting a production process from as few (costly) observations as possible. In industrial settings, complex interactions among many factors that influence a product are often regarded as a nuisance (they are often of no interest; they only complicate the process of identifying important factors, and in experiments with many factors it would not be possible or practical to identify them anyway). Hence, if you review standard texts on experimentation in industry (Box, Hunter, and Hunter, 1978; Box and Draper, 1987; Mason, Gunst, and Hess, 1989; Taguchi, 1987) you will find that they will primarily discuss designs with many factors (e.g., 16 or 32) in which interaction effects cannot be evaluated, and the primary focus of the discussion is how to derive unbiased main effect (and, perhaps, two-way interaction) estimates with a minimum number of observations. For more information, see Chapter 15 – *Experimental Design (DOE)*.

Explained Variance

The proportion of the variability in the data that is accounted for by the model (e.g., in multiple regression, ANOVA, nonlinear estimation, or neural networks).

Exploratory Data Analysis (EDA)

As opposed to traditional hypothesis testing designed to verify a priori hypotheses about relations between variables (e.g., "There is a positive correlation between the *age* of a person and his/her *risk-taking* disposition"),

exploratory data analysis (EDA) is used to identify systematic relations between variables when there are no (or not complete) a priori expectations as to the nature of those relations. In a typical exploratory data analysis process, many variables are taken into account and compared, using a variety of techniques in the search for systematic patterns. For more information, see *Exploratory Data Analysis (EDA) and Data Mining Techniques* in Chapter 12 – *Data Mining Techniques*.

Exponential Distribution

The exponential distribution function is defined as:

$$f(x) = \lambda * e^{-\lambda x}$$
$$0 \le x < \infty, \lambda > 0$$

where λ (*lambda*) is an exponential function parameter (an alternative parameterization is scale parameter $b = 1/\lambda$), and e is the base of the natural logarithm, sometimes called Euler's e (2.71...).

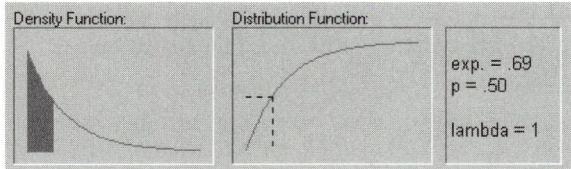

The previous illustration shows the shape of the exponential distribution when *lambda* equals 1.

Exponential Family of Distributions

A family of probability distributions with exponential terms, which includes many of the most important distributions encountered in real (neural network) problems (including the normal, or Gaussian distribution, and the *alpha* and *beta* distributions). See also, *Neural Network*.

Exponential Function

This fits to the data, an exponential function of the form: $y = b*\exp(q*x)$.

Exponentially Weighted Moving Average Line

This type of moving average can be considered to be a generalization of the simple moving average. Specifically, we could compute each data point for the plot as:

$$z_t = \lambda *x\text{-}bar_t + (1- \lambda)*z_{t-1}$$

In this formula, each point z_t is computed as λ (lambda) times the respective mean $x\text{-}bar_t$, plus one minus λ times the previous (computed) point in the plot. The parameter λ (lambda) here should assume values greater than *0* and less than *1*. You may recognize this formula as the common exponential smoothing formula. Without going into detail (see Montgomery, 1985, p. 239), this method of averaging specifies that the weight for a historically old sample means decreases geometrically as you continue to draw samples. This type of moving average line also smoothes the pattern of means across samples, and enables the engineer to detect trends more easily.

Extrapolation

Predicting the value of unknown data points by projecting a function beyond the range of known data points.

Extreme Values (in Box Plots)

Values that are far from the middle of the distribution are referred to as outliers and extreme values if they meet certain conditions.

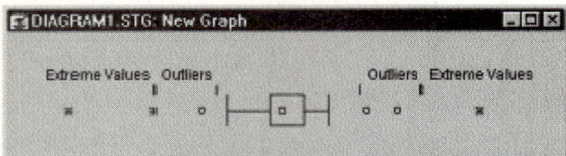

A data point is deemed to be an extreme value if the following conditions hold:

data point value > UBV + 2*o.c.*(UBV – LBV)
or
data point value < LBV – 2*o.c.*(UBV – LBV)

where *UBV* is the upper value of the box in the box plot (e.g., the mean + standard error or the 75th percentile), *LBV* is the lower value of the box in the box plot (e.g., the mean – standard error or the 25th percentile), *o.c.* is the outlier coefficient (when this coefficient equals 1.5, the extreme values are those outside the 3 box length range from the upper and lower value of the box).

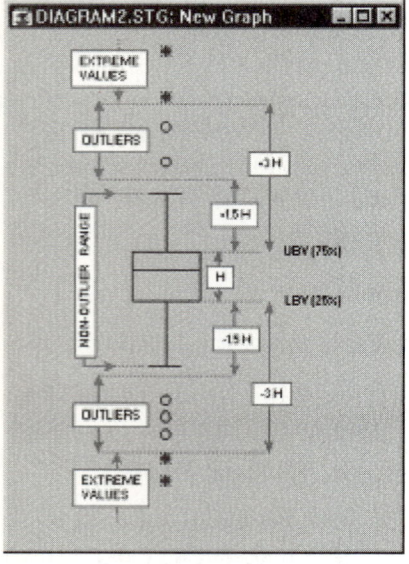

For example, the previous diagram illustrates the ranges of outliers and extremes in the classic box and whisker plot (for more information about box plots, see Tukey, 1977).

Extreme Value Distribution

The extreme value (Type I) distribution (the term first used by Lieblein, 1953) has the probability density function:

$$f(x) = 1/b * e^{-(x-a)/b} * e^{-e^{-(x-a)/b}}$$

$-\infty < x < \infty$

$b > 0$

where
a is the location parameter, *b* is the scale parameter, *e* is the base of the natural logarithm, sometimes called Euler's e (2.71...).

This distribution is also sometimes referred to as the distribution of the largest extreme. See also, *Process Analysis*.

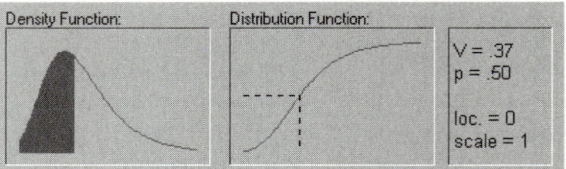

The illustration above shows the shape of the extreme value distribution when the location parameter equals 0 and the scale parameter equals 1.

F

F Distribution

The *F* distribution (for x > 0) has density function (for ν = 1, 2, ...; ω= 1, 2, ...):

$$f(x) = \{\Gamma\,[(\nu+ \omega)/2]\}/[\,\Gamma\,(\nu/2) * \Gamma\,(\omega/2)] * (\nu/\omega)^{(\nu/2)} * x^{[(\nu/2)-1]} * \{1+[(\nu/\omega)*x]\}^{[-(\nu+\omega)/2]}$$

$0 \leq x < \infty$

$\nu = 1, 2, ...,\quad \omega = 1, 2, ...$

where

ν, ω are the degrees of freedom; Γ (*gamma*) is the *Gamma* function.

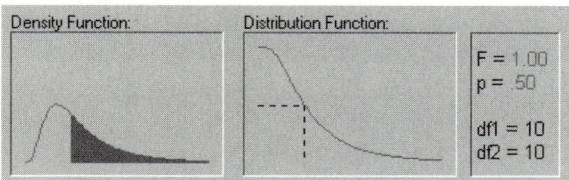

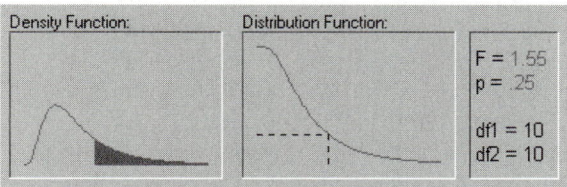

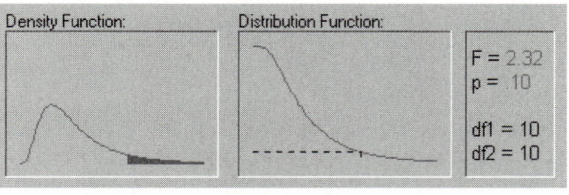

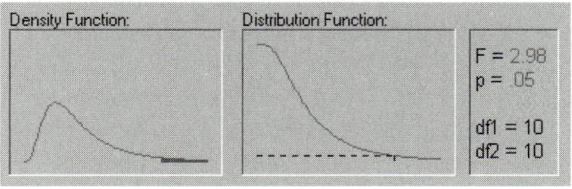

The previous illustrations show various tail areas (p-values) for an F distribution with both degrees of freedom equal to 10.

FACT

FACT is a classification tree program developed by Loh and Vanichestakul (1988) that is a precursor of the QUEST program. For discussion of the differences of FACT from other classification tree programs, see *A Brief Comparison of Classification Tree Programs* in Chapter 9 – *Classification Trees*.

Factor Analysis

The main applications of factor analytic techniques are 1) to reduce the number of variables and 2) to detect structure in the relationships between variables, that is to classify variables. Therefore, factor analysis is applied as a data reduction or (exploratory) structure detection method (the term *factor analysis* was first introduced by Thurstone, 1931).

For example, suppose we want to measure people's satisfaction with their lives. We design a satisfaction questionnaire with various items; among other things we ask our subjects how satisfied they are with their hobbies (item 1) and how intensely they are pursuing a hobby (item 2). Most likely, the responses to the two items are highly correlated with each other. Given a high correlation between the two items, we can conclude that they are quite redundant.

You can summarize the correlation between two variables in a scatterplot. A regression line can then be fitted that represents the best summary of the linear relationship between the variables. If we could define a variable that would approximate the regression line in such a plot, that variable would capture most of the essence of the two items. Subjects' single scores on that new factor, represented by the regression line, could then be used in future data analyses to represent that essence of the two items. In a sense we have reduced the two variables to one factor.

Factor analysis is an exploratory method; for information on confirmatory factor analysis, see Chapter 35 – *Structural Equation Modeling*. For more information on factor analysis, see Chapter 16 – *Factor Analysis*.

Feature Selection

One of preliminary stages in the process of data mining applicable when the data set includes more variables than could be included (or would be efficient to include) in the actual model building phase (or even in initial exploratory operations).

Feedforward Networks

Neural networks with a distinct layered structure, with all connections feeding forwards from inputs toward outputs. Sometimes used as a synonym for multilayer perceptrons.

Fishbone Chart

See *Cause-and-Effect Diagram*, page 584.

Fisher LSD

This post hoc test (or multiple comparison test) can be used to determine the significant differences between group means in an analysis of variance setting. The Fisher LSD test is considered to be one of the least conservative post hoc tests (for a detailed discussion of different post hoc tests, see Winer, Michels, & Brown (1991). For more details, see Chapter 18 – *General Linear Models*. See also, *Post hoc Comparisons*. For a discussion of statistical significance, see Chapter 1 – *Elementary Concepts in Statistics*.

Fixed Effects (in ANOVA)

The term fixed effects in the context of analysis of variance is used to denote factors in an ANOVA design with levels that are deliberately arranged by the experimenter, rather than randomly sampled from an infinite population of possible levels (those factors are called random effects). For example, if you were interested in conducting an experiment to test the hypothesis that higher temperature leads to increased aggression, you would probably expose subjects to moderate or high temperatures and then measure subsequent aggression. Temperature would be a fixed effect in this experiment, because the levels of temperature of interest to the experimenter were deliberately set, or fixed, by the experimenter.

A simple criterion for deciding whether an effect in an experiment is random or fixed is to ask how you would select (or arrange) the levels for the factor in a replication of the study. For example, if you wanted to replicate the study described in this example, you would choose the same levels of temperature from the population of levels of temperature. Thus, the factor "temperature" in this study would be a fixed factor. If instead, your interest is in how much of the variation of aggressiveness is due to temperature, you would probably expose subjects to a random sample of temperatures from the population of levels of different temperatures. Levels of temperature in the replication study would likely be different from the levels of temperature in the first study, thus temperature would be considered a random effect. See also, Chapter 3 – *ANOVA/ MANOVA* and Chapter 39 – *Variance Components and Mixed Model ANOVA/ANCOVA*.

Free Parameter

A numerical value in a structural model (see Chapter 35 – *Structural Equation Modeling*) that is part of the model, but is not fixed at any particular value by the model hypothesis. Free parameters are estimated by the program using iterative methods. Free parameters are indicated in the PATH1 language with integers placed between dashes on an arrow or a wire. For example, the following paths both have the free parameter 14.

(F1)-14->[X1]

(e1)-14-(e1)

If two different coefficients have the same free parameter number, as in the above example, both will of necessity be assigned the same numerical value. Simple equality constraints on numerical coefficients are thus imposed by assigning them the same free parameter number.

Frequency Tables (One-Way Tables)

Frequency or one-way tables represent the simplest method for analyzing categorical (nominal) data (see also, Chapter 1 – *Elementary Concepts in Statistics*). They are often used as one of the exploratory procedures to review how different categories of values are distributed in the sample. For example, in a survey of spectator interest in different sports, we could summarize the respondents' interest in watching football in a frequency table as follows:

Data: Frequency Tables (One-way Tables) (4v by 5c)				
STATISTICA BASIC STATS	FOOTBALL: "Watching Football"			
	Count	Cumulative Count	Percent	Cumulative Percent
Always: Always Interested	39	39	39	39
USUALLY: Usually Interested	16	55	16	55
SOMETIMES: Sometimes Interested	26	81	26	81
NEVER: Never Interested	19	100	19	100
Missing	0	100	0	100

The spreadsheet shows the number, proportion, and cumulative proportion of respondents who characterized their interest in watching football as either 1) *Always interested*, 2) *Usually interested*, 3) *Sometimes interested*, or 4) *Never interested*. For more information, see *Frequency Tables* in Chapter 2 – *Basic Statistics and Tables*.

Frequentist Approach

See *Bayesian Statistics (Analysis)*, page 568.

Function Minimization Algorithms

Algorithms used (e.g., in nonlinear estimation) to guide the search for the minimum of a function. For example, in the process of nonlinear estimation, the currently specified loss function is being minimized.

G

g2 Inverse

A g2 inverse is a generalized inverse of a rectangular matrix of values **A** that satisfies both:

AA`A=A  and  A`AA`=A

The *g2* inverse is used to find a solution to the normal equations in the general linear model; refer to Chapter 18 – *General Linear Models* for additional details. See also, matrix singularity and matrix inverse.

Gains Chart

The gains chart provides a visual summary of the usefulness of the information provided by one or more statistical models for predicting a binomial (categorical) outcome variable (dependent variable); for multinomial (multiple-category) outcome variables, gains charts can be computed for each category.

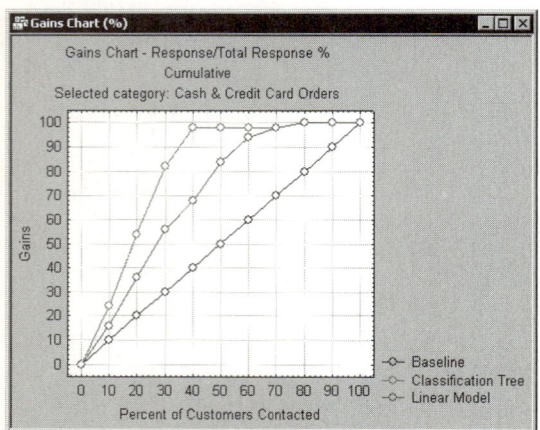

Specifically, the chart summarizes the utility that you can expect by using the respective predictive models, as compared to using baseline information only.

The gains chart is applicable to most statistical methods that compute predictions (predicted classifications) for binomial or multinomial responses. This and similar summary charts (see *Lift Chart*) are commonly used in data mining projects when the dependent or outcome variable of interest is binomial or multinomial in nature.

Example. To illustrate how the gains chart is constructed, consider this example. Suppose you have a mailing list of previous customers of your business, and you want to offer to those customers an additional service by mailing an elaborate brochure and other materials describing the service. During previous similar mail-out campaigns, you collected useful information about your customers (e.g., demographic information, previous purchasing patterns) that you could relate to the response rate, i.e., whether the respective customers responded to your mail solicitation and the type of order they placed.

Given the baseline response rate and the cost of the mail-out, sending the offer to all customers would result in a net-loss. Hence, you want to use statistical analyses to help you identify the customers who are most likely to respond. Suppose you build such a model based on the data collected in the previous mail-out campaign. You can now select only the 10 percent of the customers from the mailing lists who, according to prediction from the model, are most likely to respond. Next you can compute the number of accurately predicted responses, relative to the total number of responses in the sample; this percentage is the gain due to using the model. Put another way, of those customers likely to respond in the current sample, you can accurately identify (capture) y percent by selecting from the

customer list the top 10% who were predicted by the model with the greatest certainty to respond (where y is the gains value).

Analogous values can be computed for each percentile of the population (customers on the mailing list). You could compute separate gains values for selecting the top 20% of customers who are predicted to be among likely responders to the mail campaign, the top 30%, etc. Hence, the gains values for different percentiles can be connected by a line that will typically ascend slowly and merge with the baseline if all customers (100%) were selected.

If more than one predictive model is used, multiple gains charts can be overlaid (as shown in the illustration above) to provide a graphical summary of the utility of different models.

Gamma Coefficient

The *gamma* statistic is preferable to Spearman *R* or Kendall *tau* when the data contain many tied observations. In terms of the underlying assumptions, *gamma* is equivalent to Spearman *R* or Kendall *tau*; in terms of its interpretation and computation, it is more similar to Kendall *tau* than Spearman *R*. In short, *gamma* is also a probability; specifically, it is computed as the difference between the probability that the rank ordering of the two variables agree minus the probability that they disagree, divided by 1 minus the probability of ties. Thus, *gamma* is basically equivalent to Kendall *tau*, except that ties are explicitly taken into account. Detailed discussions of the *gamma* statistic can be found in Goodman and Kruskal (1954, 1959, 1963, 1972), Siegel (1956), and Siegel and Castellan (1988).

Gamma Distribution

The *gamma* distribution (the term first used by Weatherburn, 1946) is defined as:

$$f(x) = \{1/[b\Gamma(c)]\}*[x/b]^{c-1}*e^{-x/b}, \text{ for } 0 \leq x, c > 0$$

where Γ (*gamma*) is the *gamma* function, b is the scale parameter, c is the so-called shape parameter, and e is the base of the natural logarithm, sometimes called Euler's e (2.71...).

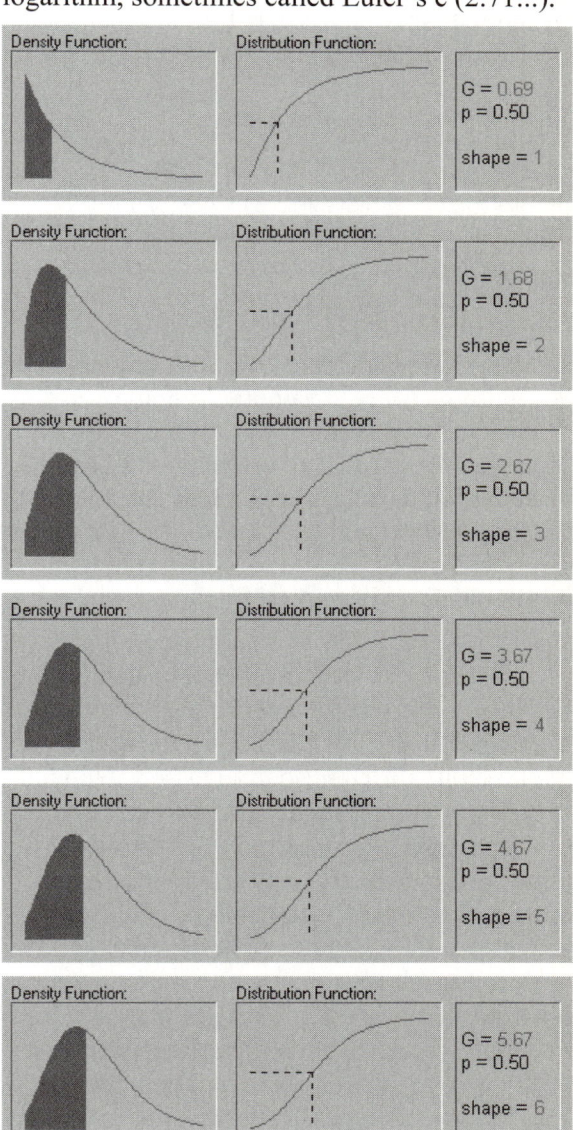

The previous illustrations show the *gamma* distribution as the shape parameter changes from 1 to 6.

Gaussian Distribution

The normal distribution, a bell-shaped function. See *Normal Distribution*.

Gauss-Newton Method

The Gauss-Newton method is a class of methods for solving nonlinear least-squares problems. In general, this method makes use of the Jacobian matrix **J** of first-order derivatives of a function F to find the vector of parameter values x that minimizes the residual sums of squares (sum of squared deviations of predicted values from observed values). An improved and efficient version of the method is the so-called Levenberg-Marquardt algorithm. For a detailed discussion of this class of methods, see Dennis & Schnabel (1983).

General ANOVA/MANOVA

The purpose of analysis of variance (ANOVA) is to test for significant differences between means by comparing (i.e., analyzing) variances. More specifically, by partitioning the total variation into different sources (associated with the different effects in the design), we are able to compare the variance due to the between-groups (or treatments) variability with that due to the within-group (treatment) variability. Under the null hypothesis (that there are no mean differences between groups or treatments in the population), the variance estimated from the within-group (treatment) variability should be about the same as the variance estimated from between-groups (treatments) variability. For more information, see Chapter 3 – *ANOVA/MANOVA*.

General Linear Model

The general linear model is a generalization of the linear regression model, such that effects can be tested 1) for categorical predictor *variables,* as well as for effects for continuous predictor variables and 2) in designs with multiple dependent variables as well as in designs with a single dependent variable. For an overview of the general linear model, see Chapter 18 – *General Linear Models*.

Generalization (in Neural Networks)

The ability of a neural network to make accurate predictions when faced with data not drawn from the original training set (but drawn from the same source as the training set). Generalization is typically achieved by dividing available training data into three subsets; one used for training the network, one used to cross verify the performance of training algorithms as they run, and one to perform a final independent test.

Generalized Additive Models

Generalized additive models [see Chapter 20 – *Generalized Additive Models (GAM)*]are generalizations of generalized linear/nonlinear models. In generalized linear/nonlinear models, the transformed dependent variable values are predicted from (is linked to) a linear combination of predictor variables; the transformation is referred to as the link function; also, different distributions can be assumed for the dependent variable values. An example of a generalized linear model is the logit regression model, where the dependent variable is assumed to be binomial, and the link function is the logit transformation. In generalized additive models, the linear function of the predictor values is replaced by an unspecified (non-parametric) function, obtained by applying a scatterplot

smoother to the scatterplot of partial residuals (for the transformed dependent variable values). See also, Hastie and Tibshirani, 1990, or Schimek, 2000.

Generalized Inverse

A generalized inverse (denoted by a superscript of -) of a rectangular matrix of values **A** is any matrix that satisfies

A⁻AA=A

A generalized inverse of a nonsingular matrix is unique and is called the regular matrix inverse. See also, *Matrix Singularity* and *Matrix Inverse*.

Generalized Linear/ Nonlinear Model

The generalized linear/nonlinear model is a generalization of the linear regression model such that nonlinear, as well as linear, effects can be tested for categorical predictor variables, as well as for continuous predictor variables, using any dependent variable whose distribution follows several special members of the exponential family of distributions (e.g., gamma, Possion, binomial, etc.), as well as for any normally-distributed dependent variable. For an overview of the generalized linear/nonlinear model, see Chapter 21 – *Generalized Linear/Nonlinear Models (GLZ)*.

Generalized Regression Neural Networks (GRNN)

A type of neural network using kernel-based approximation to perform regression. One of the so-called Bayesian networks (Speckt, 1991; Patterson, 1996; Bishop, 1995).

Genetic Algorithm

A search algorithm which locates optimal binary strings by processing an initially random population of strings using artificial mutation, crossover and selection operators, in an analogy with the process of natural selection (Goldberg, 1989). See also, *Neural Networks*.

Genetic Algorithm Input Selection

Application of a genetic algorithm to determine an optimal set of input variables by constructing binary masks which indicate which inputs to retain and which to discard (Goldberg, 1989). This method is implemented in *STATISTICA Neural Networks* and can be used as part of a model building process where variables identified as the most relevant (in *STATISTICA Neural Networks*) are then used in a traditional model building stage of the analysis (e.g., using a linear regression or nonlinear estimation method).

Geometric Distribution

The geometric distribution (the term first used by Feller, 1950) is defined as:

$$f(x) = p*(1-p)^x$$

where p is the probability that a particular event (e.g., success) will occur.

Geometric Mean

The geometric mean is a summary statistic useful when the measurement scale is not linear; it is computed as:

$$G = (x_1*x_2*...*x_n)^{1/n}$$

Where n is the sample size.

Gibbs Sampler

The Gibbs sampler is a popular method used for MCMC (Markov chain Monte Carlo) analyses. It provides an elegant way for sampling from the joint distributions of multiple variables, by applying the notion that: *to sample from a joint distribution just sample repeatedly from its one-dimensional conditionals given whatever you've seen at the time.*

For example, the values from the joint distribution of two random variables, X and Y, can be easily simulated by the *Gibbs sampler* that uses their conditional distributions rather than their joint distribution. Starting with an arbitrary choice of X and Y, X is simulated from the conditional distribution of X, given Y, and Y is simulated from conditional distribution of Y, given X. Alternating between two conditional distributions, in the subsequent steps, generates a sample from the correct joint distribution of X and Y; the approximation gets better and better as the length of the Gibbs sampler path increases.

Gini Measure of Node Impurity

According to Breiman, Friedman, Olshen, & Stone (1984), the Gini measure of node impurity at node is defined to be (pp. 28 & 38)

$$impurity(t) = \sum_{i \neq j} p(i \mid t)p(j \mid t)$$

where

$$p(j \mid t) = \frac{p(j,t)}{p(t)}$$

and

$$p(j,t) = \frac{\pi(j)N_j(t)}{N_j}$$

such that

$p(j \mid t)$ is the estimated probability that an observation belongs to group j given that it is in node t, $p(j,t)$ is the estimated probability that an observation is in group j and at node t, $p(t)$ is the estimated probability that an observation is at node t, $\frac{N(t)}{N}$, $\pi(j)$ is the prior probability for group j, $Nj(t)$ is the number of group j members at node t, and N_j is the size of group j.

Therefore, the prior probabilities play a role in every Gini Measure computation at every node. However, Breiman et al. also note that, when the prior probabilities are estimated from the data,

$$\pi(j) = \frac{N_j}{N} \Rightarrow p(j \mid t) = \frac{N_j(t)}{N(t)}$$

This fact can cause higher misclassification rates in under-represented groups.

Gompertz Distribution

The Gompertz distribution is a theoretical distribution of survival times. Gompertz (1825) proposed a probability model for human mortality, based on the assumption that the "average exhaustion of a man's power to avoid death to be such that at the end of equal infinitely small intervals of time he lost equal portions of his remaining power to oppose destruction which he had at the commencement of these intervals" (Johnson, Kotz, Blakrishnan, 1995, p. 25). The resultant hazard function:

$$r(x)=Bc^x, \quad \text{for } x \leq 0, B > 0, c \leq 1$$

is often used in survival analysis. See Johnson, Kotz, Blakrishnan (1995) for additional details.

Gradient

In structural equation modeling, the gradient is the vector of first partial derivatives of the discrepancy function with respect to the parameter values. At a local or global minimum, the discrepancy function should be at the bottom of a "valley," where all first partial derivatives are zero, so the elements of the gradient should all be near to zero when a minimum is obtained.

The elements of the gradient, by themselves, can, on occasion, be somewhat unreliable as indicators of when convergence has occurred, especially when the model fit is not good, and the discrepancy function value itself is quite large. For this reason, the gradient is not employed as a convergence criterion by this program.

Gradient Descent

Optimization techniques for nonlinear functions (e.g. the error function of a neural network as the weights are varied) which attempt to move incrementally to successively lower points in search space, in order to locate a minimum.

Gradual Permanent Impact

In time series, the gradual permanent impact pattern implies that the increase or decrease due to the intervention is gradual, and that the final permanent impact becomes evident only after some time. This type of intervention can be summarized by the expression:

$$\text{Impact}_t = \delta * \text{Impact}_{t-1} + \omega$$

(for all $t \geq$ time of impact, else = 0)

Note that this impact pattern is defined by the two parameters δ (*delta*) and ω (*omega*). If δ is near 0 (zero), the final permanent amount of impact will be evident after only a few more observations; if δ is close to 1, the final permanent amount of impact will only be evident after many more observations. As long as the d parameter is greater than 0 and less than 1 (the bounds of system stability), the impact will be gradual and result in an asymptotic change (shift) in the overall mean by the quantity:

$$\text{Asymptotic change in level} = \omega / (1 - \delta)$$

Grouping (or Coding) Variable

A grouping (or coding) variable is used to identify group membership for individual cases in the data file. Typically, the grouping variable is categorical (i.e., contains either discrete values, e.g., *1, 2, 3, ...,*

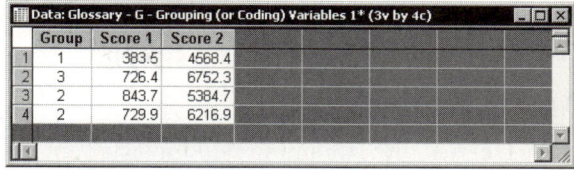

or text values, e.g., *MALE, FEMALE*)

Group	Score 1	Score 2
Male	383.5	4568.4
Female	726.4	6752.3
Female	843.7	5384.7
Male	729.9	6216.9

and the values are referred to as codes (they can be integer values or integer values with text value equivalents).

Groupware

Software intended to enable a group of users on a network to collaborate on specific projects. Groupware may provide services for communication (such as e-mail), collaborative document development, analysis, reporting, statistical data analysis, scheduling, or tracking. Documents can include text, images, or any other forms of information (e.g., multimedia). See also, *Enterprise-Wide Software Systems*, page 612.

H

Hamming Window

In time series, the Hamming window is a weighted moving average transformation used to smooth the periodogram values. In the Hamming (named after R. W. Hamming) window or Tukey-Hamming window (Blackman and Tukey, 1958), for each frequency, the weights for the weighted moving average of the periodogram values are computed as:

$$w_j = 0.54 + 0.46*cosine(\pi*j/p) \text{ (for } j=0 \text{ to } p)$$
$$w_{-j} = w_j \text{ (for } j \neq 0)$$

where $p = (m-1)/2$.

This weight function will assign the greatest weight to the observation being smoothed in the center of the window, and increasingly smaller weights to values that are further away from the center. See also, *Basic Notations and Principles*, page 512, in Chapter 38 – *Time Series/Forecasting*.

Harmonic Mean

The harmonic mean is a summary statistic used in analyses of frequency data; it is computed as:

$$H = n * 1/\Sigma(1/x_i)$$

where *n* is the sample size.

Hazard

It is often meaningful to consider the function that describes the probability of failure during a very small time increment (assuming that no failures have occurred prior to that time). This function is called the hazard function (or, sometimes, also conditional failure, intensity, or force of mortality function), and is generally defined as:

$$h(t) = f(t)/(1-F(t))$$

where *h(t)* stands for the hazard function (of time *t*), and *f(t)* and *F(t)* are the probability density and cumulative distribution functions, respectively. For additional information, see Chapter 36 – *Survival/Failure Time Analysis* or *Weibull and Reliability/Failure Time Analysis* in Chapter 32 – *Process Analysis*.

Hazard Rate

In survival analysis, the hazard rate is defined as the probability per time unit that a case that has survived to the beginning of the respective interval will fail in that interval. Specifically, it is computed as the number of failures per time units in the respective interval, divided by the average number of surviving cases at the midpoint of the interval.

Heuristic

As opposed to an algorithm (which contains a fully defined set of steps that will produce a specific outcome), heuristics are general recommendations or guides based on statistical evidence (e.g., *"quit smoking to prolong your life," "males with college education are more likely to respond positively to this advertisement than…"*) or theoretical reasoning (e.g., *"the mechanism of the vitamin X synthesis as we understand it, implies that eating Y will reduce the deficit of X"*). For more information about the concept of *heuristic*, see Kahneman, Slovic, & Tversky, 1982. See also, *Data Mining*, page 602, *Neural Networks* and *Algorithm*.

Heywood Case

A Heywood case in common factor analysis occurs when the minimum of the discrepancy function is obtained with one or more negative values as estimates for the variance of the unique variables. Such values are of course impossible. Heywood cases occur frequently when too many factors are extracted, or the sample size is too small.

Hidden Layers (in Neural Networks)

All layers of a neural network except the input and output layers. Hidden layers provide the network's nonlinear modeling capabilities.

High-Low Close

In this type of box or range plot, the serifs on the whiskers are not symmetrical but point to the left of the bar, representing the traditional stock price graph style. Note that you can change the whisker style (e.g., *Hi/Lo Left, Hi/Lo Right,* or *Whiskers*), for example:

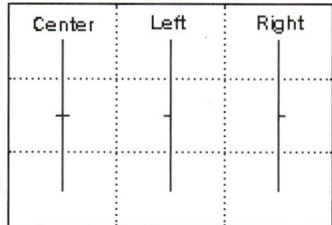

Histogram, 2D

2D histograms (the term was first used by Pearson, 1895) present a graphical representation (see the next illustration) of the frequency distribution of the selected variable(s) in which the columns are drawn over the class intervals and the heights of the columns are proportional to the class frequencies.

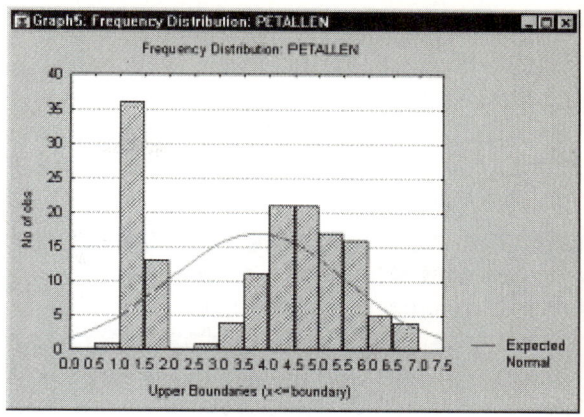

Histogram, 2D – Hanging Bars

The hanging bars histogram offers a "visual test of normality" of the distribution that helps identify the areas of the distribution where the discrepancies (between the observed and expected normal frequencies) occur. While the standard way of presenting the normal distribution fitted to an observed distribution is to overlay the best-fitting normal curve over a histogram, the hanging bars histogram does just the opposite: it suspends the bars representing the observed frequencies for consecutive ranges of values from the best-fitting normal curve.

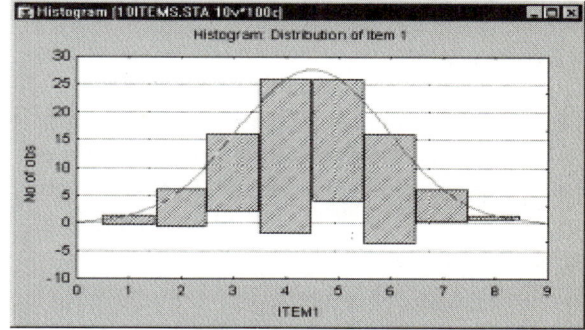

If the investigated distribution can be well approximated by the normal curve, the bottoms of all bars should form a straight, horizontal line.

Histogram, 2D – Multiple

Multiple histograms present frequency distributions of more than one variable in one 2D graph. Unlike the double-y histogram, the frequencies for all variables are plotted against the same left-Y axis.

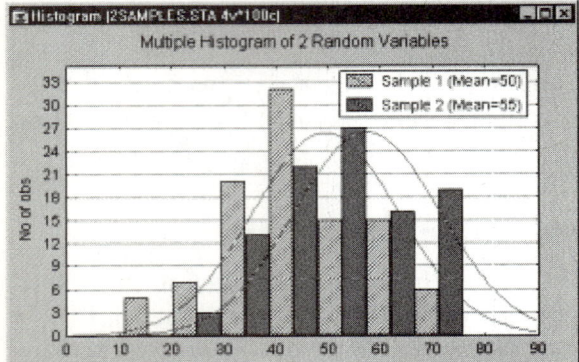

Also, the values of all examined variables are plotted against a single X axis, which facilitates comparisons between analyzed variables.

Histogram, 3D – Bivariate

Three-dimensional histograms are used to visualize crosstabulations of values in two variables. They can be considered to be a conjunction of two simple (i.e., univariate) histograms, combined such that the frequencies of co-occurrences of values on the two analyzed variables can be examined. In a most common format of this graph, a 3D bar is drawn for each cell of the crosstabulation table and the height of the bar represents the frequency of values for the respective cell of the table. Different methods of categorization can be used for each of the two variables for which the bivariate distribution is visualized (see the next illustration).

For information on smoothing 3D bivariate histograms, see *Smoothing Bivariate Distributions*, page 545.

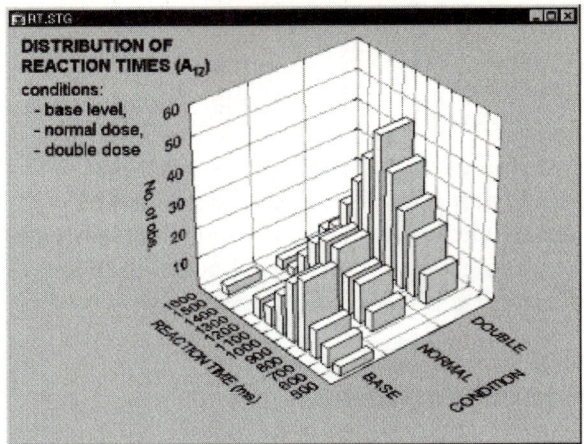

Histogram, 3D – Box Plots

This type of bivariate histogram represents the frequencies as a series of 3D bars (rectangular boxes). This is the default representation of 3D histograms. The height of each bar on the Z-axis corresponds to the frequency of the respective combination of levels for the two variables.

Histogram, 3D – Contour/Discrete

This contour plot represents a discrete projection of the 3D (smoothed) histogram.

Histogram, 3D – Contour Plot

This contour plot presents a projection of the spline-smoothed surface fit to the frequency data (see *Sequential Surface Plot, 3D*). Successive values of each series are plotted along the *X*-axis, with each successive series represented along the *Y*-axis.

Histogram, 3D – Spikes

In this type of bivariate histogram, the frequencies are represented as a series of spikes (point symbols with lines descending to the base plane). The height of each spike is

determined by the frequency for the respective combination of levels of the two variables.

Histogram, 3D – Surface Plot

In this representation of the 3D bivariate histogram, a spline-smoothed surface is fit to the frequency data.

Hollander-Proschan Test

This test compares the theoretical reliability function to the Kaplan-Meier estimate. The actual computations for this test are somewhat complex, and you may refer to Dodson (1994, Chapter 4) for a detailed description of the computational formulas. The Hollander-Proschan test is applicable to complete, single-censored, and multiple-censored data sets; however, Dodson (1994) cautions that the test may sometimes indicate a poor fit when the data are heavily single-censored. The Hollander-Proschan C statistic can be tested against the normal distribution (z).

The Hollander-Proschan test is used in Weibull and reliability/failure time analysis; see also, *Mann-Scheuer-Fertig Test* and *Anderson-Darling Test*.

Hooke-Jeeves Pattern Moves

A nonlinear estimation procedure that, at each iteration, first defines a pattern of points by moving each parameter one by one so as to optimize the current loss function. The entire pattern of points is then shifted or moved to a new location; this new location is determined by extrapolating the line from the old base point in the m dimensional parameter space to the new base point. The step sizes in this process are constantly adjusted to "zero in" on the respective optimum. This method is usually quite effective, and should be tried if both the quasi-Newton and Simplex methods fail to produce reasonable estimates.

HTM

A file name extension used to save HTML documents (see also, *HTML*).

HTML

Acronym for HyperText Markup Language. The markup language used for documents on the World Wide Web. HTML uses tags to identify elements of the document, such as text or graphics. HTML 2.0, defined by the Internet Engineering Task Force (IETF), includes features of HTML common to all Web browsers as of 1995 and was the first version of HTML widely used on the World Wide Web. Future HTML development will be carried out by the World Wide Web Consortium (W3C). HTML 3.2, the latest proposed standard, incorporates features widely implemented as of early 1996. Most Web browsers, notably Netscape Navigator and Internet Explorer, recognize HTML tags beyond those included in the present standard.

Hyperbolic Tangent (Tanh)

A symmetric S-shaped (sigmoid) function; sometimes used as an alternative to logistic functions.

Hyperplane

An N-dimensional analogy of a line or plane, which divides an $N+1$ dimensional space into two. See *Neural Networks*.

Hypersphere

An N-dimensional analogy of a circle or sphere. See *Neural Networks*.

I

Icon Plot

One of the potentially powerful general techniques of exploratory data analysis are multidimensional icon graphs. The basic idea of icon plots is to represent individual units of observation as particular graphical objects where values of variables are assigned to specific features or dimensions of the objects (usually one case = one object). The assignment is such that the overall appearance of the objects changes as a function of the configuration of values. Thus, the objects are given visual identities that are unique for configurations of values and that can be identified by the observer. Examining such icons may help to discover specific clusters of both simple relations and interactions between variables. See also, *Icon Plots*, page 547.

Icon Plot – Chernoff Faces

Chernoff faces is the most elaborate type of icon plot.

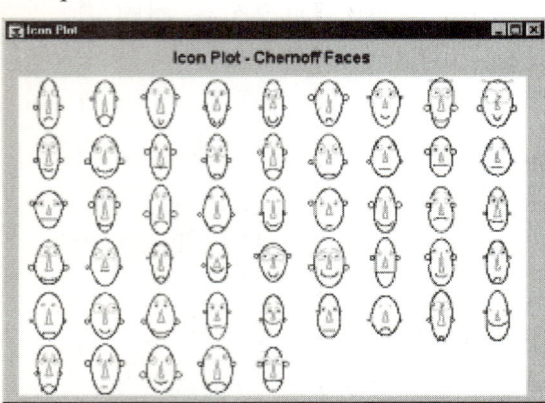

A separate face icon is drawn for each case; relative values of the selected variables for each case are assigned to shapes and sizes of individual facial features (e.g., length of nose, angle of eyebrows, width of face).

Icon Plot – Columns

In this type of icon plot, an individual column graph is plotted for each case; relative values of the selected variables for each case are represented by the height of consecutive columns.

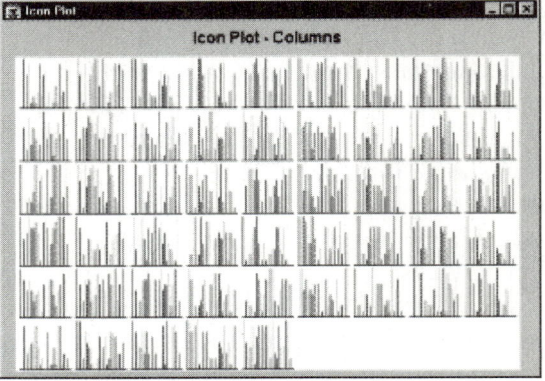

Icon Plot – Lines

In this type of icon plot, an individual line graph is plotted for each case. Relative values of the selected variables for each case are represented by the height of consecutive break points of the line above the baseline.

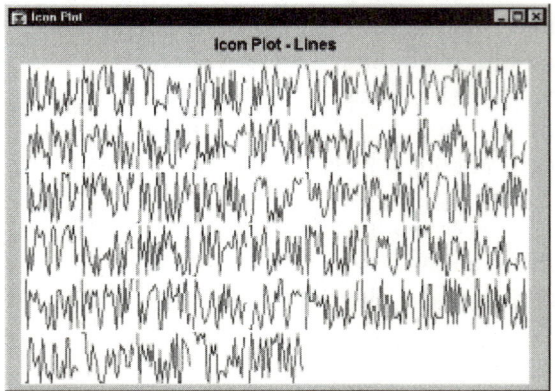

Icon Plot – Pies

In this type of icon plot, data values for each case are plotted as a pie chart (clockwise, starting at 12:00); relative values of selected variables are represented by the size of the pie slices.

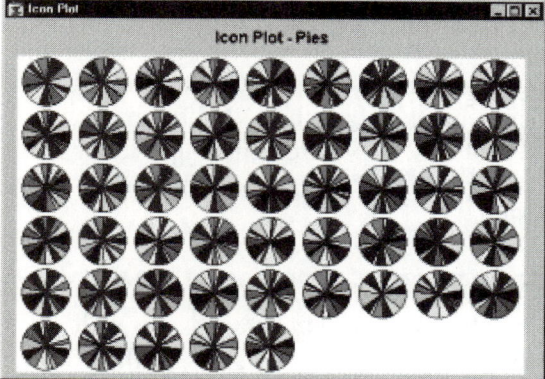

See also, *Icon Plots*, page 547.

Icon Plot – Polygons

In this type of icon plot, a separate polygon icon is plotted for each case; relative values of the selected variables for each case are represented by the distance from the center of the icon to consecutive corners of the polygon (clockwise, starting at 12:00).

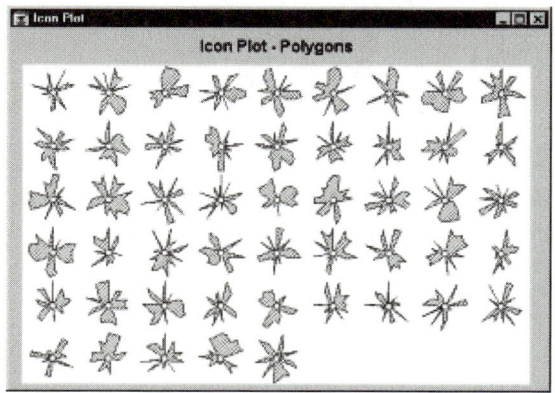

See also, *Icon Plots*, page 547.

Icon Plot – Profiles

In this type of icon plot, an individual area graph is plotted for each case; relative values of the selected variables for each case are represented by the height of consecutive peaks of the profile above the baseline.

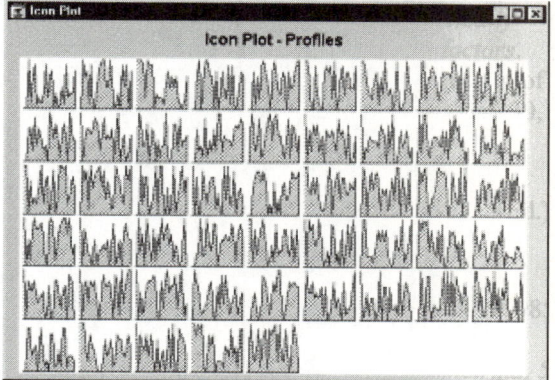

See also, *Icon Plots*, page 547.

Icon Plot – Stars

In this type of icon plot, a separate star-like icon is plotted for each case; relative values of the selected variables for each case are represented (clockwise, starting at 12:00) by the relative length of individual rays in each star. The ends of the rays are connected by a line.

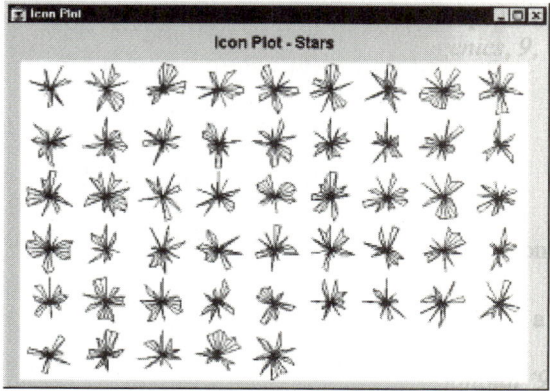

See also, *Icon Plots*, page 547.

Icon Plot – Sun Rays

In this type of icon plot, a separate sun-like icon is plotted for each case; each ray represents one of the selected variables (clockwise, starting at 12:00), and the length of the ray represents 4 standard deviations. Data values of the variables for each case are connected by a line.

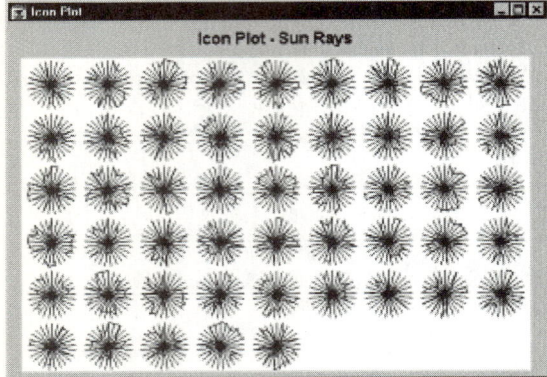

See also, *Icon Plots*, page 547.

Incremental (vs. Non-Incremental Learning Algorithms)

Methods (algorithms) for predictive data mining are also referred to as "learning" algorithms, because they derive information from the data to predict new observations. These algorithms can be divided into those that require one or perhaps two complete passes through the input data, and those that require iterative multiple access to the data to complete the estimation. The former type of algorithms are also sometimes referred to as incremental learning algorithms, because they will complete the computations necessary to fit the respective models by processing one case at a time, each time refining the solution; then, when all cases have been processed, only few additional computations are necessary to produce the final results. Non-incremental learning algorithms are those that need to process all observations in each iteration of an iterative procedure for refining a final solution. Obviously, incremental learning algorithms are usually much faster than non-incremental algorithms, and for extremely large data sets, non-incremental algorithms may not be applicable at all (without first sub-sampling).

Independent Component Analysis

Independent Component Analysis is a well established and reliable statistical method that performs signal separation. Signal separation is a frequently occurring problem and is central to Statistical Signal Processing, which has a wide range of applications in many areas of technology ranging from Audio and Image Processing to Biomedical Signal Processing, Telecommunications, and Econometrics.

Imagine being in a room with a crowd of people and two speakers giving presentations at the same time. The crowd is making comments and noises in the background. We are interested in what the speakers say and not the comments emanating from the crowd. There are two microphones at different locations, recording the speakers' voices as well as the noise coming from the crowed. Our task is to separate the voice of each speaker while ignoring the background noise.

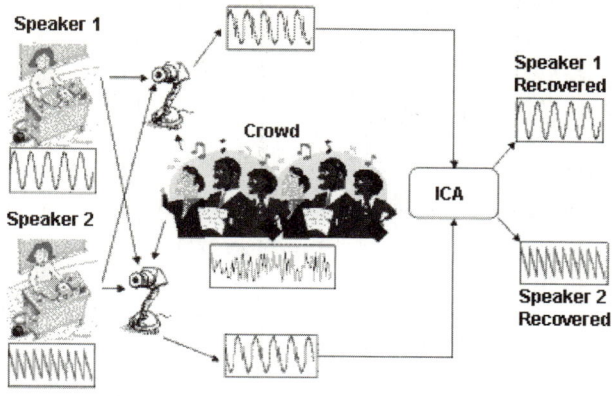

This is a classic example of the Independent Component Analysis, a well established stochastic technique. ICA can be used as a method of Blind Source Separation, meaning that it can separate independent signals from linear mixtures with virtually no prior knowledge on the signals. An example is decomposition of Electro or Magnetoencephalographic signals. In computational Neuroscience, ICA has been used for Feature Extraction, in which case it seems to adequately model the basic cortical processing of visual and auditory information. New application areas are being discovered at an increasing pace.

Inertia

The term inertia in correspondence analysis is used by analogy with the definition in applied mathematics of moment of inertia, which stands for the integral of mass times the squared distance to the centroid (e.g., Greenacre, 1984, p. 35). Inertia is defined as the total Pearson *chi*-square for a two-way frequency table divided by the total sum of all observations in the table.

In-Place Database Processing (IDP)

In-Place Database Processing (IDP) is an advanced database access technology developed at StatSoft to support high-performance, direct interface between external data sets residing on remote servers and the analytic functionality of data analysis software (such as *STATISTICA* products) residing on the client computers. The IDP technology has been developed to facilitate accessing data in large databases using a one-step process that does not necessitate creating local copies of the data set. IDP significantly increases the overall performance of data processing software; it is particularly well suited for large data mining and exploratory data analysis tasks.

The source of the IDP performance gains. The speed gains of the IDP technology are tremendous compared to accessing data in a traditional way. IDP enables the client computer to access data directly from remote databases and skip the otherwise necessary first step of importing the data and creating a local data file. The superior retrieval speed is further enhanced by IDP's multitasking architecture, which employs asynchronous and distributed processing. Specifically, IDP uses the processing resources (multiple CPUs) of the database server computers to execute the query operations, extract the requested records of data, and send them to the client computer, while the data analysis software on the client computer is simultaneously processing these records as they arrive.

Interactions

An effect of interaction occurs when a relation between (at least) two variables is modified by (at least one) other variable. In other words, the strength or the sign (direction) of a relation between (at least) two variables is different depending on the value (level) of some other variable(s). (The term interaction was first used by Fisher, 1926.) Note that the term "modified" in this context does not imply causality but represents a simple fact that depending on what subset of observations (regarding the "modifier" variable(s)) you are looking at, the relation between the other variables will be different.

For example, imagine that we have a sample of highly achievement-oriented students and another of achievement "avoiders." We now create two random halves in each sample, and give one half of each sample a challenging test, the other an easy test. We measure how hard the students work on the test. The means of this (fictitious) study are as follows:

Data: Glossary - I - Interactions (2v by 2c)	Achievement oriented	Achievement avoiders		
Challenging Test	10	5		
Easy Test	5	10		

How can we summarize these results? Is it appropriate to conclude that 1) challenging tests make students work harder, and 2) achievement-oriented students work harder than achievement avoiders? None of these statements captures the essence of this clearly systematic pattern of means. The appropriate way to summarize the result would be to say that challenging tests make only achievement-oriented students work harder, while easy tests make only achievement-avoiders work harder. In other words, the relation between the type of test and effort is positive in one group but negative in the other group. Thus, the type of achievement orientation and test difficulty interact in their effect on effort; specifically, this is an example of a two-way interaction between achievement orientation and test difficulty. (Note that statements 1 and 2 above

would describe so-called main effects.) For more information regarding interactions, see *Interaction Effects* in Chapter 3 – *ANOVA/MANOVA*.

Interpolation

Projecting a curve between known data points to infer the value of a function at points between.

Interval Scale

This scale of measurement enables you to rank order not only the items that are measured, but also to quantify and compare the sizes of differences between them (no absolute zero is required). See also, *Measurement Scales* in Chapter 1 – *Elementary Concepts in Statistics*.

Intraclass Correlation Coefficient

The value of the population intraclass correlation coefficient is a measure of the homogeneity of observations within the classes of a random factor relative to the variability of such observations between classes. It will be zero only when the estimated effect of the random factor is zero and will reach unity only when the estimated effect of error is zero, given that the total variation of the observations is greater than zero (see Hays, 1988, p. 485).

Note that the population intraclass correlation can be estimated using variance component estimation methods. For more information, see Chapter 39 – *Variance Components and Mixed-Model ANOVA/ANCOVA*.

Invariance Under a Constant Scale Factor (ICSF)

A structural model is invariant under a constant scale factor (ICSF) if model fit is not changed if all variables are multiplied by the same

constant. Most, but not all, structural models that are of practical interest are ICSF (see Chapter 35 – *Structural Equation Modeling*).

Invariance Under Change of Scale (ICS)

A structural model is invariant under change of scale if model fit is not changed by rescaling the variables, i.e., by multiplying them by scale factors (see Chapter 35 – *Structural Equation Modeling*).

Ishikawa Chart

A type of diagram used to depict the factors or variables that make up a process; named after Professor Kaoru Ishikawa of Tokyo University (e.g., see Seder, 1962), this diagram is also referred to as the Cause-and-Effect diagram. For more details, see *Cause-and-Effect Diagram*, page 584.

Isotropic Deviation Assignment

An algorithm for assigning radial unit deviations, which selects a single deviation value using a heuristic calculation based on the number of units and the volume of pattern space they occupy, with the objective of ensuring "a reasonable overlap." (Haykin, 1994). See also, Chapter 27 – *Neural Networks*.

J

Jacobian Matrix

The first-order derivative of a continuous and differentiable function F (of multiple parameters) is sometimes called the *Jacobian* matrix **J** of **F** (at some specific values of parameter vector **x**). The Jacobian matrix plays an important role in most computational algorithms for estimating parameter values for nonlinear regression problems, in particular in the Gauss-Newton and Levenberg-Marquardt algorithms; see also, Chapter 28 – *Nonlinear Estimation* for details.

Jogging Weights

Adding a small random amount to the weights in a neural network, in an attempt to escape a local optima in error space. See also, Chapter 27 – *Neural Networks*.

Johnson Curves

Johnson (1949) described a system of frequency curves that represents transformations of the standard normal curve (see Hahn and Shapiro, 1967, for details). By applying these transformations to a standard normal variable, a wide variety of non-normal distributions can be approximated, including distributions that are bounded on either one or both sides (e.g., U-shaped distributions). The advantage of this approach is that once a particular Johnson curve has been fit, the normal integral can be used to compute the expected percentage points under the respective curve. Methods for fitting Johnson curves, so as to approximate the first four moments of an empirical distribution, are described in detail in Hahn and Shapiro, 1967,

pages 199-220; and Hill, Hill, and Holder, 1976. See also, *Pearson Curves*.

Join

A join shows how data is related between two tables. When two tables contain matching values on a field, records from the two tables can be combined by defining a join. For example, suppose one table has the weight of objects with their associated part number and another table has part numbers and their associated product names. A join specifies that the two part number fields are equivalent and enables weights and product names to be related.

Joining Networks (in Neural Networks)

It is sometimes useful to be able to join two networks together to form a single composite network for a number of reasons. You might train one network to do preprocessing of data, and another to further classify the preprocessed data. Once completed, the networks can be joined together to classify raw data. Or, you might want to add a loss matrix to a classification network to make minimum cost decisions.

Note: Networks can only be joined if the number of input neurons in the second network matches the number of output neurons in the first network. The input neurons from the second network are discarded, and their fan-out weights are attached to the output neurons of the first network.

Caution: The post-processing information from the first network and the input preprocessing information from the second network are also discarded. The composite network is unlikely to make sense unless you have designed the two networks with this in mind; i.e., with no post-processing performed by the first network and no preprocessing performed by the second network.

JPEG

An acronym for Joint Photographic Experts Group, an ISO/ITU standard for storing images in compressed form using a discrete cosine transform. JPEG files offer a more economical method of saving bitmap (i.e., raster as opposed to vector/metafile) graphics than the *.bmp* format, but highly compressed images may lose some of their quality compared to *.bmp* files. See also, *PNG (Portable Network Graphics)*.

JPG

A file name extension used to save JPEG documents (see *JPEG*).

K

Kendall Tau

Kendall *tau* is a nonparametric measure of correlation defined as:

T = (# agreements - # disagreements) / total number of pairs

For small *n* (n <10), the exact probability can be calculated. The tabulated values can be found in Siegel and Castellan. However, the exact sampling distribution of *T* approaches a normal distribution very quickly with increasing *n* size. For n = 10 or more, refer to the normal distribution (Hays, 1988).

Kendall *tau* is equivalent to the Spearman *R* statistic with regard to the underlying assumptions. It is also comparable in terms of its statistical power. However, Spearman *R* and Kendall *tau* are usually not identical in magnitude because their underlying logic, as well as their computational formulas are very different. Siegel and Castellan (1988) express the relationship of the two measures in terms of the inequality:

-1 < = 3 * Kendall tau – 2 * Spearman R < = 1

More importantly, Kendall *tau* and Spearman *R* imply different interpretations. While Spearman *R* can be thought of as the regular Pearson product-moment correlation coefficient as computed from ranks, Kendall *tau* represents a probability. Specifically, it is the difference between the probability that the observed data are in the same order for the two variables versus the probability that the observed data are in different orders for the two variables. Kendall (1948, 1975), Everitt (1977), and Siegel and Castellan (1988) discuss Kendall *tau* in greater detail. Two different variants of *tau* are computed, usually called tau_b and tau_c. These measures differ only with regard as to how tied ranks are handled. In most cases these values will be fairly similar, and when discrepancies occur, it is probably always safest to interpret the lowest value.

Kernel Functions

Simple functions (typically Gaussians) that are added together, positioned at known data points, to approximate a sampled distribution (Parzen, 1962). See also, Chapter 27 – *Neural Networks*.

Kernels

A function that accepts two vectors as input and returns a scalar that represents the inner product of the vectors in some alternate dimension.

k-Means Algorithm

k-means algorithm is an algorithm to assign *k* centers to represent the clustering of *N* points (*k*<N). The points are iteratively adjusted (starting with a random sample of the *N* points) so that each of the *N* points is assigned to one of the *k* clusters, and each of the *k* clusters is the mean of its assigned points (Bishop, 1995). See also, Chapter 10 – *Cluster Analysis* and Chapter 27 – *Neural Networks*.

k-Means Algorithm (in Neural Networks)

The *k*-means algorithm (Moody and Darkin, 1989; Bishop, 1995) assigns radial centers to the first hidden layer in the network if it consists of radial units.

k-means assigns each training case to one of *k* clusters (where *k* is the number of radial units), such that each cluster is represented by the

centroid of its cases, and each case is nearer to the centroid of its cluster than to the centroids of any other cluster. It is the centroids that are copied to the radial units.

The intention is to discover a set of cluster centers which best represent the natural distribution of the training cases.

Technical details. k-means is an iterative algorithm. The clusters are first formed arbitrarily by choosing the first k cases, assigning each subsequent case to the nearest of the k, and then calculating the centroids of each cluster.

Subsequently, each case is tested to see whether the center of another cluster is closer than the center of its own cluster; if so, the case is reassigned. If cases are reassigned, the centroids are recalculated and the algorithm repeats.

Caution. There is no formal proof of convergence for this algorithm, although in practice it usually converges reasonably quickly.

k-Nearest Algorithm

An algorithm to assign deviations to radial units. Each deviation is the mean distance to the k-nearest neighbors of the point. See also, Chapter 27 – *Neural Networks*.

Kohonen Algorithm (in Neural Networks)

The Kohonen algorithm (Kohonen, 1982; Patterson, 1996; Fausett, 1994) assigns centers to a radial hidden layer by attempting to recognize clusters within the training cases. Cluster centers close to one another in pattern-space tend to be assigned to units that are close to each other in the network (topologically ordered).

The Kohonen training algorithm is the algorithm of choice for self organizing feature map networks. It can also be used to train the radial layer in other network types; specifically, radial basis function, cluster, and generalized regression neural networks.

SOFM networks are typically arranged with the radial layer laid out in two dimensions. From an initially random set of centers, the algorithm tests each training case and selects the nearest center. This center and its neighbors are then updated to be more like the training case.

Over the course of the algorithm, the learning rate (which controls the degree of adaptation of the centers to the training cases) and the size of the neighborhood are gradually reduced. In the early phases, therefore, the algorithm assigns a rough topological map, with similar clusters of cases located in certain areas of the radial layer. In later phases the topological map is fine-tuned, with individual units responding to small clusters of similar cases.

If the neighborhood is set to zero throughout, the algorithm is a simple cluster-assignment technique. It can also be used on a one-dimensional layer with or without neighborhood definition.

If class labels are available for the training cases, after Kohonen training, labels can be assigned using class labeling algorithms and Learned Vector Quantization used to improve the positions of the radial exemplars.

Technical details. The Kohonen update rule is:

$$w(t) = w(t-1) + \eta(t)(x - w(t-1))$$

x is the training case and $\eta(t)$ is the learning rate.

Kohonen Networks

Neural networks based on the topological properties of the human brain, also known as self-organizing feature maps (SOFMs) (Kohonen, 1982; Fausett, 1994; Haykin, 1994; Patterson, 1996).

Kohonen Training

An algorithm that assigns cluster centers to a radial layer by iteratively submitting training patterns to the network, and adjusting the winning (nearest) radial unit center, and its neighbors, toward the training pattern (Kohonen, 1982; Fausett, 1994; Haykin, 1994; Patterson, 1996). See also, Chapter 27 – *Neural Networks*.

Kolmogorov-Smirnov Test

The Kolmogorov-Smirnov one-sample test for normality is based on the maximum difference between the sample cumulative distribution and the hypothesized cumulative distribution. If the D statistic is significant, the hypothesis that the respective distribution is normal should be rejected. For many software programs, the probability values that are reported are based on those tabulated by Massey (1951); those probability values are valid when the mean and standard deviation of the normal distribution are known *a priori* and not estimated from the data. However, usually those parameters are computed from the actual data. In that case, the test for normality involves a complex conditional hypothesis ("how likely is it to obtain a D statistic of this magnitude or greater, contingent upon the mean and standard deviation computed from the data"), and the Lilliefors probabilities should be interpreted (Lilliefors, 1967). Note that in recent years, the Shapiro-Wilks' W test has become the preferred test of normality because of its good power

properties as compared to a wide range of alternative tests.

Kronecker Product

The Kronecker (direct) product of 2 matrices $\mathbf{A}$, with p rows and q columns, and $\mathbf{B}$, with m rows and n columns, is the matrix with pm rows and qn columns given by

$$A \otimes B = a_{ij} B$$

For example, if

$$A = \begin{bmatrix} 1 & 0 \\ 0 & 1 \\ -1 & -1 \end{bmatrix}$$

and

$$B = \begin{bmatrix} 1 \\ -1 \end{bmatrix}$$

then

$$A \otimes B = \begin{bmatrix} 1 & 0 \\ 0 & 1 \\ -1 & -1 \\ -1 & 0 \\ 0 & -1 \\ 1 & 1 \end{bmatrix}$$

Kronecker Product matrices have a number of useful properties (for a summary of these properties, see Hocking, 1985).

Kruskal-Wallis Test

The Kruskall-Wallis test is a nonparametric alternative to one-way (between-groups) ANOVA. It is used to compare three or more samples, and it tests the null hypothesis that the different samples in the comparison were drawn from the same distribution or from distributions with the same median. Thus, the interpretation of the Kruskall-Wallis test is basically similar to that of the parametric one-way ANOVA, except that it is based on ranks rather than means. For more details, see Siegel & Castellan, 1988. See also, Chapter 29 – *Nonparametric Statistics*.

Kurtosis

Kurtosis (the term first used by Pearson, 1905) measures the "peakedness" of a distribution. If the kurtosis is clearly different than 0, the distribution is either flatter or more peaked than normal; the kurtosis of the normal distribution is 0. Kurtosis is computed as:

Kurtosis = $[n*(n+1)*M_4 - 3*M_2*M_2*(n-1)]/[(n-1)*(n-2)*(n-3)*\sigma^4]$

Where M_j is equal to $\Sigma (x_i\text{-Mean}_x)^j$, n is the valid number of cases, and σ^4 is the standard deviation (*sigma*) raised to the fourth power.

See also, *Descriptive Statistics* in Chapter 2 – *Basic Statistics and Tables*.

L

Lack of Fit

For certain designs with replicates at the levels of the predictor variables, the residual sum of squares can be further partitioned into meaningful parts which are relevant for testing hypotheses. Specifically, the residual sums of squares can be partitioned into lack-of-fit and pure-error components. This involves determining the part of the residual sum of squares that can be predicted by including additional terms for the predictor variables in the model (for example, higher-order polynomial or interaction terms), and the part of the residual sum of squares that cannot be predicted by any additional terms (i.e., the sum of squares for pure error). A test of lack-of-fit for the model without the additional terms can then be performed, using the mean square pure error as the error term. This provides a more sensitive test of model fit, because the effects of the additional higher-order terms is removed from the error.

See also, *Design Matrix*, *Pure Error*, or Chapter 15 – *Experimental Design (DOE)*, Chapter 18 – *General Linear Models*, or Chapter 19 – *General Regression Models*.

Lambda Prime

Lambda is defined as the geometric sum of 1 minus the squared canonical correlation, where *lambda* is Wilks' *lambda*. The squared canonical correlation is an estimate of the common variance between two canonical variates, thus 1 minus this value is an estimate of the unexplained variance. *Lambda* is used as a test of significance for the squared canonical correlation and is distributed as *chi*-square (see below).

$$\chi^2 = [-N -1 - \{.5(p+q+1)\}] * \log_e \lambda$$

where N is the number of subjects, p is the number of variables on the right, and q is the number of variables on the left.

Laplace Distribution

The Laplace (or Double Exponential) distribution has density function:

$$f(x) = 1/(2b)*e^{-|x-a|/b} \qquad -\infty < x < \infty$$

where a is the mean of the distribution; b is the scale parameter; and e is the base of the natural logarithm, sometimes called Euler's e (2.71...).

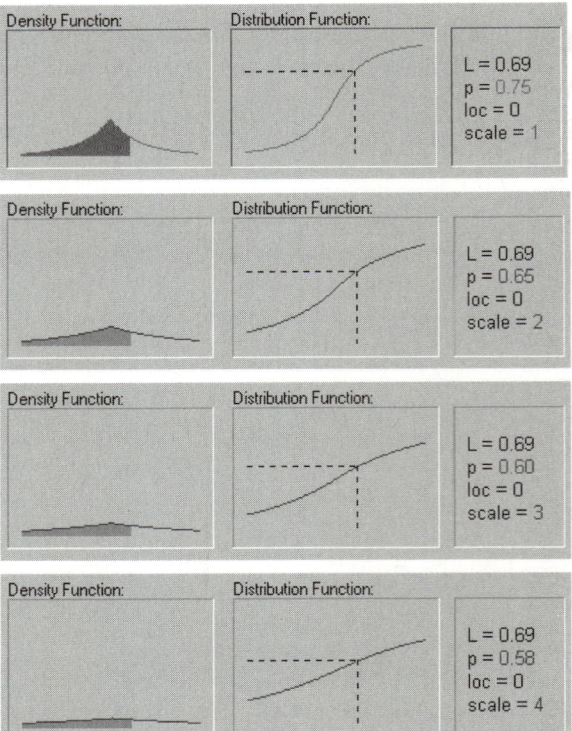

The previous illustrations show the changing shape of the Laplace distribution when the location parameter equals 0 and the scale parameter equals 1, 2, 3, and 4.

Latent Variable

A latent variable is a variable that cannot be measured directly, but is hypothesized to underlie the observed variables. An example of a latent variable is a factor in factor analysis. Latent variables in path diagrams are usually represented by a variable name enclosed in an oval or circle.

Law of Large Numbers

As the size of a random sample (from a large population) increases, the mean of that sample approximates the mean of the whole population.

Layered Compression

When layered compression is used, the main graph plotting area is reduced in size to leave space for margin graphs in the upper and right side of the display (and a miniature graph in the corner). These smaller margin graphs represent vertically and horizontally compressed images (respectively) of the main graph.

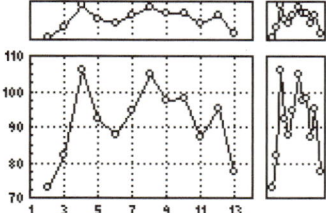

For more information on layered compression (and an additional example), see *Layered Compression*, page 546.

Learned Vector Quantization (in Neural Networks)

The learned vector quantization algorithm (LVQ) was invented by Tuevo Kohonen (Fausett, 1994; Kohonen, 1990), who also invented the self-organizing feature map.

Learned vector quantization provides a supervised version of the Kohonen training algorithm. The standard Kohonen algorithm iteratively adjusts the position of the exemplar vectors stored in the radial layer of the Kohonen network by considering only the positions of the existing vectors and of the training data. In essence, the algorithm attempts to move the exemplar vectors to positions that reflect the centers of clusters in the training data. However, the class labels of the training data cases are not taken into account. For superior classification performance, it is desirable that the exemplar vectors are adjusted, to some extent, on a per-class basis; that is, that they reflect natural clusters in each separate class. An exemplar located on a class boundary, equally close to cases of two classes, is unlikely to be of much use in distinguishing class. On the other hand, exemplars located just inside class boundaries can be extremely useful.

There are several variants of learned vector quantization. The basic version, LVQ1, is very similar to the Kohonen training algorithm. The closest exemplar to a training case is selected during training and has its position updated. However, whereas the Kohonen algorithm would move this exemplar toward the training case, LVQ1 checks whether the class label of the exemplar vector is the same as that of the training case. If it is, the exemplar is moved toward the training case; if it is not, the exemplar is moved away from the case. The more sophisticated LVQ algorithms, LVQ2.1 and LVQ3, take into account more information. They locate the nearest two exemplars to the training case. If one of these is of the right class and one the wrong class, they move the right class toward the training case and the wrong one away from it. LVQ3 also moves both exemplars toward the training case if they are both of the right class. In both LVQ2.1 and LVQ3, the concept is to move exemplars where there is some danger of misclassification.

Technical details. The basic update rule is:

$$\varpi_t = \varpi_{t-1} + \eta_t(x + \varpi_{t-1})$$

if the exemplar and training case have the same class,

$$\varpi_t = \varpi_{t-1} - \eta_t(x + \varpi_{t-1})$$

if they do not.

x is the training case, η_t is the learning rate.

In LVQ2.1, the two nearest exemplars are adjusted only if one is of the right class and one is not, and they are both "about the same" distance from the training case. The definition of "about the same" distance uses a special parameter, ε, and the formulae below:

$$\min\left(\frac{d_1}{d}, \frac{d_2}{d_1}\right) > 1 - \varepsilon$$

$$\max\left(\frac{d_1}{d}, \frac{d_2}{d_1}\right) < 1 + \varepsilon$$

In LVQ3, an alternative formula is used to ensure that the two nearest are both "about the same distance" from the training case:

$$\min\left(\frac{d_1}{d}, \frac{d_2}{d_1}\right) > (1 - \varepsilon)(1 + \varepsilon)$$

In addition, in LVQ3, if both the two nearest exemplars are of the same class as the training case, they are both moved toward the case, using a learning rate b times the standard learning rate at that epoch.

Learning Rate (in Neural Networks)

A control parameter of some training algorithms, which controls the step size when weights are iteratively adjusted. See also, Chapter 27 – *Neural Networks*.

Least Squares (2D graphs)

A curve is fitted to the *XY* coordinate data according to the distance-weighted least squares smoothing procedure (the influence of individual points decreases with the horizontal distance from the respective points on the curve).

Least Squares (3D Graphs)

A surface is fitted to the *XYZ* coordinate data according to the distance-weighted least squares smoothing procedure (the influence of individual points decreases with the horizontal distance from the respective points on the surface).

Least Squares Estimator

In the most general terms, least squares estimation is aimed at minimizing the sum of squared deviations of the observed values for the dependent variable from those predicted by the model. Technically, the least squares estimator of a parameter q is obtained by minimizing Q with respect to q where:

$$Q = \Sigma\,[Y_i - f_i(\theta)]^2$$

Note that $f_i(\theta)$ is a known function of θ, $Y_i = f_i(\theta) + \varepsilon_i$ where $i = 1$ to n, and the ε_i are random variables, and usually assumed to have expectation of 0. For more information, see Mendenhall and Sincich (1984), Bain and Engelhardt (1989), and Neter, Wasserman, and Kutner (1989). See also, Chapter 2 – *Basic Statistics and Tables*, Chapter 26 – *Multiple Linear Regression*, and Chapter 28 – *Nonlinear Estimation*.

Least Squares Means

When there are no missing cells in ANOVA designs with categorical predictor variables, the subpopulation (or marginal) means are least

square means, which are the best linear-unbiased estimates of the marginal means for the design (see, Milliken and Johnson, 1986). Tests of differences in least square means have the important property that they are invariant to the choice of the coding of effects for categorical predictor variables (e.g., the use of the sigma-restricted or the overparameterized model) and to the choice of the particular generalized inverse of the design matrix used to solve the normal equations. Thus, tests of linear combinations of least square means in general are said to not depend on the parameterization of the design.

See also, *Categorical Predictor Variable*, *Design Matrix*, *Sigma-Restricted Model*, *Overparameterized Model*, and *Generalized Inverse*; see also, Chapter 18 – *General Linear Models* or Chapter 19 – *General Regression Models*.

Levenberg-Marquardt Algorithm (in Neural Networks)

Levenberg-Marquardt (Levenberg, 1944; Marquardt, 1963; Bishop, 1995; Shepherd, 1997; Press et. al., 1992) is an advanced nonlinear optimization algorithm. It can be used to train the weights in a network just as back propagation would be. It is reputably the fastest algorithm available for such training. However, its use is restricted as follows.

Single output networks. Levenberg-Marquardt can only be used on networks with a single output unit.

Small networks. Levenberg-Marquardt has space requirements proportional to the square of the number of weights in the network. This effectively precludes its use in networks of any great size (more than a few hundred weights).

Sum-squared error function. Levenberg-Marquardt is only defined for the sum squared error function. If you select a different error function for your network, it will be ignored during Levenberg-Marquardt training. It is usually therefore only appropriate for regression networks.

Note: Just as with other iterative algorithms, Levenberg-Marquardt does not train radial units. Therefore, you can use it to optimize the non-radial layers of radial basis function networks even if there are a large number of weights in the radial layer, as those are ignored by Levenberg-Marquardt. This is significant as it is typically the radial layer that is very large in such networks.

Levenberg-Marquardt works by making the assumption that the underlying function being modeled by the neural network is linear. Based on this calculation, the minimum can be determined exactly in a single step. The calculated minimum is tested, and if the error there is lower, the algorithm moves the weights to the new point. This process is repeated iteratively on each generation. Since the linear assumption is ill founded, it can easily lead Levenberg-Marquardt to test a point that is inferior (perhaps even wildly inferior) to the current one. The clever aspect of Levenberg-Marquardt is that the determination of the new point is actually a compromise between a step in the direction of steepest descent and the above-mentioned leap. Successful steps are accepted and lead to a strengthening of the linearity assumption (which is approximately true near to a minimum). Unsuccessful steps are rejected and lead to a more cautious downhill step. Thus, Levenberg-Marquardt continuously switches its approach and can make very rapid progress.

Technical details. The Levenberg-Marquardt algorithm is designed specifically to minimize the sum-of-squares error function, using a formula that (partly) assumes that the underlying function modeled by the network is linear. Close to a minimum this assumption is approximately true, and the algorithm can make very rapid progress. Further away it may be a very poor assumption. Levenberg-Marquardt therefore compromises between the linear model and a gradient-descent approach. A move is only accepted if it improves the error, and if necessary the gradient-descent model is used with a sufficiently small step to guarantee downhill movement.

Levenberg-Marquardt uses the update formula:

$$\Delta w = -(Z^T Z + \lambda I)^{-1} Z^T \varepsilon$$

where ε is the vector of case errors, and Z is the matrix of partial derivatives of these errors with respect to the weights:

$$(Z)_{ni} \equiv \frac{\delta \varepsilon^n}{\delta w_i}$$

The first term in the Levenberg-Marquardt formula represents the linearized assumption; the second a gradient-descent step. The control parameter governs the relative influence of these two approaches. Each time Levenberg-Marquardt succeeds in lowering the error, it decreases the control parameter by a factor of 10, thus strengthening the linear assumption and attempting to jump directly to the minimum. Each time it fails to lower the error, it increases the control parameter by a factor of 10, giving more influence to the gradient descent step, and also making the step size smaller. This is guaranteed to make downhill progress at some point.

Levene and Brown-Forsythe Tests for Homogeneity of Variances (HOV)

A important assumption in analysis of variance (ANOVA and the t-test for mean differences) is that the variances in the different groups are equal (homogeneous). Two powerful and commonly used tests of this assumption are the Levene test and the Brown-Forsythe modification of this test. However, it is important to realize that 1) the homogeneity of variances assumption is usually not as crucial as other assumptions for ANOVA, in particular in the case of balanced (equal n) designs (see also, ANOVA Homogeneity of Variances and Covariances), and 2) that the tests described below are not necessarily very robust themselves (e.g., Glass and Hopkins, 1996, p. 436, call these tests "fatally flawed;" see also, the description of these tests below). If you are concerned about a violation of the HOV assumption, it is always advisable to repeat the key analyses using nonparametric methods.

Levene's test (homogeneity of variances). For each dependent variable, an analysis of variance is performed on the absolute deviations of values from the respective group means. If the Levene test is statistically significant, the hypothesis of homogeneous variances should be rejected.

Brown & Forsythe's test (homogeneity of variances). Recently, some authors (e.g., Glass and Hopkins, 1996) have called into question the power of the Levene test for unequal variances. Specifically, the absolute deviation (from the group means) scores can be expected to be highly skewed; thus, the normality assumption for the ANOVA of those absolute deviation scores is usually violated. This poses a particular problem when there is unequal n in the two (or more) groups that are to be compared. A more robust test that is very similar to the Levene test has been proposed by Brown and Forsythe (1974). Instead of performing the ANOVA on the deviations from the mean, you can perform the analysis on the deviations from the group medians. Olejnik and

Algina (1987) have shown that this test will give quite accurate error rates even when the underlying distributions for the raw scores deviate significantly from the normal distribution. However, as Glass and Hopkins (1996, p. 436) have pointed out, both the Levene test as well as the Brown-Forsythe modification suffer from what those authors call a "fatal flaw," namely, that both tests themselves rely on the homogeneity of variances assumption (of the absolute deviations from the means or medians); and hence, it is not clear how robust these tests are themselves in the presence of significant variance heterogeneity and unequal *n*.

Leverage Values

In regression, this term refers to the diagonal elements of the hat matrix $(\mathbf{X(X'X)^{-1}X'})$. A given diagonal element $(h_{(ii)})$ represents the distance between X values for the ith observation and the means of all X values. These values indicate whether or not X values for a given observation are outlying. The diagonal element is referred to as the leverage. A large leverage value indicates that the *i*th observation is distant from the center of the X observations (Neter, et al, 1985).

Life Table

The most straightforward way to describe the survival in a sample is to compute the life table. The life table technique is one of the oldest methods for analyzing survival (failure time) data (e.g., Berkson & Gage, 1950; Cutler & Ederer, 1958; Gehan, 1969; see also, Lawless, 1982, Lee, 1993). This table can be thought of as an enhanced frequency distribution table. The distribution of survival times is divided into a certain number of intervals. For each interval one can compute the number and proportion of

cases or objects that entered the respective interval alive, the number and proportion of cases that failed in the respective interval (i.e., number of terminal events, or number of cases that died), and the number of cases that were lost or censored in the respective interval. Based on those numbers and proportions, several additional statistics can be computed. Refer to Chapter 36 – *Survival/Failure Time Analysis* for additional details.

Lift Chart

The lift chart provides a visual summary of the usefulness of the information provided by one or more statistical models for predicting a binomial (categorical) outcome variable (dependent variable); for multinomial (multiple-category) outcome variables, lift charts can be computed for each category. Specifically, the chart summarizes the utility that you can expect by using the respective predictive models, as compared to using baseline information only.

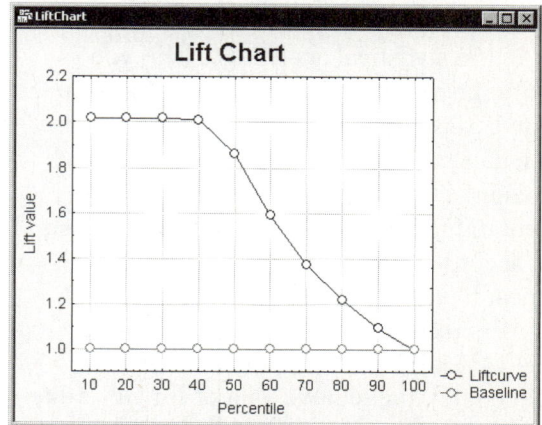

The lift chart is applicable to most statistical methods that compute predictions (predicted classifications) for binomial or multinomial responses. This and similar summary charts (see *Gains Chart*, page 620) are commonly used in data mining projects, when the dependent or outcome variable of interest is binomial or multinomial in nature.

Example. To illustrate how the lift chart is constructed, consider this example. Suppose you have a mailing list of previous customers of your business, and you want to offer to those customers an additional service by mailing an elaborate brochure and other materials describing the service. During previous similar mail-out campaigns, you collected useful information about your customers (e.g., demographic information, previous purchasing patterns) that you could relate to the response rate, i.e., whether the respective customers responded to your mail solicitation. Also, from similar prior mail-out campaigns, you were able to estimate the baseline response rate at approximately 7 percent, i.e., 7% of all customers who received a similar offer by mail responded (purchased the additional service).

Given this baseline response rate (7%) and the cost of the mail-out, sending the offer to all customers would result in a net loss. Hence, you want to use statistical analyses to help you identify the customers who are most likely to respond. Suppose you build such a model based on the data collected in the previous mail-out campaign. You can now select only the 10 percent of the customers from the mailing lists who, according to prediction from the model, are most likely to respond. If among those customers (selected by the model) the response rate is 14% percent (as opposed to the 7% baseline rate), the relative gain or lift value due to using the predictive model can be computed as 14% / 7% = 2. In other words, you were able to do twice as well as you would have done using simple random selection.

Analogous lift values can be computed for each percentile of the population (customers on the mailing list). You could compute separate lift

values for selecting the top 20% of customers who are predicted to be among likely responders to the mail campaign, the top 30%, etc. Hence, the lift values for different percentiles can be connected by a line that will typically descend slowly and merge with the baseline if all customers (100%) were selected.

If more than one predictive model is used, multiple lift charts can be overlaid (as shown in the illustration above) to provide a graphical summary of the utility of different models.

Lilliefors Test

In a Kolmogorov-Smirnov test for normality when the mean and standard deviation of the hypothesized normal distribution are not known (i.e., they are estimated from the sample data), the probability values tabulated by Massey (1951) are not valid. Instead, the so-called Lilliefors probabilities (Lilliefors, 1967) should be used in determining whether the KS difference statistic is significant.

Line Plot, 2D

In line plots, individual data points are connected by a line. Line plots provide a simple way to visually present a sequence of values. *XY* trace-type line plots can be used to display a trace (instead of a sequence).

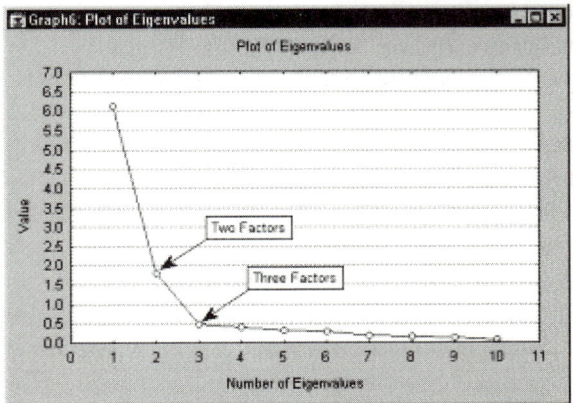

Line plots can also be used to plot continuous functions, theoretical distributions, etc.

Line Plot, 2D – Aggregated

Aggregated line plots display a sequence of means for consecutive subsets of a selected variable.

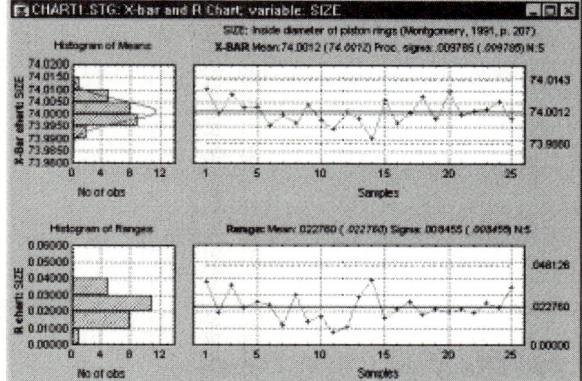

You can select the number of consecutive observations from which the mean will be calculated and if desired, the range of values in each subset will be marked by the whisker-type markers. Aggregated line plots are used to explore and present sequences of large numbers of values.

Line Plot, 2D – Case Profiles

Unlike regular line plots where the values of one variable are plotted as one line (individual data points are connected by a line), in case profile line plots, the values of the selected variables in a case (row) are plotted as one line (i.e., one line plot will be generated for each of the selected cases).

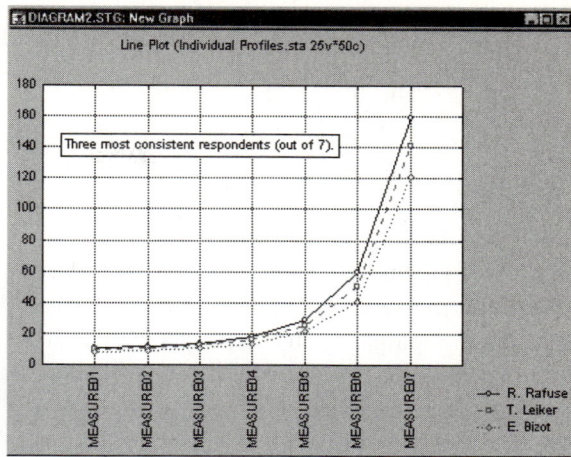

Case profile line plots provide a simple way to visually present the values in a case (e.g., test scores for several tests).

Line Plot, 2D – Double-Y

The double-Y line plot can be considered to be a combination of two separately scaled multiple line plots. One set of variables is scaled against the left-Y axis, and the others against the right-Y axis.

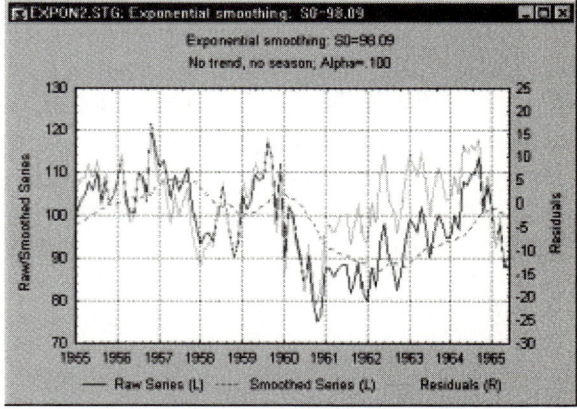

The double-Y line plot can be used to compare sequences of values of several variables by overlaying their respective line representations in a single graph. However, due to the independent scaling used for the two axes, it can facilitate comparisons between otherwise incomparable variables (i.e., variables with values in different ranges).

Line Plot, 2D – Multiple

Unlike regular line plots in which a sequence of values of one variable is represented, the multiple line plot represents multiple sequences of values (variables). A different line pattern and color is used for each of the multiple variables.

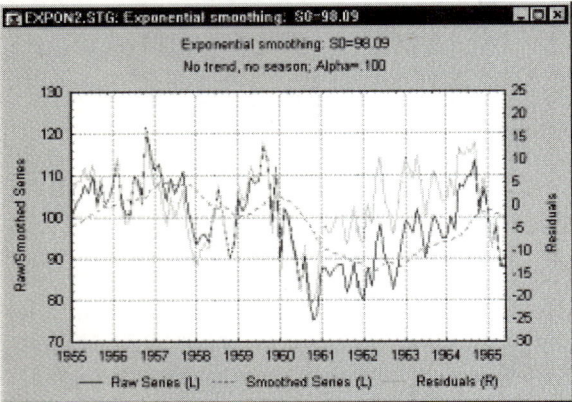

This type of line plot is used to compare sequences of values between several variables (or several functions) by overlaying them in a single graph that uses one common set of scales (e.g., comparisons between several simultaneous experimental processes, social phenomena, stock or commodity quotes, shapes of operating characteristics curves, etc.).

Line Plot, 2D – Regular

Regular line plots are used to examine and present the sequences of values (usually when the order of the presented values is meaningful).

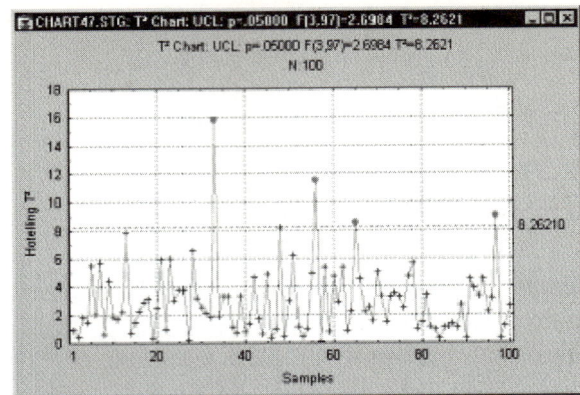

Another typical application for line sequence plots is to plot continuous functions, such as fitted functions or theoretical distributions. Note that an empty data cell (i.e., missing data) breaks the line.

Line Plot, 2D – XY Trace

In trace plots, a scatterplot of two variables is first created, and then the individual data points are connected with a line (in the order in which they are read from the data file). In this sense, trace plots visualize a trace of a sequential process (movement, change of a phenomenon over time, etc.)

Linear (2D Graphs)

A linear function (e.g., $Y = a + bX$) is fitted to the points in the 2D scatterplot.

Linear (3D Graphs)

A linear function (e.g., $Z = a + bX + cY$) is fitted to the points in the 3D scatterplot.

Linear Activation Function

A null activation function: the unit's output is identical to its activation level. See also, Chapter 27 – *Neural Networks*.

Linear Modeling

Approximation of a discriminant function or regression function using a hyperplane. See also, Chapter 27 – *Neural Networks*.

Linear Units

A unit with a linear PSP function. The unit's activation level is the weighted sum of its inputs minus the threshold, also known as a dot product or linear combination. The characteristic unit type of multilayer perceptrons. Despite the name, a linear unit may have a nonlinear activation function. See also, Chapter 27 – *Neural Networks*.

Link Function and Distribution Function

The link function in generalized linear models specifies a nonlinear transformation of the predicted values so that the distribution of predicted values is one of several special members of the exponential family of distributions (e.g., *gamma*, Possion, binomial, etc.). The link function is therefore used to model responses when a dependent variable is assumed to be nonlinearly related to the predictors.

Various link functions (see McCullagh and Nelder, 1989) are commonly used, depending on the assumed distribution of the dependent variable (y) values:

Normal, Gamma, Inverse normal, and Poisson distributions:

Identity link: $f(z) = z$

Log link: $f(z) = \log(z)$

Power link: $f(z) = z^a$, for a given a

Binomial, and Ordinal Multinomial distributions:

Logit link: $f(z) = \log(z/(1-z))$

Probit link: $f(z) = \text{invnorm}(z)$ where *invnorm* is the inverse of the standard normal cumulative distribution function.

Complementary log-log link:

$f(z) = \log(-\log(1-z))$

Loglog link: $f(z) = -\log(-\log(z))$

Multinomial distribution:

Generalized logit link:

$f(z1|z2,\ldots,zc) = \log(x1/(1-z1-\ldots-zc))$

where the model has $c+1$ categories.

For a discussion of the role of link functions, see Chapter 21 – *Generalized Linear/Nonlinear Models*.

Local Minima

Local "valleys" or minor "dents" in a loss function which, in many practical applications, will produce extremely large or small parameter estimates with very large standard errors. The Simplex method is particularly effective in avoiding such minima; therefore, this method may be particularly well suited in order to find appropriate start values for complex functions.

Logarithmic Function

This fits to the data, a logarithmic function of the form:

$y = q*[\log_n(x)] + b$

Logistic Distribution

The logistic distribution has density function:

$f(x) = (1/b)*e^{[-(x-a)/b]} * [1+e^{[-(x-a)/b]-2}\}$

where a is the mean of the distribution; b is the scale parameter; and e is the base of the natural logarithm, sometimes called Euler's e (2.71...).

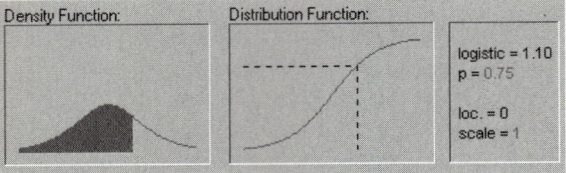

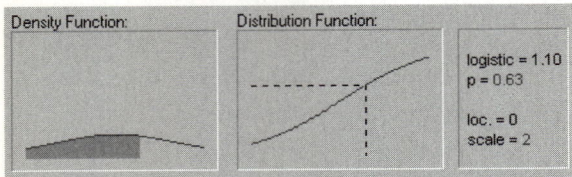

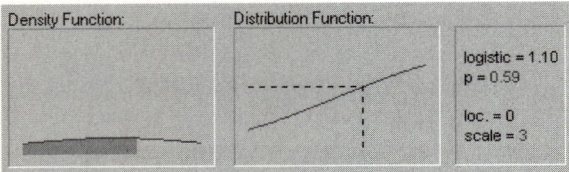

The previous illustrations show the changing shape of the logistic distribution when the location parameter equals 0 and the scale parameter equals 1, 2, and 3.

Logistic Function

An S-shaped (sigmoid) function having values in the range (0,1). See also, *Logistic Distribution*.

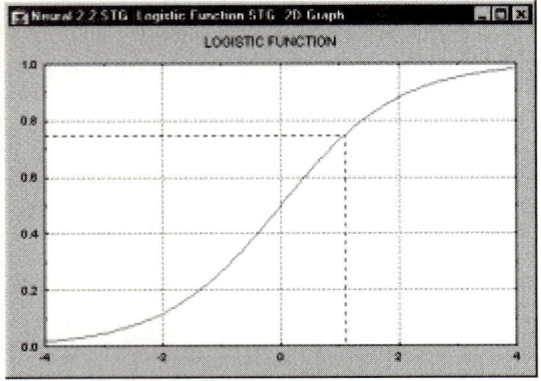

Logit Regression and Transformation

In the logit regression model, the predicted values for the dependent or response variable will never be less than (or equal to) *0*, or greater than (or equal to) *1*, regardless of the values of the independent variables; it is, therefore, commonly used to analyze binary dependent or response variables (see also, the binomial distribution). This is accomplished by applying the following regression equation (the term *logit* was first used by Berkson, 1944):

$$y = \exp(b_0 + b_1 * x_1 + \ldots + b_n * x_n) / \{1 + \exp(b_0 + b_1 * x_1 + \ldots + b_n * x_n)\}$$

Regardless of the regression coefficients or the magnitude of the *x* values, this model will always produce predicted values (predicted *y*'s) in the range of 0 to 1. The name logit stems from the fact that you can easily linearize this model via the logit transformation. Suppose we think of the binary dependent variable *y* in terms of an underlying continuous probability *p*, ranging from 0 to 1. We can then transform that probability *p* as:

$$p' = \log_e\{p/(1-p)\}$$

This transformation is referred to as the *logit* or *logistic* transformation. Note that *p'* can theoretically assume any value between minus and plus infinity. Since the logit transform solves the issue of the 0/1 boundaries for the original dependent variable (probability), we could use those (logit transformed) values in an ordinary linear regression equation. In fact, if we perform the logit transform on both sides of the logit regression equation stated earlier, we obtain the standard linear multiple regression model:

$$p' = (b_0 + b_1 * x_1 + \ldots + b_n * x_n)$$

For additional details, see Chapter 21 – *Generalized Linear/Nonlinear Models (GLZ)* or Chapter 28 – *Nonlinear Estimation*; see also, *Probit Regression and Transformation* and *Multinomial Logit and Probit Regression* for similar transformations.

Log-Linear Analysis

Log-linear analysis provides a sophisticated way of looking at crosstabulation tables (to explore the data or verify specific hypotheses), and it is sometimes considered an equivalent of ANOVA for frequency data. Specifically, it enables you to test the different factors that are used in the crosstabulation (e.g., gender, region, etc.) and their interactions for statistical significance (see Chapter 1 – *Elementary Concepts in Statistics* for a discussion of statistical significance testing). For more information, see Chapter 22 – *Log-Linear Analysis of Frequency Tables*.

Lognormal Distribution

The lognormal distribution (the term first used by Gaddum, 1945) has the probability density function:

$$f(x) = 1/[x\sigma(2\pi)^{1/2}] * \exp(-[\log(x)-\mu]^2/2\sigma^2)$$
$$0 \leq x < \infty$$
$$\mu > 0$$
$$\sigma > 0$$

where μ is the scale parameter; σ is the shape parameter; and e is the base of the natural logarithm, sometimes called Euler's e (2.71...).

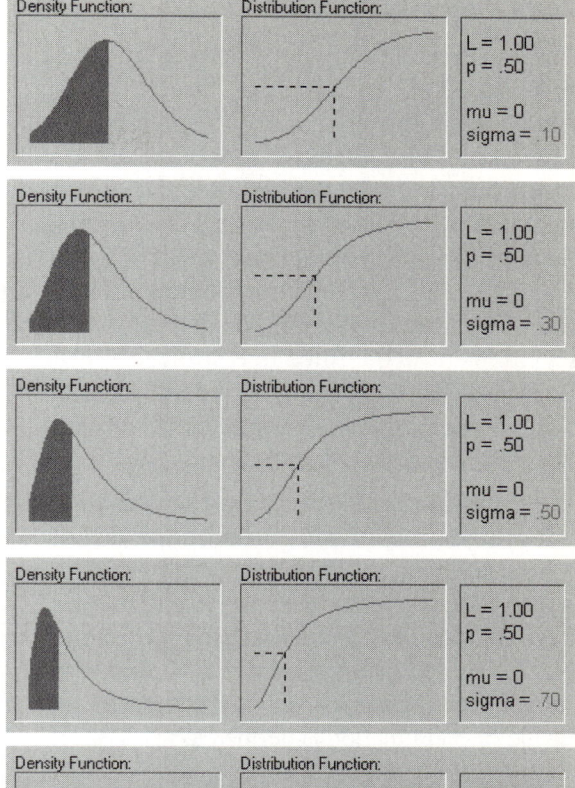

The previous illustrations show the lognormal distribution with mu equal to 0 for sigma equals .10, .30, .50, .70, and .90.

Lookahead (in Neural Networks)

For neural networks time series analysis, the number of time steps ahead of the last input variable values the output variable values should be predicted. See also, Chapter 27 – *Neural Networks*.

Loss Function

The loss function (the term loss was first used by Wald, 1939) is the function that is minimized in the process of fitting a model, and it represents a selected measure of the discrepancy between the observed data and data predicted by the fitted function. For example, in many traditional general linear model techniques, the loss function (commonly known as least squares) is the sum of squared deviations from the fitted line or plane. One of the properties (sometimes considered to be a disadvantage) of that common loss function is that it is very sensitive to outliers.

A common alternative to the least squares loss function (see above) is to maximize the likelihood or log-likelihood function (or to minimize the negative log-likelihood function; the term maximum likelihood was first used by Fisher, 1922a). These functions are typically used when fitting nonlinear models. In most general terms, the likelihood function is defined as:

$$L = F(Y, Model) = \prod_{i=1}^{n} \{p[y_i, Model\ Parameters(x_i)]\}$$

In theory, we can compute the probability (now called L, the likelihood) of the specific dependent variable values to occur in our sample, given the respective regression model.

Loss Matrix

Loss matrix is a square matrix of coefficients multiplied by a vector of class probabilities to form a vector of cost-estimates so that minimum-loss decisions can be made. See also, Chapter 27 – *Neural Networks* and *Loss Function*, page 654.

Loss Matrix (in Neural Networks)

If a network is trained so that the outputs estimate probabilities, it can be adjusted to support a loss matrix (Bishop, 1995). In simple cases, a probability estimate can be used directly; an unknown case is simply assigned to the most-probable class. Inevitably, this means that sometimes the network can be wrong (and this is unavoidable if data is noisy).

However, some mistakes can be more costly than others. For example, if diagnosing a potentially fatal illness, prescribing medication to somebody who is not actually ill may be considered a less grave error than failing to prescribe to somebody who is.

A loss matrix is a square matrix of coefficients that reflect the relative costs of various misclassifications. It is multiplied by the vector of probability estimates, resulting in a vector of cost estimates, and the case is assigned to the class with the lowest cost estimate.

Since a correct classification has zero cost, the leading diagonal of a loss matrix always contains zeros; in other positions, the coefficient in the *n*'th column and *m*'th row represents the cost of misclassifying a case that is actually in the n'th class as being in the m'th class. See also, Chapter 27 – *Neural Networks*.

LOWESS Smoothing (Robust Locally Weighted Regression)

Robust locally weighted regression is a method of smoothing 2D scatterplot data (pairs of x-y data). A local polynomial regression model is fit to each point and the points close to it. The method is also sometimes referred to as LOWESS smoothing. The smoothed data usually provide a clearer picture of the overall shape of the relationship between the x and y variables. For more information, see also, Cleveland (1979, 1985).

M

Machine Learning

Machine learning, computational learning theory, and similar terms are often used in the context of data mining to denote the application of generic model-fitting or classification algorithms for predictive data mining. Unlike traditional statistical data analysis, which is usually concerned with the estimation of population parameters by statistical inference, the emphasis in data mining (and machine learning) is usually on the accuracy of prediction (predicted classification), regardless of whether the models or techniques that are used to generate the prediction are interpretable or open to simple explanation. Good examples of this type of technique often applied to predictive data mining are neural networks or meta-learning techniques such as boosting, etc. These methods usually involve the fitting of very complex generic models, that are not related to any reasoning or theoretical understanding of underlying causal processes; instead, these techniques can be shown to generate accurate predictions or classification in crossvalidation samples.

Mahalanobis Distance

You can think of the independent variables (in a regression equation) as defining a multi-dimensional space in which each observation can be plotted. Also, you can plot a point representing the means for all independent variables. This mean point in the multidimensional space is also called the centroid. The Mahalanobis distance is the distance of a case from the centroid in the multidimensional space, defined by the correlated independent variables (if the independent

variables are uncorrelated, it is the same as the simple Euclidean distance). Thus, this measure provides an indication of whether or not an observation is an outlier with respect to the independent variable values. See also, *Standard Residual Value*, page 732, *Deleted Residual* and *Cook's Distance*.

Main Effect

As opposed to the interaction effect (in ANOVA or Regression), the main effect of a specific factor (independent variable) on dependent variable represents the effect of that independent variable averaged across levels of all other independent variables in the design.

For example, the results of the following (fictitious) experiment where performance of Achievement-oriented vs. Achievement-avoiding students on a Challenging vs. Easy Test was measured, shows a main effect of Test on performance (participants in this study did better overall on the Challenging Test - 13, than Easy Test - 11), but no main effect of the Achievement orientation (12 vs. 12):

	Achieve-ment-oriented	Achieve-ment-avoiders	Average
Challenging Test	14	12	13
Easy Test	10	12	11
Average	12	12	

Note that this concept – which is crucial in data analysis and interpretation – is commonly confused and misunderstood.

For example, common but imprecise (or even misleading) definitions of main effect state that it represents *"the effect of a factor on the dependent variable (response) measured without regard to other factors in the analysis"* (www.socialresearchmethods.net/tutorial). These definitions are incorrect in the sense that

in multivariate designs, the magnitude of any main effect is calculated controlling for the variance accounted for by other factors, thus it is not measured "without regard to other factors" but clearly with regard to those factors. Thus, main effects are clearly measured "with" regard to other factors, because the magnitude of a main effect depends on effects of those other factors.

Another common error is the assumption that the existence of a main effect of a specific factor implies somehow that this factor is not involved in interaction effects in the given design. For example, a popular Research Methods textbook (Elmes, Kantowitz, Roediger, 2006) defines main effect as *"when the effect of one independent variable is the same at all levels of another independent variable."* That statement is entirely incorrect, it defines simply a lack of interaction involving the two factors in question and has nothing to do with existence of a main effect, which can coexist with interaction effects involving the same factors in a design. For example, the following results of the (fictitious) experiment (mentioned above):

	Achieve-ment-oriented	Achieve-ment-avoiders	Average
Challenging Test	17	14	15.5
Easy Test	14	15	14.5
Average	15.5	14.5	

show both main effects of each of the factors (Achievement-oriented students perform better overall, also the Challenging Test produces better results overall), and an interaction between the two factors (Easy Test has a possibly de-motivating effect on Achievement-oriented students, but produces better results (motivates?) Achievement-avoiders).

Mallow's CP

If p regressors are selected from a set of k, Cp is defined as:

$$\sum (y - y_p)^2 / s^2 - n + 2p$$

where y_p is the predicted value of y from the p regressors; s^2 is the residual mean square after regression on the complete set of k; and n is the sample size.

The model is then chosen to give a minimum value of the criterion, or a value that is acceptably small. It is essentially a special case of Akaike Information Criterion. Mallow's CP is used in general regression models as the criterion for choosing the best subset of predictor effects when a best subset regression analysis is being performed. This measure of the quality of fit for a model tends to be less dependent (than the *R-square*) on the number of effects in the model, and hence, it tends to find the best subset that includes only the important predictors of the respective dependent variable. See *Building Models via Best Subset Regression* in Chapter 19 – *General Regression Models (GRM)* for further details.

Mann-Scheuer-Fertig Test

This test, proposed by Mann, Scheuer, and Fertig (1973), is described in detail in, for example, Dodson (1994) or Lawless (1982). The null hypothesis for this test is that the population follows the Weibull distribution with the estimated parameters. Nelson (1982) reports this test to have reasonably good power, and this test can be applied to Type II censored data. For computational details refer to Dodson (1994) or Lawless (1982); the critical values for the test statistic have been computed based on Monte Carlo studies, and have been tabulated for *n* (sample sizes) between 3 and 25; for *n* greater than 25, this test is not computed.

The Mann-Scheuer-Fertig test is used in Weibull and reliability/failure time analysis; see also, *Hollander-Proschan Test* and *Anderson-Darling Test*.

Marginal Frequencies

In a multi-way table, the values in the margins of the table are simply one-way (frequency) tables for all values in the table. They are important in that they help us to evaluate the arrangement of frequencies in individual columns or rows. The differences between the distributions of frequencies in individual rows (or columns) and in the respective margins inform us about the relationship between the crosstabulated variables. For more information on marginal frequencies, see *Crosstabulation and Stub-and-Banner Tables* in Chapter 2 – *Basic Statistics and Tables*.

Markov Chain Monte Carlo (MCMC)

The term Monte Carlo method (suggested by John von Neumann and S. M. Ulam in the 1940s) refers to simulation of processes using random numbers. The term Monte Carlo (a city long known for its gambling casinos) derived from the fact that numbers of chance (i.e., Monte Carlo simulation methods) were used in order to solve some of the integrals of the complex equations involved in the design of the first nuclear bombs (integrals of quantum dynamics). By generating large samples of random numbers from, for example, mixtures of distributions, the integrals of these (complex) distributions can be approximated from the (generated) data.

Complex equations with difficult to solve integrals are often involved in Bayesian statistics analyses. For a simple example of the MCMC method for generating bivariate normal random variables, see the definition of *Gibbs Sampler*. For a detailed discussion of MCMC methods, see Gilks, Richardson, and Spiegelhalter (1996). See also, the definition of *Bayesian Statistics (Analysis)*.

Manifest Variable

A manifest variable is a variable that is directly observable or measurable. In path analysis, diagrams used in structural modeling (see *Structural Equation Modeling and the Path Diagram*, in Chapter 35 – *Structural Equation Modeling*, manifest variables are usually represented by enclosing the variable name within a square or a rectangle.

Mass

In correspondence analysis, the term mass is used to denote the entries in the two-way table of relative frequencies (i.e., each entry is divided by the sum of all entries in the table). The results from correspondence analysis are still valid if the entries in the table are not frequencies, but some other measure of correspondence, association, similarity, confusion, etc. Since the sum of all entries in the table of relative frequencies is equal to 1.0, you could say that the table of relative frequencies shows how one unit of mass is distributed across the cells of the table. In the terminology of correspondence analysis, the row and column totals of the table of relative frequencies are called the row mass and column mass, respectively.

Matching Moments Method

This method can be employed to determine parameter estimates for a distribution (see *Quantile-Quantile Plots*, page 698, *Probability-Probability Plots*, page 692, and *Process Analysis*, page 694). The method of matching moments sets the distribution moments equal to the data moments and solves to obtain estimates for the distribution parameters. For example, for a distribution with two parameters, the first two moments of the distribution (the mean and variance of the distribution, e.g., μ and σ) would be set equal to the first two moments of the data (the sample mean and variance, respectively, e.g., the unbiased estimators x-bar and s^2, respectively) and solved for the parameter estimates. Alternatively, you could use the maximum likelihood method to estimate the parameters. For more information, see Hahn and Shapiro, 1994.

Matrix Collinearity, Multicollinearity

This term is used in the context of correlation matrices or covariance matrices, to describe the condition when one or more variables from which the respective matrix was computed are linear functions of other variables; as a consequence such matrices cannot be inverted (only the generalized inverse can be computed). See also, *Matrix Singularity*, page 660.

Matrix Ill-Conditioning

Matrix ill-conditioning is a general term used to describe a rectangular matrix of values which is unsuitable for use in a particular analysis.

This occurs perhaps most frequently in applications of linear multiple regression when the matrix of correlations for the predictors is singular and thus the regular matrix inverse cannot be computed. In some modules (i.e., in Factor Analysis) this problem is dealt with by issuing a respective warning and then artificially lowering all correlations in the correlation matrix by adding a small constant to the diagonal elements of the matrix, and then restandardizing it. This procedure will usually yield a matrix for which the regular matrix inverse can be computed.

In many applications of the general linear model and the generalized linear model, matrix singularity is not abnormal (i.e., when the overparameterized model is used to represent effects for categorical predictor variables) and is dealt with by computing a generalized inverse rather than the regular matrix inverse.

Another example of matrix ill-conditioning is intransitivity of the correlations in a correlation matrix. If in a correlation matrix variable *A* is highly positively correlated with *B*, *B* is highly positively correlated with *C*, and *A* is highly negatively correlated with *C*, this impossible pattern of correlations signals an error in the elements of the matrix. See also, *Matrix Singularity*, *Matrix Inverse*, and *Generalized Inverse*.

Matrix Inverse

The regular inverse of a rectangular matrix of values is an extension of the concept of a numeric reciprocal. For a *nonsingular matrix* **A**, its *inverse* (denoted by a superscript of -1) is the unique matrix that satisfies

$$A^{-1}AA=A$$

No such regular inverse exists for singular matrices, but generalized inverses (an infinite number of them) can be computed for any singular matrix. See also, *Matrix Singularity* and *Generalized Inverse*.

Matrix Plot

Matrix graphs summarize the relationships between several variables in a matrix of true *X-Y* plots.

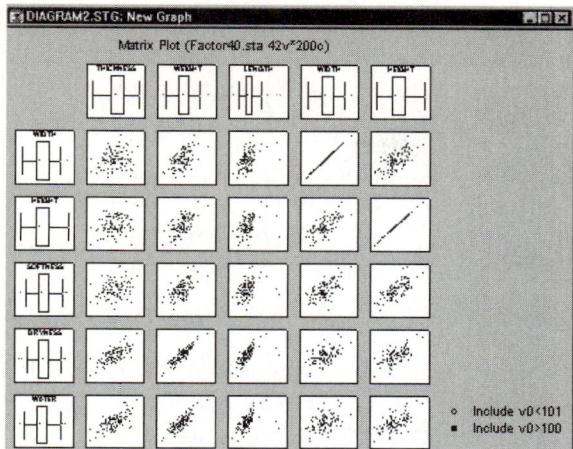

The most common type of matrix plot is the scatterplot, which can be considered to be the graphical equivalent of the correlation matrix.

Matrix Plot – Columns

In this type of matrix plot, columns represent projections of individual data points onto the X-axis (showing the distribution of the maximum values), arranged in a matrix format.

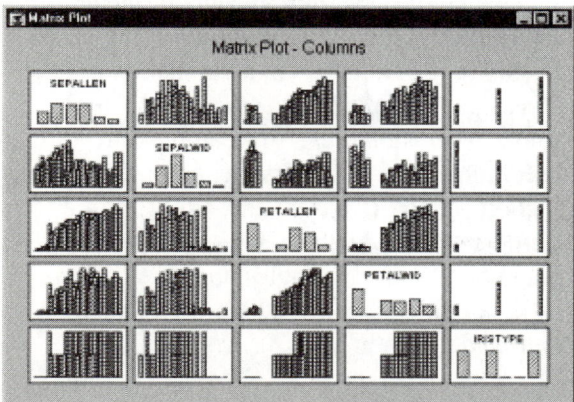

Histograms representing the distribution of each variable are displayed along the diagonal of the matrix (in square matrices, see the next illustration) or along the edges (in rectangular matrices).

Matrix Plot – Lines

In this type of matrix plot, a matrix of X-Y (i.e., nonsequential) line plots (similar to a scatterplot matrix) is produced in which individual points are connected by a line in the order of their appearance in the data file.

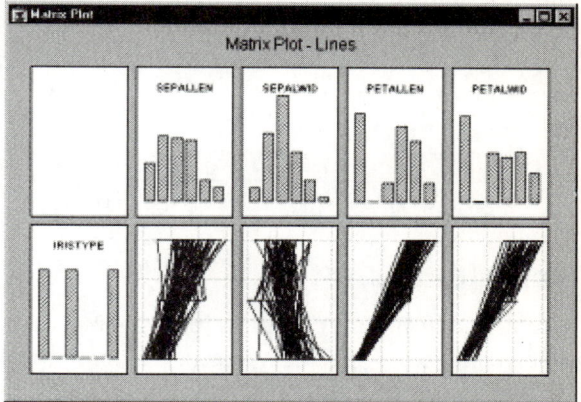

Histograms representing the distribution of each variable are displayed along the diagonal of the matrix (in square matrices) or along the edges (in rectangular matrices).

Matrix Plot – Scatterplot

In this type of matrix plot, 2D scatterplots are arranged in a matrix format (values of the column variable are used as X coordinates, values of the row variable represent the Y coordinates). Histograms representing the distribution of each variable are displayed along the diagonal of the matrix (in square matrices, see the next illustration) or along the edges (in rectangular matrices).

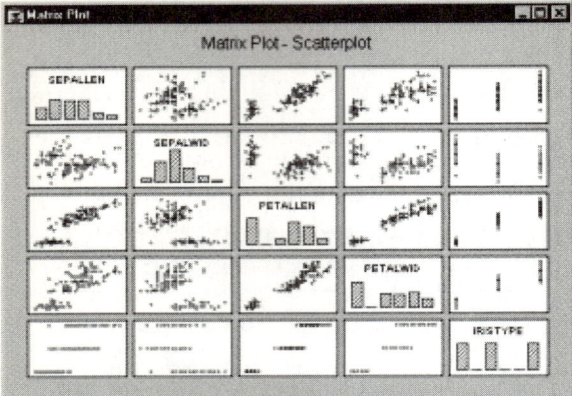

See also, *Data Reduction*.

Matrix Singularity

A rectangular matrix of values (e.g., a sums of squares and cross-products matrix) is *singular* if the elements in a column (or row) of the matrix are linearly dependent on the elements in one or more other columns (or rows) of the matrix. For example, if the elements in one column of a matrix are *1, -1, 0,* and the elements in another column of the matrix are 2, -2, 0, then the matrix is singular because 2 times each of the elements in the first column is equal to each of the respective elements in the second column. Such matrices are also said to suffer from multicollinearity problems, since one or more columns are linearly related to each other.

A unique, regular matrix inverse cannot be computed for singular matrices, but generalized inverses (an infinite number of them) can be computed for any singular matrix. See also, *Matrix Inverse*, page 658.

Matrix Rank

The column (or row) rank of a rectangular matrix of values (e.g., a sums of squares and cross-products matrix) is equal to the number of linearly independent columns (or rows) of

elements in the matrix. If there are no columns that are linearly dependent on other columns, the rank of the matrix is equal to the number of its columns and the matrix is said to have full (column) rank. If the rank is less than the number of columns, the matrix is said to have reduced (column) rank and is singular. See also, *Matrix Singularity*, page 660.

Maximum Likelihood Loss Function

An common alternative to the least squares loss function is to maximize the likelihood or log-likelihood function (or to minimize the negative log-likelihood function; the term maximum likelihood was first used by Fisher, 1922a). These functions are typically used when fitting nonlinear models. In most general terms, the likelihood function is defined as:

$$L = F(Y, Model) = \prod_{i=1}^{n} \{ p[y_i, Model\ Parameters(x_i)]\}$$

Maximum Likelihood Method

The method of maximum likelihood (the term first used by Fisher, 1922a) is a general method of estimating parameters of a population by values that maximize the *likelihood* (*L*) of a sample. The *likelihood L* of a sample of n observations $x_1, x_2, ..., x_n$, is the joint probability function $p(x_1, x_2, ..., x_n)$ when $x_1, x_2, ..., x_n$ are discrete random variables. If $x_1, x_2, ..., x_n$ are continuous random variables, then the *likelihood L* of a sample of *n* observations, $x_1, x_2, ..., x_n$, is the joint density function $f(x_1, x_2, ..., x_n)$.

Let *L* be the likelihood of a sample, where *L* is a function of the parameters $\theta_1, \theta_2, ... \theta_k$. Then the maximum likelihood estimators of $\theta_1, \theta_2, ... \theta_k$ are the values of $\theta_1, \theta_2, ... \theta_k$ that maximize *L*.

Let θ be an element of Ω. If Ω is an open interval, and if $L(\theta)$ is differentiable and assumes a maximum on W, then the MLE will be a solution

of the following equation: $(dL(\theta))/d\,\theta = 0$. For more information, see Bain and Engelhardt (1989) and Neter, Wasserman, and Kutner (1989). See also, Chapter 28 – *Nonlinear Estimation* or Chapter 39 – *Variance Components and Mixed Model ANOVA/ANCOVA*.

Maximum Unconfounding

Maximum unconfounding is an experimental design criterion that is subsidiary to the criterion of design resolution. The maximum unconfounding criterion specifies that design generators should be chosen such that the maximum number of interactions of less than or equal to the crucial order, given the resolution, are unconfounded with all other interactions of the crucial order. It is an alternative to the minimum aberration criterion for finding the best design of maximum resolution. For discussions of the role of design criteria in experimental design, see $2^{(k-p)}$ *Fractional Factorial Designs* and $2^{(k-p)}$ *Maximally Unconfounded and Minimum Aberration Designs* in Chapter 15 – *Experimental Design (Industrial DOE)*.

MD (Missing Data)

Same as missing values.

Mean

The mean is a particularly informative measure of the central tendency of the variable if it is reported along with its confidence intervals. Usually we are interested in statistics (such as the mean) from our sample only to the extent to which they are informative about the population. The larger the sample size, the more reliable its mean. The larger the variation of data values, the less reliable the mean (see also, Chapter 1 – *Elementary Concepts in Statistics*).

Mean = $(\Sigma x_i)/n$

where n is the sample size. See also, *Descriptive Statistics* in Chapter 2 – *Basic Statistics and Tables*.

Mean/SD

An algorithm (used in neural networks) to assign linear scaling coefficients for a set of numbers. The mean and standard deviation of the set are found, and scaling factors selected so that these are mapped to desired mean and standard deviation values. See also, *Neural Networks*, page 672.

Mean Substitution of Missing Data

When you select mean substitution, the missing data will be replaced by the means for the respective variables during an analysis. See also, *Casewise vs. Pairwise Deletion of Missing Data* in Chapter 2 – *Basic Statistics and Tables*.

Median

A measure of central tendency, the median (the term first used by Galton, 1882) of a sample is the value for which one-half (50%) of the observations (when ranked) will lie above that value and one-half will lie below that value. When the number of values in the sample is even, the median is computed as the average of the two middle values. See also, *Descriptive Statistics* in Chapter 2 – *Basic Statistics and Tables*.

Meta-Learning

The concept of meta-learning applies to the area of predictive data mining, to combine the predictions from multiple models. It is particularly useful when the types of models included in the project are very different. In this context, this procedure is also referred to as stacking (stacked generalization).

Suppose your data mining project includes tree classifiers, such as CART and CHAID, linear discriminant analysis (e.g., GDA), and neural networks. Each computes predicted classifications for a crossvalidation sample, from which overall goodness-of-fit statistics (e.g., misclassification rates) can be computed. Experience has shown that combining the predictions from multiple methods often yields more accurate predictions than can be derived from any one method (e.g., see Witten and Frank, 2000). The predictions from different classifiers can be used as input into a meta-learner, which will attempt to combine the predictions to create a final best predicted classification. So, for example, the predicted classifications from the tree classifiers, linear model, and the neural network classifier(s) can be used as input variables into a neural network meta-classifier, which will attempt to "learn" from the data how to combine the predictions from the different models to yield maximum classification accuracy.

You can apply meta-learners to the results from different meta-learners to create "meta-meta" learners, and so on; however, in practice such exponential increase in the amount of data processing, in order to derive an accurate prediction, will yield less and less marginal utility.

Minimax

An algorithm to assign linear scaling coefficients for a set of numbers. The minimum and maximum of the set are found, and scaling factors selected so that these are mapped to desired minimum and maximum values. See also, *Neural Networks*, page 672.

Minimum Aberration

Minimum aberration is an experimental design criterion that is subsidiary to the criterion of

design resolution. The minimum aberration design is defined as the design of maximum resolution "which minimizes the number of words in the defining relation that are of minimum length" (Fries & Hunter, 1980). Less technically, the criterion apparently operates by choosing design generators that produce the smallest number of pairs of confounded interactions of the crucial order. For example, the minimum aberration resolution IV design would have the minimum number of pairs of confounded 2-factor interactions. For discussions of the role of design criteria in experimental design, see $2^{(k-p)}$ *Fractional Factorial Designs* and $2^{(k-p)}$ *Maximally Unconfounded and Minimum Aberration Designs* in Chapter 15 – *Experimental Design (Industrial DOE)*.

Missing Data Range Plot, 2D

Use this graph to examine the pattern or distribution of missing and/or user-specified "out of range" data points in the current dataset (or in a subset of variables and cases).

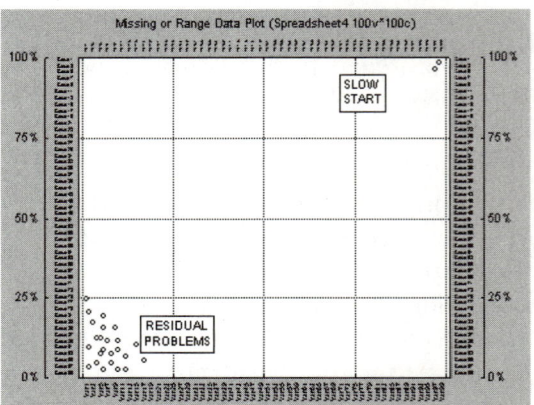

This graph is useful in exploratory data analysis to determine the extent of missing (and/or "out of range") data and whether the patterns of those data occur more or less randomly.

Missing Values

Values of variables within data sets that are not known. Although such cases that contain missing data are incomplete, they can still be used in data analysis. Various methods exist to substitute missing data (e.g., by mean substitution, various types of interpolations, and extrapolations). Also, pairwise deletion of missing data can be used. See also, *Pairwise MD Deletion*, *Casewise MD Deletion*, and *Pairwise Deletion of Missing Data vs. Mean Substitution* and *Casewise vs. Pairwise Deletion of Missing Data* in Chapter 2 – *Basic Statistics and Tables*.

Mode

A measure of central tendency, the mode (the term first used by Pearson, 1895) of a sample is the value which occurs most frequently in the sample. See also, *Descriptive Statistics* in Chapter 2 – *Basic Statistics and Tables*.

Model Profiles (in Neural Networks)

Model profiles are concise text strings indicating the architecture of networks and ensembles. A profile consists of a type code followed by a code giving the number of input and output variables and number of layers and units (networks) or members (ensembles). For time series networks, the number of steps and the lookahead factor are also given. The codes of the individual types of profiles are listed in the following table.

Code	Label
MLP	Multilayer Perceptron Network
RBF	Radial Basis Function Network
SOFM	Kohonen Self-Organizing Feature Map
Linear	Linear Network
PNN	Probabilistic Neural Network
GRNN	Generalized Regression Neural Network
PCA	Principal Components Network
Cluster	Cluster Network
Output	Output Ensemble
Conf	Confidence Ensemble

Network architecture. This is of the form I:N-N-N:O, where *I* is the number of input variable, *O* the number of output variables, and *N* the number of units in each layer.

Example. 2:4-6-3:1 indicates a network with *2* input variables, *1* output variable, *4* input neurons, *6* hidden neurons, and *3* output neurons. For a time series network, the steps factor is prepended to the profile, and signified by an "s."

Example. s10 1:10-2-1:1 indicates a time series network with steps factor (lagged input) 10.

Ensemble architecture. This is of the form I:[N]:O, where *I* is the number of input variable, *O* the number of output variables, and *N* the number of members of the ensemble.

Models for Data Mining

In the business environment, complex data mining projects may require the coordinate efforts of various experts, stakeholders, or departments throughout an entire organization. In data mining literature, various general frameworks have been proposed to serve as blueprints for how to organize the process of gathering data, analyzing data, disseminating results, implementing results, and monitoring improvements.

One such model, CRISP (Cross-Industry Standard Process for data mining) was proposed in the mid-1990s by a European consortium of companies to serve as a non-proprietary standard process model for data mining. This general approach postulates the following (perhaps not particularly controversial) general sequence of steps for data mining projects:

```
Business Understanding  ↔  Data Understanding
         ↓
Data Preparation  ↔  Modeling
                        ↓
                  Evaluation
                        ↓
                  Deployment
```

Another approach – the Six Sigma methodology – is a well-structured, data-driven methodology for eliminating defects, waste, or quality control problems of all kinds in manufacturing, service delivery, management, and other business activities. This model has recently become very popular (due to its successful implementations) in various American industries, and it appears to gain favor worldwide. It postulated a sequence of, so-called, DMAIC steps –

```
Define → Measure → Analyze → Improve → Control
```

– that grew up from the manufacturing, quality improvement, and process control traditions and is particularly well suited to production environments (including production of services, i.e., service industries).

Another framework of this kind (actually somewhat similar to Six Sigma) is the approach proposed by SAS Institute called SEMMA –

```
Sample → Explore → Modify → Model → Assess
```

– which focuses more on the technical activities typically involved in a data mining project.

All of these models are concerned with the process of how to integrate data mining methodology into an organization, how to convert data into information, how to involve important stake-holders, and how to disseminate the information in a form that can easily be converted by stake-holders into resources for strategic decision making.

Some software tools for data mining are specifically designed and documented to fit into one of these specific frameworks. See also, Chapter 12 – *Data Mining Techniques*.

Monte Carlo

A computer-intensive technique for assessing how a statistic will perform under repeated sampling. In Monte Carlo methods, the computer uses random number simulation techniques to mimic a statistical population. In the *STATISTICA* Monte Carlo procedure, the program constructs the population according to the user's prescription, and then for each Monte Carlo replication, the program:

1. Simulates a random sample from the population
2. Analyzes the sample
3. Stores the results

After many replications, the stored results will mimic the sampling distribution of the statistic. Monte Carlo techniques can provide information about sampling distributions when exact theory for the sampling distribution is not available.

Multidimensional Scaling

Multidimensional scaling can be considered to be an alternative to factor analysis (see Chapter 16 – *Factor Analysis*), and it is typically used as an exploratory method. In general, the goal of the analysis is to detect meaningful underlying dimensions that enable the researcher to explain observed similarities or dissimilarities (distances) between the investigated objects. In factor analysis, the similarities between objects (e.g., variables) are expressed in the correlation matrix. With MDS, you can analyze not only correlation matrices but also any kind of similarity or dissimilarity matrix (including sets of measures that are not internally consistent, e.g., do not follow the rule of transitivity). For more information, see Chapter 25 – *Multidimensional Scaling*.

Multilayer Perceptrons

Feedforward neural networks having linear PSP functions and (usually) nonlinear activation functions.

Multimodal Distribution

A distribution that has multiple modes (i.e., two or more peaks).

Multimodality of the distribution in a sample is often a strong indication that the distribution of the variable in population is not normal.

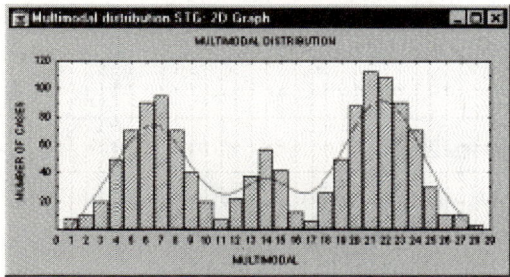

Multimodality of the distribution may provide important information about the nature of the investigated variable (i.e., the measured quality). For example, if the variable represents a reported preference or attitude, multimodality may indicate that there are several pronounced views or patterns of response in the questionnaire. Often however, the multimodality may indicate that the sample is not homogenous and the observations come in fact from two or more overlapping distributions. Sometimes, multimodality of the distribution

may indicate problems with the measurement instrument (e.g., gage calibration problems in natural sciences, or response biases in social sciences). See also, *Unimodal Distribution* and *Bimodal Distribution*.

Multinomial Distribution

The multinomial distribution arises when a response variable is categorical in nature, i.e., consists of data describing the membership of the respective cases to a particular category. For example, if a researcher recorded the outcome for the driver in accidents as "uninjured, "injury not requiring hospitalization", "injury requiring hospitalization", or "fatality," the distribution of the counts in these categories would be multinomial (see Agresti, 1996). The multinomial distribution is a generalization of the binomial distribution to more than two categories.

If the categories for the response variable can be ordered, the distribution of that variable is referred to as ordinal multinomial. For example, if in a survey the responses to a question are recorded such that respondents have to choose from the pre-arranged categories "Strongly agree," "Agree," "Neither agree nor disagree," "Disagree," and "Strongly disagree," the counts (number of respondents) that endorsed the different categories would follow an ordinal multinomial distribution (since the response categories are ordered with respect to increasing degrees of disagreement).

Specialized methods for analyzing multinomial and ordinal multinomial response variables can be found in Chapter 21 – *Generalized Linear/Nonlinear Models*.

Multinomial Logit and Probit Regression

The multinomial logit and probit regression models are extensions of the standard logit and probit regression models to the case where the dependent variable has more than two categories (e.g., not just *Pass – Fail*, but *Pass*, *Fail*, *Withdrawn*), i.e., when the dependent or response variable of interest follows a multinomial distribution rather than binomial distribution. When multinomial responses contain rank-order information, they are also called ordinal multinomial responses (see *Ordinal Multinomial Distribution*, page 678).

For additional details, see also, *Link Function and Distribution Function*, *Probit Regression and Transformation*, *Logit Regression and Transformation*, or Chapter 21 – *Generalized Linear/Nonlinear Models*.

Multi-Pattern Bar

Multi-pattern bar plots may be used to represent individual data values of the *X* variable (the same type of data as in pie charts), however, consecutive data values of the *X* variable are represented by the heights of sequential vertical bars, each of a different color and pattern (rather than as pie wedges of different widths).

Multiple Dichotomies

One possible coding scheme that can be used when more than one response is possible from a given question is to code responses using multiple dichotomies. For example, as part of a larger market survey, suppose you ask a sample of consumers to name their three favorite soft drinks. The specific item on the questionnaire may look like the following:

What are your three favorite soft drinks?
1:_____ 2:_____ 3:_____

Suppose in the above example we were only interested in *Coke*, *Pepsi*, and *Sprite*. One way to code the data in that case would be as follows:

Data: Basic Statistics Example Spreadsheet 12* (4v by 4c)				
	Coke	Pepsi	Sprite	...
Case 1	0	1	0	
Case 2	1	1	0	
Case 3	0	0	1	
...	...	...	...	

In other words, one variable was created for each soft drink, and then a value of *1* was entered into the respective variable whenever the respective drink was mentioned by the respective respondent. Note that each variable represents a dichotomy; that is, only "*1*"s and "*not 1*"s are allowed (we could have entered *1*'s and *0*'s, but to save typing we can also simply leave the *0*'s as blanks or as missing values). When tabulating these variables, we want to compute the number and percent of respondents (and responses) for each soft drink. In a sense, we compact the three variables *Coke*, *Pepsi*, and *Sprite* into a single variable (*Soft Drink*) consisting of multiple dichotomies.

For more information on multiple dichotomies, see *Multiple Responses/Dichotomies* in Chapter 2 – *Basic Statistics and Tables*.

Multiple R

The coefficient of multiple correlation (multiple R) is the positive square root of R-square (the coefficient of multiple determination, see *Residual Variance and R-Square* in Chapter 26 – *Multiple Linear Regression*). This statistic is useful in multivariate regression (i.e., multiple independent variables) when you want to describe the relationship between the variables.

Multiple Regression

The general purpose of multiple regression (the term was first used by Pearson, 1908) is to analyze the relationship between several independent or predictor variables and a dependent or criterion variable.

The computational problem that needs to be solved in multiple regression analysis is to fit a straight line (or plane in an *n*-dimensional space, where *n* is the number of independent variables) to a number of points. In the simplest case – one dependent and one independent variable – you can visualize this in a scatterplot (scatterplots are two-dimensional plots of the scores on a pair of variables). It is used as either a hypothesis testing or exploratory method. For more information, see Chapter 26 – *Multiple Linear Regression*.

Multiple Response Variables

Coding the responses to multiple response variables is necessary when more than one response is possible from a given question. For example, as part of a larger market survey, suppose you asked a sample of consumers to name their three favorite soft drinks. The specific item on the questionnaire may look like the following:

What are your three favorite soft drinks?
1:_____ 2:_____ 3:_____

Thus, the questionnaires returned to you will contain somewhere between 0 and 3 answers to this item. Also, a wide variety of soft drinks will most likely be named. One way to record the various responses would be to use three *multiple response variables* and a coding scheme for the many soft drinks. Then we could enter the respective codes (or alphanumeric labels) into the three variables, in the same way that respondents wrote them down in the questionnaire.

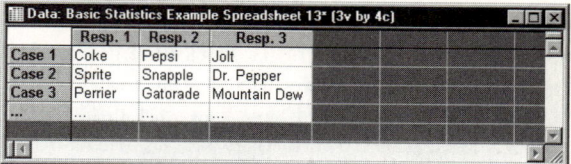

For more information, see *Multiple Responses/Dichotomies* in Chapter 2 -*Basic Statistics and Tables*.

Multiple-Response Tables

Multiple-response tables are crosstabulation tables used when the categories of interest are not mutually exclusive. Such tables can accommodate multiple response variables as well as multiple dichotomies. For more information, see *Multiple Responses/ Dichotomies* in Chapter 2 -*Basic Statistics and Tables*.

Multiplicative Season, Damped Trend

In this time series model, the simple exponential smoothing forecasts are enhanced both by a damped trend component (independently smoothed with the single parameter φ; this model is an extension of Brown's one-parameter linear model, see Gardner, 1985, p. 12-13) and a multiplicative seasonal component (smoothed with parameter δ). For example, suppose we wanted to forecast from month to month the number of households that purchase a particular consumer electronics device (e.g., VCR). Every year, the number of households that purchase a VCR will increase, however, this trend will be damped (i.e., the upward trend will slowly disappear) over time as the market becomes saturated. In addition, there will be a seasonal component, reflecting the seasonal changes in consumer demand for VCRs from month to month (demand will likely be smaller in the summer and greater during the December holidays). This seasonal component may be multiplicative, for example, sales during the December holidays may increase by factor of 1.4 (or 40%) over the average annual sales. To compute the smoothed values for the first season, initial values for the seasonal components are necessary. Also, to compute the smoothed value (forecast) for the first observation in the series, both estimates of S_0 and T_0 (initial trend) are necessary. These values are computed as:

$$T_0 = (1/\varphi)*M_k-M_1)/[(k-1)*p]$$

where φ is the smoothing parameter; k is the number of complete seasonal cycles; M_k is the mean for the last seasonal cycle; M_1 is the mean for the first seasonal cycle; p is the length of the seasonal cycle; and $S_0 = M_1-p*T_0/2$

Multiplicative Season, Exponential Trend

In this time series model, the simple exponential smoothing forecasts are enhanced both by an exponential trend component (independently smoothed with parameter γ) and a multiplicative seasonal component (smoothed with parameter c). For example, suppose we wanted to forecast the monthly revenue for a resort area. Every year, revenue may increase by a certain percentage or factor, resulting in an exponential trend in overall revenue. In addition, there could be an multiplicative seasonal component, that is, given the respective annual revenue, each year 20% of the revenue is produced during the month of December, that is, during Decembers the revenue grows by a particular (multiplicative) factor.

To compute the smoothed values for the first season, initial values for the seasonal components are necessary. Also, to compute the smoothed value (forecast) for the first

observation in the series, both estimates of S_0 and T_0 (initial trend) are necessary. By default, these values are computed as:

$$T_0 = \exp\{[\log(M_2)-\log(M_1)]/p\}$$

where M_2 is the mean for the second seasonal cycle; M_1 is the mean for the first seasonal cycle; p is the length of the seasonal cycle; and $S_0 = \exp\{\log(M_1)-p*\log(T_0)/2\}$.

Multiplicative Season, Linear Trend

In this time series model, the simple exponential smoothing forecasts are enhanced both by a linear trend component (independently smoothed with parameter γ) and a multiplicative seasonal component (smoothed with parameter γ). For example, suppose we were to predict the monthly budget for snow-removal in a community. There may be a trend component (as the community grows, there is an upward trend for the cost of snow removal from year to year). At the same time, there is obviously a seasonal component, reflecting the differential likelihood of snow during different months of the year. This seasonal component could be multiplicative, meaning that given a respective budget figure, it may increase by a factor of, for example, 1.4 during particular winter months; or it may be additive (see above), that is, a particular fixed additional amount of money is necessary during the winter months. To compute the smoothed values for the first season, initial values for the seasonal components are necessary. Also, to compute the smoothed value (forecast) for the first observation in the series, both estimates of S_0 and T_0 (initial trend) are necessary. By default, these values are computed as:

$$T_0 = (M_k-M_1)/((k-1)*p)$$

where k is the number of complete seasonal cycles; M_k is the mean for the last seasonal cycle; M_1 is the mean for the first seasonal cycle; p is the length of the seasonal cycle; and $S_0 = M_1 - T_0/2$.

Multiplicative Season, No Trend

This time series model is partially equivalent to the simple exponential smoothing model; however, in addition, each forecast is enhanced by a multiplicative component that is smoothed independently (see *Seasonal smoothing parameter δ* in Chapter 38 – *Time Series/Forecasting*). This model would, for example, be adequate when computing forecasts for monthly expected sales for a particular toy. The level of sales may be stable from year to year, or change only slowly; at the same time, there will be seasonal changes (e.g., greater sales during the December holidays), which again may change slowly from year to year. The seasonal changes may affect the sales in a multiplicative fashion, for example, depending on the respective overall level of sales, December sales may always be greater by a factor of 1.4.

Multivariate Adaptive Regression Splines (MARSplines)

Multivariate adaptive regression splines (or MARSplines) is a nonparametric regression procedure which makes no assumption about the underlying functional relationship between the dependent and independent variables. Instead MARSplines constructs this relation from a set of coefficients and basis functions that are entirely driven from the regression data. The MARSplines technique has become particularly popular in the area of data mining, because it does not assume or impose any particular type or class of relationship (e.g.,

linear, logistic, and so on) between the predictor variables and the dependent (outcome) variable of interest.

The general MARSplines model equation (see Hastie et al., 2001, equation 9.19) is given as:

$$y = f(X) = \beta_0 + \sum_{m=1}^{M} \beta_m h_m(X)$$

where the summation is over the M predictors in the model. To summarize, y is predicted as a function of the predictor variables X (and their interactions); this function consists of an intercept parameter (β_o and the weighted (by β_m sum of one or more basis functions $h_m(X)$. You may also think of this model as "selecting" a weighted sum of basis functions from the set of (a large number of) basis functions that span all values of each predictor (i.e., that set would consist of one basis function, and "knot" parameter t, for each distinct value for each predictor variable). The MARSplines algorithm then searches over the space of all inputs and predictor values (knot locations t) as well as interactions between variables. During this search an increasingly larger number of basis functions are added to the model (selected from the set of possible basis functions), to maximize an overall least squares goodness-of-fit criterion. For more information about this technique, and how it compares to other methods for nonlinear regression (or regression trees), see Hastie, Tishirani, and Friedman (2001).

Multivariate Statistical Process Control (MSPC)

Multivariate statistical process control is a methodology for simultaneously monitoring multiple inputs or variables describing a process, for the purpose of ensuring that the overall process is in control. It is an extension of simple univariate (one variable at a time) quality control.

Modern automated production processes typically measure large numbers of variables that describe the process at each stage and across multiple stages. Standard quality control charting techniques (e.g., Shewhart charts, X-bar and R charts, etc.) are applicable only to single variables. Therefore, when applied to modern production processes with hundreds of important variables that need to be monitored, the criteria typically applied to univariate charts will lead to a large number of false alarms, and in many cases nearly constant, perpetual alarms. Furthermore, this approach will ignore the inherent correlations between variables and, thus, lose important information (e.g., consider a single measure of temperature collected by one sensor drifting out of control, while 50 others stay within control, vs. a scenario where all 50 temperature readings begin to slowly drift upward; intuitively, the latter condition would be the more "significant" event).

To rectify these shortcomings, methods have been developed to monitor multiple variables simultaneously, using multivariate statistical procedures, such as Principal Component Analysis (PCA) and Partial Least Squares (PLS) methods. In short, these techniques allow users to identify a) when multiple correlated variables start to drift out of control, and b) when the fundamental relationships between variables change (so that the correlations between variables observed when the process was known to be in control are no longer applicable and valid).

A special application of MSPC is commonly found in process monitoring and quality control for industrial batch processing. Batch processes are those where goods are manufactured in "chunks" or batches, such as batches of beer, pharmaceuticals, chemicals, polymers, paint, fertilizers, cement, petroleum products, biochemicals, perfumes, or semiconductors. In those applications, one can define in-control ("good") batches; those batches can be characterized by particular maturing effects, as various measures systematically change over time (e.g., as the alcohol ferments). By building multivariate models (e.g., via PLS) describing the relationship of the various variables of interest to time (i.e., to the maturing process) for those good batches, quality control schemes can be derived to detect when a batch deviates from this known "good" multivariate pattern. For details regarding these procedures, see also Nomikos and MacGregor (1995).

N

N-in-One Encoding

For nominal variables with more than two states, the practice of representing the variable using a single unit with a range of possible values (actually implemented using minimax, explicit or none). See also, *Neural Networks*, page 672.

n Point Moving Average Line

Each point on this moving average line represents the average of the respective sample and the *n-1* number of preceding samples. Thus, this line will smooth the pattern of means across samples, enabling the quality control engineer to detect trends. You can specify the number of samples (*n*) that are to be averaged for each point in the plot. See also, *Time Series*, page 743.

Neat Scaling of Intervals

The term neat scaling is used to refer to the manner in which ranges of values are divided into intervals so that the resulting interval boundaries and steps between those boundaries are intuitive and readily interpretable (or understood).

For example, suppose you want to create a histogram for data values in the range from 1 to 10. It would be inefficient to use interval boundaries for the histogram at values such as 1.3, 3.9, 6.5, etc., i.e., to use as a minimum boundary value 1.3, and then a step size of 2.6. A much more intuitive way to divide the range of data values would be to use boundaries such as 1, 2, 3, 4, and so on, i.e., a minimum boundary at 1, with step size of 1; or you could use 2, 4, 6, etc, i.e., a minimum boundary of 2 and step size 2.

In general, *neat* in this context means that category boundaries will be round values ending either in 0, 2, or 5 (e.g., boundaries may be 0.1, 0.2, 0.3, etc.; or 50, 100, 150, etc.). To achieve this, any user-specified lower limit, upper limit, and number of categories will only be approximated.

Negative Correlation

The relationship between two variables is such that as one variable's values tend to increase, the other variable's values tend to decrease. This is represented by a negative correlation coefficient. See also, *Correlations* in Chapter 2 – *Basic Statistics and Tables*.

Negative Exponential (2D Graphs)

A curve is fitted to the *XY* coordinate data according to the negative exponentially weighted smoothing procedure (the influence of individual points decreases exponentially with the horizontal distance from the respective points on the curve).

Negative Exponential (3D Graphs)

A surface is fitted to the *XYZ* coordinate data according to the negative exponentially-weighted smoothing procedure (the influence of individual points decreases exponentially with the horizontal distance from the respective points on the surface).

Neighborhood (in Neural Networks)

In Kohonen training, a square set of units focused around the winning unit and simultaneously updated by the training algorithm.

Nested Factors

In nested designs the levels of a factor are nested (the term was first used by Ganguli,

1941) within the levels of another factor. For example, if you were to administer four different tests to four high school classes (i.e., a between-groups factor with 4 levels), and two of those four classes are in high school A, whereas the other two classes are in high school B, then the levels of the first factor (4 different tests) would be nested in the second factor (2 different high schools). See also, *ANOVA (General ANOVA/MANOVA)*.

Nested Sequence of Models

In structural equation modeling, a set of models M(1), M(2), ... M(k) form a nested sequence if model M(i) is a special case of M(i+1) for i=1 to k-1. Thus, each model in the sequence becomes increasingly more general, but includes all previous models as special cases. As an example, consider one factor, two factor, and three factor models for 10 variables. The two-factor model includes the one-factor model as a special case (simply let all the loadings on the second factor be 0). Similarly, the three-factor model contains the two and one-factor models as special cases.

Neural Networks

Neural networks are analytic techniques modeled after the (hypothesized) processes of learning in the cognitive system and the neurological functions of the brain and capable of predicting new observations (on specific variables) from other observations (on the same or other variables) after executing a process of so-called learning from existing data.

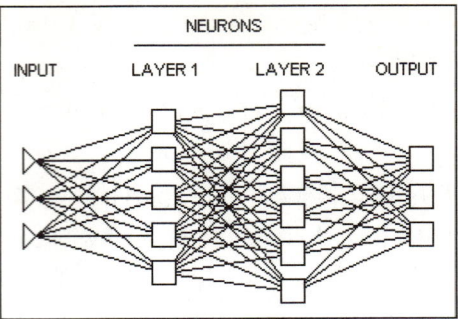

For more information, see *Neural Networks* and D*ata Mining*.

Neuron

A unit in a neural network.

Newman-Keuls Test

This post hoc test can be used to determine the significant differences between group means in an analysis of variance setting. The Newman-Keuls test, like Duncan's test, is based on the range statistic (for a detailed discussion of different post hoc tests, see Winer, Michels, & Brown (1991). For more details, see Chapter 18 – *General Linear Models*. See also, *Post hoc Comparisons*, page 687. For a discussion of statistical significance, see Chapter 1 – *Elementary Concepts in Statistics*.

NIPALS Algorithm

The standard algorithm for computing partial least squares regression components (i.e., factors) is nonlinear iterative partial least squares (NIPALS). There are many variants of the NIPALS algorithm that normalize or do not normalize certain vectors. The following algorithm, which assumes that the X and Y variables have been transformed to have means of zero, is considered to be one of most efficient NIPALS algorithms.

For each $h=1,\ldots,c$, where $A_0=X'Y$, $M_0=X'X$, $C_0=I$, and c given,

1. compute q_h, the dominant eigenvector of $A_h'A_h$

2. $w_h=G_hA_hq_h$, $w_h=w_h/||w_h||$, and store w_h into W as a column

3. $p_h=M_hw_h$, $c_h=w_h'M_hw_h$, $p_h=p_h/c_h$, and store p_h into P as a column

4. $q_h=A_h'w_h/c_h$, and store q_h into Q as a column

5. $A_{h+1}=A_h - c_hp_hq_h'$ and $B_{h+1}=M_h - c_hp_hp_h'$

6. $C_{h+1}=C_h - w_hp_h'$

The factor scores matrix **T** is then computed as $T=XW$ and the partial least squares regression coefficients B of Y on X are computed as $B=WQ$.

Noise Addition (in Neural Networks)

A practice (used in neural networks) designed to prevent overlearning during back propagation training, by adding random noise to input patterns during training (and so blurring the position of the training data). See *Neural Networks*, page 672.

Nominal Scale

This is a categorical (i.e., quantitative and not qualitative) scale of measurement where each value represents a specific category into which the variable's values fall (each category is different than the others but cannot be quantitatively compared to the others). See also, Chapter 1 – *Elementary Concepts in Statistics*.

Nominal Variables

Variables that take on one of a set of discrete values, such as *Gender*={*Male, Female*}. In neural networks, nominal output variables are used to distinguish classification tasks from regression tasks. See also, *Grouping (or Coding) Variable* and *Measurement Scales* in Chapter 1 – *Elementary Concepts in Statistics*.

Nonlinear Estimation

In the most general terms, nonlinear estimation involves finding the best fitting relationship between the values of a dependent variable and the values of a set of one or more independent variables (it is used as either a hypothesis testing or exploratory method). For example, we may want to compute the relationship between the dose of a drug and its effectiveness, the relationship between training and subsequent performance on a task, the relationship between the price of a house and the time it takes to sell it, etc. Research issues in these examples are commonly addressed by such techniques as multiple regression or analysis of variance. In fact, you can think of nonlinear estimation as a generalization of those methods. Specifically, multiple regression (and ANOVA) assumes that the relationship between the independent variable(s) and the dependent variable is linear in nature.

Nonlinear estimation leaves it up to you to specify the nature of the relationship; for example, you can specify the dependent variable to be a logarithmic function of the independent variable(s), an exponential function, a function of some complex ratio of independent measures, etc. (However, if all variables of interest are categorical in nature, or can be converted into categorical variables, you can also consider correspondence analysis as an alternative analysis technique.) For more information, see Chapter 28 – *Nonlinear Estimation*.

Non-Normal Distribution Fitting

See *Johnson Curves*.

Non-Outlier Range

The non-outlier range is the range of values in a 2D box plot, 3D sequential graph – box plot, or categorized box plots that fall below the upper outlier limit (for example, +1.5 * the height of the box) and above the lower outlier limit (for example, -1.5 * the height of the box).

Nonparametrics

Nonparametric methods were developed to be used in cases when the researcher does not know the parameters of the distribution of the variable of interest in the population (hence the name nonparametric). In more technical terms, nonparametric methods do not rely on the estimation of parameters (such as the mean or the standard deviation) describing the distribution of the variable of interest in the population. Therefore, these methods are also sometimes (and more appropriately) called parameter-free methods or distribution-free methods. For more information, see Chapter 29 – *Nonparametric Statistics*; see also, Chapter 1 – *Elementary Concepts in Statistics*.

Nonseasonal, Damped Trend

In this time series model, the simple exponential smoothing forecasts are enhanced by a damped trend component (independently smoothed with parameters γ for the trend, and φ for the damping effect). For example, suppose we want to forecast from month to month the percentage of households that own a particular consumer electronics device (e.g., a VCR). Every year, the proportion of households owning a VCR will increase; however, this trend will be damped (i.e., the upward trend will slowly disappear) over time as the market becomes saturated.

To compute the smoothed value (forecast) for the first observation in the series, both estimates of S_0 and T_0 (initial trend) are necessary. By default, these values are computed as:

$$T_0 = (1/\varphi)*(X_n-X_1)/(N-1)$$

where N is the number of cases in the series; φ is the smoothing parameter for the damped trend; and $S_0 = X_1 - T_0/2$.

Nonseasonal, Exponential Trend

In this time series model, the simple exponential smoothing forecasts are enhanced by an exponential trend component (smoothed with parameter γ). For example, suppose we want to predict the overall monthly costs of repairs to a production facility. There could be an exponential trend in the cost; that is, from year to year the costs of repairs may increase by a certain percentage or factor, resulting in a gradual exponential increase in the absolute dollar costs of repairs.

To compute the smoothed value (forecast) for the first observation in the series, both estimates of S_0 and T_0 (initial trend) are necessary. By default, these values are computed as:

$$T_0 = (X_2/X_1)$$

and

$$S_0 = X_1/T_0^{1/2}$$

Nonseasonal, Linear Trend

In this time series model, the simple exponential smoothing forecasts are enhanced by a linear trend component that is smoothed independently via the γ (gamma) parameter (see

Trend smoothing parameters in Chapter 38 – *Time Series/Forecasting*). This model is also referred to as Holt's two-parameter method. This model would, for example, be adequate when producing forecasts for spare parts inventories. The need for particular spare parts may slowly increase or decrease over time (the trend component), and the trend may slowly change as different machines etc. age or become obsolete, thus affecting the trend in the demand for spare parts for the respective machines.

In order to compute the smoothed value (forecast) for the first observation in the series, both estimates of S_0 and T_0 (initial trend) are necessary. By default, these values are computed as:

$$T_0 = (X_n-X_1)/(N-1)$$

where N is the length of the series and $S_0 = X_1 - T_0/2$.

Nonseasonal, No Trend

This time series model is equivalent to the simple exponential smoothing model. Note that, by default, the first smoothed value will be computed based on an initial S_0 value equal to the overall mean of the series.

Normal Distribution

This term was first used by Galton, 1889. The normal distribution function is determined by the following formula:

$$f(x) = \frac{1}{\sigma\sqrt{2\pi}} e^{\frac{-(x-\mu)^2}{2\sigma^2}} \quad for -\infty < x < \infty$$

where μ is the mean; σ is the standard deviation; e is the base of the natural logarithm, sometimes called Euler's e (2.71...); and π is the constant Pi (3.14...).

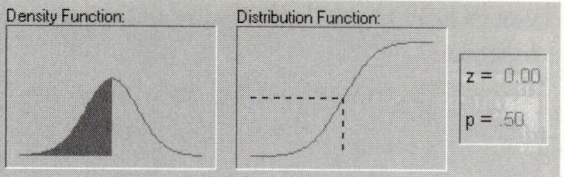
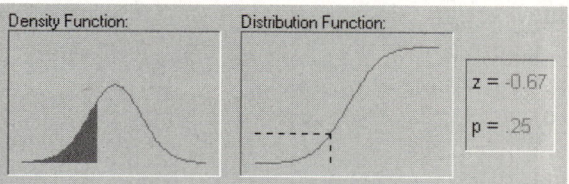
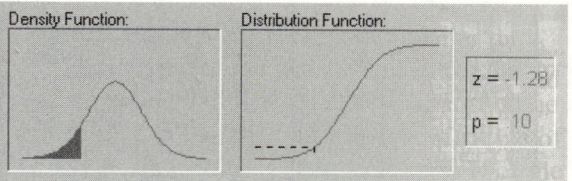
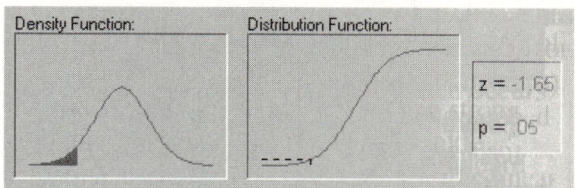
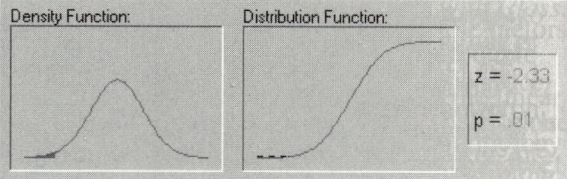

See also, *Why the Normal Distribution is Important* In Chapter 1 – *Elementary Concepts in Statistics* and the definitions of *Bivariate Normal Distribution* and *Normality Tests*.

Normal Fit

The normal/observed histogram represents the most common graphical test of normality. When you select this fit, a normal curve will be overlaid on the frequency distribution. The normal function fitted to histograms is defined as:

f(x) = NC * step * normal(x, mean, std.dev)

The normal function fitted to cumulative histograms is defined as:

f(x) = NC * inormal(x, mean, std.dev.)

where *NC* is the number of cases; *step* is the categorization step size (e.g., the integral categorization step size is 1); *normal* is the normal function; and *inormal* is the integral of the normal function. See also, *Bivariate Normal Distribution* and *Normal Distribution*.

Normality Tests

A common application for distribution fitting procedures is when you want to verify the assumption of normality before using some parametric test (see Chapter 2 – *Basic Statistics and Tables* and Chapter 29 – *Nonparametric Statistics*). A variety of statistics for testing normality are available including the Kolmogorov-Smirnov test for normality, the Shapiro-Wilks' W test, and the Lilliefors test. Additionally, you can review probability plots and normal probability plots to assess whether the data are accurately modeled by a normal distribution.

Normalization

Adjusting a series (vector) of values (typically representing a set of measurements, e.g., a variable storing heights of people, represented in inches) according to some transformation function in order to make them comparable with some specific point of reference (for example, a unit of length or a sum). For example, dividing these values by 2.54 will produce metric measurements of the height. Normalization of data is:

a. required when the incompatibility of the measurement units across variables may affect the results (e.g., in calculations based on cross products) without carrying any interpretable information, and

b. recommended whenever the final reports could benefit from expressing the results in specific meaningful/compatible units (e.g., reaction time data will be easier to interpret when converted into milliseconds from the CPU cycles of different computers that were used to measure RTs, as originally registered in a medical experiment).

Note that this term is unrelated to the term normal distribution; see also, *Standardization*, page 732.

ODBC

ODBC (Open DataBase Connectivity) is a set of conventions introduced by Microsoft that allows access to information from a wide range of databases (e.g., MS Access, Oracle) and performing queries via SQL.

Odds Ratio

The odds ratio is useful in the interpretation of the results of logistic regression (see Neter, Wasserman, and Kutner, 1989) and is computed from a 2x2 classification table which displays the predicted and observed classification of cases for a binary dependent variable:

$$(f_{11} * f_{22})/(f_{12} * f_{21})$$

where f_{ij} represents the respective frequencies in the 2x2 table.

OLAP

See *On-Line Analytic Processing*, page 677.

OLE DB

OLE DB (Object Linking and Embedding Database) is a set of conventions introduced by Microsoft that allows access to information from a wide range of databases (e.g., MS Access, Oracle). OLE DB is a database architecture that provides universal data integration over an enterprise's network, from mainframe to desktop, regardless of the data type. OLE DB is a more generalized and more efficient strategy for data access than ODBC, because it allows access to more types of data and is based on the Component Object Model (COM).

One-of-N Encoding (in Neural Networks)

Representing a nominal variable using a set of input or output units, one unit for each possible nominal value. During training, one of the units will be on and the others off. See *Neural Networks*, page 672.

One-Off (in Neural Networks)

A case typed in and submitted to the neural network as a one-off procedure (not part of a data set, and not used in training). See *Neural Networks*, page 672.

One-Sample *t*-Test

See *t-Test (for Independent and Dependent Samples)*, page 745.

One-Sided Ranges or Error Bars in Range Plots

In order to display a "one-sided" range (relative to the mid-point) or an error bar that extends in only one direction, set the respective values of the variable defining the range boundary to 0 (when the *Relative to the Mid-point* style is selected) or the mid-point (when the *absolute* style is selected).

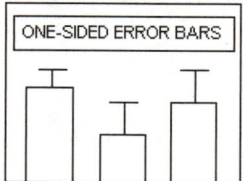

On-Line Analytic Processing (OLAP) (or Fast Analysis of Shared Multidimensional Information – FASMI)

The term On-Line Analytic Processing refers to technology that enables users of multidimensional databases to generate on-line descriptive or comparative summaries (views) of data and other analytic queries. For more information, see *On-Line Analytic Processing (OLAP)* in Chapter 12 – *Data Mining Techniques*; see also the definitions of *Data Mining* and *Data Warehousing*.

Operating Characteristic Curves, for Quality Control Charts

A common supplementary plot to standard quality control charts is the so-called operating characteristic or OC curve. One question that comes to mind when using standard variable or attribute charts is how sensitive is the current quality control procedure? Put in more specific terms, how likely is it that you will not find a sample (e.g., a mean in an X-bar chart) outside the control limits (i.e., accept the production process as in control), when, in fact, it has shifted by a certain amount? This probability is usually referred to as the b (beta) error probability, that is, the probability of erroneously accepting a process (mean, mean proportion, mean rate defectives, etc.) as being in control.

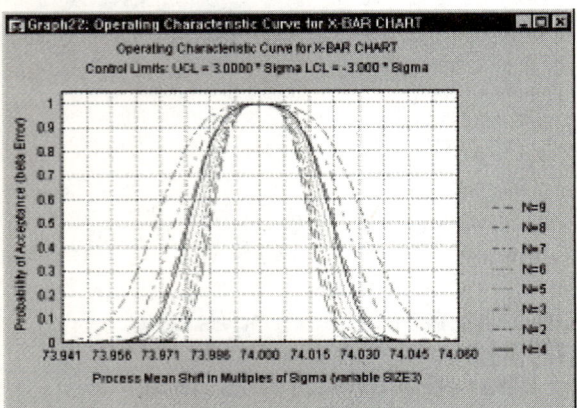

Operating characteristic curves are extremely useful for exploring the power of the quality control procedure. The actual decision concerning sample sizes should depend not only on the cost of implementing the plan (e.g., cost per item sampled), but also on the costs resulting from not detecting quality problems. The OC curve enables the engineer to estimate the probabilities of not detecting shifts of certain sizes in the production quality. For more information, see also, *Operating Characteristic Curves*, page 677.

Ordinal Multinomial Distribution

If the categories for a multinomial response variable can be ordered, the distribution of that variable is referred to as ordinal multinomial. For example, if in a survey the responses to a question are recorded such that respondents have to choose from the pre-arranged categories "Strongly agree," "Agree," "Neither agree nor disagree," "Disagree," and "Strongly disagree," then the counts (number of respondents) that endorsed the different categories would follow an ordinal multinomial distribution (since the response categories are ordered with respect to increasing degrees of disagreement).

Specialized methods for analyzing multinomial and ordinal multinomial response variables can be found in Chapter 21 – *Generalized Linear/Nonlinear Models*.

Ordinal Scale

The ordinal scale of measurement represents the ranks of a variable's values. Values measured on an ordinal scale contain information about their relationship to other values only in terms of whether they are "greater than" or "less than" other values but not in terms of "how much greater" or "how much smaller." See also, *Measurement Scales* in Chapter 1 – *Elementary Concepts in Statistics*.

Outer Arrays

In Taguchi experimental design methodology, the repeated measurements of the response variable are often taken in a systematic fashion, with the goal to manipulate noise factors. The levels of those factors are then arranged in a so-called outer array, i.e., an (orthogonal) experimental design. However, usually the repeated measurements are placed in separate columns in the data spreadsheet (i.e., each is a different variable); thus the index i (in the formulas for smaller-the-better, larger-the-better, and signed target) runs across the columns or variables in the data spreadsheet, or the levels of the factors in the outer array. See *Signal-to-Noise (S/N) Ratios* in Chapter 15 – *Experimental Design (Industrial DOE)* for more details.

Outliers

Outliers are atypical (by definition), infrequent observations; data points which do not appear to follow the characteristic distribution of the rest of the data. These may reflect genuine properties of the underlying phenomenon (variable), or be due to measurement errors or other anomalies that should not be modeled.

Because of the way in which the regression line is determined in multiple regression (especially the fact that it is based on minimizing not the sum of simple distances but the sum of squares of distances of data points from the line), outliers have a profound influence on the slope of the regression line (see the following illustrations) and consequently on the value of the correlation coefficient. A single outlier is capable of considerably changing the slope of the regression line and, consequently, the value of the correlation. Note that just one outlier can be entirely responsible for a high value of the

correlation that otherwise (without the outlier) would be close to zero. Needless to say, you should never base important conclusions on the value of the correlation coefficient alone (i.e., examining the respective scatterplot is always recommended).

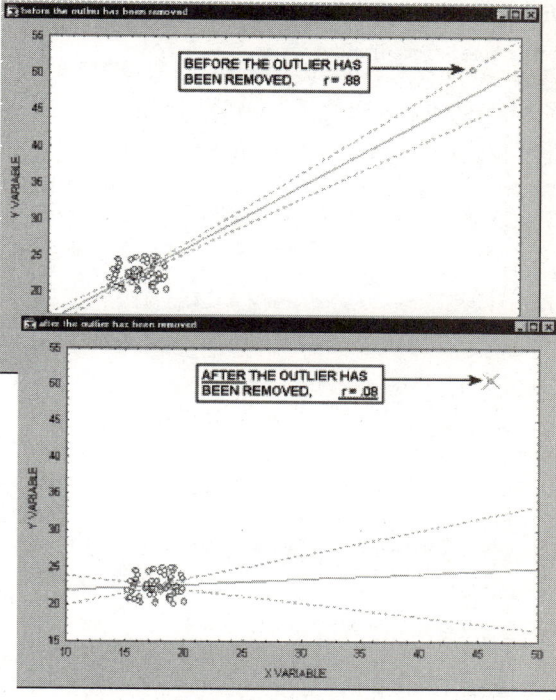

Note that if the sample size is relatively small, including or excluding specific data points that are not as clearly outliers as the one shown in the previous example may have a profound influence on the regression line (and the correlation coefficient). This is shown in the following illustrations where we call the points being excluded outliers; you may argue, however, that they are not outliers but rather extreme values.

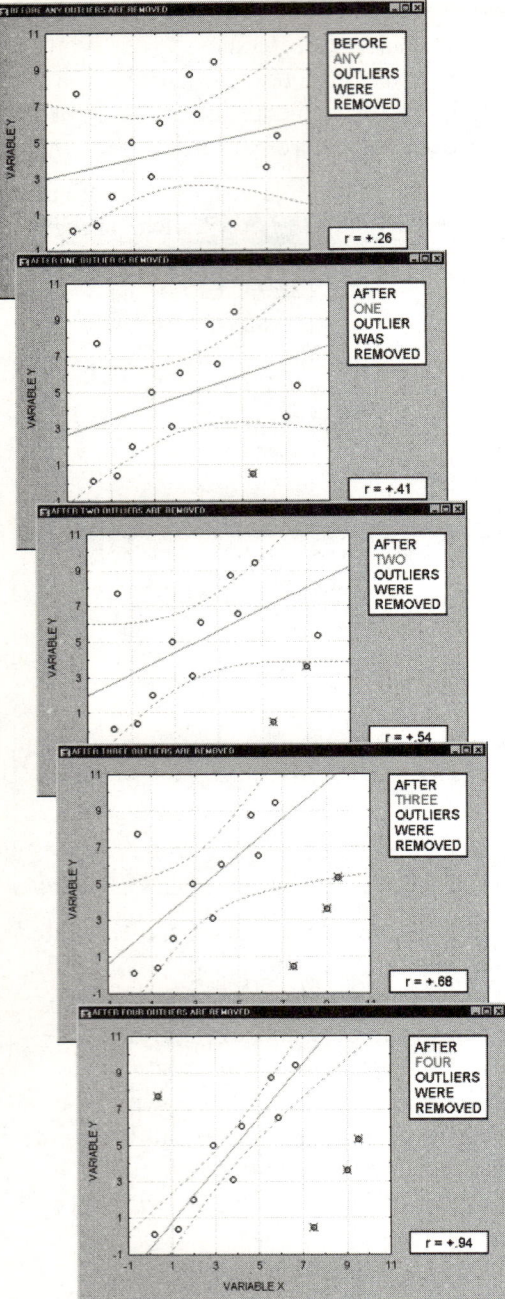

Typically, we believe that outliers represent a random error that we want to be able to control.

Needless to say, outliers may not only artificially increase the value of a correlation coefficient, but they can also decrease the value of a legitimate correlation. See also, *Ellipse*.

Outliers (in Box Plots)

Values that are far from the middle of the distribution are referred to as outliers and extreme values if they meet certain conditions.

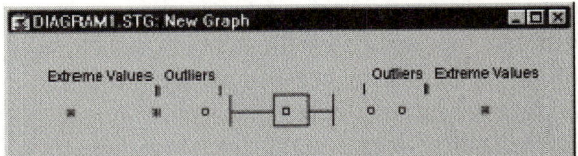

A data point is deemed to be an outlier if the following conditions hold:

data point value > UBV + *o.c.*(UBV – LBV)
or
data point value < LBV – *o.c.*(UBV – LBV)

where *UBV* is the upper value of the box in the box plot (e.g., the mean + standard error or the 75th percentile); *LBV* is the lower value of the box in the box plot (e.g., the mean – standard error or the 25th percentile); and o.c. is the outlier coefficient.

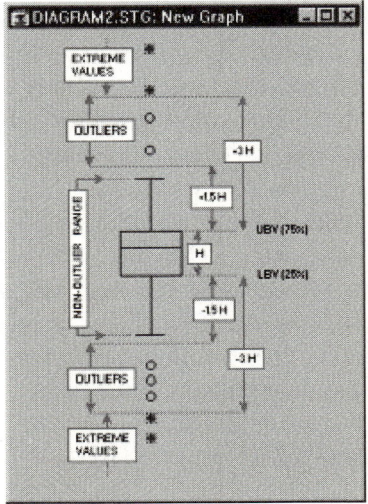

For example, the previous diagram illustrates the ranges of outliers and extremes in the classic box and whisker plot (for more information about box plots, see Tukey, 1977).

Overdispersion

The term overdispersion refers to the condition in which the variance of an observed dependent (response) variable exceeds the nominal variance, given the respective assumed distribution. This condition occurs frequently when fitting generalized linear models to categorical response variables, and the assumed distribution is binomial, multinomial, ordinal multinomial, or Poisson. When overdispersion occurs, the standard errors of the parameter estimates and related statistics (e.g., standard errors of predicted and residual statistics) must be computed taking into account the overdispersion. For details, see Agresti (1996); see also, Chapter 21 – *Generalized Linear/ Nonlinear Models*.

Overfitting

When attempting to fit a curve to a set of data points, producing a curve with high curvature that fits the data points well, but does not model the underlying function well, its shape being distorted by the noise inherent in the data. See also, *Neural Networks*, page 672.

Overlearning (in Neural Networks)

When an iterative training algorithm is run, overfitting which occurs when the algorithm is run for too long (and the network is too complex for the problem or the available quantity of data). See also, *Neural Networks*, page 672.

Overparameterized Model

An overparameterized model uses the indicator variable approach to represent effects for categorical predictor variables in general linear models and generalized linear models. To illustrate indicator variable coding, suppose that a categorical predictor variable called *Gender* has two levels (i.e., *Male* and *Female*). A separate continuous predictor variable would be coded for each group identified by the categorical predictor variable. Females might be assigned a value of 1 and males a value of 0 on a first predictor variable identifying membership in the female *Gender* group, and males would then be assigned a value of 1 and females a value of 0 on a second predictor variable identifying membership in the male *Gender* group.

Note that this method of coding for categorical predictor variables will almost always lead to design matrices with redundant columns in general linear models and generalized linear models, and thus requires a generalized inverse for solving the normal equations. As such, this method is often called the overparameterized model for representing categorical predictor variables, because it results in more columns in the design matrix than are necessary for determining the relationships of the categorical predictor variables to responses on the dependent variables. See also, the definitions of *Categorical Predictor Variable* and *Design Matrix*, or Chapter 18 – *General Linear Models (GLM)*.

P

p-Level

See *Statistical Significance (p-Level)*, page 733.

Pairwise Deletion of Missing Data vs. Mean Substitution

To avoid losing data due to casewise deletion of missing data, you can use one of two methods: 1) the so-called mean substitution of missing data (replacing all missing data in a variable by the mean of that variable) or 2) pairwise deletion of missing data. You can also use the mean substitution method to permanently remove missing data from a data set. Mean substitution offers some advantages and some disadvantages compared to pairwise deletion. Its main advantage is that it produces internally consistent sets of results (true correlation matrices). The main disadvantages are: a) mean substitution artificially decreases the variation of scores, and this decrease in individual variables is proportional to the number of missing data (i.e., the more missing data, the more perfectly average scores will be artificially added to the data set) and b) because it substitutes missing data with artificially created average data points, mean substitution may considerably change the values of correlations.

Pairwise MD Deletion

When pairwise deletion of missing data is used, cases are excluded from any calculations involving variables for which they have missing data. In the case of correlations, the correlations between each pair of variables are calculated from all cases having valid data for those two variables. See also, *Casewise vs. Pairwise Deletion of Missing Data* in Chapter 2 – *Basic*

Statistics and Tables, and *Pairwise Deletion of Missing Data vs. Mean Substitution*.

Parametric Curve

Parametric equations can be used to represent curves whose graphs are not simple functions of the type y = f(x), where *y* and *x* are represented along the vertical and horizontal axes, respecttively. Instead, the curves in the x-y plane are defined parametrically as two simultaneous functions of a parameter t that ranges over some interval (minimum, maximum). You can specify an equation y = f(t) for the y-component of the curve, and an equation x = g(t) for the *x*-component of the curve, for a specified range of parameter *t*. For example, to plot a spiral, you could specify:

y(t) = t*cos(t)   x(t) = t*sin(t)
For 0 < = t < = 12

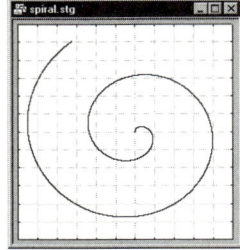

There are a wide variety of curves, from simple circles to complex shapes, that can be produced via the parametric curves facilities. Here is another example:

y(t) = (a + b)*sin(t) − b*sin((a/b + 1)*t)
x(t) = (a + b)*cos(t) − b*cos((a/b + 1)* t)

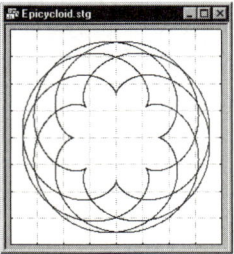

In the previous plot, *a* and *b* were set to *8* and *5*, respectively, and parameter *t* was plotted over the range from *0* to *100*. If you change *a* and *b* in the parametric equations shown above, very different shapes will emerge.

Pareto Distribution

The standard Pareto distribution has density function (for positive parameter c):

$$f(x) = c/x^{c+1} \qquad 1 \leq x, c > 0$$

where *c* is the shape parameter of the distribution.

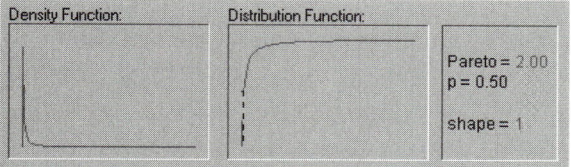

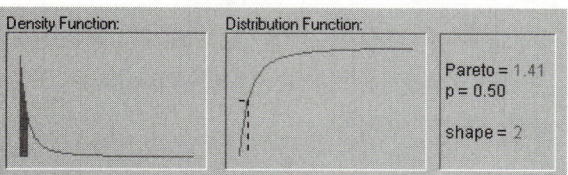

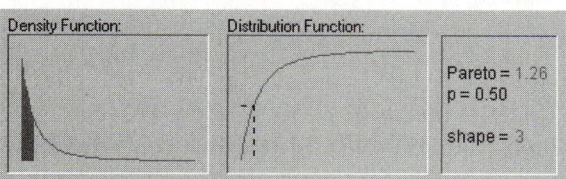

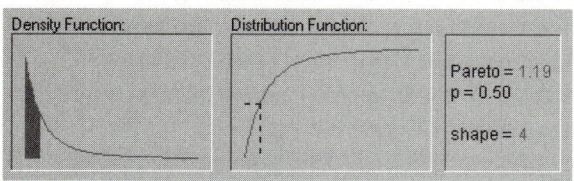

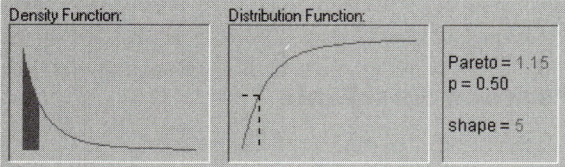

The previous illustrations show the *Pareto* distribution for the shape parameter equal to 1, 2, 3, 4, and 5.

Partial Correlation

A correlation between two variables that remains after controlling for (e.g., partialling out) one or more other variables. For example, *Hair Length* may correlate with *Height* (with taller individuals having shorter hair), however, that correlation will likely become smaller or even disappear if the influence of *Gender* is removed, since women are generally shorter and are more likely to have long hair than men. See also the definitions of *Correlation, Spurious Correlations*, and *Multiple Regression*, as well as Chapter 2 – *Basic Statistics and Tables* and Chapter 35 – *Structural Equation Modeling*.

Partial Least Squares Regression (PLS)

Partial least squares is a linear regression method that forms components (factors, or latent variables) as new independent variables (explanatory variables, or predictors) in a regression model. The components in partial least squares are determined by both the response variable(s) and the predictor variables. A regression model from partial least squares can be expected to have a smaller number of components without an appreciably smaller R-square value. For an overview of partial least squares, see Chapter 30 – *Partial Least Squares (PLS)*.

Partial Residuals

In a (linear, nonlinear, or additive) regression model with m predictors, the partial residuals for a predictor k are computed by removing from the dependent variable values the effects of all predictors i = 1, ... , m; i≠k. Scatterplots of partial residuals against predictor variables are particularly useful in generalized additive models, where they may aid in the interpretation of the (nonlinear) unique effect of the predictors in the overall model.

Parzen Window

In time series, the Parzen window is a weighted moving average transformation used to smooth the periodogram values. In the Parzen window (Parzen, 1961), for each frequency, the weights for the weighted moving average of the periodogram values are computed as:

$$w_j = 1-6*(j/p)^2 + 6*(j/p)^3 \quad \text{(for } j = 0 \text{ to } p/2)$$
$$w_j = 2*(1-j/p)^3 \quad \text{(for } j = p/2 + 1 \text{ to } p)$$
$$w_{-j} = w_j \quad \text{(for } j \neq 0)$$

where $p = (m-1)/2$

This weight function will assign the greatest weight to the observation being smoothed in the center of the window, and increasingly smaller weights to values that are further away from the center. See also, *Basic Notations and Principles*.

In neural networks, the Parzen window is an alternative name for kernel-based approximation techniques, as used in probabilistic neural networks and generalized regression neural networks (Parzen, 1962).

Pearson Correlation

The most widely used type of correlation coefficient is Pearson r (Pearson, 1896), also called linear or product-moment correlation (the term correlation was first used by Galton, 1888). Using non-technical language, you can say that the correlation coefficient determines the extent to which values of two variables are proportional to each other. The value of the correlation (i.e., correlation coefficient) does not depend on the specific measurement units used; for example, the correlation between height and weight will be identical regardless of whether inches and pounds, or centimeters and kilograms are used as measurement units. Proportional means linearly related; that is, the correlation is high if it can be approximated by a straight line (sloped upward or downward). This line is called the regression line or least squares line, because it is determined such that the sum of the squared distances of all the data points from the line is the lowest possible. Pearson correlation assumes that the two variables are measured on at least interval scales. The Pearson product moment correlation coefficient is calculated as follows:

$$r_{12} = \frac{\sum (y_{i1} - \bar{y}_1)(y_{i2} - \bar{y}_2)}{\sqrt{\sum (y_{i1} - \bar{y}_1)^2 \sum (y_{i2} - \bar{y}_2)^2}}$$

See also, *Correlations* in Chapter 2 – *Basic Statistics and Tables*.

Pearson Curves

A system of distributions proposed by Karl Pearson (e.g., see Hahn and Shapiro, 1967, pages 220-224) consists of seven solutions (of 12 originally enumerated by Pearson) to a differential equation that approximate a wide range of distributions of different shapes. Gruska, Mirkhani, and Lamberson (1989) describe in detail how the different Pearson curves can be fit to an empirical distribution. A method for computing specific Pearson percentiles is also described in Davis and Stephens (1983). See also, *Johnson Curves*.

Pearson Residuals

After fitting a generalized linear model to the data, to check the adequacy of the respective model, you usually compute various residual

statistics. The Pearson residual is computed as the raw residual $(y$-$m)$, scaled by the estimated standard deviation of y.

For a detailed discussion of residual statistics for generalized linear models, see McCullagh and Nelder (1989); see also, Chapter 21 – *Generalized Linear/Nonlinear Models*.

Penalty Functions

A constraint specified in a loss function that applies a penalty (a very large value) to the loss function when certain undesirable conditions are met. Using a penalty function enables you to control what permissible values of the parameters to be estimated that can be manipulated by the nonlinear estimation program. For more information, see *Penalty Functions, Constraining Parameters* in Chapter 28 – *Nonlinear Estimation*.

Percentiles

The percentile (this term was first used by Galton, 1885a) of a distribution of values is a number x_p such that a percentage p of the population values are less than or equal to x_p. For example, the 25th percentile (also referred to as the .25 quantile or lower quartile) of a variable is a value (x_p) such that 25% (p) of the values of the variable fall below that value.

Similarly, the 75th percentile (also referred to as the .75 quantile or upper quartile) is a value such that 75% of the values of the variable fall below that value and is calculated accordingly.

Perceptrons (in Neural Networks)

Perceptrons are a simple form of neural networks. They have no hidden layers, and can only perform linear classification tasks. Perceptrons were devised by Rosenblatt (1958), and their limitations were criticized by Minsky

and Papert (1969), leading to a loss of interest in the field. Fausett (1994) gives a good history of these early developments. A perceptron is modeled by creating a two-layer MLP network, and changing the activation function of the output layer to *Step*. The perceptron learning algorithm is modeled by using back propagation with *Momentum* 0.0 and *Shuffle* turned *Off*.

Pie Chart, 2D

Pie charts (the term first used by Haskell, 1922) are useful for representing proportions. Individual data values of the X variable are represented as the wedges of the pie.

Pie Chart, 2D – Counts

Unlike the values pie chart, this type of pie (this term was first used by Haskell, 1922) chart (sometimes called frequency pie chart) interprets data like a histogram.

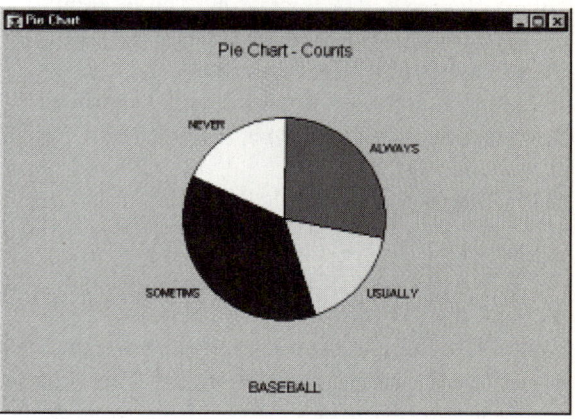

It categorizes all values of the specified variable and then displays the relative frequencies as pie slices of proportional sizes.

Pie Chart – Multi-Pattern Bar

The multi-pattern bar plot is designed to display the same type of data as the values pie chart (see *Pie Chart – Values* or *2D Histogram*),

however, the consecutive values are represented by the height of vertical bars (of different colors and patterns) and not areas of pie slices.

Their advantage over pie charts is that they may allow for more precise comparisons between presented values (e.g., small pie slices may be difficult to compare if they are not adjacent).

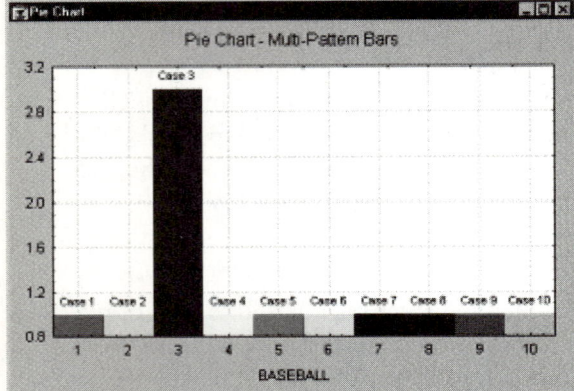

This type of graph may also have advantages over regular histograms (where one fill pattern and color is used for all columns), in cases when quick identification of specific columns is desired. See also, *Pie Chart*, page 685.

Pie Chart – Values

The sequence of values from the selected variable will be represented by consecutive slices of the pie (this term was first used by Haskell, 1922); the size of each slice will be proportional to the respective value. The values should be greater than *0* (*0*'s and negative values cannot be represented as slices of the pie). This simple type of pie chart (sometimes called data pie chart) interprets data in the most straightforward manner: one case = one slice.

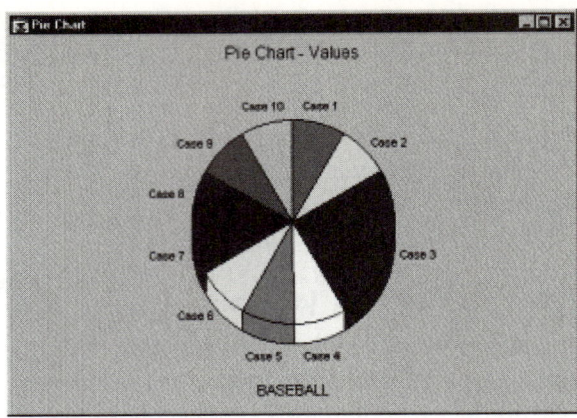

See also, *Pie Chart*, page 685.

PMML (Predictive Model Markup Language)

See *Predictive Markup Language (PMML)*, page 689.

PNG (Portable Network Graphics)

PNG (Portable Network Graphics) is a graphics file format designed to store bitmap (i.e., raster, as opposed to vector/metafile) graphic images. It was introduced to be a replacement for the patented GIF format (mostly to avoid the legal restrictions associated with the patent). A PNG image may contain a variable number of colors, including a transparent color. The size of the file depends on the number of colors used in the specific image. A compression method is used (to reduce the file size) which is highly effective if a large proportion of the image is built of the pixels with the same color attributes (e.g., it is highly effective for charts and schematic line art, but somewhat less effective than the JPEG format for photographs).

Poisson Distribution

The Poisson distribution (the term first used by Soper, 1914) is defined as:

$$f(x) = (\lambda^x * e^{-\lambda})/x!$$
for $x = 0, 1, 2, ..,\quad 0 < \lambda$

where λ (lambda) is the expected value of x (the mean), and e is the base of the natural logarithm, sometimes called Euler's e (2.71...).

Polar Coordinates

Polar coordinates (r,θ) represent the location of a point (in 2D space) by its distance (r) from a fixed point on a fixed line (polar axis) and the angle $(\theta$, in radians) from that fixed line.

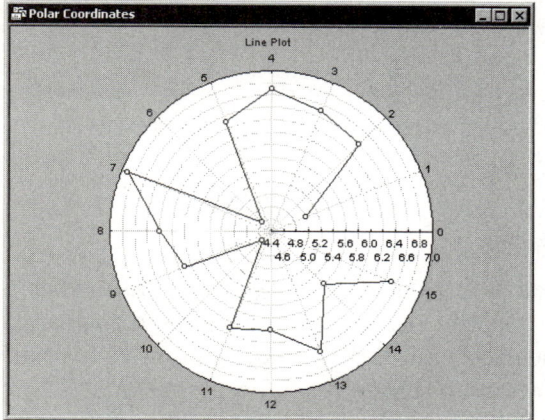

Polar plots are used to visualize functions. They also offer an intuitive way to present relations involving a variable representing direction. See also, *Cartesian Coordinates*.

Polynomial

This fits to the data, a polynomial function of the form:

$$y = b_0 + b_1x + b_2x^2 + b_3x^3 + ... + b_nx^n$$

where n is the order of the polynomial.

Fitting centered polynomial models via multiple regression. The fitting of higher-order polynomials of an independent variable with a mean not equal to zero can create difficult numerical problems. Specifically, the polynomials will be highly correlated due to the mean of the primary independent variable. With large numbers (e.g., Julian dates), this problem is very serious, and if proper protections are not put in place, can cause wrong results. The solution is to center the independent variable (sometimes, this procedures is referred to as centered polynomials), i.e., to subtract the mean, and then to compute the polynomials. See, for example, the classic text by Neter, Wasserman, & Kutner (1985, Chapter 9), for a detailed discussion of this issue (and analyses with polynomial models in general).

Positive Correlation

The relationship between two variables is such that as one variable's values tend to increase, the other variable's values also tend to increase. This is represented by a positive correlation coefficient. See also, *Correlations* in Chapter 2 – *Basic Statistics and Tables*.

Post-hoc Comparisons

Usually, after obtaining a statistically significant F test from the ANOVA, you want to know which means contributed to the effect; that is, which groups are particularly different from each other. You could of course perform a series of simple t-tests to compare all possible pairs of means. However, such a procedure would capitalize on chance. The reported probability levels would actually overestimate the statistical significance of mean differences. For example, suppose you take 20 samples of 10 random numbers each, and compute 20 means. Then, take the group (sample) with the highest mean and compare it with that of the lowest mean. The t-test for independent samples will test whether or not those two means are significantly different from each other, provided they were the only two samples taken. Post-hoc comparison techniques on the other hand, specifically take into account the fact that more than two samples were taken. They are used as either hypothesis testing or exploratory methods. For more information, see Chapter 3 – *ANOVA/MANOVA*.

Post Synaptic Potential (PSP) Function

The function applied by a unit to its inputs, weights and thresholds to form the unit's input (or activation) level. The two major PSP functions are linear (weighted sum minus threshold) and radial (scaled squared distance of weight vector from input vector). See also, *Neural Networks*, page 672.

Power Goal

The minimum power to be achieved when searching for an acceptable sample size. An acceptable sample size must yield a power greater than or equal to this value. For more information, see Chapter 31 – *Power Analysis*; see also, *Statistical Power*, page 733.

Power (Statistical)

See *Statistical Power*, page 733.

P$_{pk}$, P$_p$, P$_r$

See *Process Performance Indices* and C_{pk}, C_p, C_r. See also, *Process (Machine) Capability Analysis* in Chapter 32 – *Process Analysis*.

Prediction Interval Ellipse

Often plotted on 2D scatterplots, this interval describes the area in which a single new observation can be expected to fall with a certain probability (alpha), given that the new observation comes from a bivariate normal distribution with the parameters (means,

standard deviations, covariance) as estimated from the observed points shown in the plot.

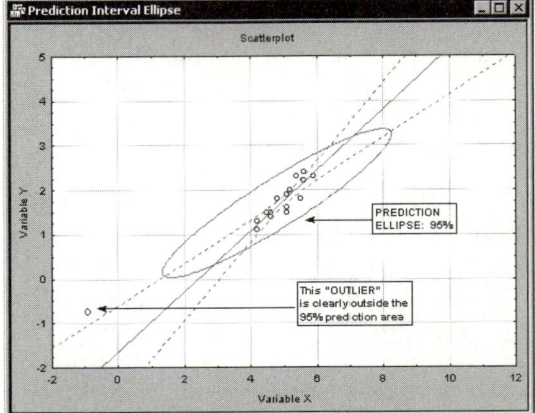

The coordinates for the ellipse are computed so that:

$$((n-p)*n)/(p*(n-1)*(n+1))*(X-X_m) \ 'S^{-1} (X-X_m) \sim F(alpha,p,n-p)$$

where n is the number of cases; p is the number of variables; i.e., p=2 in the case of the bivariate scatterplot; **X** is the vector of coordinates (pair of coordinates, since p=2); **X$_m$** is the vector of means for the p dimensions (variables) in the plot; S^{-1} is the inverse of the variance covariance matrix for the p variables; and *F(alpha,p, n-p)* is the value of F, given alpha, p, and n-p.

Note that if the number of observations in the scatterplot is small, the prediction interval may be very large. For additional information see, for example, Tracy, Young, and Mason (1992), or Montgomery 1996).

Prediction Profiles

When the results of an experiment are analyzed, the observed responses on the dependent variables can be fit to a separate prediction equation for each dependent variable (containing different coefficients but the same terms). Once these equations are constructed, predicted values for the dependent variables can be computed at any combination of levels of the predictor variables. A prediction profile for a dependent variable consists of a series of graphs, one for each independent variable, of the predicted values for the dependent variable at different levels of one independent variable, holding the levels of the other independent variables constant at specified values. Inspecting the prediction profiles for the dependent variables can show which levels of the predictor variables produce the most desirable predicted responses on the dependent variables.

For a detailed description of prediction profiles and desirability profiles, see *Profiling Predicted Responses and Response Desirability* in Chapter 15 – *Experimental Design (Industrial DOE)*.

Predictive Data Mining

The term predictive data mining is usually applied to identify data mining projects with the goal to identify a statistical or neural network model or set of models that can be used to predict some response of interest. For example, a credit card company may want to engage in predictive data mining, to derive a (trained) model or set of models (e.g., neural networks, meta-learner) that can quickly identify transactions which have a high probability of being fraudulent. Other types of data mining projects may be more exploratory in nature (e.g., to identify cluster or segments of customers), in which case drill-down descriptive and exploratory methods would be applied. Data reduction is another possible objective for data mining (e.g., to aggregate or amalgamate the information in very large data sets into useful and manageable chunks).

Predictive Mapping

One application of multiple correspondence analysis is to perform the equivalent of a multiple regression for categorical variables, by adding supplementary columns to a design matrix (see also, *Burt Table*). For example, suppose you had a design matrix containing various categorical indicators of health related behaviors (e.g., whether or not the individual smoked, exercised, etc.). You could add two columns to indicate whether or not the respective subject had or had not been ill over the past year (i.e., you could add one column *Ill* and another column *Not ill*, and enter *0*'s and *1*'s to indicate each subject's health status). If, in a simple correspondence analysis of the design matrix, you added those columns as supplementary columns to the analysis, then 1) the summary statistics for the quality of representation (see Chapter 11 – *Correspondence Analysis*) for those columns would give you an indication of how well you can explain illness as a function of the other variables in the design matrix, and 2) the display of the column points in the final coordinate system would provide an indication of the nature (e.g., direction) of the relationships between the columns in the design matrix and the column points indicating illness. This technique (adding supplementary points to a multiple correspondence analysis) is also called predictive mapping.

Predictive Model Markup Language (PMML)

PMML (Predictive Model Markup Language) is an XML-based language that allows for the efficient exchange of (trained) predictive models and shared models between different applications. A PMML document usually contains information describing fully trained or parameterized analytic models so that they can be readily deployed (applied to new cases) by another application.

Predictors

Predictors (also called independent or input variables) are variables used to predict or explain the value(s) of one or more dependent variables (also referred to as dependent or outcome variables).

PRESS Statistic

The PRESS statistic is often used in regression analyses, in order to summarize the fit of a particular model in a sample of observations that were not used to estimate the model parameters. It can simply be computed as the sums of squares of the prediction residuals for those observations. See Draper and Smith (1981); the PRESS statistic is, for example, computed in the partial least squares Model for the cross-validation (verification) samples.

Principal Components Analysis

A linear dimensionality reduction technique, which identifies orthogonal directions of maximum variance in the original data, and projects the data into a lower-dimensionality space formed of a sub-set of the highest-variance components (Bishop, 1995). See also, *Factor Analysis* and *Neural Networks*.

Prior Probabilities

Proportionate distribution of classes in the population (in a classification problem), especially where known to be different than the distribution in the training data set. Used to modify probabilistic neural network training in neural networks. See also, *Neural Networks* and *Discriminant Function Analysis*.

Probabilistic Neural Networks (PNN)

A type of neural network using kernel-based approximation to form an estimate of the probability density functions of classes in a classification problem. One of the so-called Bayesian networks (Speckt, 1990; Patterson, 1996; Bishop, 1995). See also, *Neural Networks*, page 672.

Probability Plot, Detrended

This type of graph is used to evaluate the normality of the distribution of a variable, that is, whether and to what extent the distribution of the variable follows the normal distribution. The selected variable will be plotted in a scatterplot against the values "expected from the normal distribution." This plot is constructed in the same way as the standard normal probability plot, except that before the plot is generated, the linear trend is removed. This often "spreads out" the plot, thereby enabling you to detect patterns of deviations more easily.

Probability Plot, Detrended (Categorized)

This categorized plot is constructed in the same way as the standard normal probability plot for the categorized values, except that before the plot is generated, the linear trend is removed. This often "spreads out" the plot, thereby enabling you to detect patterns of deviations more easily. For a detailed discussion of categorized graphs, see *Categorized Graphs*.

Probability Plot, Half-Normal

This type of graph is used to evaluate the normality of the distribution of a variable, that is, whether and to what extent the distribution of the variable follows the normal distribution. The

selected variable will be plotted in a scatterplot against the values "expected from the normal distribution." The half-normal probability plot is constructed in the same way as the standard normal probability plot, except that only the positive half of the normal curve is considered. Consequently, only positive normal values will be plotted on the Y-axis. This plot is used when you want to ignore the sign of the residual, that is, when you are interested mostly in the distribution of absolute residuals, regardless of the sign.

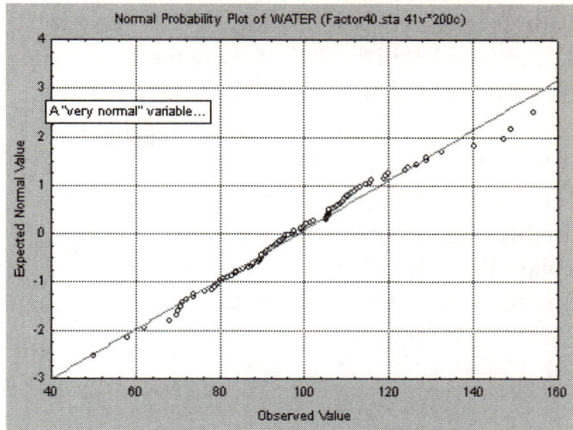

Probability Plot, Half-Normal (Categorized)

This type of graph is used to evaluate the normality of the distribution of a variable, that is, whether and to what extent the distribution of the variable follows the normal distribution. The selected variable will be plotted in a scatterplot against the values "expected from the normal distribution." The half-normal probability plot is constructed in the same way as the standard normal probability plot, except that only the positive half of the normal curve is considered. Consequently, only positive normal values will be plotted on the Y-axis. This plot is used when you want to ignore the sign of the residual, that is, when you are mostly interested in the distribution of absolute residuals, regardless of the sign.

Probability Plot, Normal

This type of graph is used to evaluate the normality of the distribution of a variable, that is, whether and to what extent the distribution of the variable follows the normal distribution. The selected variable will be plotted in a scatterplot against the values "expected from the normal distribution."

The standard normal probability plot is constructed as follows. First, the deviations from the mean (residuals) are rank ordered. From these ranks the program computes z values (i.e., standardized values of the normal distribution) based on the assumption that the data come from a normal distribution. These z values are plotted on the Y-axis in the plot. If the observed residuals (plotted on the X-axis) are normally distributed, all values should fall onto a straight line. If the residuals are not normally distributed, they will deviate from the line. Outliers may also become evident in this plot. If there is a general lack of fit, and the data seem to form a clear pattern (e.g., an S shape) around the line, the variable may have to be transformed in some way. See also, *Normal Probability Plots (Computation Note)*, page 692.

Probability Plot, Normal (Categorized)

This type of probability plot is constructed as follows. First, within each category, the values (observations) are rank ordered. From these ranks, you can compute z values (i.e., standardized values of the normal distribution) based on the assumption that the data come from a normal distribution [see *Normal Probability*

Plots (Computation Note), page 692]. These z values are plotted on the *Y*-axis in the plot. If the observed values (plotted on the *X*-axis) are normally distributed, all values should fall onto a straight line. If the values are not normally distributed, they will deviate from the line.

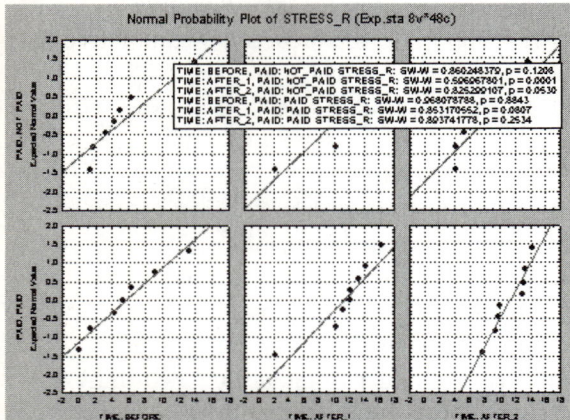

Outliers may also become evident in this plot. If there is a general lack of fit, and the data seem to form a clear pattern (e.g., an *S* shape) around the line, the variable may have to be transformed in some way (e.g., a log transformation to "pull-in" the tail of the distribution, etc.) before some statistical techniques that are affected by non-normality can be used. See also, *Categorized Graphs*.

Probability Plot – Normal (Computation Note)

The following formulas are used to convert the ranks into expected normal probability values, that is, the respective normal *z* values.

Normal probability plot. The normal probability value z_j for the *j*th value (rank) in a variable with *N* observations is computed as:

$$z_j = \Phi^{-1} [(3*j-1)/(3*N+1)]$$

where Φ^{-1} is the inverse normal cumulative distribution function (converting the normal probability *p* into the normal value *z*).

Half-normal probability plot. Here, the half-normal probability value z_j for the *j*th value (rank) in a variable with *N* observations is computed as:

$$z_j = \Phi^{-1} [3*N+3*j-1)/(6*N+1)]$$

where Φ^{-1} is again the inverse normal cumulative distribution function.

Detrended normal probability plot. In this plot each value (x_j) is standardized by subtracting the mean and dividing by the respective standard deviation (*s*). The detrended normal probability value z_j for the *j*th value (rank) in a variable with *N* observations is computed as:

$$z_j = \Phi^{-1} [(3*j-1)/(3*N+1)] - (x_j\text{-mean})/s$$

where Φ^{-1} is again the inverse normal cumulative distribution function.

Probability Plot, Normal (Multiple)

Unlike categorized normal probability plots, which produce a series of standard 2D probability plots (one for each category of cases identified by the X- and Y-axis category variables or identified by the multiple subset criteria), multiple normal probability plots enable you to evaluate the normality of the distributions of one or more variables or cases in one multiple probability plot (in order to allow for comparisons between blocks of columns or rows).

Probability-Probability Plot

In this graph, you can visually check for the fit of a theoretical distribution to the observed data visually by examining the probability-probability plot (P-P plot, also called probability plot.

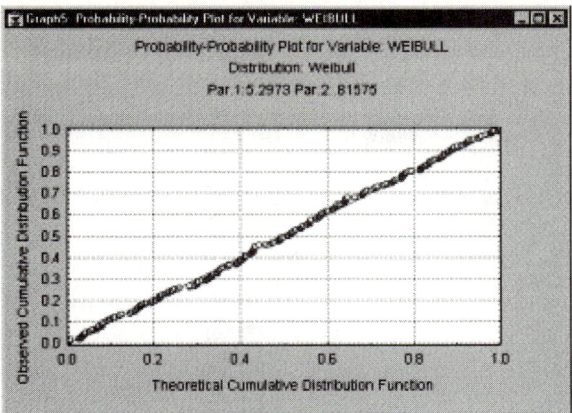

In P-P plots, the observed cumulative distribution function is plotted against the theoretical cumulative distribution function. As in the quantile-quantile plot, the values of the respective variable are first sorted into ascending order. The ith observation is plotted against one axis as i/n (i.e., the observed cumulative distribution function), and against the other axis as $F(x_{(i)})$, where $F(x_{(i)})$ stands for the value of the theoretical cumulative distribution function for the respective observation $x_{(i)}$. If the theoretical cumulative distribution approximates the observed distribution well, all points in this plot should fall onto the diagonal line.

Probability-Probability Plot (Categorized)

Visually check for the fit of a theoretical distribution to the observed data by examining each probability-probability plot (P-P plot, also called probability plot, see also, *Probability-Probability Plots*) for the respective level of the grouping variable (or user-defined subset of data). In P-P plots, the observed cumulative distribution function is plotted against the theoretical cumulative distribution function. As in the categorized Quantile-quantile plot, the values of the respective variable are first sorted into ascending order. The ith observation is plotted against one axis as i/n (i.e., the observed cumulative distribution function), and against the other axis as $F(x_{(i)})$, where $F(x_{(i)})$ stands for the value of the theoretical cumulative distribution function for the respective observation $x_{(i)}$. If the theoretical cumulative distribution approximates the observed distribution well, all points in this plot should fall onto the diagonal line. One component graph is produced for each level of the grouping variable (or user-defined subset of data) and all the component graphs are arranged in one display to allow for comparisons between the subsets of data (categories).

Probability Sampling

In probability sampling, every observation in the population from which the sample is drawn has a known probability of being selected into the sample; when that probability is the same for every observation in the population, the sample is an equal probability sample or EPSEM sample (equal probability of selection method; see Kish, 1965, for details).

EPSEM samples have certain desirable properties; for example, the simple formulas for computing means, standard deviations, and so on can be applied to estimate the respective parameters in the population.

Probit Regression and Transformation

In the probit regression model, the predicted values for the dependent variable will never be less than (or equal to) *0*, or greater than (or equal to) *1*, regardless of the values of the independent variables; it is, therefore, commonly used to analyze binary dependent or response variables (see also, *Binomial*

Distribution). This is accomplished by applying the following regression equation (the term *probit* was first used by Bliss, 1934):

$$y = NP(b_0 + b_1*X_1 \dots)$$

where *NP* stands for normal probability (space under the normal distribution; or cumulative distribution function of the normal distribution). You can easily recognize that, regardless of the regression coefficients or the magnitude of the *x* values, this model will always produce predicted values (predicted *y*'s) in the range of 0 to 1.

For additional details, see Chapter 21 – *Generalized Linear/Nonlinear Models* or Chapter 28 – *Nonlinear Estimation*; see also, *Logit Transformation and Regression* and *Multinomial Logit and Probit Regression* for similar transformations.

Process Analysis

In industrial settings, process analysis refers to a collection of analytic methods that can be used to ensure adherence of a product to quality specifications. These methods include Sampling Plans, Process (Machine) Capability Analysis, fitting measurements to Non-Normal Distributions, analysis of Gage Repeatability and Reproducibility and Weibull and Reliability/Failure Time Analysis. For more information, see Chapter 32 – *Process Analysis*.

Process Capability Indices

In industrial quality control, once a process is in control, indices are often computed to measure the quality of the items produced (and thus the capability of the process); specifically, the extent to which the items that are produced fall within allowable engineering tolerances. Given a sample of a particular size, you can estimate the standard deviation of the respective quality characteristic of interest (e.g., piston ring

diameters); you can then draw a histogram of the distribution of the characteristic of interest (piston ring diameters).

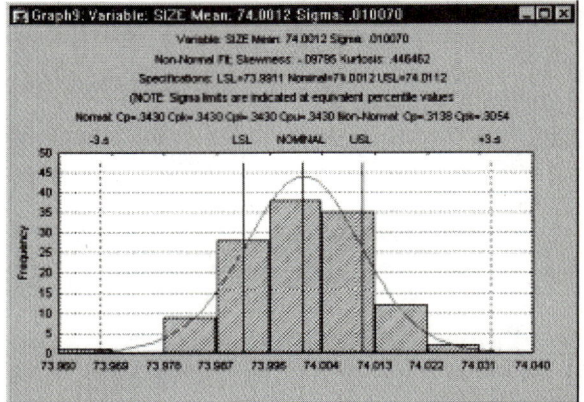

If the distribution of the quality characteristic is normal, you can make inferences concerning the proportion of items (piston rings) within specification limits (methods for non-normal distributions are also available; namely the so-called percentile method).

Common indices for process capability are C_p, C_r, and C_{pk}. All of these indices basically reflect the ratio of the engineering tolerances (process range) to the ± 3 *sigma* limits around the nominal specifications.

For computational details, see *Capability Analysis – Process Capability Indices* and *Process Performance vs. Process Capability* in Chapter 32 – *Process Analysis*.

Process Performance Indices

When monitoring a process via a quality control chart (e.g., the X-bar and R-chart; Quality Control) it is often useful to compute the process capability indices for the process. Specifically, when the data set consists of multiple samples such as data collected for the quality control chart, you can compute two different indices of variability in the data. One is the regular standard deviation for all observations, ignoring the fact that the data consist of multiple samples; the other is to estimate the process's inherent variation from the within-sample variability. For example, when plotting X-bar and R-charts, you can use the common estimator R-bar/d2 for the process sigma (e.g., see Duncan, 1974; Montgomery, 1985, 1991). Note however, that this estimator is only valid if the process is statistically stable. For a detailed discussion of the difference between the total process variation and the inherent variation refer to ASQC/AIAG reference manual (ASQC/AIAG, 1991, page 80).

When the total process variability is used in the standard capability computations, the resulting indices are usually referred to as process performance indices (as they describe the actual performance of the process; common indices are P_p, P_r, and P_{pk}), while indices computed from the inherent variation (within-sample sigma) are referred to as capability indices (since they describe the inherent capability of the process; common indices are C_p, C_r, and C_{pk}).

For additional information, see *Capability Analysis – Process Capability Indices*, and for computational details, see *Process (Machine) Capability Analysis* in Chapter 32 – *Process Analysis*.

Pruning (in Classification Trees)

Pruning a classification tree refers to the use of the automatic "right-sized" tree selection procedures developed by Breiman et. al. (1984). These procedures are designed to find, without relying on subjective judgment, the right-sized classification tree, that is, a classification tree with an appropriate number of splits and optimal predictive accuracy. The process of determining the right-sized classification tree is described in *Computational Methods* in Chapter 9 – *Classification Trees*.

Pseudo-Components

Pseudo-components are transformed values of the components (plotted in ternary graphs) where:

$$x'_i = (x_i - L_i)/(\text{Total} - L)$$

Here, x'_i stands for the ith pseudo-component, x_i stands for the original component value, L_i stands for the lower constraint (limit) for the ith component, L stands for the sum of all lower constraints (limits) for all components in the design, and *Total* stands for the mixture total. This transformation makes the coefficients for different factors comparable in size. (See Cornell, 1993, Chapter 3).

Pseudo-Inverse Algorithm

An algorithm to efficiently optimize a linear model; also known as singular value decomposition (see Bishop, 1995; Press et. al., 1992; Golub and Kahan, 1965).

Pseudo-Inverse – Singular Value Decomposition (in Neural Networks)

This algorithm uses the singular value decomposition technique to calculate the pseudo-inverse of the matrix needed to set the weights in a linear (dot product synaptic function + identity activation function) output layer, so as to find the least mean squared solution. Essentially, it guarantees to find the optimal setting for the weights in a linear layer, to minimize the RMS training set error (Bishop, 1995; Press et. al., 1992; Golub and Kahan, 1965). This is the standard least-squares optimization technique.

Linear techniques are extremely important in optimization, not least because it is possible to find an optimal solution to a linear model, something that is not guaranteed with nonlinear models, such as other types of neural networks, even if training algorithms converge.

The pseudo-inverse procedure, in addition to guaranteeing to find the absolute minimum error, is also relatively quick.

Pseudo-inverse is typically used in a number of circumstances:

- To optimize the linear output layer in a radial basis function network, subsequent to center and deviation assignment using the unsupervised algorithms.
- To optimize the output layer in a linear network.
- To fine-tune the final layer in a multilayer perceptron with a linear output layer, as used in regression problems.

Technical details. The matrix **G** is calculated, whose $i, j'th$ element is the input of the $i'th$ output unit, when the $j'th$ case is executed.

The least-squares solution is then given by:

$$w = G^+ d$$

$G^+ = (G^TG)–IG^T$ is the pseudo-inverse matrix; **w** is the weight vector into an output unit; **d** is the desired response vector (training outputs) for that output; G^+ is calculated using the singular value decomposition algorithm.

Caution. The singular value decomposition algorithm is usually numerically stable; however, occasionally a badly behaved matrix can cause it to generate mathematical errors. If this occurs, follow these steps:

Check that the training cases and (in the case of a radial basis function network) centers and deviations have been sensibly assigned.

In particular, the algorithm performs badly if radial deviations are very high (i.e., the standard

deviations of Gaussians are very small). It may be necessary to increase the number of neighbors if assigning radial deviation using K-nearest neighbors, or to increase the deviation multiplier if using Isotropic deviation assignment.

If training cases, centers and deviations are all sensible, and the algorithm still fails, use conjugate gradient descent to set the weights in the linear layer. Although typically slower than pseudo-inverse, this algorithm does not generate arithmetic errors and is guaranteed to find the minimum, as there are no local minima in this case.

Pure Error

For certain designs with replicates at the levels of the predictor variables, the residual sum of squares can be further partitioned into meaningful parts which are relevant for testing hypotheses. Specifically, the residual sums of squares can be partitioned into lack-of-fit and pure-error components. This involves determining the part of the residual sum of squares that can be predicted by including additional terms for the predictor variables in the model (for example, higher-order polynomial or interaction terms), and the part of the residual sum of squares that cannot be predicted by any additional terms (i.e., the sum of squares for pure error). A test of lack-of-fit can then be performed, using the mean square pure error as the error term. See also, *Lack of Fit* and *Design Matrix*; or see Chapter 15 – *Experimental Design (DOE)*, Chapter 18 – *General Linear Models*, or Chapter 19 – *General Regression Models*.

Quadratic

Fits a second-order polynomial function to the points in a 3D scatterplot.

Quality Control

In all production processes, the extent to which products meet quality specifications must be monitored. In the most general terms, there are two enemies of product quality: 1) deviations from target specifications, and 2) excessive variability around target specifications. During the earlier stages of developing the production process, designed experiments are often used to optimize these two quality characteristics (see the definition of *Experimental Design*); the methods discussed in Chapter 33 – *Quality Control Charts* are *on-line* or *in-process* quality control procedures to monitor an on-going production process.

The general approach to on-line quality control is straightforward: We simply extract samples of a certain size from the ongoing production process. We then produce line charts of the variability in those samples, and consider their closeness to target specifications. If a trend emerges in those lines, or if samples fall outside pre-specified limits, we declare the process to be out of control and take action to find the cause of the problem.

These types of charts are sometimes also referred to as Shewhart control charts (named after W. A. Shewhart who is generally credited as being the first to introduce these methods; see Shewhart, 1931). For more information, see Chapter 33 – *Quality Control Charts*.

Quality Control Charts, Sets of Samples In

While monitoring an ongoing process, it often becomes necessary to adjust the center line values or control limits, as those values are being refined over time. Also, you may want to compute the control limits and center line values from a *set of samples* that are known to be in control, and apply those values to all subsequent samples. Thus, each set is defined by a set of computation samples (from which various statistics are computed, e.g., *sigma*, means, etc.) and a set of application samples (to which the respective statistics, etc. are applied). Of course, the computation samples and application samples can be (and often are) not the same.

To reiterate, you may want to estimate *sigma* from a set of samples that are known to be in control (the computation set), and use that estimate for establishing control limits for all remaining and new samples (the application set).

Note that each sample must be uniquely assigned to one application set; in other words, each sample has control limits based on statistics (e.g., *sigma*) computed for one particular set. The assignment of application samples to sets proceeds in a hierarchical manner, i.e., each sample is assigned to the first set where it "fits" (where the definition of the application sample set would include the respective sample).

Quality (in Correspondence Analysis)

The term quality in the context of correspondence analysis pertains to the quality of representation of the respective row point in the coordinate system defined by the respective number of dimensions, as chosen by the user. The quality of a point is defined as the ratio of the squared distance of the point from the origin in the chosen number of dimensions, over the squared distance from the origin in the space defined by the maximum number of dimensions (remember that the metric in the typical correspondence analysis is *chi*-square). By analogy to factor analysis, the quality of a point is similar in its interpretation to the communality for a variable in factor analysis. A low quality means that the current number of dimensions does not represent well the respective row (or column).

Quantile

The quantile (this term was first used by Kendall, 1940) of a distribution of values is a number x_p such that a proportion p of the population values are less than or equal to x_p. For example, the .25 quantile (also referred to as the 25th percentile or lower quartile) of a variable is a value (x_p) such that 25% (p) of the values of the variable fall below that value.

Similarly, the .75 quantile (also referred to as the 75th percentile or upper quartile) is a value such that 75% of the values of the variable fall below that value and is calculated accordingly. See also, *Quantile-Quantile Plots*, below.

Quantile-Quantile Plot

You can visually check for the fit of a theoretical distribution to the observed data by examining the quantile-quantile (Q-Q) plot (also called quantile plot).

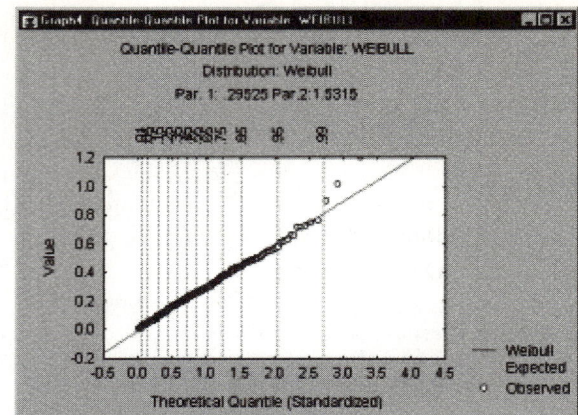

In this plot, the observed values of a variable are plotted against the theoretical quantiles. A good fit of the theoretical distribution to the observed values would be indicated by this plot if the plotted values fall onto a straight line. To produce a Q-Q plot, the program will first sort the n observed data points into ascending order, so that:

$$x_1 \leq x_2 \leq \cdots \leq x_n$$

These observed values are plotted against one axis of the graph; on the other axis the plot will show:

$$F^{-1}((i-r_{adj}) / (n+n_{adj}))$$

where i is the rank of the respective observation, r_{adj} and n_{adj} are adjustment factors (≤ 0.5), and F^{-1} denotes the inverse of the probability integral for the respective standardized distribution. The resulting plot is a scatterplot of the observed values against the (standardized) expected values, given the respective distribution. Note also that the adjustment factors r_{adj} and n_{adj} ensure that the p-value for the inverse probability integral will fall between 0 and 1, but not including 0 and 1 (see Chambers, Cleveland, Kleiner, and Tukey, 1983.

Quantile-Quantile Plot, 2D (Categorized)

In this graph, you can visually check for the fit of a theoretical distribution to the observed data by examining each quantile-quantile (or Q-Q) plot (also called quantile plot) for the respective level of the grouping variable (or user-defined subset of data).

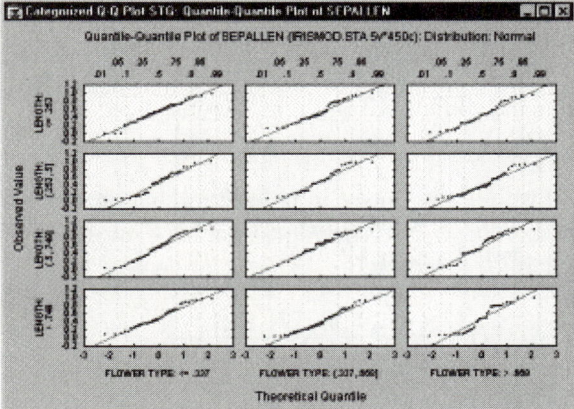

In this plot, the observed values of a variable are plotted against the theoretical quantiles. A good fit of the theoretical distribution to the observed values would be indicated by this plot if the plotted values fall onto a straight line. One component graph is produced for each level of the grouping variable(or user-defined subset of data) and all the component graphs are arranged in one display to allow for comparisons between the subsets of data (categories). (See *Quantile-Quantile Plots*, page 698, for more details on how to produce a Q-Q plot.)

Quartile

The lower and upper quartiles (this term was first used by Galton, 1882; also referred to as the .25 and .75 quantiles) are the 25th and 75th percentiles of the distribution (respectively). The 25th percentile of a variable is a value such that 25% of the values of the variable fall below that value.

Similarly, the 75th percentile is a value such that 75% of the values of the variable fall below that value and is calculated accordingly.

Quartile Range

The quartile (this term was first used by Galton, 1882) range of a variable is calculated as the value of the 75th percentile minus the value of the 25th percentile. Thus it is the width of the range about the median that includes 50% of the cases. For more information, see *Nonparametrics* , page 674.

Quasi-Newton Method (in Neural Networks)

Quasi-Newton (Bishop, 1995; Shepherd, 1997) is an advanced method of training multilayer perceptrons. It usually performs significantly better than back 0ropagation, and can be used wherever back propagation can be. It is the recommended technique for most networks with a small number of weights (less than a couple of hundred). If the network is a single output regression network and the problem has low residuals, Levenberg-Marquardt may perform better.

Quasi-Newton is a batch update algorithm: whereas back propagation adjusts the network weights after each case, Quasi-Newton works out the average gradient of the error surface across all cases before updating the weights once at the end of the epoch.

For this reason, there is no shuffle option available with Quasi-Newton, since it would clearly serve no useful function. There is also no need to select learning or momentum rates for Quasi-Newton, so it can be much easier to use than back propagation. Additive noise would destroy the assumptions made by Quasi-Newton about the shape of search space, and so is also not available.

Quasi-Newton works by exploiting the observation that, on a quadratic (i.e. parabolic) error surface, one can step directly to the minimum using the Newton step, a calculation involving the Hessian matrix (the matrix of second partial derivatives of the error surface). Any error surface is approximately quadratic "close to" a minimum. Since, unfortunately, the Hessian matrix is difficult and expensive to calculate, and anyway the Newton step is likely to be wrong on a non-quadratic surface, Quasi-Newton iteratively builds up an approximation to the inverse Hessian. The approximation at first follows the line of steepest descent, and later follows the estimated Hessian more closely.

Quasi-Newton is the most popular algorithm in nonlinear optimization, with a reputation for fast convergence. It does, however, have some drawbacks; it is rather less numerically stable than, say, conjugate gradient descent, it may be inclined to converge to local minima, and the memory requirements are proportional to the square of the number of weights in the network.

It is often beneficial to precede Quasi-Newton training with a short burst of Back Propagation (say 100 epochs), to cut down on problems with local minima.

If the network has many weights, you are advised to use Conjugate Gradient Descent instead. Conjugate Gradient Descent has memory requirements proportional only to the number of weights, not the square of the number of weights, and the training time is usually comparable with Quasi-Newton, if somewhat slower.

Technical details. Quasi-Newton is batch-based; it calculates the error gradient as the sum of the error gradients on each training case.

It maintains an approximation to the inverse Hessian matrix, called **H** below. The direction of steepest descent is called g below. The weight vector on the ith epoch is referred to as **fi** below. **H** is initialized to the identity matrix, so that the first step is in the direction g (i.e. the same direction as that chosen by Back Propagation). On each epoch, a back tracking line search is performed in the direction:

$$d = -Hg$$

Subsequently, the search direction is updated using the BFGS (Broyden-Fletcher-Goldfarb-Shanno) formula:

$$H_{i+1} = H_i + \frac{(x_{i+1} - x_i) \otimes ((x_{i+1} - x_i)}{(x_{i+1} - x_i) \bullet (\nabla f_{i+1} - \nabla f_i)} -$$
$$\frac{[H_i \bullet (\nabla f_{i+1} - \nabla f_i)] \otimes [H_i \bullet (\nabla f_{i+1} - \nabla f_i)]}{(\nabla f_{i+1} - \nabla f_i) \bullet H_i \bullet (\nabla f_{i+1} - \nabla f_i)}$$
$$+ [(\nabla f_{i+1} - \nabla f_i) \bullet H_i \bullet (\nabla f_{i+1} - \nabla f_i)]u \otimes u$$

This is guaranteed to maintain a positive-definite approximation (i.e. it will always indicate a descent direction), and to converge to the true inverse Hessian in W steps, where W is the number of weights, on a quadratic error surface. In practice, numerical errors may violate these theoretical guarantees and lead to divergence of weights or other modes of failure. In this case, run the algorithm again, or choose a different training algorithm.

Quasi-Newton Method (in Nonlinear Estimation)

The quasi-Newton method is a general optimization or search algorithm to find the minimum or maximum of some function for which the derivatives are unknown or too cumbersome to evaluate, and for which explicit solutions to compute minima and maxima do not exist. This algorithm is often used in nonlinear regression estimation. The quasi-Newton method is an efficient general algorithm that typically approximates the second-order derivatives of some (goal) function to guide the search for the minimum or maximum (e.g.., for the best parameter estimates, given the respective loss function).

QUEST

QUEST is a classification tree program developed by Loh and Shih (1997). For discussion of the differences of QUEST from other classification tree programs, see *A Brief Comparison of Classification Tree Programs* in Chapter 9 – *Classification Trees*.

Quick Propagation (in Neural Networks)

Despite the name, quick propagation (Fahlman, 1988; Patterson, 1996) is not necessarily faster than back propagation, although it may prove significantly faster for some applications.

Quick propagation also sometimes seems more inclined to instability and to getting stuck in local minima, than back propagation; these tendencies may determine whether quick propagation is more appropriate for a particular problem.

Quick propagation is a batch update algorithm: whereas back propagation adjusts the network weights after each case, quick propagation works out the average gradient of the error surface across all cases before updating the weights once at the end of the epoch.

For this reason, there is no shuffle option available with quick propagation, since it would clearly serve no useful function.

Quick propagation works by making the (typically ill-founded) assumption that the error surface is locally quadratic, with the axes of the hyper-ellipsoid error surface aligned with the weights. If this is true, the minimum of the error surface can be found after only a couple of epochs. Of course, the assumption is not generally valid, but if it is even close to true, the algorithm can converge on the minimum very rapidly.

Based on this assumption, quick propagation works as follows:

On the first epoch, the weights are adjusted using the same rule as back propagation, based upon the local gradient and the learning rate.

On subsequent epochs, the quadratic assumption is used to attempt to move directly to the minimum.

The basic quick propagation formula suffers from a number of numerical problems. First, if the error surface is not concave, the algorithm can actually go the wrong way. If the gradient changes little or not at all, the change can be extremely large, or even infinite! Finally, if a zero gradient is encountered, a weight will stop changing permanently.

Technical details. Quick propagation is batch-based; it calculates the error gradient as the sum of the error gradients on each training case.

On the first epoch, quick propagation updates weights just like back propagation.

Subsequently, weight changes are calculated using the quick propagation formula:

$$\Delta w(t) = \frac{s(t)}{s(t-1) - s(t)} \Delta w(t-1)$$

This formula is numerically unstable if $s(t)$ is very close to, equal to, or greater than $s(t-1)$. Since $s(t)$ is discovered after a move along the direction of the gradient, such conditions can only occur if the slope becomes constant, or becomes steeper (i.e., it is not concave).

In these cases, the weight update formula is:

$$\Delta w(t) = a\Delta w(t\text{-}1)$$

a is the acceleration coefficient.

If the gradient becomes zero, the weight *delta* becomes zero, and by the above formulae remains zero permanently even if the gradient subsequently changes. A conventional approach to solve this problem is to add a small factor to the weight changes calculated above. However, this approach can cause numerical instability. See also, *Neural Networks*, page 672.

Quota Sampling

Quota sampling usually refers to the process whereby a researcher attempts to match in a sample the exact makeup of the population with regard to certain demographic characteristics deemed important (such as gender, age, race, income, etc.). For example, a researcher may strive to draw a sample from a population so that the sample consists of exactly 50% males and 50% females, certain percentages of persons from particular ethnic backgrounds, etc. The purpose of this practice usually is to achieve some kind of representative sample of the underlying population.

In general, only properly drawn probability samples such as EPSEM samples will guarantee that the population to which you want to generalize is properly represented. Refer to, for example, Kish (1965) for a detailed discussion of the advantages and characteristics of probability samples (see also, *Representative Sample*, page 710, *Stratified Random Sampling*, page 734, and *Probability Sampling*, page 693).

R

Radial Basis Functions

Radial basis functions is a type of neural network employing a hidden layer of radial units and an output layer of linear units, and characterized by reasonably fast training and reasonably compact networks. Introduced by Broomhead and Lowe (1988) and Moody and Darkin (1989), they are described in most good neural network textbooks (e.g., Bishop, 1995; Haykin, 1994). See *Neural Networks*, page 672.

Radial Sampling (in Neural Networks)

Radial sampling is a simple technique to assign centers to radial units in the first hidden layer of a network by randomly sampling training cases and copying those to the centers. This is a reasonable approach if the training data are distributed in a representative manner for the problem (Lowe, 1989). The number of training cases must at least equal the number of centers to be assigned.

Random Effects (in Mixed Model ANOVA)

The term random effects in the context of analysis of variance is used to denote factors in an ANOVA design with levels that were not deliberately arranged by the experimenter (those factors are called fixed effects), but which were sampled from a population of possible samples instead. For example, if you were interested in the effect that the quality of different schools has on academic proficiency, you could select a sample of schools to estimate the amount of variance in academic proficiency (component of variance) that is attributable to differences between schools.

A simple criterion for deciding whether an effect in an experiment is random or fixed is to ask how you would select (or arrange) the levels for the respective factor in a replication of the study. For example, if you want to replicate the study described in this example, you would choose (take a sample of) different schools from the population of schools. Thus, the factor *School* in this study would be a random factor.

In contrast, if you want to compare the academic performance of boys to girls in an experiment with a fixed factor *Gender*, you would always arrange two groups: *Boys* and *Girls*. Hence, in this case (and in this case only), the same levels of the factor *Gender* would be chosen when you want to replicate the study. See also, Chapter 3 – *ANOVA/MANOVA*, and Chapter 39 – *Variance Components and Mixed Model ANOVA/ANCOVA*.

Random Sub-Sampling in Data Mining

When mining huge data sets with many millions of observations, it is neither practical nor desirable to process all cases (although efficient incremental learning algorithms exist to perform predictive data mining using all observations in the data set). For example, by properly sampling only 100 observations (from millions of observations), you can compute a very reliable estimate of the mean.

One of the rules of statistical sampling that is often not intuitively understood by untrained "observers" is the fact that the reliability and validity of results depend, among many other things, on the size of a random sample, and not on the size of the population from which it is taken. In other words, the mean estimated from 100 randomly sampled observations is as accurate (i.e., falls within the same confidence limits) regardless of whether the sample was taken from 1000 cases or 100 billion cases.

Put another way, given a certain (reasonable) degree of accuracy required, there is absolutely no need to process and include all observations in the final computations (for estimating the mean, fitting models, etc.).

Range Plot, 2D

Range plots display ranges of values or error bars related to specific data points in the form of boxes or whiskers.

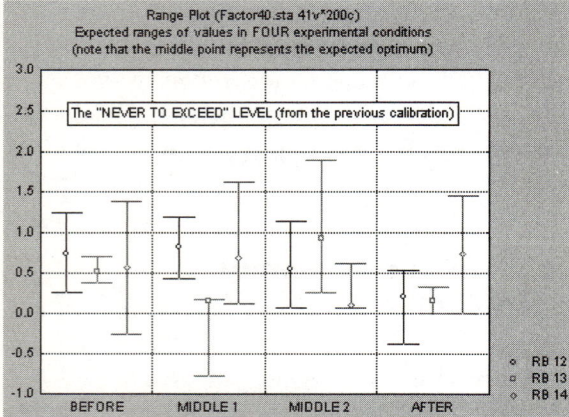

Unlike the standard box plots or means with error plots, the ranges or error bars are not calculated from data but defined by the raw values in the variables. One range or error bar is plotted for each case.

Range Plot, 2D – Error Bars

In this style of 2D range plot, the ranges or error bars are defined by the raw values in the selected variables. The midpoints are represented by point markers. One range or error bar is plotted for each case. In the simplest instance, three variables need to be selected, one representing the mid-points, one representing the upper limits, and one representing the lower limits.

Range Plot, 3D

3D range plots display ranges of values or error bars related to specific data points.

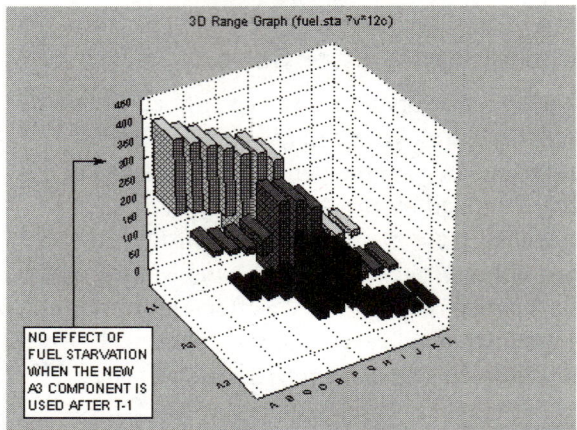

The ranges or error bars in this plot are not calculated from data (as they are in box plots) but defined by the raw values in the selected variables. One range or error bar is plotted for each case. Simple or multiple variables can be represented in the graph. Display options include simple points or error bar styles, border-style ranges, ribbons, boxes, or blocks.

Absolute vs. relative values. The range variables can be interpreted either as absolute values or values representing deviations from the midpoint.

Absolute values. When absolute values are used, the actual values of the middle point, minimum, maximum values will be plotted. For example, for a middle point value of 9, a minimum value of 1, and a maximum value of 12, the absolute range would start at 1 and end at 12, with the middle point at 9.

Relative to the Mid-Point. Use this type of variable interpretation to display the range relative to the middle point value. For the three points given in the above example, the range relative to the mid-point would start at 8 (the value that is 1 less than the mid-point of 9) and end at 21 (the value that is 12 more than the mid-point) with the mid-point at 9.

Range Plot, 3D – Error Bars

In this style of 3D sequential range plot, the error bars are not calculated from data but defined by the raw values in the selected variables. The midpoints are represented by point markers. One error bar is plotted for each case. The range variables can be interpreted either as absolute values or values representing deviations from the midpoint depending on the current setting of the *Mode* option in the graph definition dialog. Single or multiple variables can be represented in the graph.

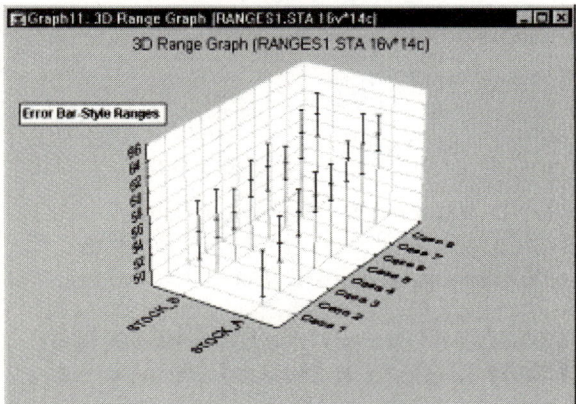

3D range plots differ from 3D box plots in that for range plots, the ranges are the values of the selected variables (e.g., one variable contains the minimum range values and another variable contains the maximum range values) while for box plots, the ranges are calculated from variable values (e.g., standard deviations, standard errors, or min-max values).

Range Plot – Boxes

In this style of range plot, the range is represented by a box (i.e., as a rectangular box

where the top of the box is the upper range and the bottom of the box is the lower range). The midpoints are represented either as point markers or horizontal lines that "cut" the box.

Range Plot – Columns

In this style of range plot, a column represents the mid-point (i.e., the top of the column is at the mid-point value) and the range (represented by whiskers) is overlaid in the column.

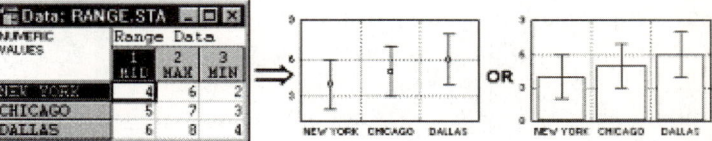

Range Plot – Whiskers

In this style of range plot (see example above), the range is represented by whiskers (i.e., as a line with a serif on both ends). The midpoints are represented by point markers.

Rank

A rank is a consecutive number assigned to a specific observation in a sample of observations sorted by their values, and thus reflecting the ordinal relation of the observation to others in the sample. Depending on the order of sorting (ascending or descending), the higher ranks represent the higher values [i.e., ascending ranks, the lowest value is assigned a rank of 1, and the highest value the "last" (highest) rank] or higher ranks represent the lower values (i.e., descending ranks, the highest value is assigned a rank of 1). See *Ordinal Scale*, page 678, and Coombs, 1950.

Rank Correlation

A rank correlation coefficient is a coefficient of correlation between two random variables that is based on the ranks of the measurements and not the actual values. For example, see *Spearman R*, *Kendall Tau*, *Gamma Coefficient*, and *Gamma Distribution*. Detailed discussions of rank correlations can be found in Hays (1981), Kendall (1948, 1975), Everitt (1977), and Siegel and Castellan (1988). See also, *Nonparametrics*, page 674.

Ratio Scale

This scale of measurement contains an absolute zero point, therefore it enables you to quantify and compare not only the sizes of differences between values, but also to interpret both values in terms of absolute measures of quantity or amount (e.g., time; 3 hours is not only 2 hours more than 1 hour, but it is also 3 times more than 1 hour). See also, *Measurement Scales* in Chapter 1 – *Elementary Concepts in Statistics*.

Raw Data Plot, 3D

With raw data plots, you can plot sequences of raw data from selected variables in a three-dimensional display.

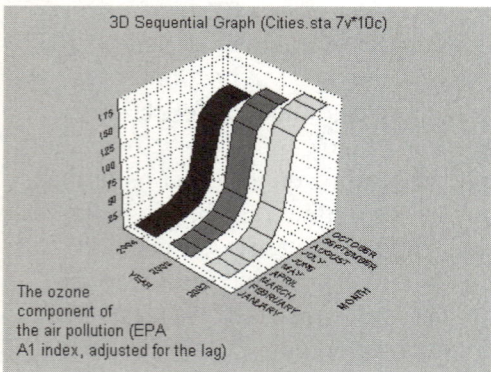

The selected variables are represented on the y-axis, the consecutive cases on the x-axis, and the values are plotted against the z-axis.

Rayleigh Distribution

The Rayleigh distribution has the probability density function:

$$f(x) = \frac{x}{b^2} e^{-x^2/(2b^2)} \quad , for\ 0 \le x < \infty,\ b > 0$$

where b is the scale parameter, e is the base of the natural logarithm, sometimes called Euler's e (2.71...). See also, *Process Analysis*, page 694.

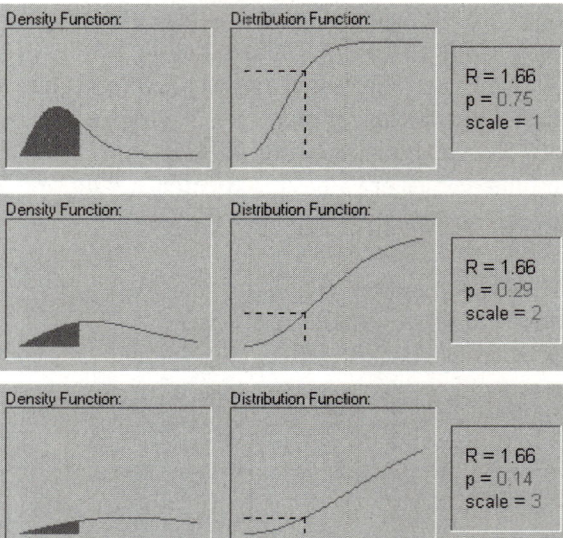

The illustrations above show the changing shape of the Rayleigh distribution when the scale parameter equals 1, 2, and 3.

Receiver Operating Characteristic (ROC) Curve (in Neural Networks)

When a neural network is used for classification, confidence levels (the Accept and Reject thresholds) determine how the neural networks assigns input cases to classes.

In the case of two-class classification problems, by default the output class is indicated by a single output neuron, with high output corresponding to one class and low output to the other. If the reject threshold is strictly less than the accept threshold, the network may include a doubt option, where it is not sure of the class if the output lies between the reject and accept thresholds.

An alternative approach is to set the accept and reject thresholds equal. In this case, as the single decision threshold is adjusted, so the classification behavior of the network changes. At one extreme all cases will be assigned to one class, and at the other extreme to the other. In between these extremes, different compromises may be found, leading to different trade-offs between the rate of erroneous assignment to each class (i.e. false-positives and false-negatives).

A Receiver Operating Characteristic (ROC) curve (Zweig, 1993) summarizes the performance of a two-class classifier across the range of possible thresholds. It plots the sensitivity (class two true positives) versus one minus the specificity (class one false negatives). An ideal classifier hugs the left side and top side of the graph, and the area under the curve is 1.0. A random classifier should achieve approximately 0.5 (a classifier with an area less than 0.5 can be improved simply by flipping the class assignment). The ROC curve is recommended for comparing classifiers, as it does not merely summarize performance at a single arbitrarily selected decision threshold, but across all possible decision thresholds.

The ROC curve can be used to select an optimum decision threshold. This threshold (which equalizes the probability of misclassification of either class; i.e. the probability of false-positives and false-negatives) can be used to automatically set confidence thresholds in classification networks with a nominal output variable with the two-state conversion function.

Regression

A category of problems where the objective is to estimate the value of a continuous output variable from some input variables. See also, *Multiple Regression*, page 666.

Regression (in Neural Networks)

In regression problems the purpose is to predict the value of a continuous output variable. Regression problems can done in neural networks using multilayer perceptrons, radial basis function networks, (Bayesian) regression networks, and linear networks.

Output scaling. Multilayer perceptrons include minimax scaling of both input and output variables. When the network is trained, shift and scale coefficients are determined for each variable, based on the minimum and maximum values in the training set, and the data is transformed by multiplying by the scale factor and adding the shift factor.

The net effect is that a 0.0 output activation level in the network is translated into the minimum value encountered in the training data, and a 1.0 activation level is translated into the maximum training data value. Consequently, the network is able to interpolate between the values represented in the training data. However, extrapolation outside the range encountered in the training set is more circumscribed. Two approaches to encoding the output are available, each of which allows a certain amount of extrapolation.

A logistic activation function is used for the output, with scaling factors determined so that the range encountered in the training set is mapped to a restricted part of the logistic functions (0,1) range (e.g. to [0.05, 0.95]. This allows a small amount of extrapolation (significant extrapolation from data is usually unjustified anyway). Using the logistic function makes training stable.

Uses an identity activation function in the final layer of the network. This supports a substantial amount of extrapolation, although not unlimited (the hidden units will saturate eventually). As a bonus, the final layer can be "fine-tuned" after iterative training using the pseudo-inverse technique. However, iterative training tends to be less stable than with a nonlinear activation function, and the learning rate must be carefully chosen to avoid weight divergence during training (i.e., less than 0.1), if using an algorithm such as back propagation.

Outliers. Regression networks can be particularly prone to problems with outlying data. The use of the sum-squared network error function means that points lying far from the others have a disproportionate influence on the position of the hyperplanes used in regression. If these points are actually anomalies (for example, spurious points generated by the failure of measuring devices) they can substantially degrade the network's performance.

One approach to this problem is to train the network, test it on the training cases, isolate those that have extremely high error values and remove them, and then to retrain the network.

If you believe the outlier is caused by a suspicious value for one of the variables in that case, you can delete that particular value, at which point the case is treated as having a missing value (see *Missing values*, below).

Another approach is to use the city-block error function. Rather than summing the squared-differences in each variable to work out an error measure, this simply sums the absolute differences. Removing the square function makes training far less sensitive to outliers.

Whereas the amount of "pull" a case has on a hyperplane is proportional to the distance of the

point from the hyperplane in the sum-squared error function, with the city block error function the pull is the same for all points, and the direction of pull simply depends on the side of the hyperplane to which the point lies. Effectively, the sum-squared error function attempts to find the mean, but the city-block error function attempts to find the median.

Missing values. It is not uncommon to come across situations where the data for some cases has some values missing; perhaps because data was unavailable, or corrupted, when gathered. In such cases, you may still need to execute a network (to get the best estimate possible given the information available) or (and this is more suspect) use the partially complete data in training because of an acute shortage of training data.

Where possible, it is usually good practice not to use variables containing a great many missing values. Cases with missing values can be excluded.

Regression Summary Statistics (in Neural Networks)

In regression problems, the purpose of the neural network is to learn a mapping from the input variables to a continuous output variable or variables.

A network is successful at regression if it makes predictions that are more accurate than a simple estimate. The simplest way to construct an estimate, given training data, is to calculate the mean of the training data, and use that mean as the predicted value for all previously unseen cases. The average expected error from this procedure is the standard deviation of the training data. The aim in using a regression network is therefore to produce an estimate that has a lower prediction error standard deviation than the training data standard deviation.

The regression statistics are:

Data mean. Average value of the target output variable.

Data S.D. Standard deviation of the target output variable.

Error mean. Average error (residual between target and actual output values) of the output variable.

Abs. E. mean. Average absolute error (difference between target and actual output values) of the output variable.

Error S.D. Standard deviation of errors for the output variable.

S.D. ratio. The error:data standard deviation ratio.

Correlation. The standard Pearson-R correlation coefficient between the predicted and observed output values.

The degree of predictive accuracy needed varies from application to application. However, generally an s.d. ratio of 0.1 or lower indicates very good regression performance.

Regular Histogram

This simple histogram is a column plot of the frequency distribution for the selected variable. See also, *Histogram, 2D*, page 627.

Regularization (in Neural Networks)

A modification to training algorithms which attempts to prevent over- or under-fitting of training data by building in a penalty factor for network complexity (typically by penalizing large weights, which correspond to networks modeling functions of high curvature) (Bishop, 1995). See also, *Neural Networks*, page 672.

Relative Function Change Criterion

The relative function change criterion is used to stop iteration when the function value is no longer changing (see Chapter 35 – *Structural Equation Modeling*). Basically, it stops iteration when the function ceases to change. The criterion is necessary because, sometimes, it is not possible to reduce the discrepancy function even when the gradient is not close to zero. This occurs, in particular, when one of the parameter estimates is at a boundary value. The true minimum, where the gradient actually is zero, includes parameter values that are not permitted (such as negative variances or correlations greater than one).

On the i'th iteration, this criterion is equal to:

$$FC_i = \left| \frac{F_i - F_{i-1}}{1 + |F_i|} \right|$$

Reliability

The reliability of a result (usually an empirical result such as a relation between variables found in a specific sample) pertains to the "representativeness" of the result found in a specific sample for the entire population. In other words, it says how probable it is that a similar result would be found if the data collection procedure (e.g., an experiment) was replicated with other samples drawn from the same population. We are rarely "ultimately" interested only in what is going on in one specific sample; we are interested in the sample only to the extent it can provide information about the population. If the study meets specific criteria (that allow us to apply the methods of statistical induction), then the reliability of a relation between variables observed in that sample can be quantitatively estimated and represented using a standard measure (technically called *p*-value or statistical significance level).

Reliability and Item Analysis

In many areas of research, the precise measurement of hypothesized processes or variables (theoretical constructs) poses a challenge by itself. For example, in psychology, the precise measurement of personality variables or attitudes is usually a necessary first step before any theories of personality or attitudes can be considered. In general, in all social sciences, unreliable measurements of people's beliefs or intentions will obviously hamper efforts to predict their behavior. The issue of precision of measurement will also come up in applied research, whenever variables are difficult to observe. For example, reliable measurement of employee performance is usually a difficult task; yet, it is obviously a necessary precursor to any performance-based compensation system.

In all of these cases, reliability & item analysis can be used to construct reliable measurement scales, to improve existing scales, and to evaluate the reliability of scales already in use. Specifically, reliability & item analysis will aid in the design and evaluation of sum scales, that is, scales that are made up of multiple individual measurements (e.g., different items, repeated measurements, different measurement devices, etc.). Reliability and item analysis provides numerous statistics that enable you to build and evaluate scales following the so-called classical testing theory model. For more information, see Chapter 34 – *Reliability/Item Analysis*.

The term reliability used in industrial statistics denotes a function describing the probability of failure (as a function of time). For a discussion of the concept of reliability as applied to

product quality (e.g., in industrial statistics), refer to *Weibull and Reliability/Failure Time Analysis* in Chapter 30 – *Process Analysis* (see also, *Gage Repeatability and Reproducibility* in the same chapter and Chapter 36 – *Survival/Failure Time Analysis*). For a comparison between these two (very different) concepts of reliability, see *Reliability*, page 709.

Reliability/Failure Time Analysis

In this context, reliability is defined as the function that describes the probability of failure (or death) of an item as a function of time. Thus, the reliability function (commonly denoted as $R(t)$) is the complement to the cumulative distribution function (i.e., $R(t)=1-F(t)$); the reliability function is also sometimes referred to as the survivorship or survival function (since it describes the probability of not failing or surviving until a certain time t; e.g., see Lee, 1992). For additional information, see *Weibull and Reliability/Failure Time Analysis* in Chapter 32 – *Process Analysis*.

Reliability of Scales

In this context, reliability is defined as the extent to which a measurement taken with a multiple-item scale (e.g., questionnaire) reflects mostly the so-called true score of the dimension that is to be measured, relative to the error. A similar notion of scale reliability is sometimes used when assessing the accuracy (and reliability) of gages or scales used in quality control charting. For additional details, refer to Chapter 34 – *Reliability/Item Analysis*, or *Gage Repeatability/Reproducibility Analysis* in Chapter 32 – *Process Analysis*.

Representative Sample

The notion of a representative sample is often misunderstood. The general intent usually is to

draw a sample from a population so that particular properties of that population can be estimated accurately from the sample. For example, political scientists may draw samples from the population of voters to predict with some certainty the outcome of an election.

In general, only properly drawn probability samples such as EPSEM samples will guarantee that the population to which you want to generalize is properly "represented." On the other hand, a generally erroneous notion is commonly expressed that, in order to achieve representativeness, it is desirable to draw a stratified sample using particular "quotas" (quota sampling) where demographic characteristics such as age, gender, race, and so on are properly "balanced," to match precisely the makeup of the underlying population. This notion is false: The precision of the estimates (such as voting margins) for a population computed from such a sample will only be enhanced, if the variables that you are attempting to match (age, gender, race...) are (strongly) related to the outcome variable of interest (e.g., voting behavior). However, in practice such *a priori* knowledge is usually elusive, and applying such quota sampling methods may yield grossly misleading results.

Refer to, for example, Kish (1965) for a detailed discussion of the advantages and characteristics of probability samples and EPSEM samples.

Resampling (in Neural Networks)

A major problem with neural networks is the generalization issue (the tendency to overfit the training data), accompanied by the difficulty in quantifying likely performance on new data.

This difficulty can be disturbing if you are accustomed to the relative security of linear modeling, where a given set of data generates a single optimal linear model. However, this security may be somewhat deceptive, and if the underlying function is not linear, the model may be very far from optimal.

In contrast, in nonlinear modeling some choice must be made about the complexity (curvature, eccentricity) of the model, and this can lead to a plethora of alternative models. Given this diversity, it is important to have ways to estimate the performance of the models on new data, and to be able to select among them.

Most work on assessing performance in neural modeling concentrates on approaches to resampling. A neural network is optimized using a training subset. Often, a separate subset (the selection subset) is used to halt training to mitigate over-learning, or to select from a number of models trained with different parameters. Then, a third subset (the test subset) is used to perform an unbiased estimation of the network's likely performance.

Although the use of a test set enables us to generate unbiased performance estimates, these estimates may exhibit high variance. Ideally, we want to repeat the training procedure a number of different times, each time using new training, selection, and test cases drawn from the population. Then, we could average the performance prediction over the different test subsets, to get a more reliable indicator of generalization performance.

In reality, we seldom have enough data to perform a number of training runs with entirely separate training, selection and test subsets. However, intuitively we might think we can do better if we train multiple networks, as when a single network is trained, only part of the data is actually involved in training. Can we find a way to use all the data in training, selection and test?

Cross validation is the simplest resampling technique. Suppose that we decide to conduct ten experiments with a given data set. We divide the data set into ten equal parts. Then, for each experiment we select one part to act as the test set. The other nine tenths of the data set are used for training and selection. When the ten experiments are finished, we can average the test set performances of the individual networks.

Cross validation has some obvious advantages. If training a single network, we would probably reserve 25% of the data for test. By using cross validation, we can reduce the individual test set size. In the most extreme version, leave-one-out cross validation, we perform a number of experiments equal to the size of the data set. On each experiment a single case is placed in the test subset, and the rest of the data is used for training. Clearly this may require a substantial number of experiments if the data set is large, but it can give you a very accurate estimate of generalization performance.

What precisely does cross validation tell us? In cross validation, each of the set of experiments should be performed with the same process parameters (same training algorithms, number of epochs, learning rates, etc.). The averaged performance measure is then an estimate of the performance on new data (drawn from the same distribution as the training data) of a single network trained using the same procedure (including the networks actually generated in the cross validation procedure).

We could select one of the cross validated networks at random and deploy it, using the estimates generated in cross validation to characterize its expected performance. However, this seems intuitively wasteful; having generated a number of networks, why not use them all? We can form the networks into an ensemble, and make predictions by averaging or voting across the resampled member networks (ensembles can also usefully combine the predictions of networks trained using different parameters, or of different architectures).

If we form an ensemble from the cross validated networks, is the performance estimate formed by averaging the test set performance of the individual networks an unbiased estimate of generalization performance? The answer is: No. The expected performance of an ensemble is not, in general, the same as the average performance of the members. Actually, the expected performance of the ensemble is at least the average performance of the members, but usually better. Thus you can use the estimate so-formed, knowing that it is conservatively biased.

Cross validation is one technique for resampling data. There are others:

Random (Monte Carlo) resampling. The subsets are randomly sampled from the available cases. Each available case is assigned to one of the three subsets.

Bootstrapping. This technique (Efron, 1979) samples a data set with replacement (i.e. a single case may be randomly sampled several times into the bootstrap set). The bootstrap can be applied any number of times, for increased accuracy. Compared with random sampling, the use of sampling with replacement can help to iron out generalization problems caused by the finite size of the data set. Breiman (1996) suggested using the bootstrap sampling technique to train multiple models for ensemble averaging (in his case the models were decision trees, but the conclusions carry over to other models), a technique he refers to as bagging.

Residual

Residuals are differences between the observed values and the corresponding values that are predicted by the model and thus they represent the variance that is not explained by the model. The better the fit of the model, the smaller the values of residuals. The ith residual (e_i) is equal to:

$$e_i = (y_i - y_i\text{-hat})$$

where y_i is the ith observed value, and y_i-hat is the corresponding predicted value.

See also, *Multiple Regression*, *Standard Residual Value*, *Mahalanobis Distance*, *Deleted Residual*, and *Cook's Distance*.

Resolution

An experimental design of resolution R is one in which no l-way interactions are confounded with any other interaction of order less than R − l. For example, in a design of resolution R equal to 5, no l = 2-way interactions are confounded with any other interaction of order less than R − l = 3, so main effects are unconfounded with each other, main effects are unconfounded with 2-way interactions, and 2-way interactions are unconfounded with each other. For discussions of the role of resolution in experimental design, see $2^{(k-p)}$ *Fractional Factorial Designs* and $2^{(k-p)}$ *Maximally Unconfounded and Minimum Aberration Designs* in Chapter 15 – *Experimental Design (Industrial DOE)*.

Response Surface

A surface plotted in three dimensions, indicating the response of one or more variable(s) (or a neural network) as two input variables are adjusted with the others held constant. See *Experimental Design (DOE)* and *Neural Networks*.

RMS (Root Mean Squared) Error

To calculate the RMS (root mean squared) error the individual errors are squared, added together, divided by the number of individual errors, and then square rooted. Gives a single number that summarizes the overall error. See *Neural Networks*, page 672.

ROC Curve

See *Receiver Operating Characteristic (ROC) Curve (in Neural Networks)*, page 706.

Root Cause Analysis

The term root cause analysis is commonly used in manufacturing to summarize the activities involved in determining the variables or factors that impact the final quality or yield of the respective processes. For example, if a particular pattern of defects emerges in the manufacture of silicon chips, engineers will pursue various methods and strategies for root cause analysis to determine the ultimate causes of those patterns of quality problems.

Root Mean Square Standardized Effect (RMSSE)

This standardized measure of effect size is used in the Analysis of Variance to characterize the overall level of population effects. It is the square root of the sum of squared standardized effects divided by the number of degrees of freedom for the effect. For example, in a 1-Way Anova, the RMSSE is calculated as

$$RMSSE = \sqrt{\frac{\sum_{j=1}^{n}\left(\frac{a_j}{\sigma}\right)^2}{j-1}} = \sqrt{\frac{S_\sigma^2}{\sigma^2}} = \frac{S_\sigma}{\sigma}$$

For more information, see Chapter 31 -*Power Analysis*.

Rosenbrock Pattern Search

This nonlinear estimation method will rotate the parameter space and align one axis with a ridge (this method is also called the method of rotating coordinates); all other axes will remain orthogonal to this axis. If the loss function is unimodal and has detectable ridges pointing toward the minimum of the function, this method will proceed with a considerable accuracy toward the minimum of the function.

Runs Tests (in Quality Control)

These tests are designed to detect patterns measurement (e.g., sample means) that may indicate that the process is out of control. In quality control charting, when a sample point (e.g., a mean in an X-bar chart) falls outside the control lines, you have reason to believe that the process may no longer be in control. In addition, you should look for systematic patterns of points (e.g., means) across samples, because such patterns may indicate that the process average has shifted. Most quality control software packages will (optionally) perform the standard set of tests for such patterns; these tests are also sometimes referred to as AT&T runs rules (see AT&T, 1959) or tests for special causes (e.g., see Nelson, 1984, 1985; Grant and Leavenworth, 1980; Shirland, 1993). The term special or assignable causes as opposed to chance or common causes was used by Shewhart to distinguish between a process that is in control, with variation due to random (chance) causes only, from a process that is out of control, with variation that is due to some non-chance or special (assignable) factors (cf. Montgomery, 1991, p. 102).

As the *sigma* control limits for quality control charts, the runs rules are based on "statistical" reasoning. For example, the probability of any sample mean in an X-bar control chart falling above the center line is equal to 0.5, provided 1) that the process is in control (i.e., that the center line value is equal to the population mean), 2) that consecutive sample means are independent (i.e., not auto-correlated), and 3) that the distribution of means follows the normal distribution. Simply stated, under those conditions there is a 50-50 chance that a mean will fall above or below the center line. Thus, the probability that two consecutive means will fall above the center line is equal to 0.5 times $0.5 = 0.25$. For additional information, see *Runs Tests*; see also, *Assignable Causes and Actions*.

S

Sampling Fraction

In probability sampling, the sampling fraction is the (known) probability with which cases in the population are selected into the sample, e.g., if you take a simple random sample with a sampling fraction of 1/10,000 from a population of 1,000,000 cases, each case would have a 1/10,000 probability of being selected into the sample, which will consist of approximately 1/10,000 * 1,000,000 = 100 observations.

Scalable Software Systems

Software (e.g., a data base management system such as MS SQL Server or Oracle) that can be expanded to meet future requirements without the need to restructure its operation (e.g., split data into smaller segments) to avoid a degradation of its performance. A scalable network enables the network administrator to add additional nodes without the need to redesign the system. An example of a non-scalable architecture is the DOS directory structure (adding files will eventually require splitting them into subdirectories). See also, *Enterprise-Wide Software Systems*.

Scaling

Altering original variable values (according to a specific function or an algorithm) into a range that meet particular criteria (e.g., positive numbers, fractions, numbers less than 10E12, numbers with a large relative variance).

Scatter Icon Plot, 2D

A scatter icon plot is a scatterplot whose point markers are icons that represent for each specific data point (defined by the scatterplot X and Y coordinates) the variability within a set of other variables chosen for the icon. Additionally, a weight variable can be selected and used to scale the entire icon representing the point. See the following illustration.

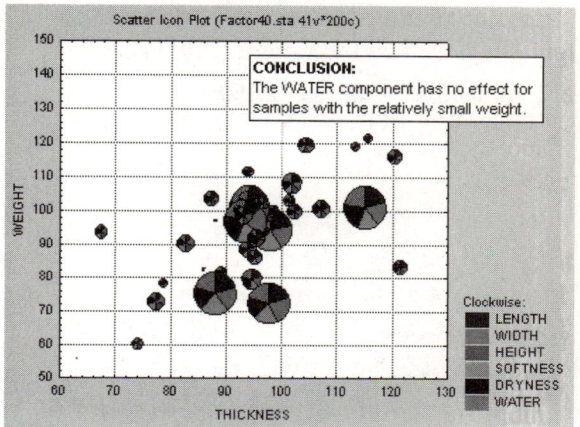

Scatterplot, 2D

The scatterplot visualizes a relation (correlation) between two variables X and Y (e.g., weight and height). Individual data points are represented in two-dimensional space (see the next illustration), where axes represent the variables (X on the horizontal axis and Y on the vertical axis).

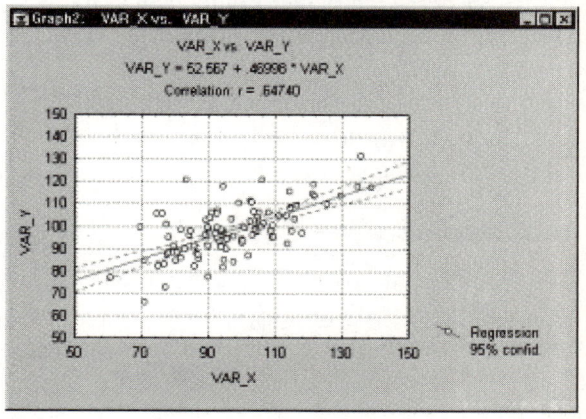

The two coordinates (X and Y) that determine the location of each point correspond to its specific values on the two variables.

Scatterplot, 2D – Bubble

In scatterplots with many overlapping data points, the relative frequencies of the number of points represented by a single plot position

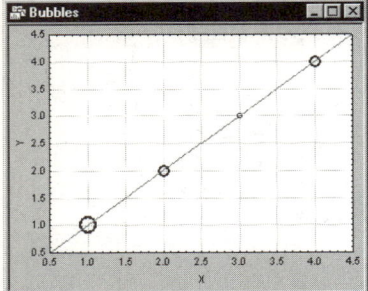

(e.g., {x, y} coordinate pair) can be indicated with "bubbles," i.e., circles of varying sizes.

Scatterplot, 2D - Categorized Ternary

The points representing the proportions of the component variables (X, Y, and Z) in a ternary graph are plotted in a 2-dimensional display for each level of the grouping variable (or user-defined subset of data). One component graph is produced for each level of the grouping variable (or user-defined subset of data) and all the component graphs are arranged in one display to allow for comparisons between the subsets of data (categories). See also, *Data Reduction*.

Scatterplot, 2D – Double-Y

This type of scatterplot can be considered to be a combination of two multiple scatterplots for one X-variable and two different sets of Y-variables.

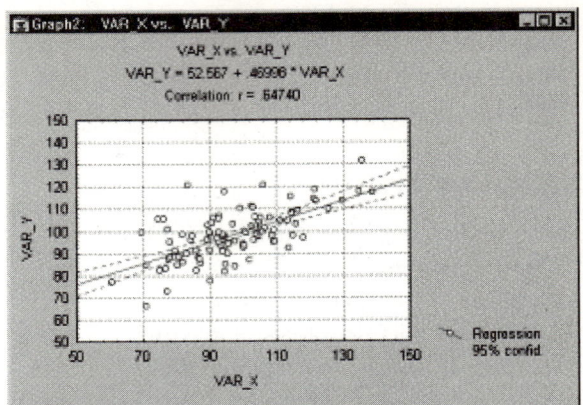

The double-Y scatterplot can be used to compare images of several correlations by overlaying them in a single graph. However, due to the independent scaling used for the two lists of variables, it can facilitate comparisons between variables with values in different ranges. See also, *Data Reduction*.

Scatterplot, 2D – Frequency

Frequency scatterplots display the frequencies of overlapping points between two variables in order to visually represent data point weight or other measurable characteristics of individual data points.

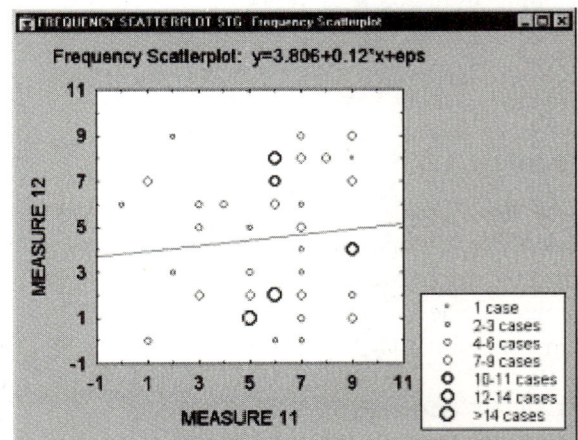

See also, *Data Reduction*.

Scatterplot, 2D – Multiple

Unlike the regular scatterplot in which one variable is represented by the horizontal axis and one by the vertical axis, the multiple scatterplot consists of multiple plots and represents multiple correlations. One variable (X) is represented by the horizontal axis, and several variables (Ys) are plotted against the vertical axis.

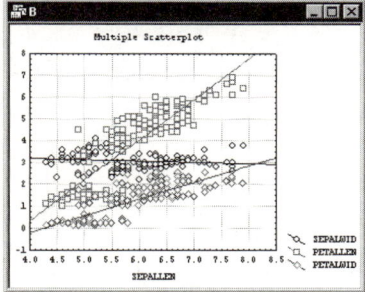

The multiple scatterplot is used to compare images of several correlations by overlaying them in a single graph that uses one common set of scales (e.g., to reveal the underlying structure of factors or dimensions in discriminant function analysis).

Scatterplot, 2D – Quantile

In quantile scatterplots, the quantiles of one variable are plotted against the quantiles of another variable in order to assess the similarity of the empirical distributions of the two variables. If the data points fall on the regression line, then it can be concluded that the two variables follow the same distribution.

Scatterplot, 2D – Regular

The regular scatterplot visualizes a relation between two variables X and Y (e.g., weight and height). Individual data points are represented by point markers in two-dimensional space, where axes represent the variables. The two coordinates

(X and Y), which determine the location of each point, correspond to its specific values on the two variables. If the two variables are strongly related, the data points form a systematic shape (e.g., a straight line or a clear curve). If the variables are not related, the points form an irregular "cloud" (see the categorized scatterplot in the following illustration for examples of both types of data sets).

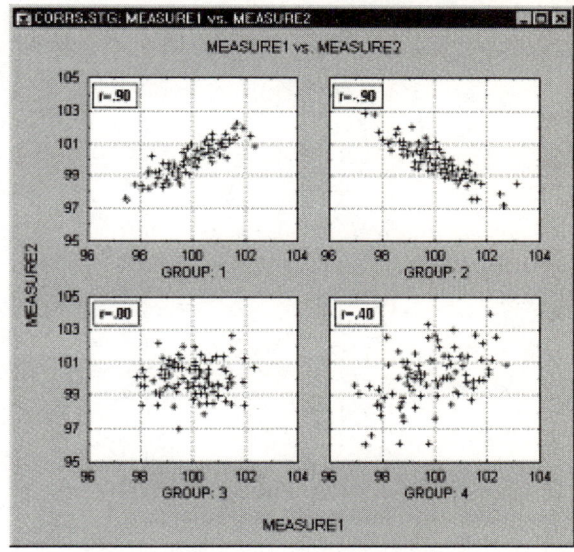

Fitting functions to scatterplot data helps identify the patterns of relations between variables (see the next illustration).

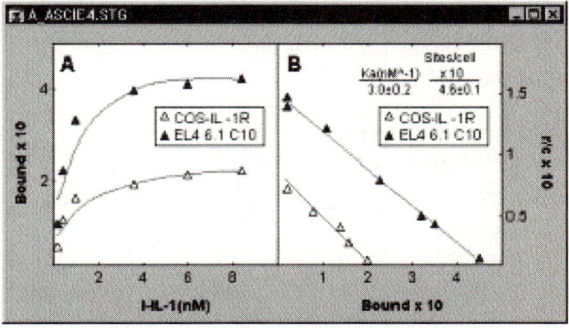

For more examples of how scatterplot data helps identify the patterns of relations between

variables, see *Outliers* and *Brushing*. See also, *Data Reduction*.

Scatterplot, 2D – Voronoi

This specialized univariate scatterplot is more an analytic technique than just a method to graphically present data. The solutions it offers help to model a variety of phenomena in natural and social sciences (e.g., Coombs, 1964; Ripley, 1981).

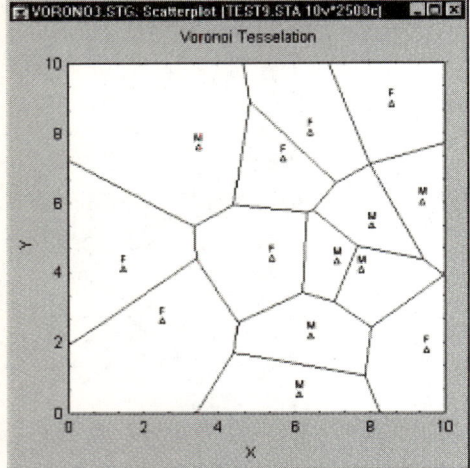

The algorithm divides the space between the individual data points represented by *XY* coordinates in 2D space. The division is such that each of the data points is surrounded by boundaries including only the area that is closer to its respective "center" data point than to any other data point.

The particular ways in which this method is used depends largely on specific research areas, however, in many of them, it is helpful to add additional dimensions to this plot by using categorization options.

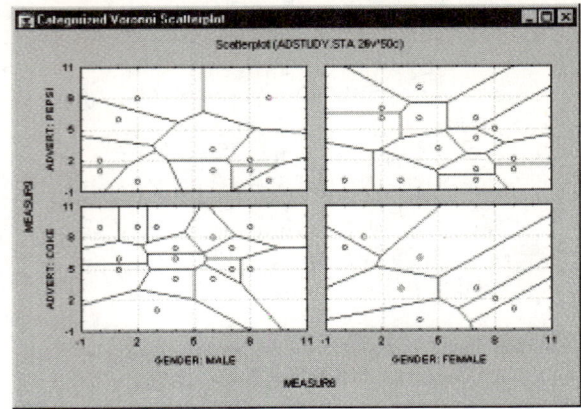

See also, *Data Reduction*.

Scatterplot, 3D

3D scatterplots visualize a relationship between three or more variables, representing the *X, Y,* and one or more *Z* (vertical) coordinates of each point in 3-dimensional space.

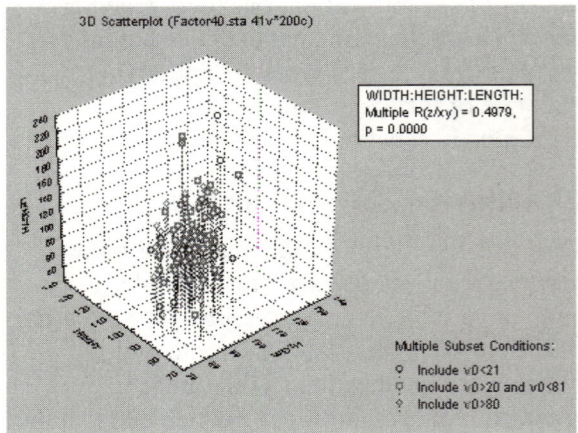

See also, *Ternary Plot, 3D – Scatterplot, Data Reduction*, and *Data Rotation (in 3D Space)*.

Scatterplot, 3D – Categorized

This type of graphs visualizes a relationship between three variables (representing the *X, Y,* and one or more *Z* [vertical] coordinates of each point in 3-dimensional space) categorized by a

grouping variable. One component graph is produced for each level of the grouping variable (or user-defined subset of data) and all the component graphs are arranged in one display to allow for comparisons between the subsets of data (categories) (see graph number 1 in the next illustration).

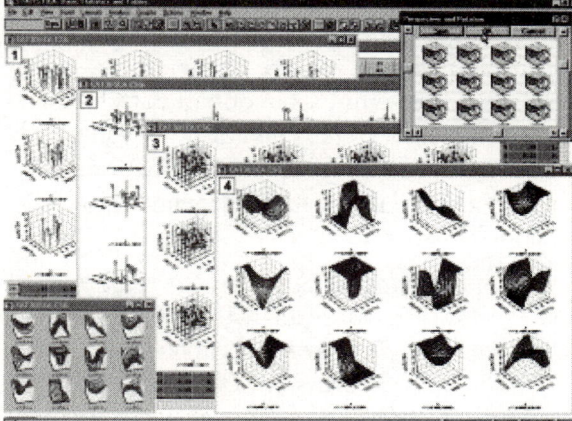

See also, the definition of *Categorized Graphs and Data Reduction*.

Scatterplot, 3D – Raw Data

An unsmoothed surface (no smoothing function is applied) is drawn through the points in the 3D scatterplot. See also, the definition of *Data Reduction*.

Scatterplot Smoothers

Various smoothing methods are used to fit a function through the points in a scatterplot to best represent (summarize) the relationship between the variables.

Scheffe's Test

This post-hoc test can be used to determine the significant differences between group means in an analysis of variance setting. Scheffe's test is considered to be one of the most conservative

post hoc tests (for a detailed discussion of different post hoc tests, see Winer, Michels, & Brown (1991). For more details, see Chapter 18 – *General Linear Models*. See also, *Post-hoc Comparisons*, page 687. For a discussion of statistical significance, see Chapter 1 – *Elementary Concepts in Statistics*.

Score Statistic

This statistic is used to evaluate the statistical significance of parameter estimates computed via maximum likelihood methods. It is also sometimes called the efficient score statistic. The test is based on the behavior of the log-likelihood function at the point where the respective parameter estimate is equal to *0.0* (zero); specifically, it uses the derivative (slope) of the log-likelihood function evaluated at the null hypothesis value of the parameter (parameter = *0.0*). While this test is not as accurate as explicit likelihood-ratio test statistics based on the ratio of the likelihoods of the model that includes the parameter of interest, over the likelihood of the model that does not, its computation is usually much faster. It is therefore the preferred method for evaluating the statistical significance of parameter estimates in stepwise or best-subset model building methods. An alternative statistic is the Wald statistic.

Scree Plot, Scree Test

The eigenvalues for successive factors can be displayed in a simple line plot. Cattell (1966) proposed that this scree plot can be used to graphically determine the optimal number of factors to retain.

The scree test involves finding the place where the smooth decrease of eigenvalues appears to level off to the right of the plot. To the right of this point, presumably, you find only factorial scree – scree is the geological term referring to the debris that collects on the lower part of a rocky slope. Thus, no more than the number of factors to the left of this point should be retained.

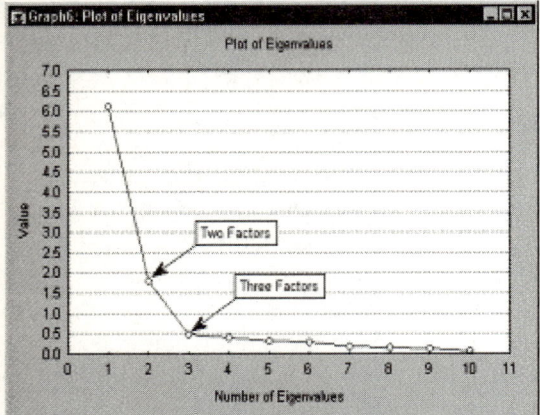

For more information on procedures for determining the optimal number of factors to retain, see *Reviewing the results of a principal components analysis* in Chapter 16 – *Factor Analysis* and *How Many Dimensions to Specify* in Chapter 25 – *Multidimensional Scaling*.

S.D. Ratio

In a regression problem, the ratio of the prediction error standard deviation to the original output data standard deviation. A lower S.D. ratio indicates a better prediction. This is equivalent to one minus the explained variance of the model. See *Multiple Regression*, page 666, and *Neural Networks*, page 672.

Semi-Partial (or Part) Correlation

The semi-partial or part correlation is similar to the partial correlation statistic. Like the partial correlation, it is a measure of the correlation between two variables that remains after controlling for (i.e., partialling out) the effects of one or more other predictor variables.

However, while the squared partial correlation between a predictor X_1 and a response variable Y can be interpreted as the proportion of (unique) variance accounted for by X_1, in the presence of other predictors $X_2, \ldots, X_k$, relative to the residual or unexplained variance that cannot be accounted for by $X_2, \ldots, X_k$, the squared semi-partial or part correlation is the proportion of (unique) variance accounted for by the predictor X_1, relative to the total variance of Y. Thus, the semi-partial or part correlation is a better indicator of the "practical relevance" of a predictor, because it is scaled to (i.e., relative to) the total variability in the dependent (response) variable.

See also, *Correlation*, *Spurious Correlations*, and *Partial Correlation*, as well as Chapter 2 – *Basic Statistics and Tables*; Chapter 26 – *Multiple Linear Regression*; Chapter 18 – *General Linear Models*; Chapter 19 – *General Regression Models*; and Chapter 35 – *Structural Equation Modeling*.

SEMMA

An acronym for Sample-Explore-Modify-Model-Assess, but denotes a framework for implementation of data miner techniques in a business environment. See *Models for Data Mining*, page 663. See also, Chapter 12 – *Data Mining Techniques*.

Sensitivity Analysis (in Neural Networks)

A sensitivity analysis indicates which input variables are considered most important by that particular neural network. Sensitivity analysis can be used purely for informative purposes, or to perform input pruning.

Sensitivity analysis can give important insights into the usefulness of individual variables. It

often identifies variables that can be safely ignored in subsequent analyses, and key variables that must always be retained. However, it must be deployed with some care, for reasons that are explained below.

Input variables are not, in general, independent; that is, there are interdependencies between variables. Sensitivity analysis rates variables according to the deterioration in modeling performance that occurs if that variable is no longer available to the model. In so doing, it assigns a single rating value to each variable. However, the interdependence between variables means that no scheme of single ratings per variable can ever reflect the subtlety of the true situation.

Consider, for example, the case in which two input variables encode the same information (they might even be copies of the same variable). A particular model might depend wholly on one, wholly on the other, or on some arbitrary combination of them. Then sensitivity analysis produces an arbitrary relative sensitivity to them. Moreover, if either is eliminated the model may compensate adequately because the other still provides the key information. It may therefore rate the variables as of low sensitivity, even though they might encode key information. Similarly, a variable that encodes relatively unimportant information, but is the only variable to do so, may have higher sensitivity than any number of variables that mutually encode more important information.

There may be interdependent variables that are useful only if included as a set. If the entire set is included in a model, they can be accorded significant sensitivity, but this does not reveal the interdependency. Worse, if only part of the interdependent set is included, their sensitivity will be zero, as they carry no discernable information.

In summary, sensitivity analysis does not rate the usefulness of variables in modeling in a reliable or absolute manner. You must be cautious in the conclusions you draw about the importance of variables. Nonetheless, in practice it is extremely useful. If a number of models are studied, it is often possible to identify key variables that are always of high sensitivity, others that are always of low sensitivity, and ambiguous variables that change ratings and probably carry mutually redundant information.

How does sensitivity analysis work? Each input variable is treated in turn as if it were "unavailable" (Hunter, 2000). There is a missing value substitution procedure, which is used to allow predictions to be made in the absence of values for one or more inputs. To define the sensitivity of a particular variable, v, we first run the network on a set of test cases, and accumulate the network error. We then run the network again using the same cases, but this time replacing the observed values of v with the value estimated by the missing value procedure, and again accumulate the network error.

Given that we have effectively removed some information that presumably the network uses (i.e. one of its input variables), we would reasonably expect some deterioration in error to occur. The basic measure of sensitivity is the ratio of the error with missing value substitution to the original error. The more sensitive the network is to a particular input, the greater the deterioration we can expect, and therefore the greater the ratio.

If the ratio is one or lower, making the variable unavailable either has no effect on the performance of the network, or actually enhances it.

Once sensitivities have been calculated for all variables, they can be ranked in order.

Sequential Contour Plot, 3D

This contour plot presents a 2-dimensional projection of the spline-smoothed surface fit to the data (see *Sequential Surface Plot, 3D*, page 723). Successive values of each series are plotted along the X-axis, with each successive series represented along the Y-axis.

Sequential/Stacked Plot, 2D

In this type of graph, the sequence of values from each variable is stacked on one another. See the next illustration, which shows a column version of this plot.

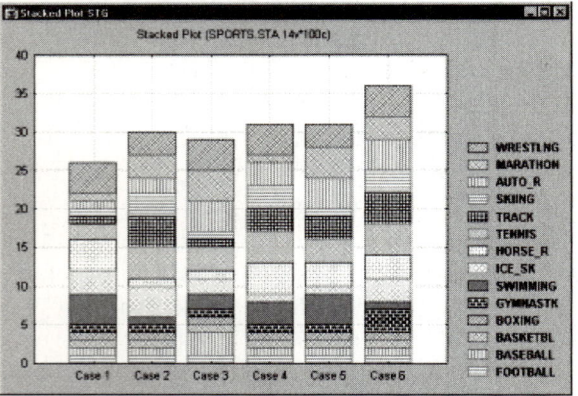

Sequential/Stacked Plot, 2D – Area

The sequence of values from each variable will be represented by consecutive areas stacked on one another in this type of graph.

Sequential/Stacked Plot, 2D – Column

The sequence of values from each variable will be represented by consecutive segments of

vertical columns stacked on one another in this type of graph.

Sequential/Stacked Plot, 2D – Lines

The sequence of values from each variable will be represented by consecutive lines stacked on one another in this type of graph.

Sequential/Stacked Plot, 2D – Mixed Lines

In this type of graph, the sequences of values of variables selected from one list will be represented by consecutive areas stacked on one another while the sequences of values of variables selected from a second list will be represented by consecutive *lines* stacked on one another (over the area representing the last variable from the first list).

Sequential/Stacked Plot, 2D – Mixed Step

In this type of graph, the sequences of values of variables selected from one list will be represented by consecutive step areas stacked on one another while the sequences of values of variables selected from a second list will be represented by consecutive step lines stacked on one another (over the step area representing the last variable from the first list).

Sequential/Stacked Plot, 2D – Step

The sequence of values from each selected variable will be represented by consecutive step lines stacked on one another in this type of graph.

Sequential/Stacked Plot, 2D – Step Area

The sequence of values from each selected variable will be represented by consecutive step areas stacked on one another in this type of graph.

Sequential Surface Plot, 3D

In this sequential plot, a spline-smoothed surface is fit to each data point. Successive values of each series are plotted along the X-axis, with each successive series represented along the Y-axis.

Shapiro-Wilks' W Test

The Shapiro-Wilks' W test is used in testing for normality. If the W statistic is significant, the hypothesis that the respective distribution is normal should be rejected. The Shapiro-Wilks' W test is the preferred test of normality because of its good power properties as compared to a wide range of alternative tests (Shapiro, Wilk, & Chen, 1968). Some software programs implement an extension to the test described by Royston (1982), which allows it to be applied to large samples (with up to 2,000 observations). See also, *Kolmogorov-Smirnov Test* and *Lilliefors Test*.

Shewhart Control Charts

This is a standard graphical tool widely used in statistical quality control. The general approach to quality control charting is straightforward: You extract samples of a certain size from the ongoing production process. You then produce line charts of the variability in those samples, and consider their closeness to target specifications. If a trend emerges in those lines, or if samples fall outside pre-specified limits, the process is declared to be out of control and the operator will take action to find the cause of the problem. These types of charts are sometimes also referred to as Shewhart control charts (named after W. A. Shewhart who is generally credited as being the first to introduce these methods; see Shewhart, 1931). For additional information, see Chapter 33 – *Quality Control Charts* and the definition of *Assignable Causes and Actions*.

Short Run Control Charts

The short run quality control chart, for short production runs, plots transformations of the observations of variables or attributes for multiple parts, each of which constitutes a distinct run on the same chart. The transformations rescale the variable values of interest such that they are of comparable magnitudes across the different short production runs (or parts). The control limits computed for those transformed values can then be applied to determine if the production process is in control, to monitor continuing production, and to establish procedures for continuous quality improvement.

Shuffle, Back Propagation (in Neural Networks)

Presenting training cases in a random order on each epoch, to prevent various undesirable effects that can otherwise occur (such as oscillation and convergence to local minima). See *Neural Networks*, page 672.

Shuffle Data (in Neural Networks)

Randomly assigning cases to the training and verification sets, so that these are (as far as possible) statistically unbiased. See *Neural Networks*, page 672.

Sigma Restricted Model

A *sigma*-restricted model uses the sigma-restricted coding to represent effects for categorical predictor variables in general linear models and generalized linear models. To illustrate the sigma-restricted coding, suppose that a categorical predictor variable called Gender has two levels (i.e., male and female). Cases in the two groups would be assigned values of 1 or -1, respectively, on the coded predictor variable, so that if the regression coefficient for the variable is positive, the group coded as 1 on the predictor variable will have a higher predicted value (i.e., a higher group mean) on the dependent variable, and if the regression coefficient is negative, the group coded as -1 on the predictor variable will have a higher predicted value on the dependent variable. This coding strategy is aptly called the sigma-restricted parameterization, because the values used to represent group membership (1 and -1) sum to zero. See also, *Categorical Predictor Variables, Design Matrix*, and Chapter 18 – *General Linear Models*.

Sigmoid Function

An S-shaped curve, with a near-linear central response and saturating limits. See also, *Logistic Function* and *Hyperbolic Tangent (Tanh)*.

Signal Detection Theory (SDT)

Signal detection theory (SDT) is an application of statistical decision theory used to detect a signal embedded in noise. SDT is used in psychophysical studies of detection, recognition, and discrimination, and in other areas such as medical research, weather forecasting, survey research, and marketing research.

A general approach to estimating the parameters of the signal detection model is via the use of the generalized linear model. For example, DeCarlo (1998) shows how signal detection models based on different underlying distributions can easily be considered by using the generalized linear model with different link functions.

For a discussion of the generalized linear/nonlinear model and the link functions it uses, see Chapter 21 – *Generalized Linear/Nonlinear Models*.

Simple Random Sampling (SRS)

Simple random sampling is a type of probability sampling where observations are randomly selected from a population with a known probability or sampling fraction. Typically, you begin with a list of N observations that comprises the entire population from which you want to extract a simple random sample (e.g., a list of registered voters); you can then generate k random case numbers (without replacement) in the range from 1 to N, and select the respective cases into the final sample (with a sampling fraction or known selection probability of k/N).

Refer to, for example, Kish (1965) for a detailed discussion of the advantages and characteristics of probability samples and EPSEM samples.

Simplex Algorithm

A nonlinear estimation algorithm that does not rely on the computation or estimation of the derivatives of the loss function. Instead, at each iteration the function will be evaluated at m+1 points in the m dimensional parameter space. For example, in two dimensions (i.e., when there are two parameters to be estimated), the program will evaluate the function at three points around the current optimum. These three points would define a triangle; in more than two dimensions, the figure produced by these points is called a simplex.

SIMPLS Algorithm

An alternative estimation method for partial least squares regression components is the SIMPLS algorithm (de Jong, 1993), which can be described as follows.

For each $h=1,\ldots,c$, where $A_0=X'Y$, $M_0=X'X$, $C_0=I$, and c given,

1. compute q_h, the dominant eigenvector of $A_h{}'A_h$

2. $w_h=A_hq_h$, $c_h=w_h{}'M_hw_h$, $w_h=w_h/sqrt(c_h)$, and store w_h into W as a column

3. $p_h=M_hw_h$, and store p_h into P as a column

4. $q_h=A_h{}'w_h$, and store q_h into Q as a column

5. $v_h=C_hp_h$, and $v_h=v_h/||v_h||$

6. $C_{h+1}=C_h - v_hv_h{}'$ and $M_{h+1}=M_h - p_hp_h{}'$

7. $A_{h+1}=C_hA_h$

Similarly to NIPALS, the T of SIMPLS is computed as $T=XW$ and B for the regression of Y on X is computed as $B=WQ'$.

Single and Multiple Censoring

There are situations in which censoring can occur at different times (multiple censoring), or only at a particular point in time (single censoring). Consider an example experiment where we start with 100 light bulbs, and terminate the experiment after a certain amount of time. If the experiment is terminated at a particular point in time, a single point of censoring exists and the data set is said to be single-censored. However, in biomedical research multiple censoring often exists, for example, when patients are discharged from a hospital after different amounts (times) of treatment, and the researcher knows that the patient survived up to those (differential) points of censoring.

Data sets with censored observations can be analyzed via survival analysis or via Weibull and reliability/failure time analysis. See also, *Type I and II Censoring* and *Left and Right Censoring*.

Singular Value Decomposition

An efficient algorithm for optimizing a linear model. See also, *Pseudo-Inverse Algorithm*, page 695.

Six Sigma (DMAIC)

Six sigma is a well-structured, data-driven methodology for eliminating defects, waste, or quality control problems of all kinds in manufacturing, service delivery, management, and other business activities. Six sigma methodology is based on the combination of well-established statistical quality control techniques, simple and advanced data analysis methods, and the systematic training of all personnel at every level in the organization involved in the activity or process targeted by six sigma.

Six sigma methodology and management strategies provide an overall framework for organizing company wide quality control efforts. These methods have recently become very popular, due to numerous success stories from major US-based as well as international corporations. For reviews of six sigma strategies, refer to Harry and Schroeder (2000), or Pyzdek (2001).

These are organized into the categories of activities that make up the six sigma effort: Define (D), Measure (M), Analyze (A), Improve (I), Control (C); or DMAIC.

Define. The define phase is concerned with the definition of project goals and boundaries, and the identification of issues that need to be addressed to achieve the higher sigma level.

Measure. The goal of the measure phase is to gather information about the current situation, to obtain baseline data on current process performance, and to identify problem areas.

Analyze. The goal of the analyze phase is to identify the root cause(s) of quality problems, and to confirm those causes using the appropriate data analysis tools.

Improve. The goal of the improve phase is to implement solutions that address the problems (root causes) identified during the previous (analyze) phase.

Control. The goal of the control phase is to evaluate and monitor the results of the previous phase (improve).

Six Sigma Process

A six sigma process is one that can be expected to produce only 3.4 defects per one million opportunities. The concept of the six sigma process is important in six sigma quality improvement programs. The idea can best be summarized with the following graphs.

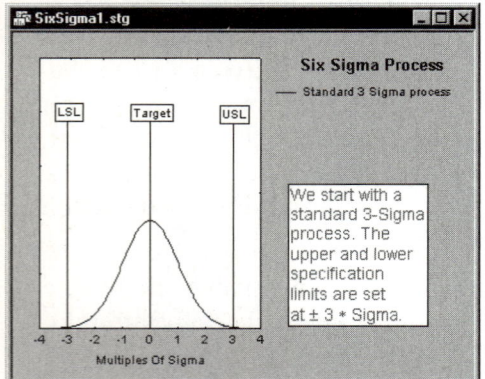

The term Six Sigma derives from the goal to achieve a process variation, so that ± 6 * sigma (the estimate of the population standard deviation) will "fit" inside the lower and upper specification limits for the process. In that case, even if the process mean shifts by 1.5 * *sigma* in one direction (e.g., to +1.5 *sigma* in the direction of the upper specification limit), the process will still produce very few defects.

For example, suppose we expressed the area above the upper specification limit in terms of one million opportunities to produce defects. The 6 * *sigma* process shifted upward by 1.5 * *sigma* will only produce 3.4 defects (i.e., parts or cases greater than the upper specification limit) per one million opportunities.

Shift. An ongoing process that at some point was centered will shift over time. Motorola, in their implementation of Six Sigma strategies, determined that it is reasonable to assume that a process will shift over time by approximately 1.5 * *sigma* (see, for example, Harry and Schroeder, 2000). Hence, most standard Six Sigma calculators will be based on a 1.5 * *sigma* shift.

One-sided vs. two-sided limits. In the illustration shown above the area outside the upper specification limit (greater than USL) is defined as one million opportunities to produce defects. Of course, in many cases any outcomes (e.g., parts) that are produced that fall below the specification limit can be equally defective. In this case, you may want to consider the lower tail of the respective (shifted) normal distribution as well. However, in practice, you usually ignore the lower tail of the normal curve because 1) in many cases, the process naturally has one-sided specification limits (e.g., very low delay times are not really a defect, only very long times; very few customer complaints are not a problem, only very many, etc.), and 2) when a 6 * sigma process has been achieved, the area under the normal curve below the lower specification limit is negligible.

Yield. The previous illustration focuses on the number of defects that a process produces. The number of non-defects can be considered the Yield of the process. Six Sigma calculators will compute the number of defects per million

opportunities (DPMO) as well as the yield, expressed as the percent of the area under the normal curve that falls below the upper specification limit (in the illustration above).

Skewness

Skewness (this term was first used by Pearson, 1895) measures the deviation of the distribution from symmetry. If the skewness is clearly different from 0, that distribution is asymmetrical, while normal distributions are perfectly symmetrical.

$$\text{Skewness} = \frac{nM_3}{(n-1)(n-2)\sigma^3}$$

where M_3 is equal to $\Sigma\,(x_i\text{-Mean}_x)^3$; σ^3 is the standard deviation (*sigma*) raised to the third power; and n is the valid number of cases.

See also, *Descriptive Statistics* in Chapter 2 – *Basic Statistics and Tables*.

Smoothing

Smoothing techniques can be used in two different situations. Smoothing techniques for 3D bivariate histograms enable you to fit surfaces to 3D representations of bivariate frequency data. Thus, every 3D histogram can be turned into a smoothed surface providing a sensitive method for revealing non-salient overall patterns of data and/or identifying patterns to use in developing quantitative models of the investigated phenomenon.

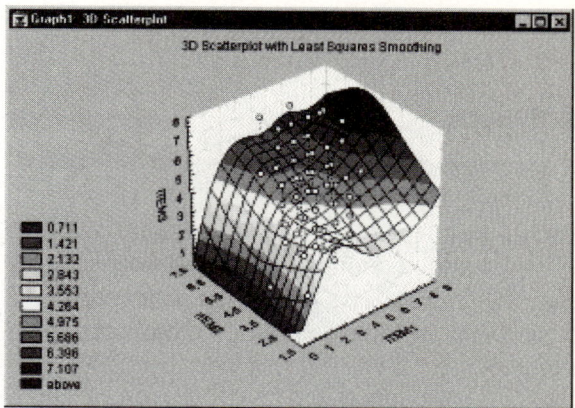

In time series analysis, the general purpose of smoothing techniques is to bring out the major patterns or trends in a time series, while de-emphasizing minor fluctuations (random noise). Visually, as a result of smoothing, a jagged line pattern should be transformed into a smooth curve.

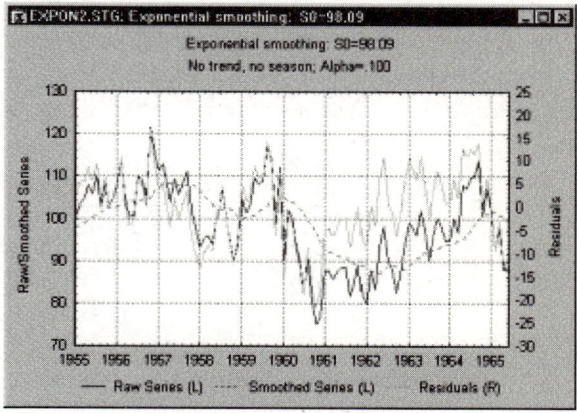

See also, *Exploratory Data Analysis and Data Mining Techniques* in Chapter 12 – *Data Mining Techniques*, and the definition of *Smoothing Bivariate Distributions*.

SOFM (Self-Organizing Feature Map; Kohonen Networks)

Neural networks based on the topological properties of the human brain, also known as

Kohonen Networks (Kohonen, 1982; Fausett, 1994; Haykin, 1994; Patterson, 1996).

Softmax

A specialized activation function for one-of-N encoded classification networks. Performs a normalized exponential (i.e. the outputs add up to 1). In combination with the cross entropy error function, allows multilayer perceptron networks to be modified for class probability estimation (Bishop, 1995; Bridle, 1990). See, *Neural Networks*, page 672.

Space Plot, 3D

This type of graph offers a distinctive means of representing 3D scatterplot data through the use of a separate X-Y plane positioned at a user-selectable level of the vertical Z-axis (which "sticks up" through the middle of the plane).

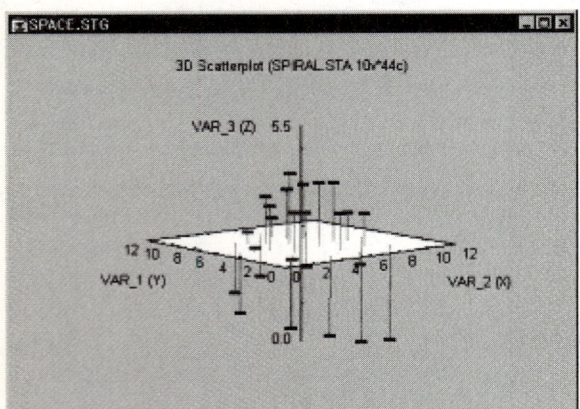

The space plots specific layout may facilitate exploratory examination of specific types of three-dimensional data. It is recommended to assign variables to axes such that the variable that is most likely to discriminate between patterns of relation among the other two is specified as Z. See also, *Data Rotation (in 3D Space)*.

Space Plot, 3D – Categorized

This type of graph offers a distinctive means of representing 3D scatterplot data through the use of a separate *X-Y plane* positioned at a user-selectable level of the vertical *Z-axis* (which "sticks up" through the middle of the plane). The level of the X-Y plane can be adjusted in order to divide the X-Y-Z space into meaningful parts (e.g., featuring different patterns of the relation between the three variables) (see graph number 2, above). See also, *Categorized Graphs*.

Spearman R

Spearman *R* can be thought of as the regular Pearson product-moment correlation coefficient (Pearson *r*); that is, in terms of the proportion of variability accounted for, except that Spearman *R* is computed from ranks. Spearman *R* assumes that the variables under consideration were measured on at least an *ordinal* (rank order) scale; that is, the individual observations (cases) can be ranked into two ordered series. Detailed discussions of the Spearman *R* statistic, its power and efficiency can be found in Gibbons (1985), Hays (1981), McNemar (1969), Siegel (1956), Siegel and Castellan (1988), Kendall (1948), Olds (1949), or Hotelling and Pabst (1936).

Spectral Plot

The original application of this type of plot was in the context of spectral analysis in order to investigate the behavior of non-stationary time series. On the horizontal axes, you can plot the frequency of the spectrum against consecutive time intervals, and indicate on the Z-axis the spectral densities at each interval (see for example, Shumway, 1988, page 82).

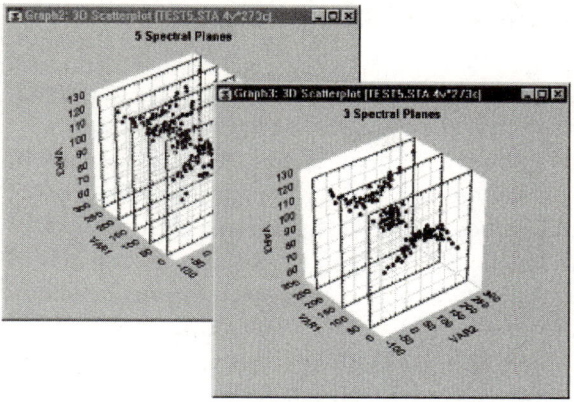

Spectral plots have clear advantages over the regular 3D scatterplots when you are interested in examining how a relationship between two variables changes across the levels of a third variable, as is shown in the next illustration.

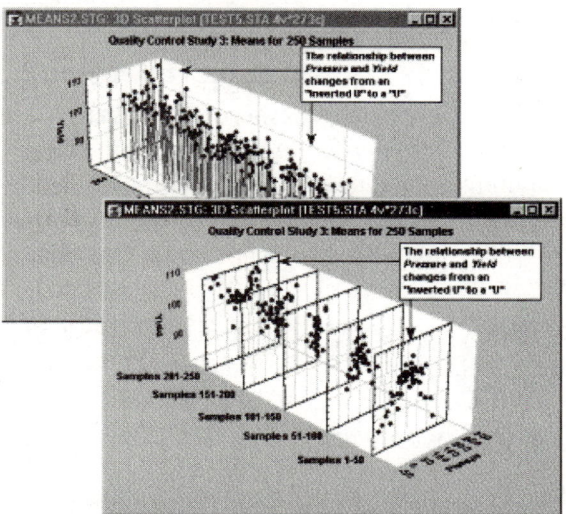

The advantage of spectral plots over the regular 3D scatterplots is well illustrated in the comparison of the two displays of the same data set shown in the previous illustration.

The spectral plot makes it easier to see that the relationship between *Pressure* and *Yield* changes from an "inverted U" to a "U." See also, *Data Rotation (in 3D Space)*.

Spectral Plot, 3D (Categorized)

This type of graph produces multiple spectral plots (for subsets of data determined by the selected categorization method) arranged in one display to allow for comparisons between the subsets of data. Values of variables X and Z are interpreted as the X- and Z-*axis* coordinates of each point, respectively; values of variable Y are clustered into equally-spaced values, corresponding to the locations of the consecutive spectral planes (see graph number 3, above). See also, *Categorized Graphs*.

Spikes (3D Graphs)

In this type of graph, individual values of one or more series of data are represented along the X-axis as a series of spikes (point symbols with lines descending to the base plane). Each series to be plotted is spaced along the Y-axis. The height of each spike is determined by the respective value of each series.

Spline (2D Graphs)

A curve is fitted to the XY coordinate data using the bicubic spline smoothing procedure.

Spline (3D Graphs)

A surface is fitted to the XYZ coordinate data using the bicubic spline smoothing procedure.

Split Selection (for Classification Trees)

Split selection for classification trees refers to the process of selecting the splits on the predictor variables that are used to predict membership in the classes of the dependent variable for the cases or objects in the analysis. Given the hierarchical nature of classification trees, these splits are selected one at time, starting with the split at the root node, and continuing with splits of resulting child nodes until splitting stops, and the child nodes that have not been split become terminal nodes.

The split selection process is described in *Computational Methods in* Chapter 9 – *Classification Trees*.

Spurious Correlations

Correlations that are not due to causal relationships between the respective variables but are due mostly to the influences of one or more "other" variables. For example, there is a correlation between the total amount of losses in a fire and the number of firemen that were putting out the fire; however, what this correlation does not indicate is that if you call fewer firemen then you would lower the losses. There is a third variable (the initial size of the fire) that influences both the amount of losses and the number of firemen. If you control for this variable (e.g., consider only fires of a fixed size), the correlation will either disappear or perhaps even change its sign. The main problem with spurious correlations is that we typically do not know what the hidden agent is. However, in cases when we know where to look, we can use partial correlations that control for (i.e., partial out) the influence of specified variables. See also, *Correlation*, *Multiple Regression*, and *Partial Correlation*, as well as Chapter 2 – *Basic Statistics and Tables*; and Chapter 35 – *Structural Equation Modeling*.

SQL

SQL (Structured Query Language) enables you to query an outside data source about the data it contains. You can use a SQL statement in order to specify the desired tables, fields, rows, etc. to return as data. For information on SQL syntax, consult an SQL manual.

Square Root of the Signal to Noise Ratio (f)

This standardized measure of effect size is used in analysis of variance to characterize the overall level of population effects, and is very similar to the RMSSE. It is the square root of the sum of squared standardized effects divided by the number of effects. For example, in a 1-Way ANOVA, with J groups, f is calculated as:

$$f = \sqrt{\frac{\sum_{j=1}^{n} \left(\frac{a_j}{\sigma}\right)^2}{j}} = \sqrt{\frac{\sigma_\sigma^2}{\sigma^2}} = \frac{\sigma_\sigma}{\sigma}$$

For more information see Chapter 31 – *Power Analysis*.

Stacked Generalization

See *Stacking*, below.

Stacking (Stacked Generalization)

The concept of stacking (short for Stacked Generalization) applies to the area of predictive data mining, to combine the predictions from multiple models. It is particularly useful when the types of models included in the project are very different.

Suppose your data mining project includes tree classifiers, such as CART and CHAID, linear discriminant analysis (e.g., see Chapter 17 – *General Discriminant Analysis*), and neural networks. Each computes predicted classifications for a crossvalidation sample, from which overall goodness-of-fit statistics (e.g., misclassification rates) can be computed. Experience has shown that combining the predictions from multiple methods often yields more accurate predictions than can be derived from any one method (e.g., see Witten and Frank, 2000). In stacking, the predictions from different classifiers are used as input into a meta-learner, which attempts to combine the predictions to create a final best predicted classification. So, for example, the predicted classifications from the tree classifiers, linear model, and the neural network classifier(s) can be used as input variables into a neural network meta-classifier, which will attempt to "learn" from the data how to combine the predictions from the different models to yield maximum classification accuracy.

Other methods for combining the prediction from multiple models or methods (e.g., from multiple data sets used for learning) are boosting and bagging (voting).

Standard Deviation

The standard deviation (this term was first used by Pearson, 1894) is a commonly used measure of variation. The standard deviation of a population of values is computed as:

$$\sigma = \sqrt{\frac{\sum (x_i - \mu)^2}{N}}$$

where μ is the population mean and N is the population size.

The sample estimate of the population standard deviation is computed as:

$$s = \sqrt{\frac{\sum (x_i - \bar{x})^2}{n-1}}$$

where $\bar{x}$ is the sample mean and n is the sample size.

See also, *Descriptive Statistics* in Chapter 2 – *Basic Statistics and Tables*.

Standard Error

The *standard error* (this term was first used by Yule, 1897) is the standard deviation of a mean and is computed as:

$$std.\,err. = \sqrt{\frac{s^2}{n}}$$

Where s^2 is the sample variance and n is the sample size.

Standard Error of the Mean

The standard error of the mean (first used by Yule, 1897) is the theoretical standard deviation of all sample means of size n drawn from a population and depends on both the population variance (sigma) and the sample size (n) as indicated below:

$$\sigma_{\bar{x}} = \sqrt{\frac{\sigma^2}{n}}$$

where σ^2 is the population variance and n is the sample size.

Since the population variance is typically unknown, the best estimate for the standard error of the mean is then calculated as:

$$s_{\bar{x}} = \sqrt{\frac{s^2}{n}}$$

where s^2 is the sample variance (our best estimate of the population variance) and n is the sample size.

See also, *Descriptive Statistics* in Chapter 2 – *Basic Statistics and Tables*.

Standard Error of the Proportion

This is the standard deviation of the distribution of the sample proportion over repeated samples. If the population proportion is π, and the sample size is N, the standard error of the proportion when sampling from an infinite population is:

$$\sigma_p = \sqrt{\frac{\pi(1-\pi)}{N}}$$

For more information see Chapter 31 – *Power Analysis*.

Standard Residual Value

This is the standardized residual value (observed minus predicted divided by the square root of the residual mean square). See also, *Mahalanobis Distance*, *Deleted Residual*, and *Cook's Distance*.

Standardization

While in everyday language the term standardization means converting to a common standard or making something conform to a standard (its meaning is similar to the term normalization in data analysis, see *Normalization*, page 676), in statistics, this term has a very specific meaning and refers to the transformation of data by subtracting each value from some reference value (typically a sample mean) and diving it by the standard deviation (typically a sample SD). This important transformation will bring all values (regardless of their distributions and original units of measurement) to compatible units from a distribution with a mean of 0 and a standard deviation of 1. This transformation has a wide variety of applications because it makes the distributions of values easy to compare across variables and/or subsets. If applied to the input data, standardization also makes the results of a variety of statistical techniques entirely independent of the ranges of values or the units of measurements (see the discussion of these issues in Chapter 1 – *Elementary Concepts in Statistics*, Chapter 2 – *Basic Statistics*, Chapter 16 – *Factor Analysis*, Chapter 26 – *Multiple Linear Regression*, and others).

Standardized DFFITS

This is another measure of impact of the respective case on the regression equation. The formula for *standardized DFFITS* is

$$SDFIT_i = \frac{DFFIT_i}{s_i \sqrt{\tilde{h}_i}}$$

where h_i is the leverage for the ith case,

and

$$\tilde{h}_i = \frac{1}{N + h_i}$$

See also, *DFFITS*, *Studentized Residuals*, and *Studentized Deleted Residuals*. For more information, see Hocking (1996) and Ryan (1997).

Standardized Effect (E$_s$)

A statistical effect expressed in convenient standardized units. For example, the standardized effect in a 2-sample *t*-test is the difference between the two means, divided by the standard deviation, i.e.:

$E_s = (\mu_1 - \mu_2)/\sigma$

In a 1-sample *t*-test, the formula is:

$E_s = (\mu - \mu_0)/\sigma$

In a dependent sample *t*-test, the formula is:

$E_s = (\mu_1 - \mu_2)/sqrt(\sigma_1^2 + \sigma_2^2 - 2\rho\sigma_1\sigma_2)$

For planned contrasts, the formula is:

$E_s = (\Sigma\ c_j\ \mu_j)/\sigma = \Psi/\sigma$

For more information, see Chapter 31 – *Power Analysis*.

Stationary Series (in Time Series)

In time series analysis, a stationary series has a constant mean, variance, and autocorrelation through time (i.e., seasonal dependencies have been removed via differencing).

Statistical Power

The probability of rejecting a false statistical null hypothesis. For more information, see Chapter 31 – *Power Analysis*.

Statistical Process Control (SPC)

The term statistical process control (SPC) is typically used in context of manufacturing processes (although it may also pertain to services and other activities), and it denotes statistical methods used to monitor and improve the quality of the respective operations. By gathering information about the various stages of the process and performing statistical analysis on that information, the SPC engineer is able to take necessary action (often preventive) to ensure that the overall process stays in-control and to allow the product to meet all desired specifications. SPC involves monitoring processes, identifying problem areas, recommending methods to reduce variation and verifying that they work, optimizing the process, assessing the reliability of parts, and other analytic operations. SPC uses such basic statistical quality control methods as quality control charts (Sheward, Pareto, and others), capability analysis, gage repeatability/reproducibility analysis, and reliability analysis. However, also specialized experimental methods (DOE) and other advanced statistical techniques are often part of global SPC systems. Important components of effective, modern SPC systems are real-time access to data and facilities to document and respond to incoming QC data on-line, efficient central QC data warehousing, and groupware facilities allowing QC engineers to share data and reports (see *Enterprise SPC*). See also, *Quality Control* and *Process Analysis*. For more information on process control systems, see the ASQC/AIAG's *Fundamental Statistical Process Control Reference Manual* (1991).

Statistical Significance (*p*-Level)

The statistical significance of a result is an estimated measure of the degree to which it is "true" (in the sense of "representative of the population"). More technically, the value of the *p*-level represents a decreasing index of the reliability of a result. The higher the *p*-level, the less we can believe that the observed relation between variables in the sample is a reliable indicator of the relation between the respective variables in the population. Specifically, the p-level represents the probability of error that is involved in accepting our observed result as valid, that is, as "representative of the population." For example, the *p*-level of .05 (i.e., 1/20) indicates that there is a 5% probability that the relation between the variables found in our sample is a "fluke." In other words, assuming that in the population there was no relation between those variables whatsoever, and we were repeating experiments such as ours one after another, we could expect that approximately in every 20 replications of the experiment there would be one in which the relation between the variables in question would be equal or stronger than in ours. In many areas of research, the *p*-level of .05 is customarily treated as a border-line acceptable error level. See also, Chapter 1 – *Elementary Concepts in Statistics*.

Steepest Descent Iterations

When initial values for the parameters are far from the ultimate minimum, the approximate Hessian used in the Gauss-Newton procedure may fail to yield a proper step direction during iteration. In this case, the program may iterate into a region of the parameter space from which recovery (i.e., successful iteration to the true minimum point) is not possible. One option offered by structural equation modeling is to precede the Gauss-Newton procedure with a few iterations utilizing the "method of steepest descent." In the steepest descent approach, values of the parameter vector **q** on each iteration are obtained as:

$$\hat{\theta}_{k+1} = \hat{\theta}_k + \lambda_k g_k$$

In simple terms, what this means is that the Hessian is not used to help find the direction for the next step. Instead, only the first derivative information in the gradient is used. Inserting a few steepest descent iterations may help in situations where the iterative routine "gets lost" after only a few iterations.

Steps

Repetitions of a particular analytic or computational operation or procedure.

Stepwise Regression

A model-building technique that finds subsets of predictor variables that most adequately predict responses on a dependent variable by linear (or nonlinear) regression, given the specified criteria for adequacy of model fit.

For an overview of stepwise regression and model fit criteria see Chapter 19 – *General Regression Models* or Chapter 26 – *Multiple Linear Regression*; for nonlinear stepwise and best subset regression, see Chapter 21 – *Generalized Linear/Nonlinear Models*.

Stiffness Parameter (in Fitting Options)

The stiffness parameter determines the degree to which the fitted curve depends on local configurations of the analyzed values. The lower the coefficient, the more the shape of the curve is influenced by individual data points (i.e., the curve "bends" more to accommodate individual values and subsets of values).

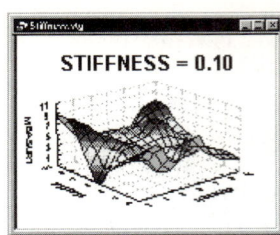 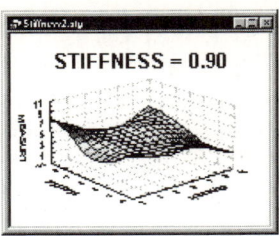

The range of the stiffness parameters is $0 < s < 1$. Large values of the parameter produce smoother curves that adequately represent the overall pattern in the data set at the expense of local details. See also, McLain, 1974.

Stopping Conditions

During an iterative process (e.g., fitting, searching, training), the conditions which must be true for the process to stop. (For example, in neural networks, the stopping conditions include the maximum number of epochs, target error performance and the minimum error improvement thresholds.

Stopping Rule (in Classification Trees)

The stopping rule for a classification tree refers to the criteria that are used for determining the right-sized classification tree, that is, a classification tree with an appropriate number of splits and optimal predictive accuracy. The process of determining the right-sized classification tree is described in *Computational Methods* in Chapter 9 – *Classification Trees*.

Stratified Random Sampling

In general, random sampling is the process of randomly selecting observations from a population, to create a subsample that represents the observations in that population (see Kish, 1965; see also, *Probability Sampling*, *Simple Random Sampling*, and *EPSEM Samples*; see also, *Representative Sample*, page

710, for a brief exploration of this often misunderstood notion). In stratified sampling, you usually apply specific (identical or different) sampling fractions to different groups (strata) in the population to draw the sample.

Over-sampling particular strata to over-represent rare events. In some predictive data mining applications it is often necessary to apply stratified sampling to systematically over-sample (apply a greater sampling fraction) to particular rare events of interest. For example, in catalog retailing the response rate to particular catalog offers can be below 1%, and when analyzing historical data (from prior campaigns) to build a model for targeting potential customers more successfully, it is desirable to over-sample past respondents (i.e., the rare respondents who ordered from the catalog); you can then apply the various model building techniques for classification (see *Data Mining*) to a sample consisting of approximately 50% responders and 50% non-responders. Otherwise, if you were to draw a simple random sample for the analysis (with 1% of responders), practically all model building techniques would likely predict a simple no-response for all cases, and would be (trivially) correct in 99% of the cases.

Stub and Banner Tables

Stub-and-banner tables (or banner tables) are essentially two-way tables, except that two lists of categorical variables (instead of just two individual variables) are crosstabulated. For more information, see *Stub and Banner Tables* in Chapter 2 – *Basic Statistics and Tables*.

Student's *t* Distribution

The student's *t* distribution has density function (for $\nu = 1, 2, ...$):

$$f(x) = \frac{\Gamma\left(\dfrac{\nu+1}{2}\right)}{\Gamma\left(\dfrac{\nu}{2}\right)} \frac{1}{\sqrt{\nu\pi}} \left(1 + \frac{x^2}{\nu}\right)^{-\frac{\nu+1}{2}}$$

where ν is the degrees of freedom; Γ (*gamma*) is the *Gamma* function, and π is the constant Pi (3.14...). The following illustrations show various tail areas (p-values) for a Student's t distribution with 15 degrees of freedom.

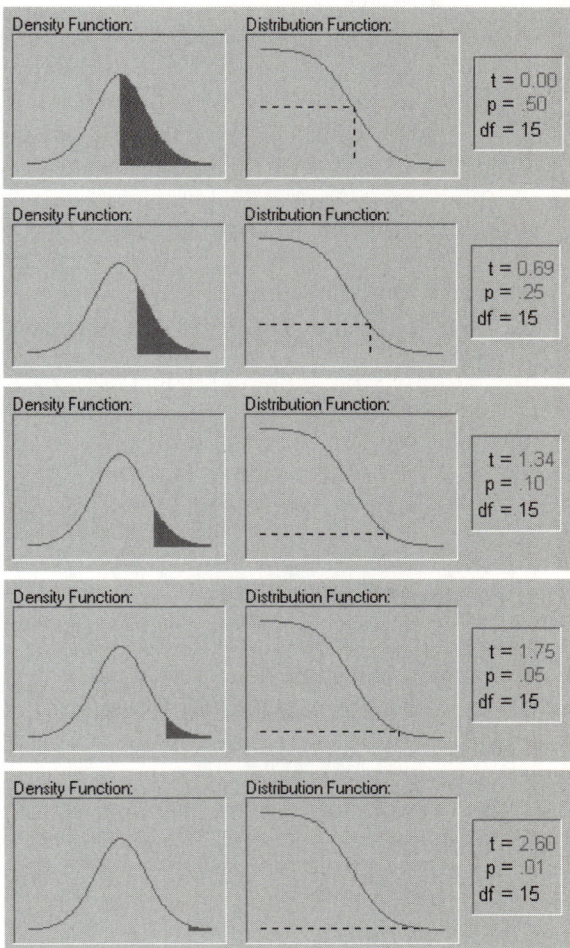

Studentized Deleted Residuals

In addition to standardized residuals several methods (including studentized residuals, studentized deleted residuals, DFFITS, and standardized DFFITS) are available for detecting outlying values (observations with extreme values on the set of predictor variables or the dependent variable). The formula for studentized deleted residuals is given by

$$\text{SDRESID}_i = \frac{DRESID_i}{s_{(i)}}$$

for

$$\text{DRESID} = \frac{e_i}{1-\tilde{h}_i}$$

and where

$$s_{(i)} = \frac{1}{\sqrt{(C-p-1)}} \sqrt{\frac{(C-p)s^2}{1-h_i} - \text{DRESID}_i^2}$$

e_i is the error for the ith case, h_i is the leverage for the ith case, p is the number of coefficients in the model
and

$$\tilde{h}_i = \frac{1}{N} + h_i$$

For more information see Hocking (1996) and Ryan (1997).

Studentized Residuals

In addition to standardized residuals several methods (including studentized residuals, studentized deleted residuals, DFFITS, and standardized DFFITS) are available for detecting outlying values (observations with extreme values on the set of predictor variables or the dependent variable). The formula for studentized residuals is:

$$\text{SRES}_i = \frac{\frac{e_i}{s}}{\sqrt{1-\tilde{h}_i}}$$

where e_i is the error for the ith case, h_i is the leverage for the ith case, and $\tilde{h}_i = 1/N + h_i$. For

more information see Hocking (1996) and Ryan (1997).

Sweeping

The sweeping transformation of matrices is commonly used to efficiently perform stepwise multiple regression (see Dempster, 1969, Jennrich, 1977) or similar analyses; a modified version of this transformation is also used to compute the *g2 generalized inverse*. The forward sweeping transformation for a column k can be summarized in the following four steps (where the *e*'s refer to the elements of a symmetric matrix):

1.　$e_{ij} = e_{ij} - e_{jk} * e_{kj} / e_{kk}$ for i<>k, j<>k
2.　$e_{kj} = e_{kj} / e_{kk}$
3.　$e_{ik} = e_{ik} / e_{kk}$
4.　$e_{kk} = -1 / e_{kk}$

The reverse sweeping operation reverses the changes affected by these transformations. The sweeping operator is used extensively in general linear models, multiple regression, and similar techniques.

Sum-Squared Error Function

An error function composed by squaring the difference between sets of target and actual values, and adding these together (see also, *Loss Function*).

Supervised Learning (in Neural Networks)

Training algorithms that adjust the weights in a neural network by executing the network on input cases with known outputs, and using the error between the actual and target outputs to adjust the weights. See *Neural Networks*, page 672.

Support Value (in Association Rules)

When applying (in data or text mining) algorithms for deriving association rules of the

general form *If Body then Head* (e.g., *If (Car=Porsche and Age<20) then (Risk=High and Insurance=High)*), the Support value is computed as the joint probability (relative frequency of co-occurrence) of the *Body* and *Head* of each association rule.

Suppressor Variable

A suppressor variable (in multiple regression) has zero (or close to zero) correlation with the criterion but is correlated with one or more of the predictor variables, and therefore, it will suppress irrelevant variance of independent variables. For example, you are trying to predict the times of runners in a 40-meter dash. Your predictors are *Height* and *Weight* of the runner. Now, assume that *Height* is not correlated with *Time*, but *Weight* is. Also assume that *Weight* and *Height* are correlated. If *Height* is a suppressor variable, it will suppress, or control for, irrelevant variance (i.e., variance that is shared with the predictor and not the criterion), thus increasing the partial correlation. This can be viewed as ridding the analysis of noise.

Let t = Time, h = Height, w – Weight, r_{th} = 0.0, r_{tw} = 0.5, and r_{hw} = 0.6.

Weight in this instance accounts for 25% ($R_{tw}**2 = 0.5**2$) of the variability of *Time*. However, if *Height* is included in the model, an additional 14% of the variability of *Time* is accounted for even though *Height* is not correlated with *Time* (see below):

$$R_{t.hw}**2 = 0.5**2/(1 - 0.6**2) = 0.39$$

For more information, refer to Pedhazur, 1982.

Surface Plot, 3D

With a surface plot, a surface is fitted to the data (variables corresponding to sets of XYZ coordinates).

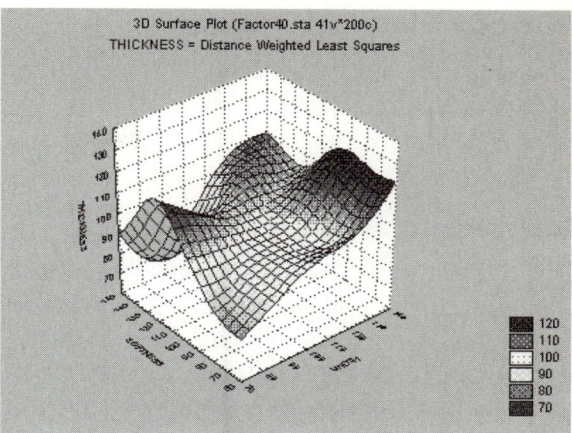

The specified surface is fit to the values of the variable represented by the z-axis and is shaded in colors corresponding to the z-axis values.

Surface Plot, 3D (Categorized)

In this type of graph, a surface (defined by a smoothing technique or user-defined mathematical expression) is fitted to the categorized data (variables corresponding to sets of *XYZ* coordinates, for subsets of data determined by the selected categorization method) arranged in one display to allow for comparisons between the subsets of data (categories).

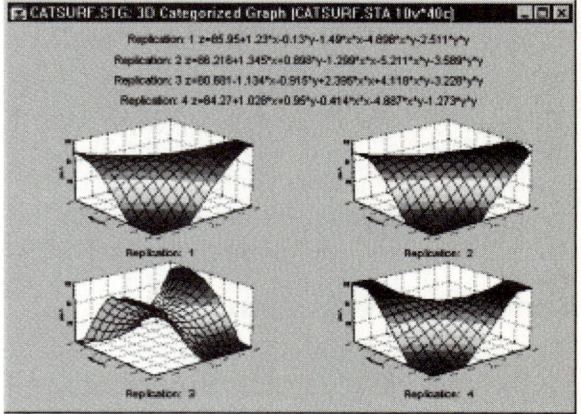

See also, *Categorized Graphs*.

Surface Plot – Raw Data

This sequential plot fits a spline-smoothed surface to each data point. Successive values of each series are plotted along the X-axis, with each successive series represented along the Y-axis.

Survival Analysis

Survival analysis (exploratory and hypothesis testing) techniques include descriptive methods for estimating the distribution of survival times from a sample, methods for comparing survival in two or more groups, and techniques for fitting linear or nonlinear regression models to survival data. A defining characteristic of survival time data is that they usually include so-called censored observations, e.g., observations that survived to a certain point in time, and then dropped out from the study (e.g., patients who are discharged from a hospital). Instead of discarding such observations from the data analysis all together (i.e., unnecessarily lose potentially useful information) survival analysis techniques can accommodate censored observations and use them in statistical significance testing and model fitting. Typical survival analysis methods include life table, survival distribution, and Kaplan-Meier survival function estimation, and additional techniques for comparing the survival in two or more groups. Finally, survival analysis includes the use of regression models for estimating the relationship of (multiple) continuous variables to survival times. For more information, see Chapter 36 – *Survival/Failure Time Analysis*.

Survivorship Function

The survivorship function [commonly denoted as R(t)] is the complement to the cumulative distribution function [i.e., R(t)=1-F(t)]; the survivorship function is also referred to as the reliability or survival function (since it

describes the probability of not failing or of surviving until a certain time *t*; e.g., see Lee, 1992). For additional information, see Chapter 36 – *Survival/Failure Time Analysis*, or *Weibull and Reliability/Failure Time Analysis* in Chapter 32 – *Process Analysis*.

Symmetric Matrix

A matrix is symmetric if the transpose of the matrix is itself (i.e., $\mathbf{A} = \mathbf{A}'$). In other words, the lower triangle of the square matrix is a "mirror image" of the upper triangle with 1s on the diagonal (see the next illustration).

$$\begin{bmatrix} 1 & 2 & 3 & 4 \\ 2 & 1 & 5 & 6 \\ 3 & 5 & 1 & 7 \\ 4 & 6 & 7 & 1 \end{bmatrix}$$

Symmetrical Distribution

If you split the distribution in half at its mean (or median), the distribution of values would be a mirror image about this central point. See also, *Descriptive Statistics* in Chapter 2 – *Basic Statistics and Tables*.

Synaptic Functions (in Neural Networks)

Dot product. Dot product units perform a weighted sum of their inputs, minus the threshold value. In vector terminology, this is the dot product of the weight vector with the input vector, plus a bias value. Dot product units have equal output values along hyperplanes in pattern space. They attempt to perform classification by dividing pattern space into sections using intersecting hyperplanes.

Radial. Radial units calculate the square of the distance between the two points in N dimensional space (where N is the number of inputs) represented by the input pattern vector and the unit's weight vector. Radial units have equal output values lying on hyperspheres in pattern space. They attempt to perform classification by measuring the distance of normalized cases from exemplar points in pattern space (the exemplars being stored by the units). The squared distance is multiplied by the threshold (which is, therefore, actually a deviation value in radial units) to produce the post synaptic value of the unit (which is then passed to the unit's activation function). Dot product units are used in multilayer perceptron and linear networks, and in the final layers of radial basis function, PNN, and GRNN networks. Radial units are used in the second layer of Kohonen, radial basis function, clustering, and probabilistic and generalized regression networks. They are not used in any other layers of any standard network architecture.

Division. This is specially designed for use in generalized regression networks, and should not be employed elsewhere. It expects one incoming weight to equal +1, one to equal -1, and the others to equal zero. The post-synaptic value is the +1 input divided by the -1 input.

T

Tapering

The so-called process of split-cosine-bell tapering in time series is a recommended transformation of the series prior to the spectrum analysis. It usually leads to a reduction of leakage in the periodogram. The rational for this transformation is explained in detail in Bloomfield (1976, p. 80-94). In essence, a proportion (p) of the data at the beginning and at the end of the series is transformed via multiplication by the weights:

$w_t = 0.5*\{1-\cos[\pi*(t - 0.5)/m]\}$
(for t=0 to m-1)
$w_t = 0.5*\{1-\cos[\pi*(N - t + 0.5)/m]\}$
(for t=N-m to N-1)

where m is chosen so that $2*m/N$ is equal to the proportion of data to be tapered (p).

Ternary Plot, 2D – Scatterplot

In this type of graph, the triangular coordinate systems are used to plot 3 (or more) variables (the components X, Y, and Z) in 2 dimensions.

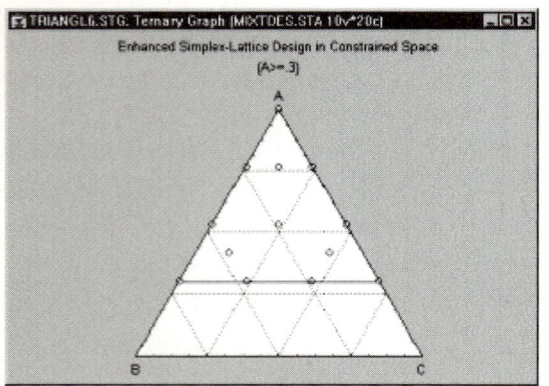

In the previous illustration, the points representing the proportions of the component variables (X, Y, and Z) are plotted. See also, *Data Reduction*.

Ternary Plot, 3D

A ternary plot can be used to examine relations between four or more dimensions where three of those dimensions represent components of a mixture (i.e., the relations between them are constrained such that the values of the three variables add up to the same constant).

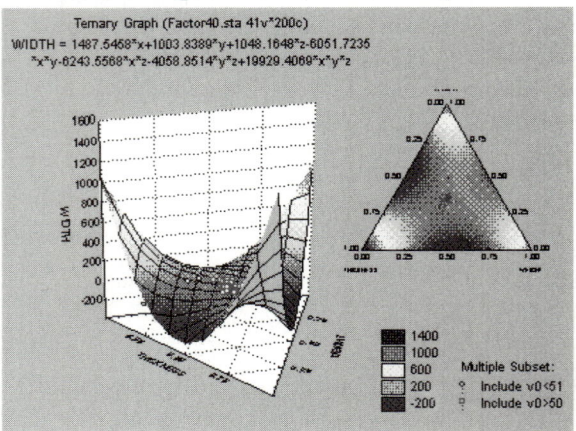

One typical application of this graph is when the measured response(s) from an experiment depends on the relative proportions of three components (e.g., three different chemicals) which are varied in order to determine an optimal combination of those components (e.g., in mixture designs).

Ternary Plot, 3D - Contour (Categorized)

This 3D categorized plot projects a 3-dimensional surface onto a 2-dimensional plane as area contour plots for each level of the grouping variable. One component graph is produced for each level of the grouping variable (or user-defined subset of data) and all the component graphs are arranged in one display

to allow for comparisons between the subsets of data (categories).

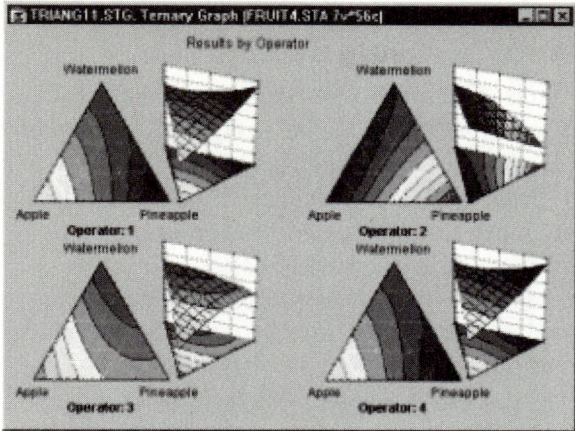

See also, *Categorized Graphs*.

Ternary Plot, 3D – Contour/Areas

In this type of ternary graph, the 3-dimensional surface (fitted to a four-coordinate data set) is projected onto a 2-dimensional plane as an area contour (see the next illustration).

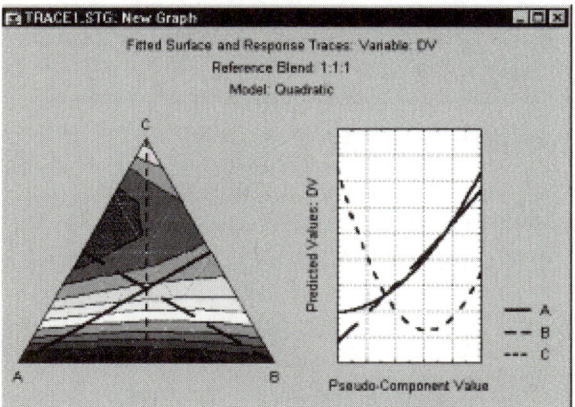

Ternary Plot, 3D – Contour/Lines

In this type of ternary graph, the 3-dimensional surface (fitted to a four-coordinate data set) is projected onto a 2-dimensional plane as a line contour (see the next illustration).

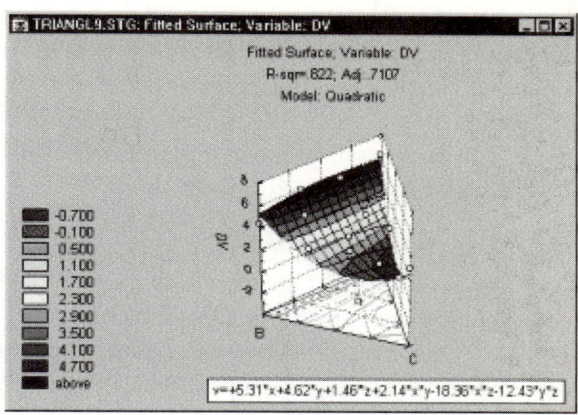

Ternary Plot, 3D – Deviation

With this type of ternary graph, you can examine the relations between four or more dimensions (*X, Y, Z,* and *V1, V2,* etc.) as deviations from a specified base-level of the *V-axis* where three of those dimensions (*X, Y,* and *Z*) represent components of a mixture (i.e., the relations between them are constrained such that the values of the three variables add up to the same constant for each case).

Ternary Plot, 3D – Scatterplot

In this type of ternary graph, the triangular coordinate systems are used to plot four (or more) variables (the components *X, Y,* and *Z,* and the responses *V1, V2,* etc.) in three dimensions (ternary 3D scatterplots or surface plots). Here, the responses (*V1, V2,* etc.) associated with the proportions of the component variables (*X, Y,* and *Z*) in a ternary graph are plotted as the heights of the points. See also, *Data Reduction*.

Ternary Plot, 3D - Scatterplot (Categorized)

The responses associated with the proportions of the component variables (*X, Y,* and *Z*) in a ternary graph are plotted in a 3-dimensional display for each level of the grouping variable (or user-defined subset of data). One component graph is produced for each level of the grouping variable (or user-defined subset of data) and all the component graphs are arranged in one display to allow for comparisons between the subsets of data (categories). See also, *Data Reduction*.

Ternary Plot, 3D – Space

This type of ternary graph offers a distinctive method of representing 3D scatterplot data through the use of an *X-Y-Z* plane (defined via a triangular coordinate system) positioned at a user-selectable level of the vertical *V-axis* (which "sticks up" through the middle of the plane). The level of the *X-Y-Z* plane can be adjusted in order to divide the *X-Y-Z* space into meaningful parts (e.g., featuring different patterns of the relation between the three variables).

Ternary Plot, 3D - Space (Categorized)

In this type of ternary graph, 3D scatterplot data are represented through the use of an *X-Y-Z* plane (defined via a triangular coordinate system) positioned at a user-selectable level of the vertical *V-axis* (which "sticks up" through the middle of the plane) and categorized by each level of the grouping variable (or user-defined subset of data). One component graph is produced for each level of the grouping variable (or user-defined subset of data) and all the component graphs are arranged in one display to allow for comparisons between the subsets of data (categories).

The level of the *X-Y-Z* plane can be adjusted in order to divide the *X-Y-Z-V* space into meaningful parts (e.g., featuring different patterns of the relation between the three variables).

Ternary Plot, 3D – Surface

In a 3D ternary surface plot, a surface is fit to a four-coordinate data set in a 3-dimensional

ternary graph using a first, second, full cubic, or special cubic polynomial fit.

Ternary Plot, 3D – Surface (Categorized)

A surface is fit to a four-coordinate data set in this 3-dimensional ternary graph categorized by each level of the grouping variable (or user-defined subset of data). One component graph is produced for each level of the grouping variable (or user-defined subset of data). All the component graphs are arranged in one display to allow for comparisons between the subsets of data (categories). See the next illustration.

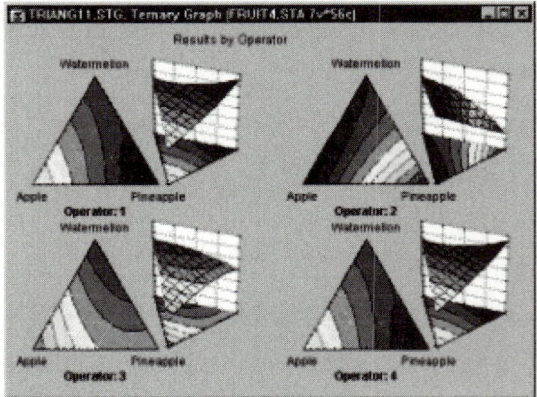

Ternary Plot, 3D – Trace

With this type of ternary graph, you can examine the relations between four or more dimensions (X, Y, Z, and V1, V2, etc.) as a 3D trace plot where three of those dimensions (X, Y, and Z) represent components of a mixture (i.e., the relations between them is constrained such that the values of the three variables add up to the same constant for each case). Data points in this graph are positioned as in regular 3D scatterplots, however, individual data points are connected with a line (in the order in which they were read from the data file), visualizing a "trace" of sequential values.

StatSoft®
Copyright © StatSoft, 2006

Ternary Plot, 3D - Trace (Categorized)

With this type of ternary graph, you can examine the relations between four or more dimensions (*X, Y, Z,* and *V1, V2,* etc.) as a 3D trace plot where three of those dimensions (*X, Y,* and *Z*) represent components of a mixture (i.e., the relations between them is constrained such that the values of the three variables add up to the same constant for each case). Data points in this graph are positioned as in regular 3D scatterplots, however, individual data points are connected with a line (in the order in which they were read from the data file), visualizing a trace of sequential values.

Text Mining

While data mining is typically concerned with the detection of patterns in numeric data, very often, important (e.g., critical to business) information is stored in the form of text. Unlike numeric data, text is often amorphous, and difficult to deal with (e.g., email messages, open-ended comments on a questionnaire or suggestion form, patients' descriptions of their symptoms, searches of written historical records, etc.). Text mining generally consists of the analysis of (multiple) text documents by extracting key phrases, concepts, etc. and the preparation of the text processed in that manner for further analyses with numeric data mining techniques (e.g., to determine co-occurrences of concepts, key phrases, names, addresses, product names, etc.).

A typical (first) goal in data mining is feature extraction, i.e., the identification of the terms and concepts most frequently used in the input documents; a second goal typically is to discover any associations between features (e.g., associations between symptoms as described by patients). Hence, a first step to text

mining usually consists of "coding" the information in the input text; as a second step various methods such as Association Rules algorithms may be applied to determine relations between features.

THAID

THAID is a classification trees program developed by Morgan & Messenger (1973) that performs multi-level splits when computing classification trees. For discussion of the differences of *THAID* from other classification tree programs, see *A Brief Comparison of Classification Tree Programs* in Chapter 9 – *Classification Trees*.

Threshold

A criterion value (sometimes arbitrarily established) that is used to determine if particular conditions are met or a point separating conditions. (In neural networks, a value subtracted from the weighted sum in a linear PSP unit to produce the activation level. In radial units, the threshold is actually treated as a deviation.)

Time-Dependent Covariates

Time-dependent covariates occur when the effect of the covariate on survival is dependent on time (i.e., the conditional hazard at each point in time is a function of the covariate and time).

Time Series

A *time series* is a sequence of measurements, typically taken at successive points in time. Time series analysis includes a broad spectrum of exploratory and hypothesis testing methods that have two main goals: a) identifying the nature of the phenomenon represented by the sequence of observations, and b) forecasting (predicting future values of the time series variable). Both of these goals require that the pattern of observed time series data is identified and more or less formally described. Once the pattern is established, we can interpret and integrate it with other data (i.e., use it in our theory of the investigated phenomenon, e.g., seasonal commodity prices). Regardless of the depth of our understanding and the validity of our interpretation (theory) of the phenomenon, we can extrapolate the identified pattern to predict future events. For more information, see Chapter 38 – *Time Series/Forecasting*.

Time Series (in Neural Networks)

Many important problems can be classified as time series problems; the objective is to predict the value of some (typically continuous) variable, giving previous values of that and/or other variables (Bishop, 1995).

Tolerance (in Multiple Regression)

The *tolerance* of a variable is defined as 1 minus the squared multiple correlation of this variable with all other independent variables in the regression equation. Therefore, the smaller the tolerance of a variable, the more redundant is its contribution to the regression (i.e., it is redundant with the contribution of other independent variables). If the tolerance of any of the variables in the regression equation is equal to zero (or very close to zero), the regression equation cannot be evaluated (the matrix is said to be ill-conditioned, and it cannot be inverted).

Topological Map

The radial layer of a Kohonen network, with units laid out in two-dimensions, and trained so that inter-related clusters tend to be situated close together in the layer. Used for cluster analysis (Kohonen, 1982; Fausett, 1994;

Haykin, 1994; Patterson, 1996). See *Neural Networks*, page 672.

Trace Plot, 3D

As in 3D scatterplots, each data point in trace plots is represented by its location in 3D space as determined by the values of the variables selected as X, Y, and Z (and interpreted as the X, Y, and Z axis coordinates). The data points are then connected sequentially (in the order encountered in the data file) with a line to form a "trace" of a sequential process (e.g., movement, change of a phenomenon over time, etc.).

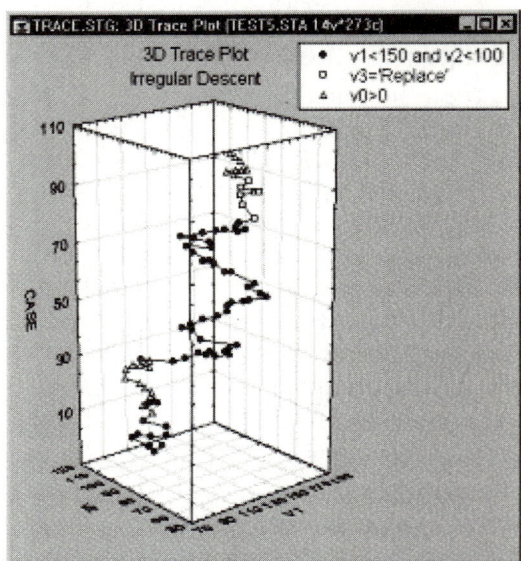

A good metaphor of the information that is best represented in a trace plot is that of the trajectory of an object in 3-dimensional space.

Training/Testing/ Classification Accuracy

A measure of how well a model is trained to predict the training/testing data.

Trimmed Means

For certain graphs (e.g., 2D box plots, 3D box plots, categorized box plots), an option is available to trim the extreme values from the distribution of values of a variable. For example, you can trim (i.e., remove) the lowest 5% and the highest 5% from the distribution of values. The mean of the trimmed distribution of values is referred to as a trimmed mean (this term was first used by Tukey, 1962).

t-Test (for Independent and Dependent Samples)

The *t*-test is the most commonly used method to evaluate the differences in means between two groups. The groups can be independent (e.g., blood pressure of patients who were given a drug vs. a control group who received a placebo) or dependent (e.g., blood pressure of patients "before" vs. "after" they received a drug). Theoretically, the *t*-test can be used even if the sample sizes are very small (e.g., as small as 10; some researchers claim that even smaller *n*'s are possible), as long as the variables are approximately normally distributed and the variation of scores in the two groups is not reliably different (see also, Chapter 1 – *Elementary Concepts in Statistics*).

Dependent samples test. The *t*-test for dependent samples can be used to analyze designs in which the within-group variation (normally contributing to the error of the measurement) can be easily identified and excluded from the analysis. Specifically, if the two groups of measurements to be compared are based on the same sample of observation units (e.g., subjects) that were tested twice (e.g., before and after a treatment), then a considerable part of the within-group variation in both groups of scores can be attributed to the initial individual differences between the

observations and, thus, accounted for (i.e., subtracted from the error). This, in turn, increases the sensitivity of the design.

One-sample test. In so-called one-sample *t*-test, the observed mean (from a single sample) is compared to an expected (or reference) mean of the population (e.g., some theoretical mean), and the variation in the population is estimated based on the variation in the observed sample. See Hays, 1988. See also, *t-Test for Independent Samples* and *t-Test for Dependent Samples* in Chapter 2 – *Basic Statistics and Tables*.

Tukey HSD

This post-hoc test (or multiple comparison test) can be used to determine the significant differences between group means in an analysis of variance setting. The Tukey HSD is generally more conservative than the Fisher LSD test but less conservative than Scheffe's test (for a detailed discussion of different post hoc tests, see Winer, Michels, & Brown (1991). For more details, see Chapter 18 – *General Linear Models*. See also, *Post-hoc Comparisons*, page 687. For a discussion of statistical significance, see Chapter 1 – *Elementary Concepts in Statistics*.

Tukey Window

In time series, the Tukey window is a weighted moving average transformation used to smooth the periodogram values. In the Tukey (Blackman and Tukey, 1958) or Tukey-Hanning window (named after Julius Von Hann), for each frequency, the weights for the weighted moving average of the periodogram values are computed as:

$w_j = 0.5 + 0.5*\cos(\pi*j/p)$, (for j=0 to p)
$w_{-j} = w_j$, (for $j \neq 0$)

where $p = (m-1)/2x$.

This weight function will assign the greatest weight to the observation being smoothed in the center of the window, and increasingly smaller weights to values that are further away from the center. See also, *Spectrum Analysis – Basic Notations and Principles*.

Two-State (in Neural Networks)

An encoding technique for nominal variables with only two values, where the nominal variable is represented by a single input or output unit, either set or cleared. See *Neural Networks*, page 672.

Type I and II Censoring

So-called type I censoring describes the situation when a test is terminated at a particular point in time, so that the remaining items are only known not to have failed up to that time (e.g., we start with 100 light bulbs, and terminate the experiment after a certain amount of time). In this case, the censoring time is often fixed, and the number of items failing is a random variable. In type II censoring, the experiment would be continued until a fixed proportion of items have failed (e.g., we stop the experiment after exactly 50 light bulbs have failed). In this case, the number of items failing is fixed, and time is the random variable. Data sets with censored observations can be analyzed via survival analysis or via Weibull and reliability/failure time analysis. See also, *Left and Right Censoring*, and *Single and Multiple Censoring*.

Type I Error (Alpha)

The probability of incorrectly rejecting the null hypothesis. For more information, see Chapter 31 – *Power Analysis*.

Type I, II, III (IV, V) Sums of Squares

When there are missing cells in a factorial ANOVA design, there is ambiguity regarding the specific comparisons between the (population, or least-squares) cell means that constitute the main effects and interactions of interest. Chapter 18 – *General Linear Models* discusses the methods commonly labeled Type I, II, III, and IV sums of squares as well as methods for testing effects in incomplete designs, that are widely used in other areas (and traditions) of research.

Type V sums of squares. We propose the term Type V sums of squares to denote the approach that is widely used in industrial experimentation, to analyze fractional factorial designs; these types of designs are discussed in detail in $2^{(k-p)}$ *Fractional Factorial Designs* in Chapter 15 – *Experimental Design (DOE)*. In effect, for those effects for which tests are performed all population marginal means (least squares means) are estimable.

Type VI sums of squares. We propose the term Type VI sums of squares to denote the approach that is often used in programs that only implement the *sigma* restricted model (as opposed to programs such as *STATISTICA*'s *General Linear Models*, which offers a choice between the *sigma* restricted and overparameterized). This approach is identical to what is described as the effective hypothesis method in Hocking (1996). For additional details, see *Six Types of Sums of Squares* in Chapter 18 – *General Linear Models*.

U

Unequal N HSD

This post hoc test can be used to determine the significant differences between group means in an analysis of variance setting. The *Unequal N HSD* test is a modification of the Tukey HSD test, and it provides a reasonable test of differences in group means if group n's are not too discrepant (for a detailed discussion of different post hoc tests, see Winer, Michels, & Brown (1991). For more details, see Chapter 18 – *General Linear Models*. See also, *Post-hoc Comparisons*, page 687. For a discussion of statistical significance, see Chapter 1 – *Elementary Concepts in Statistics*.

Uniform Distribution

The discrete Uniform distribution (the term first used by Uspensky, 1937) has density function:

$$f(x) = 1/N \qquad x = 1, 2, ..., N$$

The continuous Uniform distribution has density function:

$$f(x) = 1/(b-a) \qquad a < x < b$$

where *a* is the lower limit of the interval from which points will be selected, *b* is the upper limit of the interval from which points will be selected.

Unimodal Distribution

A distribution that has only one mode. A typical example is the normal distribution that happens to be also symmetrical but many unimodal distributions are not symmetrical (e.g., typically the distribution of income is not symmetrical but "left-skewed"; see skewness). See also, bimodal distribution, multimodal distribution.

Unit Penalty

In several search algorithms, a penalty factor which is multiplied by the number of units in the network and added to the error of the network, when comparing the performance of the network with others. This has the effect of selecting smaller networks at the expense of larger ones. See also, *Penalty Functions*, page 685.

Unit Types (in Neural Networks)

Units in the input layer are extremely simple: they simply hold an output value, which they pass onto units in the second layer. Input units do no processing. Input units have their synaptic function set to Dot Product, and their activation function set to Identity by default; actually these functions are ignored in input units.

Each hidden or output unit has a number of incoming connections from units in the preceding layer (the fan-in): one for each unit in the preceding layer. Each unit also has a threshold value.

The outputs of the units in the preceding layer, the weights on the associated connections, and the threshold value are fed through the unit's synaptic function (post synaptic potential function) to produce a single value (the unit's input value).

The input value is passed through the unit's activation function to produce a single output value, also known as the activation level of the unit.

Unsupervised Learning (in Neural Networks)

Of the following unsupervised learning algorithms, all except principal components

analysis are concerned with assignment of radial unit centers and deviations.

Unsupervised learning algorithms require a data set that includes typical input variable values. Observed output variable values are not required. If output variable values are present in the data set, they are simply ignored.

Center Assignment. Kohonen Algorithm, Radial Sampling, K-Means Algorithm.

Deviation Assignment. Explicit Deviation Assignment, Isotropic Deviation Assignment, K-Nearest Neighbor.

Principal Components Analysis. Principal Components Analysis.

Unweighted Means

If the cell frequencies in a multi-factor ANOVA design are unequal, the unweighted means (for levels of a factor) are calculated from the means of sub-groups without weighting, that is, without adjusting for the differences between the sub-group frequencies.

V

Variables – Independent vs. Dependent

The terms dependent and independent variable apply mostly to experimental research where some variables are manipulated, and in this sense they are independent from the initial reaction patterns, features, intentions, etc. of the subjects. Some other variables are expected to be dependent on the manipulation or experimental conditions. That is to say, they depend on what the subject will do in response. Independent variables are those that are manipulated whereas dependent variables are only measured or registered.

Somewhat contrary to the nature of this distinction, these terms are also used in studies where we do not literally manipulate independent variables, but only assign subjects to experimental groups based on some preexisting properties of the subjects. For example, if in an experiment, males are compared with females regarding their white cell count (*WCC*), *Gender* could be called the independent variable and *WCC* the dependent variable.

For more information, see *Dependent vs. Independent Variables* in Chapter 1 – *Elementary Concepts in Statistics*.

Variance

The variance (this term was first used by Fisher, 1918a) of a *population* of values is computed as:

$$\sigma^2 = \Sigma\ (x_i - \mu)^2 / N$$

where μ is the population mean, and N is the population size.

The unbiased *sample* estimate of the population variance is computed as:

$$s^2 = \Sigma\ (x_i - xbar)^2 / n-1$$

where *xbar* is the sample mean, and n is the sample size.

See also, *Descriptive Statistics* in Chapter 2 – *Basic Statistics and Tables*.

Variance Components (in Mixed Model ANOVA)

The term variance components is used in the context of experimental designs with random effects, to denote the estimate of the (amount of) variance that can be attributed to those effects. For example, if you were interested in the effect that the quality of different schools has on academic proficiency, you could select a sample of schools to estimate the amount of variance in academic proficiency (component of variance) that is attributable to differences between schools. See also, *Methods for Analysis of Variance* in Chapter 3 – *ANOVA/MANOVA*

Variance Inflation Factor (VIF)

The diagonal elements of the inverse correlation matrix (i.e., -1 times the diagonal elements of the sweep matrix) for variables that are in the equation are also sometimes called variance inflation factors (VIF; e.g., see Neter, Wasserman, Kutner, 1985). This terminology denotes the fact that the variances of the standardized regression coefficients can be computed as the product of the residual variance (for the correlation transformed model) times the respective diagonal elements of the inverse correlation matrix. If the predictor variables are uncorrelated, the diagonal elements of the inverse correlation matrix are equal to 1.0; thus, for correlated predictors, these elements represent an "inflation factor" for the variance of the regression coefficients, due to the redundancy of the predictors. See also, Chapter 26 – *Multiple Linear Regression*.

V-Fold Cross-Validation

In v-fold cross-validation, repeated (v) random samples are drawn from the data for the analysis, and the respective model or prediction method, etc. is then applied to compute predicted values, classifications, etc. Typically, summary indices of the accuracy of the prediction are computed over the v replications; thus, this technique enables the analyst to evaluate the overall accuracy of the respective prediction model or method in repeatedly drawn random samples. This method is customarily used in tree classification and regression.

Voronoi Tessellation Graph

The Voronoi tessellation graph plots values of two variables *X* and *Y* in a scatterplot, and then divides the space between individual data points into regions such that the boundaries surrounding each data point enclose an area that is closer to that data point than to any other neighboring points.

This specialized plot is more an analytic technique than just a method to graphically present data. The solutions it offers help to model a variety of phenomena in natural and social sciences (e.g., Coombs, 1964; Ripley, 1981).

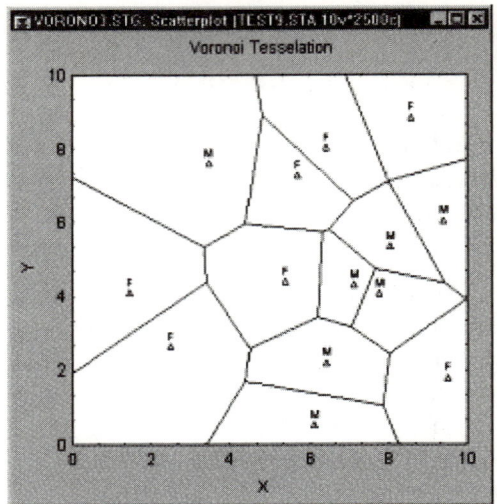

The particular ways in which this method is used depends largely on specific research areas, however, in many of them, it is helpful to add additional dimensions to this plot by using categorization options (as shown in the next illustration).

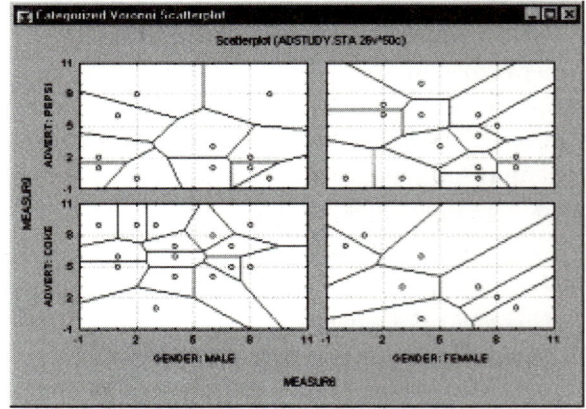

See also, *Data Reduction*.

Voting

See *Bagging*.

Wald Statistic

The results spreadsheet with the parameter estimates for the Cox proportional hazard regression model includes the so-called Wald statistic, and the *p* level for that statistic. This statistic is a test of significance of the regression coefficient; it is based on the asymptotic normality property of maximum likelihood estimates, and is computed as:

$$W = \beta * 1/Var(\beta) * \beta$$

In this formula, β stands for the parameter estimates, and Var(β) stands for the asymptotic variance of the parameter estimates. The Wald statistic is tested against the *chi*-square distribution.

Weibull Distribution

The Weibull distribution (Weibull, 1939, 1951; see also, Lieblein, 1955) has density function (for positive parameters *b, c,* and $\bar\theta$):

$$f(x) = c/b*[(x-\theta)/b]^{c-1} * e^{\wedge}\{-[(x-\theta)/b]^c\}$$
$$\theta < x, \; b > 0, \; c > 0$$

where *b* is the scale parameter of the distribution, *c* is the shape parameter of the distribution, θ is the location parameter of the distribution, and e is the base of the natural logarithm, sometimes called Euler's e (2.71...).

The following illustrations show the Weibull distribution as the shape parameter increases (.5, 1, 2, 3, 4, 5, and 10).

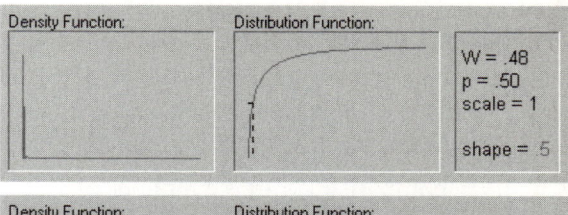

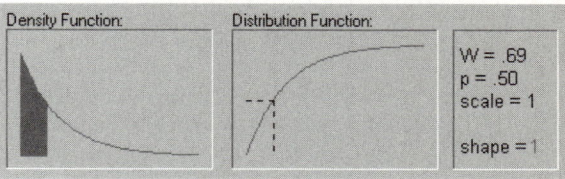

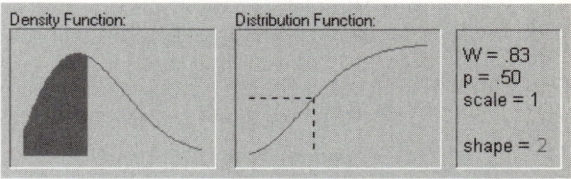

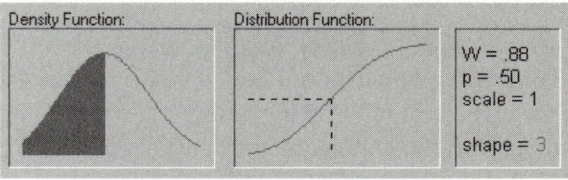

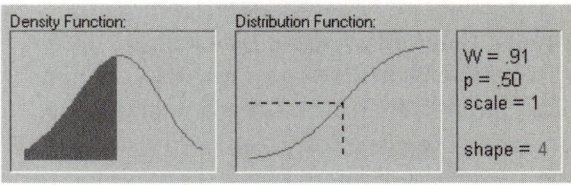

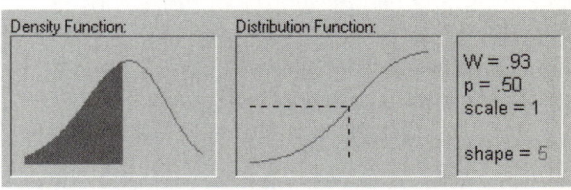

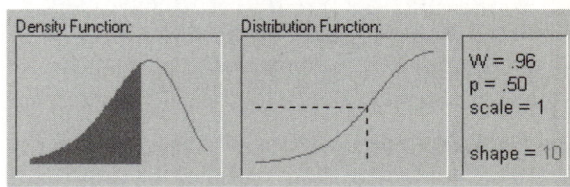

Weigend Weight Regularization (in Neural Networks)

A common problem in neural network training (particularly of multilayer perceptrons) is over-fitting. A network with a large number of weights in comparison with the number of training cases available can achieve a low training error by modeling a function that fits the training data well despite failing to capture the underlying model. An over-fitted model typically has high curvature, as the function is contorted to pass through the points, modeling any noise in addition to the underlying data.

There are several approaches in neural network training to deal with the over-fitting problem (Bishop, 1995). These approaches are listed below.

Select a neural network with just enough units to model the underlying function. The problem with this approach is determining the correct number of units, which is problem-dependent.

Add some noise to the training cases during training (altering the noise on each case each epoch): this "blurs" the position of the training data, and forces the network to model a smoothed version of the data.

Stop training (see *Stopping Conditions*, page 734) when the selection error begins to rise, even if the training error continues to fall. This event is a sure sign that the network is beginning to over-fit the data.

Use a regularization technique, which explicitly penalizes networks with large curvature, thus encouraging the development of a smoother model.

The last technique mentioned is regularization, and this section describes Weigend weight regularization (Weigend et. al., 1991).

A multilayer perceptron model with sigmoid (logistic or hyperbolic tangent) activation functions has higher curvature if the weights are larger. You can see this by considering the shape of the sigmoid curve: if you just look at a small part of the central section, around the value 0.0, it is "nearly linear," and so a network with very small weights will model a "nearly linear" function, which has low curvature. As an aside, note that during training the weights are first set to small values (corresponding to a low curvature function), and then (at least some of them) diverge. One way to promote low curvature therefore is to encourage smaller weights.

Weigend weight regularization does this by adding an extra term to the error function, which penalizes larger weights. Hence the network tends to develop the larger weights that it needs to model the problem, and the others are driven toward zero. The technique can be used with any multilayer perceptron training algorithms (back propagation, conjugate gradient descent, Quasi-Newton Method, quick propagation, and Delta-bar-Delta) apart from Levenberg-Marquardt, which makes its own assumptions about the error function.

The technique is commonly referred to as Weigend weight elimination, as it is possible, once weights become very small, to simply remove them from the network. This is an extremely useful technique for developing models with a "sensible" number of hidden units, and for selecting input variables.

Once a model has been trained with Weigend regularization and excess inputs and hidden units removed, it can be further trained with Weigend regularization turned off, to "sharpen up" the final solution.

Weigend regularization can also be very helpful in that it tends to prevent models from becoming over-fitted.

Note: When using Weigend regularization, the error on the progress graph includes the Weigend penalty factor. If you compare a network trained with Weigend to one without, you may get a false impression that the Weigend-trained network is under-performing. To compare such networks, view the error reported in the summary statistics on the model list (this does not include the Weigend error term).

Technical details. The Weigend error penalty is given by:

$$\lambda \sum_i \frac{w_i^2}{w_0 + w_i^2}$$

where λ is the Regularization coefficient, w_i is each of the weights, and w_o is the Scale coefficient.

The error penalty is added to the error calculated by the network's error function during training, and its derivative is added to the weight's derivative. However, the penalty is ignored when running a network.

The regularization coefficient is usually manipulated to adjust the selective pressure to prune units. The relationship between this coefficient and the number of active units is roughly logarithmic, so the coefficient is typically altered over a wide range (0.01-0.0001, say).

The scale coefficient defines what is a "large" value to the algorithm. The default setting of 1.0 is usually reasonable, and it is seldom altered.

A feature of the Weigend error penalty is that it does not just penalize larger weights. It also prefers to tolerate an uneven mix of some large and some small weights, as opposed to a number of medium-sized weights. It is this property that allows it to "eliminate" weights.

Weighted Least Squares (in Regression)

In some cases it is desirable to apply differential weights to the observations in a regression analysis, and to compute so-called weighted least squares regression estimates. This method is commonly applied when the variances of the residuals are not constant over the range of the independent variable values. In that case, you can apply the inverse values of the variances for the residuals as weights and compute weighted least squares estimates. (In practice, these variances are usually not known, however, they are often proportional to the values of the independent variable(s), and this proportionality can be exploited to compute appropriate case weights.) Neter, Wasserman, and Kutner (1985) describe an example of such an analysis.

Wilcoxon Test

The Wilcoxon test is a nonparametric alternative to t-test for dependent samples. It is designed to test a hypothesis about the location (median) of a population distribution. It often involves the use of matched pairs, for example, "before" and "after" data, in which case it tests for a median difference of zero.

This procedure assumes that the variables under consideration were measured on a scale that allows the rank ordering of observations based on each variable (i.e., ordinal scale) and that allows rank ordering of the differences between variables (this type of scale is sometimes referred to as an ordered metric scale, see Coombs, 1950). For more details, see Siegel & Castellan, 1988. See also, Chapter 29 – *Nonparametric Statistics*.

Win Frequencies (in Neural Networks)

In a Kohonen network, the number of times that each radial unit is the winner when the data set is executed. Units that win frequently represent cluster centers in the topological map. See also, Chapter 27 – *Neural Networks*.

Wire

A wire is a line, usually curved, used in a path diagram to represent variances and covariances of exogenous variables.

X

X-11 Census Method II Seasonal Adjustment

The general idea of seasonal decomposition and adjustment are discussed in the context of the Census I seasonal adjustment method [Seasonal Decomposition (Census I)]. The Census method II (2) is an extension and refinement of the simple adjustment method. Over the years, different versions of the Census method II evolved at the Census Bureau; the method that has become most popular and is used most widely in government and business is the so-called X-11 variant of the Census method II (see Hiskin, Young, & Musgrave, 1967). Subsequently, the term X-11 has become synonymous with this refined version of the Census method II. In addition to the documentation that can be obtained from the Census Bureau, a detailed summary of this method is also provided in Makridakis, Wheelwright, and McGee (1983) and Makridakis and Wheelwright (1989). For more information, see *Classical Seasonal Decomposition (Census Method 1)*.

XML (Extensible Markup Language)

XML (Extensible Markup Language) is a specification language developed by the World Wide Web Consortium (W3C). XML is a language standard designed especially for Web documents to enable programmers to create their own customized tags, thus enabling the definition, transmission, validation, and interpretation of data between applications and between organizations. A special version of .XML is PMML.

Y

Yates Corrected Chi-square

The approximation of the *chi*-square statistic in small 2 x 2 tables can be improved by reducing the absolute value of differences between expected and observed frequencies by 0.5 before squaring (Yates' correction). This correction, which makes the estimation more conservative, is usually applied when the table contains only small observed frequencies, so that some expected frequencies become less than 10 (for further discussion of this correction, see Conover, 1974; Everitt, 1977; Hays, 1988; Kendall & Stuart, 1979; and Mantel, 1974).

Z

Z Distribution (Standard Normal)

The Z distribution (or standard normal distribution) function is determined by the following formula:

$$f(x) = \frac{1}{\sqrt{2\pi}} e^{\frac{-x^2}{2}}$$

$-\infty < x < \infty$

where e is the base of the natural logarithm, sometimes called Euler's e (2.71...) and π is the constant Pi (3.14...).

Note that this distribution is simply a normal distribution where the mean is zero and the standard deviation is one. The Z distribution is commonly used in hypothesis testing for large samples or in situations where the standard deviation is known.

DISTRIBUTION TABLES

Compared to probability calculators, the traditional format of distribution tables such as those presented in this chapter has the advantage of showing many values simultaneously and, thus, enables you to examine and quickly explore ranges of probabilities.

Note that all table values were calculated using the distribution facilities in *STATISTICA*, and they were verified against other published tables.

Standard Normal (Z) Table

The standard normal distribution is used in various hypothesis tests including tests on single means, the difference between two means, and proportions. The standard normal distribution has a mean of 0 and a standard deviation of 1. The following illustrations show various (left) tail areas for this distribution.

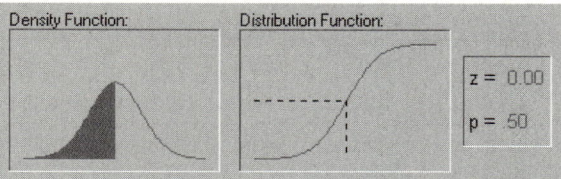

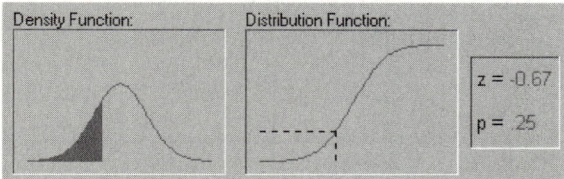

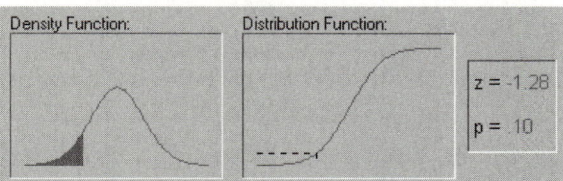

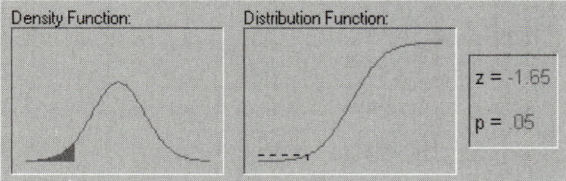

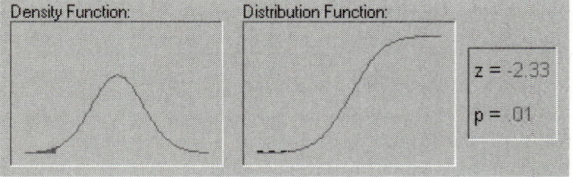

For more information on the normal distribution as it is used in statistical testing, see Chapter 1 – *Elementary Concepts in Statistics*. See also, *Normal Distribution*, page 675.

As shown in the next illustration, the values inside the given table represent the areas under the standard normal curve for values between 0 and the relative z-score. For example, to determine the area under the curve between 0 and 2.36, look in the intersecting cell for the row labeled 2.30 and the column labeled 0.06. The area under the curve is .4909. To determine the area between 0 and a negative value, look in the intersecting cell of the row and column which sums to the absolute value of the number in question. For example, the area under the curve between -1.3 and 0 is equal to the area under the curve between 1.3 and 0, so look at the cell on the 1.3 row and the 0.00 column (the area is 0.4032).

Area between 0 and z

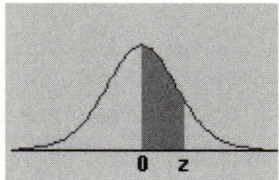

Normal (Z) Distribution

	0.00	0.01	0.02	0.03	0.04	0.05	0.06	0.07	0.08	0.09
0.0	0.0000	0.0040	0.0080	0.0120	0.0160	0.0199	0.0239	0.0279	0.0319	0.0359
0.1	0.0398	0.0438	0.0478	0.0517	0.0557	0.0596	0.0636	0.0675	0.0714	0.0753
0.2	0.0793	0.0832	0.0871	0.0910	0.0948	0.0987	0.1026	0.1064	0.1103	0.1141
0.3	0.1179	0.1217	0.1255	0.1293	0.1331	0.1368	0.1406	0.1443	0.1480	0.1517
0.4	0.1554	0.1591	0.1628	0.1664	0.1700	0.1736	0.1772	0.1808	0.1844	0.1879
0.5	0.1915	0.1950	0.1985	0.2019	0.2054	0.2088	0.2123	0.2157	0.2190	0.2224
0.6	0.2257	0.2291	0.2324	0.2357	0.2389	0.2422	0.2454	0.2486	0.2517	0.2549
0.7	0.2580	0.2611	0.2642	0.2673	0.2704	0.2734	0.2764	0.2794	0.2823	0.2852
0.8	0.2881	0.2910	0.2939	0.2967	0.2995	0.3023	0.3051	0.3078	0.3106	0.3133
0.9	0.3159	0.3186	0.3212	0.3238	0.3264	0.3289	0.3315	0.3340	0.3365	0.3389
1.0	0.3413	0.3438	0.3461	0.3485	0.3508	0.3531	0.3554	0.3577	0.3599	0.3621
1.1	0.3643	0.3665	0.3686	0.3708	0.3729	0.3749	0.3770	0.3790	0.3810	0.3830
1.2	0.3849	0.3869	0.3888	0.3907	0.3925	0.3944	0.3962	0.3980	0.3997	0.4015
1.3	0.4032	0.4049	0.4066	0.4082	0.4099	0.4115	0.4131	0.4147	0.4162	0.4177
1.4	0.4192	0.4207	0.4222	0.4236	0.4251	0.4265	0.4279	0.4292	0.4306	0.4319
1.5	0.4332	0.4345	0.4357	0.4370	0.4382	0.4394	0.4406	0.4418	0.4429	0.4441
1.6	0.4452	0.4463	0.4474	0.4484	0.4495	0.4505	0.4515	0.4525	0.4535	0.4545
1.7	0.4554	0.4564	0.4573	0.4582	0.4591	0.4599	0.4608	0.4616	0.4625	0.4633
1.8	0.4641	0.4649	0.4656	0.4664	0.4671	0.4678	0.4686	0.4693	0.4699	0.4706
1.9	0.4713	0.4719	0.4726	0.4732	0.4738	0.4744	0.4750	0.4756	0.4761	0.4767
2.0	0.4772	0.4778	0.4783	0.4788	0.4793	0.4798	0.4803	0.4808	0.4812	0.4817
2.1	0.4821	0.4826	0.4830	0.4834	0.4838	0.4842	0.4846	0.4850	0.4854	0.4857
2.2	0.4861	0.4864	0.4868	0.4871	0.4875	0.4878	0.4881	0.4884	0.4887	0.4890
2.3	0.4893	0.4896	0.4898	0.4901	0.4904	0.4906	0.4909	0.4911	0.4913	0.4916
2.4	0.4918	0.4920	0.4922	0.4925	0.4927	0.4929	0.4931	0.4932	0.4934	0.4936
2.5	0.4938	0.4940	0.4941	0.4943	0.4945	0.4946	0.4948	0.4949	0.4951	0.4952
2.6	0.4953	0.4955	0.4956	0.4957	0.4959	0.4960	0.4961	0.4962	0.4963	0.4964
2.7	0.4965	0.4966	0.4967	0.4968	0.4969	0.4970	0.4971	0.4972	0.4973	0.4974
2.8	0.4974	0.4975	0.4976	0.4977	0.4977	0.4978	0.4979	0.4979	0.4980	0.4981
2.9	0.4981	0.4982	0.4982	0.4983	0.4984	0.4984	0.4985	0.4985	0.4986	0.4986
3.0	0.4987	0.4987	0.4987	0.4988	0.4988	0.4989	0.4989	0.4989	0.4990	0.4990

Student's *t* Table

The shape of the Student's *t* distribution is determined by the degrees of freedom. As shown in the following illustrations, its shape changes as the degrees of freedom increases.

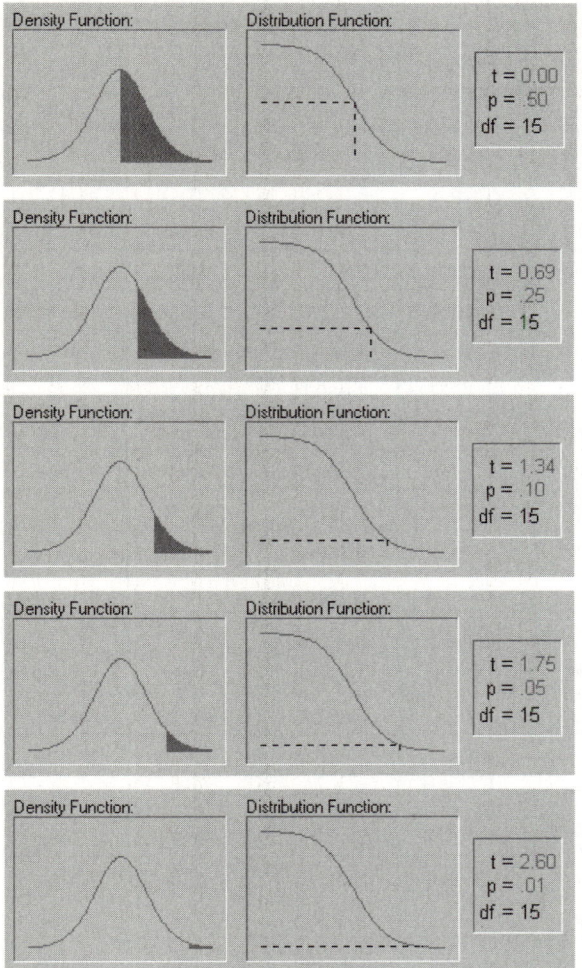

For more information on how this distribution is used in hypothesis testing, see *t*-test for independent samples and *t*-test for dependent samples in Chapter 2 – *Basic Statistics and Tables*. See also, Student's *t* Distribution. As indicated by the chart in the next illustration, the areas given at the top of this table are the right tail areas for the t-value inside the table. To determine the 0.05 critical value from the t-distribution with 6 degrees of freedom, look in the 0.05 column at the 6 row: $t_{(.05,6)} = 1.943180$.

t table with right tail probabilities

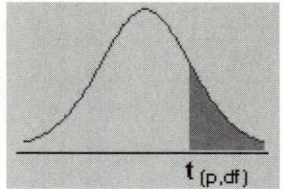

Student's *t* Distribution

df\p	0.40	0.25	0.10	0.05	0.025	0.01	0.005	0.0005
1	0.324920	1.000000	3.077684	6.313752	12.70620	31.82052	63.65674	636.6192
2	0.288675	0.816497	1.885618	2.919986	4.30265	6.96456	9.92484	31.5991
3	0.276671	0.764892	1.637744	2.353363	3.18245	4.54070	5.84091	12.9240
4	0.270722	0.740697	1.533206	2.131847	2.77645	3.74695	4.60409	8.6103
5	0.267181	0.726687	1.475884	2.015048	2.57058	3.36493	4.03214	6.8688
6	0.264835	0.717558	1.439756	1.943180	2.44691	3.14267	3.70743	5.9588
7	0.263167	0.711142	1.414924	1.894579	2.36462	2.99795	3.49948	5.4079
8	0.261921	0.706387	1.396815	1.859548	2.30600	2.89646	3.35539	5.0413
9	0.260955	0.702722	1.383029	1.833113	2.26216	2.82144	3.24984	4.7809
10	0.260185	0.699812	1.372184	1.812461	2.22814	2.76377	3.16927	4.5869
11	0.259556	0.697445	1.363430	1.795885	2.20099	2.71808	3.10581	4.4370
12	0.259033	0.695483	1.356217	1.782288	2.17881	2.68100	3.05454	4.3178
13	0.258591	0.693829	1.350171	1.770933	2.16037	2.65031	3.01228	4.2208
14	0.258213	0.692417	1.345030	1.761310	2.14479	2.62449	2.97684	4.1405
15	0.257885	0.691197	1.340606	1.753050	2.13145	2.60248	2.94671	4.0728
16	0.257599	0.690132	1.336757	1.745884	2.11991	2.58349	2.92078	4.0150
17	0.257347	0.689195	1.333379	1.739607	2.10982	2.56693	2.89823	3.9651
18	0.257123	0.688364	1.330391	1.734064	2.10092	2.55238	2.87844	3.9216
19	0.256923	0.687621	1.327728	1.729133	2.09302	2.53948	2.86093	3.8834
20	0.256743	0.686954	1.325341	1.724718	2.08596	2.52798	2.84534	3.8495
21	0.256580	0.686352	1.323188	1.720743	2.07961	2.51765	2.83136	3.8193
22	0.256432	0.685805	1.321237	1.717144	2.07387	2.50832	2.81876	3.7921
23	0.256297	0.685306	1.319460	1.713872	2.06866	2.49987	2.80734	3.7676
24	0.256173	0.684850	1.317836	1.710882	2.06390	2.49216	2.79694	3.7454
25	0.256060	0.684430	1.316345	1.708141	2.05954	2.48511	2.78744	3.7251
26	0.255955	0.684043	1.314972	1.705618	2.05553	2.47863	2.77871	3.7066
27	0.255858	0.683685	1.313703	1.703288	2.05183	2.47266	2.77068	3.6896
28	0.255768	0.683353	1.312527	1.701131	2.04841	2.46714	2.76326	3.6739
29	0.255684	0.683044	1.311434	1.699127	2.04523	2.46202	2.75639	3.6594
30	0.255605	0.682756	1.310415	1.697261	2.04227	2.45726	2.75000	3.6460
inf	0.253347	0.674490	1.281552	1.644854	1.95996	2.32635	2.57583	3.2905

Chi-Square Table

As with the Student's *t*-distribution, the *chi*-square distribution's shape is determined by its degrees of freedom.

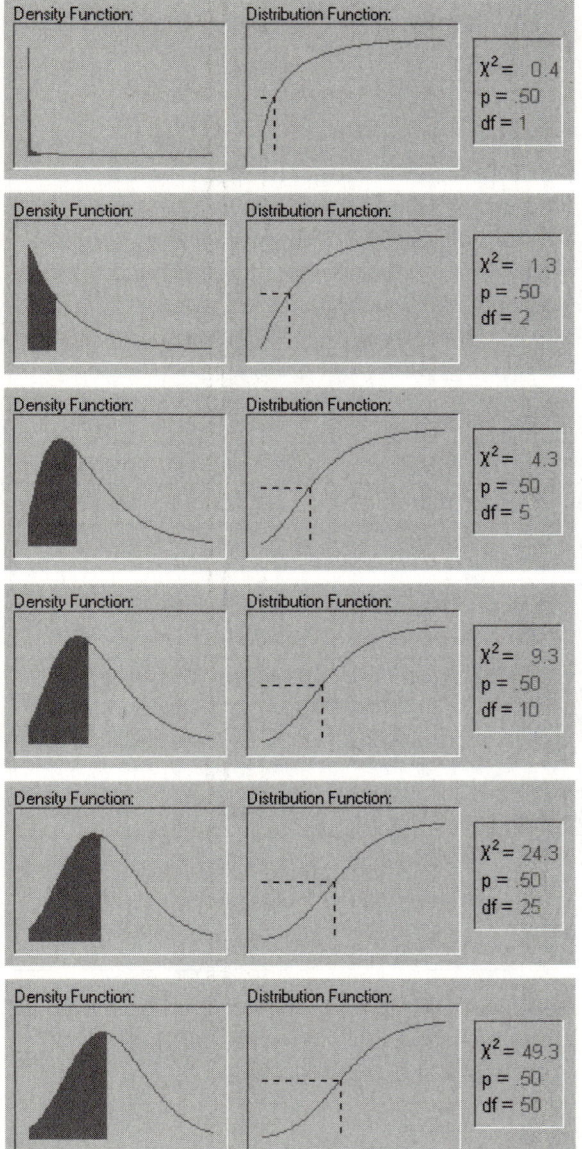

The previous illustrations show the shape of the *chi-square* distribution as the degrees of freedom increase (1, 2, 5, 10, 25 and 50).

For examples of tests of hypothesis that use the *chi*-square distribution, see *Statistics in Crosstabulation Tables* in Chapter 2 – *Basic Statistics and Tables* and Chapter 28 – *Nonlinear Estimation*. See also, *Chi-square Distribution* in Chapter 14 – *Distribution Fitting*. As shown in the illustration the next illustration, the values inside this table are critical values of the *chi*-square distribution with the corresponding degrees of freedom. To determine the value from a *chi*-square distribution (with a specific degree of freedom) which has a given area above it, go to the given area column and the desired degree of freedom row. For example, the .25 critical value for a *chi*-square with 4 degrees of freedom is 5.38527. This means that the area to the right of 5.38527 in a *chi*-square distribution with 4 degrees of freedom is .25.

Right tail areas for the Chi-square distribution

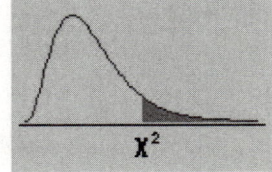

Chi-Square Distribution

df\area	.995	.990	.975	.950	.900	.750
1	0.00004	0.00016	0.00098	0.00393	0.01579	0.10153
2	0.01003	0.02010	0.05064	0.10259	0.21072	0.57536
3	0.07172	0.11483	0.21580	0.35185	0.58437	1.21253
4	0.20699	0.29711	0.48442	0.71072	1.06362	1.92256
5	0.41174	0.55430	0.83121	1.14548	1.61031	2.67460
6	0.67573	0.87209	1.23734	1.63538	2.20413	3.45460
7	0.98926	1.23904	1.68987	2.16735	2.83311	4.25485
8	1.34441	1.64650	2.17973	2.73264	3.48954	5.07064
9	1.73493	2.08790	2.70039	3.32511	4.16816	5.89883
10	2.15586	2.55821	3.24697	3.94030	4.86518	6.73720
11	2.60322	3.05348	3.81575	4.57481	5.57778	7.58414
12	3.07382	3.57057	4.40379	5.22603	6.30380	8.43842
13	3.56503	4.10692	5.00875	5.89186	7.04150	9.29907
14	4.07467	4.66043	5.62873	6.57063	7.78953	10.16531
15	4.60092	5.22935	6.26214	7.26094	8.54676	11.03654
16	5.14221	5.81221	6.90766	7.96165	9.31224	11.91222
17	5.69722	6.40776	7.56419	8.67176	10.08519	12.79193
18	6.26480	7.01491	8.23075	9.39046	10.86494	13.67529
19	6.84397	7.63273	8.90652	10.11701	11.65091	14.56200
20	7.43384	8.26040	9.59078	10.85081	12.44261	15.45177
21	8.03365	8.89720	10.28290	11.59131	13.23960	16.34438
22	8.64272	9.54249	10.98232	12.33801	14.04149	17.23962
23	9.26042	10.19572	11.68855	13.09051	14.84796	18.13730
24	9.88623	10.85636	12.40115	13.84843	15.65868	19.03725
25	10.51965	11.52398	13.11972	14.61141	16.47341	19.93934
26	11.16024	12.19815	13.84390	15.37916	17.29188	20.84343
27	11.80759	12.87850	14.57338	16.15140	18.11390	21.74940
28	12.46134	13.56471	15.30786	16.92788	18.93924	22.65716
29	13.12115	14.25645	16.04707	17.70837	19.76774	23.56659
30	13.78672	14.95346	16.79077	18.49266	20.59923	24.47761

Chi-Square Distribution (continued)

df\area	.500	.250	.100	.050	.025	.010	.005
1	0.45494	1.32330	2.70554	3.84146	5.02389	6.63490	7.87944
2	1.38629	2.77259	4.60517	5.99146	7.37776	9.21034	10.59663
3	2.36597	4.10834	6.25139	7.81473	9.34840	11.34487	12.83816
4	3.35669	5.38527	7.77944	9.48773	11.14329	13.27670	14.86026
5	4.35146	6.62568	9.23636	11.07050	12.83250	15.08627	16.74960
6	5.34812	7.84080	10.64464	12.59159	14.44938	16.81189	18.54758
7	6.34581	9.03715	12.01704	14.06714	16.01276	18.47531	20.27774
8	7.34412	10.21885	13.36157	15.50731	17.53455	20.09024	21.95495
9	8.34283	11.38875	14.68366	16.91898	19.02277	21.66599	23.58935
10	9.34182	12.54886	15.98718	18.30704	20.48318	23.20925	25.18818
11	10.34100	13.70069	17.27501	19.67514	21.92005	24.72497	26.75685
12	11.34032	14.84540	18.54935	21.02607	23.33666	26.21697	28.29952
13	12.33976	15.98391	19.81193	22.36203	24.73560	27.68825	29.81947
14	13.33927	17.11693	21.06414	23.68479	26.11895	29.14124	31.31935
15	14.33886	18.24509	22.30713	24.99579	27.48839	30.57791	32.80132
16	15.33850	19.36886	23.54183	26.29623	28.84535	31.99993	34.26719
17	16.33818	20.48868	24.76904	27.58711	30.19101	33.40866	35.71847
18	17.33790	21.60489	25.98942	28.86930	31.52638	34.80531	37.15645
19	18.33765	22.71781	27.20357	30.14353	32.85233	36.19087	38.58226
20	19.33743	23.82769	28.41198	31.41043	34.16961	37.56623	39.99685
21	20.33723	24.93478	29.61509	32.67057	35.47888	38.93217	41.40106
22	21.33704	26.03927	30.81328	33.92444	36.78071	40.28936	42.79565
23	22.33688	27.14134	32.00690	35.17246	38.07563	41.63840	44.18128
24	23.33673	28.24115	33.19624	36.41503	39.36408	42.97982	45.55851
25	24.33659	29.33885	34.38159	37.65248	40.64647	44.31410	46.92789
26	25.33646	30.43457	35.56317	38.88514	41.92317	45.64168	48.28988
27	26.33634	31.52841	36.74122	40.11327	43.19451	46.96294	49.64492
28	27.33623	32.62049	37.91592	41.33714	44.46079	48.27824	50.99338
29	28.33613	33.71091	39.08747	42.55697	45.72229	49.58788	52.33562
30	29.33603	34.79974	40.25602	43.77297	46.97924	50.89218	53.67196

F Distribution Tables

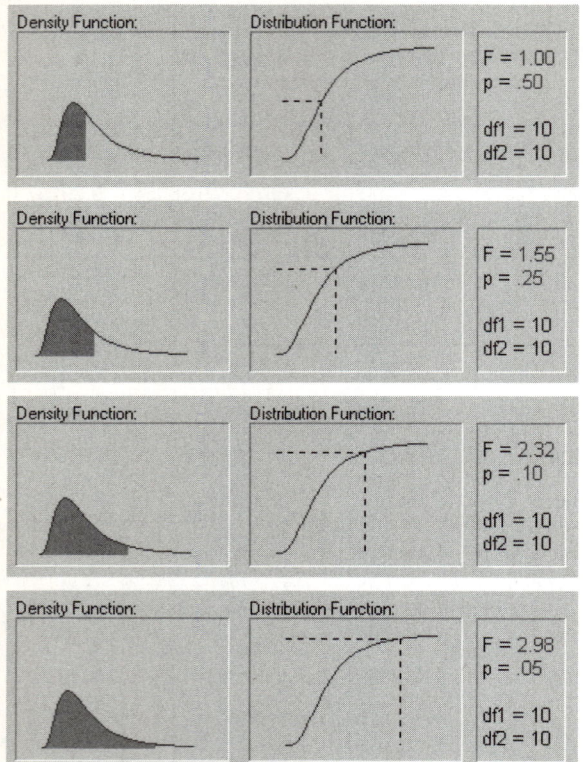

The F distribution is a right-skewed distribution used most commonly in Analysis of Variance (see Chapter 3 – *ANOVA/MANOVA*). The F distribution is a ratio of two *chi-square* distributions, and a specific F distribution is denoted by the degrees of freedom for the numerator *chi*-square and the degrees of freedom for the denominator *chi*-square. An example of the $F_{(10,10)}$ distribution is shown in the animation above. When referencing the F distribution, the numerator degrees of freedom are always given first, as switching the order of degrees of freedom changes the distribution (e.g., $F_{(10,12)}$ does not equal $F_{(12,10)}$). For the four F tables below, the rows represent denominator degrees of freedom and the columns represent

numerator degrees of freedom. The right tail area is given in the name of the table. For example, to determine the .05 critical value for an F distribution with 10 and 12 degrees of freedom, look in the 10 column (numerator) and 12 row (denominator) of the F table for alpha=.05. $F_{(.05, 10, 12)} = 2.7534$.

See F Table for alpha=.10 on page 769.

See F Table for alpha=.05 on page 771.

See F Table for alpha=.025 on page 773.

See F Table for alpha=.01 on page 775.

F Table for alpha = .10

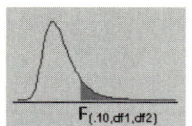

$F_{(.10, df1, df2)}$

df2/df1	1	2	3	4	5	6	7	8	9	10
1	39.86346	49.50000	53.59324	55.83296	57.24008	58.20442	58.90595	59.43898	59.85759	60.19498
2	8.52632	9.00000	9.16179	9.24342	9.29263	9.32553	9.34908	9.36677	9.38054	9.39157
3	5.53832	5.46238	5.39077	5.34264	5.30916	5.28473	5.26619	5.25167	5.24000	5.23041
4	4.54477	4.32456	4.19086	4.10725	4.05058	4.00975	3.97897	3.95494	3.93567	3.91988
5	4.06042	3.77972	3.61948	3.52020	3.45298	3.40451	3.36790	3.33928	3.31628	3.29740
6	3.77595	3.46330	3.28876	3.18076	3.10751	3.05455	3.01446	2.98304	2.95774	2.93693
7	3.58943	3.25744	3.07407	2.96053	2.88334	2.82739	2.78493	2.75158	2.72468	2.70251
8	3.45792	3.11312	2.92380	2.80643	2.72645	2.66833	2.62413	2.58935	2.56124	2.53804
9	3.36030	3.00645	2.81286	2.69268	2.61061	2.55086	2.50531	2.46941	2.44034	2.41632
10	3.28502	2.92447	2.72767	2.60534	2.52164	2.46058	2.41397	2.37715	2.34731	2.32260
11	3.22520	2.85951	2.66023	2.53619	2.45118	2.38907	2.34157	2.30400	2.27350	2.24823
12	3.17655	2.80680	2.60552	2.48010	2.39402	2.33102	2.28278	2.24457	2.21352	2.18776
13	3.13621	2.76317	2.56027	2.43371	2.34672	2.28298	2.23410	2.19535	2.16382	2.13763
14	3.10221	2.72647	2.52222	2.39469	2.30694	2.24256	2.19313	2.15390	2.12195	2.09540
15	3.07319	2.69517	2.48979	2.36143	2.27302	2.20808	2.15818	2.11853	2.08621	2.05932
16	3.04811	2.66817	2.46181	2.33274	2.24376	2.17833	2.12800	2.08798	2.05533	2.02815
17	3.02623	2.64464	2.43743	2.30775	2.21825	2.15239	2.10169	2.06134	2.02839	2.00094
18	3.00698	2.62395	2.41601	2.28577	2.19583	2.12958	2.07854	2.03789	2.00467	1.97698
19	2.98990	2.60561	2.39702	2.26630	2.17596	2.10936	2.05802	2.01710	1.98364	1.95573
20	2.97465	2.58925	2.38009	2.24893	2.15823	2.09132	2.03970	1.99853	1.96485	1.93674
21	2.96096	2.57457	2.36489	2.23334	2.14231	2.07512	2.02325	1.98186	1.94797	1.91967
22	2.94858	2.56131	2.35117	2.21927	2.12794	2.06050	2.00840	1.96680	1.93273	1.90425
23	2.93736	2.54929	2.33873	2.20651	2.11491	2.04723	1.99492	1.95312	1.91888	1.89025
24	2.92712	2.53833	2.32739	2.19488	2.10303	2.03513	1.98263	1.94066	1.90625	1.87748
25	2.91774	2.52831	2.31702	2.18424	2.09216	2.02406	1.97138	1.92925	1.89469	1.86578
26	2.90913	2.51910	2.30749	2.17447	2.08218	2.01389	1.96104	1.91876	1.88407	1.85503
27	2.90119	2.51061	2.29871	2.16546	2.07298	2.00452	1.95151	1.90909	1.87427	1.84511
28	2.89385	2.50276	2.29060	2.15714	2.06447	1.99585	1.94270	1.90014	1.86520	1.83593
29	2.88703	2.49548	2.28307	2.14941	2.05658	1.98781	1.93452	1.89184	1.85679	1.82741
30	2.88069	2.48872	2.27607	2.14223	2.04925	1.98033	1.92692	1.88412	1.84896	1.81949
40	2.83535	2.44037	2.22609	2.09095	1.99682	1.92688	1.87252	1.82886	1.79290	1.76269
60	2.79107	2.39325	2.17741	2.04099	1.94571	1.87472	1.81939	1.77483	1.73802	1.70701
120	2.74781	2.34734	2.12999	1.99230	1.89587	1.82381	1.76748	1.72196	1.68425	1.65238
inf	2.70554	2.30259	2.08380	1.94486	1.84727	1.77411	1.71672	1.67020	1.63152	1.59872

F Table for alpha = .10 (continued)

df2/df1	12	15	20	24	30	40	60	120	INF
1	60.70521	61.22034	61.74029	62.00205	62.26497	62.52905	62.79428	63.06064	63.32812
2	9.40813	9.42471	9.44131	9.44962	9.45793	9.46624	9.47456	9.48289	9.49122
3	5.21562	5.20031	5.18448	5.17636	5.16811	5.15972	5.15119	5.14251	5.13370
4	3.89553	3.87036	3.84434	3.83099	3.81742	3.80361	3.78957	3.77527	3.76073
5	3.26824	3.23801	3.20665	3.19052	3.17408	3.15732	3.14023	3.12279	3.10500
6	2.90472	2.87122	2.83634	2.81834	2.79996	2.78117	2.76195	2.74229	2.72216
7	2.66811	2.63223	2.59473	2.57533	2.55546	2.53510	2.51422	2.49279	2.47079
8	2.50196	2.46422	2.42464	2.40410	2.38302	2.36136	2.33910	2.31618	2.29257
9	2.37888	2.33962	2.29832	2.27683	2.25472	2.23196	2.20849	2.18427	2.15923
10	2.28405	2.24351	2.20074	2.17843	2.15543	2.13169	2.10716	2.08176	2.05542
11	2.20873	2.16709	2.12305	2.10001	2.07621	2.05161	2.02612	1.99965	1.97211
12	2.14744	2.10485	2.05968	2.03599	2.01149	1.98610	1.95973	1.93228	1.90361
13	2.09659	2.05316	2.00698	1.98272	1.95757	1.93147	1.90429	1.87591	1.84620
14	2.05371	2.00953	1.96245	1.93766	1.91193	1.88516	1.85723	1.82800	1.79728
15	2.01707	1.97222	1.92431	1.89904	1.87277	1.84539	1.81676	1.78672	1.75505
16	1.98539	1.93992	1.89127	1.86556	1.83879	1.81084	1.78156	1.75075	1.71817
17	1.95772	1.91169	1.86236	1.83624	1.80901	1.78053	1.75063	1.71909	1.68564
18	1.93334	1.88681	1.83685	1.81035	1.78269	1.75371	1.72322	1.69099	1.65671
19	1.91170	1.86471	1.81416	1.78731	1.75924	1.72979	1.69876	1.66587	1.63077
20	1.89236	1.84494	1.79384	1.76667	1.73822	1.70833	1.67678	1.64326	1.60738
21	1.87497	1.82715	1.77555	1.74807	1.71927	1.68896	1.65691	1.62278	1.58615
22	1.85925	1.81106	1.75899	1.73122	1.70208	1.67138	1.63885	1.60415	1.56678
23	1.84497	1.79643	1.74392	1.71588	1.68643	1.65535	1.62237	1.58711	1.54903
24	1.83194	1.78308	1.73015	1.70185	1.67210	1.64067	1.60726	1.57146	1.53270
25	1.82000	1.77083	1.71752	1.68898	1.65895	1.62718	1.59335	1.55703	1.51760
26	1.80902	1.75957	1.70589	1.67712	1.64682	1.61472	1.58050	1.54368	1.50360
27	1.79889	1.74917	1.69514	1.66616	1.63560	1.60320	1.56859	1.53129	1.49057
28	1.78951	1.73954	1.68519	1.65600	1.62519	1.59250	1.55753	1.51976	1.47841
29	1.78081	1.73060	1.67593	1.64655	1.61551	1.58253	1.54721	1.50899	1.46704
30	1.77270	1.72227	1.66731	1.63774	1.60648	1.57323	1.53757	1.49891	1.45636
40	1.71456	1.66241	1.60515	1.57411	1.54108	1.50562	1.46716	1.42476	1.37691
60	1.65743	1.60337	1.54349	1.51072	1.47554	1.43734	1.39520	1.34757	1.29146
120	1.60120	1.54500	1.48207	1.44723	1.40938	1.36760	1.32034	1.26457	1.19256
inf	1.54578	1.48714	1.42060	1.38318	1.34187	1.29513	1.23995	1.16860	1.00000

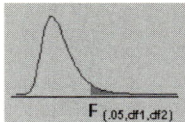

F Table for alpha = .05

$F_{(.05, df1, df2)}$

df2/df1	1	2	3	4	5	6	7	8	9
1	161.4476	199.5000	215.7073	224.5832	230.1619	233.9860	236.7684	238.8827	240.543
2	18.5128	19.0000	19.1643	19.2468	19.2964	19.3295	19.3532	19.3710	19.3848
3	10.1280	9.5521	9.2766	9.1172	9.0135	8.9406	8.8867	8.8452	8.8123
4	7.7086	6.9443	6.5914	6.3882	6.2561	6.1631	6.0942	6.0410	5.9988
5	6.6079	5.7861	5.4095	5.1922	5.0503	4.9503	4.8759	4.8183	4.7725
6	5.9874	5.1433	4.7571	4.5337	4.3874	4.2839	4.2067	4.1468	4.0990
7	5.5914	4.7374	4.3468	4.1203	3.9715	3.8660	3.7870	3.7257	3.6767
8	5.3177	4.4590	4.0662	3.8379	3.6875	3.5806	3.5005	3.4381	3.3881
9	5.1174	4.2565	3.8625	3.6331	3.4817	3.3738	3.2927	3.2296	3.1789
10	4.9646	4.1028	3.7083	3.4780	3.3258	3.2172	3.1355	3.0717	3.0204
11	4.8443	3.9823	3.5874	3.3567	3.2039	3.0946	3.0123	2.9480	2.8962
12	4.7472	3.8853	3.4903	3.2592	3.1059	2.9961	2.9134	2.8486	2.7964
13	4.6672	3.8056	3.4105	3.1791	3.0254	2.9153	2.8321	2.7669	2.7144
14	4.6001	3.7389	3.3439	3.1122	2.9582	2.8477	2.7642	2.6987	2.6458
15	4.5431	3.6823	3.2874	3.0556	2.9013	2.7905	2.7066	2.6408	2.5876
16	4.4940	3.6337	3.2389	3.0069	2.8524	2.7413	2.6572	2.5911	2.5377
17	4.4513	3.5915	3.1968	2.9647	2.8100	2.6987	2.6143	2.5480	2.4943
18	4.4139	3.5546	3.1599	2.9277	2.7729	2.6613	2.5767	2.5102	2.4563
19	4.3807	3.5219	3.1274	2.8951	2.7401	2.6283	2.5435	2.4768	2.4227
20	4.3512	3.4928	3.0984	2.8661	2.7109	2.5990	2.5140	2.4471	2.3928
21	4.3248	3.4668	3.0725	2.8401	2.6848	2.5727	2.4876	2.4205	2.3660
22	4.3009	3.4434	3.0491	2.8167	2.6613	2.5491	2.4638	2.3965	2.3419
23	4.2793	3.4221	3.0280	2.7955	2.6400	2.5277	2.4422	2.3748	2.3201
24	4.2597	3.4028	3.0088	2.7763	2.6207	2.5082	2.4226	2.3551	2.3002
25	4.2417	3.3852	2.9912	2.7587	2.6030	2.4904	2.4047	2.3371	2.2821
26	4.2252	3.3690	2.9752	2.7426	2.5868	2.4741	2.3883	2.3205	2.2655
27	4.2100	3.3541	2.9604	2.7278	2.5719	2.4591	2.3732	2.3053	2.2501
28	4.1960	3.3404	2.9467	2.7141	2.5581	2.4453	2.3593	2.2913	2.2360
29	4.1830	3.3277	2.9340	2.7014	2.5454	2.4324	2.3463	2.2783	2.2229
30	4.1709	3.3158	2.9223	2.6896	2.5336	2.4205	2.3343	2.2662	2.2107
40	4.0847	3.2317	2.8387	2.6060	2.4495	2.3359	2.2490	2.1802	2.1240
60	4.0012	3.1504	2.7581	2.5252	2.3683	2.2541	2.1665	2.0970	2.0401
120	3.9201	3.0718	2.6802	2.4472	2.2899	2.1750	2.0868	2.0164	1.9588
inf	3.8415	2.9957	2.6049	2.3719	2.2141	2.0986	2.0096	1.9384	1.8799

F Table for alpha = .05 (continued)

df2/df1	10	12	15	20	24	30	40	60	120	INF
1	241.8817	243.9060	245.9499	248.0131	249.0518	250.0951	251.1432	252.1957	253.2529	254.3144
2	19.3959	19.4125	19.4291	19.4458	19.4541	19.4624	19.4707	19.4791	19.4874	19.4957
3	8.7855	8.7446	8.7029	8.6602	8.6385	8.6166	8.5944	8.5720	8.5494	8.5264
4	5.9644	5.9117	5.8578	5.8025	5.7744	5.7459	5.7170	5.6877	5.6581	5.6281
5	4.7351	4.6777	4.6188	4.5581	4.5272	4.4957	4.4638	4.4314	4.3985	4.3650
6	4.0600	3.9999	3.9381	3.8742	3.8415	3.8082	3.7743	3.7398	3.7047	3.6689
7	3.6365	3.5747	3.5107	3.4445	3.4105	3.3758	3.3404	3.3043	3.2674	3.2298
8	3.3472	3.2839	3.2184	3.1503	3.1152	3.0794	3.0428	3.0053	2.9669	2.9276
9	3.1373	3.0729	3.0061	2.9365	2.9005	2.8637	2.8259	2.7872	2.7475	2.7067
10	2.9782	2.9130	2.8450	2.7740	2.7372	2.6996	2.6609	2.6211	2.5801	2.5379
11	2.8536	2.7876	2.7186	2.6464	2.6090	2.5705	2.5309	2.4901	2.4480	2.4045
12	2.7534	2.6866	2.6169	2.5436	2.5055	2.4663	2.4259	2.3842	2.3410	2.2962
13	2.6710	2.6037	2.5331	2.4589	2.4202	2.3803	2.3392	2.2966	2.2524	2.2064
14	2.6022	2.5342	2.4630	2.3879	2.3487	2.3082	2.2664	2.2229	2.1778	2.1307
15	2.5437	2.4753	2.4034	2.3275	2.2878	2.2468	2.2043	2.1601	2.1141	2.0658
16	2.4935	2.4247	2.3522	2.2756	2.2354	2.1938	2.1507	2.1058	2.0589	2.0096
17	2.4499	2.3807	2.3077	2.2304	2.1898	2.1477	2.1040	2.0584	2.0107	1.9604
18	2.4117	2.3421	2.2686	2.1906	2.1497	2.1071	2.0629	2.0166	1.9681	1.9168
19	2.3779	2.3080	2.2341	2.1555	2.1141	2.0712	2.0264	1.9795	1.9302	1.8780
20	2.3479	2.2776	2.2033	2.1242	2.0825	2.0391	1.9938	1.9464	1.8963	1.8432
21	2.3210	2.2504	2.1757	2.0960	2.0540	2.0102	1.9645	1.9165	1.8657	1.8117
22	2.2967	2.2258	2.1508	2.0707	2.0283	1.9842	1.9380	1.8894	1.8380	1.7831
23	2.2747	2.2036	2.1282	2.0476	2.0050	1.9605	1.9139	1.8648	1.8128	1.7570
24	2.2547	2.1834	2.1077	2.0267	1.9838	1.9390	1.8920	1.8424	1.7896	1.7330
25	2.2365	2.1649	2.0889	2.0075	1.9643	1.9192	1.8718	1.8217	1.7684	1.7110
26	2.2197	2.1479	2.0716	1.9898	1.9464	1.9010	1.8533	1.8027	1.7488	1.6906
27	2.2043	2.1323	2.0558	1.9736	1.9299	1.8842	1.8361	1.7851	1.7306	1.6717
28	2.1900	2.1179	2.0411	1.9586	1.9147	1.8687	1.8203	1.7689	1.7138	1.6541
29	2.1768	2.1045	2.0275	1.9446	1.9005	1.8543	1.8055	1.7537	1.6981	1.6376
30	2.1646	2.0921	2.0148	1.9317	1.8874	1.8409	1.7918	1.7396	1.6835	1.6223
40	2.0772	2.0035	1.9245	1.8389	1.7929	1.7444	1.6928	1.6373	1.5766	1.5089
60	1.9926	1.9174	1.8364	1.7480	1.7001	1.6491	1.5943	1.5343	1.4673	1.3893
120	1.9105	1.8337	1.7505	1.6587	1.6084	1.5543	1.4952	1.4290	1.3519	1.2539
inf	1.8307	1.7522	1.6664	1.5705	1.5173	1.4591	1.3940	1.3180	1.2214	1.000

F Table for alpha = .025

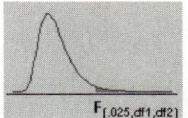

$F_{(.025, df1, df2)}$

df2/df1	1	2	3	4	5	6	7	8	9
1	647.7890	799.5000	864.1630	899.5833	921.8479	937.1111	948.2169	956.6562	963.2846
2	38.5063	39.0000	39.1655	39.2484	39.2982	39.3315	39.3552	39.3730	39.3869
3	17.4434	16.0441	15.4392	15.1010	14.8848	14.7347	14.6244	14.5399	14.4731
4	12.2179	10.6491	9.9792	9.6045	9.3645	9.1973	9.0741	8.9796	8.9047
5	10.0070	8.4336	7.7636	7.3879	7.1464	6.9777	6.8531	6.7572	6.6811
6	8.8131	7.2599	6.5988	6.2272	5.9876	5.8198	5.6955	5.5996	5.5234
7	8.0727	6.5415	5.8898	5.5226	5.2852	5.1186	4.9949	4.8993	4.8232
8	7.5709	6.0595	5.4160	5.0526	4.8173	4.6517	4.5286	4.4333	4.3572
9	7.2093	5.7147	5.0781	4.7181	4.4844	4.3197	4.1970	4.1020	4.0260
10	6.9367	5.4564	4.8256	4.4683	4.2361	4.0721	3.9498	3.8549	3.7790
11	6.7241	5.2559	4.6300	4.2751	4.0440	3.8807	3.7586	3.6638	3.5879
12	6.5538	5.0959	4.4742	4.1212	3.8911	3.7283	3.6065	3.5118	3.4358
13	6.4143	4.9653	4.3472	3.9959	3.7667	3.6043	3.4827	3.3880	3.3120
14	6.2979	4.8567	4.2417	3.8919	3.6634	3.5014	3.3799	3.2853	3.2093
15	6.1995	4.7650	4.1528	3.8043	3.5764	3.4147	3.2934	3.1987	3.1227
16	6.1151	4.6867	4.0768	3.7294	3.5021	3.3406	3.2194	3.1248	3.0488
17	6.0420	4.6189	4.0112	3.6648	3.4379	3.2767	3.1556	3.0610	2.9849
18	5.9781	4.5597	3.9539	3.6083	3.3820	3.2209	3.0999	3.0053	2.9291
19	5.9216	4.5075	3.9034	3.5587	3.3327	3.1718	3.0509	2.9563	2.8801
20	5.8715	4.4613	3.8587	3.5147	3.2891	3.1283	3.0074	2.9128	2.8365
21	5.8266	4.4199	3.8188	3.4754	3.2501	3.0895	2.9686	2.8740	2.7977
22	5.7863	4.3828	3.7829	3.4401	3.2151	3.0546	2.9338	2.8392	2.7628
23	5.7498	4.3492	3.7505	3.4083	3.1835	3.0232	2.9023	2.8077	2.7313
24	5.7166	4.3187	3.7211	3.3794	3.1548	2.9946	2.8738	2.7791	2.7027
25	5.6864	4.2909	3.6943	3.3530	3.1287	2.9685	2.8478	2.7531	2.6766
26	5.6586	4.2655	3.6697	3.3289	3.1048	2.9447	2.8240	2.7293	2.6528
27	5.6331	4.2421	3.6472	3.3067	3.0828	2.9228	2.8021	2.7074	2.6309
28	5.6096	4.2205	3.6264	3.2863	3.0626	2.9027	2.7820	2.6872	2.6106
29	5.5878	4.2006	3.6072	3.2674	3.0438	2.8840	2.7633	2.6686	2.5919
30	5.5675	4.1821	3.5894	3.2499	3.0265	2.8667	2.7460	2.6513	2.5746
40	5.4239	4.0510	3.4633	3.1261	2.9037	2.7444	2.6238	2.5289	2.4519
60	5.2856	3.9253	3.3425	3.0077	2.7863	2.6274	2.5068	2.4117	2.3344
120	5.1523	3.8046	3.2269	2.8943	2.6740	2.5154	2.3948	2.2994	2.2217
inf	5.0239	3.6889	3.1161	2.7858	2.5665	2.4082	2.2875	2.1918	2.1136

F Table for alpha = .025 (continued)

df2/df1	10	12	15	20	24	30	40	60	120	INF
1	968.6274	976.7079	984.8668	993.1028	997.2492	1001.414	1005.598	1009.800	1014.020	1018.258
2	39.3980	39.4146	39.4313	39.4479	39.4562	39.465	39.473	39.481	39.490	39.498
3	14.4189	14.3366	14.2527	14.1674	14.1241	14.081	14.037	13.992	13.947	13.902
4	8.8439	8.7512	8.6565	8.5599	8.5109	8.461	8.411	8.360	8.309	8.257
5	6.6192	6.5245	6.4277	6.3286	6.2780	6.227	6.175	6.123	6.069	6.015
6	5.4613	5.3662	5.2687	5.1684	5.1172	5.065	5.012	4.959	4.904	4.849
7	4.7611	4.6658	4.5678	4.4667	4.4150	4.362	4.309	4.254	4.199	4.142
8	4.2951	4.1997	4.1012	3.9995	3.9472	3.894	3.840	3.784	3.728	3.670
9	3.9639	3.8682	3.7694	3.6669	3.6142	3.560	3.505	3.449	3.392	3.333
10	3.7168	3.6209	3.5217	3.4185	3.3654	3.311	3.255	3.198	3.140	3.080
11	3.5257	3.4296	3.3299	3.2261	3.1725	3.118	3.061	3.004	2.944	2.883
12	3.3736	3.2773	3.1772	3.0728	3.0187	2.963	2.906	2.848	2.787	2.725
13	3.2497	3.1532	3.0527	2.9477	2.8932	2.837	2.780	2.720	2.659	2.595
14	3.1469	3.0502	2.9493	2.8437	2.7888	2.732	2.674	2.614	2.552	2.487
15	3.0602	2.9633	2.8621	2.7559	2.7006	2.644	2.585	2.524	2.461	2.395
16	2.9862	2.8890	2.7875	2.6808	2.6252	2.568	2.509	2.447	2.383	2.316
17	2.9222	2.8249	2.7230	2.6158	2.5598	2.502	2.442	2.380	2.315	2.247
18	2.8664	2.7689	2.6667	2.5590	2.5027	2.445	2.384	2.321	2.256	2.187
19	2.8172	2.7196	2.6171	2.5089	2.4523	2.394	2.333	2.270	2.203	2.133
20	2.7737	2.6758	2.5731	2.4645	2.4076	2.349	2.287	2.223	2.156	2.085
21	2.7348	2.6368	2.5338	2.4247	2.3675	2.308	2.246	2.182	2.114	2.042
22	2.6998	2.6017	2.4984	2.3890	2.3315	2.272	2.210	2.145	2.076	2.003
23	2.6682	2.5699	2.4665	2.3567	2.2989	2.239	2.176	2.111	2.041	1.968
24	2.6396	2.5411	2.4374	2.3273	2.2693	2.209	2.146	2.080	2.010	1.935
25	2.6135	2.5149	2.4110	2.3005	2.2422	2.182	2.118	2.052	1.981	1.906
26	2.5896	2.4908	2.3867	2.2759	2.2174	2.157	2.093	2.026	1.954	1.878
27	2.5676	2.4688	2.3644	2.2533	2.1946	2.133	2.069	2.002	1.930	1.853
28	2.5473	2.4484	2.3438	2.2324	2.1735	2.112	2.048	1.980	1.907	1.829
29	2.5286	2.4295	2.3248	2.2131	2.1540	2.092	2.028	1.959	1.886	1.807
30	2.5112	2.4120	2.3072	2.1952	2.1359	2.074	2.009	1.940	1.866	1.787
40	2.3882	2.2882	2.1819	2.0677	2.0069	1.943	1.875	1.803	1.724	1.637
60	2.2702	2.1692	2.0613	1.9445	1.8817	1.815	1.744	1.667	1.581	1.482
120	2.1570	2.0548	1.9450	1.8249	1.7597	1.690	1.614	1.530	1.433	1.310
inf	2.0483	1.9447	1.8326	1.7085	1.6402	1.566	1.484	1.388	1.268	1.000

F Table for alpha = .01

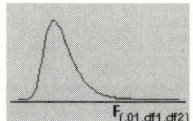

$F_{(.01, df1, df2)}$

df2/df1	1	2	3	4	5	6	7	8	9
1	4052.181	4999.500	5403.352	5624.583	5763.650	5858.986	5928.356	5981.070	6022.473
2	98.503	99.000	99.166	99.249	99.299	99.333	99.356	99.374	99.388
3	34.116	30.817	29.457	28.710	28.237	27.911	27.672	27.489	27.345
4	21.198	18.000	16.694	15.977	15.522	15.207	14.976	14.799	14.659
5	16.258	13.274	12.060	11.392	10.967	10.672	10.456	10.289	10.158
6	13.745	10.925	9.780	9.148	8.746	8.466	8.260	8.102	7.976
7	12.246	9.547	8.451	7.847	7.460	7.191	6.993	6.840	6.719
8	11.259	8.649	7.591	7.006	6.632	6.371	6.178	6.029	5.911
9	10.561	8.022	6.992	6.422	6.057	5.802	5.613	5.467	5.351
10	10.044	7.559	6.552	5.994	5.636	5.386	5.200	5.057	4.942
11	9.646	7.206	6.217	5.668	5.316	5.069	4.886	4.744	4.632
12	9.330	6.927	5.953	5.412	5.064	4.821	4.640	4.499	4.388
13	9.074	6.701	5.739	5.205	4.862	4.620	4.441	4.302	4.191
14	8.862	6.515	5.564	5.035	4.695	4.456	4.278	4.140	4.030
15	8.683	6.359	5.417	4.893	4.556	4.318	4.142	4.004	3.895
16	8.531	6.226	5.292	4.773	4.437	4.202	4.026	3.890	3.780
17	8.400	6.112	5.185	4.669	4.336	4.102	3.927	3.791	3.682
18	8.285	6.013	5.092	4.579	4.248	4.015	3.841	3.705	3.597
19	8.185	5.926	5.010	4.500	4.171	3.939	3.765	3.631	3.523
20	8.096	5.849	4.938	4.431	4.103	3.871	3.699	3.564	3.457
21	8.017	5.780	4.874	4.369	4.042	3.812	3.640	3.506	3.398
22	7.945	5.719	4.817	4.313	3.988	3.758	3.587	3.453	3.346
23	7.881	5.664	4.765	4.264	3.939	3.710	3.539	3.406	3.299
24	7.823	5.614	4.718	4.218	3.895	3.667	3.496	3.363	3.256
25	7.770	5.568	4.675	4.177	3.855	3.627	3.457	3.324	3.217
26	7.721	5.526	4.637	4.140	3.818	3.591	3.421	3.288	3.182
27	7.677	5.488	4.601	4.106	3.785	3.558	3.388	3.256	3.149
28	7.636	5.453	4.568	4.074	3.754	3.528	3.358	3.226	3.120
29	7.598	5.420	4.538	4.045	3.725	3.499	3.330	3.198	3.092
30	7.562	5.390	4.510	4.018	3.699	3.473	3.304	3.173	3.067
40	7.314	5.179	4.313	3.828	3.514	3.291	3.124	2.993	2.888
60	7.077	4.977	4.126	3.649	3.339	3.119	2.953	2.823	2.718
120	6.851	4.787	3.949	3.480	3.174	2.956	2.792	2.663	2.559
inf	6.635	4.605	3.782	3.319	3.017	2.802	2.639	2.511	2.407

F Table for alpha = .01 (continued)

df2/df1	10	12	15	20	24	30	40	60	120	INF
1	6055.847	6106.321	6157.285	6208.730	6234.631	6260.649	6286.782	6313.030	6339.391	6365.864
2	99.399	99.416	99.433	99.449	99.458	99.466	99.474	99.482	99.491	99.499
3	27.229	27.052	26.872	26.690	26.598	26.505	26.411	26.316	26.221	26.125
4	14.546	14.374	14.198	14.020	13.929	13.838	13.745	13.652	13.558	13.463
5	10.051	9.888	9.722	9.553	9.466	9.379	9.291	9.202	9.112	9.020
6	7.874	7.718	7.559	7.396	7.313	7.229	7.143	7.057	6.969	6.880
7	6.620	6.469	6.314	6.155	6.074	5.992	5.908	5.824	5.737	5.650
8	5.814	5.667	5.515	5.359	5.279	5.198	5.116	5.032	4.946	4.859
9	5.257	5.111	4.962	4.808	4.729	4.649	4.567	4.483	4.398	4.311
10	4.849	4.706	4.558	4.405	4.327	4.247	4.165	4.082	3.996	3.909
11	4.539	4.397	4.251	4.099	4.021	3.941	3.860	3.776	3.690	3.602
12	4.296	4.155	4.010	3.858	3.780	3.701	3.619	3.535	3.449	3.361
13	4.100	3.960	3.815	3.665	3.587	3.507	3.425	3.341	3.255	3.165
14	3.939	3.800	3.656	3.505	3.427	3.348	3.266	3.181	3.094	3.004
15	3.805	3.666	3.522	3.372	3.294	3.214	3.132	3.047	2.959	2.868
16	3.691	3.553	3.409	3.259	3.181	3.101	3.018	2.933	2.845	2.753
17	3.593	3.455	3.312	3.162	3.084	3.003	2.920	2.835	2.746	2.653
18	3.508	3.371	3.227	3.077	2.999	2.919	2.835	2.749	2.660	2.566
19	3.434	3.297	3.153	3.003	2.925	2.844	2.761	2.674	2.584	2.489
20	3.368	3.231	3.088	2.938	2.859	2.778	2.695	2.608	2.517	2.421
21	3.310	3.173	3.030	2.880	2.801	2.720	2.636	2.548	2.457	2.360
22	3.258	3.121	2.978	2.827	2.749	2.667	2.583	2.495	2.403	2.305
23	3.211	3.074	2.931	2.781	2.702	2.620	2.535	2.447	2.354	2.256
24	3.168	3.032	2.889	2.738	2.659	2.577	2.492	2.403	2.310	2.211
25	3.129	2.993	2.850	2.699	2.620	2.538	2.453	2.364	2.270	2.169
26	3.094	2.958	2.815	2.664	2.585	2.503	2.417	2.327	2.233	2.131
27	3.062	2.926	2.783	2.632	2.552	2.470	2.384	2.294	2.198	2.097
28	3.032	2.896	2.753	2.602	2.522	2.440	2.354	2.263	2.167	2.064
29	3.005	2.868	2.726	2.574	2.495	2.412	2.325	2.234	2.138	2.034
30	2.979	2.843	2.700	2.549	2.469	2.386	2.299	2.208	2.111	2.006
40	2.801	2.665	2.522	2.369	2.288	2.203	2.114	2.019	1.917	1.805
60	2.632	2.496	2.352	2.198	2.115	2.028	1.936	1.836	1.726	1.601
120	2.472	2.336	2.192	2.035	1.950	1.860	1.763	1.656	1.533	1.381
inf	2.321	2.185	2.039	1.878	1.791	1.696	1.592	1.473	1.325	1.000

REFERENCES CITED

Abraham, B., & Ledolter, J. (1983). *Statistical methods for forecasting*. New York: Wiley.

Adorno, T. W., Frenkel-Brunswik, E., Levinson, D. J., & Sanford, R. N. (1950). *The authoritarian personality*. New York: Harper.

Agrawal, R., Imielinski, T., & Swami, A. (1993). Mining association rules between sets of items in large databases. *Proceedings of the 1993 ACM SIGMOD Conference*, Washington, DC.

Agrawal, R. & Srikant, R. (1994). Fast algorithms for mining association rules. *Proceedings of the 20th VLDB Conference*. Santiago, Chile.

Agresti, Alan (1996). *An Introduction to Categorical Data Analysis*. New York: Wiley.

Akaike, H. (1973). Information theory and an extension of the maximum likelihood principle. In B. N. Petrov and F. Csaki (Eds.), *Second International Symposium on Information Theory*. Budapest: Akademiai Kiado.

Akaike, H. (1983). Information measures and model selection. *Bulletin of the International Statistical Institute: Proceedings of the 44th Session, Volume 1*. Pages 277-290.

Aldrich, J. H., & Nelson, F. D. (1984). *Linear probability, logit, and probit models*. Beverly Hills, CA: Sage Publications.

Almon, S. (1965). The distributed lag between capital appropriations and expenditures. *Econometrica, 33*, 178-196.

Altman, M. *Software review*: JMP, V. 4.02. At http://lilt.ilstu.edu/gmklass/ITPnews/fall00/MAltmanf00.htm

American Supplier Institute (1984-1988). *Proceedings of Supplier Symposia on Taguchi Methods*. (April, 1984; November, 1984 October, 1985; October, 1986; October, 1987; October, 1988), Dearborn, MI: American Supplier Institute.

Anderson, O. D. (1976). *Time series analysis and forecasting*. London: Butterworths.

Anderson, S. B., & Maier, M. H. (1963). 34,000 pupils and how they grew. *Journal of Teacher Education, 14*, 212-216.

Anderson, T. W. (1958). *An introduction to multivariate statistical analysis*. New York: Wiley.

Anderson, T. W. (1984). *An introduction to multivariate statistical analysis* (2nd ed.). New York: Wiley.

Anderson, T. W., & Rubin, H. (1956). Statistical inference in factor analysis. *Proceedings of the Third Berkeley Symposium on Mathematical Statistics and Probability*. Berkeley: The University of California Press.

Andrews, D. F. (1972). Plots of high-dimensional data. *Biometrics, 28*, 125-136.

ASQC/AIAG (1990). *Measurement systems analysis reference manual*. Troy, MI: AIAG.

ASQC/AIAG (1991). *Fundamental statistical process control reference manual*. Troy, MI: AIAG.

AT&T (1956). *Statistical quality control handbook, Select code 700-444*. Indianapolis, AT&T Technologies.

Auble, D. (1953). Extended tables for the Mann-Whitney statistic. *Bulletin of the Institute of Educational Research, Indiana University, 1*, No. 2.

Bagozzi, R. P., & Yi, Y. (1989). On the use of structural equation models in experimental design. *Journal of Marketing Research, 26*, 271-284.

Bagozzi, R. P., Yi, Y., & Singh, S. (1991). On the use of structural equation models in experimental designs: Two extensions. *International Journal of Research in Marketing, 8,* 125-140.

Bailey, A. L. (1931). The analysis of covariance. *Journal of the American Statistical Association, 26,* 424-435.

Bails, D. G., & Peppers, L. C. (1982). *Business fluctuations: Forecasting techniques and applications.* Englewood Cliffs, NJ: Prentice-Hall.

Bain, L. J. (1978). *Statistical analysis of reliability and life-testing models.* New York: Decker.

Bain, L. J. and Engelhart, M. (1989) *Introduction to Probability and Mathematical Statistics.* Kent, MA: PWS.

Baird, J. C. (1970). *Psychophysical analysis of visual space.* New York: Pergamon Press.

Baird, J. C., & Noma, E. (1978). *Fundamentals of scaling and psychophysics.* New York: Wiley.

Barcikowski, R., & Stevens, J. P. (1975). A Monte Carlo study of the stability of canonical correlations, canonical weights, and canonical variate-variable correlations. *Multivariate Behavioral Research, 10,* 353-364.

Barker, T. B. (1986). Quality engineering by design: Taguchi's philosophy. *Quality Progress, 19,* 32-42.

Barlow, R. E., & Proschan, F. (1975). *Statistical theory of reliability and life testing.* New York: Holt, Rinehart, & Winston.

Barlow, R. E., Marshall, A. W., & Proschan, F. (1963). Properties of probability distributions with monotone hazard rate. *Annals of Mathematical Statistics, 34,* 375-389.

Barnard, G. A. (1959). Control charts and stochastic processes. *Journal of the Royal Statistical Society, Ser. B, 21,* 239.

Bartholomew, D. J. (1984). The foundations of factor analysis. *Biometrika, 71,* 221-232.

Bates, D. M., & Watts, D. G. (1988). *Nonlinear regression analysis and its applications.* New York: Wiley.

Bayne, C. K., & Rubin, I. B. (1986). *Practical experimental designs and optimization methods for chemists.* Deerfield Beach, FL: VCH Publishers.

Becker, R. A., Denby, L., McGill, R., & Wilks, A. R. (1986). Datacryptanalysis: A case study. *Proceedings of the Section on Statistical Graphics, American Statistical Association,* 92-97.

Bellman, R. (1961). *Adaptive Control Processes: A Guided Tour.* Princeton University Press.

Belsley, D. A., Kuh, E., and Welsch, R. E. (1980). *Regression Diagnostics.* New York: Wiley.

Bendat, J. S. (1990). *Nonlinear system analysis and identification from random data.* New York: Wiley.

Bentler, P. M, & Bonett, D. G. (1980). Significance tests and goodness of fit in the analysis of covariance structures. *Psychological Bulletin, 88,* 588-606.

Bentler, P. M. (1986). Structural modeling and Psychometrika: A historical perspective on growth and achievements. *Psychometrika, 51,* 35-51.

Bentler, P. M. (1989). *EQS Structural equations program manual.* Los Angeles, CA: BMDP Statistical Software.

Bentler, P. M., & Weeks, D. G. (1979). Interrelations among models for the analysis of moment structures. *Multivariate Behavioral Research, 14,* 169-185.

Bentler, P. M., & Weeks, D. G. (1980). Linear structural equations with latent variables. *Psychometrika, 45,* 289-308.

Benzécri, J. P. (1973). *L'Analyse des Données: T. 2, I' Analyse des correspondances.* Paris: Dunod.

Bergeron, B. (2002). *Essentials of CRM: A guide to customer relationship management.* NY: Wiley.

Berkson, J. (1944). Application of the Logistic Function to Bio-Assay. *Journal of the American Statistical Association, 39,* 357-365.

Berkson, J., & Gage, R. R. (1950). The calculation of survival rates for cancer. *Proceedings of Staff Meetings, Mayo Clinic, 25,* 250.

Berry, M., J., A., & Linoff, G., S., (2000). *Mastering data mining*. New York: Wiley.

Bhote, K. R. (1988). *World class quality*. New York: AMA Membership Publications.

Binns, B., & Clark, N. (1986). The graphic designer's use of visual syntax. *Proceedings of the Section on Statistical Graphics, American Statistical Association*, 36-41.

Birnbaum, Z. W. (1952). Numerical tabulation of the distribution of Kolmogorov's statistic for finite sample values. *Journal of the American Statistical Association*, 47, 425-441.

Birnbaum, Z. W. (1953). Distribution-free tests of fit for continuous distribution functions. *Annals of Mathematical Statistics*, 24, 1-8.

Bishop, C. (1995). *Neural Networks for Pattern Recognition*. Oxford: University Press.

Bishop, Y. M. M., Fienberg, S. E., & Holland, P. W. (1975). *Discrete multivariate analysis*. Cambridge, MA: MIT Press.

Bjorck, A. (1967). Solving linear least squares problems by Gram-Schmidt orthonormalization. *Bit*, 7, 1-21.

Blackman, R. B., & Tukey, J. (1958). *The measurement of power spectral from the point of view of communication engineering*. New York: Dover.

Blackwelder, R. A. (1966). *Taxonomy: A text and reference book*. New York: Wiley.

Blalock, H. M. (1972). *Social statistics* (2nd ed.). New York:McGraw-Hill

Bliss, C. I. (1934). The method of probits. *Science*, 79, 38-39.

Bloomfield, P. (1976). *Fourier analysis of time series: An introduction*. New York: Wiley.

Bock, R. D. (1963). Programming univariate and multivariate analysis of variance. *Technometrics*, 5, 95-117.

Bock, R. D. (1975). *Multivariate statistical methods in behavioral research*. New York: McGraw-Hill.

Bolch, B.W., & Huang, C. J. (1974). *Multivariate statistical methods for business and economics*. Englewood Cliffs, NJ: Prentice-Hall.

Bollen, K. A. (1989). *Structural equations with latent variables*. New York: John Wiley & Sons.

Borg, I., & Lingoes, J. (1987). *Multidimensional similarity structure analysis*. New York: Springer.

Borg, I., & Shye, S. (in press). *Facet Theory*. Newbury Park: Sage.

Bouland, H. and Kamp, Y. (1988). Auto-association by multilayer perceptrons and singular value decomposition. *Biological Cybernetics 59*, 291-294.

Bowker, A. G. (1948). A test for symmetry in contingency tables. *Journal of the American Statistical Association*, 43, 572-574.

Bowley, A. L. (1897). Relations between the accuracy of an average and that of its constituent parts. *Journal of the Royal Statistical Society*, 60, 855-866.

Bowley, A. L. (1907). *Elements of Statistics*. London: P. S. King and Son.

Box, G. E. P. (1953). Non-normality and tests on variances. *Biometrika*, 40, 318-335.

Box, G. E. P. (1954a). Some theorems on quadratic forms applied in the study of analysis of variance problems: I. Effect of inequality of variances in the one-way classification. *Annals of Mathematical Statistics*, 25, 290-302.

Box, G. E. P. (1954b). Some theorems on quadratic forms applied in the study of analysis of variance problems: II. Effect of inequality of variances and of correlation of errors in the two-way classification. *Annals of Mathematical Statistics*, 25, 484-498.

Box, G. E. P., & Anderson, S. L. (1955). Permutation theory in the derivation of robust criteria and the study of departures from assumptions. *Journal of the Royal Statistical Society*, 17, 1-34.

Box, G. E. P., & Behnken, D. W. (1960). Some new three level designs for the study of quantitative variables. *Technometrics*, 2, 455-475.

Box, G. E. P., & Cox, D. R. (1964). An analysis of transformations. *Journal of the Royal Statistical Society*, 26, 211-253.

Box, G. E. P., & Cox, D. R. (1964). An analysis of transformations. *Journal of the Royal Statistical Society, B26*, 211-234.

Box, G. E. P., & Draper, N. R. (1987). *Empirical model-building and response surfaces.* New York: Wiley.

Box, G. E. P., & Jenkins, G. M. (1970). *Time series analysis.* San Francisco: Holden Day.

Box, G. E. P., & Jenkins, G. M. (1976). *Time series analysis: Forecasting and control.* San Francisco: Holden-Day.

Box, G. E. P., & Muller, M. E. (1958). A note on the generation of random normal deviates. *Annals of Mathematical Statistics*, 29, 610=611. Inversion method (Muller, 1959).

Box, G. E. P., & Tidwell, P. W. (1962). Transformation of the independent variables. *Technometrics, 4*, 531-550.

Box, G. E. P., & Wilson, K. B. (1951). On the experimental attainment of optimum conditions. *Journal of the Royal Statistical Society*, Ser. B, *13*, 1-45.

Box, G. E. P., Hunter, W. G., & Hunter, S. J. (1978). *Statistics for experimenters: An introduction to design, data analysis, and model building.* New York: Wiley.

Braha, D. (ed) (2001). *Data Mining for design and manufacturing: Methods and Applications.* Dordrecht, The Netherlands: Kluwer Academic Publishers.

Breiman, L., Friedman, J. H., Olshen, R. A., & Stone, C. J. (1984). *Classification and regression trees.* Monterey, CA: Wadsworth & Brooks/Cole Advanced Books & Software.

Breiman, L. (2001). *Random Forests*, Technical Report, Statistics Department, University of California, Berkeley.

Brenner, J. L., et al. (1968). Difference equations in forecasting formulas. *Management Science, 14*, 141-159.

Brent, R. F. (1973). *Algorithms for minimization without derivatives.* Englewood Cliffs, NJ: Prentice-Hall.

Breslow, N. E. (1970). A generalized Kruskal-Wallis test for comparing K samples subject to unequal pattern of censorship. *Biometrika*, *57*, 579-594.

Breslow, N. E. (1974). Covariance analysis of censored survival data. *Biometrics*, *30*, 89-99.

Bridle, J.S. (1990). Probabilistic interpretation of feedforward classification network outputs, with relationships to statistical pattern recognition. In F. Fogelman Soulie and J. Herault (Eds.), *Neurocomputing: Algorithms, Architectures and Applications*, 227-236. New York: Springer-Verlag.

Brigham, E. O. (1974). *The fast Fourier transform.* Englewood Cliffs, NJ: Prentice-Hall.

Brillinger, D. R. (1975). *Time series: Data analysis and theory.* New York: Holt, Rinehart. & Winston.

Broomhead, D.S. and Lowe, D. (1988). Multivariable functional interpolation and adaptive networks. *Complex Systems 2*, 321-355.

Brown, D. T. (1959). A note on approximations to discrete probability distributions. *Information and Control*, *2*, 386-392.

Brown, M. B., & Forsythe, A. B. (1974). Robust tests for the equality of variances. *Journal of the American Statistical Association*, *69*, 264-267.

Brown, R. G. (1959). *Statistical forecasting for inventory control.* New York: McGraw-Hill.

Browne, M. W. (1968). A comparison of factor analytic techniques. *Psychometrika*, *33*, 267-334.

Browne, M. W. (1974). Generalized least squares estimators in the analysis of covariance structures. *South African Statistical Journal*, *8*, 1-24.

Browne, M. W. (1982). Covariance Structures. In D. M. Hawkins (Ed.) *Topics in Applied Multivariate Analysis.* Cambridge, MA: Cambridge University Press.

Browne, M. W. (1984). Asymptotically distribution free methods for the analysis of covariance structures. *British Journal of Mathematical and Statistical Psychology, 37,* 62-83.

Browne, M. W., & Cudeck, R. (1990). Single sample cross-validation indices for covariance structures. *Multivariate Behavioral Research, 24,* 445-455.

Browne, M. W., & Cudeck, R. (1992). Alternative ways of assessing model fit. In K. A. Bollen and J. S. Long (Eds.), *Testing structural equation models.* Beverly Hills, CA: Sage.

Browne, M. W., & DuToit, S. H. C. (1982). AUFIT (Version 1). A computer programme for the automated fitting of nonstandard models for means and covariances. *Research Finding WS-27.* Pretoria, South Africa: Human Sciences Research Council.

Browne, M. W., & DuToit, S. H. C. (1987). Automated fitting of nonstandard models. *Report WS-39.* Pretoria, South Africa: Human Sciences Research Council.

Browne, M. W., & DuToit, S. H. C. (1992). Automated fitting of nonstandard models. *Multivariate Behavioral Research, 27,* 269-300.

Browne, M. W., & Mels, G. (1992). *RAMONA User's Guide.* The Ohio State University: Department of Psychology.

Browne, M. W., & Shapiro, A. (1989). Invariance of covariance structures under groups of transformations. *Research Report 89/4.* Pretoria, South Africa: University of South Africa Department of Statistics.

Browne, M. W., & Shapiro, A. (1991). Invariance of covariance structures under groups of transformations. *Metrika, 38,* 335-345.

Browne, M.W., & Cudeck, R. (1993). Alternative ways of assessing model fit. In K. A. Bollen & J. S. Long, (Eds.), *Testing structural equation models.* Beverly Hills, CA: Sage.

Brownlee, K. A. (1960). *Statistical Theory and Methodology in Science and Engineering.* New York: John Wiley.

Buffa, E. S. (1972). *Operations management: Problems and models* (3rd. ed.). New York: Wiley.

Buja, A., & Tukey, P. A. (Eds.) (1991). *Computing and Graphics in Statistics.* New York: Springer-Verlag.

Buja, A., Fowlkes, E. B., Keramidas, E. M., Kettenring, J. R., Lee, J. C., Swayne, D. F., & Tukey, P. A. (1986). Discovering features of multivariate data through statistical graphics. *Proceedings of the Section on Statistical Graphics, American Statistical Association,* 98-103.

Burman, J. P. (1979). Seasonal adjustment - a survey. *Forecasting, Studies in Management Science, 12,* 45-57.

Burns, L. S., & Harman, A. J. (1966). *The complex metropolis, Part V of profile of the Los Angeles metropolis: Its people and its homes.* Los Angeles: University of Chicago Press.

Burt, C. (1950). The factorial analysis of qualitative data. *British Journal of Psychology, 3,* 166-185.

Campbell D. T., & Fiske, D. W. (1959). Convergent and discriminant validation by the multitrait-multimethod matrix. *Psychological Bulletin, 56,* 81-105

Carling, A. (1992). *Introducing Neural Networks.* Wilmslow, UK: Sigma Press.

Carmines, E. G., & Zeller, R. A. (1980). *Reliability and validity assessment.* Beverly Hills, CA: Sage Publications.

Carrol, J. D., Green, P. E., and Schaffer, C. M. (1986). Interpoint distance comparisons in correspondence analysis. *Journal of Marketing Research, 23,* 271-280.

Carroll, J. D., & Wish, M. (1974). Multidimensional perceptual models and measurement methods. In E. C. Carterette and M. P. Friedman (Eds.), *Handbook of perception.* (Vol. 2, pp. 391-447). New York: Academic Press.

Cattell, R. B. (1966). The scree test for the number of factors. *Multivariate Behavioral Research, 1,* 245-276.

Cattell, R. B., & Jaspers, J. A. (1967). A general plasmode for factor analytic exercises and research. *Multivariate Behavioral Research Monographs.*

Chambers, J. M., Cleveland, W. S., Kleiner, B., & Tukey, P. A. (1983). *Graphical methods for data analysis*. Bellmont, CA: Wadsworth.

Chan, L. K., Cheng, S. W., & Spring, F. (1988). A new measure of process capability: Cpm. *Journal of Quality Technology, 20*, 162-175.

Chen, J. (1992). Some results on 2(nk) fractional factorial designs and search for minimum aberration designs. *Annals of Statistics, 20*, 2124-2141.

Chen, J., & Wu, C. F. J. (1991). Some results on s(nk) fractional factorial designs with minimum aberration or optimal moments. *Annals of Statistics, 19*, 1028-1041.

Chen, J., Sun, D. X., & Wu, C. F. J. (1993). A catalog of two-level and three-level fractional factorial designs with small runs. *International Statistical Review, 61*, 131-145.

Chernoff, H. (1973). The use of faces to represent points in k-dimensional space graphically. *Journal of American Statistical Association, 68*, 361-368.

Christ, C. (1966). *Econometric models and methods*. New York: Wiley.

Clarke, G. M., & Cooke, D. (1978). *A basic course in statistics*. London: Edward Arnold.

Clements, J. A. (1989). Process capability calculations for non-normal distributions. *Quality Progress*. September, 95-100.

Cleveland, W. S. (1979). Robust locally weighted regression and smoothing scatterplots. *Journal of the American Statistical Association, 74*, 829-836.

Cleveland, W. S. (1984). Graphs in scientific publiccations. *The American Statistician, 38*, 270-280.

Cleveland, W. S. (1985). *The elements of graphing data*. Monterey, CA: Wadsworth.

Cleveland, W. S. (1993). *Visualizing data*. Murray Hill, NJ: AT&T.

Cleveland, W. S., Harris, C. S., & McGill, R. (1982). Judgements of circle sizes on statistical maps. *Journal of the American Statistical Association, 77*, 541-547.

Cliff, N. (1983). Some cautions concerning the application of causal modeling methods. *Multivariate Behavioral Research, 18*, 115-126.

Cochran, W. G. (1950). The comparison of percentages in matched samples. *Biometrika, 37*, 256-266.

Cohen, J. (1977). *Statistical power analysis for the behavioral sciences. (*Rev. ed.). New York: Academic Press.

Cohen, J. (1983). *Statistical power analysis for the behavioral sciences. (2nd Ed.)*. Mahwah, NJ: Lawrence Erlbaum Associates.

Cohen, J. (1994). The earth is round ($p < .05$). *American Psychologist, 49,* 9971003.

Cole, D. A., Maxwell, S. E., Arvey, R., & Salas, E. (1993). Multivariate group comparisons of variable systems: MANOVA and structural equation modeling. *Psychological Bulletin, 114,* 174-184.

Connor, W. S., & Young, S. (1984). Fractional factorial experiment designs for experiments with factors at two and three levels. In R. A. McLean & V. L. Anderson (Eds.), *Applied factorial and fractional designs*. New York: Marcel Dekker.

Connor, W. S., & Zelen, M. (1984). Fractional factorial experiment designs for factors at three levels. In R. A. McLean & V. L. Anderson (Eds.), *Applied factorial and fractional designs*. New York: Marcel Dekker.

Conover, W. J. (1974). Some reasons for not using the Yates continuity correction on 2 x 2 contingency tables. *Journal of the American Statistical Association, 69*, 374-376.

Conover, W. J., Johnson, M. E., & Johnson, M. M. (1981). A comparative study of tests for homogeneity of variances with applications to the outer continental shelf bidding data. *Technometrics, 23*, 357-361.

Cook, R. D. (1977). Detection of influential observations in linear regression. *Technometrics, 19*, 15-18.

Cook, R. D., & Nachtsheim, C. J. (1980). A comparison of algorithms for constructing exact D-optimal designs. *Technometrics, 22*, 315-324.

Cook, R. D., & Weisberg, S. (1982). *Residuals and Influence in Regression.* (Monographs on statistics and applied probability). New York: Chapman and Hall.

Cooke, D., Craven, A. H., & Clarke, G. M. (1982). *Basic statistical computing.* London: Edward Arnold.

Cooley, J. W., & Tukey, J. W. (1965). An algorithm for the machine computation of complex Fourier series. *Mathematics of Computation, 19*, 297-301.

Cooley, W. W., & Lohnes, P. R. (1971). *Multivariate data analysis.* New York: Wiley.

Cooley, W. W., & Lohnes, P. R. (1976). *Evaluation research in education.* New York: Wiley.

Coombs, C. H. (1950). Psychological scaling without a unit of measurement. *Psychological Review, 57*, 145-158.

Coombs, C. H. (1964). *A theory of data.* New York: Wiley.

Corballis, M. C., & Traub, R. E. (1970). Longitudinal factor analysis. *Psychometrika, 35*, 79-98.

Corbeil, R. R., & Searle, S. R. (1976). Restricted maximum likelihood (REML) estimation of variance components in the mixed model. *Technometrics, 18*, 31-38.

Cormack, R. M. (1971). A review of classification. *Journal of the Royal Statistical Society, 134*, 321-367.

Cornell, J. A. (1990a). *How to run mixture experiments for product quality.* Milwaukee, Wisconsin: ASQC.

Cornell, J. A. (1990b). *Experiments with mixtures: designs, models, and the analysis of mixture data* (2nd ed.). New York: Wiley.

Cox, D. R. (1957). Note on grouping. *Journal of the American Statistical Association, 52*, 543-547.

Cox, D. R. (1959). The analysis of exponentially distributed lifetimes with two types of failures.

Journal of the Royal Statistical Society, 21, 411-421.

Cox, D. R. (1964). Some applications of exponential ordered scores. *Journal of the Royal Statistical Society, 26*, 103-110.

Cox, D. R. (1970). *The analysis of binary data.* New York: Halsted Press.

Cox, D. R. (1972). Regression models and life tables. *Journal of the Royal Statistical Society, 34*, 187-220.

Cox, D. R., & Oakes, D. (1984). *Analysis of survival data.* New York: Chapman & Hall.

Cramer, H. (1946). *Mathematical methods in statistics.* Princeton, NJ: Princeton University Press.

Cristianini, N., & Shawe-Taylor, J. (2000). *Introduction to support vector machines and other kernel-based learning methods.* Cambridge, UK: Cambridge University Press.

Crowley, J., & Hu, M. (1977). Covariance analysis of heart transplant survival data. *Journal of the American Statistical Association, 72*, 27-36.

Cudeck, R. (1989). Analysis of correlation matrices using covariance structure models. *Psychological Bulletin, 105*, 317-327.

Cudeck, R., & Browne, M. W. (1983). Cross-validation of covariance structures. *Multivariate Behavioral Research, 18*, 147-167.

Cutler, S. J., & Ederer, F. (1958). Maximum utilization of the life table method in analyzing survival. *Journal of Chronic Diseases, 8*, 699-712.

Dahlquist, G., & Bjorck, A. (1974). *Numerical Methods.* Englewood Cliffs, NJ: Prentice-Hall.

Daniel, C. (1976). *Applications of statistics to industrial experimentation.* New York: Wiley.

Daniell, P. J. (1946). Discussion on symposium on autocorrelation in time series. *Journal of the Royal Statistical Society*, Suppl. *8*, 88-90.

Daniels, H. E. (1939). The estimation of components of variance. *Supplement to the Journal of the Royal Statistical Society, 6*, 186-197.

Darlington, R. B. (1990). *Regression and linear models.* New York: McGraw-Hill.

Darlington, R. B., Weinberg, S., & Walberg, H. (1973). Canonical variate analysis and related techniques. *Review of Educational Research*, *43*, 433-454.

DataMyte (1992). *DataMyte handbook.* Minnetonka, MN.

David, H. A. (1995). First (?) occurrence of common terms in mathematical statistics. *The American Statistician*, *49*, 121-133.

Davies, P. M., & Coxon, A. P. M. (1982). *Key texts in multidimensional scaling.* Exeter, NH: Heinemann Educational Books.

Davis, C. S., & Stephens, M. A. Approximate percentage points using Pearson curves. *Applied Statistics, 32*, 322-327.

De Boor, C. (1978). *A practical guide to splines.* New York: Springer-Verlag.

DeCarlo, L. T. (1998). Signal detection theory and generalized linear models, *Psychological Methods*, 186-200.

De Gruitjer, P. N. M., & Van Der Kamp, L. J. T. (Eds.). (1976). *Advances in psychological and educational measurement.* New York: Wiley.

de Jong, S (1993) SIMPLS: An Alternative Approach to Partial Least Squares Regression, *Chemometrics and Intelligent Laboratory Systems*, 18, 251-263

de Jong, S and Kiers, H. (1992) Principal Covariates regression, *Chemometrics and Intelligent Laboratory Systems*, 14, 155-164

Deming, S. N., & Morgan, S. L. (1993). *Experimental design: A chemometric approach.* (2nd ed.). Amsterdam, The Netherlands: Elsevier Science Publishers B.V.

Deming, W. E., & Stephan, F. F. (1940). The sampling procedure of the 1940 population census. *Journal of the American Statistical Association*, *35*, 615-630.

Dempster, A. P. (1969). *Elements of Continuous Multivariate Analysis.* San Francisco: Addison-Wesley.

Dempster, A. P., Laird, N. M., & Rubin, D. B. (1977). Maximum likelihood from incomplete data via the EM algorithm. *Journal of the Royal Statistical Society*, *39*, 1-38.

Dennis, J. E., & Schnabel, R. B. (1983). *Numerical methods for unconstrained optimization and nonlinear equations.* Englewood Cliffs, NJ: Prentice Hall.

Derringer, G., & Suich, R. (1980). Simultaneous optimization of several response variables. *Journal of Quality Technology*, *12*, 214-219.

Devroye, L. (1986). *Non-uniform random variate generation.* New York: Springer.

Diamond, W. J. (1981). *Practical experimental design.* Belmont, CA: Wadsworth.

Dijkstra, T. K. (1990). Some properties of estimated scale invariant covariance structures. *Psychometrika*, *55*, 327-336.

Dinneen, L. C., & Blakesley, B. C. (1973). A generator for the sampling distribution of the Mann Whitney U statistic. *Applied Statistics*, *22*, 269-273.

Dixon, W. J. (1954). Power under normality of several nonparametric tests. *Annals of Mathematical Statistics*, *25*, 610-614.

Dixon, W. J., & Massey, F. J. (1983). *Introduction to statistical analysis* (4th ed.). New York: McGraw-Hill.

Dobson, A. J. (1990). *An introduction to generalized linear models.* New York: Chapman & Hall.

Dodd, B. (1979). Lip reading in infants: Attention to speech presented in- and out-of-synchrony. *Cognitive Psychology*, *11*, 478-484.

Dodge, Y. (1985). *Analysis of experiments with missing data.* New York: Wiley.

Dodge, Y., Fedorov, V. V., & Wynn, H. P. (1988). *Optimal design and analysis of experiments.* New York: North-Holland.

Dodson, B. (1994). *Weibull analysis.* Milwaukee, Wisconsin: ASQC.

Doyle, P. (1973). The use of Automatic Interaction Detection and similar search procedures. *Operational Research Quarterly*, *24*, 465-467.

Duncan, A. J. (1974). *Quality control and industrial statistics*. Homewood, IL: Richard D. Irwin.

Duncan, O. D., Haller, A. O., & Portes, A. (1968). Peer influence on aspiration: a reinterpretation. *American Journal of Sociology*, *74*, 119-137.

Dunnett, C. W. (1955). A multiple comparison procedure for comparing several treatments with a control. *Journal of the American Statistical Association*, *50*, 1096-1121.

Durbin, J. (1970). Testing for serial correlation in least-squares regression when some of the regressors are lagged dependent variables. *Econometrica*, *38*, 410-421.

Durbin, J., & Watson, G. S. (1951). Testing for serial correlations in least squares regression. II. *Biometrika*, *38*, 159-178.

Dykstra, O. Jr. (1971). The augmentation of experimental data to maximize |X'X|. *Technometrics*, *13*, 682-688.

Eason, E. D., & Fenton, R. G. (1974). A comparison of numerical optimization methods for engineering design. *ASME Paper 73-DET-17*.

Eastment, H. T., & Krzanowski, W. J. (1982, February) Cross-validatory choice of the number of components from a principal components analysis, *Technometrics, Vol. 24, No. 1*.

Edelstein, H., A. (1999). *Introduction to data mining and knowledge discovery (3rd ed)*. Potomac, MD: Two Crows Corp.

Edgeworth, F. Y. (1885). Methods of statistics. In *Jubilee Volume, Royal Statistical Society*, 181-217.

Efron, B. (1982). *The jackknife, the bootstrap, and other resampling plans*. Philadelphia, Pa. Society for Industrial and Applied Mathematics.

Eisenhart, C. (1947). The assumptions underlying the analysis of variance. *Biometrics*, *3*, 1-21.

Elandt-Johnson, R. C., & Johnson, N. L. (1980). *Survival models and data analysis*. New York: Wiley.

Elliott, D. F., & Rao, K. R. (1982). *Fast transforms: Algorithms, analyses, applications*. New York: Academic Press.

Elmes, D. G., Kantowitz, B. H., & Roediger, H. C. (2006). *Research Methods in Psychology*. Belmont, CA: Thompson.

Elsner, J. B., Lehmiller, G. S., & Kimberlain, T. B. (1996). Objective classification of Atlantic hurricanes. *Journal of Climate*, *9*, 2880-2889.

Enslein, K., Ralston, A., & Wilf, H. S. (1977). *Statistical methods for digital computers*. New York: Wiley.

Euler, L. (1782). Recherches sur une nouvelle espece de quarres magiques. *Verhandelingen uitgegeven door het zeeuwsch Genootschap der Wetenschappen te Vlissingen*, *9*, 85-239. (Reproduced in *Leonhardi Euleri Opera Omnia*. Sub auspiciis societatis scientiarium naturalium helveticae, 1st series, *7*, 291-392.)

Evans, M., Hastings, N., & Peacock, B. (1993). *Statistical Distributions*. New York: Wiley.

Everitt, B. S. (1977). *The analysis of contingency tables*. London: Chapman & Hall.

Everitt, B. S. (1984). *An introduction to latent variable models*. London: Chapman and Hall.

Ewan, W. D. (1963). When and how to use Cusum charts. *Technometrics*, *5*, 1-32.

Fahlman, S.E. (1988). Faster-learning variations on back-propagation: an empirical study. In D. Touretzky, G.E. Hinton and T.J. Sejnowski (Eds.), *Proceedings of the 1988 Connectionist Models Summer School*, 38-51. San Mateo, CA: Morgan Kaufmann.

Fausett, L. (1994). *Fundamentals of Neural Networks*. New York: Prentice Hall.

Fayyad, U. M., Piatetsky-Shapiro, G., Smyth, P., & Uthurusamy, R. (Eds.). (1996). *Advances in Knowledge Discovery and Data Mining*. Cambridge, MA: The MIT Press.

Fayyad, U. S., & Uthurusamy, R. (Eds.) (1994). *Knowledge Discovery in Databases; Papers from the 1994 AAAI Workshop*. Menlo Park, CA: AAAI Press.

Feigl, P., & Zelen, M. (1965). Estimation of exponential survival probabilities with concomitant information. *Biometrics, 21*, 826-838.

Feller, W. (1948). On the Kolmogorov-Smirnov limit theorems for empirical distributions. *Annals of Mathematical Statistics, 19*, 177-189.

Fetter, R. B. (1967). *The quality control system*. Homewood, IL: Richard D. Irwin.

Fienberg, S. E. (1977). *The analysis of cross-classified categorical data*. Cambridge, MA: MIT Press.

Finn, J. D. (1974). *A general model for multivariate analysis*. New York: Holt, Rinehart & Winston.

Finn, J. D. (1977). Multivariate analysis of variance and covariance. In K. Enslein, A. Ralston, and H. S. Wilf (Eds.), *Statistical methods for digital computers. Vol. III.* (pp. 203-264). New York: Wiley.

Finney, D. J. (1944). The application of probit analysis to the results of mental tests. *Psychometrika, 9*, 31-39.

Finney, D. J. (1971). *Probit analysis*. Cambridge, MA: Cambridge University Press.

Firmin, R. (2002). Advanced time series modeling for semiconductor process control: The fab as a time machine. In Mackulak, G. T., Fowler, J. W., & Schomig, A. (eds.). *Proceedings of the International Conference on Modeling and Analysis of Semiconductor Manufacturing (MASM 2002)*.

Fisher, R. A. (1918). The correlation between relatives on the supposition of Mendelian inheritance. *Transactions of the Royal Society of Edinbrugh, 52*, 399-433.

Fisher, R. A. (1922). On the interpretation of *Chi-square* from contingency tables, and the calculation of *p*. *Journal of the Royal Statistical Society, 85*, 87-94.

Fisher, R. A. (1922). On the mathematical foundations of theoretical statistics. *Philosophical Transactions of the Royal Society of London*, Ser. A, *222*, 309-368.

Fisher, R. A. (1926). The arrangement of field experiments. *Journal of the Ministry of Agriculture of Great Britain, 33*, 503-513.

Fisher, R. A. (1928). The general sampling distribution of the multiple correlation coefficient. *Proceedings of the Royal Society of London*, Ser. A, *121*, 654-673.

Fisher, R. A. (1935). *The Design of Experiments*. Edinburgh: Oliver and Boyd.

Fisher, R. A. (1936). *Statistical Methods for Research Workers (6th ed.)*. Edinburgh: Oliver and Boyd.

Fisher, R. A. (1936). The use of multiple measurements in taxonomic problems. *Annals of Eugenics, 7*, 179-188.

Fisher, R. A. (1938). The mathematics of experimentation. *Nature, 142*, 442-443.

Fisher, R. A., & Yates, F. (1934). The 6 x 6 Latin squares. *Proceedings of the Cambridge Philosophical Society, 30*, 492-507.

Fisher, R. A., & Yates, F. (1938). *Statistical Tables for Biological, Agricultural and Medical Research*. London: Oliver and Boyd.

Fleishman, A. E. (1980). Confidence intervals for correlation ratios. *Educational and Psychological Measurement, 40*, 659670.

Fletcher, R. (1969). *Optimization*. New York: Academic Press.

Fletcher, R., & Powell, M. J. D. (1963). A rapidly convergent descent method for minimization. *Computer Journal, 6*, 163-168.

Fletcher, R., & Reeves, C. M. (1964). Function minimization by conjugate gradients. *Computer Journal, 7*, 149-154.

Fomby, T.B., Hill, R.C., & Johnson, S.R. (1984). *Advanced econometric methods*. New York: Springer-Verlag.

Ford Motor Company, Ltd. & GEDAS (1991). *Test examples for SPC software*.

Fouladi, R. T. (1991). *A comprehensive examination of procedures for testing the significance of a correlation matrix and its elements*. Unpublished master's thesis, University of British Columbia, Vancouver, British Columbia, Canada.

Franklin, M. F. (1984). Constructing tables of minimum aberration p(nm) designs. *Technometrics, 26*, 225-232.

Fraser, C., & McDonald, R. P. (1988). COSAN: Covariance structure analysis. *Multivariate Behavioral Research, 23*, 263-265.

Freedman, L. S. (1982). Tables of the number of patients required in clinical trials using the logrank test. *Statistics in Medicine, 1,* 121129.

Friedman, J. (1991). Multivariate adaptive regression splines (with discussion), *Annals of Statistics, 19*, 1-141.

Friedman, J. H. (1993). Estimating functions of mixed ordinal and categorical variables using adaptive splines. in S. Morgenthaler, E. Ronchetti, & W. A. Stahel (Eds.) (1993, p. 73-113). *New directions in statistical data analysis and robustness*. Berlin: Birkhäuser Verlag.

Friedman, J. H. (1999a). Greedy function approximation: A gradient boosting machine. *IMS 1999 Reitz Lecture*.

Friedman, J. H. (1999b). *Stochastic gradient boosting*. Stanford University.

Friedman, M. (1937). The use of ranks to avoid the assumption of normality implicit in the analysis of variance. *Journal of the American Statistical Association, 32*, 675-701.

Friedman, M. (1940). A comparison of alternative tests of significance for the problem of *m* rankings. *Annals of Mathematical Statistics, 11*, 86-92.

Fries, A., & Hunter, W. G. (1980). Minimum aberration 2 (kp) designs. *Technometrics, 22*, 601-608.

Frost, P. A. (1975). Some properties of the Almon lag technique when one searches for degree of polynomial and lag. *Journal of the American Statistical Association, 70*, 606-612.

Fuller, W. A. (1976). *Introduction to statistical time series*. New York: Wiley.

Gaddum, J. H. (1945). Lognormal distributions. *Nature, 156*, 463-466.

Gale, N., & Halperin, W. C. (1982). A case for better graphics: The unclassed choropleth map. *The American Statistician, 36*, 330-336.

Galil, Z., & Kiefer, J. (1980). Time- and space-saving computer methods, related to Mitchell's DETMAX, for finding D-optimum designs. *Technometrics, 22*, 301-313.

Galton, F. (1882). Report of the anthropometric committee. In *Report of the 51st Meeting of the British Association for the Advancement of Science, 1881*, 245-260.

Galton, F. (1885). Section H. Anthropology. Opening address by Francis Galton. *Nature, 32*, 507- 510.

Galton, F. (1885). Some results of the anthropometric laboratory. *Journal of the Anthropological Institute, 14*, 275-287.

Galton, F. (1888). Co-relations and their measurement. *Proceedings of the Royal Society of London, 45*, 135-145.

Galton, F. (1889). *Natural Inheritance*. London: Macmillan.

Ganguli, M. (1941). A note on nested sampling. *Sankhya, 5*, 449-452.

Gara, M. A., & Rosenberg, S. (1979). The identification of persons as supersets and subsets in free-response personality descriptions. *Journal of Personality and Social Psychology, 37*, 2161-2170.

Gara, M. A., & Rosenberg, S. (1981). Linguistic factors in implicit personality theory. *Journal of Personality and Social Psychology, 41*, 450-457.

Gardner, E. S., Jr. (1985). Exponential smoothing: The state of the art. *Journal of Forecasting, 4*, 1-28.

Garthwaite, P. H. (1994) An Interpretation of Partial Least Squares, *Journal of the American Statistical Association*, 89 NO. 425, 122-127.

Garvin, D. A. (1987). Competing on the eight dimensions of quality. *Harvard Business Review*, November/December, 101-109.

Gatsonis, C., & Sampson, A. R. (1989). Multiple correlation: exact power and sample size calculations. *Psychological Bulletin, 106*, 516524.

Gbur, E., Lynch, M., & Weidman, L. (1986). An analysis of nine rating criteria on 329 U. S. metropolitan areas. *Proceedings of the Section on Statistical Graphics, American Statistical Association*, 104-109.

Gedye, R. (1968). *A manager's guide to quality and reliability.* New York: Wiley.

Gehan, E. A. (1965a). A generalized Wilcoxon test for comparing arbitrarily singly-censored samples. *Biometrika, 52*, 203-223.

Gehan, E. A. (1965b). A generalized two-sample Wilcoxon test for doubly-censored data. *Biometrika, 52*, 650-653.

Gehan, E. A., & Siddiqui, M. M. (1973). Simple regression methods for survival time studies. *Journal of the American Statistical Association, 68*, 848-856.

Gehan, E. A., & Thomas, D. G. (1969). The performance of some two sample tests in small samples with and without censoring. *Biometrika, 56*, 127-132.

Geladi, P. and Kowalski, B. R. (1986) Partial Least Squares Regression: A Tutorial, *Analytica Chimica Acta*, 185, 1-17.

Gerald, C. F., & Wheatley, P. O. (1989). *Applied numerical analysis* (4th ed.). Reading, MA: Addison Wesley.

Gibbons, J. D. (1976). *Nonparametric methods for quantitative analysis.* New York: Holt, Rinehart, & Winston.

Gibbons, J. D. (1985). *Nonparametric statistical inference* (2nd ed.). New York: Marcel Dekker.

Gifi, A. (1981). *Nonlinear multivariate analysis.* Department of Data Theory, The University of Leiden. The Netherlands.

Gifi, A. (1990). *Nonlinear multivariate analysis.* New York: Wiley.

Gill, P. E., & Murray, W. (1972). Quasi-Newton methods for unconstrained optimization. *Journal of the Institute of Mathematics and its Applications, 9*, 91-108.

Gill, P. E., & Murray, W. (1974). *Numerical methods for constrained optimization.* New York: Academic Press.

Gini, C. (1911). Considerazioni sulle probabilita a posteriori e applicazioni al rapporto dei sessi nelle nascite umane. *Studi Economico-Giuridici della Universita de Cagliari*, Anno III, 133-171.

Glass, G V., & Hopkins, K. D. (1996). *Statistical methods in education and psychology.* Needham Heights, MA: Allyn & Bacon.

Glass, G. V., & Stanley, J. (1970). *Statistical methods in education and Psychology.* Englewood Cliffs, NJ: Prentice-Hall.

Glasser, M. (1967). Exponential survival with covariance. *Journal of the American Statistical Association, 62*, 561-568.

Gnanadesikan, R., Roy, S., & Srivastava, J. (1971). *Analysis and design of certain quantitative multiresponse experiments.* Oxford: Pergamon Press, Ltd.

Goldberg, D. E. (1989). *Genetic Algorithms.* Reading, MA: Addison Wesley.

Golub, G. and Kahan, W. (1965). Calculating the singular values and pseudo-inverse of a matrix. *SIAM Numerical Analysis, B 2 (2)*, 205-224.

Golub, G. H. and van Loan, C. F. (1996) *Matrix Computations*, The Johns Hopkins University Press

Golub, G. H., & van Loan, C. F. (1983). *Matrix computations.* Baltimore: Johns Hopkins University Press.

Gompertz, B. (1825). On the nature of the function expressive of the law of human mortality.

Philosophical Transactions of the Royal Society of London, Series A, 115, 513-580.

Goodman, L. A. (1954). Kolmogorov-Smirnov tests for psychological research. *Psychological Bulletin*, *51*, 160-168.

Goodman, L. A. (1971). The analysis of multidimensional contingency tables: Stepwise procedures and direct estimation methods for models building for multiple classification. *Technometrics*, *13*, 33-61.

Goodman, L. A., & Kruskal, W. H. (1954). Measures of association for cross-classifications. *Journal of the American Statistical Association*, *49*, 732-764.

Goodman, L. A., & Kruskal, W. H. (1959). Measures of association for cross-classifications II: Further discussion and references. *Journal of the American Statistical Association*, *54*, 123-163.

Goodman, L. A., & Kruskal, W. H. (1963). Measures of association for cross-classifications III: Approximate sampling theory. *Journal of the American Statistical Association*, *58*, 310-364.

Goodman, L .A., & Kruskal, W. H. (1972). Measures of association for cross-classifications IV: Simplification of asymptotic variances. *Journal of the American Statistical Association*, *67*, 415-421.

Goodnight, J. H. (1980). Tests of hypotheses in fixed effects linear models. *Communications in Statistics*, *A9*, 167-180.

Gorman, R.P., & Sejnowski, T.J. (1988). Analysis of hidden units in a layered network trained to classify sonar targets. *Neural Networks 1 (1)*, 75-89.

Grant, E. L., & Leavenworth, R. S. (1980). *Statistical quality control* (5th ed.). New York: McGraw-Hill.

Green, P. E., & Carmone, F. J. (1970). *Multidimensional scaling and related techniques in marketing analysis*. Boston: Allyn & Bacon.

Green, P. J. & Silverman, B. W. (1994) *Nonparametric regression and generalized linear models: A roughness penalty approach*. New York: Chapman & Hall.

Greenacre, M. J. & Hastie, T. (1987). The geometric interpretation of correspondence analysis. *Journal of the American Statistical Association*, *82*, 437-447.

Greenacre, M. J. (1984). *Theory and applications of correspondence analysis*. New York: Academic Press.

Greenacre, M. J. (1988). Correspondence analysis of multivariate categorical data by weighted least-squares. *Biometrica*, *75*, 457-467.

Greenhouse, S. W., & Geisser, S. (1958). Extension of Box's results on the use of the F distribution in multivariate analysis. *Annals of Mathematical Statistics*, *29*, 95-112.

Greenhouse, S. W., & Geisser, S. (1959). On methods in the analysis of profile data. *Psychometrika*, *24*, 95-112.

Grizzle, J. E. (1965). The two-period change-over design and its use in clinical trials. *Biometrics*, *21*, 467-480.

Gross, A. J., & Clark, V. A. (1975). *Survival distributions: Reliability applications in the medical sciences*. New York: Wiley.

Gruska, G. F., Mirkhani, K., & Lamberson, L. R. (1989). *Non-Normal data Analysis*. Garden City, MI: Multiface Publishing.

Gruvaeus, G., & Wainer, H. (1972). Two additions to hierarchical cluster analysis. *The British Journal of Mathematical and Statistical Psychology*, *25*, 200-206.

Guttman, L. (1954). A new approach to factor analysis: the radex. In P. F. Lazarsfeld (Ed.), *Mathematical thinking in the social sciences*. New York: Columbia University Press.

Guttman, L. (1968). A general nonmetric technique for finding the smallest coordinate space for a configuration of points. *Pyrometrical*, *33*, 469-506.

Guttman, L. B. (1977). What is not what in statistics. *The Statistician, 26*, 81107.

Haberman, S. J. (1972). Loglinear fit for contingency tables. *Applied Statistics*, *21*, 218-225.

Haberman, S. J. (1974). *The analysis of frequency data*. Chicago: University of Chicago Press.

Hahn, G. J., & Shapiro, S. S. (1967). *Statistical models in engineering*. New York: Wiley.

Hahn, G. J. & Meeker, W. Q. (1991). *Statistical intervals: A guide for practitioners*. New York: Wiley.

Hakstian, A. R., Rogers, W. D., & Cattell, R. B. (1982). The behavior of numbers of factors rules with simulated data. *Multivariate Behavioral Research, 17*, 193-219.

Hald, A. (1949). Maximum likelihood estimation of the parameters of a normal distribution which is truncated at a known point. *Skandinavisk Aktuarietidskrift, 1949*, 119-134.

Hald, A. (1952). *Statistical theory with engineering applications*. New York: Wiley.

Han, J., Kamber, M. (2000). *Data mining: Concepts and Techniques*. New York: Morgan-Kaufman.

Han, J., Lakshmanan, L. V. S., & Pei, J. (2001). Scalable frequent-pattern mining methods: An overview. In T. Fawcett (Ed.). *KDD 2001: Tutorial Notes of the Seventh ACM SIGKDD International Conference on Knowledge Discovery and Data Mining*. New York: The Association for Computing Machinery.

Harlow, L. L., Mulaik, S. A., & Steiger, J. H. (Eds.) (1997). *What if there were no significance tests*. Mahwah, NJ: Lawrence Erlbaum Associates.

Harman, H. H. (1967). *Modern factor analysis*. Chicago: University of Chicago Press.

Harris, R. J. (1976). The invalidity of partitioned U tests in canonical correlation and multivariate analysis of variance. *Multivariate Behavioral Research, 11*, 353-365.

Harrison, D. & Rubinfield, D. L. (1978). Hedonic prices and the demand for clean air. *Journal of Environmental Economics and Management, 5*, 81-102.

Harry, M.J. & Schroeder, R. (2000). *Six Sigma: The breakthrough management strategy revolu-tionizing the world's top corporations*. New York: Doubleday.

Hart, K. M., & Hart, R. F. (1989). *Quantitative methods for quality improvement*. Milwaukee, WI: ASQC Quality Press.

Hartigan, J. A. (1975). *Clustering algorithms*. New York: Wiley.

Hartigan, J. A. & Wong, M. A. (1978). Algorithm 136. A *k*-means clustering algorithm. *Applied Statistics, 28*, 100.

Hartley, H. O. (1959). Smallest composite designs for quadratic response surfaces. *Biometrics, 15*, 611-624.

Harville, D. A. (1977). Maximum likelihood approaches to variance component estimation and to related problems. *Journal of the American Statistical Association, 72*, 320-340.

Haskell, A. C. (1922). *Graphic Charts in Business*. New York: Codex.

Hastie, T. J. & Tibshirani, R. J. (1990). *Generalized Additive Models*. New York: Chapman & Hall.

Hastie, T., Tibshirani, R., & Friedman, J. H. (2001). *The elements of statistical learning : Data mining, inference, and prediction*. New York: Springer.

Haviland, R. P. (1964). *Engineering reliability and long life design*. Princeton, NJ: Van Nostrand.

Hayduk, L. A. (1987). *Structural equation modelling with LISREL: Essentials and advances*. Baltimore: The Johns Hopkins University Press.

Haykin, S. (1994). *Neural Networks: A Comprehensive Foundation*. New York: Macmillan Publishing.

Haykin, S. (1994). *Neural Networks: A Comprehensive Foundation*. New York: Macmillan College Publishing.

Hays, W. L. (1981). *Statistics* (3rd ed.). New York: CBS College Publishing.

Hays, W. L. (1988). *Statistics* (4th ed.). New York: CBS College Publishing.

Heiberger, R. M. (1989). *Computation for the analysis of designed experiments*. New York: Wiley.

Hemmerle, W. J., & Hartley, H., O. (1973). Computing maximum likelihood estimates for the mixed A.O.V. model using the W transformation. *Technometrics*, *15*, 819-831.

Henley, E. J., & Kumamoto, H. (1980). *Reliability engineering and risk assessment*. New York: Prentice-Hall.

Hettmansperger, T. P. (1984). *Statistical inference based on ranks*. New York: Wiley.

Hibbs, D. (1974). Problems of statistical estimation and causal inference in dynamic time series models. In H. Costner (Ed.), *Sociological Methodology 1973/1974* (pp. 252-308). San Francisco: Jossey-Bass.

Hill, I. D., Hill, R., & Holder, R. L. (1976). Fitting Johnson curves by moments. *Applied Statistics*. *25*, 190-192.

Hilton, T. L. (1969). *Growth study annotated bibliography*. Princeton, NJ: Educational Testing Service Progress Report 69-11.

Hochberg, J., & Krantz, D. H. (1986). Perceptual properties of statistical graphs. *Proceedings of the Section on Statistical Graphics, American Statistical Association*, 29-35.

Hocking, R. R. (1985). *The analysis of linear models*. Monterey, CA: Brooks/Cole.

Hocking, R. R. (1996). *Methods and Applications of Linear Models. Regression and the Analysis of Variance*. New York: Wiley.

Hocking, R. R., & Speed, F. M. (1975). A full rank analysis of some linear model problems. *Journal of the American Statistical Association*, *70*, 707-712.

Hoerl, A. E. (1962). Application of ridge analysis to regression problems. *Chemical Engineering Progress*, *58*, 54-59.

Hoerl, A. E., & Kennard, R. W. (1970). Ridge regression: Applications to nonorthogonal problems. *Technometrics*, *12*, 69-82.

Hoff, J. C. (1983). *A practical guide to Box-Jenkins forecasting*. London: Lifetime Learning Publications.

Hoffman, D. L. & Franke, G. R. (1986). Correspondence analysis: Graphical representation of categorical data in marketing research. *Journal of Marketing Research*, *13*, 213-227.

Hogg, R. V., & Craig, A. T. (1970). *Introduction to mathematical statistics*. New York: Macmillan.

Holzinger, K. J., & Swineford, F. (1939). *A study in factor analysis: The stability of a bi-factor solution*. University of Chicago: Supplementary Educational Monographs, No. 48.

Hooke, R., & Jeeves, T. A. (1961). Direct search solution of numerical and statistical problems. *Journal of the Association for Computing Machinery*, *8*, 212-229.

Hoskuldsson, A. (1988). PLS regression methods, *Journal of Chemometrics, Vol. 2*, 211-228.

Hosmer, D. W and Lemeshow, S. (1989), *Applied Logistic Regression*, John Wiley & Sons, Inc.

Hotelling, H. (1947). Multivariate quality control. In Eisenhart, Hastay, and Wallis (Eds.), *Techniques of Statistical Analysis*. New York: McGraw-Hill.

Hotelling, H., & Pabst, M. R. (1936). Rank correlation and tests of significance involving no assumption of normality. *Annals of Mathematical Statistics*, *7*, 29-43.

Hoyer, W., & Ellis, W. C. (1996). A graphical exploration of SPC. *Quality Progress*, *29*, 65-73.

Hsu, P. L. (1938). Contributions to the theory of Student's *t* test as applied to the problem of two samples. *Statistical Research Memoirs*, *2*, 1-24.

Huba, G. J., & Harlow, L. L. (1987). Robust structural equation models: implications for developmental psychology. *Child Development*, *58*, 147-166.

Huberty, C. J. (1975). Discriminant analysis. *Review of Educational Research*, *45*, 543-598.

Hunter, A., Kennedy, L., Henry, J., & Ferguson, R.I. (2000). Application of Neural Networks and Sensitivity Analysis to improved prediction of Trauma Survival. *Computer Methods and Algorithms in Biomedicine* 62, 11-19.

Huynh, H., & Feldt, L. S. (1970). Conditions under which mean square ratios in repeated measures designs have exact *F*-distributions. *Journal of the American Statistical Association, 65*, 1582-1589.

Ireland, C. T., & Kullback, S. (1968). Contingency tables with given marginals. *Biometrika, 55*, 179-188.

Jaccard, J., Weber, J., & Lundmark, J. (1975). A multitrait-multimethod factor analysis of four attitude assessment procedures. *Journal of Experimental Social Psychology, 11*, 149-154.

Jacobs, D. A. H. (Ed.). (1977). *The state of the art in numerical analysis.* London: Academic Press.

Jacobs, R.A. (1988). Increased Rates of Convergence Through Learning Rate Adaptation. *Neural Networks 1 (4)*, 295-307.

Jacoby, S. L. S., Kowalik, J. S., & Pizzo, J. T. (1972). *Iterative methods for nonlinear optimization problems.* Englewood Cliffs, NJ: Prentice-Hall.

Jambu, M. (1978). Classification automatique pour l'analyse des donnees. Dunod.

Jambu, M. (1991). Exploratory and multivariate data analysis. Academic Press.

James, L. R., Mulaik, S. A., & Brett, J. M. (1982). *Causal analysis. Assumptions, models, and data.* Beverly Hills, CA: Sage Publications.

Jardine, N., & Sibson, R. (1971). *Mathematical taxonomy.* New York: Wiley.

Jastrow, J. (1892). On the judgment of angles and position of lines. *American Journal of Psychology, 5*, 214-248.

Jenkins, G. M., & Watts, D. G. (1968). *Spectral analysis and its applications.* San Francisco: Holden-Day.

Jennrich, R. I. (1970). An asymptotic test for the equality of two correlation matrices. *Journal of the American Statistical Association, 65*, 904-912.

Jennrich, R. I. (1977). Stepwise regression. In K. Enslein, A. Ralston, & H.S. Wilf (Eds.), *Statistical methods for digital computers*. New York: Wiley.

Jennrich, R. I., & Moore, R. H. (1975). Maximum likelihood estimation by means of nonlinear least squares. *Proceedings of the Statistical Computing Section*, American Statistical Association, 57-65.

Jennrich, R. I., & Sampson, P. F. (1968). Application of stepwise regression to nonlinear estimation. *Technometrics, 10*, 63-72.

Jennrich, R. I., & Sampson, P. F. (1976). Newton-Raphson and related algorithms for maximum likelihood variance component estimation. *Technometrics, 18*, 11-17.

Jennrich, R. I., & Schuchter, M. D. (1986). Unbalanced repeated-measures models with structured covariance matrices. *Biometrics, 42*, 805-820.

Jennrich. R. I. (1977). Stepwise discriminant analysis. In K. Enslein, A. Ralston, & H.S. Wilf (Eds.), *Statistical methods for digital computers*. New York: Wiley.

Johnson, L. W., & Ries, R. D. (1982). *Numerical Analysis* (2nd ed.). Reading, MA: Addison Wesley.

Johnson, N. L. (1961). A simple theoretical approach to cumulative sum control charts. *Journal of the American Statistical Association, 56*, 83-92.

Johnson, N. L. (1965). Tables to facilitate fitting *SU* frequency curves. *Biometrika, 52*, 547.

Johnson, N. L., & Kotz, S. (1970). *Continuous univariate distributions, Vol I and II.* New York: Wiley.

Johnson, N. L., Kotz, S., Blakrishnan, N. (1995). *Continuous univariate distributions: Volume II.* (2nd Ed). NY: Wiley.

Johnson, N. L., & Leone, F. C. (1962). Cumulative sum control charts - mathematical principles applied to their construction and use. *Industrial Quality Control, 18*, 15-21.

Johnson, N. L., Nixon, E., & Amos, D. E. (1963). Table of percentage points of pearson curves. *Biometrika, 50*, 459.

Johnson, N. L., Nixon, E., Amos, D. E., & Pearson, E. S. (1963). Table of percentage points of Pearson curves for given 1 and 2, expressed in standard measure. *Biometrika*, 50, 459-498.

Johnson, P. (1987). *SPC for short runs: A programmed instruction workbook*. Southfield, MI: Perry Johnson.

Johnson, S. C. (1967). Hierarchical clustering schemes. *Psychometrika*, *32*, 241-254.

Johnston, J. (1972). *Econometric methods*. New York: McGraw-Hill.

Jöreskog, K. G. (1973). A general model for estimating a linear structural equation system. In A. S. Goldberger and O. D. Duncan (Eds.), *Structural Equation Models in the Social Sciences*. New York: Seminar Press.

Jöreskog, K. G. (1974). Analyzing psychological data by structural analysis of covariance matrices. In D. H. Krantz, R. C. Atkinson, R. D. Luce, and P. Suppes (Eds.), *Contemporary Developments in Mathematical Psychology, Vol. II*. New York: W. H. Freeman and Company.

Jöreskog, K. G. (1978). Structural analysis of covariance and correlation matrices. *Psychometrika*, *43*, 443-477.

Jöreskog, K. G., & Lawley, D. N. (1968). New methods in maximum likelihood factor analysis. *British Journal of Mathematical and Statistical Psychology*, *21*, 85-96.

Jöreskog, K. G., & Sörbom, D. (1979). *Advances in factor analysis and structural equation models*. Cambridge, MA: Abt Books.

Jöreskog, K. G., & Sörbom, D. (1982). Recent developments in structural equation modeling. *Journal of Marketing Research*, *19*, 404-416.

Jöreskog, K. G., & Sörbom, D. (1984). *Lisrel VI. Analysis of linear structural relationships by maximum likelihood, instrumental variables, and least squares methods*. Mooresville, Indiana: Scientific Software.

Jöreskog, K. G., & Sörbom, D. (1989). *Lisrel 7. A guide to the program and applications*. Chicago, Illinois: SPSS Inc.

Judge, G. G., Griffith, W. E., Hill, R. C., Luetkepohl, H., & Lee, T. S. (1985). *The theory and practice of econometrics*. New York: Wiley.

Juran, J. M. (1960). Pareto, Lorenz, Cournot, Bernnouli, Juran and others. *Industrial Quality Control*, *17*, 25.

Juran, J. M. (1962). *Quality control handbook*. New York: McGraw-Hill.

Juran, J. M., & Gryna, F. M. (1970). *Quality planning and analysis*. New York: McGraw-Hill.

Juran, J. M., & Gryna, F. M. (1980). *Quality planning and analysis* (2nd ed.). New York: McGraw-Hill.

Juran, J. M., & Gryna, F. M. (1988). *Juran's quality control handbook* (4th ed.). New York: McGraw-Hill.

Kachigan, S. K. (1986). *Statistical analysis: An interdisciplinary introduction to univariate & multivariate methods*. New York: Redius Press.

Kackar, R. M. (1985). Off-line quality control, parameter design, and the Taguchi method. *Journal of Quality Technology*, *17*, 176-188.

Kackar, R. M. (1986). Taguchi's quality philosophy: Analysis and commentary. *Quality Progress*, *19*, 21-29.

Kahneman, D., Slovic, P., & Tversky, A. (1982). *Judgment under uncertainty: Heuristics and biases*. New York: Cambridge University Press.

Kaiser, H. F. (1958). The varimax criterion for analytic rotation in factor analysis. *Pyrometrical*, *23*, 187-200.

Kaiser, H. F. (1960). The application of electronic computers to factor analysis. *Educational and Psychological Measurement*, *20*, 141-151.

Kalbfleisch, J. D., & Prentice, R. L. (1980). *The statisticcal analysis of failure time data*. New York: Wiley.

Kane, V. E. (1986). Process capability indices. *Journal of Quality Technology*, *18*, 41-52.

Kaplan, E. L., & Meier, P. (1958). Nonparametric estimation from incomplete observations. *Journal of the American Statistical Association*, *53*, 457-481.

Karsten, K. G., (1925). *Charts and graphs*. New York: Prentice-Hall.

Kass, G. V. (1980). An exploratory technique for investigating large quantities of categorical data. *Applied Statistics, 29*, 119-127.

Keats, J. B., & Lawrence, F. P. (1997). Weibull maximum likelihood parameter estimates with censored data. *Journal of Quality Technology*, 29, 105-110.

Keeves, J. P. (1972). *Educational environment and student achievement*. Melbourne: Australian Council for Educational Research.

Kendall, M. G. (1940). Note on the distribution of quantiles for large samples. *Supplement of the Journal of the Royal Statistical Society*, 7, 83-85.

Kendall, M. G. (1948). *Rank correlation methods*. (1st ed.). London: Griffin.

Kendall, M. G. (1975). *Rank correlation methods* (4th ed.). London: Griffin.

Kendall, M. G. (1984). *Time Series*. New York: Oxford University Press.

Kendall, M., & Ord, J. K. (1990). *Time series* (3rd ed.). London: Griffin.

Kendall, M., & Stuart, A. (1977). *The advanced theory of statistics. (Vol. 1)*. New York: MacMillan.

Kendall, M., & Stuart, A. (1979). *The advanced theory of statistics* (Vol. 2). New York: Hafner.

Kennedy, A. D., & Gehan, E. A. (1971). Computerized simple regression methods for survival time studies. *Computer Programs in Biomedicine, 1*, 235-244.

Kennedy, W. J., & Gentle, J. E. (1980). *Statistical computing*. New York: Marcel Dekker, Inc.

Kenny, D. A. (1979). *Correlation and causality*. New York: Wiley.

Keppel, G. (1973). *Design and analysis: A researcher's handbook*. Engelwood Cliffs, NJ: Prentice-Hall.

Keppel, G. (1982). *Design and analysis: A researcher's handbook* (2nd ed.). Engelwood Cliffs, NJ: Prentice-Hall.

Keselman, H. J., Rogan, J. C., Mendoza, J. L., & Breen, L. L. (1980). Testing the validity conditions for repeated measures F tests. *Psychological Bulletin*, 87, 479-481.

Khuri, A. I., & Cornell, J. A. (1987). *Response surfaces: Designs and analyses*. New York: Marcel Dekker, Inc.

Kiefer, J., & Wolfowitz, J. (1960). The equivalence of two extremum problems. *Canadian Journal of Mathematics*, *12*, 363-366.

Kim, J. O., & Mueller, C. W. (1978a). *Factor analysis: Statistical methods and practical issues*. Beverly Hills, CA: Sage Publications.

Kim, J. O., & Mueller, C. W. (1978b). *Introduction to factor analysis: What it is and how to do it*. Beverly Hills, CA: Sage Publications.

Kirk, D. B. (1973). On the numerical approximation of the bivariate normal (tetrachoric) correlation coefficient. *Psychometrika*, *38*, 259-268.

Kirk, R. E. (1968). *Experimental design: Procedures for the behavioral sciences*. (1st ed.). Monterey, CA: Brooks/Cole.

Kirk, R. E. (1982). *Experimental design: Procedures for the behavioral sciences*. (2nd ed.). Monterey, CA: Brooks/Cole.

Kirk, R. E. (1995). *Experimental design: Procedures for the behavioral sciences*. Pacific Grove, CA: Brooks-Cole.

Kirkpatrick, S., Gelatt, C.D. and Vecchi, M.P. (1983). Optimization by simulated annealing. *Science 220 (4598)*, 671-680.

Kish, L. (1965). *Survey sampling*. New York: Wiley.

Kivenson, G. (1971). *Durability and reliability in engineering design*. New York: Hayden.

Klecka, W. R. (1980). *Discriminant analysis*. Beverly Hills, CA: Sage.

Klein, L. R. (1974). *A textbook of econometrics*. Englewood Cliffs, NJ: Prentice-Hall.

Kleinbaum, D. G. (1996). *Survival analysis: A self-learning text.* New York: Springer-Verlag.

Kline, P. (1979). *Psychometrics and psychology.* London: Academic Press.

Kline, P. (1986). *A handbook of test construction.* New York: Methuen.

Kmenta, J. (1971). *Elements of econometrics.* New York: Macmillan.

Knuth, Donald E. (1981). *Seminumerical algorithms.* 2nd ed., Vol 2 of: *The art of computer programming.* Reading, Mass.: Addison-Wesley.

Kohonen, T. (1982). Self-organized formation of topologically correct feature maps. *Biological Cybernetics, 43,* 59-69.

Kohonen, T. (1990). Improved versions of learning vector quantization. *International Joint Conference on Neural Networks 1,* 545-550. San Diego, CA.

Kolata, G. (1984). The proper display of data. *Science, 226,* 156-157.

Kolmogorov, A. (1941). Confidence limits for an unknown distribution function. *Annals of Mathematical Statistics, 12,* 461-463.

Korin, B. P. (1969). On testing the equality of k covariance matrices. *Biometrika, 56,* 216-218.

Kramer, M.A. (1991). Nonlinear principal components analysis using autoassociative neural networks. *AIChe Journal 37 (2),* 233-243.

Kruskal, J. B. (1964). Nonmetric multidimensional scaling: A numerical method. *Pyrometrical, 29,* 1-27, 115-129.

Kruskal, J. B., & Wish, M. (1978). *Multidimensional scaling.* Beverly Hills, CA: Sage Publications.

Kruskal, W. H. (1952). A nonparametric test for the several sample problem. *Annals of Mathematical Statistics, 23,* 525-540.

Kruskal, W. H. (1975). Visions of maps and graphs. In J. Kavaliunas (Ed.), *Auto-carto II, proceedings of the international symposium on computer assisted cartography.* Washington, DC: U. S. Bureau of the Census and American Congress on Survey and Mapping.

Kruskal, W. H., & Wallis, W. A. (1952). Use of ranks in one-criterion variance analysis. *Journal of the American Statistical Association, 47,* 583-621.

Ku, H. H., & Kullback, S. (1968). Interaction in multidimensional contingency tables: An information theoretic approach. *J. Res. Nat. Bur. Standards Sect. B,* 72, 159-199.

Ku, H. H., Varner, R. N., & Kullback, S. (1971). Analysis of multidimensional contingency tables. *Journal of the American Statistical Association,* 66, 55-64.

Kullback, S. (1959). *Information theory and statistics.* New York: Wiley.

Kvålseth, T. O. (1985). Cautionary note about R2. *The American Statistician, 39,* 279-285.

Lagakos, S. W., & Kuhns, M. H. (1978). Maximum likelihood estimation for censored exponential survival data with covariates. *Applied Statistics,* 27, 190-197.

Lakatos, E., & Lan, K. K. G. (1992). A comparison of sample size methods for the logrank statistic. *Statistics in Medicine, 11,* 179191.

Lance, G. N., & Williams, W. T. (1966). A general theory of classificatory sorting strategies. *Computer Journal, 9,* 373.

Lance, G. N., & Williams, W. T. (1966). Computer programs for hierarchical polythetic classification ("symmetry analysis"). *Computer Journal, 9,* 60.

Larsen, W. A., & McCleary, S. J. (1972). The use of partial residual plots in regression analysis. *Technometrics, 14,* 781-790.

Lawless, J. F. (1982). *Statistical models and methods for lifetime data.* New York: Wiley.

Lawley, D. N., & Maxwell, A. E. (1971). *Factor analysis as a statistical method.* New York: American Elsevier.

Lawley, D. N., & Maxwell, A. E. (1971). *Factor analysis as a statistical method* (2nd. ed.). London: Butterworth & Company.

Lebart, L., Morineau, A., and Tabard, N. (1977). *Techniques de la description statistique.* Paris: Dunod.

Lebart, L., Morineau, A., and Warwick, K., M. (1984). *Multivariate descriptive statistical analysis: Correspondence analysis and related techniques for large matrices*. New York: Wiley.

Lee, E. T. (1980). *Statistical methods for survival data analysis*. Belmont, CA: Lifetime Learning.

Lee, E. T., & Desu, M. M. (1972). A computer program for comparing *K* samples with right-censored data. *Computer Programs in Biomedicine*, *2*, 315-321.

Lee, E. T., Desu, M. M., & Gehan, E. A. (1975). A Monte-Carlo study of the power of some two-sample tests. *Biometrika*, *62*, 425-532.

Lee, S., & Hershberger, S. (1990). A simple rule for generating equivalent models in covariance structure modeling. *Multivariate Behavioral Research*, *25*, 313-334.

Lee, Y. S. (1972). Tables of upper percentage points of the multiple correlation coefficient. *Biometrika*, *59*, 175189.

Legendre, A. M. (1805). *Nouvelles Methodes pour la Determination des Orbites des Cometes*. Paris: F. Didot.

Lehmann, E. L. (1975). *Nonparametrics: Statistical methods based on ranks*. San Francisco: Holden-Day.

Lennox, B., Hiden, H. G., Montague, G. A., Kornfeld, G., & Goulding, P. R. Application of multivariate statistical process control to batch operations, Elsevier, Computers and Chemical Engineering 24 (2000) 291-296.

Levenberg, K. (1944). A method for the solution of certain nonlinear problems in least squares. *Quarterly Journal of Applied Mathematics II (2)*, 164-168.

Lewicki, P., Hill, T., & Czyzewska, M. (1992). Nonconscious acquisition of information. *American Psychologist*, 47, 796-801.

Lieblein, J. (1953). On the exact evaluation of the variances and covariances of order statistics in samples form the extreme-value distribution. *Annals of Mathematical Statistics*, *24*, 282-287.

Lieblein, J. (1955). On moments of order statistics from the Weibull distribution. *Annals of Mathematical Statistics*, *26*, 330-333.

Lilliefors, H. W. (1967). On the Kolmogorov-Smirnov test for normality with mean and variance unknown. *Journal of the American Statistical Association, 64*, 399-402.

Lim, T.-S., Loh, W.-Y., & Shih, Y.-S. (1997). An empirical comparison of decision trees and other classification methods. *Technical Report 979*, Department of Statistics, University of Wisconsin, Madison.

Lindeman, R. H., Merenda, P. F., & Gold, R. (1980). *Introduction to bivariate and multivariate analysis*. New York: Scott, Foresman, & Co.

Lindman, H. R. (1974). *Analysis of variance in complex experimental designs*. San Francisco: W. H. Freeman & Co.

Linfoot, E. H. (1957). An informational measure of correlation. *Information and Control*, *1*, 50-55.

Linn, R. L. (1968). A Monte Carlo approach to the number of factors problem. *Psychometrika*, *33*, 37-71.

Lipson, C., & Sheth, N. C. (1973). *Statistical design and analysis of engineering experiments*. New York: McGraw-Hill.

Lloyd, D. K., & Lipow, M. (1977). *Reliability: Management, methods, and mathematics*. New York: McGraw-Hill.

Loehlin, J. C. (1987). *Latent variable models: An introduction to latent, path, and structural analysis*. Hillsdale, NJ: Erlbaum.

Loh, W.-Y, & Shih, Y.-S. (1997). Split selection methods for classification trees. *Statistica Sinica*, *7*, 815-840.

Loh, W.-Y., & Vanichestakul, N. (1988). Tree-structured classification via generalized discriminant analysis (with discussion). *Journal of the American Statistical Association*, *83*, 715-728.

Long, J. S. (1983a). *Confirmatory factor analysis*. Beverly Hills: Sage.

Long, J. S. (1983b). *Covariance structure models: An introduction to LISREL.* Beverly Hills: Sage.

Longley, J. W. (1967). An appraisal of least squares programs for the electronic computer from the point of view of the user. *Journal of the American Statistical Association, 62,* 819-831.

Longley, J. W. (1984). *Least squares computations using orthogonalization methods.* New York: Marcel Dekker.

Lord, F. M. (1957). A significance test for the hypothesis that two variables measure the same trait except for errors of measurement. *Psychometrika, 22,* 207-220.

Lorenz, M. O. (1904). Methods of measuring the concentration of wealth. *American Statistical Association Publication, 9,* 209-219.

Lowe, D. (1989). Adaptive radial basis function non-linearities, and the problem of generalisation. *First IEEE International Conference on Artificial Neural Networks,* 171-175, London, UK.

Lucas, J. M. (1976). The design and use of cumulative sum quality control schemes. *Journal of Quality Technology, 8,* 45-70.

Lucas, J. M. (1982). Combined Shewhart-CUSUM quality control schemes. *Journal of Quality Technology, 14,* 89-93.

MacCallum, R. C., Browne, M. W., & Sugawara, H. M. (1996). Power analysis and determination of sample size for covariance structure modeling. *Psychological Methods, 1,* 130149.

MacGregor, J. F. & Kourti, T. (1995). Statistical Process Control of Multivariate Processes, *Control Eng. Practice,* Vol. 3, No. 3, 403-414.

Maddala, G. S. (1977) *Econometrics.* New York: McGraw-Hill.

Maiti, S. S., & Mukherjee, B. N. (1990). A note on the distributional properties of the Jöreskog-Sörbom fit indices. *Psychometrika, 55,* 721-726.

Makridakis, S. G. (1983). Empirical evidence versus personal experience. *Journal of Forecasting, 2,* 295-306.

Makridakis, S. G. (1990). *Forecasting, planning, and strategy for the 21st century.* London: Free Press.

Makridakis, S. G., & Wheelwright, S. C. (1978). *Interactive forecasting: Univariate and multivariate methods* (2nd ed.). San Francisco, CA: Holden-Day.

Makridakis, S. G., & Wheelwright, S. C. (1989). *Forecasting methods for management* (5th ed.). New York: Wiley.

Makridakis, S. G., Wheelwright, S. C., & McGee, V. E. (1983). *Forecasting: Methods and applications* (2nd ed.). New York: Wiley.

Makridakis, S., Andersen, A., Carbone, R., Fildes, R., Hibon, M., Lewandowski, R., Newton, J., Parzen, R., & Winkler, R. (1982). The accuracy of extrapolation (time series) methods: Results of a forecasting competition. *Journal of Forecasting, 1,* 11-153.

Malinvaud, E. (1970). *Statistical methods of econometrics.* Amsterdam: North-Holland Publishing Co.

Mandel, B. J. (1969). The regression control chart. *Journal of Quality Technology, 1,* 3-10.

Mann, H. B., & Whitney, D. R. (1947). On a test of whether one of two random variables is stochastically larger than the other. *Annals of Mathematical Statistics, 18,* 50-60.

Mann, N. R., Schafer, R. E., & Singpurwalla, N. D. (1974). *Methods for statistical analysis of reliability and life data.* New York: Wiley.

Mann, N. R., Scheuer, R. M, & Fertig, K. W. (1973). A new goodness of fit test for the two-parameter Weibull or extreme value distribution. *Communications in Statistics, 2,* 383-400.

Manning, C. D., & Schutze, H. (2002). *Foundations of statistical natural language processing* (5th printing). Cambridge, MA: MIT Press.

Mantel, N. (1966). Evaluation of survival data and two new rank order statistics arising in its consideration. *Cancer Chemotherapy Reports, 50,* 163-170.

Mantel, N. (1967). Ranking procedures for arbitrarily restricted observations. *Biometrics, 23*, 65-78.

Mantel, N. (1974). Comment and suggestion on the Yates continuity correction. *Journal of the American Statistical Association, 69*, 378-380.

Mantel, N., & Haenszel, W. (1959). Statistical aspects of the analysis of data from retrospective studies of disease. *Journal of the National Cancer Institute, 22*, 719-748.

Marascuilo, L. A., & McSweeney, M. (1977). *Nonparametric and distribution free methods for the social sciences*. Monterey, CA: Brooks/Cole.

Marple, S. L., Jr. (1987). *Digital spectral analysis*. Englewood Cliffs, NJ: Prentice-Hall.

Marquardt, D.W. (1963). An algorithm for least-squares estimation of nonlinear parameters. *Journal of the Society of Industrial and Applied Mathematics 11 (2)*, 431-441.

Marsaglia, G. (1962). Random variables and computers. In J. Kozenik (Ed.), *Information theory, statistical decision functions, random processes: Transactions of the third Prague Conference.* Prague: Czechoslovak Academy of Sciences.

Marsaglia, G. (1996). *DIEHARD: A battery of tests of randomness*. At http://stat.fsu.edu/~geo/diehard.html.

Mason, R. L., Gunst, R. F., & Hess, J. L. (1989). *Statistical design and analysis of experiments with applications to engineering and science.* New York: Wiley.

Massey, F. J., Jr. (1951). The Kolmogorov-Smirnov test for goodness of fit. *Journal of the American Statistical Association, 46*, 68-78.

Masters (1995). *Neural, Novel, and Hybrid Algorithms for Time Series Predictions*. New York: Wiley.

Matsueda, R. L., & Bielby, W. T. (1986). Statistical power in covariance structure models. In N. B. Tuma (Ed.), *Sociological methodology.* Washington, DC: American Sociological Association.

McArdle, J. J. (1978). A structural view of structural models. Paper presented at the *Winter Workshop on Latent Structure Models Applied to Developmental Data, University of Denver, December, 1978.*

McArdle, J. J., & McDonald, R. P. (1984). Some algebraic properties of the Reticular Action Model for moment structures. *British Journal of Mathematical and Statistical Psychology, 37*, 234-251.

McCleary, R., & Hay, R. A. (1980). *Applied time series analysis for the social sciences.* Beverly Hills, CA: Sage Publications.

McCullagh, P. & Nelder, J. A. (1989). *Generalized linear models* (2nd Ed.). New York: Chapman & Hall.

McCullough, B. D. (1998). Assessing the reliability of statistical software: Part I. *The American Statistician, 52*, 358-366.

McCullough, B. D. (1999). Assessing the reliability of statistical software: Part II. *The American Statistician, 53*, 149-159.

McDonald, R. P. (1980). A simple comprehensive model for the analysis of covariance structures. *British Journal of Mathematical and Statistical Psychology, 31*, 59-72.

McDonald, R. P. (1989). An index of goodness-of-fit based on noncentrality. *Journal of Classification, 6*, 97-103.

McDonald, R. P., & Hartmann, W. M. (1992). A procedure for obtaining initial value estimates in the RAM model. *Multivariate Behavioral Research, 27*, 57-76.

McDonald, R. P., & Mulaik, S. A. (1979). Determinacy of common factors: A nontechnical review. *Psychological Bulletin, 86*, 297-306.

McDowall, D., McCleary, R., Meidinger, E. E., & Hay, R. A. (1980). *Interrupted time series analysis.* Beverly Hills, CA: Sage Publications.

McKenzie, E. (1984). General exponential smoothing and the equivalent ARMA process. *Journal of Forecasting, 3*, 333-344.

McKenzie, E. (1985). Comments on 'Exponential smoothing: The state of the art' by E. S. Gardner, Jr. *Journal of Forecasting, 4,* 32-36.

McLachlan, G. J. (1992). *Discriminant analysis and statistical pattern recognition.* New York: Wiley.

McLain, D. H. (1974). Drawing contours from arbitrary data points. *The Computer Journal, 17,* 318-324.

McLean, R. A., & Anderson, V. L. (1984). *Applied factorial and fractional designs.* New York: Marcel Dekker.

McLeod, A. I., & Sales, P. R. H. (1983). An algorithm for approximate likelihood calculation of ARMA and seasonal ARMA models. *Applied Statistics,* 211-223 (Algorithm AS).

McNemar, Q. (1947). Note on the sampling error of the difference between correlated proportions or percentages. *Psychometrika, 12,* 153-157.

McNemar, Q. (1969). *Psychological statistics* (4th ed.). New York: Wiley.

Melard, G. (1984). A fast algorithm for the exact likelihood of autoregressive-moving average models. *Applied Statistics, 33,* 104-119.

Mels, G. (1989). *A general system for path analysis with latent variables.* M. S. Thesis: Department of Statistics, University of South Africa.

Mendoza, J. L., Markos, V. H., & Gonter, R. (1978). A new perspective on sequential testing procedures in canonical analysis: A Monte Carlo evaluation. *Multivariate Behavioral Research, 13,* 371-382.

Meredith, W. (1964). Canonical correlation with fallible data. *Psychometrika, 29,* 55-65.

Miettinnen, O. S. (1968). The matched pairs design in the case of all-or-none responses. *Biometrics, 24,* 339352.

Miller, R. (1981). *Survival analysis.* New York: Wiley.

Milligan, G. W. (1980). An examination of the effect of six types of error perturbation on fifteen clustering algorithms. *Psychometrika, 45,* 325-342.

Milliken, G. A., & Johnson, D. E. (1984). *Analysis of messy data: Vol. I. Designed experiments.* New York: Van Nostrand Reinhold, Co.

Milliken, G. A., & Johnson, D. E. (1992). *Analysis of messy data: Vol. I. Designed experiments.* New York: Chapman & Hall.

Minsky, M.L. and Papert, S.A. (1969). *Perceptrons.* Cambridge, MA: MIT Press.

Mitchell, T. J. (1974a). Computer construction of "D-optimal" first-order designs. *Technometrics, 16,* 211-220.

Mitchell, T. J. (1974b). An algorithm for the construction of "D-optimal" experimental designs. *Technometrics, 16,* 203-210.

Mittag, H. J. (1993). *Qualitätsregelkarten.* München/Wien: Hanser Verlag.

Mittag, H. J., & Rinne, H. (1993). *Statistical methods of quality assurance.* London/New York: Chapman & Hall.

Monro, D. M. (1975). Complex discrete fast Fourier transform. *Applied Statistics, 24,* 153-160.

Monro, D. M., & Branch, J. L. (1976). The chirp discrete Fourier transform of general length. *Applied Statistics, 26,* 351-361.

Montgomery, D. C. (1976). *Design and analysis of experiments.* New York: Wiley.

Montgomery, D. C. (1985). *Statistical quality control.* New York: Wiley.

Montgomery, D. C. (1991) *Design and analysis of experiments* (3rd ed.). New York: Wiley.

Montgomery, D. C. (1996). *Introduction to Statistical Quality Control* (3rd Edition). New York: Wiley.

Montgomery, D. C. (1996). *Statistical quality control* (3rd. Edition). New York: Wiley.

Montgomery, D. C., & Wadsworth, H. M. (1972). Some techniques for multivariate quality control applications. *Technical Conference Transactions.* Washington, DC: American Society for Quality Control.

Montgomery, D. C., Johnson, L. A., & Gardiner, J. S. (1990). *Forecasting and time series analysis* (2nd ed.). New York: McGraw-Hill.

Mood, A. M. (1954). *Introduction to the theory of statistics.* New York: McGraw Hill.

Moody, J. and Darkin, C.J. (1989). Fast learning in networks of locally-tuned processing units. *Neural Computation 1 (2)*, 281-294.

Moré, J. J., (1977). *The Levenberg-Marquardt Algorithm: Implementation and Theory.* In G.A. Watson, (ed.), *Lecture Notes in Mathematics 630,* p. 105-116. Berlin: Springer-Verlag.

Morgan, J. N., & Messenger, R. C. (1973). THAID: A sequential analysis program for the analysis of nominal scale dependent variables. *Technical report,* Institute of Social Research, University of Michigan, Ann Arbor.

Morgan, J. N., & Sonquist, J. A. (1963). Problems in the analysis of survey data, and a proposal. *Journal of the American Statistical Association, 58,* 415-434.

Morris, M., & Thisted, R. A. (1986). Sources of error in graphical perception: A critique and an experiment. *Proceedings of the Section on Statistical Graphics, American Statistical Association,* 43-48.

Morrison, A. S., Black, M. M., Lowe, C. R., MacMahon, B., & Yuasa, S. (1973). Some international differences in histology and survival in breast cancer. *International Journal of Cancer, 11,* 261-267.

Morrison, D. (1967). *Multivariate statistical methods.* New York: McGraw-Hill.

Morrison, D. F. (1990). *Multivariate statistical methods.* (3rd Ed.). New York: McGraw-Hill.

Moses, L. E. (1952). Nonparametric statistics for psychological research. *Psychological Bulletin, 49,* 122-143.

Mulaik, S. A. (1972). *The foundations of factor analysis.* New York: McGraw Hill.

Muller, M. E. (1959). A comparison of methods for generating normal deviates on digital computers. *Journal for the ACM, 6,* 376-383.

Murphy, K. R., & Myors, B. (1998). *Statistical power analysis: A simple general model for traditional and modern hypothesis tests.* Mahwah, NJ: Lawrence Erlbaum Associates.

Muth, J. F. (1960). Optimal properties of exponentially weighted forecasts. *Journal of the American Statistical Association, 55,* 299-306.

Nachtsheim, C. J. (1979). *Contributions to optimal experimental design.* Ph.D. thesis, Department of Applied Statistics, University of Minnesota.

Nachtsheim, C. J. (1987). Tools for computer-aided design of experiments. *Journal of Quality Technology, 19,* 132-160.

Nelder, J. A., & Mead, R. (1965). A Simplex method for function minimization. *Computer Journal, 7,* 308-313.

Nelson, L. (1984). The Shewhart control chart - tests for special causes. *Journal of Quality Technology, 15,* 237-239.

Nelson, L. (1985). Interpreting Shewhart X-bar control charts. *Journal of Quality Technology, 17,* 114-116.

Nelson, P. R. C., Taylor, P. A., & MacGregor, J. F. (1996). Missing data methods in PCA and PLS: Score calculations with incomplete observations, elsevier, chemometrics and intelligent laboratory systems, 35, 45-65.

Nelson, W. (1982). *Applied life data analysis.* New York: Wiley.

Nelson, W. (1990). *Accelerated testing: Statistical models, test plans, and data analysis.* New York: Wiley.

Neter, J., Wasserman, W., & Kutner, M. H. (1985). *Applied linear statistical models: Regression, analysis of variance, and experimental designs.* Homewood, IL: Irwin.

Neter, J., Wasserman, W., & Kutner, M. H. (1989). *Applied linear regression models* (2nd ed.). Homewood, IL: Irwin.

Newcombe, Robert G. (1998). Two-sided confidence intervals for the single proportion: comparison of seven methods. *Statistics in Medicine, 17,* 857872.

Neyman, J., & Pearson, E. S. (1931). On the problem of *k* samples. *Bulletin de l'Academie Polonaise des Sciences et Lettres,* Ser. A, 460-481.

Neyman, J., & Pearson, E. S. (1933). On the problem of the most efficient tests of statistical hypothesis. *Philosophical Transactions of the Royal Society of London,* Ser. A, *231,* 289-337.

Nisbett, R. E., Fong, G. F., Lehman, D. R., & Cheng, P. W. (1987). Teaching reasoning. *Science, 238,* 625-631.

Nomikos, P., & MacGregor, J. F. (1995, February). Multivariate SPC Charts for Monitoring Batch Processes, *Technometrics, Volume 37, Number 1.*

Noori, H. (1989). The Taguchi methods: Achieving design and output quality. *The Academy of Management Executive, 3,* 322-326.

Nunally, J. C. (1970). *Introduction to psychological measurement.* New York: McGraw-Hill.

Nunnally, J. C. (1978). *Psychometric theory.* New York: McGraw-Hill.

Nussbaumer, H. J. (1982). *Fast Fourier transforms and convolution algorithms* (2nd ed.). New York: Springer-Verlag.

O'Brien, R. G., & Kaiser, M. K. (1985). MANOVA method for analyzing repeated measures designs: An extensive primer. *Psychological Bulletin, 97,* 316-333.

Okunade, A. A., Chang, C. F., & Evans, R. D. (1993). Comparative analysis of regression output summary statistics in common statistical packages. *The American Statistician, 47,* 298-303.

Olds, E. G. (1949). The 5% significance levels for sums of squares of rank differences and a correction. *Annals of Mathematical Statistics, 20,* 117-118.

Olejnik, S. F., & Algina, J. (1987). Type I error rates and power estimates of selected parametric and nonparametric tests of scale. *Journal of Educational Statistics, 12,* 45-61.

Olson, C. L. (1976). On choosing a test statistic in multivariate analysis of variance. *Psychological Bulletin, 83,* 579-586.

O'Neill, R. (1971). Function minimization using a Simplex procedure. *Applied Statistics, 3,* 79-88.

Ostle, B., & Malone, L. C. (1988). *Statistics in research: Basic concepts and techniques for research workers* (4th ed.). Ames, IA: Iowa State Press.

Ostrom, C. W. (1978). *Time series analysis: Regression techniques.* Beverly Hills, CA: Sage Publications.

O'Sullivan, F. (1985). Comments on some aspects of the spline smoothing approach to nonparametric regression curve fitting. *Journal of the Royal Statistical Society,* B, 47, 39-40.

Overall, J. E., & Speigel, D. K. (1969). Concerning least squares analysis of experimental data. *Psychological Bulletin, 83,* 579-586.

Page, E. S. (1954). Continuous inspection schemes. *Biometrics, 41,* 100-114.

Page, E. S. (1961). Cumulative sum charts. *Technometrics, 3,* 1-9.

Palumbo, F. A., & Strugala, E. S. (1945). Fraction defective of battery adapter used in handie-talkie. *Industrial Quality Control, November,* 68.

Pande, P.S., Neuman, R. P., Cavanagh, R. R. (2000). *The Six Sigma way: How GE, Motorola, and other top companies are honing their performance.* New York: McGraw.

Pankratz, A. (1983). *Forecasting with univariate Box-Jenkins models: Concepts and cases.* New York: Wiley.

Parker, D.B. (1985). *Learning logic. Technical Report TR-47,* Cambridge, MA: MIT Center for Research in Computational Economics and Management Science.

Parzen, E. (1961). Mathematical considerations in the estimation of spectra: Comments on the discussion of Messers, Tukey, and Goodman. *Technometrics, 3,* 167-190; 232-234.

Parzen, E. (1962). On estimation of a probability density function and mode. *Annals of Mathematical Statistics 33*, 1065-1076.

Patil, K. D. (1975). Cochran's Q test: Exact distribution. *Journal of the American Statistical Association*, *70*, 186-189.

Patterson, D. (1996). *Artificial Neural Networks.* Singapore: Prentice Hall.

Peace, G. S. (1993). *Taguchi methods: A hands-on approach.* Milwaukee, Wisconsin: ASQC.

Pearson, E. S., and Hartley, H. O. (1972). *Biometrika tables for statisticians, Vol II.* Cambridge: Cambridge University Press.

Pearson, K. (1894). Contributions to the mathematical theory of evolution. *Philosophical Transactions of the Royal Society of London*, Ser. A, *185*, 71-110.

Pearson, K. (1895). Skew variation in homogeneous material. *Philosophical Transactions of the Royal Society of London*, Ser. A, *186*, 343-414.

Pearson, K. (1896). Regression, heredity, and panmixia. *Philosophical Transactions of the Royal Society of London*, Ser. A, *187*, 253-318.

Pearson, K. (1900). On the criterion that a given system of deviations from the probable in the case of a correlated system of variables is such that it can be reasonably supposed to have arisen from random sampling. *Philosophical Magazine*, 5th Ser., *50*, 157-175.

Pearson, K. (1904). On the theory of contingency and its relation to association and normal correlation. *Drapers' Company Research Memoirs*, Biometric Ser. I.

Pearson, K. (1905). Das Fehlergesetz und seine Verallgemeinerungen durch Fechner und Pearson. A Rejoinder. *Biometrika*, *4*, 169-212.

Pearson, K. (1908). On the generalized probable error in multiple normal correlation. *Biometrika*, *6*, 59-68.

Pearson, K., (Ed.). (1968). *Tables of incomplete beta functions* (2nd ed.). Cambridge, MA: Cambridge University Press.

Pedhazur, E. J. (1973). *Multiple regression in behavioral research*. New York: Holt, Rinehart, & Winston.

Pedhazur, E. J. (1982). *Multiple regression in behavioral research* (2nd ed.). New York: Holt, Rinehart, & Winston.

Peressini, A. L., Sullivan, F. E., & Uhl, J. J., Jr. (1988). *The mathematics of nonlinear programming.* New York: Springer.

Peto, R., & Peto, J. (1972). Asymptotically efficient rank invariant procedures. *Journal of the Royal Statistical Society*, *135*, 185-207.

Phadke, M. S. (1989). *Quality engineering using robust design.* Englewood Cliffs, NJ: Prentice-Hall.

Phatak, A., Reilly, P. M., and Penlidis, A. (1993) An Approach to Interval Estimation in Partial Least Squares Regression, *Analytica Chimica Acta*, 277, 495-501

Piatetsky-Shapiro, G. (Ed.) (1993). *Proceedings of AAAI-93 Workshop on Knowledge Discovery in Databases.* Menlo Park, CA: AAAI Press.

Piepel, G. F. (1988). Programs for generating extreme vertices and centroids of linearly constrained experimental regions. *Journal of Quality Technology, 20*, 125-139.

Piepel, G. F., & Cornell, J. A. (1994). Mixture experiment approaches: Examples, discussion, and recommendations. *Journal of Quality Technology, 26*, 177-196.

Pigou, A. C. (1920). *Economics of Welfare*. London: Macmillan.

Pike, M. C. (1966). A method of analysis of certain class of experiments in carcinogenesis. *Biometrics, 22*, 142-161.

Pillai, K. C. S. (1965). On the distribution of the largest characteristic root of a matrix in multivariate analysis. *Biometrika*, *52*, 405-414.

Plackett, R. L., & Burman, J. P. (1946). The design of optimum multifactorial experiments. *Biometrika*, *34*, 255-272.

Polya, G. (1920). Uber den zentralen Grenzwertsatz der Wahrscheinlichkeitsrechnung und das Momentenproblem. *Mathematische Zeitschrift, 8*, 171-181.

Porebski, O. R. (1966). Discriminatory and canonical analysis of technical college data. *British Journal of Mathematical and Statistical Psychology, 19*, 215-236.

Powell, M. J. D. (1964). An efficient method for finding the minimum of a function of several variables without calculating derivatives. *Computer Journal, 7*, 155-162.

Pregibon, D. (1997). Data Mining. *Statistical Computing and Graphics, 7*, 8.

Prentice, R. (1973). Exponential survivals with censoring and explanatory variables. *Biometrika, 60*, 279-288.

Press, W. H., Flannery, B. P., Teukolsky, S. A., Vetterling, W. T. (1992). *Numerical recipies (2nd Edition)*. New York: Cambridge University Press.

Press, W.H., Teukolsky, S.A., Vetterling, W.T. and Flannery, B.P. (1992). *Numerical Recipes in C: The Art of Scientific Computing (Second ed.)*. Cambridge University Press.

Press, William, H., Flannery, B. P., Teukolsky, S. A., Vetterling, W. T. (1986). *Numerical recipies*. New York: Cambridge University Press.

Priestley, M. B. (1981). *Spectral analysis and time series*. New York: Academic Press.

Pyzdek, T. (1989). *What every engineer should know about quality control*. New York: Marcel Dekker.

Pyzdek, T. (1999). *Quality Engineering Handbook* (Quality and Reliability, 57). New York: Marcel Dekker.

Pyzdek, T. (2000). *The Handbook for Quality Management*. New York: Quality Publishing.

Pyzdek, T. (2001). *The Six Sigma Handbook: A complete guide for greenbelts, blackbelts, and managers at all levels*. New York: McGraw.

Qazaz, Cazhaow. (1996). *Bayesian Error Bars for Regression*, Ph.D. thesis, Aston University, England

Quinlan. (1992). *C4.5: Programs for Machine Learning*, Morgan Kaufmann

Quinlan, J.R., & Cameron-Jones, R.M. (1995). Oversearching and layered search in empirical learning. *Proceedings of the 14th International Joint Conference on Artificial Intelligence, Montreal* (Vol. 2). Morgan Kaufman, 1019-10244.

Ralston, A., & Wilf, H.S. (Eds.). (1960). *Mathematical methods for digital computers*. New York: Wiley.

Ralston, A., & Wilf, H.S. (Eds.). (1967). *Mathematical methods for digital computers* (Vol. II). New York: Wiley.

Randles, R. H., & Wolfe, D. A. (1979). *Introduction to the theory of nonparametric statistics*. New York: Wiley.

Rannar, S., Lindgren, F., Geladi, P, and Wold, S. (1994) A PLS Kernel Algorithm for Data Sets with Many Variables and Fewer Objects. Part 1: Theory and Algorithm, *Journal of Chemometrics, 8*, 111-125.

Rao, C. R. (1951). An asymptotic expansion of the distribution of Wilks' criterion. *Bulletin of the International Statistical Institute, 33*, 177-181.

Rao, C. R. (1952). *Advanced statistical methods in biometric research*. New York: Wiley.

Rao, C. R. (1965). *Linear statistical inference and its applications*. New York: Wiley.

Rath & Strong (2000). *Rath & Strong's Six Sigma pocket guide*. Lexington, MA: Rath & Strong, Inc.

Rhoades, H. M., & Overall, J. E. (1982). A sample size correction for Pearson chi-square in 2 x 2 contingency tables. *Psychological Bulletin, 91*, 418-423.

Rinne, H., & Mittag, H. J. (1995). *Statistische Methoden der Qualitätssicherung (3rd. edition)*. München/Wien: Hanser Verlag.

Ripley, B. D. (1981). *Spacial statistics*. New York: Wiley.

Ripley, B. D. (1996). *Pattern recognition and neural networks*. Cambridge: Cambridge University Press.

Rodriguez, R. N. (1992). Recent developments in process capability analysis. *Journal of Quality Technology, 24*, 176-187.

Rogan, J. C., Keselman, J. J., & Mendoza, J. L. (1979). Analysis of repeated measurements. *British Journal of Mathematical and Statistical Psychology, 32*, 269-286.

Rosenberg, S. (1977). New approaches to the analysis of personal constructs in person perception. In A. Landfield (Ed.), *Nebraska symposium on motivation* (Vol. 24). Lincoln, NE: University of Nebraska Press.

Rosenberg, S., & Sedlak, A. (1972). Structural representations of implicit personality theory. In L. Berkowitz (Ed.). *Advances in experimental social psychology* (Vol. 6). New York: Academic Press.

Rosenblatt, F. (1958). The Perceptron: A probabilistic model for information storage and organization in the brain. *Psychological Review 65*, 386-408.

Roskam, E. E., & Lingoes, J. C. (1970). MINISSA-I: A Fortran IV program for the smallest space analysis of square symmetric matrices. *Behavioral Science, 15*, 204-205.

Ross, P. J. (1988). *Taguchi techniques for quality engineering: Loss function, orthogonal experiments, parameter, and tolerance design.* Milwaukee, Wisconsin: ASQC.

Roy, J. (1958). Step-down procedure in multivariate analysis. *Annals of Mathematical Statistics, 29*, 1177-1187.

Roy, J. (1967). *Some aspects of multivariate analysis.* New York: Wiley.

Roy, R. (1990). *A primer on the Taguchi method.* Milwaukee, Wisconsin: ASQC.

Royston, J. P. (1982). An extension of Shapiro and Wilks' W test for normality to large samples. *Applied Statistics, 31*, 115-124.

Rozeboom, W. W. (1979). Ridge regression: Bonanza or beguilement? *Psychological Bulletin, 86*, 242-249.

Rozeboom, W. W. (1988). Factor indeterminacy: the saga continues. *British Journal of Mathematical and Statistical Psychology, 41*, 209-226.

Rubinstein, L.V., Gail, M. H., & Santner, T. J. (1981). Planning the duration of a comparative clinical trial with loss to follow-up and a period of continued observation. *Journal of Chronic Diseases, 34,* 469479.

Rud, O., P. (2001). *Data mining cookbook: Modeling data for marketing, risk, and customer relationship management.* NY: Wiley.

Rumelhart, D.E. and McClelland, J. (eds.) (1986). *Parallel Distributed Processing, Vol 1.* Cambridge, MA: MIT Press.

Rumelhart, D.E., Hinton, G.E. and Williams, R.J. (1986). Learning internal representations by error propagation. In D.E. Rumelhart, J.L. McClelland (Eds.), *Parallel Distributed Processing, Vol 1.* Cambridge, MA: MIT Press.

Runyon, R. P., & Haber, A. (1976). *Fundamentals of behavioral statistics.* Reading, MA: Addison-Wesley.

Ryan, T. P. (1989). *Statistical methods for quality improvement.* New York: Wiley.

Ryan, T. P. (1997). *Modern Regression Methods.* New York: Wiley.

Sandler, G. H. (1963). *System reliability engineering.* Englewood Cliffs, NJ: Prentice-Hall.

SAS Institute, Inc. (1982). *SAS user's guide: Statistics, 1982 Edition.* Cary, NC: SAS Institute, Inc.

Satorra, A., & Saris, W. E. (1985). Power of the likelihood ratio test in covariance structure analysis. *Psychometrika, 50*, 83-90.

Saxena, K. M. L., & Alam, K. (1982). Estimation of the noncentrality parameter of a chi squared distribution. *Annals of Statistics, 10*, 1012-1016.

Scheffé, H. (1953). A method for judging all possible contrasts in the analysis of variance. *Biometrica, 40*, 87-104.

Scheffé, H. (1959). *The analysis of variance.* New York: Wiley.

Scheffé, H. (1963). The simplex-centroid design for experiments with mixtures. *Journal of the Royal Statistical Society, B25*, 235-263.

Scheffé, H., & Tukey, J. W. (1944). A formula for sample sizes for population tolerance limits. *Annals of Mathematical Statistics, 15*, 217.

Scheines, R. (1994). Causation, indistinguishability, and regression. In F. Faulbaum, (Ed.), *SoftStat '93. Advances in statistical software 4*. Stuttgart: Gustav Fischer Verlag.

Schiffman, S. S., Reynolds, M. L., & Young, F. W. (1981). *Introduction to multidimensional scaling: Theory, methods, and applications*. New York: Academic Press.

Schimek, M. G. (2000). *Smoothing and regression: Approaches, computations, and application*. New York: Wiley.

Schmidt, F. L., & Hunter, J. E. (1997). Eight common but false objections to the discontinuation of significance testing in the analysis of research data. In Harlow, L. L., Mulaik, S. A., & Steiger, J. H. (Eds.), *What if there were no significance tests*. Mahwah, NJ: Lawrence Erlbaum Associates.

Schmidt, P., & Muller, E. N. (1978). The problem of multicollinearity in a multistage causal alienation model: A comparison of ordinary least squares, maximum-likelihood and ridge estimators. *Quality and Quantity, 12*, 267-297.

Schmidt, P., & Sickles, R. (1975). On the efficiency of the Almon lag technique. *International Economic Review, 16*, 792-795.

Schmidt, P., & Waud, R. N. (1973). The Almon lag technique and the monetary versus fiscal policy debate. *Journal of the American Statistical Association, 68*, 11-19.

Schnabel, R. B., Koontz, J. E., and Weiss, B. E. (1985). A modular system of algorithms for unconstrained minimization. *ACM Transactions on Mathematical Software, 11*, 419-440.

Schneider, H. (1986). *Truncated and censored samples from normal distributions*. New York: Marcel Dekker.

Schneider, H., & Barker, G.P. (1973). *Matrices and linear algebra* (2nd ed.). New York: Dover Publications.

Schönemann, P. H., & Steiger, J. H. (1976). Regression component analysis. *British Journal of Mathematical and Statistical Psychology, 29*, 175-189.

Schrock, E. M. (1957). *Quality control and statistical methods*. New York: Reinhold Publishing.

Schwarz, G. (1978). Estimating the dimension of a model. *Annals of Statistics, 6*, 461-464.

Scott, D. W. (1979). On optimal and data-based histograms. *Biometrika, 66*, 605-610.

Searle, S. R. (1987). *Linear models for unbalanced data*. New York: Wiley.

Searle, S. R., Casella, G., & McCullock, C. E. (1992). *Variance components*. New York: Wiley.

Searle, S., R., Speed., F., M., & Milliken, G. A. (1980). The population marginal means in the linear model: An alternative to least squares means. *The American Statistician, 34*, 216-221.

Seber, G. A. F., & Wild, C. J. (1989). *Nonlinear regression*. New York: Wiley.

Sebestyen, G. S. (1962). *Decision making processes in pattern recognition*. New York: Macmillan.

Seder, L. A. (1962). *Quality improvement*. In J. M. Juran. *Quality control handbook*. New York: McGraw-Hill.

Sen, P. K., & Puri, M. L. (1968). On a class of multivariate multisample rank order tests, II: Test for homogeneity of dispersion matrices. *Sankhya, 30*, 1-22.

Serlin, R. A., & Lapsley, D. K. (1993). Rational appraisal of psychological research and the good-enough principle. In G. Keren & C. Lewis (Eds.), *A handbook for data analysis in the behavioral sciences: Methodological issues* (pp. 199-228). Hillsdale, NJ: Lawrence Erlbaum Associates.

Serlin. R. A., & Lapsley, D. K. (1985). Rationality in psychological research: The good-enough principle. *American Psychologist, 40*, 7383.

Shapiro, A., & Browne, M. W. (1983). On the investigation of local identifiability: A counter example. *Psychometrika, 48*, 303-304.

Shapiro, S. S., Wilk, M. B., & Chen, H. J. (1968). A comparative study of various tests of normality. *Journal of the American Statistical Association, 63*, 1343-1372.

Shepherd, A. J. (1997). *Second-Order Methods for Neural Networks*. New York: Springer.

Shewhart, W. A. (1931). *Economic control of quality of manufactured product*. New York: D. Van Nostrand.

Shewhart, W. A. (1939). *Statistical method from the viewpoint of quality*. Washington, DC: The Graduate School Department of Agriculture.

Shirland, L. E. (1993). *Statistical quality control with microcomputer applications*. New York: Wiley.

Shiskin, J., Young, A. H., & Musgrave, J. C. (1967). *The X-11 variant of the census method II seasonal adjustment program*. (Technical paper no. 15). Bureau of the Census.

Shumway, R. H. (1988). *Applied statistical time series analysis*. Englewood Cliffs, NJ: Prentice Hall.

Siegel, A. E. (1956). Film-mediated fantasy aggression and strength of aggressive drive. *Child Development, 27*, 365-378.

Siegel, S. (1956). *Nonparametric statistics for the behavioral sciences*. New York: McGraw-Hill.

Siegel, S., & Castellan, N. J. (1988). *Nonparametric statistics for the behavioral sciences* (2nd ed.) New York: McGraw-Hill.

Simkin, D., & Hastie, R. (1986). Towards an information processing view of graph perception. *Proceedings of the Section on Statistical Graphics, American Statistical Association*, 11-20.

Sinha, S. K., & Kale, B. K. (1980). *Life testing and reliability estimation*. New York: Halstead.

Smirnov, N. V. (1948). Table for estimating the goodness of fit of empirical distributions. *Annals of Mathematical Statistics, 19*, 279-281.

Smith, D. J. (1972). *Reliability engineering*. New York: Barnes & Noble.

Smith, K. (1953). Distribution-free statistical methods and the concept of power efficiency. In L. Festinger and D. Katz (Eds.), *Research methods in the behavioral sciences* (pp. 536-577). New York: Dryden.

Sneath, P. H. A., & Sokal, R. R. (1973). *Numerical taxonomy*. San Francisco: W. H. Freeman & Co.

Snee, R. D. (1975). Experimental designs for quadratic models in constrained mixture spaces. *Technometrics, 17*, 149-159.

Snee, R. D. (1979). Experimental designs for mixture systems with multi-component constraints. *Communications in Statistics - Theory and Methods, A8(4)*, 303-326.

Snee, R. D. (1985). Computer-aided design of experiments - some practical experiences. *Journal of Quality Technology, 17*, 222-236.

Snee, R. D. (1986). An alternative approach to fitting models when re-expression of the response is useful. *Journal of Quality Technology, 18*, 211-225.

Sokal, R. R., & Mitchener, C. D. (1958). A statistical method for evaluating systematic relationships. *University of Kansas Science Bulletin, 38*, 1409.

Sokal, R. R., & Sneath, P. H. A. (1963). *Principles of numerical taxonomy*. San Francisco: W. H. Freeman & Co.

Soper, H. E. (1914). Tables of Poisson's exponential binomial limit. *Biometrika, 10*, 25-35.

Spearman, C. (1904). "General intelligence," objectively determined and measured. *American Journal of Psychology, 15*, 201-293.

Speckt, D.F. (1990). Probabilistic Neural Networks. *Neural Networks 3 (1)*, 109-118.

Speckt, D.F. (1991). A Generalized Regression Neural Network. *IEEE Transactions on Neural Networks 2 (6)*, 568-576.

Spirtes, P., Glymour, C., & Scheines, R. (1993). *Causation, prediction, and search*. Lecture Notes in Statistics, V. 81. New York: Springer-Verlag.

Spjotvoll, E., & Stoline, M. R. (1973). An extension of the *T*-method of multiple comparison to include the cases with unequal sample sizes. *Journal of the American Statistical Association, 68*, 976-978.

Springer, M. D. (1979). *The algebra of random variables*. New York: Wiley.

Spruill, M. C. (1986). Computation of the maximum likelihood estimate of a noncentrality parameter. *Journal of Multivariate Analysis, 18*, 216-224.

StatSoft, Inc. (2001) *STATISTICA System Reference*

Steiger, J. H. (1979). Factor indeterminacy in the 1930's and in the 1970's; some interesting parallels. *Psychometrika, 44*, 157-167.

Steiger, J. H. (1980a). Tests for comparing elements of a correlation matrix. *Psychological Bulletin, 87*, 245-251.

Steiger, J. H. (1980b). Testing pattern hypotheses on correlation matrices: Alternative statistics and some empirical results. *Multivariate Behavioral Research, 15*, 335-352.

Steiger, J. H. (1988). Aspects of person-machine communication in structural modeling of correlations and covariances. *Multivariate Behavioral Research, 23*, 281-290.

Steiger, J. H. (1989). *EzPATH: A supplementary module for SYSTAT and SYGRAPH*. Evanston, IL: SYSTAT, Inc.

Steiger, J. H. (1990). Some additional thoughts on components and factors. *Multivariate Behavioral Research, 25*, 41-45.

Steiger, J. H., & Browne, M. W. (1984). The comparison of interdependent correlations between optimal linear composites. *Psychometrika, 49*, 11-24.

Steiger, J. H., & Fouladi, R. T. (1992). *R2:* A computer program for interval estimation, power calculation, and hypothesis testing for the squared multiple correlation. *Behavior Research Methods, Instruments, and Computers, 4,* 581582.

Steiger, J. H., & Fouladi, R. T. (1997). Noncentrality interval estimation and the evaluation of statistical models. In Harlow, L. L., Mulaik, S. A., & Steiger, J. H. (Eds.), *What if there were no significance tests*. Mahwah, NJ: Lawrence Erlbaum Associates.

Steiger, J. H., & Hakstian, A. R. (1982). The asymptotic distribution of elements of a correlation matrix: Theory and application. *British Journal of Mathematical and Statistical Psychology, 35*, 208-215.

Steiger, J. H., & Lind, J. C. (1980). *Statistically-based tests for the number of common factors*. Paper presented at the annual Spring Meeting of the Psychometric Society in Iowa City. May 30, 1980.

Steiger, J. H., & Schönemann, P. H. (1978). A history of factor indeterminacy. In S. Shye, (Ed.), *Theory Construction and Data Analysis in the Social Sciences*. San Francisco: Jossey-Bass.

Steiger, J. H., Shapiro, A., & Browne, M. W. (1985). On the multivariate asymptotic distribution of sequential chi-square statistics. *Psychometrika, 50*, 253-264.

Stelzl, I. (1986). Changing causal relationships without changing the fit: Some rules for generating equivalent LISREL models. *Multivariate Behavioral Research, 21*, 309-331.

Stenger, F. (1973). Integration formula based on the trapezoid formula. *Journal of the Institute of Mathematics and Applications, 12*, 103-114.

Stevens, J. (1986). *Applied multivariate statistics for the social sciences*. Hillsdale, NJ: Erlbaum.

Stevens, W. L. (1939). Distribution of groups in a sequence of alternatives. *Annals of Eugenics, 9*, 10-17.

Stewart, D. K., & Love, W. A. (1968). A general canonical correlation index. *Psychological Bulletin, 70*, 160-163.

Steyer, R. (1992). *Theorie causale regressionsmodelle* [Theory of causal regression models]. Stuttgart: Gustav Fischer Verlag.

Steyer, R. (1994). Principles of causal modeling: a summary of its mathematical foundations and practical steps. In F. Faulbaum, (Ed.), *SoftStat '93. Advances in statistical software 4*. Stuttgart: Gustav Fischer Verlag.

Stone, M. and Brooks, R. J. (1990) Continuum Regression: Cross-validated Sequentially Constructed Prediction Embracing Ordinary Least Squares, Partial Least Squares, and Principal Components Regression, *Journal of Royal Statistical Society*, 52, No. 2, 237-269.

Student (1908). The probable error of a mean. *Biometrika*, *6*, 1-25.

Swallow, W. H., & Monahan, J. F. (1984). Monte Carlo comparison of ANOVA, MIVQUE, REML, and ML estimators of variance components. *Technometrics*, *26*, 47-57.

Taguchi, G. (1987). *Jikken keikakuho* (3rd ed., Vol I & II). Tokyo: Maruzen. English translation edited by D. Clausing. *System of experimental design*. New York: UNIPUB/Kraus International

Taguchi, G., & Jugulum, R. (2002). *The Mahalanobis-Taguchi strategy*. New York, NY: Wiley.

Tanaka, J. S., & Huba, G. J. (1985). A fit index for covariance structure models under arbitrary GLS estimation. *British Journal of Mathematical and Statistical Psychology*, *38*, 197-201.

Tanaka, J. S., & Huba, G. J. (1989). A general coefficient of determination for covariance structure models under arbitrary GLS estimation. *British Journal of Mathematical and Statistical Psychology*, *42*, 233-239.

Tatsuoka, M. M. (1970). *Discriminant analysis*. Champaign, IL: Institute for Personality and Ability Testing.

Tatsuoka, M. M. (1971). *Multivariate analysis*. New York: Wiley.

Tatsuoka, M. M. (1976). Discriminant analysis. In P. M. Bentler, D. J. Lettieri, and G. A. Austin (Eds.), *Data analysis strategies and designs for substance abuse research*. Washington, DC: U.S. Government Printing Office.

Taylor, D. J., & Muller, K. E. (1995). Computing confidence bounds for power and sample size of the general linear univariate model. *The American Statistician, 49,* 4347.

Thorndike, R. L., & Hagen, E. P. (1977). *Measurement and evaluation in psychology and education*. New York: Wiley.

Thurstone, L. L. (1931). Multiple factor analysis. *Psychological Review*, *38*, 406-427.

Thurstone, L. L. (1947). *Multiple factor analysis*. Chicago: University of Chicago Press.

Timm, N. H. (1975). *Multivariate analysis with applications in education and psychology*. Monterey, CA: Brooks/Cole.

Timm, N. H., & Carlson, J. (1973). *Multivariate analysis of non-orthogonal experimental designs using a multivariate full rank model*. Paper presented at the American Statistical Association Meeting, New York.

Timm, N. H., & Carlson, J. (1975). Analysis of variance through full rank models. *Multivariate behavioral research monographs*, No. *75-1*.

Tippett, L. H. C. (1925). On the extreme individuals and the range of samples taken from a normal population. *Biometrika*, 17, 364-387.

Tracey, N. D., Young, J., C., & Mason, R. L. (1992). Multivariate control charts for individual observations. *Journal of Quality Technology*, *2*, 88-95.

Tribus, M., & Sconyi, G. (1989). An alternative view of the Taguchi approach. *Quality Progress*, *22*, 46-48.

Trivedi, P. K., & Pagan, A. R. (1979). Polynomial distributed lags: A unified treatment. *Economic Studies Quarterly*, *30*, 37-49.

Tryon, R. C. (1939). *Cluster Analysis*. Ann Arbor, MI: Edwards Brothers.

Tucker, L. R., Koopman, R. F., & Linn, R. L. (1969). Evaluation of factor analytic research procedures by means of simulated correlation matrices. *Psychometrika*, *34*, 421-459.

Tufte, E. R. (1983). *The visual display of quantitative information*. Cheshire, CT: Graphics Press.

Tufte, E. R. (1990). *Envisioning information*. Cheshire, CT: Graphics Press.

Tukey, J. W. (1953). *The problem of multiple comparisons*. Unpublished manuscript, Princeton University.

Tukey, J. W. (1962). The future of data analysis. *Annals of Mathematical Statistics, 33*, 1-67.

Tukey, J. W. (1967). An introduction to the calculations of numerical spectrum analysis. In B. Harris (Ed.), *Spectral analysis of time series*. New York: Wiley.

Tukey, J. W. (1972). Some graphic and semigraphic displays. In *Statistical Papers in Honor of George W. Snedecor*; ed. T. A. Bancroft, Arnes, IA: Iowa State University Press, 293-316.

Tukey, J. W. (1977). *Exploratory data analysis*. Reading, MA: Addison-Wesley.

Tukey, J. W. (1984). *The collected works of John W. Tukey*. Monterey, CA: Wadsworth.

Tukey, P. A. (1986). A data analyst's view of statistical plots. *Proceedings of the Section on Statistical Graphics, American Statistical Association*, 21-28.

Tukey, P. A., & Tukey, J. W. (1981). Graphical display of data sets in 3 or more dimensions. In V. Barnett (Ed.), *Interpreting multivariate data*. Chichester, U.K.: Wiley.

Upsensky, J. V. (1937). *Introduction to Mathematical Probability*. New York: McGraw-Hill.

Vale, C. D., & Maurelli, V. A. (1983). Simulating multivariate non-normal distributions. *Psychometrika, 48*, 465-471.

Vandaele, W. (1983). *Applied time series and Box-Jenkins models*. New York: Academic Press.

Vaughn, R. C. (1974). *Quality control*. Ames, IA: Iowa State Press.

Velicer, W. F., & Jackson, D. N. (1990). Component analysis vs. factor analysis: some issues in selecting an appropriate procedure. *Multivariate Behavioral Research, 25*, 1-28.

Velleman, P. F., & Hoaglin, D. C. (1981). *Applications, basics, and computing of exploratory data analysis*. Belmont, CA: Duxbury Press.

Von Mises, R. (1941). Grundlagen der Wahrscheinlichkeitsrechnung. *Mathematische Zeitschrift, 5*, 52-99.

Wainer, H. (1995). Visual revelations. *Chance, 8*, 48-54.

Wald, A. (1939). Contributions to the theory of statistical estimation and testing hypotheses. *Annals of Mathematical Statistics, 10*, 299-326.

Wald, A. (1945). Sequential tests of statistical hypotheses. *Annals of Mathematical Statistics, 16*, 117-186.

Wald, A. (1947). *Sequential analysis*. New York: Wiley.

Walker, J. S. (1991). *Fast Fourier transforms*. Boca Raton, FL: CRC Press.

Wallis, K. F. (1974). Seasonal adjustment and relations between variables. *Journal of the American Statistical Association, 69*, 18-31.

Wang, C. M., & Gugel, H. W. (1986). High-performance graphics for exploring multivariate data. *Proceedings of the Section on Statistical Graphics, American Statistical Association*, 60-65.

Ward, J. H. (1963). Hierarchical grouping to optimize an objective function. *Journal of the American Statistical Association, 58*, 236.

Warner B. & Misra, M. (1996). Understanding Neural Networks as Statistical Tools. *The American Statistician, 50*, 284-293.

Weatherburn, C. E. (1946). *A First Course in Mathematical Statistics*. Cambridge: Cambridge University Press.

Wei, W. W. (1989). *Time series analysis: Univariate and multivariate methods*. New York: Addison-Wesley.

Weibull, W. (1951). A statistical distribution function of wide applicability. *Journal of Applied Mechanics, September*.

Weibull, W., (1939). A statistical theory of the strength of materials. *Ing. Velenskaps Akad. Handl., 151*, 1-45.

Weigend, A.S., Rumelhart, D.E. and Huberman, B.A. (1991). Generalization by weight-elimination with application to forecasting. In R.P. Lippmann, J.E. Moody and D.S. Touretzky (Eds.) *Advances in Neural Information Processing Systems 3*, 875-882. San Mateo, CA: Morgan Kaufmann.

Weiss, S. M., & Indurkhya, N. (1997). *Predictive data mining: A practical guide*. New York: Morgan-Kaufman.

Welch, B. L. (1938). The significance of the differences between two means when the population variances are unequal. *Biometrika*, *29*, 350-362.

Welstead, S. T. (1994). *Neural network and fuzzy logic applications in C/C++*. New York: Wiley.

Werbos, P.J. (1974). *Beyond regression: new tools for prediction and analysis in the behavioural sciences*. Ph.D. thesis, Harvard University, Boston, MA.

Wescott, M. E. (1947). Attribute charts in quality control. *Conference Papers, First Annual Convention of the American Society for Quality Control*. Chicago: John S. Swift Co.

Westphal, C., Blaxton, T. (1998). *Data mining solutions*. New York: Wiley.

Wheaton, B., Múthen, B., Alwin, D., & Summers G. (1977). Assessing reliability and stability in panel models. In D. R. Heise (Ed.), *Sociological Methodology*. New York: Wiley.

Wheeler, D. J., & Chambers, D.S. (1986). *Understanding statistical process control*. Knoxville, TN: Statistical Process Controls, Inc.

Wherry, R. J. (1984). *Contributions to correlational analysis*. New York: Academic Press.

Whitney, D. R. (1948). *A comparison of the power of non-parametric tests and tests based on the normal distribution under non-normal alternatives*. Unpublished doctoral dissertation, Ohio State University.

Whitney, D. R. (1951). A bivariate extension of the U statistic. *Annals of Mathematical Statistics*, *22*, 274-282.

Widrow, B., and Hoff Jr., M.E. (1960). Adaptive switching circuits. *IRE WESCON Convention Record*, 96-104.

Wiggins, J. S., Steiger, J. H., and Gaelick, L. (1981). Evaluating circumplexity in models of personality. *Multivariate Behavioral Research*, *16*, 263-289.

Wilcoxon, F. (1945). Individual comparisons by ranking methods. *Biometrica Bulletin*, *1*, 80-83.

Wilcoxon, F. (1947). Probability tables for individual comparisons by ranking methods. *Biometrics*, *3*, 119-122.

Wilcoxon, F. (1949). *Some rapid approximate statistical procedures*. Stamford, CT: American Cyanamid Co.

Wilde, D. J., & Beightler, C. S. (1967). *Foundations of optimization*. Englewood Cliffs, NJ: Prentice-Hall.

Wilks, S. S. (1943). *Mathematical Statistics*. Princeton, NJ: Princeton University Press.

Wilks, S. S. (1946). *Mathematical statistics*. Princeton, NJ: Princeton University Press.

Williams, W. T., Lance, G. N., Dale, M. B., & Clifford, H. T. (1971). Controversy concerning the criteria for taxonometric strategies. *Computer Journal*, *14*, 162.

Wilson, G. A., & Martin, S. A. (1983). An empirical comparison of two methods of testing the significance of a correlation matrix. *Educational and Psychological Measurement*, *43*, 11-14.

Winer, B. J. (1962). *Statistical principles in experimental design*. New York: McGraw-Hill.

Winer, B. J. (1971). *Statistical principles in experimental design* (2nd ed.). New York: McGraw Hill.

Winer, B. J., Brown, D. R., Michels, K. M. (1991). *Statistical principals in experimental design. (3rd ed.)*. New York: McGraw-Hill.

Witten, I., H., & Frank, E. (2000). *Data Mining: Practical Machine Learning Tools and Techniques*. New York: Morgan Kaufmann.

Wold, H., (1966). Estimation of principal components and related models by iterative least squares. *Multivariate Analysis* (Ed., P.R. Krishnaiah), Academic Press, NY, pp. 391-420.

Wold, H., (1975). Path models with latent variables: The NIPALS approach. *Quantitative Sociology: International perspectives on mathematical and statistical model building* (Ed.s, H.M. Blalock et al.). Academic Press, NY, 307-357.

Wold, S. (1978, November) Cross-validatory estimation of the number of components in factor and principal components models, *Technometrics, Vol. 20, No. 4.*

Wold, S., Geladi, P., Esbensen, K., & Ohman, J. (1987). Multi-way Principal components and PLS-analysis, J*ournal of Chemometrics, Vol. 1*, 41-56.

Wold, S., Kettaneh, N., Friden, H., & Holmberg, A. (1998). Modeling and diagnosis of batch processes and analogous kinetic experiments, *Chemometrics Intel. Lab. Syst., 44*, 331-340.

Wolfowitz, J. (1942). Additive partition functions and a class of statistical hypotheses. *Annals of Mathematical Statistics, 13*, 247-279.

Wolynetz, M. S. (1979a). Maximum likelihood estimation from confined and censored normal data. *Applied Statistics, 28*, 185-195.

Wolynetz, M. S. (1979b). Maximum likelihood estimation in a linear model from confined and censored normal data. *Applied Statistics, 28*, 195-206.

Wonnacott, R. J., & Wonnacot, T. H. (1970). *Econometrics*. New York: Wiley.

Woodward, J. A., & Overall, J. E. (1975). Multivariate analysis of variance by multiple regression methods. *Psychological Bulletin, 82*, 21-32.

Woodward, J. A., & Overall, J. E. (1976). Calculation of power of the *F* test. *Educational and Psychological Measurement, 36*, 165-168.

Woodward, J. A., Bonett, D. G., & Brecht, M. L. (1990). *Introduction to linear models and experimental design*. New York: Harcourt, Brace, Jovanovich.

Woodward, J. A., Douglas, G. B., & Brecht, M. L. (1990). *Introduction to linear models and experimental design*. New York: Academic Press.

Yates, F. (1933). The principles of orthogonality and confounding in replicated experiments. *Journal of Agricultural Science, 23*, 108-145.

Yates, F. (1937). *The Design and Analysis of Factorial Experiments*. Imperial Bureau of Soil Science, Technical Communication No. 35, Harpenden.

Yokoyama, Y., & Taguchi, G. (1975). *Business data analysis: Experimental regression analysis*. Tokyo: Maruzen.

Youden, W. J., & Zimmerman, P. W. (1936). Field trials with fiber pots. *Contributions from Boyce Thompson Institute, 8*, 317-331.

Young, F. W, & Hamer, R. M. (1987). *Multidimensional scaling: History, theory, and applications*. Hillsdale, NJ: Erlbaum

Young, F. W., Kent, D. P., & Kuhfeld, W. F. (1986). Visuals: Software for dynamic hyper-dimensional graphics. *Proceedings of the Section on Statistical Graphics, American Statistical Association*, 69-74.

Younger, M. S. (1985). *A first course in linear regression* (2nd ed.). Boston: Duxbury Press.

Yuen, C. K., & Fraser, D. (1979). *Digital spectral analysis*. Melbourne: CSIRO/Pitman.

Yule, G. U. (1897). On the theory of correlation. *Journal of the Royal Statistical Society, 60*, 812-854.

Yule, G. U. (1907). On the theory of correlation for any number of variables treated by a new system of notation. *Proceedings of the Royal Society*, Ser. A, *79*, 182-193.

Yule, G. U. (1911). *An Introduction to the Theory of Statistics*. London: Griffin.

Zippin, C., & Armitage, P. (1966). Use of concomitant variables and incomplete survival information in the estimation of an exponential survival parameter. *Biometrics, 22*, 665-672.

Zupan, J. (1982). *Clustering of large data sets*. New York: Research Studies Press.

Zweig, M.H., & Campbell, G. (1993). Receiver-Operating Characteristic (ROC) Plots: A Fundamental Evaluation Tool in Clinical Medicine. *Clin. Chem 39 (4)*, pp. 561-577.

Zwick, W. R., & Velicer, W. F. (1986). Comparison of five rules for determining the number of components to retain. *Psychological Bulletin*, *99*, 432-442.

INDEX

2

3

A